ISBN 978-0-266-04309-6
PIBN 10955890

The Canadian Entomologist

VOLUME XLIV.

1912.

EDITED BY

DR. E. M. WALKER,

Biological Department,

UNIVERSITY OF TORONTO, TORONTO.

Editor Emeritus: REV. C. J. S. BETHUNE.

ONTARIO AGRICULTURAL COLLEGE, GUELPH, ONT.

London, Ontario:

The London Printing and Lithographing Company Limited.

1912.

LIST OF CONTRIBUTORS TO VOL. XLIV.

ABBOTT, JAMES F.	ST. LOUIS, MO.
ALDRICH, PROF. J. M., University of Idaho	MOSCOW, IDAHO.
ALEXANDER, CHARLES P., Cornell University	ITHACA, N. Y.
BACK, E. A., Virginia Agricultural Station	BLACKSBURG, VA.
BANKS, NATHAN	EAST FALLS CHURCH, VA.
BARNES, DR. WILLIAM	DECATUR, ILL.
BLATCHLEY, PROF. W. S	INDIANAPOLIS, IND.
BOWDITCH, FRED C	BROOKLINE, MASS.
BRAUN, MISS ANNETTE F., University of Cincinnati	CINCINNATI, OHIO.
BUENO, J. R. DE LA TORRE	WHITE PLAINS, N. Y.
CAESAR, LAWSON, Ont. Agric. College	GUELPH, ONT.
CHAMBERLIN, RALPH V., University of Pennsylvania	PHILADELPHIA, PA.
CLEMENCE, VICTOR L	PASADENA, CAL.
COCKERELL, PROF. T. D. A., University of Colorado	BOULDER, COL.
COCKERELL, MRS W. P	BOULDER, COL.
CRAWFORD, J. C., U. S. National Museum	WASHINGTON, D. C.
CROSBY, PROF. CYRUS R., Cornell University	ITHACA, N. Y.
DOD, F. H. WOLLEY	MILLARVILLE, ALTA.
EHRMANN, GEORGE A	PITTSBURG, PA.
EVANS, JOHN D	TRENTON, ONT.
FALL, H. C	PASADENA, CAL.
FELT, DR. E. P.	ALBANY, N. Y.
FROST, C. A	SOUTH FRAMINGHAM, MASS.
FYLES, REV. THOMAS W.	OTTAWA, ONT.
GIBSON, ARTHUR, Central Experimental Farm	OTTAWA, ONT.
GIRAULT, A. ARSÈNE	BRISBANE, AUSTRALIA.
HEWITT, DR. C. GORDON, Central Experimental Farm	OTTAWA, ONT.
HOOD, J. DOUGLAS, U. S. Biological Survey	WASHINGTON, D. C.
LEONARD, M. D., Cornell University	ITHACA, N. Y.
LYMAN, HENRY H.	MONTREAL, Que.
McDERMOTT, F. ALEX., Hygienic Laboratory	WASHINGTON, D. C.
McDUNNOUGH, DR. J. H	DECATUR, ILL.
MACGILLIVRAY, PROF. ALEX. D., University of Illinois	URBANA, ILL.
PEARSALL, RICHARD F	BROOKLYN, N. Y.
PRATT, F. C. (deceased) Bureau of Entomology	WASHINGTON, D. C.
PROUT, LOUIS B.	LONDON, ENGLAND.
ROHWER, S. A., Bureau of Entomology	WASHINGTON, D. C.
ROOT, F. M	OBERLIN, O.
SCHAEFFER, CHAS., Brooklyn Institute	BROOKLYN, N. Y.
SMITH, HARLAN I.	OTTAWA, ONT.
SPOONER, C. S., Georgia State Board of Entomology	ATLANTA, GA.
SWAINE, J. M., Central Experimental Farm	OTTAWA, ONT.
SWETT, LOUIS W	BOSTON, MASS.
THOMPSON, W. R.	LONDON, ONT.
TOTHILL, J. D.	FREDERICTON, N. B.
TOWNSEND, PROF. C. H. TYLER	LIMA, PERU.
VAN DUZEE, E. P., Grosvenor Public Library	BUFFALO, N. Y.
WALKER, DR. E. M., University of Toronto	TORONTO, ONT.
WALLIS, J. B	WINNIPEG, MAN.
WEBSTER, PROF. F. M., Bureau of Entomology	WASHINGTON, D. C.
WILSON, H. F., Oregon Agricultural College	CORVALLIS, OR.
WINN, ALBERT F.	WESTMOUNT, P. Q.
WOLCOTT, A. B.	CHICAGO, ILL.

The Canadian Entomologist.

VOL. XLIV. LONDON, JANUARY, 1912. No. 1

SYSTEMATIC NOTES ON NORTH AMERICAN TACHINIDÆ.*

BY JOHN D. TOTHILL, DIVISION OF ENTOMOLOGY, OTTAWA.

The following is the first of a series of articles on Tachinidæ which the writer hopes to publish in this journal from time to time. They will be of a strictly systematic nature and emphasis will be laid upon Canadian species. This present article contains a description of a new Canadian species of *Winthemia* together with a key for the separation of the North American species of the genus known at the present time; it also contains suggestions for improvement at two difficult points in Mr. Coquillett's generally excellent key to the genera in his "Revision of the N. American Tachinidæ."

Winthemia Desv.

In the course of a study of the parasites of the Spruce Budworm (*Tortrix fumiferana* Clemens) in Canada by the Division of Entomology a new species of *Winthemia* was reared in considerable numbers. A description of this species, which is here named *W. fumiferanæ* after its host, together with a key for the separation of the North American species of the genus follows.

The genus *Winthemia* Desv. ("Essai sur les Myodaires," p. 173, 1830), is represented in North America by four known species. One of these, *W. quadripustulata* Fab., is an exceedingly variable species. The writer has examined the types of *W. obscura* Coq. and *W. antennalis* Coq. and there seems to be little doubt but that they are both good species. For the privilege of examining these types and for numerous other courtesies while at the United States National Museum the writer is indebted to the authorities of that institution and more particularly to Mr. Frederick Knab.

The four species may be separated as follows :—

1. With 3 sternopleural bristles.......................................2.
   With 2 sternopleural bristles.......................................3

*Contributions from the Division of Entomology, Ottawa.

2. With 3 postsutural bristles, 3rd joint of antenna nearly 3 times as long as 2nd; arista thickened almost to middle........... *obscura* Coq.

With 4 postsutural bristles, 3rd joint of antenna 1¼ times as long as 2nd; arista thickened on basal ⅓ ....*fumiferanæ*, n. sp.

3. Palpi, scutellum and apex of abdomen black......... *antennalis* Coq.

Palpi, scutellum and apex of abdomen yellowish. *quadripustulata* Fab.

Winthemia fumiferanæ, n. sp.

Black species with palpi, scutellum and usually antennæ and sides of 2nd and 3rd segments of the abdomen yellowish; thorax, abdomen and legs lightly dusted with whitish pollen; 4 postsutural and 3 sternopleural bristles; hind legs ciliate. Length, 7-9 mm.

Head ¾ times as broad as long; front in male ⅓ width of eye, in female equal to width of eye; frontal vitta in female 1½ times width of parafrontal plate (measurements taken immediately anterior to ocellar triangle); parafacials at narrowest point ¼ as wide as facial plate at widest point; genæ 1/6 eye height; antennæ reaching to lowest ¼ of face, 3rd joint in both sexes 1¼ times length of 2nd; arista thickened on basal ⅓, 2nd joint 1½ times as long as broad. Colour of head except eyes whitish pollinose on black ground; frontal vitta black; frontalia, parafacialia, facial plate, occiput and genæ whitish pollinose on black ground; antennæ varying from black to yellowish in all its segments, palpi yellow. A strong pair of proclinate ocellar bristles in both sexes; females with two pair of, males without orbital bristles; frontal bristles to, or slightly beyond base of second antennal joint; parafacials with numerous fine black hairs on upper ⅔; bristles of facialia on lower ¼ only.

Thorax including the pleuræ light grey pollinose on a black ground, the pollen being somewhat irregularly distributed; scutellum blackish at base and yellowish at apex, the extent of yellowish area varying in different specimens. Four pairs of postsutural dorsocentral bristles (in one specimen only 3 pairs), 3 pairs of postacrostichals and three sternopleurals, the latter usually strong but the lower one absent on one side in one specimen; scutellum with three strong pairs of marginal, and a pair of cruciate apical and several weak discal bristles.

Legs black; coxæ and femora strongly, tibiæ and tarsi faintly, whitish pollinose; middle tibiæ with 2 or 3 bristles on front side near the middle;

hind tibiæ ciliate but with two longer bristles near the middle; tarsal claws and pulvilli considerably longer in the male than in the female.

Wings hyaline, becoming somewhat fuscous toward base. Vein M_{4+5} with one to five bristles near base; the medium cross vein quite distinctly S shaped; there is no appendage at bend of R_{1+2}; the anterior end of medium cross vein is situate at 1/3 distance from the bend of R_{1+2} to radio medial cross vein.

Abdomen black and polished on the whole of first segment and on posterior margins of the other segments; the narrow anterior margins of segments 2, 3 and 4 white pollinose; the median fascia irregularly white pollinose on black ground; sides of segments 2 and 3 sometimes yellowish. A pair of median marginal macrochætæ on segments 1 and 2, a row of very long marginals on segment 3; no discal bristles on segments 2 and 3; all the segments are thickly covered with rather long fine hairs, which, especially medially, are erect and not proclinate; fourth segment covered on disc with fine bristles about 3/4 length of marginal macrochætæ on segment 3.

Described from 18 males and 18 females bred in the Division from the Spruce Budworm (*Tortrix fumiferana* Clemens). The localities are as follows: Two males and one female from Maniwaki, Province of Quebec; 16 males and 17 females from Duncans, British Columbia, Canada. The adults issued from both larvæ and pupæ, but principally the latter, of the host. Type female from Duncans, B.C., and 33 co-types deposited with Division of Entomology, Experimental Farms, Ottawa; 2 co-types a male and female from Duncans, B. C., deposited in the United States National Museum, Washington, D. C.

Amobia distincta Towns., and *Senotainia trilineata* V. & W.

In a recent attempt by the writer to determine with the aid of Coquillett's "Revision" some Tachinids that have since proved to *Senotainia trilineata* V. & W., considerable difficulty was experienced in deciding whether the species was the above mentioned or *Amobia distincta* Towns.; moreover reference to the original description did not materially facilitate the determination. From an examination of a large series of both species in the United States National Museum it was found that they are abundantly distinct and that the generic separation is fully justified. The following is a table, which it is hoped may prove useful, of some of the more obvious differences between the two species :—

Amobia distincta Towns.	*Senotainia trilineata* V. & W.
1. Radiomedial cross vein far before tip of R_{2+3}.	1. Radiomedial cross vein at or close to tip of R_{2+3}.
2. Palpi black.	2. Palpi yellow.
3. Parafacials at narrowest point at least 1.5 times length of 3rd antennal joint.	3. Parafacials at narrowest point about equal to length of 3rd antennal joint.
4. The three black thoracic vittæ, broad and conspicuous.	4. The three or four black thoracic vittæ narrow and inconspicuous.
5. The abdominal markings (three rows of black triangles on yellowish gray ground) very distinct even without lens, especially in male.	5. The abdominal markings not all distinct.
6. Female with piercing ovipositor.	6. Female without piercing oviposter.

Tachinophyto variabilis Coq., and *floridensis* Towns.

Tachynophyto Towns., Trans. Amer. Ent. Soc., Vol. 19, p. 130, 1892, generic synonymy.

Pseudomyothria Towns., 1892, loc. cit.

Methypostena Towns., 1908, Tax of Musc. Flies.

Lixophaga Towns., 1908, Tax of Musc. Flies.

Hypostena of authors (non Meig).

The above synonymy is pointed out by Mr. D. W. Coquillett in his recent and valuable paper "The Type Species of North American Genera of Diptera," p. 611.

In the "Revision," page 62, key section No. 7, two species of the above genus are separated, namely, *variabilis* Coq., and *floridensis* Towns. The key reads as follows :—

"7. Third segment of abdomen pollinose on at least the basal two-thirds, the pollen yellowish, abdomen subopaque ;
length, 4–9 mm . . . . *variabilis* Coq.
"Third segment at most pollinose on the basal third, the pollen white, abdomen subshining ; length, 4–9 mm . . . *floridensis* Towns."

The characters made use of are purely colorational and since the publication of Coquillett's valuable "Revision" larger series of the two species have been accumulated which clearly demonstrate that such characters, at least in *variabilis*, are subject to great variation. A recent examination of the types and of the series both at the Gipsy Moth Parasite Laboratory and at the United States National Museum by the writer brings out two points, i.e., that the species are abundantly distinct and that the pollinosity on the third segment of the abdomen in *variabilis* varies all the way from the typical condition to the condition met with in typical *floridensis*. The following conspicuous structural differences will serve to separate the species :—

T variabilis Coq.	*T. floridensis* Tn.
1. Third joint of antennæ 3.5 to 4 times length of 2nd.	1. Third joint of antennæ 2 to 2.5 times length of 2nd.
2. Costal spine very inconspicuous.	2. Costal spine strongly developed and very conspicuous.

(To be continued.)

NOTES ON THE PARASITIC HYMENOPTERA.

BY A. A. GIRAULT, BRISBANE, AUSTRALIA.

Superfamily Chalcidoidea.
Family Encyrtidæ.
Subfamily Encyrtinæ.
Tribe Arrhenophagini.
Genus *Rhopoideus* Howard.

1. *Rhopoideus fuscus*, new species.

Dr. C. Gordon Hewitt, Dominion Entomologist, Ottawa, Canada, has sent me among other things eight specimens of an Encyrtine bearing acute edentate mandibles, which agree well with the genus *Rhopoideus* Howard. This species, however, has but 9-jointed antennæ, counting a very short, almost imperceptible ring-joint; its antennal club is solid. Now Ashmead gives as a diagnostic character of the genus in question 10-jointed antennæ (the funicle 5-jointed, no ring-joint mentioned), which would imply at least a 2-jointed antennal club. The original description of *Rhopoideus* leaves one in doubt as to the total number of antennal joints, the only statement made concerning them being to the effect that the funicle is 5-jointed. Nevertheless, this Canadian species agrees so well with the generic description, even to the possible hosts, except in the antennæ, that we have reason to question Ashmead's statement concerning the latter. With this species the funicle is 5-jointed, the first three joints "small and narrow, each rather broader than long, 4 and 5 broader and longer and as broad as long," as described for the type species, except that in each case here the joints are longer than wide. The antennæ of this species appear to agree in general form with those of the type of *Rhopoideus*.*

Female.—Length, 1.95 mm. Rather long and slender, the body flattened or depressed.

General colour uniformly brown, but the abdomen somewhat paler, the brown emphasized along the caudal margins of the segments (making at least four transverse brown stripes across the abdomen, which, however, are not conspicuous). Antennæ concolorous; legs somewhat lighter, with some yellow, the tarsi pallid yellowish, the distal tarsal joint clouded. Eyes dark. Wings hyaline, with the exception of a slight cloud of fuscous under the stigmal vein and just out a slight distance from the base, and also sometimes slightly touched with fuscous along the caudal margin irregularly, proximad and along the oblique hairless line at either margin of it. Trochanters and bulbs of the antennæ pallid.

Mesoscutum and mesoscutellum polygonally sculptured, as if covered with flat scales, both bearing a few, sparse, short setæ; the concave face finely lined with circular lines (concentric about the rather deep and large, crescentic scrobicular cavity in about the centre of the face); carina of vertex present; tarsi 5-jointed, the joints short yet longer than wide; tibial spurs single, the cephalic spur curved and forked at tip; caudal femora somewhat thickened, legs otherwise slender or usual. Fore wings with an oblique, hairless line running proximo-caudad from the origin of the stigmal vein. Mandibles short, claw-shaped, acute and edentate at tip. Submarginal vein long and slender, the costal cell rather wide, the marginal vein a mere rounded point where the submarginal touches the cephalic margin, the postmarginal vein absent, the stigmal vein distinct, moderate in length, with a slender neck. Fore wing (including the costal cell) densely, finely ciliated, the blade ample and wide, only about twice longer than broad, the marginal cilia short, becoming noticeable only at apex and disto-caudad, where they are moderately short. Caudal wings densely ciliate discally, rather short. Parapsidal furrows absent. Abdomen

*Dr. L. O. Howard has very kindly examined the type of his *Rhopoideus citrinus* for me, and tells me in a letter dated August 8, 1911, that the antennal club of that species is solid, hence the antennæ 8-jointed (excluding any question of a ring-joint). Ashmead's diagnosis of the genus is therefore wrong.

longer than the thorax, cylindrical but pointed at the apex, the valves of the ovipositor slightly extruded. Ocelli in a slightly curved line. Scutellum peltate, angular, as wide as long or nearly. Cephalic aspect of head nearly quadrate.

Antennæ 9-jointed; scape long and slender, slightly thickened in the middle, the bulb rather long, both together over twice longer than the pedicel; the latter obconic, rather long, over twice longer than wide; a very flat, short ring-joint, which has the shape of a mushroom; funicle joints 1–3 short, 4 and 5 longer and wider; 1 and 2 subequal, each slightly longer than wide; 3 of same width but slightly longer; 4 a third longer and broader than 3; and 5 a third longer and broader than funicle joint 4; all much shorter than the pedicel, which is subequal in length to the combined lengths of the first three funicle joints; club solid, long and cylindrical, obtusely pointed, not quite as long as the funicle. Pubescence of antenna short, not dense or conspicuous.

(From eight specimens, 2/3-inch objective, 1-inch optic, Bausch and Lomb.)

Male.—Unknown.

Described from eight specimens mounted singly on slides, received for identification from Dr. C. Gordon Hewitt, as noted above, each slide labelled "from spruce budworm material, Province of Quebec," and respectively, "Maniwaki, 27, VI, 11," and "Montcalm, 6, VII, 11," two females, two slides (homotypes in Canada); "Chicoutimi, 3, VII, 11," and "St. Gabriel de Brandon, 3, VII, 11," two females, two slides co-types, as noted below); "Chicoutimi, 3, VII, 11," and "Montcalm, 6, VII, 11," two females, two slides (types); and "Chicoutimi, 3, VII, 11, two females, two slides (homotypes in collection Illinois State Laboratory of Natural History). The supposed host is *Tortrix fumiferana* Clemens, but a coccid is indicated instead. Other coccid parasites, some noted beyond, were reared from the same host material.

Habitat.—Dominion of Canada—Quebec (Chicoutimi, Maniwaki, Montcalm and St. Gabriel de Brandon).

Types.—Cat. No. *14,206*, United States National Museum, Washington, D. C., the two females as indicated above. *Co-types:* Accession No. *45,080*, Illinois State Laboratory of Natural History, Urbana, U. S. A., the two females, two slides as indicated above. *Homotypes:* The two females as above indicated, in the collections of the Division of Entomology, Central Experimental Farm, Ottawa, Canada, and the two indicated as being in the collections of the Illinois State Laboratory of Natural History (Accession No. *45,085*).

This species evidently differs considerably from the type species, *citrinus* Howard; it must be considerably larger and more slender, the colour is brown, not light orange; the mesonotum is differently sculptured, namely, polygonally, not finely, transversely lined; the sheaths of the ovipositor not nearly black at tip but concolorous; the joints of the funicle somewhat longer, and the antennal club shorter, not as long as the funicle, even; and no oblique hairless line on the fore wing is noted for the type species, nor a ring-joint in the antennæ.

Family Pteromalidæ.
Subfamily Sphegigasterinæ
Tribe Sphegigasterini.
Genus *Urios* Girault MS.

1. *Urios vestali* Girault.

This nearly wingless species, which was described in the Journal of the New York Entomological Society (December, 1911), was captured by Mr. A. G. Vestal at the Devil's Hole, near Havana, Illinois. It was found in an ant's nest (*Pheidole vinelandica* Forel.), April 1, 1911. The nest of the ant was in sandy soil, in a bunch grass area. Mr. Vestal stated that, casually, he was unable to distinguish the pteromalid from the ants. In other words, it closely mimics the host ant.

Family Eulophidæ.
Subfamily Entedoninæ.
Tribe Omphalini.
Genus *Astichus* Foerster.

1. *Astichus bimaculatipennis*, new species.

Normal position.

Female.—Length, 1.85 mm. Funicle not ringed with white, scutellum not smooth, and without grooved lines, parapsidal furrows not very distinct, cilia of wings not in rows, dense; wings with two maculæ. Species large for the genus.

General colour metallic green, the head and pronotum metallic bluish, the face æneous just dorsad, purplish just ventrad, of the insertion of the antennæ; the dorsum of the abdomen, except the metallic green proximal segment, dark, purplish black; metanotum with æneous reflections; scape pallid dusky; thoracic pleura and coxæ metallic dark bluish, the femora the same, pallid at apical end; trochanters dark; tibiæ and tarsi pallid, the apical tarsal joint dark or black; flagellum dusky; ventum of thorax and abdomen dark, purplish black. Tegulæ dark. Wings hyaline, venation dusky, the fore wing with two dusky blotches along the cephalic or costal margin, the first or proximal one at the junction of the submarginal and marginal veins, rounded and about one-half the size of the apical one, which is situated at the stigmal vein, and is more irregular in outline. Eyes chestnut red; ocelli ruby red.

Head flat from lateral aspect, the occipital margin acute; front broad, concave, vertex narrow, broader laterad; eyes lateral, oval, covering a little over a half of the lateral aspect of the head, the malar space present; antennæ inserted, about on an imaginary line drawn between the ventral ends of the eyes, the scape not reaching to the vertex; lateral ocelli on the narrow vertex at the occipital margin, distant from the eyes, but farther apart from each other than each is from the eye margin, dorsal, an imaginary line connecting them convex; the cephalic ocellus barely visible from dorsal aspect, cephalic, forming a flat triangle with the others, and situated in the cephalic aspect of the vertex front, against the acute occipital margin at the meson and closer to the lateral ocelli than they are to each other. Head delicately shagreened, its surface not as coarse as the surface of the eyes; the entire thorax dorsad moderately, coarsely, polygonally reticulated, the parapsidal furrows mere impressions, inconspicuous, not well defined grooves, and from some aspects seen only caudad; pronotum visible from dorsal aspect, about one-third the length of the mesocutum, the caudal margin of the latter, between the advanced axillæ, convex; scutellum rounded, normal, convex, without grooved lines; mesopostscutellum not large, crescentic, sculptured like the scutellum; metathoracic spiracle margined, distinct, short and broadly oval; metathorax slightly more delicately sculptured than the scutellum and scutum of the mesothorax, and with a delicate median carina, and two others, on each side of the meson, both curved and running caudolaterad from the caudal margin of the mesopostscutellum; of these two lateral carinæ, the more laterad or cephalic one is the shorter. Coxæ and the thoracic pleura sculptured similarly to the metanotum, the cephalic coxæ less so; caudal coxæ enlarged, subtriquetrous. Abdomen conic-ovate, but very slightly produced or convex ventrad, longer than the head and thorax combined and than the wings; very delicately reticulated.

Fore and hind wings densely ciliate discally, the cilia short and close; marginal cilia of both wings short and close, longer on the hind wings. Marginal vein of fore wings subequal to or slightly shorter than the submarginal vein, the postmarginal vein equal to half the length of the marginal, and nearly twice the length of the stigmal vein, which is bifurcate at apex.

Antennæ filiform, not very long, the scape cylindrical and of moderate length, the funicle 4-jointed, the club 3-jointed, and with a single ring-joint. Pedicel small, obconic, about a third of the length of the long first funicle joint; ring-joint minute; first funicle joint cylindrical, the longest antennal joint, wider than the pedicel and a fourth longer than the following joint; funicle joints 2 and 3 subequal, cylindrical oval, joint 2 slightly longer and slightly narrower than joint 3, and both distinctly shorter and broader than joint 1 of the funicle; funicle joint 4 subquadrate, of about the same width, but only about two-thirds of the length of joint 3; proximal club joint large, half the length of the club, but distinctly smaller than the apical funicle joint; the intermediate club joint smaller, conical, not much larger than the pedicel, about a little over half the size of the proximal club joint; the apical or discal joint minute, nipple- or spur-like. Funicle and club hispid, with white hairs, of which there are about three transverse rows on the first funicle joint, and two rows on each of the following funicle joints and the proximal club joint.

(From a single specimen, ⅔-inch objective, 2-inch optic, Bausch and Lomb.)

Male.—Unknown.

Described from a single female specimen received for identification from Mr. R. L. Webster, Iowa College of Agriculture and Mechanic Arts, Ames, Iowa, who reared it as a probable hyperparasite of *Alceris minuta* Robinson. (Bull. No. 102, Iowa State College of Agriculture and Mechanic Arts Experiment Station, Ames, Iowa, p. 210.)

Habitat.—United States, Ames, Iowa.

Type.—Accession No *40,290*, Illinois State Laboratory of Natural History, Urbana, one female on a tag, plus one balsam slide with antennæ.

Subfamily Aphelininæ.

Tribe Aphelinini.

Genus *Physcus* Howard.

1. *Physcus varicornis* Howard.

I desire to record the occurrence of this coccid parasite in some Canadian localities. Dr. Hewitt recently sent me four female specimens on slides labelled "From spruce budworm material, Maniwaki, Montcalm and St. Gabriel de Brandon, Province of Quebec, 2, 3 and 6, July, 1911."

In the original description of this genus (Howard, Bull. No. 1, technical series, Division of Entomology, U. S. Department of Agriculture, 1895, p. 43), a statement is made to the effect that the "second and third funicle joints" are "subequal and each longer than joint 1." Later (Id., Bull. No. 12, technical series, Bureau of Entomology, U. S. Department of Agriculture, 1907, p. 72), this statement is used as a diagnostic character in a table of genera to the Aphelinini. The character varies. In the four specimens noted above, three have the joint as described, but the fourth specimen has the first funicle joint slightly longer than either joint 2 or 3. In some Illinois specimens I have noted the same variation, sometimes all three joints equal, sometimes the second shortest, and in others as described originally. The matter is of no great importance, since the table mentioned can do without the line containing the statement about the funicle joints. The variation itself, however, is a rather peculiar one, and important from the standpoint of specific characters.

Genus *Prospaltella* Ashmead.

1. *Prospaltella aurantii* (Howard).

This widely distributed parasite of coccids has lately been received from Dr. Hewitt from several localities in Canada, which I think should be recorded in this connection. There were seven females on six slides labelled "From spruce budworm material, Chicoutimi, St. Gabriel de Brandon and Maniwaki, Province of Quebec, July 2 and 3, 1911." They evidently originated from some coccid concealed in the host material. All of the specimens were compared with the type, and are homotypes, therefore. The fore wings in this species seem to have a tendency to be very slightly clouded out to the end of the marginal vein from base, but this cloudiness is so slight that one cannot always be sure that it is real. From the collections of the United States Department of Agriculture I have a series of six females on a slide, with a number of other coccid parasites labelled "1725. *Aspidiotus* on common wild shrub on streams, California and Cuautla, Morelos, Mexico, July 1, '97. Koebele."

I have also seen two other series of this species from Mexico, on slides from the same collection, which evidently form a distinct race of *aurantii*, and which I thought at first would certainly prove to be specifically distinct. This form differs in having distinctly broader fore wings (about from 19 to 21 lines across the widest portion of the blade, *aurantii* bearing only about 15), their longest marginal cilia less than a third of the greatest width (in the type form over a third), and the antennæ differing in that each joint of the funicle is longer than the one preceding, the third longest; whereas in the type form the second joint is longest. I have no doubt but that these forms grade into each other. The specimens should be recorded. They are: Three females on a slide with *Signiphora* labelled "1745. *Aspidiotus* on soft wooded fibrous tree. Cordoba, Mexico, 17, 7, '97, Koebele"; and thirteen females on a single slide, with several species of *Signiphora* (*mexicana* Ashmead, *flavopalliata* Ashmead, and *townsendi* Ashmead), together with an *Ablerus*, labelled "1768. *Aspidiotus* on Hibiscus, Cuautla, Morelos, Mexico, May 29, '97, Koebele." These last specimens varied in colour, most of them having the abdomen wholly black-brown instead of brownish yellow.

FURTHER NOTES ON DIABROTICA.
No. II.

BY FRED. C. BOWDITCH, BROOKLINE, MASS.

(Continued from Vol. XLIII, page 417.)

D. boucardi, nov. sp.

Head, thorax and scutel smooth shining black; antennæ and feet fuscous yellow; elytra bright purple, lateral margin obsoletely viridescent, with two transverse depressions and also humeral and lateral submedian impressions. Length 7 mm.

One example, Panama, in the Boucard collection of the Tring material; Belongs in sec. D, near *coccinea* Baly. The palpi are the colour of the antennæ; head with a deep frontal puncture; antennæ more than half as long as body, 2nd joint short, 4 much longer than 3. Thorax elongate, sinuate and sharply angled behind with a deep transverse depression, occupying the rear half; elytra somewhat dilated at the rear; the 1st transverse impression is at the rear of the anterior third, the 2nd is much the larger of the two and occupies the middle of the elytra, the two connected by a depression along the suture; the humeral depression is slightly curved inwards and ends about the beginning of the middle third the lateral depression is broad and submedian. The depressions give a much swollen or torous effect to the rear of the elytra. The relative length of the antennal joints might place this form in sec. 1, but the general appearance is such that I have put it next *coccinea* Baly; the elytra are sparsely diffusely punctulate, body beneath black.

D. tæniolescens, nov. sp.

Head and thorax rufous yellow; antennæ and legs yellow; scutel dark rufous; elytra pale yellow, each with 4 elongate black spots, a humeral and subbasal median, and two behind the middle, in the rear of the anterior ones, giving the appearance of 2 interrupted black vittæ, body beneath black. Length, 5 mm.

One example, Callanga, Peru.

A very well marked form coming next to *tæniolata* Gahan, from which it differs in the wholly flavous antennæ and spotted in place of vittate elytra. The thorax is finely punctulate and has a deep oblique fovea on each side and a third round one at the rear; the punctures of the elytra are moderate and arranged in obsolete striæ; the subbasal spots do not attain the margin and are a little short of ⅓ the length of the elytra; the two rear spots end just over the convexity; the exterior angles of all the spots are rounded; the general appearance on each side of the rear spots on each elytron, is that of an oblong black patch bisected by a narrow yellow stripe.

D. 4-signata, nov. sp. (Jac. in litt.).

Head black; antennæ black, extreme base and joints 8–11 flavous, except extreme tip of 11 and base of 8 which are black. Thorax flavous-rufous, wider than long, smooth, obsoletely trifoveate; elytra flavous, rather coarsely and thickly punctate, each side with 5 short viridicyaneous streaks, a basal median, a humeral, a lateral and two small postmedian spots, placed on the convexity longitudinally behind the first two anterior streaks. Body below and legs testaceous, tibiæ and tarsi fuscous. Length, 6 mm.

Two examples, Marcapata, Peru.

Very close apparently to *humeralis* Gahan. The antennæ, however, seem to differ and the markings are viridicyaneous in place of nigrocyaneous. The species has been distributed with the manuscript name *4-signata* Jac. and two co-types have been sent me by Messrs. Staudinger & Bang-Haas. In my two examples the inner basal marks are not joined to the outer as in *humeralis*, but I have one or two examples which are puzzling to place in either species, and I feel a little doubtful as to the specific limits of either form.

The antennæ are a little more than half as long as the body, the basal joints piceous below, the punctures of the elytra are fairly well arranged in striæ on the disk and obsoletely biseriate.

D. subangulata, nov. sp. (Jac. in litt.).

Head black with a large triangular fovea; antennæ three-fourths the length of the body, black, with the four or five basal joints fuscous and the three last (extreme tip of the eleventh excepted) white. Thorax transverse, rufous, depressed, bifoveate (third obsolete) ; scutel black ; elytra moderately coarsely punctate, black, a large oblong basilar spot, a median transverse fascia, not attaining either the margin or suture, and a large quadrate apical spot in each elytron, whitish flavous. Legs yellow with black or piceous tibiæ and tarsi. Body below yellow with breast black. Length, 8 mm.

Seven examples, Marcapata, Peru.

Very variable in the light elytral marking, as noted hereafter. The third joint of the antennæ is ½ longer than the second, the fourth shorter than the preceding two, the colour of the basal joints varies somewhat in the amount of black, some being much darker than others ; the thorax is much broader than long, with a few fine punctures scattered over the surface ; the two side fovea are distinct and the third just before the scutel is obsolete ; the sides are widely depressed and moderately sinuate, the surface dull shining ; elytra are moderately dilated behind, only slightly depressed behind the scutel. What I consider the typical light spot marking varies at the base so that the basal spot and median fascia may unite and the black band between becomes a curved lunule from the shoulder, running towards, but not reaching, the suture ; also the apical spot may unite with the median fascia, leaving the band between as a more or less well defined spot. In one example the spots are so suffused as to indicate that specimens occur which are wholly light flavous, except the margins. The colour of the tibiæ and tarsi seem to vary from black to light piceous according to the predominance of the black markings of the elytra.

This species has been distributed with the manuscript name *sub-angulata* Jac. I have received 4 co-types from Messrs. Staudinger & Bang-Haas.

D. inconspicua, nov. sp. (Jac. in litt.).

Entirely pallid testaceous, with black eyes; antennæ fuscous in the middle, mandibles dark. Head with a deep frontal fovea, antennæ nearly three-fourths as long as body, 3rd joint nearly as long as 4. Thorax nearly as broad as long, strongly obliquely bifoveate and somewhat depressed behind ; elytra slightly dilated to the rear, a faint piceous spot on the shoulder, and with punctures thick, moderate and obsoletely arranged in rows. Length, 6 mm.

Two examples, Callanga, Peru.

This species has been distributed with the manuscript name *incon-spicua* Jac. My two examples were sent me as co-types by Messrs. Staudinger & Bang-Haas. The species is analogous to many in the latter part of sec. 1. The elytra are not plicate.

D. guyanensis, nov. sp.

Head shining black ; mouth parts flavous ; antennæ piceous at base, fuscous in middle, last four joints flavous, extreme tip of 11 dark. Thorax rufous, shining, lightly bifoveate, depressed ; scutel black ; elytra strongly plicate, flavous, with a short common sutural and a long humeral stripe black ; punctation thick and subrugose. Body below flavous, breast black, legs flavous.

Var. with short submedian, subsutural black stripe or spot.

Three examples, Br. Guiana type form, 1 var. do., and 1 Pachitea, Peru. Length, 6–6½ mm.

The antennæ are long and slim, reaching nearly three-fourths the length of the body ; the thorax is very slightly sinuate behind, rounded in front, almost square, and noticeably polished ; the sutural stripe is of medium width, a little less than one-third the length of the elytra, hardly narrowed behind ; the humeral stripe stops just round the bend of the convexity, and is evidently sometimes interrupted at a little behind the middle. Both stripes attain the base. This species resembles the forms in sec. 1, division O.

D. bertonii, nov. sp.

Head rufous ; labrum and spots on the vertex piceous ; antennæ black, joints 8–9 flavous with bases and tips dark. Thorax rufous, rather coarsely punctate, deeply trifoveate; scutel piceous rufous; elytra thickly punctate, shining black, the margins, the suture (very narrowly behind) and a median band dilated at the sides and suture, flavous; beneath, thorax red, body black, legs flavous with tibiæ and tarsi and apex of femora above, and coxæ black. Length, 6 mm.

Type 1 example, Puerto Bertoni Alto Parana, Paraguay, sent me by Mr. Schrotky, also a specimen in the Jacoby collection labelled "1 A.A. 28 In Ocynis."

Should be placed near *borrei* Baly, but the punctuation and flavous markings very different; the thorax is much broader than long, slightly sinuate behind. the fovea connected by a well marked sulcus; the punctuation of the elytra is rather coarse and thick, becoming obsolete behind; the black spots do not attain the basal margin (as in *borrei*) and there is a well marked sulcus on the sutural side of the shoulder; and the elytra are very obsoletely shortly plicate; the sutural yellow stripe is narrowed behind to a mere line just separating the black spots on the elytra; all the spots have the angles rounded.

D. thammii, nov. sp.

Head and thorax rufous; antennæ fuscous, tip of last joint dark. Thorax wider than long, trifoveate, the lateral ones oblique and deep; scutel rufous or piceous; elytra thickly punctate, dull shiny black, the lateral margin slightly dilated at the apex and a median transverse fascia more or less dilated up and down the suture, flavous. Beneath, thorax red, body black, legs yellow, tarsi more or less piceous. Length, 5–5½ mm.

Type, Marcapata, Peru, in 2nd Jacoby collection; also Pachitea, Peru; single example in the 1st Jacoby collection, Chanchomayo, Peru (Thamm) labelled "*concula* Er. ♀?" Twelve examples in all.

Very like *bertonii* supra and near *borrei* Baly. The antennæ ♂ are ¾ as long as body, joint 3 and 4 nearly equal; the thorax is broadly margined and slightly sinuate behind and the surface is sparsely evidently punctate. The black of the elytra varies considerably; in what I consider typical there is a solid basilar fascia; but specimens occur where the fascia is divided into two distinct rounded subbasal spots, leaving a wide sutural yellow stripe and a narrow basilar stripe, flavous, all the flavous marks connected. In this form also the rear fascia is narrowly subdivided by the narrow yellow suture. The elytra are shortly obsoletely plicate.

NEW SPECIES AND GENERA OF NORTH AMERICAN LEPIDOPTERA.

BY WM. BARNES, M.D., AND J. MCDUNNOUGH, PH.D., DECATUR, ILL.

Family Noctuidæ.

Heliothis ætheria, sp. nov.

Head and thorax clothed with olivaceous hairs; primaries dark olive green, in most cases entirely suffused with rich purplish as far as the subterminal line; a blue spot on costa near base, often extending along costa to t. a. line; this latter pale blue, rather broad, strongly and evenly convex; t. p. line narrower, slightly defined by blue, especially on costal and inner margins, perpendicular to costa and well beyond reniform to a point opposite base of same, then strongly incurved to below reniform and again straight to inner margin; median area largely filled with pale yellowish, leaving a narrow costal border and a larger patch on inner margin of the ground colour, and containing a large dark quadrate reniform more or less scaled with purplish, the upper portion of which tends to suffuse with costal border; s. t. line marked by the difference in shade between the purplish subterminal area and the narrow olive terminal portion, slightly waved, on the whole parallel to outer margin; fringes concolorous. Secondaries black, with an irregular pale yellow median band not reaching inner margin, and much constricted in central portion, or even broken into two spots; fringes whitish. Beneath, primaries black, with a broad, sharply defined pale yellow median band containing a large black discal spot corresponding to reniform, terminal area at costa suffused with whitish; secondaries as above, but costal area pale yellowish and median band broader, entirely enclosing black discal spot; costal half of terminal area suffused with whitish; fringes pale, darker at apex of primaries. Expanse, 25 mm.

Habitat: Redington, Ariz, 10 ♂s, 14 ♀s. Types, collection Barnes.

The species is closely related to *sueta* Grt., which, however, lacks the blue shading of the Arizona form; the ground colour is quite variable, at times all traces of the pink suffusion being lost.

Schinia velutina, sp. nov.

Head, thorax, abdomen and wings white, very slightly suffused with a pale ochreous. At first glance apparently immaculate, by holding in certain lights the maculation of primaries is distinctly visible as satiny white lines; t. a. line strongly outwardly oblique to just below cubital

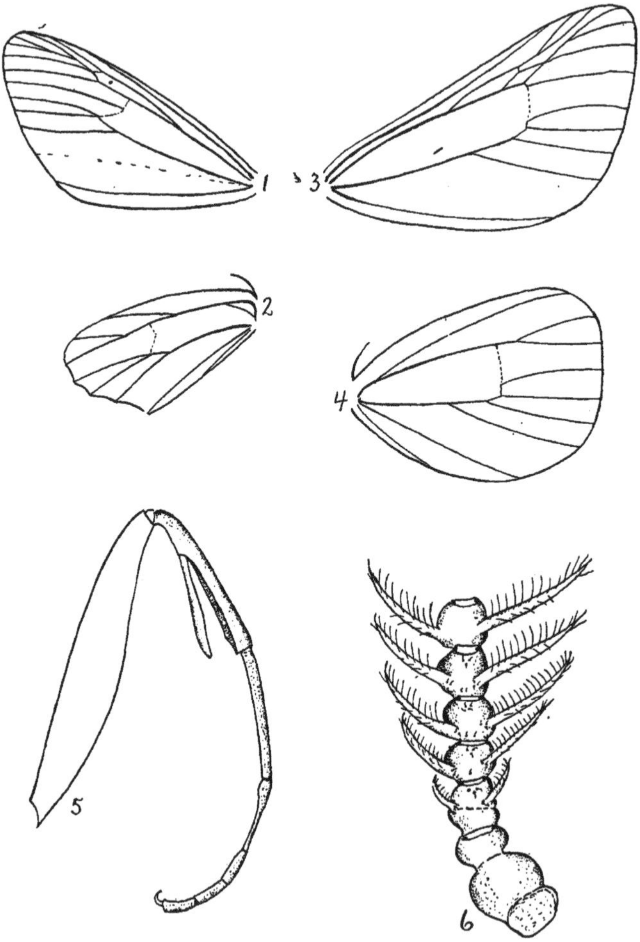

FIG. 1.—Generic features of *Grossbeckia* and *Friesia*.

1. Venation of fore wing of *Grossbeckia semimaculata*.
2. " of hind wing of " "
3. " of fore wing of *Friesia anormalis*.
4. " of hind wing of " "
5. Fore leg of *Friesia anormalis*.
6. Basal portion of antennæ of *Friesia anormalis* (much enlarged).

vein, then just as strongly oblique inwardly to inner margin ; reniform indistinctly visible as a satiny white patch ; t. p. line well excurved around reniform, thence parallel to outer margin to a point on inner margin a little more than two-thirds from base ; s. t. line rather indistinct, slightly waved, approached to t. p. line on vein Cu_2. Secondaries immaculate ; all fringes white. Beneath, white, with a very prominent discocellular dusky spot on primaries. Expanse, 25 mm.

Habitat : Eureka, Ut. (Spalding), 2 ♂s. Type, collection Barnes.

The fore tibiæ possess one long curved claw on inner side, and a small claw with strong spine above it on outer side.

Grotella parvipuncta, sp. nov.

Head, thorax and primaries creamy white, latter with only faint traces of black dots, consisting of one in the median fold near base of wing, an oblique postmedian row of three parallel to outer margin, the upper just above the origin of vein 5, the second in submedian fold below vein 2, and the third on inner margin ; occasionally a trace of a second dot on inner margin about two-fifths from base. Primaries smoky brown, with pale fringes, darker in ♀ than in ♂. Beneath, primaries deep smoky, with ochreous costal margin and pale fringes ; secondaries whitish, at times slightly smoky, immaculate. Expanse, 23 mm.

Habitat : Ft. Wingate, N. M.; Deming, N. M., 2 ♂s, 3 ♀. Types, collection Barnes.

The species is close to *dis* Grt., which it resembles in the almost obsolete maculation ; it is, however, slightly smaller, the primaries are creamy white and not chalky white as in *dis*, the secondaries are paler brown on upper side, and lack the faint median band and discal dot on under side, which is present in seven specimens of the true *dis* examined by us. It is probably confused with this species in collections, but we have recently had specimens compared with the type of *dis* in the Snow collection by Mr. F. X. Williams, and he agrees with us that the two species are distinct. Hampson's figure of *dis* probably represents that species.

Grotella soror, sp. nov.

Head and thorax white, abdomen ochreous brown ; primaries very pale ochreous, white along inner margin ; two brown spots on costa in basal third, the outer one larger and oblique, forming the commencement of a broken antemedial line, the continuation of which is formed by a perpendicular brown dash between cubital and anal veins, and a dot on inner margin about two-fifths from base of wing; a diffuse brown spot just beyond the middle of costa and a faint dot at end of cell; a subterminal line of brown dots, very evenly outcurved to submedian fold, terminating in a straight dash, perpendicular to inner margin but not quite attaining same; a large apical brown blotch, between which and subterminal line are two small brown dashes, placed vertically, the upper resting on costa; two terminal black patches at extremity of veins 2 and 3; fringes broadly checkered with dark brown; outer margin pale. Secondaries whitish, largely suffused with smoky, and with a broad dark brown marginal band, narrowing towards anal angle; a faint discal dot and pale fringes. Beneath, primaries smoky brown, fringes checkered; secondaries as above.

Habitat: Redington, Ariz., 1 ♀. Type, collection Barnes.

Closely related to *binda* Barnes; differs in the more even subterminal line, the presence of an apical brown patch, and patches at termination of veins 2 and 3, and the fact that the brown checkering of the fringes does *not* extend backward on the terminal area of the wing itself as in *binda*; the secondaries are darker, with more distinct marginal band.

Eriopyga dubia, sp. nov.

Palpi with the third joint longer and less porrect than is generally the case in this genus; head and thorax gray; primaries dark gray, very suffused and slightly shiny in appearance, and with all the maculation indistinct; t. a. line black, single, inclined outwardly, lunulate, preceded by a pale shade; basal area before t. a. line largely suffused with blackish shading; t. p. line excurved around cell, then parallel to outer margin, single, black, crenulate, mostly very indistinct; orbicular a pale, oval, indistinct mark, situated near t. a. line; reniform a black shade more or less hidden by the dark median shade, which is angled slightly below same; s. t. line not recognizable; a terminal black line broken by yellow dots at termination of veins; fringes concolorous. Secondaries smoky; fringes with an ochreous basal line, followed by dark line, beyond which the fringes are lighter. Beneath smoky, with an indistinct postmedian line and discal dot to both wings. Expanse, 20–25 mm.

Habitat: Redington, Palmerlee, Ariz., 10 ♂s, 1 ♀. Types, collection Barnes.

The antennæ in both sexes are ciliate, and the species is quite delicate, more like a *Cerma* species in general appearance; the hairy eyes would preclude this association, however. Considerable variation in size exists in the specimens before us.

Eriopyga antennata, sp. nov.

Antennæ of ♂ very strongly bipectinate; of ♀ slightly ciliate; head and thorax clothed with a mixture of gray and red-brown hair and scales; primaries deep brown or purple-brown, the distinctness of the maculation variable; basal line slightly marked on costa; t. a. line geminate, black, filled with a pale shade of the ground colour, inner line indistinct, slightly outwardly oblique, with an outcurve in submedian fold and another below vein 1; orbicular when present small, round, outlined in dark, filled with ground colour; reniform indistinctly outlined, outer portion filled with yellow above, which is a single white dot and below it two, remainder filled with ground colour; no trace of claviform; t. p. line indistinct, geminate, the outer line being reduced to a series of venular dots, well exserted around cell and slightly incurved in submedian fold; median shade very faint; subterminal space slightly paler than rest of wings; s. t. line pale yellow, rather broken, defined by a dark preceding shade, angled outwardly below apex of wing, incurved slightly opposite cell and in submedian fold; a very faint black broken terminal line and an ochreous line at base of the dark fringes. Secondaries whitish, hyaline, strongly suffused with smoky in all but the basal portion; with small discal dot. Beneath, primaries smoky, sprinkled outwardly with ochreous, with small discal dot and traces of a postmedian line on costa; secondaries whitish, sprinkled along costa and outer margin with ochreous, a small discal dot and broken postmedian line; a faint broken dark terminal line to both wings. Expanse, ♂ 25 mm.; ♀ 28 mm.

Habitat: Redington, Ariz., 4 ♂s, 3 ♀. Types, collection Barnes.

A variable species. The markings of the reniform tend to obsolescence, and only to well marked specimens is the above description applicable; sometimes the white dots are absent, in other specimens the yellow patch is greatly reduced as well, but a careful examination will usually show sufficient of the typical maculation to avoid confusion with other species. The male antennæ are more strongly pectinate than in other species we have seen.

Eriopyga gigantoides, sp. nov.

♀.—Palpi outwardly black-brown, a few ochreous hairs at tip of second joint; front with a strong tuft of dark-brown hairs sprinkled with ochreous ones; tegulæ and thorax rather lighter brown; primaries purplish-brown, rather shiny, the basal portion of wing to t. a. line sometimes shaded considerably with blackish; in such specimens the basal line is not visible; in lighter forms it may be distinguished as a geminate black mark on costa; t. a. line black, geminate, the outer line most distinct, filled with the ground colour, slightly outwardly oblique and rather evenly crenulate; orbicular and reniform obsolete; t. p. line black, geminate, crenulate, the inner line only distinct, filled with rather paler shade than the ground colour, strongly outcurved just below costa, then parallel to outer margin, forming an outward angle on anal vein; a strong black median shade, the most prominent feature of the maculation, extends obliquely outwards from costa to below position of reniform and close to t. p. line, where it is sharply angled; its course is then sinuate to middle of inner margin; s. t. line indistinct, at times almost wanting, pale, angled below apex of wing, then rather evenly sinuate and close to outer margin; a terminal series of black dots; fringes dusky, with ochreous basal line. Secondaries smoky, with incomplete dark terminal line; fringes somewhat lighter, with ochreous basal line. Beneath, primaries smoky, shaded with ochreous along costa and outer margin, with a rather rigid dark postmedian line, slightly curved at costa, a slight discal mark, and prominent terminal row of dark dots; secondaries shiny whitish, sprinkled in costal half with ochreous and black; a dark discal spot, crenulate postmedial line, indistinct towards inner margin, and terminal row of dots. Expanse, 32 mm.

Habitat: White Mts., Ariz., 3 ♀s. Type, collection Barnes.

Allied to *gigas* Sm., of which we possess co-types: differs in the much smoother and darker appearance, the narrower wings, and the lack of orbicular and reniform.

(To be continued.)

THE ENTOMOLOGICAL SOCIETY OF ONTARIO.

The forty-eighth annual meeting of the Entomological Society of Ontario was held at the Ontario Agricultural College, Guelph, on Thursday and Friday, Nov. 23rd and 24th. During the day meetings the chair was occupied by the president, Dr. E. M. Walker, and during the evening meeting by President Creelman, of the college.

Among those present were Messrs. H. H. Lyman and A. F. Winn, Montreal; Dr. C. Gordon Hewitt and Mr. Arthur Gibson, Ottawa; Prof. J. M. Swaine, Macdonald College, St. Anne's, P.Q.; Mr. J. D. Evans, Trenton; Dr. E. M. Walker and Mr. J. B. Williams, Toronto; President Creelman, Professors Bethune, Zavitz, Jarvis, Hutt, Howitt, Messrs. Pettit, Cæsar, McCubbin and Baker, of the staff, and a number of students of the Ontario Agricultural College and Macdonald Institute.

On Thursday morning a meeting of the Council was held, at which the report of the proceedings of the Society during the past year was drawn up and various matters of interest to its members were discussed In acceptance of an invitation from Dr. Hewitt, it was decided to hold the next annual meeting at the Central Experimental Farm, Ottawa, the exact date to be decided upon later. Prof. J. H. Comstock of Cornell University, and Dr. E. P. Felt, State Entomologist of New York, were York, were elected Honorary Members of the Society. Mr. E. Baynes Reed, Meteorological Station, Victoria, B. C., was elected a Life Member.

In the afternoon the proceedings commenced with the reading of the reports of the following directors on the insects of the year in their respective districts: Mr. A. Gibson, Ottawa; Mr. C. E. Grant, Orillia; Mr. A. Cosens, Toronto, and Mr. R. C. Treherne, Grimsby. Dr. Hewitt then gave an account of the work of the Division of Entomology, which showed that, with a much increased and most efficient staff, gratifying progress was being made along many lines of entomological work, particularly in the establishment of field stations in various parts of the country, in the campaign against the Brown-tail Moth and in the study of the parasites of the Larch Saw-fly and Spruce Bud-worm. Mr. Cæsar then read an extended and valuable paper on the insects of the year in Ontario, which was discussed at considerable length by many of those present. This was followed by a paper by Dr. Fyles, "Notes on the Season of 1911," after which the reports of the Montreal and Toronto branches and of the Treasurer, Curator and Librarian of the Society were read and adopted.

In the evening a public meeting was held in the Massey Hall Auditorium, which, considering the inclemency of the weather, was fairly well attended by students of the college and visitors from the town, as well as by members of the Society.

President Creelman, who occupied the chair, opened the meeting with a short address of welcome in his usual cordial manner. Dr. William Riley, of Cornell University, who was to have been the speaker of the evening, was unfortunately prevented by illness from being present, but his place was ably filled by Dr. Hewitt, whose address on "Insect Scourges of Mankind" was listened to with great interest and attention by those present. He gave a very thorough account of various diseases, the germs of which are carried from one patient to another through the agency of insects, dwelling especially upon the Sleeping Sickness and other tropical diseases caused by trypanosomes and transmitted by Tse-tse flies, and on malaria and yellow fever, which are transmitted only by particular species of mosquitoes. The address was illustrated by many excellent lantern slides.

On the following morning the members spent a pleasant hour in the Biological Museum, where many interesting specimens were exhibited by those present. At 10.30 o'clock the proceedings were resumed in the Biological Lecture Hall, the president, Dr. E. M. Walker, opening the meeting with the reading of the presidential address, which dealt with the entomological field in Canada at the present time and the directions along which progress in this science may be expected in in the near future.

In the afternoon the following papers were read: "Some Forest Insects from De Grassi Point, Lake Simcoe," by Dr. Walker; "Thrips Affecting Cereals," by Dr. Hewitt; "The Stream," by Dr. Fyles; "Blister Beetles," by Mr. Arthur Gibson; "A Parasite of Hepialus Thule," by Mr. A. F. Winn; "Common Ipidæ of Eastern Canada," by Prof. J. M. Swaine; "Insect Migrations in Manitoba," by Mr. Norman Criddle; "The Catalogue of Canadian Insects," by Dr. Hewitt; "Entomological Record for 1911," by Mr. Gibson; and "Notes on Hepialus hyperboreus," by Mr. Horace Dawson.

The election of officers for the ensuing year resulted as follows:

President—Dr. Edmund M. Walker, Lecturer in Zoology, University of Toronto.

Vice-President—Dr. C. Gordon Hewitt, Dominion Entomologist, Central Experimental Farm, Ottawa.

Secretary-Treasurer—Mr. A. W. Baker, B. S. A., Demonstrator in Entomology, O. A. College, Guelph.

Curator—Mr. Lawson Cæsar, B. A., B. S. A., Lecturer in Entomology and Plant Diseases, O. A. College.

Librarian—Rev. C. J. S. Bethune, M. A., D. C. L., F. R. S. C., Professor of Entomology and Zoology, O. A. College.

Directors—Division No. 1, Mr. Arthur Gibson, Div. of Entomology, Central Experimental Farm, Ottawa; Division No. 2, Mr. C. E. Grant, Orillia; Division No. 3, Mr. A. Cosens, Parkdale Collegiate Institute, Toronto; Division No. 4, Mr. C. W. Nash, East Toronto; Division No. 5, Mr. R. S. Duncan, Port Hope; Division No. 6, Mr. R. S. Hamilton, Collegiate Institute, Galt; Division No. 7, Mr. W. A. Ross, Jordan Harbour.

Delegate to the Royal Society—Prof. J. M. Swaine, Macdonald College, P. Q.

Auditors—Prof J. E. Howitt and Mr. W. A. McCubbin, Ontario Agricultural College.

NOTES ON THE LIFEHISTORY OF *NEPTICULA SLINGERLANDELLA* KEARFOTT (TINEIDÆ).

BY C. R. CROSBY, ITHACA, N. Y.

The following notes on the life history of the plum leaf-miner are compiled from the notes of the late Professor M. Slingerland, supplemented by observations by the writer :—

The plum leaf-miner is a new fruit pest which was brought to Professor Slingerland's attention in the fall of 1907 by C. M. Hooker & Sons, of Rochester, N. Y., who stated that it had been present in their plum and prune orchards for a number of years and had been gradually increasing in numbers. The mines were so abundant that the trees were partially defoliated and the size and quality of the crop injured.

We have not been able to find this miner in other orchards, and with the possible exception of apple no other food plant is known. That it may occasionally attack apple is quite probable. While examining some old apple trees in a neglected orchard, about a quarter of a mile from the Hooker orchard, on July 7, 1911, the writer found that mines very closely resembling those of the plum leaf-miner were abundant in the leaves of the water sprouts growing at the base of several trees. Infested leaves were brought to the insectary, but the larvæ left the mines while in transit and constructed cocoons indistinguishable from those of the plum leaf-miner. The identity of this apple leaf-miner cannot be settled definitely until the moths are reared next spring.

In the Hooker orchard the plum leaf-miner has shown a decided preference for certain varieties. German and Italian prunes are most severely infested ; French and Shropshire Damsons are less subject to attack, although some years ago the former variety was badly infested ; Diamond, Bradshaw, Lombard and Rheinclaude are nearly immune.

LIFE HISTORY.

The moth.—The adult of the plum leaf-miner is a small bronzy black moth having an expanse of 1/7 to 1/5 inch. The fore wings are crossed by a shining white band on the outer third and the head bears a conspicuous orange tuft. These moths emerge from cocoons at or near the surface of the ground during the daytime in the latter part of May and in early June. During the day they remain quietly on the bark of the trunk and larger branches, none being found on the leaves. Several hundred moths are often found on a single tree ; when disturbed they suddenly take flight and most of them settle on the opposite side of the tree. They gradually decrease in numbers and about the middle of June disappear.

The egg.—The act of egg-laying has not been observed, but probably takes place in the evening or at night as the moths are rarely seen on the leaves during the day. The eggs are attached to the under surface of the leaf, usually at the forks of the more prominent veins. The egg is about .3 mm. long by .2 mm. wide, oval in outline, flattened where attached to the leaf and dome-shaped in profile. The green of the leaf shows through the transparent egg-shell, making it a difficult object to find. They are most easily located by holding a leaf at an angle in the sun so the light will strike it obliquely when the eggs will be seen as minute glistening dots. The exact time required for the hatching of the egg has not been determined, but it cannot be far from two weeks. On June 2, 1908, an examination of the orchard showed that a great number of eggs had been laid; on June 9 no eggs had hatched, and on June 18 hatching had just nicely begun.

The larva.—In hatching the larva eats its way out of the egg-shell on the under side next to the leaf and enters the leaf directly without coming out on the surface. Wnen full grown the larva is about 1/6 inch in length, greenish white in colour, with the head light brown; the contents of the alimentary canal show through the semitransparent body wall as a greenish or brownish stripe. The larva is legless and only slightly flattened; the constrictions between the segments are rather deep but obtuse; the surface of the body is smooth and clothed with dense, very short, microscopic hairs interspersed with a few larger ones.

The mine.—After entering the leaf directly from the under side of the egg the young larva eats out a narrow linear burrow or mine an inch or less in length, leaving the outer layers of the leaf intact. This portion of the mine usually follows a tortuous course but may be nearly straight. The larva next enlarges its mine into an irregular ovate blotch about one-half inch in length. In the linear portion of the mine the excrement is left as a blackish streak extending along the centre of the burrow; in the blotch mine it forms a broad irregular band along the centre, but does not extend to the tip. The outer leaf layers overlying the mines turn brownish or yellowish; the upper layer seems to be thinner than the lower and the mines are more conspicuous when viewed from above. There are ten or a dozen mines in a single leaf.

The cocoon.—When full grown the larva leaves the mine through a cut in the upper surface of the leaf, falls to the ground and there constructs a small flattened brownish cocoon in cracks in the soil, under loose stones, or between the base of the tree and the surrounding soil. When the ground is undisturbed they are rarely found more than an inch below the surface. Sod furnishes ideal winter quarters for the cocoons. The cocoon is light brownish in colour, broadly oval and moderately arched; it is about 2½ mm. long by 1½ to 2 mm. wide, and is usually slightly wider at one end. It is surrounded by a thin flange formed by the closely united edges of the two valves of which the cocoon is composed. The cocoons are held in place by a few strands of silk. The time at which the larvæ become mature and construct their cocoons varies considerably with the season. On July 6, 1911, about one-half the larvæ had left the mines; on July 21, 1908, and July 19, 1909, a few larvæ were still present in the mines.

The pupa.—After forming the cocoon the larva apparently does not transform at once; a cocoon opened August 4, 1908, contained a larva. The winter, however, is passed in the pupal stage. On October 10, 1911, the writer opened a number of freshly gathered cocoons and found that all the larvæ had transformed to pupæ.

The pupa is about 2 mm. in length, ovate, pointed behind and somewhat flattened. The ventral surface is brownish yellow, the dorsum greenish. The eyes are dark coloured and the orange tuft on the head of the moth shows through the pupal skin. On the dorsum of the abdomen there are six transverse interrupted rows of short brownish spines. On each side of the dorsum there is a longitudinal row of wart-like protuberances, each bearing a colourless spine. The anterior spines are very short and they gradually increase in length towards the tip of the body. When about to transform to the adult the pupa works itself partly out of the end of the cocoon, probably by the aid of these spines. The empty pupa skin is left protruding from the cocoon.

Parasites.—No parasites were observed infesting the plum leaf-miner until May 11, 1911, when a cocoon was found containing the larva of a Chalcis-fly. The larva is 1.4 mm. long, smooth, whitish in colour and rounded at both ends. On June 2, 1911, two adults of the parasite were found in a vial containing cocoons of the moth. They had emerged through a smooth, round hole in the side of the cocoon. In the fall of 1911 the parasites had increased in numbers so that nearly one-half of the cocoons examined were infested. So far only three adults, all males, have been reared. They are small, four-winged flies, metallic green in colour and about 1½ mm. in length. This species has recently been described as *Derostenus salutaris* Crosby.

GEOMETRIDÆ AS YET UNDESCRIBED.

BY RICHARD F. PEARSALL, BROOKLYN, N. Y.

(Continued from page 253, Vol. XLIII.)

Eupithecia vaporata, n. sp.

Expanse, 13–15 mm. Palpi moderate, rather heavily scaled, dark brown. Antennæ slender, gray, barred faintly with dark brown, shortly ciliate beneath. All above gray, mixed with dusky brown, scaled, the front and first two segments of the abdomen being paler, the latter without tufts. Fore wings narrow, extended at apex, are crossed by a number of ill-defined hair lines composed of the darker scales. The intra- and extra-discal lines appear double and heavier, the former obliterating the discal dot, which it reaches at a sharp angle from costa, and thence with a strongly basal trend to inner margin; the latter, with less of an angle below costa, becomes slightly heavier opposite cell, and wavy to inner margin, nearly parallel to intradiscal. The basal line is obsolete. Beyond extradiscal the usual geminate pale line is present, not clear, but faintly margined outside with a fine hair line of dark scales. These cloud the subterminal space, which is without definite markings. Marginal line broad, black, broken at veins. Fringes rather long, gray. Hind wings with lines as on primaries, but fainter, except at inner margin, the extra-discal being heaviest, with strong outward curve around discal dot, which is a mere pin point. Marginal line and fringes as on primaries. Beneath the body and wings are somewhat paler and more glossy, the lines on wings reproduced as above, and all except the marginal line are fainter.

Types, ♂ and ♀, from San Diego, California (Ricksecker), were taken at light, May 16, 1910, and co-types in a series of fourteen in both sexes, from the same locality, are in author's collection. Also a co-type ♀ from collection of Geo. H. Field, San Diego, Cal., taken by him April 22, 1910, is in possession of Mr. J. A. Grossbeck.

This species is smaller even than *huachuca* Gros., less distinct in markings, and paler in ground colour.

Eupithecia scabrogata, n. sp.

Expanse, 22 mm. Of the same size, and much resembling *subapicata* Guen. Palpi short, stout, loosely scaled, dark brown. Front and vertex rough, with a mixture of dark and pale brown scales. Antennæ dark brown, flattened, slender, ciliate in ♂, almost bare in ♀. Thorax black centrally, crossed from base to base of wings by a broad dusky white band, with front and sides dark brown. Abdomen above dark brown, the segments pale ventrally, except the second, which has an irregular black patch above. Beneath generally paler. Ground colour of primaries pale yellowish brown, covered with dark brown and black scales, more thickly massed along costal region, at apex, subterminally at anal angle, and broadly along inner margin below cell to base. This leaves a central patch of yellowish brown, clear of dark scales, from a point at base, broadening out over cell, especially clear about the small, round, black discal dot, thence in an irregular patch, narrowing rapidly, in an upward slant to margin a little below apex. The basal, median and extradiscal are pale geminate cross lines, indicated chiefly at costal and inner margins, the latter a little better defined. The subterminal white line, indicated very faintly, ending in an irregular whitish patch, between veins 2 and 3 at anal angle. Marginal line black, cut with white opposite veins. Fringes long, dusky, having a central dark line, with pale line at base. Veins black-scaled, broken at cross lines, especially beneath cell, and on veins 1 to 4. Secondaries dusky. Inner margin broadly sprinkled with dark scales, cut by the beginnings of pale cross lines, which fade out at centre, the extradiscal pale band being wider and more distinctly outlined. Discal dot very small, dusky. Fringes as on primaries. Beneath dusky, silken, the lines above faintly indicated in dusky dots across wings. Discal dots present on all wings, small, dusky. Thorax beneath and femora darker.

The type, a ♀ from the Hy. Edwards collection, is labeled California, without date, and will rest eventually in the Amer. Muesum of Nat. History, N. Y. City, or so soon as the author can obtain a duplicate. The specimen was inadvertently given to him several years ago as *subapicata*. The only other specimen I have seen is a male, submitted to me by Mr. J. A. Grossbeck, from Dr. Barnes' collection, which I have made a co-type. It is labeled Redington, Arizona, also without date, but taken in 1910, and differs only in being rather more strongly and clearly marked than the type. *Subapicata* is taken in December and January, and I doubt not this species appears also about that time.

Genus *Eucymatoge* Hub.

Eucymatoge penumbrata, n. sp.

Expanse, 22 mm. Of same size and shape as *Eup. scabrogata* just described, and might easily be confused with it. Palpi long, moderately stout, dark brown, almost black. Front covered with an even mixture of dark brown and paler scales, the latter showing more abundantly over vertex, on collar and patagiæ. Antennæ sordid white, ringed with dark brown, simple in both sexes. Above, the ground colour of all wings is a soiled brownish white, overlaid with dark rich brown, mingled with black scales. Thorax above black centrally, has a conspicuous line of pure scales crossing it between wing bases, with front and scutellar region dark brown. Abdomen paler brown, broadly ringed with darker on second segment, without dorsal tufts. The dark brown basal and extradiscal lines cross the primaries, but many hair lines along inner margin in basal and central spaces quickly fade out. The basal composed of three parallel hair lines, the two outer being heavier and black, includes a space, paler than the ground colour, traversed centrally by a fine hair line of brown. It starts from costa one-third out, makes a long sharp angle almost to discal spot, thence backward nearly straight, to inner margin, one-fourth from base. In the male co-type this line is suffused with black scales. The extradiscal two-thirds out runs straight across costa, makes a sharp outward angle below it, and thence nearly parallel to outer margin, slightly waved as it reaches the inner, a little more than two-thirds out. A narrow indistinct geminate pale line borders this outwardly. A pale area occupies the central portion of the wing, with an extension across extradiscal below costa toward apex, and an isolated spot between veins 3 and 4 on subterminal space. The large oblique linear discal spots are rich brown, conspicuous, and are surrounded by a ruddy brown suffusion, which appears again on the apical prolongation, and more faintly on the spot on subterminal space, particularly in the female type. Subterminal space darker, especially opposite cell, where a cluster of black scales starts within extradiscal, extends across it to margin, and upward toward apex. It is traversed centrally by an indistinct volute white line. Terminal line not well marked. Fringes paler than ground colour. Secondaries darkened with brown scales along outer and inner margins, are almost devoid of them centrally and along costa. The geminate pale line and the basal lines are outlined at inner margin, the former traceable nearly to costa. Discal dots round, dark brown. Terminal line black, broken at veins. Fringes pale, cut with black opposite veins. Beneath, all wings pale brownish ashen, darkened apically, the lines as above faintly showing on primaries, and on secondaries, dotted on veins. Discal dots on primaries are large, oblique, linear, black, on secondaries round, black.

Marginal lines black, present on all wings. Fringes paler than above Thorax, abdomen and legs ashen, with dark scales sprinkled heavily on femora and fore legs, and on abdomen toward tip.

The type, a female, has long been a unique in the collection of Mr. W. H. Broadwell, who has kindly allowed me to retain it, and bears the label, Palmerlee, Arizona. The single male co-type is from the collection of Dr. Barnes, submitted to me by Mr. J. A. Grossbeck, and was taken at Redington, Arizona, January 1, 1910.

(To be continued.)

ON *MERRAGATA LACUNIFERA* BERG.

BY J. R. DE LA TORRE BUENO, WHITE PLAINS, NEW YORK.

In 1879, in his "*Hemiptera Argentina*,"[1] Carlos Berg described a new Lygæid genus, *Lipogomphus*, placing it near *geocoris*, which contained a new species, *lacuniferus*,[2] so called because of the white corial *lacuna* bounded by the thickened brown veins. This was founded on three specimens of an unknown bug taken in Buenos Aires by himself, in company with the Argentine Entomologist, Enrique Lynch. Subsequent study caused him in 1884[3] to place his new genus near *Hebrus* Curtis (now sometimes *Nœogeus* Laporte). No further reference appears to have been made to this species, except its enumeration by Lethierry and Sévérin,[4] till 1898, when Champion[5] referred the genus to *Merragata* Buchanan White, whose type, *Merragata hebroides* F. B. White, is from Mexico.

To my good friend, Rev. Longinos Navas, the learned Spanish Jesuit, I owe the possession of four examples of this very interesting form, which were secured at Montevideo, Uruguay—a second locality for the species. These specimens agree very well with the original description, although, being carded and slightly mutilated, the discrepancy in the number of joints in the hind tarsi, which Champion points out, could not be determined without further mutilating them. The number and proportion of the rostal joints in the individuals before me does not agree with the

1. P. 286.
2. P. 287.
3. Add. et Em. Hem. Arg., pp. 116-117.
4. 1896, Catalogue Général des Hémiptères, III, 52.
5. Biologia Centrali Americana Het. II, 198 (Aug., 1898).

description, but as in other more easily demonstrable particulars subsequently pointed out, especially in the relative lengths of the antennal segments, there is no material divergence, this is perhaps attributable to an error of observation on the part of the describer. It differs from the Central American species noted by Champion in having the 2nd, 3rd and 4th antennal joints subequal, the first joint being the shortest and stoutest and somewhat curved, and in the bifid scutellar apex, in which last character it resembles the figures of *Hebrus major* Champ. and *H. hirsutus* Champ.[6] In fact, in regard to the latter species, it would not surprise me at all to find it eventually transferred to *Merragata*, especially since in the unique type the antennæ were broken, but were *assumed* to be five-jointed, a somewhat risky proceeding in view of the fact that the generic difference lies in this character.

Nothing appears to be known as to the habits of the genus. I secured *M. hebroides* B. White in a ditch draining into the Canal de la Viga in Mexico City, in April of 1910, but made no further note than that it was taken by dredging in grasses growing into the water at the edge.

The recognized species of *Merragata* may be separated by the following key, based on Champion's, in the "Biologia Centrali Americana."

1. (2) Scutellum bifid at apex ; antennal joints 2 to 4 subequal, joint 1 shortest, stoutest .... 1. *Lacunifera* Berg.
2. (1) Scutellum blunt, *not* bifid at apex.
3. (2) Antennal joints 1 to 3 subequal, 4 rather stout and fusiform........................... 2. *hebroides* F. B. White.
4. (3) Antennal joint 3 slender and very much longer than 2 ; 4 slender and subfusiform.
5. (6) Pronotum deeply constricted at the sides. 3 .... *Leucosticta* Champion.
6. (5) Pronotum moderately constricted .... 4. *Brevis* Champion.

Næogeus (or *Hebrus*) and *Merragata* look extremely like *Microvelia* in the Gerridæ, but the *apical* tarsal claws at once serve to distinguish them from the last named, in which they are subapical and set in a cleft in the tarsus. The two genera of *Næogeidæ* (= Hebridæ) are thus distinguished :

Antennæ 4-jointed............................ *Merragata* F. B. White.
Antennæ 5-jointed................ *Næogeus* Laporte (= *Hebrus* Curtis).

6. 1898, Biol. Cent. Am., Het. II, pl. VIII, figs. 1 and 2.

Mailed January 10th, 1912.

The Canadian Entomologist.

Vol. XLIV. LONDON, FEBRUARY, 1912. No. 2

FURTHER NOTES ON ALBERTA LEPIDOPTERA.

BY F. H. WOLLEY DOD, MILLARVILLE, ALTA.

(Continued from Vol. XLIII, page 399.)

278. *Ufeus plicatus* Grt.—I have not seen the type of this species, but what I have as such is the *plicatus* of the British Museum and most other collections that I have seen. It differs principally from what I hold as *satyricus*, probably also correctly, in being redder, having larger wings, with more acute apices, the transverse lines narrower and less diffuse, the t. a. deeply dentate rather than curved. There are other distinctive characters, but these seem the most reliable and most obvious. Of *barometricus* Goosens, I know nothing beyond the mere reference given in Dyar's Catalogue. *Hulstii* Smith was described in Ann. N. Y. Acad. Sci., XVIII, p. 99, Jan., 1908, from two males from Stockton, Utah, and Black Hills, Wyoming. The type is the Stockton specimen, whence I have a pair, the female of which I have compared with it. The description states that it is "perhaps nearest to *satyricus* in type of maculation, but differs obviously in colour, in the absence of all trace of ordinary spots, and in the immaculate under side." In my Stockton male the discoidal spots are practically obsolete, in the female they are very distinctly marked. The under sides are very pale, but not quite immaculate. These obviously merge into my series of Calgary *plicatus*, and if mixed they would be inseparable without the labels. I would suggest that Prof. Smith's comparing *hulstii* to *satyricus* was a slip. The species is a decided rarity here.

279. *U. satyricus* Grt.—I have never found this in any numbers, though it is much less rare than the preceding. It is extremely variable in the quantity and distribution of black and dark brown scales.

280. *Agrotiphila incognita* Smith.—This species is not in my collection, but I have seen the two male types from Laggan, though one of them is only labelled "B. C." in error. A male is in the British Museum, taken by Mrs. Nicholl in 1907, on Brobokton Creek, in the mountains far north of Laggan.

281. *A. maculata* Smith.—Though less rare than the preceding, this seems never to have been taken in any numbers. I have taken a few myself at Laggan, and Mrs. Nicholl has taken a few there and on Mt. Athabasca, and near Lake O'Hara, on the British Columbian side of the divide. My dates are all between July 16th and 27th. It is an above-timber species, occurring between 7,000 feet and the summits, though I have not been on any above 9,000 feet.

283. *Mamestra mystica* Smith.—In my note on this species in Vol. XXXVII, p. 151, line 5, for "The palest *discalis* and the darkest *mystica*," read, "the darkest *discalis* and the palest *mystica*." I overlooked the slip in the proofs. As to the distinctness of these two, there can be no doubt. In colour *discalis* is pale blue-gray, *mystica* lacks the bluish tint and is browner. They are also distinguishable on the characters previously pointed out. In colour and ornamentation *mystica* is really nearer *nimbosa*, and occasional specimens are indistinguishable. I had almost decided that they were forms of one species, when I discovered slight antennal differences, which may, however, prove to intergrade, though I have not both forms from the same locality. In my males of *nimbosa* from Montreal; Milwaukee Co, Wis.; and Vancouver Island, the antennæ are ciliate and bristled, with the joints scarcely marked. Some Pacific coast specimens have the ground colour very clean, with the brown irrorations very much reduced, though so far I have found nothing else about them to suggest distinctness of species. In my *mystica*, from Miniota, Man.; Alberta; and Windermere, B. C., the male antennæ are minutely serrate, fasciculate and bristled, the bristle appearing to be longest in Miniota, and shortest in Windermere specimens. In some of those from Miniota, however, the joints are scarcely marked, and the character may fail as distinctive. Sir George Hampson places *mystica* and *nimbosa*, with *rogenhoferi*, in a different group from *discalis* and *imbrifera* on antennal characters, as having them ciliate only. He has *mystica* from the type locality, Winnipeg, and I have not, though I have seen the type, and know the species well. *Discalis* has male antennæ serrate-fasciculate, but the serrations are not more prominent than in most of my *mystica*, and the bristle seems to be lacking. The type of *nimbosa* is a male in the British Museum from Trenton Falls, New York.

284. *M. imbrifera* Grt.—I have seen the type of this species in the British Museum, a female, which, according to the Catalogue, comes from

Trenton Falls, N. Y., the same locality as *nimbosa*. The male antennæ are minutely serrate-fasciculate, and strongly bristled. The serrations are less coarse though more distinct than in *nimbosa*, *mystica* or *discalis*, and the bristle longer than in the two former, *discalis* having none. I have it from Montreal; Biddeford, Maine; several Manitoba localities, and Red Deer River, near Gleichen, Alta. It seems much less common here in the hills, though I took a couple in the Upper Columbia valley, near Windermere, B. C. *Rogenhoferi* Möschler, as catalogued by Sir George Hampson, has male antennæ ciliate only. He had no specimens under the name in the collection, but figures as such a male from "W. Manitoba" in Prof. Smith's collection. This specimen I have examined. The label is, I think, in Mr. Hanham's writing, and the "W" probably stands for Winnipeg. When I saw it, it had a small piece only of one antenna, which my notes say were "ciliate, with joints little marked." It is something distinct from *imbrifera*, which I have often received under the name, and a species unfamiliar to me. Prof. Smith, in his Monograph of Mamestra, states that he has examined the male type from Labrador in Mr. Möschler's collection, and that "the antennal joints are distinctly serrated, and furnished with bristly tufts." (Pr. U. S. N. M., XIV, 204, 1891.) This leaves some doubt as to the correctness of the identity of the Rutger's college specimen.

284. The single specimen which I recorded under this number as *juncimacula* is probably a variation of *purpurissata* Grt. It is, however, extremely like Holland's figure of *juncimacula*, stated in the text to occur in Colorado, which is therefore presumably the locality of the specimen figured. Sir George Hampson's figure of a Colorado female is much more like the form described by Dr. Dyar from Kaslo as var. *crydina* (CAN. ENT., XXXVI, 32, 1904). Hampson lists *crydina* as a synonym of *purpurissata*, but had no Kaslo specimens in the collection. Prof. Smith, in Journ. N. Y. Ent. Soc., XV, 152, 1907, claims that *crydina* is a good species. I thought that might be so at one time, but after studying more material I find that the Kaslo form, as stated under the description, intergrades with eastern *purpurissata*, which is the predominating form at Calgary. The type of *juncimacula* Smith, is a male in the Washington collection, bearing no locality label. Neither, by the way, is the labelled "type" in Prof. Smith's handwriting, though bearing the Museum red type label. I have not seen the description, but the form appears to have

been described from the mountains of Colorado as a *purpurissata*, and subsequently, in Ent. News, IX, 241, Dec., 1898, separated as a species. My only Colorado specimen is a female from Durango, and looks like an obscure *purpurissata* merely. Vancouver Island specimens are paler and more distinctly maculate than any others that I have. I believe that *crydina* is merely a strongly marked form of *purpurissata*, and *juncimacula* is very doubtfully distinct.

287. *M. columbia* Smith.—I have seen two specimens of this form marked "type," both males labelled "Ft. Calgary, B. C.," one in the Neumœgen collection, and the other at Washington. The description refers to both male and female types, which may be an error. In 1884, when Capt. Geddes collected the specimens, Calgary was merely a Northwest Mounted Police fort. The "B. C." error I have repeatedly corrected. Closer acquaintance has brought me to look upon this as a local race of *meditata* Grt. The majority of Calgary specimens are considerably paler than *meditata* from the Eastern States, and tinged with reddish rather than brown. Specimens from Cartwright, Man., and Redvers, Sask., include obvious intergrades, as well as specimens inseparable from some in both eastern and Calgary series, except in being smaller, as is usual with Manitoba and Saskatchewan races. *Determinata* Smith is a Colorado form very closely allied to these, with darker central band, and rather conspicuous discoidal spots, those in *meditata* and *columbia* being usually rather obscure, and sometimes scarcely discernible. Sir George Hampson separates *determinata* from the other two in the tables on the character of the orbicular being concave anteriorly. This is a variable character in my *columbia* series, in which I do not suspect two species. I have only a single Colorado male in my collection, from Colorado Springs, and a few of my local specimens come very near it. Prof. Smith has a good series from California.

288. *M. cervina* Smith.—I do not feel at all confident that this is distinct from *lustralis*, of which the type is a Wisconsin female in the British Museum. The eastern form does not appear to be very common, and I have not the material to enable me to form a definite opinion. The character by which Hampson separates *lustralis* from *cervina* in the table is the presence in the former of a black mark preceding the white patch near the anal angle in submedian fold. In his description, however, the mark is called brown. A brown mark is faintly discernible here in some of my local series of *cervina*. It is rather more evident in my one *lustralis*

from an unknown locality, and two from Sudbury, Ontario, which are the only eastern specimens which I possess. However, I see no differences that I should suspect of being specific. Cartwright and Miniota specimens in my collection are alike, and probably more typical *cervina* than those from Alberta, being a little smaller and darker. *Tæniocampa suffusa* Smith, type, is an Arizona female in the Washington Museum, and appeared to me to be a pale *lustralis*, and is referred to that species by its author in his Check List.

289. *M. segregata* Smith.

290. *var. gussata* Smith.—I am convinced that these two are the same species. Dr. Dyar suggests in the Kootenai List that *gussata* is only a variety of *segregata*, and I agree with him. *Gussata* is less highly coloured than *segregata*, and has more black markings. In my former notes I stated that Sir George Hampson considered the two to be the same species. That was his opinion expressed in a letter to me about that time. Before publishing, however, he altered his opinion, as he places *segregata* in the genus *Polia*, and *gussata* in *Hyssia*, separating them in the Catalogue by 136 pages, and figures a Calgary specimen under each name. Prof. Smith, in Journ. N. Y. Ent. Soc., XV, 156–7, Sept., 1907, closely analyzes Hampson's descriptions of *Polia* and *Hyssia* and points ont that there is no tangible difference except a very doubtful one of abdominal tufting. At the same time, he makes no suggestion that *segregata* and *gussata* are the same.

291. *M. negussa* Smith.—A series received from Redvers, Sask., from Mr. Croker, is very constant, which fact, in this genus, I accept as evidence in favour of distinctness from *segregata*, which the form resembles in almost every respect, only entirely lacking all black markings. Hampson places it in *Polia*, and figures a Calgary specimen as *plicata*, of which he makes it a synonym. The figure is bad, and too contrasting He mentions in his description that the discoidal spots are defined by black, which is not the case in any of my Calgary or Redvers specimens. The male type of *plicata* from Glenwood Springs, Colo., has the spots outlined in black, and a black basal streak, agreeing in these respects with my only specimen from that locality, a female. *Negussa* is also slightly smaller, but whether really distinct I will not at present venture to suggest.

292. *M. neoterica* Smith.—I have now seen the types of this species from Winnipeg, and have a similar series in my collection from Cartwright. This form is small and rather dull and even in colour. Walker's type of *detracta* is a male in the British Museum from Trenton Falls, N.Y., where is also Grote's *claviplena* from Evans Centre. These two are certainly one species, and I cannot see that *neoterica* is anything but a local variation of it. Typical *detracta* is larger, more olivaceous, and usually far less even in colour. Calgary specimens are intermediate in average size, but nearest the eastern form in colour. Calgary and eastern specimens can be found exactly alike, but usually the former are paler. Dr. Dyar in the Kootenai List refers *neoterica* as a race of *detracta*, and mentions that the Kaslo form differs slightly from either, being dull and even like *neoterica* and large, like *detracta*. I have some from Provo, Utah, which are most like the Kaslo form, but paler. The relative difference in size between the sexes at Calgary and in the east, does not appear to be constant, the females seeming to average a trifle smaller than the males where the species occurs.

294. *M. meodana* Smith.—(Journ. N.Y. Ent. Soc., XVIII, 95, June, 1910.) This is the name which Prof. Smith has given to what I had listed as *liquida*, and he made a Calgary male and female type, and co-types from Calgary ; Pullman, Wash ; Yellowstone Park, Wyo.; Arrowhead Lake, B.C.; and Denver, Colo. He says : "The species has been confused with *liquida* Grt., which is a much more contrastingly mottled form occurring in Washington, and probably over a similar range. *Liquida*, as described, and as figured by Hampson, has narrower, more pointed primaries, and while the type of maculation is similar, *meodana* is neatly and quietly ornamented, while *liquida* is strongly contrasted and showy." In Ent News., XXI, 398, Nov., 1910, I commented upon the forms, expressing a doubt as to their distinctness as species. I have little to add to that. Vancouver Island specimens in my collection are a bit brighter than typical *meodana*, which I look upon as variation rather than a species.

297. *M. nevadæ* Grt.—One of my Calgary specimens I have compared with the type, a female (not male as stated in the Catalogue), in the British Museum, from the Sierra Nevada, California. Banff and Kaslo specimens are similar. Sir George Hampson makes *canadensis* Smith a synonym, as had previously been suggested by Dr. Dyar in the Kootenai

List. Prof. Smith, in Journ. N. Y. Ent. Soc., XV, 153, Sept., 1907, takes exception to this view, but suggests that they may be races only. He states that *nevadæ* is much brighter, more contrasting, and broader winged than *canadensis*. The latter was described from a unique male from the Province of New Brunswick, and the type, which I have not seen, is probably in the Thaxter collection in the Museum of Comparative Zoology at Cambridge, Mass. Dr. Dyar says that specimens from Wisconsin and from Kaslo, B. C., in the Washington Museum, are alike. I saw them there, and have no note that they differed. They stood, by the way, under *canadensis*, whilst Calgary specimens did duty for *nevadæ*. The *canadensis* of Prof. Smith's collection was a badly worn male from Winnipeg, which I should call about typical *nevadæ*. Last winter I examined a specimen from Hymers, Ontario, belonging to Mr. Winn, which I thought might be typical *canadensis*, as it almost entirely lacked the red shades of *nevadæ*, though doubtfully distinct therefrom. But, according to the description, red shades exist in *canadensis*. At present I have no evidence in favour of distinctness, though it requires more material to permit of a fair judgment.

298. *M. invalida* Smith.—I have not taken this species here for some years, but it seems to be of more frequent occurrence at Banff, whence I have a few. I have no males in my collection, and I notice that an absence of that sex is complained of under the description, made from specimens from Sierra Nevada and Placer Co., Calif. I have seen a type at Rutger's College, another at Washington, and three are in the Henry Edwards collection, though I overlooked these. My Alberta specimens appear to be the same species. I have examined the type of Walker's *cristifera* in the British Museum, a worn specimen from St. Martin's Falls, Albany River, Hudson's Bay Territory. It is the specimen figured by Hampson, but most of the pale shades shown in the figure merely denote the worn condition of the specimen. He makes *lubens* Grt., "ab. 1," and *rufula* Morr., a synonym. Of the latter I know nothing, but *lubens*, of which the female type from New York is in the Museum also, is easily distinct, as pointed out originally by Grote in CAN. ENT., XXVI, 141-146, 1894, and latterly by Prof. Smith, who has suggested that *cristifera* may be prior to his *invalida*. I know nothing against the suggestion, and were it not that the worn condition of *cristifera* type leaves an element of doubt, I should say it was certainly correct.

NEW COLEOPTERA CHIEFLY FROM THE SOUTHWEST.—V.

BY H. C. FALL, PASADENA, CAL.

The new species herein described have, with a single exception, come to hand during the past year (1911) and seem worthy of prompt publication.

Quedius compransor, n. sp.

Robust, head and prothorax black, elytra and abdomen dark rufous, the latter dusky toward the base. Head including the mandibles (♂) slightly longer than wide, gradually wider posteriorly; eyes small, not at all prominent, distant from the nuchal constriction by about 2½ times their longest diameter; a large setigerous puncture at the base of the antennæ, one at the upper margin of the eye, and two others posteriorly in a transverse line and fully twice as far from the eye as from the nuchal constriction; front without punctures. Labrum bilobed. Antennæ rather stout, filiform, but little longer than the head, joints 4–10 subsimilar and a little wider than long. Prothorax about ¼ wider than the head, 1/5 wider than long and evidently wider than the elytra at base, and equal to the width of the latter posteriorly; narrowed in front, sides rounding into the base with but feeble evidence of hind angles; disk entirely without punctures, margin evidently but not strongly explanate posteriorly. Scutellum impunctate. Elytra subequal in length to the prothorax, punctuation fine and rather close throughout. Abdomen similarly but slightly less closely and evenly punctate. Head beneath with a few fine scattered punctures, lateral carina broadly interrupted. Hind tibiæ spinulose.

Length 9–11 mm.; width 2.5–3.2 mm.

Manhattan, Kansas.

Described from three males sent me by Mr. Knaus, who writes that they were taken Jan. 6, from the burrow of a "pocket gopher."

By Horn's table this interesting species would fall with *spelaeus*, to which it is allied by the small eyes and explanate side margins of the thorax, this latter character being however less marked than in *spelaeus*. It differs from *spelaeus* in its stouter form, colour, ovate head (parallel at sides in *spelaeus*), with the infraorbital carina obliterated except toward its extremities, and the absence of the usual discal series of punctures near the front of the pronotum, the marginal punctures only being present. The surface of the head and pronotum appears to the eye to be smooth and polished, but as in most species of the genus is really strigillate with a system of exceedingly fine wavy lines with sparse very minute feebly impressed punctures, a little more evident on the head.

This is the first species to be described from our fauna without discal pronotal punctures. It is not possible to assert that their absence is constant, although completely wanting in the three specimens before me.

Tritoma tenebrosa, n. sp.

Very similar in size, form and colour to *unicolor*; broadly ovate, black, mouth, antennal stem, tarsi and tip of last ventral segment dark rufous or rufopiceous, upper surface very finely alutaceous throughout and rather dull. Head closely, distinctly but not coarsely punctate. Prothorax finely rather sparsely punctulate; punctures of elytral series moderately strong and close, nearly as in *unicolor*; intervals minutely sparsely punctulate. Body beneath dull, rather finely sparsely punctate, the ventral segments more closely so.

Length 4.8 mm.; width 2.75 mm.

Southern Pines, N.C. (Rev. A. H. Manee).

The resemblance to *unicolor* is very close in all respects except the lustre and sculpture of the pronotum, which in the latter species is strongly shining without alutaceous sculpture and with the punctuation relatively very coarse. *Angulata* is more nearly in agreement with *tenebrosa* in punctuation and surface lustre, but it is distinctly smaller, with finer less closely punctured elytral striæ and red legs.

Agrilus strigicollis, n. sp.

Form moderately stout, about as in *pensus* and *obolinus*, moderately shining, æneous, prothorax somewhat cupreous, beneath cupreo-aeneous. Antennæ barely attaining the middle of the prothorax, serrate from the fifth joint, which is a little longer than wide, the following ones wider than long. Front broadly and rather deeply concave in superior half, the concavity confluent with a smaller post-clypeal impression, coarsely closely punctate, the punctures uniting in part to form short rugæ. Prothorax a little wider than long, sides nearly straight and parallel, narrowed only at the anterior angles; median line rather deeply impressed throughout, the impression broader behind; surface coarsely transversely strigose in wavy lines at the middle, the strigæ becoming longitudinal laterally; hind angles not carinate though with an obtuse elevation in the position of the usual carina, within which is a small basal impression, and another larger at the middle of the outer margin. Scutellum impressed and without transverse carina. Elytra scarcely sinuate behind the humeri, gradually narrowed from about the middle, apices separately rounded and finely serrulate, surface rather coarsely imbricate, disk a little flattened at middle, pubescence very short, sparse and recurved, evenly distributed. Prosternum rather densely punctuate and with short recurved pubescence, lobe truncate and feebly emarginate ; intercoxal process rather broad and seemingly obtuse at apex. Abdomen moderately punctate, last segment with a small emargination at apex ; pygidium without projecting carina. Claws deeply cleft, the apices of the inner portions nearly in contact.

Length 9 mm.; width 2.35 mm.

The type is a female from the Huachuca Mts. of Arizona, collected and given me by Mr. Carl R. Coolidge.

This species is at once separable from any of our previously described forms by the combination of antennal and ungual character, no other species with the serration of the antennæ beginning with the fifth joint having the long inflexed claw tooth.

Diphyllostoma Fall.

The discovery of a second species of this remarkable Lucanid genus is, like the first, due to Mr. Ralph Hopping, of Kaweah, California. Of the specimens sent Mr. Hopping writes : "The Lucanid seems to have different habits from *fimbriata*, flying about 10 a.m. and in the pines at 6,600 ft. elevation, whereas *fimbriata* seems to be a night flier at 1,000 ft. A reference to the original description of *fimbriata* shows that at least one specimen of that species was taken in flight shortly after noon and it is doubtful if this distinction is more than incidental ; the difference in altitude however is probably of more significance.

The new form agrees very closely in all essentials and most details with *fimbriata* and it is only necessary to refer the student to the description of the latter (CAN. ENT., 1901, p. 289), and state the differences.

D. nigricollis.

Form slightly narrower than in *fimbriata*, the prothorax a little smaller and black, the elytra piceo-testaceous, in *fimbriata* dark brown or castaneous and concolorous throughout ; mandibular process less strongly emarginate ; prothorax distinctly more finely and sparsely punctate, the elytra similarly but less deeply sculptured than in *fimbriata*. Tarsi a little longer and more slender, the joints more than three times as long as wide, while in *fimbriata* they are less than three times as long as wide.

Length 6½-8 mm.

Described from ten examples—all ♂s—taken at Huckleberry Meadow, Fresno Co., California, July 15 and Aug. 1 ; elevation, 6,600 ft.

Cremastochilus quadratus, n. sp.

Black, subopaque, above with very sparse short brownish erect or suberect hairs which become on the pronotal disk distinctly squamiform, varying from two to three times as long as wide ; hairs beneath sparse and very short, stiff and setiform. Mentum deeply and regularly cupuliform, the margin entire. Head as in *schaumii* and *westwoodi*. Prothorax nearly one-half wider than long, widest across the hind angles which are not at all retracted, sides very broadly and just visibly sinuate before the hind angles, arcuately narrowed in front, the apex 3/5 as wide as the base ; front angles foveate, hind angles rectangular, triangularly smooth above, not limited within by an impression ; disk broadly convex, median line impressed, punctures coarse and shallow, dense at sides, well separated toward the middle. Elytra moderately flattened, rather more so than in *westwoodi*, sculpture as in the latter species. Pygidium coarsely cribrate punctate. Body beneath coarsely moderately closely punctate. Tibiæ distinctly less broad than in *westwoodi* ; front tarsi short, passing the apex of the tibiæ by only the terminal joint, or slightly more ; middle tarsi subequal in length to the tibiæ ; hind tarsi a little shorter than the tibiæ. All the tarsal joints are concavely compressed laterally, more strongly so basally, so that when viewed from above the joints appear much narrower at base.

Length 12.5–14 mm.: width 5–5.8 mm.

Described from three examples sent by Mr. Junius Henderson, of the University of Colorado, who took them at Ft. Mojave on the Colorado River in Western Arizona, March 16, 1911.

As indicated in the descrption, this insect is most nearly related to *C. westwoodi*, to which the student would be led by attempting to identify it by Horn's table of the genus. It differs markedly from that and other allied species, however, by the thorax not being narrowed behind ; the pronotum is also more coarsely and less closely punctured toward the middle, the erect hairs are here more truly scales, the pygidium is more coarsely punctured, the tibiæ less stout, the front tarsi shorter and the mentum more deeply concave. The peculiar concave compression of the tarsal joints is not closely approached by any other species known to me.

Lachnosterna carolina, n. sp.

Moderately elongate, cylindrical, entirely rather pale rufo-testaceous, surface moderately shining. Clypeus broadly feebly emarginate, moderately reflexed, surface closely punctate, the front a little less densely so.

Prothorax fully twice as wide as long from a vertical view point, sides parallel posteriorly, accurately narrowed in front, margin entire, surface moderately closely, not coarsely, punctate. Elytra as closely and somewhat more coarsely punctate than the prothorax, costæ faint. Pygidium vaguely finely punctate and with a tendency to become longitudinally wrinkled. Metasternum closely punctate, hairs short and not dense. Abdomen finely sparsely punctate, nearly smooth at middle. Last joint of maxillary palpi fusiform ovate, slightly impressed.

Length 14–15 mm.; width 7½–8½ mm.

Male.—Antenna 10-jointed; club a little shorter than the stem; abdomen slightly flattened at middle, penultimate segment faintly sinuate at middle and with a slightly roughened arcuate impression which anteriorly attains the middle of the segment, and is about twice as wide as long; last segment with a shallow subrectangular emargination, the apical limiting angles not produced or acute, the bottom of the emargination feebly roughened on its extreme edge; surface of the segment with a transverse polished fovea occupying the entire length; inner spur of hind tibia short, varying from $^1/_{10}$ to ¼ the length of the long and slender outer spur.

This species is very closely allied to *ephilida*, which it resembles perfectly in all the more obvious characters. The latter, however, has a slightly longer antennal club, the abdomen in the male is distinctly channeled or concave at middle, the penultimate segment more evidently roughened posteriorly, the last segment more deeply emarginate, the lateral lobes more prominent, the posterior border of the emargination more widely and strongly roughened, the genitalia quite different, though of a similar type.

Described from five examples—all males—taken at Southern Pines, N.C., by Rev. A. H. Manee, the dates of capture ranging from June 14 to July 15.

Microphotus rinconis, n. sp.

Oblong, prothorax testaceous, the disk rather broadly infuscate, elytra fuscous, under surface and appendages testaceous. Antennæ (♂) 8 or 9 jointed. Prothorax about ¼ wider than long, sides parallel posteriorly, arcuately narrowed in anterior half, the apex subangularly rounded; surface dull, coarsely reticulate punctate in front, somewhat less so behind, especially on the convex median portion of the disk, the latter neither channeled nor carinate. Elytra a little more than twice as long as the prothorax, subparallel, rather coarsely but vaguely punctate, costæ variable in distinctness.

Length 6.1–6.6 mm.; width 2–2.4 mm.

Rincon Mts , Southern Arizona (Beyer).

M. octarthrus, n. sp.

Nearly similar to the preceding, the elytra and median parts of the prothorax fuscotestaceous. Antennæ 8-jointed. Prothorax but slightly wider than long, sides feebly obliquely convergent behind. Elytra a little shorter with sides more evidently arcuate in some specimens, the punctuation dense, rather coarse and better defined than in the preceding.

Length 4.75–5.3 mm.; width 2 mm.

Rincon Mts., Arizona (Byer).

Var. *pecosensis*, n. var.

Under this name I include as a variety or race of the above a series of four examples from Pecos, New Mexico, taken by Prof. Cockerell. They differ from the typical form by their larger size (5.5–6.6 mm.), relatively longer elytra, somewhat larger eyes and slightly less stout antennæ. One specimen is anomalous in its shorter elytra and is probably aberrant; it has, however, deterred me from describing this as a distinct species. This species or subspecies was recorded as *angustatus* in the New Mexico List, following LeConte's determination of Colorado specimens, which are probably the same thing.

M. decarthrus, n. sp.

Elongate, parallel, prothorax distinctly, elytra moderately shining, colour as in the preceding species. Antennæ 10-jointed. Prothorax slightly wider than long, sides a little more convergent behind, apex narrowly subtruncate at middle, surface shining, the punctures of different sizes, but as a whole finer, shallower and distinctly separated; median line channeled posteriorly. Elytra narrow, parallel, more than three times as long as the prothorax, punctuation close but vague and rather fine; costæ evident.

Length 6.6 mm.; width 2.2 mm.

Chiricahua Mts., Southern Arizona. A single specimen collected and given me by Mr. V. L. Clemence, of Pasadena, California.

The five species of *Microphotus* known to me may be easily separated by the following table, the characters of course pertaining to the males only :—

Elytra suboval, distinctly rounded on the sides .... *dilatatus*.
Elytra parallel, sides straight or but little rounded.
Prothorax not narrowed behind.
Prothorax semielliptical, widest at extreme base; elytra 2⅔ to 4 times as long as the prothorax, brownish testaceous frequently with a pinkish tinge .... *angustatus*.
Prothorax with sides parallel posteriorly; elytra 2⅔ to 3 times as long as the prothorax, fuscous in colour .... *rinconis*.
Prothorax obliquely narrowed behind, widest at about the middle.
Antennæ 8-jointed, prothorax more coarsely and densely reticulate punctate, less shining, the median line carinate or subcarinate posteriorly .... *octarthrus*.
Autennæ 10-jointed, prothorax more sparsely punctate and shining, median line sulcate posteriorly...................... *decarthrus*.

So far as known, *dilatatus* is confined to the Cape region of Lower California. *Angustatus* occurs in and to the west of the Sierras from Southern California to Oregon.

There is a confusing disagreement in published references to *Microphotus* as to the number or antennal joints. LeConte, in the original diagnosis of the genus (based on *dilatatus*), describes these organs as 11-jointed. In his subsequent description of *angustatus* they are said to be 9-jointed. Later, in his synopsis of the Lampyridæ, the number of joints is given in the generic table as nine in the male and eight in the female, but in the remarks upon the genus which follow on the same page the males are said to have 10 and the females 9-jointed antennæ. The small subulate appendage to the terminal joint was evidently counted by LeConte in the original desciption, but not afterwards. So far as my material goes *decarthrus* alone has 10-jointed antennæ; of my three examples of *dilatatus* one has three organs 9-jointed, another 8-jointed, the outer joints being lacking in the third. In both specimens of *rinconis* the antennæ are evidently 8-jointed when viewed from the front, but there is a more or less complete division of the sixth joint on the lower and posterior sides so that viewed from that position they appear to be 9-jointed. In the eight examples of *octarthrus* the antennæ are uniformly 8-jointed, while in *angustatus* they are as constantly 9-jointed. The following measurements in millimeters of the length and width of the prothorax and the length of the elytra exhibit considerable variation, but the deduced ratios are in most cases sufficiently different to be distinctive. No measurements of width of elytra are recorded, the tendency to warp, curl and separate when dry rendering them unreliable for comparative purposes.

	Length of Prothorax	Width of Prothorax	Ratio of Length to Width of Prothorax	Length of Elytra	Length of Elytra in Terms of Prothorax
dilatatus	2.30	2.95	.75	5.80	2.52
"	2.38	3 00	.76	6.40	2.69
"	1.78	2.25	.79	4.75	2.67
rinconis	1.80	2.20	.82	4.83	2.68
"	1.51	1.82	.83	4.60	3.05
octarthrus	1.50	1.53	.98	5.10	3 40
"	1.40	1.50	.93	4.75	3.39
"	1.50	1.60	.94	5.00	3.33
"	1.55	1.70	.91	4.00	2.58
var. *pecosensis*	1.50	1.62	.93	3.40	2.27
"	1.60	1.70	.94	3.55	2.22
"	1.40	1.52	.92	3.35	2 39
"	1.60	1.60	1.00	3 70	2.31
decarthrus	1.60	1.70	.94	5.00	3.13
angustatus	1.50	1.76	.85	5.95	3.97
"	1.50	1.75	.85	5.50	3.67
"	1.88	2.20	.85	7.00	3.72
"	1.34	1.50	.89	4.75	3.54
"	1.60	2.00	.80	6.25	3.91

Ammodonus granosus, n. sp.

Broadly oval, moderately convex, dull black, densely clothed above with appressed ash coloured scales varied with brownish, and with numerous short subrecumbent squamiform hairs which on the elytra are arranged subserially in great part; beneath with sparse appressed narrow scales or scale-like hairs, side margins of the body fimbriate with short feebly clavate squamiform hairs. Head and prothorax with numerous naked granules which are separated by about their own diameters on the head and anterior parts of the prothorax, a little less close toward the base of the latter. The prevailing colour of the scales is ashy, feebly nubilously varied with pale brown at the middle of the basal and apical parts of the pronotum; elytra with a uniformly slightly brownish shade along the suture, exterior to which is a fuscous basal spot and an irregular transverse median spot, and behind the latter and nearer to the suture than to the side margin, an elongate oblong spot of same colour.

Length 5 mm.; width 2¾ mm.

Rincon Mts., Southern Arizona.

Three examples of this interesting species were taken by Mr. G. Beyer, from one of which the above description is drawn. It is evidently closely related to *A. fossor*, but differs notably in its conspicuously granulose head and pronotum, slightly wider head, less transverse prothorax—$^{3}/_{5}$ as long as wide—(about ½ as long as wide in *fossor)* more pronounced elytral markings and stouter front tibiæ with broader apical process. In *fossor* there are a few small granules on the head and pronotum, but these are discernible with difficulty, being nearly or quite concealed by the vestiture in all specimens I have seen.

I find it impossible from description to distinguish between *Ammodonus* and the genus *Scaptes* as defined in the "Biologia." *Scaptes tropicus* Kirsch, widely distributed over the central portions of the American continent and the adjacent islands, must be closely allied to the present species, and perhaps still more closely to *fossor*.

Supplementary Note on Microphotus.

Since sending the MS. of the present article to the publisher, Mr. A. B. Wolcott, of the Field Museum, of Chicago, has called my attention to some remarks on *Microphotus*, including the description of a new species, by Ernest Oliver, in the Revue Scientifique du Bourbonnais et du Centre de la France—1911, No. 3, p. 79. The author calls attention to the discrepancies in LeConte's writings as to the number of antennal joints, which I have alluded to above, and says that in all ♂s seen by him—excepting the new species about to be described—the antennæ are 9-jointed. This new species has 10-jointed antennæ and is described as follows:—

"*M. robustus*, n. sp.—Pallide testaceus, elongatus, antennis decemarticulatis; prothorace supra caput rugose et profunde punctato, lateribus leviter attenuatis, antice rotundato, basi vix sinuato, angulis rectis, carinate, pallide testaceo, macule parva basali rubescente; scutello testaceo, triangulari; elytris elongatis, subparallelis, fuscis, rugosis, obsolete costulatis, prothorace haud latioribus, apicem versus attenuatis; ♀ ignota. Long. 12 mill.—San Diego.

"Bien distinct des autres espèces par sa taille beaucoup plus grande, ses antennes de 10 articles, son prothorax court, atténué, à sommet bien arrondi, ses élytres acuminés, plus long que l'abdomen et un peu déhiscents à partir de la moitié de leur longeur, etc."

Where San Diego is we are not informed, but presumably in California. The size is much greater than in any species of the genus known to me, being nearly double that of *decarthrus*, which alone agrees with *robustus* in the number of antennal joints.

NOTES ON THE CHALCIDOID *TRICHAPORUS* FOERSTER OF THE FAMILY EULOPHIDÆ, WITH DESCRIPTION OF ONE NEW NORTH AMERICAN FORM FROM ILLINOIS.

BY A. ARSÈNE GIRAULT, BRISBANE, AUSTRALIA.

History and Description.

Arnold Foerster in 1856, in his "Hymenopterologische Studien," designated as follows a group of generic rank called *Trichaporus*, which had no species named in connection with it. Quoting the table of genera given under his family Tetrastichoidæ we find the following omitting those portions of the table having no relevancy :

"*a.* Das Schildchen ohne Furchen.
b. Fühler scheinbar dreigliedrig . . . . *Triphasius m.*[a])
bb. Fühler deutlich mehrgliedrig.
c. Flügel ohne ramus stigmaticus . . . . *Anoxus m.**)
cc. Flügel mit einem ramus stigmaticus.
d. Der ganze Flugelrand mit langen Wimperhaaren besetzt . . . . *Pterothrix* Westw.
dd. Der Vorderrand des Flügels ohne längere Wimperhaare . . . . *Trichaporus m.***)
aa. Das Schildchen mit Furchen versehen.
e. Der Schaft übermässig verdickt (♂).
f. . . . . *Ceranisus* Walk.
ff. . . . . *Baryscapus m.****)
ee. Der Schaft nicht übermässig verdickt.
g. . . . .(. . . . .).
gg. . . . . *Hyperteles m.*†)
. . . . *Tetrastichus* Hal."
pp. 83–84.

In the last paragraph of the next page (p. 85), Foerster stated in regard to the group *Trichaporus* : "Eine gleiche Bewandtniss hat es mit der Gattung Trichaporus. Von Pterothrix wird sie in gleicher Weise durch achtgliedrige Fühler beim ♂ und ♀ geschieden. Dazu kommt, dass die siebengliedrigen Fühler des ♂ von Pterothrix mit langen Haaren

a. The footnotes are omitted, with one exception—A.A.G.

**)Trichaporus von θρίξ, τριχός, ἡ und ἄπορος ον, arm, dürftig Im Vergleich zu Pterothrix erscheint der Flügel arm an Wimperhaaren.

bekleidet sind, grade so wie bei den ♂ von Tetrastichus, bei Trichaporus aber sind sie ganz kurz und gleichförmig behaart. Der Flügel weicht ebenfalls von Pterothrix ab, indem er gleich Anozus am Vorderrande nur einen kurzen Haarsaum hat. Der deutliche ramus stigmaticus gibt aber auf der anderen Seite wieder ein gutes Unterscheidungsmerkmal der Gattung Anozus gegenüber ab."

Hence the group was originally defined as tetrastichines, having 8-jointed antennæ in both sexes, without a ring joint, and with uniform short hairs, the scutellum without furrows, the fore wings with a stigmal vein, but without long cilia on the cephalic margin.

Foerster gave nothing more concerning the genus; no species was mentioned as belonging to it; under the code it is therefore without status. Notwithstanding this, Taschenberg (1866) recognized the group, as did also de Dalla Torre (1898), the latter, however, placing it among the "Genera Sedis Incertæ" of the subfamily Tetrastichinæ, with the comment "Species exstat."

In 1904 Ashmead took the name and applied it to a group of his own species and one of Philippi's (1873), still quoting Foerster as responsible for the name, and stating that the type was unknown but giving a wholly different definition of the genus. Several years earlier Ashmead (1900) removed *Euderus columbianus* Ashmead to *Trichaporus* Foerster, thereby recognizing the latter. This species can not become the type of the genus, since Ashmead in 1904 defined the genus with characters which *columbianus* does not possess. *Trichaporus* Foerster, 1856, being non-existent, the group *Trichoporus* defined by Ashmead in 1904, and referred to Foerster, 1856, should become a *genus novum* Foerster without designated type. For the latter purpose I select *Trichoporus melleus* Ashmead, 1904, being one of the species upon which the definition of the genus was evidently based, and the first one described by Ashmead in 1904. *(Exurus) Trichoporus colliguayæ* (Philippi, 1873) is the first species listed by Ashmead in 1904, but this is not selected as the type of the genus because of the fact that it may not have been actually seen by him, and its reference to this genus is, I believe, somewhat doubtful. I retain the original spelling of Foerster—*Trichaporus*.

The genus has no synonyms, unless *Exurus* Philippi, 1873, should prove to be such. It is true that Ashmead (1904, p. 374) designated *Euderus* Thomson *(sic)* (1878, p. 276) to be a synonym of *Trichaporus* Foerster, 1856. But in the first place the latter was non-existent, and secondly, Thomson never described a genus called *Euderus*, but distinctly (in the place cited) quotes Haliday as being responsible for the group, in fact as he was. *Euderus* Haliday has little in common with *Trichaporus* Foerster.

The characters of the new genus, extracted from the key of the Tetrastichini as given by Ashmead (1904, pp. 348–349), are as follows: Tetrastichines with a sessile cylindrical abdomen as long as, or longer than, the thorax, and convex above (dorsad), a slender marginal vein in the hind wings, the mesonotum without a median groove, the pronotum not conical, the antennæ 9- or 10-jointed with one or two ring-joints, the scutellum with four (or two?) longitudinal grooved lines, the metanotum usually punctate, and the head, thorax and abdomen punctate or shagreened; the fore wings with short marginal fringes, the hind wings not acutely pointed at apex and the segments of the abdomen subequal in length. The genus is closely related to *Tetrastichodes* Ashmead on the one hand and to *Syntomosphyrum* Foerster on the other.

Ashmead always spelled the name *Trichoporus* instead of *Trichaporus*; as stated, I have adopted the latter as being correct.

The genus as it now stands contains six species, two of which (*colliguayæ* and *columbianus*), however, are more or less doubtfully placed. I have been unable to gain access to the types of any of the species.

Host Relations.

The habits of the parasites of this genus are not well known. In fact, in no case is there a definite record of the host of any one species, and but three of the species have been in any way connected with hosts. *Trichaporus columbianus* (Ashmead) is stated by Smith (1900) to live in cecidomyid galls; *colliguayæ* (Philippi) was reared in large numbers from a gall on *Colliguaya odorifera* Molina in Chile, but under circumstances which would necessitate confirmation of this gall-forming habit; it is possible that the gall was cecidomyid, and the species parasitic on the latter; *æneoviridis* was reared under conditions which make it impossible to decide whether it is parasitic on an ichneumonid, a syrphid or a larva of a lasiocampid moth. Nothing definite can therefore be stated in regard to the host relations of the genus.

Distribution of the Genus.

The species of the genus *Trichaporus* as now known are restricted to the Western Hemisphere, and the majority of the species belong to South America; *melleus* Ashmead, *viridicyaneus* Ashmead, *Persimilis* Ashmead are known from Brazil only; *colliguayæ* (Philippi) from Chile. The North American species are: *columbianus* (Ashmead) [Florida, District of Columbia, New Jersey]; and *æneoviridis* Girault (Illinois). Of the six species of the genus, four are South American and two North American, and the genus as a whole is distributed between the meridians of 40.6° and 89.2° west longitude, and between the parallels of about 40° south and 42° north latitude.

(To be continued.)

NEW SPECIES AND GENERA OF NORTH AMERICAN LEPIDOPTERA.

BY WM. BARNES, M.D., AND J. H. MCDUNNOUGH, PH.D., DECATUR, ILL.

(Continued from page 22.)

Leucania suavis, sp. nov.

Head and thorax clothed with olivaceous hair; primaries straw-coloured, slightly sprinkled with black atoms, especially along inner margin; a dark shade extends from base of wing above cubital vein and along vein 5 to outer margin; veins in outer portion of wing finely lined on both sides with dusky; an indistinct oblique row of black dots across the wing beyond the cell, not attaining costa; a faint row of black marginal dots, mostly incomplete; fringes whitish. Secondaries deep smoky in ♂, fringes pale, cut by a dark line; in ♀ smoky, but much lighter than in ♂, an incomplete row of terminal dots and pale fringes without dark line. Beneath primaries smoky, outer margin and a ray extending outwards from discocellular vein pale straw-colour; secondaries pale, slightly suffused with smoky; a small discocellular spot and incomplete row of terminal dots on each wing. Expanse, 31 mm.

Habitat: White Mts., Ariz., 1 ♂, 6 ♀s. Type, collection Barnes.

Our single ♂ specimen shows a black dot at the inception of vein 2 of primaries, and another below it on anal vein; these are lacking in the ♀s. The species may easily be separated from all other N. Am. species of the genus *Leucania*, as defined by Hampson, by the fact that there are no black lines in the interspaces of the veins in the terminal area.

Trachea cara, sp. nov.

Palpi blackish outwardly; head and thorax clothed with an admixture of reddish-ochreous and black scales; an indistinct black transverse line on tegulæ and a rather more distinct black line before upper margin of patagia. Primaries purplish-red, suffused with ochreous; a black basal dash extending to t. a. line; basal line only indicated by a slight dark mark on costa, surrounded by diffuse ochreous shading; t. a. line indistinct in costal half of wing, indicated by two spots on costa, below basal dash distinct, geminate, black, filled with ochreous, and bent inwards to inner margin near base; beyond t. a. line considerable ochreous shading, especially along inner margin; orbicular round or slightly oval, outlined partially in black, with ochreous annulus and smoky central portion; claviform a slight black arrow mark below orbicular, not extending back to t. a. line and preceded by ochreous shading; reniform large, constricted centrally, the lower portion considerably broader than the upper, outlined in black, with dark centre; a slight dark median shade angled at reniform; t. p. line indistinct in costal portion, geminate, black, the inner line most distinct, filled with ochreous, outcurved around reniform, almost touching same at base, from which point it is evenly oblique and slightly lunate to inner margin; space between it and reniform shaded with ochreous; several pale dots beyond on costa; subterminal space even purplish-red, with little ochreous shading; s. t. line pale, wavy, crossed by two black sharply defined lines above and below vein 5, reaching from outer margin almost to t. p. line; terminal space with less reddish than remainder of wing, crossed by black line below vein 7, and with faint black mark on vein 2, neither of these crossing subterminal line; a terminal series of small black lunules; fringes dusky, streaked with ochreous opposite veins; secondaries smoky, with an incomplete dark terminal line; fringes smoky, with slight pinkish tinge, cut indistinctly by a darker line. Beneath smoky, with slight pinkish tinge, traces of a medial line on primaries mostly confined to costal area, distinct medial line and discal dot on secondaries. Expanse, 32 mm.

Habitat: Eureka, Ut.; Provo, Ut., 2 ♂s, 1 ♀. Type, collection Barnes.

Very similar in maculation to *T. adnixa* Grt., but lacking the blackish mark in subterminal area below vein 2, which is mentioned by Hampson (Cat. Lep. Het., VII, 187) in his description, and is also present in a coloured drawing of the type in the Tepper collection, which we possess. We have several specimens from Vanc. Is., B. C., which we take to be *adnixa*, and which are generally darker in ground colour, with a more prominent light patch beyond reniform; the black lines on each side of vein 5 are also not so clearly cut in the B. C. specimens, tending to become suffused with each other, and the s. t. line is more prominent and distinct.

Hadenella cervoides, sp. nov.

Palpi outwardly dark brown, scaled with white at base, upturned, third joint short, porrect, antennæ ciliate; front and thorax closely scaled with brown and pale scales; divided scale tuft on metathorax; primaries brown, ordinary lines wanting, two black spots on costa above orbicular and reniform indicating their position, and a faint pale shade-line beyond reniform giving the approximate course of t. p. line; orbicular and reniform small, outlined in white, former round, latter kidney-shaped and open towards costa; faint terminal row of black dots, preceded by much more distinct pale ones, from the inferior one of which a slight black dash extends inwardly; fringes long, dusky, cut by a darker line. Secondaries smoky; fringes whitish, cut by a broad dark shade; beneath smoky brown, secondaries white at base and inner margin. Expanse, 25 mm.

Habitat: Redington, Ariz, 1 ♂. Type, collection Barnes.

The generic reference is doubtful; the front has a small truncate prominence, with raised edges and slight central process, but as the abdomen is devoid of squammation we are unable to tell whether tufts are present or not; in general appearance it fits in very well with *pergentilis*. We thought at first this might be *Fotella notalis* Grt., but as far as can be judged by Grote's rather meagre description, combined with Hampson's remarks, this is a larger species (34 mm.), without orbicular, and with a pale terminal border.

Perigea orta, sp. nov.

Palpi ochreous, sides of 2nd joint and 3rd joint dark brown; front ochreous, shaded posteriorly with dark brown; base of tegulæ ochreous, bordered with a dark line; remainder of head and thorax clothed with an admixture of reddish, ochreous and dark brown scales; abdomen yellow-brown, with darker tufting; primaries dark brown, with a distinct reddish tinge; maculation indistinct; basal line represented by two dark streaks on costa, with intermediate space filled with olive; t. a. line geminate, inner line obsolete, filled with olive shading, slightly oblique in course, dentate, a small inward angle below costa, prominent ones in the cell and on vein 1; orbicular small, round, partially outlined in black, filled with olive; claviform, when present, a small blackish blotch resting on t. a. line, and occasionally filled with olive; reniform large, the lower portion filled with a prominent quadrate white patch shaded inferiorly with black, the upper portion filled with several irregular olive spots and dashes, separated from each other by dusky shades; above reniform on costa a small olive dot; t. p. line indistinct, broken, represented by a series of olive spots, shaded inwardly more or less distinctly with black sagittate marks, and followed by a row of minute white dots on the veins; in course parallel to outer margin, slightly incurved in submedian fold; s. t. line usually very indistinct, marked by the difference in shade between the dark subterminal and the ochreous shaded terminal spaces, irregular, incurved opposite cell, dentate on veins 2–4, occasionally preceded by black sagittate marks, most prominent in costal half; terminal area usually but slightly lighter than subterminal portion; at times rather heavily streaked with ochreous; a dark terminal line, broken by yellow points opposite the veins; fringes dark, rayed with ochreous or olive opposite the veins. Secondaries entirely smoky, with broken dark terminal line. Beneath primaries smoky, costa and outer margin ochreous, shaded with pinkish; secondaries whitish, sprinkled with smoky; a more or less evident discal spot and postmedian line on both wings. Expanse, ♂ 23 mm.; ♀ 28.5 mm.

Habitat: Gila Co., Ariz. (2 ♂s, 5 ♀s); Redington, Ariz. (1 ♂, 2 ♀s); Santa Catalina Mts., Ariz. (1 ♀). Types, collection Barnes.

The species bears considerable resemblance to *vecors* Gn., is, however, much smaller, lighter in appearance, and differs in the marking of the reniform, as well as in other minor details; it shows considerable variation as regards the distinctness of the subterminal line and the shading in terminal space. It is possibly Mexican, but we can find nothing in Hampson's work that agrees with it.

Oligia (Hadena) tonsa ab. *fasciata*, ab. nov.

Maculation as in *tonsa* Grt., or *subjuncta* Sm.; the ground colour of the wings, however, is white, streaked slightly with blackish; a broad red-brown band stretches across the median area of wing, bordered inwardly by the t. a. line, outwardly in the upper portion by the inner margin of the reniform, in the lower portion by the curved t. p. line; in the basal area of wing, near inner margin, two short black streaks, and a black dash across the median band as in *tonsa;* orbicular and reniform white, former very prominent against the dark surrounding area, latter with a slight yellowish outer shading between it and t. p. line. Secondaries deep smoky. Expanse, 22 mm.

Habitat: Eureka, Ut., 1 ♂. Type collection, Barnes.

This very striking form we received, along with a number of ordinary *tonsa*, from Mr. T. Spalding. As it agrees exactly in the course of the lines and general maculation with these specimens, we prefer to regard it for the present as an aberration, although it may prove to be a good species.

Athetis (Caradrina) mona, sp. nov.

Palpi outwardly black, 3rd joint pale ochreous; head and thorax gray, latter slightly paler; primaries very even dark gray-brown, with a sprinkling of black scales; t. a. line fine, black, slightly broken, originating from a black spot on costa; wavy and somewhat outwardly inclined; orbicular a small dark spot; reniform large, concave towards apex of wing, the concavity outlined in yellow, basal half outlined with white dots, 4–5 in number, central portion very slightly darker than rest of wing; on costa above reniform a dark spot; a faint dark median shade; t. p. line faint, crenulate, evenly sinuate; s. t. line barely visible, pale ochreous, irregular; terminal area slightly darker than remainder of wing; indistinct dark broken terminal line bordered outwardly with paler. Secondaries white, with broad outward dusky suffusion and dark terminal line, fringes pale, cut by a dark line near base. Beneath primaries wholly smoky, with faint discal dot and traces of postmedian line; secondaries white, sprinkled with brown along costa and outer margin, with distinct discal dot. Expanse, 22 mm.

Habitat: Witch Creek, San Diego Co., Calif., 1 ♀. Type, collection Barnes.

A species resembling certain forms of the European *selini* rather than any American species known to us; *multifera* Wlk. is probably its closest ally.

Papaipema errans, sp. nov.

Head and thorax purplish-brown. sprinkled slightly with white; tegulæ tipped with white; primaries purple-brown, sprinkled with white and suffused with golden-yellow, which is particularly prominent along inner margin to t. p. line and in terminal space; all maculation dull and indistinct; t. a. line only distinguishable as a fine dark line crossing the yellowish area near inner margin; above this two oval dark shades represent claviform and orbicular; a dark median shade with prominent outward angle on cubital vein, inwardly oblique from below reniform to near t. a. line; t. p. line fine, rigid, inclined slightly outward from reniform to inner margin, separating gradually from median shade, and approaching subterminal line; beyond it on costa several pale dots; reniform an obscure, dark, figure-of-eight shade; subterminal area lighter and more evenly purplish than remainder of wing; s. t. line marked by difference of shade between subterminal and terminal areas, shaded inwardly with smoky brown; terminal area with golden tinge; fringes dark. Secondaries smoky, paler basally, with obscure discal mark; beneath smoky, with discal dots and obscure postmedian lines on both wings. Expanse, 26 mm.

Habitat: White Mts., Ariz., 1 ♂. Type, collection Barnes.

Related to *unimoda* Sm., but the t. p. line is not lunulate, and is distinctly bent outward towards inner margin.

Too late to avoid publication we learn that in the foregoing article on "New Species and Genera of Lepidoptera" we have in two instances created a synonym. Our species *Hudenella cervoides* proves to be *Caradrina fragosa* Grt.; Dr. Barnes has just recently compared the two types. Our new genus and species *Friesia anormalis* is Grote's *Prosoparia perfuscaria*, placed at present in the *Geometridæ*. Mr. J. A. Grossbeck has sent us a specimen compared with the type, remarking at the same time that it is a Noctuid; we are glad to find our opinion supported by such a good authority, and trust that our figures of the structural features may serve to elucidate and augment Grote's very meagre and inadequate description.—J. H. McDunnough.

FURTHER NOTES ON DIABROTICA.
No. III.

BY FRED. C. BOWDITCH, BROOKLINE, MASS.

(Continued from page 16.)

D. quadrinotata, nov. sp.

Head black; antennæ long, black, joints 9–11 flavous, except extreme tip of last. Thorax wider than long, flavous, shining, finely sparsely punctulate, trifoveate, nearly straight on sides, angles all acute; scutel black; elytra somewhat dilated at rear, thickly finely punctate, light pale flavous, each elytron with two small black spots, a humeral or subhumeral and a submedian. Body beneath black, legs black, femora flavous, under side of thorax flavous. Length, 8 mm.

Two examples, Peru, green label (Marcapata ?).

The antennæ are nearly as long as the body (♂), or shorter than the body (♀); joint 3 nearly as long as 4; head with a deep fovea. The thoracic fovea are well marked and connected by a well-marked depression, and one example has a median piceous mark at the rear. The elytra are much wider than the thorax, convex, very thickly punctured, almost obsoletely granular, the black spots relatively small but very conspicuous. Somewhat similar in shape and appearance to *7-punctata* Jac. from Parada, Mex.

D. bicincta, nov. sp. (Jac. in litt.).

Head black; antennæ black, last 3 joints (♂?), or 4 joints (♀?), flavous, except tip of 11. Thorax punctured, flavous, trifoveate; scutel black; elytra thickly and rather coarsely punctate, flavous, a broad basilar and somewhat narrower submedian fascia dull black, neither quite attaining the margin. Thorax beneath and mesosternum flavous, rest of body black, legs black with yellow femora. Length, 4½–5 mm.

Five examples, Marcapata, Peru.

This form has been distributed with the manuscript name *bicincta* Jac. Four co-types have been sent me by Messrs. Staudinger & Bang-Haas. It is a pretty little species, and easily recognized by the two black fasciæ. The antennæ are long and slender, and nearly equal to the length of the body in the ♂. The apparent difference in the number of white joints may be significant of similar variation in other so-called species. The elytra are depressed near the scutel and not at all plicate; in one example the suture is very narrowly black between the bands. One example has the rear band abbreviated into two spots, the anterior band much narrowed, and joint 8 of the antennæ half black and half white.

D. parambaensis, nov. sp.

Head rufous; front in the ♂ profoundly excavate concave, in the ♀ convex; antennæ flavous, fuscous at tip, joint 2 very short, 3 several times longer than 2 in the ♂, stout, *curved*, and a trifle longer than 4; in the ♀ relatively shorter, equal to 4 and *both joints* cylindrical. Thorax rufous, polished, *distinctly punctulate*, deeply bifoveate and obsoletely depressed before the scutel, which is black; elytra shining black, punctured subseriately on the disk, becoming obsolete behind; the tip moderately, and a narrow median fascia bright, yellow. Thorax below red and black, body black, edges of the segments narrowly flavous, legs bright yellow, extreme tip of tarsi fuscous.

Var. A with small yellow subbasal median dot.

Six examples of typical form, "Paramba 3500 iv; '97," dry season (Rosenburg); 14 examples from Cachabé seem the same, though smaller, and with the upper flavous marks occasionally joined along the margin. Length, 5½–7 mm.

Seems close to *excelsa* Baly (which I have not seen), but that species is said to have the thorax *impunctate* and the elytra obsoletely "elevato-vittatis," and the length is 9½ mm. Curiously enough, the description of *excelsa* does not mention the shape of the third antennal joint, except inferentially in the statement, "the fourth cylindrical, not curved." The curved third joint in the ♂ of *parambaensis* is very marked, and allies it closely to the Central American forms, *lepida* Say and *variabilis* Jac. It seems to indicate a tendency towards the dilated joint of the ♂ of *Ceratoma*. All the forms in Baly's paper, sec. K, with concave front in the ♂, are represented in my collection, with the exception of *excelsa* Baly. The forms *imitans* Jac. (type in my collection) and *deliciosa* Baly have a very strong tendency to run together.

In *parambaensis*, the entire front is occupied in the ♂ by the concavity, which is very deep; the antennæ are about half as long as the body, and if the small second joint is bent to a particular angle it pushes up a supplemental hinge, which appears at first sight like a small joint. The thorax is wider than long, sinuate and angled behind; the lateral fovea oblique and deep, and a distinct antescutellar depression is present; punctuation fine, but perfectly distinct; elytra with the usual shape of species of this section, quite strongly punctate, subseriately on the disk, becoming obsolete behind, transversely depressed behind the scutel and subplicate. The Cachabé specimens, as a rule, are smaller and with smoother elytra. Apparently a common form. The species of this section K need large series to determine the species, and even then it is difficult.

D. stuarti, nov. sp.

Head yellow; mouth black; antennæ yellow, tip of last joint dark. Thorax yellow, bifoveate; scutel blackish rufous; elytra thickly and rather coarsely punctate, yellow, with the rear part nearly to the middle semi-shining blue-black. Beneath yellow, with metasternum and abdomen black; legs yellow, tibiæ and tarsi and apex of femora dark fuscous.

Two examples, San Augustin, Mapiri, 3,500 feet, Sept. 1896 (M. Stuart). Length, 10 mm.

Belongs to sec. M, and should be placed near *dimidiata* Baly. Head with front distinctly carinate, antennæ reaching to just within the black apex of the elytra, with joints 3 and 4 nearly equal; thorax transverse, margined, all the angles acute, sparsely, finely and evenly punctate. The elytra are slightly dilated at rear, just a trifle more than in *prodiga* Er., and not as much as in *dimidiata* Baly. The rear tibiæ are rather darker than the others. The only one of its large allies having a black scutel is *prodiga* Er.

D. haenschi, nov sp.

Head rufous flavous; mouth-parts piceous; front carinate; antennæ black, with extreme base rufous. Thorax transverse, shiny, rufous flavous bifoveate, depressed, with all the angles prominent and a few scattered punctures; scutel rufous; elytra dilated behind, thickly and almost rugosely punctured, especially behind; yellow, with the rear half black. Below, thorax and mesosternum yellow, remainder black, legs black, with base of femora rufous. Length, 5½–6 mm.

Type, Balzabampa. Ecuador (R. Haensch); also Sn. Inez, Ecuador.

Belongs to sec. M, and comes nearest to *atriventris* Jac., from Ecuador, but is easily distinguished by the black legs.

The antennæ are long and slender, and reach nearly to the tip of the elytra. The Sn. Inez example is much less rugosely punctured than the type.

D. marcapa, nov. sp.

Head, thorax, antennæ, scutel, rear half of the elytra, body beneath and legs black, with a faint tinge of green on the thorax; anterior half of the elytra bright rufous; thorax with three deep fovea, and distinctly though sparsely punctate; elytra thickly and coarsely punctured, becoming obsolete at the rear. Length, 7 mm.

One example, Peru, green label (Callanga ?), Jacoby collection.

This species should be placed in sec. M, though the form is more like that of some of the species of sec. L. Head with carinate front and hairy, especially in front of the eyes; thorax strongly transverse, margined, with oblique lateral fovea, the rear round and just in front of the scutel; the elytra only slightly dilated behind. The antennæ are about three-fourths the length of the elytra, and joint 3 is not quite as long as 4. The extreme base of the femora is piceous. The tibiæ are noticeably covered with sericeous hairs.

D. cyaneo-maculata, nov. sp. (Jac. in litt.).

Head and thorax yellow; antennæ, except the extreme base, black. Thorax bifoveate and depressed; scutel yellow; elytra yellow, slightly dilated at the rear thickly, coarsely and subseriately punctured, with a large cyaneous blue spot occupying the whole apical third, leaving the extreme margin narrowly flavous to the tip. Body beneath paler yellow, legs yellow, with tibiæ, tarsi and upper surface of apex of femora black. Length, 5½ mm.

Five examples, Callanga, Peru.

This species has been distributed with the manuscript name *cyaneo-maculata* Jac. Of what purport to be three co-types sent me by Messrs. Staudinger & Bang-Haas, two belong to a different genus. The antennæ attain the apical spot; joint 3 a trifle shorter than 4. The thorax is finely but obviously punctulate, while the elytra show here and there fragments of smooth lines between the punctures. In one of the specimens the apical spot includes the side margin, in the typical form it does not.

D. cyaneo-plagiata, nov. sp.

Head yellow; mouth-parts dark; antennæ black, piceous at base, joints 9–11 flavous, with extreme tip of last dark. Thorax flavous, bifoveate, a sublateral dark streak on each side, not attaining either edge; scutel flavous; elytra flavous rufous, dilated behind, coarsely thickly punctured, with a common apical cyaneous blue spot, which does not attain either the side or apex. Body beneath flavous, legs flavous, with apex of femora and tibiæ and tarsi black. Length, 4½ mm.

One example, Peru, green label (Callanga ?).

Superficially like *cyaneo-maculata*, but the light joints of the antennæ and the black streaks of the thorax at once separate it. The elytral spot is smaller; the elytral punctuation is coarse, subseriate in the disk, and becomes obsolete behind and at the sides.

A CORRECTION.

From the description of *Gnorimoschema septentrionella*, in the December number, p. 422, some words have been left out. The description of the hind wing of the insect should read: *Hind wing* dark grey—dries with a gloss. Fringe lengthening towards the body to 3½ millimetres at the longest part, light brown.—T. W. F.

WASHINGTON MEETING OF THE ENTOMOLOGICAL SOCIETY OF AMERICA.

The sixth annual meeting of the Entomological Society of America was held in Room 376 of the new U. S. National Museum Building on Tuesday and Wednesday, December 26 and 27.

The following papers were read:—

Herbert Osborn.—Faunistic studies in entomology.

E. P. Felt.—Numerals as aids in classification.

E. S. Tucker.—Studies of insects bred and collected from the American mistletoe. Presented by Andrew Rutherford.

H. C. Severin.—The influence of temperature on the moulting of the walking-stick, *Diapheromera femorata*. (Title only.)

R. Matheson and C. R. Crosby.—Notes on aquatic Hymenoptera. Presented by C. R. Crosby.

Ann H. Morgan.—Photographs illustrating the life histories of May-flies.

H. Y. Tsou.—The Chinese wax-scale, *Erecerus pe-la*.

A. D. MacGillivray.—The lacinia in the maxilla of the Hymenoptera.

Lucy Wright Smith.—Glycogen in insects, especially in the nervous system and the eyes.

J. A. Nelson.—Note on an abnormal queen bee.

J. Chester Bradley.—The designation of the venation of the hymenopterous wing.

Ann H. Morgan.—Homologies in the wing-veins of May-flies.

A. D. MacGillivray.—The pupal wings of *Hepialus thule*.

J. Chester Bradley.—The wing venation of chaloid flies.

F. M. Webster.—Our present educational system in relation to the training of economic entomologists.

C. W. Johnson.—The use of colour in designating types and varieties.

Leonard Haseman.—Entomological work in Missouri.

Herbert Osborn.—A problem in the flight of insects.

E. P. Felt.—The biology of Miastor and Oligarces.

P. P. Calvert.—Seasonal collecting in Costa Rica.

W. L. W. Field.—Hybrid butterflies of the genera *Basilarchia*.

The following papers were read by title only, because of the expiration of the time allowed for the reading of papers :—

O. A. Johannsen.—Cocoon making by *Bucculatrix canadensisella*.

J. G. Needham.—Some adaptive features of myrmeleonid venation.

E. H. Strickland.—The Pezomachini of North America.

Z. P. Metcalf.—Homologies of the wing-veins of Homoptera Auchenorhynchi.

The following officers were elected for 1912 :—

President.—S. A. Forbes.

First Vice-President.—A. D. Hopkins.

Second Vice-President.—C. P. Gillette.

Secretary-Treasurer.—A. D. MacGillivray.

Additional Members of Executive Committee.—J. H. Comstock, John B. Smith, Henry Skinner, Herbert Osborn, E. D. Ball, P. P. Calvert.

Member of Committee on Nomenclature for three years.—H. T. Fernald.

The Society adjourned, to meet with the American Association for the Advancement of Science, at Cleveland, Ohio, January, 1913.

ALEX. D. MACGILLIVRAY, Secretary-Treasurer.

FINAL REPORT OF THE JAMES FLETCHER MEMORIAL COMMITTEE OF THE OTTAWA FIELD NATURALISTS' CLUB.

The Memorial Fountain, erected on the Central Experimental Farm, was unveiled on July 19th, 1910. Several hundreds of people were present at the ceremony, including some distinguished visitors from a distance. Official representatives of the Royal Society of Canada, the Entomological Society of Ontario and the Ottawa Field Naturalists' Club, were present, and took a prominent part in the proceedings. The Fountain, including the medallion, is the work of Dr. R. Tait McKenzie, of the University of Pennsylvania, Philadelphia, U. S. A.

The Memorial portrait, which is the work of Mr. Franklyn Brownell, R.C.A., was unveiled at an evening meeting of the Ottawa Field Naturalists' Club on January 9th, 1912. It is an exceedingly good likeness of the late Dr. Fletcher, and, as most satisfactory arrangements have been made with the Municipal Library Board and the Librarian of the Carnegie Library, the portrait will be hung in a prominent place in this latter building.

CASH STATEMENT.

Receipts.

Total amount paid by subscribers...............		$1838 85
Bank interest..............................		22 61
		$1861 46

Expenditure.

Cost of Memorial Fountain....................	$1500 00	
Cost of Portrait, including frame................	225 00	
Miscellaneous expenses : printing envelopes, receipt forms, postage, travelling, etc................	136 46	
		$1861 46

On behalf of the Committee,

ARTHUR GIBSON, Secretary-Treasurer.

Ottawa, January 29th, 1912.

ADDITIONS TO THE LIBRARY.

The Librarian of the Entomological Society of Ontario has much pleasure in acknowledging the receipt of the following publications, a present from the Trustees of the British Museum, London, England : "Monograph of the Culicidæ of the World," by F. V. Theobald, Vols. 3 and 5 ; "Synonymic Catalogue of Orthoptera," by W. F. Kirby, Vol. III; "Illustrations of Lepidoptera," Parts 6 to 9, 4 vols., quarto, illustrated with beautiful coloured plates.

These books form a very welcome addition to the library.

M. PAUL NOEL, Directeur du Laboratoire Regional, d'Entomologie Agricole de la Seine-Inferieure, Route de Neufchatel, 41, Rouen, France, desiring to publish a work on the properties which certain female insects possess of being able to attract the males from a great distance, will be very grateful to entomologists who would be willing to give him any well-authenticated facts relating to this attraction. He would send in return some of his entomological publications and the work in question immediately after it is printed.

Mailed February, 9th, 1912.

The Canadian Entomologist.

VOL. XLIV. LONDON, MARCH, 1912. No. 3

NOTES ON GEOPHILOIDEA FROM IOWA AND SOME NEIGHBOURING STATES.

BY RALPH V. CHAMBERLIN.

University of Pennsylvania, Philadelphia.

During several weeks in June and July of 1910 I had opportunity for making collections of chilopods in the district indicated by the title of this paper. Unfortunately, the season was unusually dry in these States, particularly in Michigan and Wisconsin, and, as a result, unfavourable for securing an abundance of material. The members of the Geophiloidea seemed especially difficult to uncover; but among the species obtained are several of exceptional interest, two representing new genera, for which it seems necessary to erect a new family. The families of the Geophiloidea now recognized as occurring in the United States, east of the Rocky Mts., may be separated as follows :

a. Mandibles with a dentate lamella.
  b. Mandibles with a single pectinate lamella ; antennæ filiform or somewhat clavate....................Family *Schendylidæ.*
  bb. Mandibles with several pectinate lamellæ ; antennæ flattened, attenuated distad..................Family *Himantariidæ.*

aa. Mandibles with no dentate lamella ; with a single pectinate lamella.
  b. Labrum fused for a short distance at middle ; antennæ flattened, at least narrowly elliptic in cross-section, attenuated distad...............................Family *Sogonidæ.*
  bb. Labrum entirely free ; antennæ cylindrical, filiform or a little clavate.
    c. Median piece of labrum extending along and, at least in part, fused with the lateral ; at middle of free edge with two much larger and more strongly-chitinized teeth directed more or less ventrad............Family *Soniphilidæ*, fam. nov.
    cc. Three divisions of labrum distinct ; without two such larger and ventrally-directed teeth......................Family *Geophilidæ.*

Of these families, representatives of the Geophilidæ and Soniphilidæ alone were secured in the region covered by this paper. However, the Schendylidæ is represented, *Escaryus urbicus* (Meinert), having been taken in Minnesota, and the same species having been found by the writer to be quite common in New York State. The family Himantariidæ is represented in Texas and Mississippi by a species of Haplophilus, and by at least one of the genus Gosiphilus, *G. laticeps* (Wood). These genera may be found to range into the present section. The family Sogonidæ is at present known to be distributed in Texas (*Timpina texana* Chamberlin, a form with but five joints to the anal legs), and in South Carolina and Tennessee (*Sogona minima* Chamberlin). On the Pacific Coast occur several families not found east of the Rockies.

Family Geophilidæ.
Subfamily Geophilinæ.
Genus Geophilus Leach.

Geophilus rubens Say.

Syn. Geophilus cephalicus Wood.
Geophilus lævis Wood.
Geophilus okolonæ Bollman.

Localities.—DeWitt, Mongona and Boone, Iowa; Franklin Grove, Ill.; Saunder's, Mich.

This is a very common species in Indiana, Ohio and more Eastern States. The form described by Bollman from Arkansas agrees perfectly with this species, excepting that the number of pairs of legs is higher than usual in northern specimens. This, however, is in line with a tendency shown by many other species for the number of legs to show an increase in going from the north to the south or from high elevations to low. It is one of the commonest forms in this district. California specimens also frequently have a larger number of legs.

Genus Arenophilus Chamberlin.

Arenophilus bipuncticeps (Wood).

Syn. Geophilus attenuatus Bollman (but not certainly of Say).
Geophilus georgianus and latro Meinert.
Schendyla perforata McNeill.

Localities.—Mongona, Boone, DeWitt, Tama, Marshalltown, Iowa; Fremont, Neb.; Peoria, Ill.; Janesville, Wis.

This is by far the most abundant species. It ranges as a common form through the greater part of the United States in and east of the

Mississippi Valley. At Mongona (June 22), and Marshalltown (June 24), Iowa, and at Sterling, Ill. (June 26.), females were taken with recently-laid eggs.

Genus Pachymerium Koch.

Pachymerium ferrugineum Koch.

Syn. Geophilus foveatus (McNeill).

Localities.—DeWitt, Iowa; Peoria, Ill.; Devil's Lake and Fond du Lac, Wisc.

At Fond du Lac (July 6), the species was found in great abundance among the stones at a river's edge, partly grown individuals being common, and a considerable number of females being found with bodies still coiled about their recently-hatched young.

This is a species widespread in the Eastern United States, as it is in Europe. The specimens secured are similar in size to Austrian specimens, most being under 25 mm. in length.

Subfamily Chilenophilinæ

Genus Taiyuna Chamberlin.

Taiyuna opita, sp. nov.

Proportionately robust; attenuated strongly caudad, and also decidedly but less strongly cephalad. Sparsely clothed throughout with long bristles.

Head with corners rounded; sides convexly curving; caudal margin straight; anterior margin extended forward from corners to middle, and a little incurved at median line; longer than wide in ratio, 19:16, and five times longer than exposed portion of basal plate. Prebasal plate absent. Basal plate overlapped in front by the cephalic, and behind by the first dorsal plate; free portion wider than median length in about ratio 34:7. Antennæ short; articles moderate and short, the ultimate equal in length to the two preceding taken together.

Claws of prehensorial feet when closed reaching the distal end of the first antennal article. Claw at base with a subcylindric, apically truncate, tooth; prefemur also with a strongly chitinized tooth at distal end; the intermediate joints also each with a distinct, conical and well-chitinized tooth. Prosternum unarmed; its anterior median margin nearly straight, not excised; chitinous lines not evident; suture parallel with margin; wider than long in ratio 39:35, longer than prefemur in ratio 7:4. Dorsum weakly bisulcate, also with a more median pair of fine sulci. Anterior præscuta short, being of moderate length in the middle region, and then again shortening caudad. Spiracles all circular, the first greatly exceeding the second in size. First pair of legs much reduced; anterior pairs more robust than the caudal, not shorter. Anterior ventral plates with a rather weak median sulcus, most plates plane; pores not detected. Last ventral plate moderately wide; margins straight, the lateral moderately converging caudad. Coxopleuræ with about four pores in a row under edge of plate, and four or five free on the sides, well separated from each other. Anal legs longer and more crassate than the penult; without claws. Pairs of legs (in female) 41.

Length of female 15 mm.; width .9 mm.

Localities.—Posers and Kimball's, Mich.

Genus Gnathomerium Ribaut.

Gnathomerium umbraticum (McNeill).

Syn. Gnathomerium americanum Ribaut.

Locality.—Manitou, Colorado.

This seems to be a southern species, occurring widely and abundantly throughout the Southern States. In favourable seasons it may be found to be not rare in the present region, as Bollman reports it as common in Indiana.

Subfamily Linoteniinæ.

Genus Linotenia Koch.

Linotenia chionophila (Wood).

Localities.—Devil's Lake and Ashland, Wis.

Many specimens were taken at the former locality under leaves and stones about the margin of the lake. This species is boreal, being abundant, comparatively, in Alaska and adjacent islands. It was first described from specimens taken at Fort Simpson on the Red River of the North. It is very close to *Linotenia acuminata* (Leach) of Europe, and may have to be merged with it.

Linotenia fulva (Sager).

Localities.—Mongona, Boone, DeWitt and Marshalltown, Iowa; Franklin Grove, Ill.; Sterling, Ill.

Very much the commonest Linotenia in the Northern United States, and one of the commonest members of the entire order.

Family Soniphilidæ, fam. nov.

Genus Soniphilus, gen. nov.

Labrum free; the median part firmly fused to the lateral, at least at ends; edge of median portion directed ventrad and bearing a number of very stout teeth, which extend directly ventrad (the figure accompanying suggests a bedding of these teeth somewhat caudad, which does not exist); of these teeth the two median are clearly largest, the others decreasing from median to outermost; lateral portions with edge bearing a few spinous processes much more weakly chitinized than the teeth of middle portion. (See pl. 1, fig. 3.) Mandibles with a single pectinate lamella; no dentate lamella. Both branches of first maxillæ set off by a suture; the outer branch biarticulate, entirely without lappets or with a single short, conical one on outer edge of base; coxæ completely fused at mesal line. Coxæ of second maxillæ fused at middle; palpi short, bearing a simple claw of normal size.

Chitinous lines of prosternum strongly developed. Prehensorial feet with joints all unarmed; claws when closed not attaining front margin of head. Frontal plate not discrete. Prebasal plate absent. At least the anterior sterna with caudal margin strongly chitinized in a sharp edge or blade-like form, which fits into a transverse groove in anterior margin of succeeding plate. (See fig. 5.) Pores not detected. Dorsal plates bisulcate. Last ventral plates very wide. Anal legs six-jointed, ending in claws.

Type.—*Soniphilus embius*, sp. nov.

Soniphilus secundus Chamb., a Californian species, also belongs here.

Soniphilus embius, sp. nov.

Slender, attenuated cephalad and caudad; body very sparsely provided with short straight hairs, the head with longer ones.

Yellowish-white, the anterior region more strongly yellow or lemon colored; head with prosternum and prehensorial feet pale reddish brown; antennæ yellowish white.

Head widest over caudal portion, the sides from middle caudad but very slightly converging, the sides in front of middle nearly straight and clearly converging; anterior margin with middle part straight, transverse, on each side a little oblique, extending a little caudad in running from middle to lateral cornea, straight. Frontal suture absent. Prebasal plate absent. Basal plate four times as wide as its median length, a little wider than cephalic plate (24:23). Antennæ filiform, of moderate length; articles longer than wide, decreasing in length distad to the penult, the ultimate about equal in length to the two preceding taken together.

Claws of prehensorial feet when closed not attaining the anterior margin of head, short, the inner free margin of prefemur very short or almost obliterated ; claw within a very small conical tooth at base, other articles unarmed. Prosternum with chitinous lines well developed ; two submedian longitudinal sulci ; anterior margin unarmed, weakly angularly depressed from sides to median line ; much wider than long (14:9), longer than greatest length of prefemur nearly in ratio 9:5. Dorsum weakly bisulcate. Prescuta of middle region moderate or short, not much decreasing in length cephalad and caudad. Spiracles all circular, relatively large, the first considerably larger than the second, the others gradually decreasing from the second caudad. Legs of first pair decidedly shorter and more slender than the second. Ventral pores not detected. Plates of anterior portion of body with a transverse groove along cephalic edge, which is protected by a flange-like extension on the ventral side ; into this groove fits the well-chitinized, extended blade-like caudal edge of the preceding plate in each case. Last ventral plate very wide, strongly narrowed caudad, the lateral margins a little incised below middle ; caudal margin straight. Coxo-pleuræ each with a single free isolated pore of small size, and two larger pits covered by the edge of the plate. (See fig. 4.) Anal legs longer and slightly stouter than the penult (female). Pairs of legs (female) 43.

Length, 13 mm.

Localities.—DeWitt, Iowa.

The type is a single female, which was taken with her recently-laid eggs.

Genus Poaphilus, gen. nov.

Agreeing in general with Soniphilus, as described above, but readily distinguished in having the joints of the prehensorial feet dentate within and its claws extending much beyond the front margin of the head. The last ventral plate is narrow or but moderate in width, not very wide, as in the preceding genus

Type.—*Poaphilus kewinus*, sp. nov.

Aside from the species here described, a second one from New Mexico is also known.

Poaphilus kewinus, sp. nov.

Body very small, strongly attenuated cephalad and caudad.

Antennæ and legs pale yellow ; body light yellowish brown ; head with prosternum and prehensorial feet light reddish brown.

Head much longer than wide (11:8) ; ten or eleven times as long as the very short basal plate ; relatively narrow, leaving sides of prehensorial feet exposed for entire length; caudal margin truncate, sides weakly bowed outward from end to end, the anterior margin rounded on each side, mesally incised. Frontal plate not discrete. Prebasal plate absent. Basal plate greatly abbreviated, the exposed portion eight times as wide as long.

Antennæ filiform, as compared with body length rather long; articles moderately long, decreasing distad, the ultimate a little longer than the two preceding together; bristles very long, distad, becoming shorter and denser as usual.

Claws of prehensorial feet when closed attaining distal end of first antennal article; claw armed at base with an acute conical tooth, prefemur with a low, conical and subdentiform protrusion on mesal surface, other joints unarmed. Prosternum wider than long in ratio 20:17; longer than the prefemur in the ratio 17:10, nearly; chitinous lines distinct. Dorsal plates bisulcate; also with a weak median sulcus. All prescuta short. All spiracles circular, the first larger than the second. First pair of legs shorter and much more slender than the second; anterior pairs shorter and thicker than those of posterior portion of body. Last ventral plate moderately wide, narrowed caudad, the margins nearly straight, the caudal slightly excised. Coxopleural pores four, small, two of these covered or partly covered by the edge of the last ventral plate and the other two free. Anal legs longer and thicker than the penult, ending in a long slender claw. Pairs of legs in female, 37.

Length, 6.5 mm.

Locality.—Marshalltown, Iowa.

The type, as with the preceding species, is a single female which was taken—her eggs were very few in number.

EXPLANATION OF PLATE.

Soniphilus embius, gen. et sp. nov.

Fig. 1.—Dorsal view of anterior portion.

Fig. 2.—Ventral view of anterior portion.

Fig. 3.—Labrum, ventral aspect. (The teeth of median portion normally extend directly ventrad; the figure shows them extending caudo-ventrad, this resulting from depression by the cover-glass).

Fig. 4.—Ventral view of posterior portion.

Fig. 5.—Ninth and tenth ventral plates.

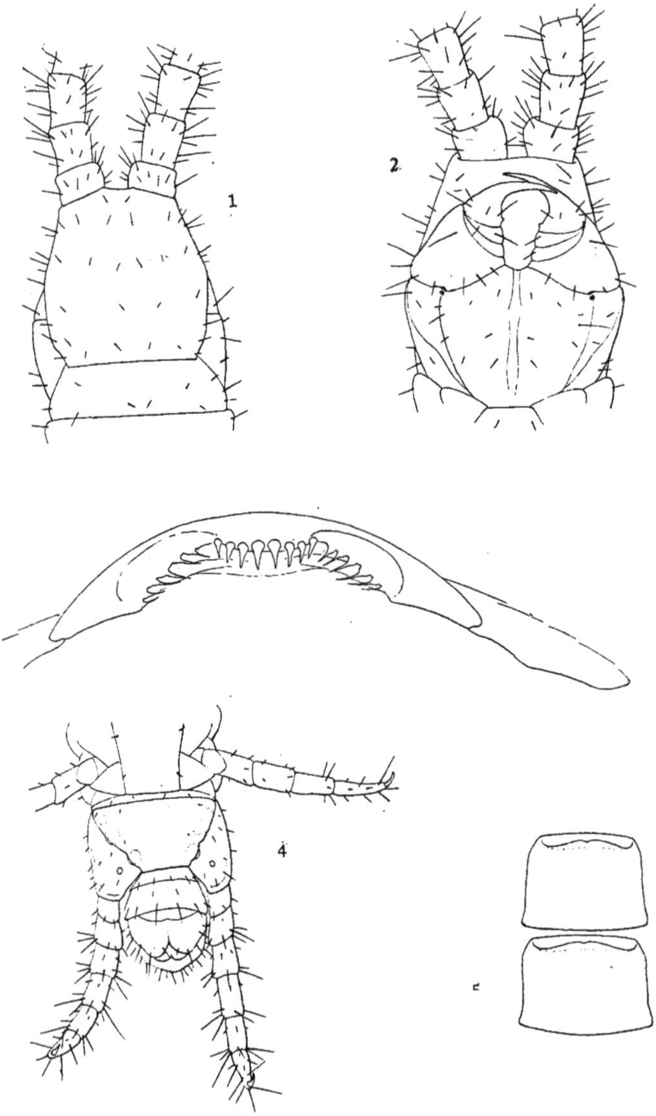

SONIPHILUS EMBIUS, GEN. ET SP. NOV.

THE LIGHT-EMISSION OF AMERICAN LAMPYRIDÆ: NOTES AND CORRECTIONS ON FORMER PAPERS.

BY F. ALEX. MCDERMOTT, WASHINGTON, D. C.

The author wishes to make the following corrections in and additions to his former papers on "The Light-Emission of American Lampyridæ" in this journal:

Vol. 42 (1910), p. 360.—Modify lines 13 to 9 from bottom to read:

"The *consanguineus* emits two such flashes, separated by an interval of about a second, followed by a longer interval before the next two; sometimes the double flash is followed by a residual phosphoresence, as in *pyralis*. The *angulata* usually emits a single flash, much shorter and more sudden than that of *pyralis*, being in this regard like that of *scintillans*, but more greenish in colour than the light of the latter insect." (The twinkling light ascribed to *angulata* was no doubt that of a male *Lecontea lucifera* Melsh., its somewhat larger and very similar relative.)

P. 363.—Delete note at foot of page, as this paper proved to have no bearing on the immediate subject.

Vol. 43 (1911), p. 404.—After line 4, *Photuris* has been observed mating only rarely; upon one occasion a pair of these insects were observed to meet when flying low in almost directly opposite directions, and to alight on the ground and couple; this occurred in a little patch of woods where there were very few other fireflies of any species near. Both were flashing rapidly as they flew toward each other.

P. 405, line 17 from bottom, after "p. 142."—Rennie (Insect Miscellanies, Lond., 1831, pp. 222–232) cites some observations on *Lampyris noctiluca*, which, however, are on the whole opposed to the theory of the sexual significance of the photogenicity.

Line 11 from bottom.—Olivier (Compt. Rend. Assn. Fr. Av. Sci., 1909, Sess. 37, pp. 573–580; 1er. Cong. Internat. d'Entomol., Brux, Aug., 1910, pp. 273–382), has also made some observations along the same line as Gorham.

Line 4 from bottom.—For "Avesbury" read "Avebury." (Sir John Lubbock.)

Rennie (supra) also notes the tendency of *Lampyris noctiluca* ♂ to fly into lighted rooms.

P. 406, line 8.—Before "light" insert "ordinary."

NOTES ON THE CHALCIDOID *TRICHAPORUS* FOERSTER OF THE FAMILY EULOPHIDÆ, WITH DESCRIPTION OF ONE NEW NORTH AMERICAN FORM FROM ILLINOIS.

BY A. ARSÈNE GIRAULT, BRISBANE, AUSTRALIA.

(Continued from page 52.)

Family Eulophidæ.

Subfamily Tetrastichinæ.

Tribe Tetrastichini.

Trichaporus Foerster, novum Ashmead, 1904.

(*Type*: *Trichaporus melleus* Ashmead.)

1. *Trichaporus melleus* Ashmead.

Ashmead, 1904, p. 512.

"Trichoporus melleus, sp. nov.

"Female: Length, 1.8 mm. Honey yellow, punctate, the eyes brown, the abdomen with a blackish spot on each side near the middle, the scape and legs pale yellowish; flagellum long, filiform, hairy; wings hyaline, the veins pale yellowish. The abdomen is cylindrical, pointed at apex, and as long as the head and thorax united.

"Male: Length, 1.4 mm. Agrees in colour with the female except that the blackish spots near the middle of the abdomen unite and form a transverse band, while the veins in the front wings are brownish. The flagellum is long, and the hairs are much longer than in the female.

"Brazil: Santarem; Chapada."

Type in the Carnegie Museum, Pittsburgh, Pennsylvania.

2. *Trichaporus viridicyaneus* Ashmead.

Ashmead, 1904, p. 512.

"Trichoporus viridicyaneus, sp. nov.

"Female: Length, 2–2.6 mm. Metallic bluish green to blue, punctate; scape, trochanters, apices of all femora, and all tibiæ and tarsi, except the last joint, pale yellowish; flagellum brownish yellow, pubescent; wings hyaline, the veins yellowish. The abdomen is long, cylindrical, twice as long as the thorax, pubescent, the first and second body segments about equal, shorter than the third, the first segment longer than the third, the sixth and seventh short, the seventh conical.

"Male: Length, 1.4–1.5 mm. Agrees well with the female, except in the usual sexual differences and in a slight difference in the colour of the antennæ and legs. The flagellum is darker, with longer hairs, and with only *one* ring-joint, while the front and middle femora are dusky only at base. The abdomen is cylindrical, a little longer than the head and thorax united.

"Brazil: Chapada, in April. Fourteen females, six male specimens."

Types in the Carnegie Museum, Pittsburgh, Pennsylvania.

3. *Trichaporus persimilis* Ashmead.

Ashmead, 1904, p. 512.

"Trichoporus persimilis, sp. nov.

"Female: Length, 2.8 mm. Metallic brown-black, punctate, the abdomen brown beneath; flagellum brown, hairy; scape, pedicel and legs, including the coxæ, honey yellow, the femora more or less dusky or brownish, especially basally; otherwise it is very similar to *T. viridicyaneus* except that the first body segment of the abdomen is twice the length of the second.

"Brazil: Chapada, in April. Two specimens."

Types in the Carnegie Museum, Pittsburgh, Pennsylvania.

4. *Trichaporus æneoviridis*, species nova.

Normal position.

Female: Length, 1.8 mm. Average, moderate for the family.

General colour bright metallic green, with a brassy sheen, the scutellum with a purplish hue in certain lights. Legs pallid yellow, including the apices of the coxæ; tips of apical tarsal joints dusky; antennæ dark, indefinite in colour, the scape paler yellowish, the pedicel and ring-joint slightly paler; venation indefinite, dusky yellowish. Eyes and ocelli dark garnet, the latter moderately large, in a flat isosceles triangle on the vertex, the distance between each about the same as the distance between each lateral ocellus and the margin of the eye.

Head (cephalic aspect) bilobed, longer than wide (dorsal aspect), as wide as the greatest width of the thorax, five times wider than long (cephalo-caudad), squamosely reticulated, the cephalic ocellus at the extreme median apex of the vertex (cephalad), one-third wider across the eyes (cephalo-caudad) than at the median line of the vertex, which is narrowed. Eyes ovate, half the length of the genæ, their surface much rougher than that of the head. Antennæ inserted below (ventrad) the middle of the face, slightly above (dorsad) the ventral ends of the eyes. Genal sulcus broad, distinct.

Mesothorax, including axillæ, scutum, scutellum and parapsides, strongly shagreened, or squamosely reticulated, the pleura less so, polygonally sculptured; dorsum of the metathorax roughly reticulated, or punctate, the median carina moderately strong. Abdomen closely, squamosely reticulated. Scutellum with four conspicuous longitudinal furrows, two on each side of the meson, the lateral ones barely visible from the direct dorsal aspect and at the lateral margin of the sclerite in the dorso-lateral aspect. Parapsidal furrows conspicuous, deep; mesoscutum with a slight carina along the median line. Abdomen ovate, the segments subequal, segment 2 longest, 3 shorter than 4, 5 and 6, widest at segment 4, and about equal in length to the head and thorax combined; caudal margins of the segments straight, or slightly concaved at the meson. Tarsi 4-jointed.

Fore wings hyaline, the marginal fringes short, the discal cilia uniform; postmarginal vein absent, the marginal vein about a fourth longer than the submarginal, the latter broken. Hind wings uniformly ciliate discally. Stigmal vein of fore wings moderate in length, clavate, and with an uncus.

Antennæ 9-jointed; scape about equal in length to the pedicel and first funicle joint combined, inserted not much below the middle of the face; pedicel subconic, over one-third shorter than funicle joint 1; ring-joint inconspicuous but evident; funicle joints 1 and 3 subequal, cylindrical oval, funicle 2 slightly longer, nearly twice the length of the pedicel; the three club joints decreasing in size, the basal joint longer than the pedicel, nearly a third shorter than funicle 3 and a fourth longer than joint 2 of the club; the apical joint conical, smaller than the pedicel, and ending in a short spine-like projection. Antennæ bearing stiff curved bristles, the flagellum longitudinally carinate. Ovipositor not exserted. Mandibles fuscous, tridentate, symmetrical, the outer and next tooth conspicuous, the inner or third tooth one-half shorter than the second, barely defined, its inner (nasal) margin obliquely concaved.

From 33 specimens, ⅔-inch objective, 1-inch optic, Bausch and Lomb.

Male.—Length 1.5 mm., smaller, more slender. The same. Abdomen cylindrical; antennæ slightly lighter in colour, more hairy, the setæ longer, 10-jointed, the same, but the funicle 4-jointed, the first joint distinctly smaller than the others, the ring-joint minute, subobsolete, and the two basal joints of the club subequal in length; funicle joints 2 and 3 equal, longer than 4. Mandibles the same.

From 4 specimens, the same magnification.

Described from four males, thirty-five females, reared in the insectary of the State Entomologist, Urbana, Illinois, May 27, 1908, from, supposedly, a single larva of *Malacosoma americana* (Harris) on apple twigs. On May 18 this supposed lasiocampid larva was dissected and found to contain the pupæ of this species, which emerged as adults on the date mentioned. The host was one of several larvæ sent to the insectary, together with several incidental puparia of a species of *Syrphus*, all of them, with this single exception, being parasitized by a solitary Ichneumonoid. The nature of its parasitism is, therefore, unknown, the host being either a dipteron of the family Syrphidæ, a hymenopteron of the family Ichneumonidæ or the lasiocampid. It is apparently a primary parasite of the latter.

It differs from all other species in the genus in being comparatively smooth, the head, thorax and abdomen squamosely reticulated, not punctate, agreeing somewhat in this respect with *Syntomosphyrum*, Foerster.

Habitat.—Centralia, Illinois (L. M. Smith).

Type.—Accession No. 37,543, Illinois State Laboratory of Natural History, Urbana, Illinois, 3 females, tag-mounted, female antenna in xylol-balsam (1 slide). Cotype No. 12,200, United States National Museum, Washington, D. C., 2 females on tags.

5. *Trichaporus colliguayæ* (Philippi).

Exurus colliguayæ Philippi, 1873, pp. 296–298, taf. 1, figs. 1, 1a, 1e.

Exurus colliguayæ Philippi, de Dalla Torre, 1898, p. 159.

Trichaporus colliguaya (Philippi), Ashmead, 1904, p. 512.

"1. *Gallen des Colliguai.* Taf. 1, fig. 1a-e.

"Man sieht sehr häufig Gallen am untern Theil der Kätzchen des Colliguai, *Colliguaya odorifera* Molina, eines Strauches aus der Familie der Euphorbiaceen, der in den mittleren Provinzen Chile's gemein ist, und dessen Holz beim Brennen angenehm riecht. Fig. 1 zeigt eine solche Galle in natürlicher Grösse. Die unteren zwei Drittel der Axe des Kätchens sind gewaltig aufgetrieben, im Gestalt einer länglich eiförmigen, etwas unregelmässigen Knolle, und tragen auf ihrer Oberfläche noch die schuppenartigen Deckblätter, auf denen die Staubgefässe entspringen, welche mehr oder weniger vollständig entwickelt sind. Schneidet man die Galle durch, so sieht man im Innern derselben eine unregelmassige, von 2–3 mill. dicken Wänden eingeschlossene Höhle, in welcher zahlreiche Maden sitzen. Im Anfang sind die Gallen gelbgrün, spater mehr roth, zuletzt, wenn sie beginnen trocken zu werden, braun. Sie sind von mässiger Consistenz, und milchen beim Durchschneiden weniger als die übrigen Theile der Pflanze. Die Maden zeigen nichts Auffallendes; man sieht deutlich mit dem Kopf dreizehn Ringe; sie zeigen keine Spur von Füssen und Augen, etc. Ich sammelte eine Menge dieser Gallen, und that sie in ein grosses Einmacheglas, um zu sehen, was sich daraus entwickeln würde, vergass aber über andern Geschäften meine Gallen, bis ich nach längerer Zeit in dem Glase viele hundert kleine Pteromalinen, so wie ein Paar grössere einer zweiten Art, aber kein einziges Exemplar einer Gallwespe oder Fliege fand. Ich muss daher glauben, dass die erst erwahnte Pteromaline die Gallen hervorbringt, und die zweite im Larvenzustand die Maden der ersten auffrisst.

"Die Gallen bildende Pteromaline scheint mir ein besonderes Genus bilden zu müssen, und ich habe sie *Exurus Colliguayæ* genannt (ξεονὸς "was einen Schwanz bildet, spitz zulauft). Der Körper ist 3 mill. lang, die Flügelgespannung beträgt 7 mill. Das vollkommene Insekt ist ganz schwarz und glänzend, bis auf die untere Hälfte der Schenkel, die Schienen und Tarsen, welche schalgelb sind. Der Kopf ist quer; seine drei Punktaugen liegen in einer graden Linie zwischen* den Netzaugen, wie man deutlich an der Puppe wahrnimmt, da deren Körper heller gefärbt ist; am vollkommenen Insekt sind sie schwer zu sehen. Die Fühler entspringen in der Höhe des untern Augenrandes, sind gekniet und nur so lang, das sie, zürückgeschlagen, bis etwas über den Ursprung der Flügel reichen würden, und sind nach dem Geschlecht verschieden, beim Männchen nämlich federbuschartig lang behaart und siebengliedrig, beim Weibchen sehr kurz behaart und sechsgliedrig. Das erste Glied ist keulenförmig und ziemlich dick; es reicht bis an den Scheitel und ist auf der Oberseite schwach behaart, sonst kahl; das zweite Glied ist verkehrt kegelförmig und etwa ein Drittel so lang; das dritte Glied ist beim Männchen an der Basis verdickt, zwei Drittel so lang wie das erste, und ähnlich ist das vierte, fünfte und sechste, nur nimmt ihre Dicke allmählich ab; das siebente ist etwas länger als das vorhergehende, im ganzen walzenformig, in der Mitte etwas dicker. Das dritte, vierte, fünfte, sechste Glied haben am Grunde einen Wirtel längerer Haare, das siebente ist überall gleichmässig und ziemlich lang behaart. Beim Weibchen sind die beiden ersten Glieder der Fühler ziemlich wie beim Männchen, aber alle folgenden sind walzenförmig und überall gleichmässig behaart; das dritte Glied scheint aus der Verschmelzung von zwei Gliedern entstanden zu sein. Mundtheile habe ich am vollkommenen Insekt nicht bemerkt, obgleich ich an der Puppe an der Stelle zwei braune Punkte gesehen habe, die ich fur die Mandibeln halten möchte.

"Der Hinterleib ist nicht gestielt, verlängert, allmählich zugespitzt, und ist die Spitze beim Weibchen länger; er ist auf der Bauchseite gekielt

*Beginning p. 297.

auf dem Rucken (wenigstens bei trocknen Exemplaren) concav, mit anliegenden, kurzen Härchen bekleidet, aber doch sehr glänzend, so dass man nur schwer erkennen kann, dass er aus sieben Gliedern besteht; an der Puppe ist dies leichter. Die Brust ist fast ganz kahl; der Rucken der Vorderbrust ist klein, kaum so lang wie der Kopf; die Mittelbrust ist ziemlich gross; das Schildchen deutlich, sonst durch nichts ausgezeichnet; die Hinterbrust sanft abschussig. Die Vorderflügel sind dadurch ausgezeichnet, dass der erste und einzige Nerv den Vorderrand selbst bildet bis zu zwei Dritteln der Länge, wo er einen stielformigen Ast nach hinten und aussen schickt. Die Hinterflügel haben an der Basis keinen Lappen. Die Beine sind lang und schlank, durch nichts Besonderes ausgezeichnet. Die Hüfte ist ziemlich dick; es sind zwei kleine Trochanter vorhanden; der Schenkel ist schlank, in der Mitte mässig verdickt, schwach und kurz behaart; die Schiene ist ziemlich walzenförmig,† und tragt am Ende einen kurzen Dorn. Die Tarsen sind kurzen als die Schienen, bedeutend dunner, walzenförmig; die einzelnen Glieder sind schwer zu unterscheiden; es sind ihrer fünf; das erste Glied ist das längste, das vierte und fünfte sind zusammen kaum länger als das dritte. Sehr lang sind die beiden Haffenlappen, fast länger als das fünfte Glied, während die Klauen sehr klein sind, nämlich kaum halb so lang wie das funfte Glied dick ist.

"Da ich mich sehr wenig mit dem Studium der Hymenopteren, namentlich der kleineren, beschäftigt habe, so muss ich es anderen Entomologen überlassen, zu entscheiden, welches die genauere Stellung dieses Insektes im System ist; in den wenigen, einschlägigen Büchern, die mir zu Gebote stehen, habe ich, wie gesagt, kein Genus finden konnen, in welches ich dasselbe hätte einordnen können.

"In der Abbildung auf Taf. 1 ist fig. 1d eine Galle, von aussen gesehen; fig. 1e dieselbe aufgeschnitten; 1 das weibliche Insekt, vergrössert; die darunter stehenden Linien geben die Grösse an; 1a ist ein stark vergrösserter Fühler des Männchens, 1b des Weibchens, 1c der Tarsus." Pp. 296-298.

The form of this species, as shown in the figure, is similar in general to that of *Trichaporus*, but the tarsi are 5-jointed, the antenna 6 jointed in the female, the scutellum small and apparently without grooves, the parapsidal furrows apparently absent, the body not punctate, and other characters, which make its present position questionable. I think, however, that Ashmead has placed it as well as circumstances allow. I

†Beginning p. 298.

do not know of existing specimens, but there are probably some in the Naturhistorische Museum, St. Yago, Chile.

6. *Trichaporus columbianus* (Ashmead).

Euderus columbiana Ashmead, 1888, pp. 104-105.

Euderus columbianus Ashmead, de Dalla Torre, 1888, p. 6.

Trichoporus columbianus Ashmead, 1900, p. 561.

"*Euderus* Haliday.

"(14) *Euderus columbiana*, n. sp.

"♀. Length, 10 inch. Dull brown, or bronzy-green, its whole surface, including the abdomen, strongly, confluently punctate. Head transverse, not wider than the posterior part of the mesothorax, and with only a slight antennal groove in front. Antennæ about as long as the thorax,‡ eight-jointed; scape slender, yellowish brown; flagellum dark brown, about twice as long as the scape, pubescent, the pedicel shorter than the funicle joint, the latter joint the longest, about twice as long as wide, the following joints being not much longer than wide, sub-moniliform. Thorax: collar transverse, rounded before; mesothorax with parapsidal grooves well defined; scutellum longer than wide, without grooves, rounded behind, sides parallel. Abdomen conic-ovate, cylindric, one-third longer than head and thorax together, the segments of nearly equal length. Legs dark brown, trochanters, knees, fore and middle tibiæ, and all the tarsi honey-yellow, hind tibiæ dusky in the middle. Wings hyaline, fringed with short ciliæ; the veins brown, the marginal is twice the length of the submarginal, the stigmal short, while the postmarginal is wanting.

"Hab.—Florida and District of Columbia."

I have been unable to connect, directly, *Euderus columbianus* Ashmead with *Trichaporus columbianus* Ashmead listed in Smith's (1900) Catalogue of the Insects of New Jersey, but as I cannot, in addition, find the original description of the latter, conclude that they are synonymic and that Ashmead intended the former species.

In characters, the species does not agree with either genus (*Euderus* Haliday or *Trichaporus*), as now limited (but does with the definition of Foerster, 1856), and can hardly belong to the Tetrastichini as limited by Ashmead (1904), the scutellum having no grooves. From the description of the species quoted in foregoing, being an eulophid with a long marginal vein, short submarginal and stigmal veins, without a postmarginal vein and grooves on the scutellum, with 8-jointed antennæ and complete parapsidal

‡Ending p. 104.

furrows (implied) sessile abdomen, the species falls near the omphalinine genus, *Closterocerus* Westwood. As I have not seen the species, however, I think that Ashmead's later determination should be accepted for the present, and so I have included it here.*

Smith (1900), gives the following note concerning this species: "Lives in Cecidomyid galls, widely distributed (Ashm.)." The species occurs in Florida, District of Columbia and New Jersey.

The types are probably in the United States National Museum, Washington, D. C.

Table of Species.

This table is constructed from the literature, and caution should therefore be exercised in identifying species by its aid alone. It forms merely an index to the species included within the group.

Females.

A. Species metallic bronze-greenish to bluish or brownish.

a. Dull brown or bronzy-green, confluently punctate.

Flagellum dark brown; scutellum without grooves; legs dark brown and honey-yellow..........*columbianus* (Ashmead).

b. Metallic brown-black.

Flagellum brown; scutellum with grooves; legs and coxæ honey-yellow, with some brownish on femora..*persimilis* Ashmead.

c. Metallic blue-green, punctate.

Flagellum brown-yellow; scutellum with grooves; tibiæ and tarsi pallid yellow, femora and coxæ green..........................*viridicyaneus* Ashmead.

d. Metallic, shining, bright green, brassy, squamosely reticulated.

Flagellum dusky, neutral; scutellum with four grooved lines; legs uniformly pallid yellow, the coxæ metallic green............................*æneoviridis* Girault.

B. Species shining black.

a. Flagellum black; legs yellow and black; scutellum without grooves.............................*colliguayæ* (Philippi).

C. Species honey-yellow.

a. Flagellum yellow; punctate; legs pallid yellow.*melleus* Ashmead.

D. Species bright metallic green, brassy; squamosely reticulated.

a. Metathorax punctate; scutellum with four grooved lines; legs

*It may be that *Trichaporus* Foerster, with Ashmead's *columbianus* as type, could be resurrected, while the group of Ashmeadian species now forming the genus, as here proposed, including the new species, could be renamed.

pallid yellow, the coxæ metallic green; flagellum dusky, the scape and pedicel paler................*æneoviridis* Girault.

Literature referred to.

1856. Foerster, Arnold.—Hymenopterologische Studien, Aachen, II, pp. 83, 84, 85.

1866. Taschenberg, E. L.—Die Hymenopteren Deutschlands nach ihren Gattungen und theilweise nach ihren Arten als Wegweiser für angehende Hymenopterologen, *etc.* Leipzig, p. 109.
Trichaporus Foerster.—Date of publication not given; preface dated August, 1865, Halle.

1867. Kirchner, Leopold—Catalogus hymenopterorum europæ, Vindobonæ, p. 186, No. 700.
"700.—G. Trichaporus Förster, Hym. Stud. II 85. 1, Tr. Sp.?"

1872. Walker, Francis.—Notes on Chalcidiæ, London, Part VI, pp. 104-105.
Translation of Foerster (1856).

1873. Philippi, Rudolph Amandus.—Chilenische Insekten beschrieben von Dr. R. A. Philippi. (Stettinger) Entomologischer Zeitung. (Herausgegeben von dem entomologischen Vereine zu Stettin), Stettin, XXXIV, pp 296-298, Taf: I, figs. 1, 1a—1e.

1888. Ashmead, William Harris.—Descriptions of some new North American Chalcididæ. CAN. ENT., London, Ontario, XX, pp. 104-105.

1898. De Dalla Torre, Carl G.—Catalogus Hymenopterorum hujusque descriptorum systematicus et synonymicus, Lipsiæ, V, pp. 27, 159.

1900. Ashmead, William Harris, in John Bernhard Smith.—Insects of New Jersey. A list of species occurring in New Jersey, *etc.* Supplement, 27th Annual Rep. State Board of Agr., Trenton, p. 561.
Trichoporus columbianus Ashmead.

1900. Smith, John Bernhard.—*Vide* Ashmead, 1900.

1904. Ashmead, William Harris.—Classification of the Chalcid flies or the superfamily Chalcidoidea, with descriptions of new species, *etc.* Memoirs of the Carnegie Museum, Pittsburgh, Pennsylvania, I, No. 4 (Publications of the Carnegie Museum, Serial No. 21), pp. 348-350, 392, 512.

1907. Schenied́eknecht, Otto.—Die Hymenopteren Mitteleuropas, *etc.*, Jena, pp. 489, 490.
Same as Ashmead (1904); table of genera.

1909. Idem.—Genera Insectorum (dirigés par P. Wytsman), Bruxelles, 97 me fascicule, Family Chalcididæ. pp. 427, 464, 465, 468.

Table to the genus as in Ashmead (1904); brief diagnosis of the genus, listing *colliguayæ*, *melleus*, *persimilis* and *viridicyaneus*, *Euderus columbianus* (p. 427).

(See also Kieffer, bionomic note on *colliguayæ*, Révista Chilena de Historia Natural. Organo del Museo de Valpaiso, VII, p. 111.)

NEW AFRICAN *TIPULIDÆ*.

BY C. P. ALEXANDER, ITHACA, N. Y.

The following species were given by Mr. Chas. W. Howard to Prof. Needham, and later turned over to me for examination. There were four specimens, representing three species, of which two are herein characterized as new. Mr. Howard's remark, that "the species were as thick as gnats," is interesting.

Styringomyia howardi, n. sp.

Holotype.— ♂, brown and gray; length, 5 25 mm; width, 4.75 mm.

Mouthparts dark brownish black; palpi, first segment very short; second segment large, oval, brown, apical third black; third more slender, brown, apical two-thirds black; terminal segment about as thick as the penultimate. Antennæ: first segment elongated, gray; second oval, enlarged at the distal end, remaining segments oval, gradually becoming more elongated to the tip; segments with a short pubescence and long irregular hairs, which are scarcely verticillate; first segment gray, second dark brown at tip, yellowish at base; remaining segments pale brownish yellow, the hairs darker; ommatidia large, coarse, black; front, vertex, genæ and occiput gray, with stout, scattered black bristles.

Pronotum large and prominent, showing an unusually generalized condition; the scutellum U-shaped, encircling the cephalic margin of the mesothoracic præscutum, with about three prominent bristles on the lateral margin; the scutum is narrower, running to an obtuse point cephalad, with a group of bristles along the lateral margin. Mesonotum: præscutum with a row of bristles along each side of the median line and a row along the lateral margin, this row incurving near the cephalic margin of the sclerite; scutum with four bristles on each half; the scutellum with a bristle on either side of the median line; postscutum and metanotum unarmed. Pronotum brown, pale apically, with an inverted U-shaped pale mark on the scutum; mesonotum præscutum, middle line pale, remainder brown; scutum grayish brown, yellow along the cephalic margin passing around the black bristle; scutellum yellow medially, brown laterad of the bristle, postscutum brown; metanotum brown; sterna yellow; epimera and episterna reddish brown, forming a narrow longitudinal band.

Halteres pale brown, subapically darker brown; tip yellow. Legs short and stout, thickly covered with appressed hairs; coxæ short, cylindrical, in the fore leg about as long as the trochanter; in the middle leg shorter than the prominent trochanter; in the hind leg prominent, much exceeding the shorter and narrower trochanter. Femora rather short, slender proximally, soon thickening so as to become almost clavate distally; the fore femora have stout, long hairs, which are scattered irregularly amongst the appressed hairs, becoming very numerous near the apical portion of the lower surface of the segment. Tibiæ slender throughout, tibiæ and metatarsi with a few prominent hairs regularly disposed; the other tarsal segments with a single hair at the tip. The fore femora are as long as the succeeding segments combined; the hind legs are longer than the others. Fore legs lacking (in the holotype); middle leg, coxæ and trochanter light yellow; femora yellow, with a medial and subapical brown band; tibiæ yellow, with a dark band before the middle and at the tip; tarsi yellow-tipped with dark brown; fifth segment and claws dark brown. Hind legs, coxæ, trochanters and femora as in the fore leg; tibiæ and tarsi yellow, excepting the last tarsal segment, which is darker.

Abdomen with numerous scattered hairs, yellow; the apical margins of the segments brown.

Wings with a faint yellow tinge; costal border and radial veins yellow; remaining veins darker; a dark suffusion around cross-vein *r-m*, at the union of M_3 with M_{1+2} and along the basal deflection of Cu_1. Venation (see fig. 2): S_c short, approximated with R basally; its tip opposite the origin of R_s; R short, the tip of R_1 before the middle of the

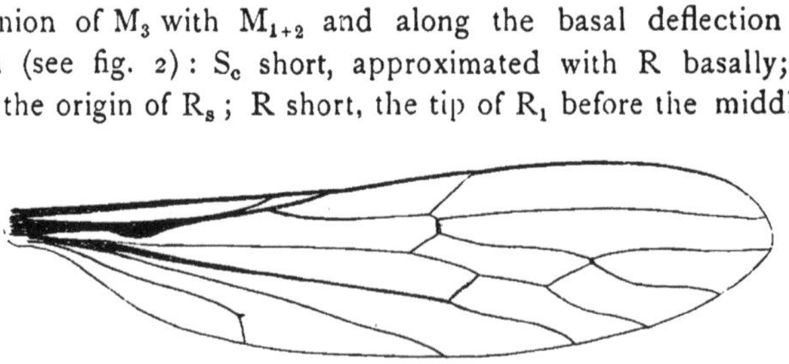

FIG. 2.—*Styringomyia howardi*, holotype.

wing, the sector originating a short distance back from the tip; R_s straight, rather long; R_{2+3} very short, oblique; deflection of R_{4+5} very short, scarcely equal to the *r-m* cross-vein; R_{4+5} long. M forks anterior to the fork of R_s; deflection of M_{1+2} rather long; M_3 in a line with M, strongly deflected cephalad toward M_{1+2}, nearly, if not quite, obliterating the cross-vein *m*. Basal deflection of Cu_1 under the middle of cell 1st M_2. First anal fused with Cu at extreme base; 2nd anal strongly curved at tip with a spur at the curve, which may be a remnant of a forked anal.

Paratype.—♂. This specimen is much darker than the type; the first six antennal segments are dark, remainder yellowish; thoracic dorsum dark brown, where it is light brown in the type; yellow of abdomen replaced by dark brownish gray, etc. This is but an extreme in colour.

This species is remarkably similar to the species mentioned by Osten Sacken (Mon. Dipt. N. Am., IV, p. 102, 103). The main differences are in the venation, the elongated cell 1st M_2 and incurved second anal with a spur at the curve being peculiar to *S. howardi.*

Holotype.—♂, Queliniani, Zambesi R., Dec. 20, '08; coll. Mr. C. W. Howard.

Paratype.—♂, with the type.

The only species described from Africa is *S. cornigera* Speiser (Dipt. aus Deutschland Afrikanischen Kolonieen, p. 130-132, fig. 1*). This insect differs so remarkably from the remaining species of the genus, which otherwise form a homogenous compact group, that I propose to set it off in a new subgenus.

Neostyringomyia, subgen. n.

Char.—Radius long, its tip beyond the middle of the wing; R_s remarkably shortened, no longer than the *r-m* cross-vein; R_{2+3} sinuate, leaving cell R_1 very different in shape from that which obtains in the subgenus *Styringomyia;* cross-vein *m* long and prominent; basal fusion of Cu and 1st A very long; prothorax narrow, scarcely one-fourth as wide as the head; above the antennæ a short, bent spatulate horn.

Type.—*S. cornigera*, Speis.

Cornigera is obviously of more recent derivation than the members of the subgenus *Styringomyia*, and its venation is almost normal; the retreat of R_{2+3} toward the base of the wing may give a hint to the manner in which the remarkable venation of *Toxorhina* came about, perhaps by the fusion of R_{2+3} with some other vein, such as R_1.

A species was described from the Pacific Islands by Grimshaw in 1901, as *S. didyma* (Fauna *hawaiiensis*, Vol. 3, pt. 1 (Dipt.), pl. 1, figs. 14–16), from Honolulu, Oahu De Meijere, in his recent paper, "Studien

*Berl. Ent. Zeitschr., 52 (1907).

uber Süd-ostasiat. Dipteren, V,"† records the species from much farther west (Batavia, Java, etc.). *Styringomyia didyma* belongs to the typical subgenus, and is extremely similar to the fossil species described by Löew** and Osten Sacken, as well as to the species under consideration. All of the species of the subgenus *Styringomyia,* as here limited, are very similar to one another in venation, and the coloration is inclined to be variable. *S. didyma* differs from the new species as follows: The wings are shorter in *didyma;* R_{4+5} is in a direct line with R_8, whereas there is a deflection at the origin of R_{4+5} in *S. howardi. Didyma* has no spur at the curve of 2nd anal. The coloration of the thorax of the two species is different. The male genitalia of the species have not been studied critically, and must furnish the ultimate criterion. It is, of course, possible that when further collections are made, intermediate stations for the genus will be discovered, and then it may be proved that *S. howardi* is merely a variant of *S. didyma.* However, I prefer to describe it as distinct at present.

In the end of Vol. III of the Monograph, p. VII, Osten Sacken came forward with the surprising intelligence that the genus *Styringomyia* still existed. He says: "During my passage through Stockholm in 1872, I made the interesting discovery that the genus, besides its occurence in amber and copal, is found living in Africa. I saw several specimens among the unnamed Diptera from Caffraria (from Wahlberg's voyage) in the Stockholm Museum. The species was apparently different from that included in the copal, which I possess." Later, in "Studies on Tipulidæ,"* he states, "This singular genus, originally described from specimens included in copal from Zanzibar, and also in amber, has been discovered since as still living in South Africa. In the museum in Stockholm I have seen recent specimens brought from Caffraria by Wahlberg."

Despite Prof. Speiser's statement (l.c., p. 132), that Osten Sacken probably referred to *Elephantomyia wahlbergi* Bergr., when he made the last-quoted statement, I have no doubt but that Osten Sacken saw specimens of a true *Styringomyia* in Stockholm; an error of this calibre was not customary with Osten Sacken.

Mongoma zambesiæ, n. sp.

Holotype.— ♀, brown; length, 5.75 mm.; width, 5.5 mm.

Rostrum and palpi dark brown; antennæ, first two segments dark

†Tijschr. voor Entomol., April, 1911, p. 40.
**Loew, H. Dipterol Beiträge, I, p. 7, with f. (1847).
*Berl. Ent. Zeitschr., Bd. XXXI, 1887; Heft., II, pp. 185, 186.

brown, third light brown, remainder lacking. Front, vertex, genæ and occiput dark brown.

Thorax: Mesothoracic præscutum strongly produced cephalad, entirely covering the pronotum; cervical sclerite elongated, prominent; transverse suture scarcely V-shaped; mesothoracic præscutum, dark brown anteriorly, posteriorly with a pale brown median line, which extends back across the scutum, remainder of thoracic dorsum dark brown. Sterna, episterna and epimera brownish yellow; halteres pale; legs long, dull brown, at the joints somewhat darker; no processes on the fore femora, as described for *M. fragillima* and *M. curtipennis.*

Abdomen uniform brown.

Wings hyaline, costal margin yellow, stigma rather indistinct. Venation (see fig. 3), Sc very long, as in all members of the genus; R long, cross-vein *r* near its tip. R_s gently arcuated, forking far before the tip

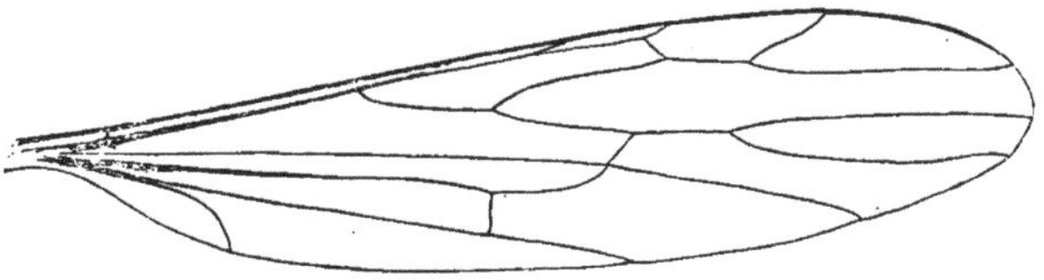

FIG 3 —*Mangonia zambeziæ*, holotype.

of Sc_1 and in a line with R_{4+5}; the cross-vein *r* far before the fork of R_{2+3}; R_2 short, oblique; R_3 long, in a line with R_{2+3}. R_{4+5} fusing with M_{1+2} to form the proximo-anterior border of cell M_2, thus obliterating the *r-m* cross-vein. M forks at the lower corner of cell M_2, M_{1+2} departing cephalad, fusing with R_{4+5} for a distance and finally separating, free at the margin; M_3 in a line with M. Cu short, its fork far back, the free position of Cu_1 very long, fusing with M_3 at the fork of M, and continuing to the margin so fused. Cu_2 fuses with 1st A far back from the wing-margin, so that 1st A + Cu_2 is over twice the length of the free portion of Cu_2 alone. 2nd A is very short, suggesting the condition found in *Petaurista.*

Holotype.—♀, Queliniani, Zambesi R., Dec. 20, '08; Mr. C. W. Howard.

The genus *Mongoma*, of which ten species have been described, has a world-wide distribution in the tropics; two species have been described from the West Indies, five species from the East Indies and Australia, and three species from Africa. The genus is distinguished by the excessive length of Sc, the obliteration of the *radio-medial* cross-vein by the long fusion of R_{4+5} with M_{1+2}, and the decided tendency of Cu_2 to fuse with 1st A.

The West Indian species (*manca* and *pallida* Will., Dipt. St. Vincent, p. 291–293, figs. 6, 7, of *pallida*) and possibly *M. albitarsia* Dol. (E. Ind.), also, which I have not seen, are the most generalized members of the genus, in that Cu_2 and 1st A are distinct to the wing-margin. The intermediate group, containing *trentepohlii* Wied. (see Wiedemann, Aussereur. Zweifl. Insekt., I, 551; 18, tab. VIb, fig. 12 ; a better figure in De Meijere, Tijd. voor Ent., 1911, pl. IV, fig. 42), *fragillima* Westw. (see Westwood, Trans. Ent. Soc. Lond., 1881, pl. 17, fig. 1 ; also Needham, 23rd Rept., N. Y. St. Ent., pl. 21, fig. 6), and *exornata* Berg. (Bergr., Entomol. Tidskrift, 1888, opp. p. 130, fig. 3), has Cu_2 fused with 1st A for a short distance back from the tip (Cu_2+ 1st A less than one-half Cu_2). A third stage in the specialization of this part occurs in *M. pennijes* O. S. (E. Ind). (See De Meijere, l.c., pl. IV, fig. 39.) The maximum of specialization, as far as I know, occurs in the present species, where the fusion of Cu_2 with 1st A is notable, and suggests the condition obtained in the families *Empididæ* and *Dolichopodidæ*.

Of the three described African species, *M. zambesiæ*, comes closest, apparently, to *exornata*. *M. fragillima* (and probably *M. curtipennis* also, according to Speiser, who compares it with *fragillima*), has vein M_3 separating from Cu_1, and continuing distinct to the wing-margin ; both of these species possess a curious spur-like structure at the base of the fore femora, which does not occur in *M. zambesiæ*.

I have a ♂ of *M. exornata* Bergr., taken at Queliniani, Zambesi R., Dec. 20, '08, in which the fore legs are lacking, and I am unable to state whether or not this structure occurs there. *M. exornata* has been recorded from Delagoa Bay, Portuguese East Africa ; Caffraria, E. Cape Colony, and Amani, German E. Africa. It is apparently widely distributed throughout Eastern Africa.

ON THE OCCURRENCE OF A EUROPEAN SPECIES OF MYMARIDÆ IN NORTH AMERICA.

BY A. ARSÈNE GIRAULT, BRISBANE, AUSTRALIA.

Up to the present I have been successful in finding but a single species of the family Mymaridæ, common to Europe and North America. This species is *Anaphes pratensis* Foerster, which I have captured in Illinois, and of whose characteristics I write of in a paper on Chalcidoidea, to be published soon in Germany ; the species is recorded from America in another paper, to appear in the Journal of the New York Entomological Society. The identification of the species is based on comparison with specimens found in the collections of the United States Museum, labelled as from France, the specific label in the handwriting of Ashmead. The evidence of the establishment of the identification is but presumptive, yet even if wrong, it is still true that we have specimens of a species common to both continents, whatever the name of the species may be. Only the specimens in the National Museum bear witness that it is *pratensis*, and their origin is not known. Nevertheless, Ashmead must have had good reason for so labelling them. For the present. identification must hold.

As I state elsewhere, the species is allied to both *iole* Girault and *nigrellus* Girault, and in this statement the species *hercules* Girault should have been included also; these are all American forms. From both *nigrellus* and *hercules* Foerster's species may be distinguished readily by reason of the fact that the marginal cilia of the fore wing at apex are distinctly longer (by over a third, they are about two-thirds the greatest width of the fore wings). There are a number of minute discal cilia scattered under the venation of the posterior wing, the fore wings are less regularly and uniformly fumated, and the proximal tarsal joints of all legs are longer. Its other characteristics, as compared with those of the American species mentioned above, are given in the papers referred to in this connection. The posterior wing bears two lines of discal cilia along each edge, the inner line of the two out some distance from the edge, toward the mid-longitudinal line of the blade.

In addition to the specimen of *pratensis*, recorded elsewhere, as having been captured in Illinois, I have since seen the following specimens, kindly sent to me by Mr. H. L. Viereck, and belonging to the Connecticut Agricultural College: Two slides bearing respectively a single male and female specimen (one pair in all), and each the label, "New Haven, Ct., 10 May, 1904. H. L. Viereck, *Taraxacum officinale*." In the United States the species occurs in Illinois (Urbana), and Connecticut (New Haven). The Connecticut specimens have been returned to Mr. Viereck.

While on this topic, it is meet to mention the possible identities of several other American forms with those of Europe. A species recently described as *Gonatocerus brunneus* Girault may possibly be *Gonatocerus flavus* Walker (so called), and my *(Stephanodes) Polynema psecas* is very similar, and possibly identical with *Polynema enockii* (Girault), a species which Enock described as *Stephanodes elegans (Stephanodes* equals *Polynema; elegans* preoccupied in *Polynema)*. I have considered them distinct, however, as they seem so. Still they must be considered but questionably valid until a better opportunity is afforded for comparing them.

NEW SPECIES AND GENERA OF NORTH AMERICAN LEPIDOPTERA.

BY WM. BARNES, M.D., AND J. H. MCDUNNOUGH, PH.D., DECATUR, ILL.

(Continued from page 57.)

Amolita delicata, sp. nov.

♂.—Head, thorax and abdomen pale gray; primaries very pale ochreous, suffused in the basal half and along costa with grayish, and finely sprinkled with black scales; faint traces of an oblique ochreous dash from apex to end of cell, caused by a lack of black scaling at this point; two minute black points may or may not be present at end of cell, close together; veins more or less marked with ochreous; fringes concolorous; secondaries slightly smoky, with traces of a dark terminal line and a smoky line cutting the white fringes.

♀.—Very similar to the ♂; the oblique apical dash better defined, due to darker marginal shading on both sides; frequently traces of a dark shade in the cell; veins light ochreous, giving a distinct strigate appearance to outer area of wing; secondaries pure white, with slight sprinkling of dark scales along costa and outer margin; fringes white with dark basal line. Beneath primaries of ♂ smoky, secondaries whitish, sprinkled with smoky especially along costa, and with faint discal dot; in ♀ primaries are much lighter than in ♂ and the discal dot of secondaries is wanting, Expanse, ♂, 25 mm.; ♀, 29 mm.

Habitat: White Mts., Ariz., 9 ♂s, 7 ♀s. Types, collection Barnes.

Vein 8 of secondaries arises from about the middle of the cell and not from the base as in Hampson's definition of the genus *Amolita*. As, however, this is also the case with *roseola* Sm., which is retained in the genus, we place it here rather than in *Doerriesa* Staud., in which it would fall according to Hampson's tables. The ♂ antennæ are laminate. The species varies somewhat as regards the black sprinkling, several specimens being almost uniformly pale ochreous, whilst others are distinctly sprinkled, with the veins showing clearly.

Amolita fratercula, sp. nov.

Primaries: ground colour pale ochreous suffused with gray, and rather evenly shaded with smoky brown; the most prominent feature is an oblique dash of the ground colour which extends from a point on outer margin just below costa inwards to the cell and is shaded superiorly with smoky brown, which shade extends more or less distinctly through the cell to the base of wing, leaving the cubital vein as a fine ochreous line distinct to the discocellular vein. In the ♂ two very faint dark dots are visible at the end of cell; fringes concolorous. Secondaries in the ♂ deep smoky with a pale line at the base of the dusky fringes; in ♀ slightly smoky with pale fringes, cut by a slightly darker line. Beneath primaries smoky, lighter outwardly; secondaries lighter, sprinkled with smoky brown. Expanse, ♂ 24 mm., ♀ 31 mm.

Habitat: ♂, Palmerlee, Ariz.; ♀, White Mts., Ariz.; 1 ♂, 1 ♀. Types, collection Barnes.

The species is closely related to *delicata*, but is in general much darker and lacks the strigate appearance of this species, due to the fact that the veins in the outer area of primaries are not visible; the dusky secondaries in both sexes first led us to separate it. The apical ochreous dash is also not direct from the apex of the wing but from a point on the outer margin below apex; the palpi are distinctly longer than in *delicata*.

Redingtonia. gen. nov.—(Type *R. alba*, sp. nov.)

Palpi short, upturned, third joint porrect; proboscis well developed, front with a pointed corneous prominence, its lower edge produced to a trilobate plate with corneous plate below it; head and thorax clothed with rough hair, intermingled with scales; anterior tibia unarmed; posterior tibiæ clothed with long hair, without spines; primaries with broad cell, vein Cu_2 from well before lower angle, veins Cu_1, M_3 and M_2 from around lower angle, M_1 from just below upper angle, areole present, veins R_3 and R_4 stalked, from apex of areole with R_5, R_2 from areole, R_1 from middle of cell. Secondaries with M_2 obsolescent from below middle of discocellular, R and M_1 from apex of cell, M_3 and Cu_1 from lower angle.

The extraordinary frontal protuberance, which may be compared to that of *Azenia*, with an extra pointed prominence added dorsally, as well as the rough hairy squammation, sufficiently characterize this genus. It falls near *Azenia* Grt., according to Hampson's tables (Lep. Het., Vol. IX).

R. alba, sp. nov.

Front and abdomen pale ochreous. Head, thorax, and wings pure white, immaculate. Beneath primaries rather smoky, secondaries white. Expanse, 29 mm.

Habitat: Redington, Ariz., 2 ♀s. Type, collection Barnes.

Genus Homolagoa, gen. nov.—(Type *H. grotelliformis*, sp. nov.)

Palpi upturned, 3rd joint long, pointed, smoothly scaled; antennæ ciliate, ocelli present; thorax clothed rather roughly with hair and scales; abdomen of ♀ with a thick tuft of hairs at extremity; tibiæ unarmed, front with a prominent wart-like conical tubercle and a slight infra-clypeal plate; primaries with well rounded outer margin, vein R_1 from middle of cell, R_2 from upper angle of areole, R_3 and R_4 on long stalk, from apex of areole with R_5, M_1 from below upper angle of cell, M_2 and M_3 close together from above lower angle of cell, Cu_1 from lower angle, Cu_2 from beyond centre of cell. Secondaries with Sc. joined to cell at base, R and M_1 slightly stalked, M_2 curved downwards at base from well above lower angle of cell, M_3 and Cu_1 connate from lower angle, Cu_2 from beyond centre of cell.

The presence of a well developed vein M_2 on secondaries would place the genus in the family *Erastrianæ* of Hampson. Apparently its position would be somewhere near *Exyra* Grt. The frontal structure and the abdominal tuft of the ♀, similar to that found in *Lagoa* and certain Liparid species, render the genus easily recognizable.

H. grotelliformis, sp. nov.

Palpi blackish; head, thorax and primaries white; abdomen white with the segmental divisions banded with black; maculation of primaries much as in certain *Grotella* species; a black dot on costa at base; a transverse subbasal band of three black dots, one on costa, one on inner margin and the middle one equidistant from both; a transverse median band of 5 black dots slightly curved inward at costa consisting of a dot on costa, two vertically placed dots at end of cell, a dot below vein 2 and another on inner margin; fringes white, slightly tipped with black in costal portion; secondaries smoky, paler basally, with faint trace of a dark antemedial line; fringes white. Beneath primaries dark smoky brown with white fringes, secondaries white with discal dot and dot about middle of costa. Expanse, 22 mm.

Habitat: Redington, Ariz; Palmerlee, Ariz., 1 ♂, 2 ♀s. Types, collection Barnes.

Taruche areloides, sp. nov.

Head, front and palpi dark purple-brown, tegulæ, thorax and abdomen cream-coloured; primaries with basal third as far as inner margin of orbicular cream-coloured, remainder of wing deep purple-brown, shaded broadly at anal angle with lighter shades; basal line geminate, gray-green, extending half across wing; t. a. line geminate, gray-green, the lines broader at costa, angled inwardly in the cell, incurved on submedian fold; orbicular and reniform small, oval, outlined with black and filled with blue-black scales, the former usually entirely within the dark area of wing, occasionally with the inner edge just projecting into the white area; on the costa just beyond reniform a large white quadrate patch from the base of which the geminate t. p. line arises and bends sharply inward below reniform and orbicular, almost reaching the margin of the dark area of wing; from a point below the orbicular it turns towards the inner margin, forming two lunulate marks, the upper being the larger; the space beyond the t. p. line is almost entirely filled with bluish purple; s. t. line indistinct, marked with creamy at costa and in central area, incurved at vein 2; a broken terminal dark purple-brown line; fringes bluish purple, cut with white opposite cell and between veins 2 and 3. Secondaries whitish with narrow smoky border in ♂, almost entirely smoky in ♀; fringes pale. Beneath, primaries smoky with the white patch of upper side marked in ochreous. Secondaries suffused with pale smoky brown, with a discal spot and indistinct postmedian line angled sharply opposite the cell. Expanse, 27 mm.

Habitat: White Mts., Ariz., 3 ♂s, 5 ♀s. Types, collection Barnes.

Closely related to *areli* Stkr.; differs in the much larger size of the white patch and the fact that the orbicular is not contained within the light area of wing.

(To be continued.)

THE OLDEST AMERICAN HOMOPTEROUS INSECT.

BY T. D. A. COCKERELL. UNIVERSITY OF COLORADO.

With very few exceptions, the cretaceous strata of North America, so rich in various organic remains, have failed to yield insects. A cockroach from the Judith River Beds in Montana has been described as *Stantoniella cretacea* (Handlirsch). A Protoblattoid from the Kootanie of Montana is called *Lygobius knowltoni* Mitchell. Beetle remains named *Archiorhynchus angusticollis* Heer, *Curculiopsis cretacea* (Heer), and *Elytrulum multipunctatum* (Heer), are from the lower cretaceous of Greenland, while one from the Pierre formation of Manitoba is named *Hylobiites cretaceus* Scudder. Egg-masses from the Laramie Beds of Colorado are called *Corydalites fecundus* Scudder. Considering the enormous time represented by the cretaceous, and the richness of the flora, it is certain that there must have existed a succession of insect-faunæ including innumerable types, almost all of which are now unknown to us. This is particularly unfortunate, because during this period the modern families of insects must have been in course of evolution. Tertiary insects we have in abundance, but they are not old enough to afford much clue to the history of living groups; early mesozoic fossils, so far as found in this country, represent the least specialized of modern orders. In other parts of the world, cretaceous insects are also extremely scarce; of Homoptera, excepting some very dubious gall-like objects on *Eucalyptus* leaves, there is only a single species, the cicadid *Hylæoneura lignei* Lameere and Severian, from Belgium. The first American cretaceous Homopteron has just been found by Mr. Terry Duce in the Pierre formation at Lesser's brickyard, Boulder, Colorado. There is no doubt about the formation, as the specimen is in the same piece of rock as the characteristic mollusc *Scaphites nodosus* Owen. The formation is marine, but it was evidently laid down close to land, and the insect doubtless fell or was washed into the sea.

Petropteron mirandum, n. g., n. sp.

A tegmen or upper wing, the part preserved 7½ mm. long, the actual length probably about 9½; width near the middle about 4½; shape subtriangular, broadly widening apically; veins strong, reddish brown, membrane apparently strong, no markings of any kind; venation as shown in the figure, the interpretation given being scarcely open to doubt, with the possible exception of the first anal, which may be in

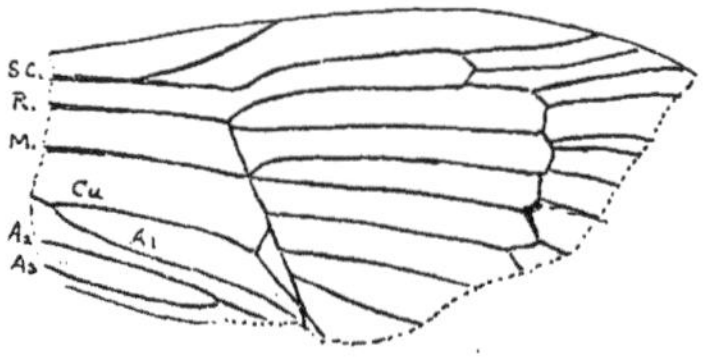

FIG. 4.—*Petropteron mirandum*, n. sp.

reality the inferior branch of the cubitus; there is no sign of a free first anal. There are two series of gradate veins, the inner placed somewhat as in *Dicranotropis*, the outer much as in the eocene genus *Eofulgorella*, and many living forms. The closed anal cell is normal for many Homoptera, and is exactly as in the European cretaceous *Hylæoneura*. The lower branch of the subcosta, although bulging in the direction of the radius near the beginning of the first series of gradate veins, is not connected with it by any cross-vein at this point. The triangular cell in the branches of the cubitus, contiguous with the first gradate series, finds a parallel in Kirkaldy's "restored" figure of *Aneono*. The basal union of cubitus and first anal is as in *Scolypopa*.

I suppose the insect to be a Fulgorid, and this possibility is supported by the occurrence of Fulgoridæ in the older Purbeck Beds of

England. It is quite possible, however, that it belongs to an extinct family.

The name *Petropteron* is in allusion to the Pierre formation.

P. S.—On renewed minute examination, I feel sure I see traces of the end of a free first anal. There seems to be a longitudinal fold or distortion which makes it impossible to follow it any distance toward the base.

BOOK NOTICE.

CONTRIBUTIONS TO THE NATURAL HISTORY OF THE LEPIDOPTERA OF NORTH AMERICA, by Dr. William Barnes and Dr. J. H. McDunnough, Decatur, Ill.

Under the above title have appeared the first two parts of a new and much needed work on the Lepidoptera of North America which will meet with the heartiest commendations from all corners of the continent.

Prior to the publication by Dr. Holland of "The Butterfly Book" and "The Moth Book," there were but few entomologists who could afford to possess, or were fortunate enough to have access to, the rare and expensive separate works and long sets of volumes of periodicals in which to look for figures and descriptions. These two books, and particularly the plates, for the text is of necessity very limited, have proved of the greatest help to beginners and the more advanced as well, and many of us ventured to hope that the time might soon come when a reliable figure of every known North American species would be available. Now after a wait of eight years, our expectations begin to be realized, as the "Contributions" are exactly what we most needed, namely, a series of monographic reviews of families or smaller groups, giving descriptions of all the species, references to the more important literature, placing the generic names on a more stable basis, and last, but not least, photographic figures of each and every species.

Part I deals with "The Cossidæ of North America" and consists of 35 pages of text and seven plates (three of structure and four of imagos) and an index. Part II is entitled "The Lasiocampid Genus Gloveria and Its Allies"—17 pages of text, one plate of venation and three plates of imagos and an index, and covers a little wider range of territory, species from Mexico and Central America being included.

The size of page conforms with Dr. Holland's books, the text is well printed on excellent paper and the illustrations are all on plate paper. Much care has evidently been taken to secure accurate reproductions and in many cases the actual types are shown. We quote the authors' remarks from the introductory chapter of Part I: "Owing to the relative rarity of many of the smaller species from the south and south-west, very few of our North American species have ever been figured. It has therefore seemed advisable to us to illustrate as fully as possible. In many instances we have been enabled to present a figure of the type specimen; in all other cases the specimen used for figuring purposes has been compared with the type either by ourselves or some competent authority." Some of the types referred to are in rather dilapidated condition and it is well that they have been photographed so that their appearance may be retained in a more permanent manner and it is to be hoped that the under surface has also been preserved photographically, although no under sides are shown in any of the figures. Probably this is because, in these groups, the markings of the under side are not of much value in differentiating between species.

While we have nothing but praise for this work, it is our duty to mention the slight and almost inevitable typographical errors, which have caught our eye. In Part I, page 33, lines 12 and 8 from foot of page, the genus Prionoxystus is spelled without the second o.; and also on line 12 from foot for *robinæ* read *robiniæ*.

In Part II, in the explanation of Plates II and IV, the word forma is printed in italics, making it appear to be part of the specific names, while the text indicates that they are aberrations.

We hope that these parts will shortly be followed by many others dealing with groups badly in need of elucidation. The price of Part I is $1.50 and of Part II $1.00, and they are obtainable from the authors.

Since writing the above, Part III has come to hand, entitled "Revision of the Megathymidæ," 43 pages. Price, $1.25.

There has been much confusion in identification of the species of these "giant skippers," and also concerning the two sexes of several species, as well as through publication of wrong figures—and the seven half-tone plates of the butterflies, and of their structure, combined with the carefully-prepared text, should enable anyone to correctly identify the specimens they may be fortunate enough to acquire.

Other parts to follow in the near future will deal with "A List of Types in the Barnes' Collection," "Illustrations of Typical and Rare Specimens," and "Fifty New Species, Fully Illustrated."

A. F. WINN.

Mailed March 9th, 1912.

John B. Smith

The Canadian Entomologist.

VOL. XLIV. LONDON, APRIL, 1912. NO. 4

OBITUARY.

JOHN BERNHARDT SMITH.

It is with profound regret that we have to record the death, from Bright's disease, of Dr. John Bernhardt Smith, Professor of Entomology at Rutgers College, New Brunswick, N.J., Entomologist to the New Jersey Agricultural Experiment Station, and State Entomologist of New Jersey, which occurred at his home during the morning of March 12, 1912.

Dr. Smith was born in New York City on November 21, 1858, so he died at a comparatively early age. It is a coincidence that the late Dr. James Fletcher and the one we now mourn, who were such close friends, should be called away at about the same age. Dr. Smith's early education was received at the Public Schools. He practised law from 1880 to 1884, but his heart was not in such work, and during this latter year he was appointed as a special agent to the United States Department of Agriculture, which position he held until 1886, when he was made Assistant-Curator of Insects in the United States National Museum. Here he remained until 1889, when he was appointed Professor of Entomology at Rutgers College and Entomologist to the New Jersey Agricultural Experiment Station. In 1894, he also received the title of State Entomologist of New Jersey. During the years 1882 to 1890 he was the editor of Entomologica Americana. For several years he was also editor of the "Bulletin of the Brooklyn Entomological Society."

Dr. Smith was an extremely busy man, one who in every way served his state and country as few men have. A man of wide experience and deep study he has, in his published works, left behind him a monument of knowledge which will last for all time and which will undoubtedly serve as a guide for many future students of entomology. While in the Museum at Washington, he published some very valuable monographic works, namely, "A Monograph of the Sphingidæ of America, North of Mexico," "A Revision of the Lepidopterous Family Saturniidæ;" and "Preliminary Catalogue of the Arctiidæ of Temperate North America." Bulletin No. 44 of the U. S. N. M., pp. 1-424, "A Catalogue, Bibliographical and Synonymical, of the species of moths of the Lepidopterous Superfamily

Noctuidæ, found in Boreal America, with critical notes," was prepared by him and appeared in 1893. This is indispensable to students of these insects, as are also his many "Contributions toward a Monograph of the Noctuidæ of Boreal America." His best scientific work was undoubtedly in this family, of which he was our leading American authority. It is impossible to mention here the many articles which he published, in revising genera, describing new species (of which he created many hundreds), etc. The first paper he published in THE CANADIAN ENTOMOLOGIST appeared in Volume XIV. Since that date he has been one of our most valued contributors. Articles from his pen have been published in 24 different volumes of this journal. A bibliography of his systematic papers would fill many pages; it is to be hoped that such will soon be prepared.

In 1891, Dr. Smith published a "List of the Lepidoptera of Boreal America," which was used generally by lepidopterists. This check list was revised and re-published in 1903. "Explanation of Terms Used in Entomology" was prepared by him and appeared in 1906. His "Catalogue of the Insects of New Jersey," the third edition of which recently appeared, is an extremely useful publication, and the only one of its kind which has been published by any state in the United States.

Other important works, of a popular nature, written by Dr. Smith, are "Economic Entomology," published in 1896, which is a valuable book for students of entomology, farmers, etc., and "Our Insect Friends and Enemies," which appeared in 1909. This latter treats of insects in relation to man, to other animals, to one another, and to plants, and in it there is also a chapter on the war against insects.

As an economic entomologist few men in the world were his equal. His series of annual reports, the first of which was included in the Tenth Annual Report of the New Jersey Agricultural Experiment Station, 1889, and the last, that for 1910, which was published in 1911, together with the many economic bulletins which he prepared, form a valuable source of reference concerning injurious insects, particularly those occurring within the State of New Jersey. The very successful work he did on the control of mosquitoes has been commented upon widely. His special report, published in 1904, 482 pp., upon the mosquitoes occurring within the State of New Jersey, in which is included an account of the different species, their habits, life-history, economic treatment, etc., is an extremely valuable contribution and shows the remarkable capability of the man in dealing with problems of such magnitude. Further accounts of this mosquito work are given at considerable length in his annual reports, since the above dates.

At meetings of farmers, horticulturists, etc, and those of scientific societies, which he was closely identified with, his lectures and helpful talks will be much missed. He received honours from many societies, among which may be mentioned that of Fellow of the American Association for the Advancement of Science, Fellow of the New York Academy of Sciences, Fellow of the Entomological Society of America, Honourary Member of the Entomological Society of Ontario, Honourary Member of the Newark Entomological Society, Corresponding Member of the Entomological Society of Washington, and Corresponding Member of the Ottawa Field-Naturalists' Club. He also had active membership in the Association of Economic Entomologists, Society for the Promotion of Agricultural Science, Brooklyn Entomological Society, Philadelphia Feldman Collecting Social, Brooklyn Institute, Washington Academy, and New Jersey State Microscopical Society.

In 1891, Rutgers College conferred upon him the honourary degree of Doctor of Science.

Like many busy men, he always found time to help others ; in his death, we in Canada have lost a true and valued friend. To-day there is a gap in our ranks which it will indeed be difficnlt to fill.

To Mrs. Smith and the two grown-up children who survive him, we extend our deepest sympathy. ARTHUR GIBSON.

GEOMETRIDÆ AS YET UNDESCRIBED.

BY RICHARD F. PEARSALL, BROOKLYN, N. Y.

(Continued from page 31.)

Stamnodes ululata, n. sp.

Expanse, 30 mm. Palpi moderate, extending well beyond the bulging front, pink, rough scaled, the last joint clay-yellow. Front dusky clay. Antennæ clay-yellow, heavily dusted with black scales above. The base of fore legs in front, the collar and bases of patagiæ are deeply roseate. Body clay-yellow, except a white cloud covering the scutellar region and base of abdomen, the latter sparingly sprinkled with roseate scales toward apex. Wings broad and thin in texture, of an even, pale, glistening clay-yellow, a little paler beneath. The primaries along costa, and broadly at apex, are sprinkled with roseate scales. The costa at one-fourth and one-half from the base is crossed by a pale bar, and at three-fourths out, a semilucent pale band, after crossing costa, curves outward around cell, and is lost at middle of wing. Secondaries with a faint roseate hue at outer apical margin, without markings of any kind. All wings above and below without discal dots or marginal lines. Fringes rather long, pink, sprinkled towards apices with black scales, which tend to gather at end of veins. Beneath, the primaries, over cell to extradiscal luteous band, outlining this rather sharply, and beyond this band a well-defined apical triangle reaching to centre of outer margin, are heavily dusted with roseate scales, the rest of the surface being entirely clear. Secondaries, without markings of any kind, are evenly and finely dusted with roseate.

Type, one female taken at La Puerta, Calif., Oct. 15th, 1911, is in the author's collection. The immaculate under side of secondaries is quite unusual.

Parexcelsa, nov. gen.

Palpi very small. Tongue obsolete. Front flat, finely shagreened, opaque, margined above between the orbits by a fine polished ridge, the edge apparently of the projecting vertex, which is slightly elevated between the pedicellate bases of the antennæ, these, together with the whole upper surface, being black and highly polished. Antennæ long, bipectinate to apex, where they are shortened so abruptly as to give a blunt appearance. Thorax and abdomen untufted. Legs normal, the hind tibiæ not swollen, without hair-pencil, and having two pairs of spurs. All wings somewhat extended, the outer margins rounded and slightly crenulate, without basal foveæ. Fore wings 12 veins, with 8 and 9 stalked on 7 and with each other, all others free. No accessory cell. Hind wings with costal margin concave, vein 8 parallelling cell for a short distance at or before centre, and showing a vestige of vein 5 toward outer margin; 3 and 4 are widely separate, and 6 and 7 from point.

Type.—*Parexcelsa ultraria* Pears.

This genus, by the faint trace of vein 5, shows a Hydriomenid tendency, but is, in my judgment, best retained among the Ennominæ, next to *Hulstina* Dyar.

Parexcelsa ultraria, n. sp.

Expanse, 28–30 mm. Palpi very short, hairy; pale gray. Antennæ stout, the shaft white, with pectinations dark gray. Front covered with a velvety pile of livid gray. Head, thorax above and primaries above silver-gray, with a mixture of scattered black scales, frosted with white in part. Abdomen and secondaries a dusky silver-gray. Head and thorax rough scaled ; a tuft overhanging front between antennæ. Primaries are marked with jet-black lines, one of which crosses costa at a sharp angle about one-fourth out. Another runs from centre of wing-base along vein beneath cell to its end, and a third from point of cell to apex of wing, broken at its middle by a sharp angle. Below this line to inner margin a series of long sharp points, outlined by a fine black hair line, rest on the veins, their bases joined about centre of wing into an irregular patch. The two lower points are shorter and broader than the rest. Another long point reaches backward through centre of cell nearly to its base, one beneath the black line along its lower margin, and a third, short and broad, between this and inner margin. The included space within these points, and the patch at centre, is a livid gray, almost free from the frosting of white scales, which cover the rest of the surface, forming a snow-white patch above and bordering the black line apex. A short black dash, ending in a cluster of black scales at margin, between the ends of veins. Fringes dusky gray, long and silken.. Secondaries without markings. Fringes as on primaries. No discal dots. A fine black marginal line. Beneath ashen, dusky along costa of primaries, and outwardly on all wings, which are bordered with a fine black marginal line. Fringes as above. No discal dots. Body beneath and legs pale ash-gray, sprinkled with black atoms.

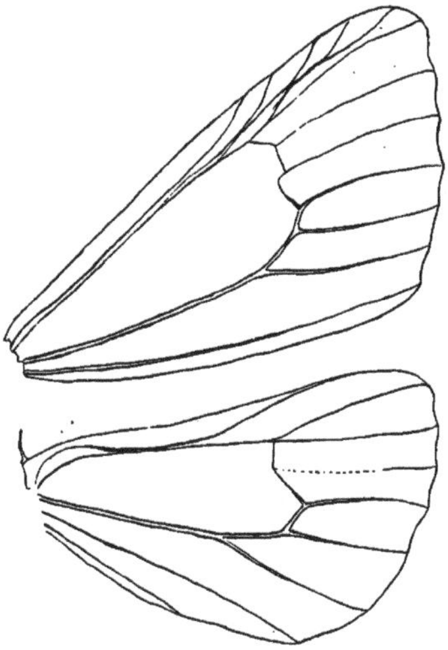

FIG. 5.—*Parexcelsa ultraria*, venation.

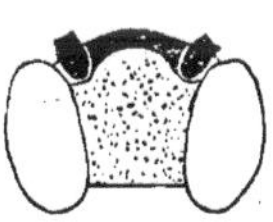

FIG. 6.—*P. ultraria*, front view of head.

Type, ♂ from San Diego, Calif., taken Oct. 9, 1910, and thirteen male co-types, Oct. 28, 1911, are in author's collection. The female is unknown to me.

NOTES ON *MELITÆA ALMA* STRECKER.

BY VICTOR L. CLEMENCE, PASADENA, CALIF.

In a recent number of the CANADIAN ENTOMOLOGIST, Karl R. Coolidge published an article entitled "Melitæa alma and Its Synonymy." Since the publication of the above article I have added a considerable number of specimens to that group in my collection, with the intention of verifying Coolidge's classification. I have also received specimens of *M. alma*, *M. fulvia* and *M. cyneas* from Dr. Barnes, which have been compared with the types, and which agree with my own series.

I have *M. alma* from Chiricahua Mts., South Arizona; Santa Catalina Mts., Arizona, and Fort Wingate, N. M.

M. fulvia from Fort Wingate, N. M., and Santa Catalina Mts., Ariz.

M. cyneas from Chiricahua Mts. and the Huachuca Mts., Ariz.

All my *fulvia* males are constant, and show very little variation. A few of the females show a tendency to the *alma* form, which is also the case in my *cyneas*. There is no doubt in my mind that typical *alma* is a variety which occurs occasionally in both *fulvia* and *cyneas*, more often in the females. I have taken two female *alma* in the Chiricahua Mts. flying with *cyneas* males. Out of a series of twenty *fulvia* from Fort Wingate three of the females approach the *alma* form. Dr. J. McDunnough, with Dr. Barnes, says: "Many of the *fulvia* females show a tendency towards becoming yellow, but most of our males are very constant in this respect." I have not heard of any locality where the *alma* form predominates, but on the other hand there was not one *cyneas* among the *fulvia* from Fort Wingate, and I have never seen a *fulvia* either in the Chiricahua Mts. or Huachuca Mts., where *cynéas* is common.

I believe *fulvia* and *cyneas* bear the same relationship to each other that *leanira* does to *wrighti*, *fulvia* being the more northern form occurring in Colorado, New Mexico and Arizona, while *cyneas* is the more southern form occurring from S. Arizona to S. Mexico. The fact that *alma* occurs in the same loçalities as both *fulvia* and *cyneas* leads me to think that it was the original form occupying the whole general region, and that owing to geographical surroundings each of the others has become permanent and has gradually taken the place of the parent form, which still is occasionally found among both *fulvia* and *cyneas*, the latter becoming a geographical subspecies. According to priority I should give the following classification:

Melitæa alma Strecker.
Sub.-sp. " *alma fulvia* Edwards.
" " *alma cyneas* Godman and Salvin.

The accompanying plate shows the three typical forms.

MELITAEA ALMA (♀) STRECKER.

M. ALMA CYNEAS (♂, ♀) GOD.-SALV.

M. ALMA FULVIA (♂, ♀) EDWARDS.

THE TRANSMISSION OF TYPHUS FEVER BY LICE.

In a recent paper* some interesting experiments are recorded by Drs. T. Goldberger and T. F. Anderson, which indicate that not only the body louse *(Pediculus vestimenti)* but the head louse *(P. capitis)* also may transmit the virus of typhus fever. These authors had previously shown that Brill's disease, which appears to be endemic in New York City, is identical with the typhus fever of Mexico, which, accordingly, may be identical with the European typhus fever. Evidence of the ability of lice to transmit typhus fever has been previously adduced by several investigators. In 1909, Nicolle, Comte and Conseil demonstrated that body lice *(P. vestimenti)*, which had been allowed to feed upon an infected bonnet monkey *(M. sinicus)*, were able to transmit typhus fever to two other monkeys, somewhere between the first and seventh day after feeding. In the following year, Ricketts and Wilder, who were working in Mexico, reported the successful transmission of the virus of typhus fever by *P. vestimenti* from man to monkey and from monkey to monkey. They were also able to infect a monkey by intradermal inoculation with the abdominal contents of infected lice, and similar experiments were successfully carried out by Wilder in 1911. Drs. Goldberger and Anderson commenced their work in 1909. They have confirmed the results of previous workers in regard to the body louse *(P. vestimenti)* and have also shown that the head louse *(P. capitis)* is able to transmit Mexican typhus fever from man to monkey by the subcutaneous injection of a saline suspension of crushed and infected head lice and almost certainly by its bite. The typhus virus is able to retain its virulence in the body of the head louse for twenty to twenty-four hours. The authors' conclusions are as follows:

1. The body louse *(P. vestimenti)* may become infected with typhus. The virus is contained in the body of the infected louse and is transmissible by subcutaneous injection of the crushed insect or its bite.

2. The head louse *(P. capitis)* may become infected with virus. The virus is contained in the body of the infected louse and may be transmitted by cutaneous injection of the crushed insect, and, we believe, also by its bite.

These results are of great interest to the entomologist. One by one our most common insects affecting man have been shown to be important factors in the transmission of disease; the house fly carries typhoid and

*"The Transmission of Typhus Fever, with Special Reference to Transmission by the Head Louse *(Pediculus capitis)*." *"Public Health Reports"* of the U.S. Public Health and Marine Hospital Service, Washington. W.C. 27, No. 9, 1st March, 1912, pp. 297-307.

certain other infectious diseases; the flea carries the plague bacillus; the bed bug has been shown to be transmitting agent of the causative organisms of the serious tropical Black Fever or Kala Azar, and the louse transmits typhus fever. That all the insects directly attendant upon man's person are disease carriers is not a pleasant fact for contemplation!

C. GORDON HEWITT.

TO THE EDITOR OF THE CANADIAN ENTOMOLOGIST:

In your journal of October, 1908, pp. 370–373, I published an account of the attempt made by Hendel to revolutionize the nomenclature of Diptera by introducing generic names from an obscure early paper of Meigen's, which were published without any described species being associated with them—in other words, without types.

Mr. Hendel based his action at the time on his interpretation of the rules of nomenclature of the International Zoological Congress, expressing great regret at the overturning of names, but protesting that the rules compelled it; later, in Wiener Entomologische Zeitung, XXVIII, 33–36, 1909, he took up my argument from the rules themselves, and endeavoured to show that I had not interpreted them correctly. So far, his action was as if forced by these rules. It was interesting, indeed, to find (W. E. Z., XXX, 89–92, 1911), that he has revolted against the rules commission of the I. Z. C., on a minor problem, the mode of designation of types, and refuses to follow them. I cannot help but regret that he did not revolt sooner, so as to spare us the trouble about Meigen's 1800 paper. I think he is perfectly right in his present contention, which relates to point g under Article 30, as amended at the Boston meeting, 1907. But my present purpose is merely to show the embarrassment of a too sweeping acceptance of any rules of nomenclature.

American dipterists have shown a commendable disposition to sit tight during this nomenclatural flurry, and already the worst seems past. On the general question of the validity of a genus without a type, I have noticed two expressions recently that are of interest. One is by S. A. Rohwer, in Technical Bulletin, No. 20, Bureau of Entomology, p. 70. He was fixing the types of saw-fly genera, and used the following language: "In this paper a genus is considered to be without standing until it contains a species; and genera which were founded without species take the first species placed in them as the type, and date from the time when that species was placed in them." If this rule were followed, Meigen's 1800 genera would date from 1908. The other case I found in the CANADIAN ENTOMOLOGIST itself, 1912, p. 50, where Mr. Girault is discussing the genus Trichaporus, and says: "No species was mentioned as belonging to it; under the code it is therefore without status."

J. M. ALDRICH, Moscow, Ida.

REMARKS ON *GNYPETA* THOMS. (*STAPHYLINIDÆ* COL.).

BY A. FENYES, PASADENA, CAL.

This rather feebly-characterized Aleocharine genus has the following diagnosis :

Tarsi 4–5–5-jointed ; antennæ 11-jointed ; maxillary palpi 4-jointed ; labial palpi 3-jointed ; ligula bifid ; genæ simple. Hind tarsi with joint 1 not longer than 2 and 3 together. Prosternum membranous under the front coxæ ; mesosternal process obtuse at tip ; middle coxæ separated. The first free ventral segment of the abdomen transversely sulcate at base.

Gnypeta differs from *Atheta* Thoms. by the sulcate ventrites, from *Tachyusa* Er. by the obtuse mesosternal process, and from *Myrmecopora* Saulcy by the comparatively short first joint of the hind tarsi.

In the males the sixth ventrite is produced, and at tip rounded ; in the females less produced than in the males, and at tip sinuate.

The following descriptions of nearctic forms have been published :

1. *nigrella* Lec., n. sp., N. Amer. Col., I, 1863, 29 (Tachyusa).
2. *baltifera* Lec., ibid., 29 (Tachyusa).
3. *laticollis* Csy., Bull. Cal. Ac. Sc., I, 1885, 287 (Falagria).
4. *experta* Csy., ibid., 300 (Tachyusa).
5. *linearis* Csy., ibid., 301 (Tachyusa).
6. *harfordi* Csy., ibid., 304 (Tachyusa).
7. *crebrepunctata* Csy., Bull. Cal. Ac. Sc., II, 1886, 203 (Tachyusa).
8. *atrolucens* Csy., Ann. N. Y. Ac. Sc., VII, 1893, 346.
9. *lucens* Brnhr., Deutsch. Ent. Ztschr., 1905, 254.
10. *helenæ* Csy., Tr. Ac. Sc. St. Louis, XVI, 1906, 193.
11. *deserticola* Csy., ibid., 193.
12. *punctulata* Csy., ibid., 194.
13. *ventralis* Csy., ibid., 194.
14. *floridana* Csy., ibid., 195.
15. *bockiana* Csy., ibid., 195.
16. *manitobæ* Csy., ibid., 196.
17. *brevicornis* Csy., ibid., 196.
18. *incrassata* Csy., ibid., 198.
19. *leviventris* Csy., ibid., 198.
20. *oregona* Csy., ibid, 199.
21. *impressiceps* Csy., ibid., 199.
22. *curtipennis* Csy., ibid., 201.
23. *abducens* Csy., ibid., 201.
24. *shastana* Csy., ibid., 202.

25. *majuscula* Csy., ibid., 217 (Euliusa).
26. *sparsella* Csy., ibid., 217 (Euliusa).
27. *elsinorica* Csy., ibid , 218 (Euliusa).
28. *transversa* Csy., ibid., 218 (Euliusa).
29. *mollis* Csy., ibid., 219 (Euliusa).
30. *pimalis* Csy., ibid., 220 (Euliusa).
31. *citrina* Csy., ibid., 220 (Euliusa).
32. *wickhami* Csy., Mem. Col , II, 1911, 166.
33. *brunnescens* Csy., ibid., 167.
34. *boulderensis* Csy., ibid., 167.
35. *oblata* Csy., ibid., 168.
36. *pallidipes* Csy., ibid., 168.
37. *uteana* Csy,, ibid., 169.
38. *modica* Csy., ibid., 170.
39. *sensilis* Csy., ibid., 170.
40. *limatula* Csy , ibid., 170 (Euliusa).

REMARKS.

1. *G. nigrella* Lec.—"*Tachyusa nigrella.* Elongate, black, shining, with delicate ashy pubescence, very finely punctulate; thorax obsoletely canaliculate, before the base transversely slightly foveate, a little shorter than broad; abdomen in front slightly narrowed; segments 1–3 transversely deeply impressed; legs blackish-pitchy. Length, .12. Middle and Western States; common."

There is in the Leconte collection at Cambridge, Mass., a specimen labelled *nigrella*, and there is a pink disk on the pin under the specimen; the pink disk meaning: Middle States (Ohio, Penna., West N. Y.). Below the above specimen, without any name, but probably referred by Leconte to *nigrella*, is another specimen with a pink *and* yellow disk, the latter disk meaning: Western States (N.-W. Va., Ky., N.-W. Tenn., S. Ill.), and, finally, there is another row with three specimens, all with pink disks.

This species is well characterized amongst the eastern nearctic Gnypetas by the prothorax being simply foveate at base in the female and lacking the geminate punctures; the impression becomes longer in the male, but is not traceable beyond the basal half of the prothorax. It is the most common eastern species in our fauna, not closely related to any other from the northern hemisphere; it is known to me from the following localities: Penna. (Pittsburg); Mass. (Chicopee, Framingham); N. Y. and Md. (Baltimore).

2. *baltifera* Lec.—"*Tachyusa baltifera.* Less elongate, blackish-pitchy, shining, finely punctulate, with delicate pubescence; thorax, elytra

and segments 1-3 of the abdomen piceo-testaceous; thorax canaliculate behind, before the base transversely impressed, a little shorter than broad; abdomen slightly narrowed towards the base; joints 1-3 deeply impressed transversely; antennæ and legs piceo-testaceous. Length, .10. One specimen, Coney Island, near New York. Less elongate than the other species, with fine punctures, especially of the thorax, less dense and more distinct than in the two preceding species." (The two preceding species are *cavicollis* and *nigrella*.)

If my notes are correct, the specimen in the Leconte collection from Coney Island is labelled "*balteata* Lec." This undoubtedly good species is recognizable amongst the eastern forms by the longitudinally, broadly, entirely excavated prothorax of the male; the first joint of the hind tarsi is fully as long as joints 2 and 3 together, the species thus appearing as a connecting link between *Gnypeta* and *Tachyusa*. Not closely related to any other species of the northern hemisphere; known to me from the following localities: Mass. (Tyngsboro, Framingham); Penna. (Jeanette); Ill. (Algonquin); Maine and New York.

3. *laticollis* Csy.—The most common species of *Gnypeta* in California, recognizable at once by the open middle acetabula. It is known to me from Los Gatos, Pacific Grove, Nordhoff, S. Juan Capistrano, Oceanside, Lakeside, Foster, S. Diego, Victorville, S. Bernardino, Elsinore, Riverside, Pomona, Azusa, Pasadena, Sierra Madre, Mt. Wilson, Yuma, all in California. I have also some specimens from El Paso, Texas, which do not appear to be different from the California specimens.

4. *experta* Csy.—Recognized by the male characters. The head in this sex being broadly, almost entirely concave, and the prothorax entirely concave in the middle. The females are somewhat different from the males, especially in habitus, with apparently shorter antennæ and a trifle broader abdomen; they are consequently not easily recognized, unless taken in company of males.

5. *linearis* Csy.—I fail to find any reliable characters which would separate *linearis* from *experta*, and propose to unite these two forms under the latter name.

6. *harfordi* Csy.—In this species the male prothorax is obsoletely depressed in about the basal two-thirds, the depression being shallow and not as sharply limited as in *experta*. I believe I interpret this species correctly. I have seen specimens from Cole, Applegate, Nordhoff, Pasadena, Lakeside and Foster, all in California.

7. *crebrepunctata* Csy.—Apparently a good species, differing from our other western forms chiefly by the much less shining surface and by the

coarse punctuation, also by the shorter and outwardly more incrassate antennæ. I have specimens from California (S. Francisco, S. Barbara, Pacific Grove and Point Reyes); also from Oregon (Newport), the latter being co-types of *G. nigerrima* Brnhr. i. litt. My Pacific Grove specimens were taken near the ocean, on a moist, thickly-incrusted salt flat, under cow manure; and I have no doubt that *crebrepunctata* is exclusively a seashore-inhabiting species.

8. *atrolucens* Csy.—Closely related to (if not identical with), *Gnypeta cærulea* Shalb. from Northern Europe, differing chiefly from the latter by the more shining integuments. Occurs in the mountainous regions of the Eastern United States (Catskill Mts. in N. Y.; Mt. Washington in N. H.). My 5 specimens from Mt. Washington are, curiously enough, all males, and I have in my collection another male from Vermont. The type of *atrolucens* is said to be a female.

9. *lucens* Brnhr.—A synonym of *laticollis* Csy.

10. *helenæ* Csy.—A good species of the *carbonaria-ripicola* group, with a well-marked, bifoveate antebasal transverse impression on the prothorax. In the male the head impressed, the bottom of the impression impunctate, the prothorax not modified, the 6th ventrite produced and, at the tip, moderately rounded. A species of wide distribution. I have before me specimens from Montana (Kalispell); N. Dakota (Williston); Colorado (Buena Vista); Nevada (Reno); Arizona (Williams), and California (Occidental, Deer Park Springs, Tahoe City and Tallac). My Tahoe City specimens were taken in a swampy place; they appeared to feed on dead tadpoles; other specimens from the same locality were taken during the evening flight.

11. *deserticola* Csy.—Smaller, with paler legs than *helenæ* Csy., but otherwise scarcely different, and probably a synonym of the latter. I have not seen this form.

12. *punctulata* Csy.—Still smaller than *deserticola*, placed at present also as a synonym of *helenæ* Csy. I refer to this form a specimen of the Fall cabinet (from Pomona, Cal.); another specimen from Pasadena, Cal., is more shining, with more sparsely punctate abdomen, but otherwise not different.

13. *ventralis* Csy.—One specimen in my collection from Arizona (Williams), agrees fairly well with the original description, but appears to be quite similar to *majuscula* Csy. also, judging from the description; no opinion can be pronounced about the status of this species until more material is accumulated in our collections.

14. *floridana* Csy.—Can be distinguished by the strongly-dilated abdomen and the very long antennal joint. I have only one male before me, from Enterprise in Florida; it makes the impression of belonging to a good species, possibly from the Central American fauna.

15. *bockiana* Csy.—Can be distinguished from *nigrella* Lec. by the presence of two foveolæ in the transverse basal impression of the prothorax, and by the absence of punctuation in the transverse impressions of the first three tergites. Of this form I have only males before me, one each from Arkansas (Little Rock); Tennessee (Nashville), and Texas (Vaco). Casey does not mention the sex of the type, which is from Missouri (St. Louis).

16. *manitobæ* Csy.—Unknown to me; possibly a more northern form of *bockiana* Csy.

17. *brevicornis*, Csy.—A specimen in my cabinet (collected by Wickham, and labelled "*brevicornis*"), from British Columbia, makes the impression of being a good species; it has the prothorax bifoveolate at base, and can be recognized by the rather short antennæ and the pale colour.

18. *incrassata* Csy.—Five specimens from Montana (Kalispell) in my collection make the impression of a good species, recognizable by the long antennæ and the unusually long third antennal joint.

19. *leviventris* Csy.—Is a close relative of *incrassata*, and may prove to be conspecific with the latter. I have seen nine specimens from the type locality (Ojai Valley in California), they have shorter antennæ than my specimens of *incrassata*, but do not seem to present other characters of specific value.

20. *oregona* Csy.—A synonym of *helenæ* Csy. (vide Casey, Mem. Col., II, 1911, 167).

21. *impressiceps*, Csy.—Unknown to me, possibly a synonym of *laticollis* Csy.

22. *curtipennis* Csy.; 23. *abducens* Csy.; 24. *shastana* Csy.; are, I believe, synonyms of *harfordi* Csy. I have specimens from British Columbia (Duncans), Washington (Baring) and California (Cole) before me, which, after careful study, must be referred to *harfordi* Csy., yet exhibit some slight differences in the general form of the body and in the length of the elytra.

25. *majuscula* Csy.—Unknown to me; perhaps a synonym of *laticollis* Csy.

26. *sparsella* Csy.; 27. *elsinorica* Csy.; 28. *transversa* Csy.; 29. *mollis* Csy.; 30. *pimalis* Csy., and 31. *citrina* Csy., are all synonyms of *laticollis* Csy. My large series of Southern Californian specimens, while showing

variations in size, shape, colour and sculpture, represents undoubtedly one species only, and must be referred to *laticollis* Csy. I have five specimens of *pimalis* Csy. before me (three from the author himself, two from Wickham), none of them specifically different from *laticollis*.

32. *wickhami* Csy.—Possibly a synonym of *helenæ* Csy.

33. *brunnescens* Csy.—Without much doubt identical with *nigrella* Lec. I have two specimens in my collection from New York, which can be referred to *brunnescens*, and which I cannot separate satisfactorily from *nigrella*.

34. *boulderensis* Csy.—Probably a synonym of *helenæ* Csy.

35. *oblata* Csy.—Probably also a synonym ot *helenæ* Csy.

36. *pallidipes* Csy.—Possibly a synonym of *harfordi* Csy.

37. *uteana* Csy.—Unknown to me.

38. *modica* Csy.—Probably a synonym of *laticollis* Csy.

39. *sensilis* Csy.—Is *experta* Csy.

40. *limatula* Csy.—Is a synonym of *laticollis* Csy.

The following list of nearctic species (with synonym) is offered tentatively:

1. *nigrella* Lec.
   brunnescens Csy.

2. *baltifera* Lec.

3. *laticollis* Csy.
   lucens Brnhr.
   ventralis Csy.
   impressiceps Csy.
   majuscula Csy.
   sparsella Csy.
   elsinorica Csy.
   transversa Csy.
   mollis Csy.
   pimalis Csy.
   citrina Csy.
   modica Csy.
   limatula Csy.

4. *experta* Csy.
   linearis Csy.
   sensilis Csy.

5. *harfordi* Csy.
   curtipennis Csy.
   abducens Csy.
   shastana Csy.
   pallidipes Csy.

6. *crebrepunctata* Csy.

7. *atrolucens* Csy.

8. *helenæ* Csy.
   deserticola Csy.
   punctulata Csy.
   oregona Csy.
   wickhami Csy.
   boulderensis Csy.
   oblata Csy.

9. *floridana* Csy.

10. *bockiana* Csy.
    manitobæ Csy.

11. *brevicornis* Csy.

12. *incrassata* Csy.
    leviventris Csy.

13. *uteana* Csy.

The accumulation of more material in our collections will probably still reduce the above number of species; and I venture to express my belief that *bockiana* may prove to be conspecific with *helenæ*, *brevicornis* and *incrassata* with *harfordi*, and *uteana* with *helenæ*. This reduced number of forms can be tabulated roughly as follows:

I.—EASTERN FORMS:

1. Abdomen strongly dilated in the middle.............*floridana* Csy.
   Abdomen at most only moderately dilated in the middle........2.
2. Prothorax broadly excavated in the male.............*baltifera* Lec.
   Prothorax not excavated in the male....................3.
3. Colour bluish-black............................*atrolucens* Csy.
   Colour black.....................................4.
4. Prothorax bifoveolate at base......................*bockiana* Csy.
   Prothorax not bifoveolate........................*nigrella* Lec.

II.—WESTERN FORMS:

1. Middle acetabula open behind......................*laticollis* Csy.
   Middle acetabula entirely closed..........................2.
2. Prothorax broadly, entirely concave in the male........*experta* Csy.
   Prothorax never broadly and entirely concave in the male.......3.
3. Prothorax not bifoveolate in the basal impression......*harfordi* Csy.
   Prothorax bifoveolate in the basal impression..................4.
4. More or less shining..............................*helenæ* Csy.
   Rather opaque.........................*crebrepunctata* Csy.

The described Central and South American forms are:

1. *dissimilis* Shp., Biol. Centr. Amer. Col., I, 2, 1883, 173 (Homalota).
2. *nigricans* Shp., ibid., 227.
3. *fragilis* Shp., ibid., 227.
4. *mexicana* Shp., ibid., 228.
5. *boliviana* Brnhr., Bull. Soc. Ent. Ital., LX, 1908, 247.

1. *G. dissimilis* Shp. is tentatively placed here in the genus *Gnypeta*; it is described from a single specimen.

2. *nigricans* Shp.—In our fauna *bockiana* Csy. seems to be the nearest relative of this form.

3. *fragilis* Shp.—A good species, having no relatives in the nearctic fauna; recognizable by the pale eleventh antennal joint and the impunctate basal impressions of the first tergites.

4. *mexicana* Shp.--Quite likely identical with *nigricans* Shp.

5. *boliviana* Brnhr.—Apparently without relatives in the nearctic fauna.

There is another species in the neotropical fauna, described by Sharp under the name *Rechota impressa* (Biol. Centr. Amer. Col., I, 2, 1883, 228, 229). *Rechota* cannot be separated from *Gnypeta*, the only distinguishing feature being the truncation of the meso- and metasternal processes; in *impressa* Shp. the middle coxal cavities are open behind, and, in the male, the prothorax is broadly impressed, very much in the same way as in our *baltifera* and *experta*. In our fauna *laticollis* Csy. is somewhat similar in habitus to *impressa*, but lacks the modification of the male prothorax.

The palæarctic fauna contains the following described forms:

1. *carbonaria* Mannh., Prec. Brachel., 1830, 75.
2. *cærulea* Sahlb., Ins. Fenn., I, 1834, 351.
3. *velata* Er., Kaef. Mk. Brdbg., I, 1837, 319.
4. *ripicola* Kiesw., Stett. Ent. Ztg., V, 1844, 317.
5. *canaliculata* J. Sahlb., Sv. Ak. Handl., XVII, 1880, No. 4, 84.
6. *cavicollis* J. Sahlb., ibid,, 85.
7. *ænescens* J. Sahlb., ibid., 85.

1. *G. carbonaria* Mannh.—Represented in our fauna apparently by *G. helenæ* Csy.

2. *cærulea* Sahlb.—Our *atrolucens* Csy. may prove to be conspecific with *cærulea*; they are, at any rate, very closely related to each other.

3. *velata* Er.—Without relatives in our country.

4. *ripicola* Kiesw.—Very near to *carbonaria* Mannh.

5. *canaliculata* J. Sahlb., and 6. *cavicollis* J. Sahlb.—Quite likely conspecific; represented in our fauna by the very closely related *G. experta* Csy.

7. *ænescens* J. Sahlb.—Apparently without close relatives in our fauna.

There is one species described from the Indo-Oriental fauna, *elegans* Brnhr., Deutsch. Ent. Ztschr., 1902, 22, from Ceylon; one species from the Australian fauna, *fulgida* Fvl., Ann. Mus. Civ. Genova, XIII, 1878, 583, and two species from the Æthiopian fauna, 1. *angulicollis* Fvl., Rev d'Entom., XXVI, 1907, 58, and 2. *pulchricornis* Fvl., ibid., 58, both from English East Africa.

A NEW TYPE OF CORIXIDÆ (RAMPHOCORIXA BALANODIS, N. GEN., ET SP.) WITH AN ACCOUNT OF ITS LIFE HISTORY.[1]

BY JAMES FRANCIS ABBOTT, ST. LOUIS, MO.

Our knowledge of the developmental history of the water bugs is very incomplete. In the early days of embryology, Corixa was studied by Metschnikoff,[2] and Brandt,[3] and others with especial reference to the germ layers, the revolution of the embryo, etc. Leon Dufour[4] had previously described the eggs of the two European species, *Arctocorisa striata* (L) and *heiroglyphica* (Duf.).

The only account of the metamorphosis of any member of the group that I have been able to find is that of F. Buchanan White,[5] who, in addition to describing the egg of *Corixa nigrolineata* (= *Arctocerisa fabricii*), also described the first moult, remarking that the tarsus of the third pair of legs is but one-jointed. "At this stage," he says, "they died"—a result which apparently has been obtained by all who have attempted similar observations since. Indeed the rearing of both Notonecta and Corixa seems attended with unusual difficulties,[6] although I believe that by the use of mosquito larvæ for food, success has been attained with the former.

The writer has succeeded in carrying a species of Corixid through the whole series of moults from egg to imago, and since the critical study of the larger groups of Hemiptera is greatly hampered by our ignorance of the developmental stages, it seems worth while to describe the various instars in some detail.

The present species, which appears to be undescribed, has the remarkable habit of attaching its eggs to the carapace of the crayfish, some individuals of which were found almost completely covered by hundreds of tiny eggs. As the writer intends later to discuss this habit in detail, it will be merely alluded to here. The egg-bearing crayfish were captured in a small clear-water pond near Columbia, Mo., the early part of July and were isolated in small aquaria. All the eggs were in the same advanced stage of development, with the red eye spots showing through the shell, and they began to hatch July 8th (1910).

1. From the Zoological Laboratory of Washington University.
2. Zeit. wiss. Zool., XVI, 1866, 422-436, Taf., 26 and 27a.
3. Mem. Acad. Imp. Sci. St. Petersb., XIII, 1869.
4. Recherches sur les Hémiptères, 1833.
5. "Notes on Corixa," Ent. Mon. Mag., X, 1873, 60-63.
6. Cf. Bueno, CAN. ENTOM., XXXVII, 390, 1905.

The mistake was made of attempting to rear the nymphs in small jars. Whether on account of lack of sufficient oxygen or of appropriate food, or too high a temperature or some other unknown cause, the greater number of nymphs perished as soon as hatched.

The remnant were transferred to a large aquarium used for breeding mussels. This was a zinc-lined tank about two feet deep and with a superficial area of thirty or thirty-five square feet, with a layer of soft mud in the bottom and an overflow arrangement by means of which a quiet but constant stream of fresh water was kept circulating through the tank. There were a number of mussels in the mud and several crayfish. A few water weeds supplied shelter for smaller organisms, of which a large Ostracod was the most plentiful. I observed several of the older nymphs feeding on the Ostracods, and it is possible that the absence of some similar food caused the individuals in the separate jars to die.

The newly-hatched nymphs are very active and, as a rule, keep close to the bottom. They are negatively phototropic until the fourth or fifth instar and this condition, which keeps them in the shadows, aided by their great transparency, is doubtless of much value in enabling them to escape their enemies. The bulk of the eggs hatched July 8. The first moult (second instar) occurred about July 16th, the second about July 24th, the third, July 31-Aug. 3, the fourth, August 10th, and the imagos appeared about August 18th. From the third instar on, the mortality was high. In the morning, numbers would be found on the surface of the water near the edge held by a bubble of air, the buoyancy of which they were unable to overcome and, unless assisted, they perished in this way. It seems probable that they are most active at night as they were rarely seen to dart to the surface frequently, except on dull, dark days.

The Egg.

Length about 9 mm. Breadth about 4 mm. Shape elongate-oval, bilaterally rather than axially symmetrical, i.e., one side nearly straight, the opposite strongly curved. (See Fig. 1.) Colour grayish yellow (later stages only were observed); the surface ornamented with a delicate tracery in the form of interlocking hexagons like a honeycomb or the facets of a compound eye. The egg is fastened in a sort of shallow cup which is of a leathery texture and dark brown in colour. The distal end through which the nymph emerges, is provided with six to eight short lobes arranged in a circle. The appearance of the whole egg is much like that of a minute Grantia sponge.

Dufour described the eggs of *striata* and *hieroglyphica* as acuminate at the free end and placed on a pad. White speaks of the eggs he describes as pyriform and attached at the broader end. He does not mention the pad or cup, nor does Heidemann,[7] of *Corixa mercenaria*. It would be of interest to discover if there is a difference in this regard between different species of Corixids or whether in some cases the pad or cup has merely escaped observation.

First Instar.

Length about 1.15 mm. Width about .55 mm. General appearance of adult, but wider in proportion to length. (Fig. 2.) Head about three times as wide as long (dorsal aspect); distance from vertex to tip of beak about equal to the width between eyes (ventral aspect). Eyes prominent and conspicuous, deeply pigmented, facets relatively large. The beak is apparently four-jointed, rather broad and conical. The black tips of the mandibles and maxillæ project slightly between the two halves. The former are somewhat shorter than the latter, curved, with minute serrations at the tips, and may be seen to extend into the head apparently up to the level of the eyes.

The antennæ are two-jointed, inserted far down toward the beak, the last joint about ⅓ the interorbital width in length. Tarsi all one-jointed. Those of first leg when at rest, curved over beak as in imago. First tarsi triangular in section, about ⅓ as long as those of third leg, 3½ times as long as broad, oblong-triangular, broadly rounded above, the comb of bristles prominent. (Fig. 2a.) Tibia of second leg 3/5 the length of tarsus and squarish in section with the anterior angles armed each with a row of short bristles. Intermediate tarsus nearly 8 times as long as broad, with a ventral row of long bristles and several rows of much shorter ones; tarsal claws weak, variable in length. Third leg sparsely bristled, tarsal joint slightly longer than the tibia or the femur, which are subequal. Body a little less than twice as long as broad, the posterior angles not so truncate as in later instars, provided and armed each with a half dozen rather long bristles. Lateral margin of body with bristles on posterior half only.

7. Proc. Ent. Soc., Wash., XIII, 1911, p. 140, Pl. XII, fig. 7.

The tracheal system is comparatively simple, consisting of two longitudinal trunks sending off laterals in each abdominal segment and one stout branch to each leg. Anterior branches supply the brain and the eyes.

Second Instar.

A marked increase in size is noticeable, the length being now about 1.9 mm. and the width about .9, roughly one-half as much. Head strongly convex, the frontal margin with a row of rather long bristles, longest in the middle, shorter toward the eyes. Posterior border deeply sinuate or arcuate.

Prothorax about as long as mesothorax, the two together a trifle longer than metathorax; the contour of the two together forming a narrow oval. Posterior margin of metathorax straight, anterior margin concave; its median length about equal to that of head. Abdomen truncate, seven-jointed, last joint about ½ as wide as first joint, terminated by two groups of rather long setæ at the angles.

Tarsi all one-jointed. First tarsus fringed with moderately long setæ, about equal to tibia in length. Second legs; tarsus equal to tibia, both together about as long as femur. Third legs with femora but slightly flattened, tarsus nearly as long as femur and tibia together, clothed with setæ, these longest at the joint, becoming much shorter distally. Colour very transparent. A median grayish line on thorax.

Third Instar. (Fig. 3.)

Length 2 mm. Width 1 mm. Head as before. Eyes a little more than 1/5 the head-width in width. The wing-pads first appear; about 3/5 the length of thorax, sparsely hairy. Thorax ½ as long as wide. Abdomen as before, fringed on the sides by rather long setæ, the posterior angles with conspicuous tufts. Ventral surface sparsely pilose.

Tarsi all one-jointed. The whole first leg about equal in length to the femur of second leg. Tarsus about three times as long as broad, terminated by a sharp spine. *Second legs* slender; tarsal claws as long as tarsus, other joints as in third leg, all feebly setose. *Third legs;* tarsus 1⅓ times the tibia, the latter equal to femur. Tibia and femur together about equal to femur of second leg. Abdomen strongly truncate.

Fourth Instar.

Length 3 mm. Width 1.2 mm. Very much more pigmented and less transparent than previous instars. Posterior margin of head, posterior angles of eyes, and posterior margin of thorax fuscous. Anterior margin of thorax and inner edge of wing pads with rather dense brownish-black hairs. These together with the pigmented posterior margin of the thorax form a square; a median patch of brown hair joining the band on the anterior margin. General surface of thorax smoky brown with narrow median clear line, and a paler transverse band in the middle. Head pale brown with a darker shading on vertex. Whole dorsal surface of thorax and abdomen sparsely hairy, the abdominal segments faintly indicated by transverse brown stripes. A median longitudinal white stripe ⅓ the body-width in diameter runs the length of the dorsal surface of the abdomen. Within this is a series of large pale brown blotches, one on each segment, the third and fourth of these with a distinct crescent of chestnut brown,[8] marginal third of abdomen smoky, fringed with cilia, but these less conspicuous because of the general hairiness of the body. The wing-pads hardly extend beyond the thorax.

Tarsi all one-jointed. *First legs* as before. *Second leg* with femur as long as width of head, equal to tibia and tarsus together. Claws 1/5 longer than tarsus. *Third leg* with tarsus equal to width of head, feathered with dense hairs. Antennæ ½ the length of tarsus of first leg. Interorbital space ⅔ the width of head, and equal to ¾ the length from vertex to tip of beak.

Fifth Instar. (Fig. 4).

Length 3.8 mm. Width 1.4 mm. Dorsal marking as in previous instar, but more intensified. The two median dark brown marks of third and fourth abdominal terga oblong surrounded by a larger oblong of smoky brown. Hairy covering of wing-pads and thorax conspicuous, the median patch of the anterior border extending more

8. These conspicuous markings are found on the dorsal surface of older nymphs of all species of Corixids that I have examined. I have considered them of glandular nat re, and they are so considered by Kunckel d'Herculais (Comptes Rendus, cxx, p. 1,002, 1895), who remarks that the dorsal position of the "scent glands" differentiates the Corixids from Nepa and Notonecta, and put them phylogenetically nearer the Cimicids. J. Gulde, however, in an elaborate monograph published later ("Die Dorsaldrüsen der Larven der Hemiptera Heteroptera" Ber. Senckenb. Ges., 1902, p. 85–136), describes the dorsal glands in all the various families of Rhynchota, including the aquatic families, and denies the presence of such glan s in any waterbugs. The Corixids examined were *Corixa geoffroyi* (Leach), *Arctocorisa linnei* (Fieb.) and *Cymatia coleoptrata* (L). He claims that the conspicuous markings are merely the site of the insertion of certain abdominal muscles. It would seem worth while to investigate the matter further.

than ⅓ the length of thorax down the median axis. Wing-pads extend half way to third abdominal segment. Beak brownish, with short pubescence. Legs pure white, antennæ no larger than before, but fringed with short cilia. Tarsi of first two legs one-jointed; those of third leg two-jointed, otherwise legs as before.

In comparing the various larval stages one is struck by marked increase in the size of the eyes relative to the size of the head as development proceeds. Another point is of great theoretical interest. As is well known, there exists throughout the group an extraordinary sexual dimorphism, such that the uninitiated might be led to class males and females of the same species in different families, so great is the dissimilarity in structure. It is of interest to note that the larval stages up to the last instar, with respect to those structures (palæ, frontal fovea, asymmetry of abdominal segments, etc.), that exhibit this dimorphism, are *entirely of the female type.*[9] The writer has dissected the much larger *Arctocorisa harrisii* Uhl. during the last moults, and has found the same thing to be true. A specimen in the fifth instar just ready to moult may easily be "shelled out" of its cuticle and, if a male, the irregular arrangement of the abdominal segments will be found fully developed, but entirely concealed by the regular and symmetrical arrangement, characteristic of the females and larvæ.

DESCRIPTION OF THE IMAGO.

Ramphocorixa balanodis, n. gen. et sp.

Colour.—Head yellowish, tegmina pale silvery grayish, almost iridescent in the female, darker in the male, the characteristic vermiculate or banded markings usual in the group nearly obsolete. Pronotum grayish or smoky brown, suffused with darker in the male. Rostrum pale yellowish. Tergum, legs and whole ventral surface of female pure white. Dorsum of male black, except the lateral margins, which are pale, the ventral surface white, except for two broad almost black oblong bands on either side, each nearly ⅓ the body-width in width, parallel to but not quite reaching the lateral margin and extending over sternites 3, 4 and 5. Genital segments pale in both sexes. A tiny reddish spot on the outer surface of posterior coxæ, next the distal joint. The hairs of the limbs tinged with yellow. Anterior and posterior margins of pronotum fuscous, the former line sinuate. Surface of pronotum otherwise with three complete pale brown lines, little, if at all, arched, and two shorter ones alternating. Clavus nearly transparent, margined with brown, about one-third

9. F. G. Smith, Quar. J. Mic. Sci., 1909, pp. 54, 577; 1910, 55.

its area adjacent to the scutellum, immaculate. A few complete markings beyond the middle. Corial lines pale smoky or grayish brown, confusedly interrupted or obsolescent, fusing to form three delicate vermiculated longitudinal stripes, these continue upon the membrane. Corium clothed with sparse, fine depressed whitish hairs.

Pronotum and anterior half of clavus rastrate. The tegmina are semi-hyaline, the colour of the dorsum showing through, on account of which the male appears darker than the female. *Pronotum lenticular*, 2½ times as wide as long, its posterior margin evenly rounded and not produced, a small area of scutellum visible between it and the clavus. Head emarginate behind, the lateral angles (with the eyes) acute and slightly produced. Interorbital space about equal to posterior width of eye. Posterior margin of eye touching occipital margin, except for a short distance at the inner angle. Two parallel rows of punctures on either side of the vertex. Intermediate tarsi ½ the length of tibia, the latter 3/5 the length of femur. Posterior femora and tibia subequal, a little more than ½ the tarsi in length. Metaxyphus small, short, triangular.

Sexual characters.—Male: Head acuminate, strongly carinate, about ⅓ longer than pronotum. *Fovea acorn-shaped*, broad and deep, occupying the entire space between the eyes and reaching from the labrum to the acute termination of the carina mentioned. Foveal surface clothed with fine depressed whitish hairs. *Palæ shiny ivory-white, very irregular in shape. (cf. Fig. 6.) Lower edge entire, slightly concave; upper surface flat, deeply incised about midway the length, so as almost to cut the pala into two joints.* Viewed from the inner surface the outline suggests somewhat the head of a bird of prey. Inner surface with a row of 23 dark brown "pegs"; the first nine following the curve of the upper margin, then the line arching downward to the limit of the cleft. Tip of pala with a single long, serrated spur, a row of short spines along the lower inner edge, a row of longer ones along lower outer edge. The posterior upper margin of the pala projects slightly over the tibia in a flattened spur. Tibia a little less than half the pala in length. Femur with a large stridular area composed of fine spines set in rows. *Asymmetry dextral. Strigil very minute*, .05 mm. long and 1/5 as wide as long, crescentric in shape, lying in a small membranous projection of the 6th tergite, in the antero-posterior axis, with about 18–20 transverse striæ. Fifth, sixth and seventh tergites divided, fourth deeply cleft.

Female: Venter evenly rounded, front plane with a *small circular depressed fovea* between the lower inner angles of the eyes. Palæ oblong-cultrate, lower edge straight or slightly incurved, upper edge straight to the middle, thence truncate to the tip, where there is a short retrorse spine. Tibia same width as palæ and ½ as long.

Length 5 mm.–5½ mm. Boone Co. and St. Louis Co., Mo. July and November.

This species appears to resemble *Corixa (Arctocorisa ?) acuminata* Uhl., but the structure of the male palæ, which are quite unlike those of any other species in the group, together with the shape of the head in the male, the minute strigil, and the short lenticular pronotum, sharply sets it off from other species. The presence of a frontal fovea in the female is also extraordinary, and together with the points mentioned above seems to warrant separating the species from its congeners in a new genus, for which the name *RAMPHOCORIXA* is proposed, and of which the following may stand as a diagnosis :

RAMPHOCORIXA, n. gen.—Allied to Arctocorisa Wallen., from which it differs in the form of the male palæ, strigil and shape of head. Differs from Glænocorisa Thoms. in the absence of bristles among the palar pegs. Pronotum lenticular rastrate. Head of male sharply acuminate, with fovea acorn-shaped, ♂ palæ dorsally, deeply cleft, much longer than tibia, terminated by a long serrated spine; femur with a large stridular area of minute spines. Strigil minute. Fifth, sixth and seventh tergites divided in the male. Asymmetry of male dextral. Female palæ cultrate with a short retrorse terminal spine; face of ♀ foveate.

EXPLANATION OF PLATE IV.

Fig. 1.—Egg of *Ramphocorixa balanodis*, × 34. The dorsal cup is affixed to the carapace of the crayfish.

Fig. 2.—First instar, ventral aspect, × 82. A = the pala or first tarsus, × 240.

Fig. 3.—Third instar, dorsal aspect, × 24, showing the beginning of the wing-pads. The setæ of the legs and body are omitted.

Fig. 4.—Fifth instar, dorsal aspect, × 10. The wing-pads have grown beyond the thorax and are covered with downy hair. Cilia of abdomen and legs omitted.

Fig. 5.—Frontal aspect, head of male, × 20.

Fig 6.—Pala of male, × 51, viewed from inner upper angle. F = femur ; T = tibia ; P = Pala or tarsus ; A = row of pegs ; B = stridular area ; Q = diagrammatic section of pala marked X.

Fig. 7.—Pala of female, × 68.

Fig. 8.—Antenna × 68.

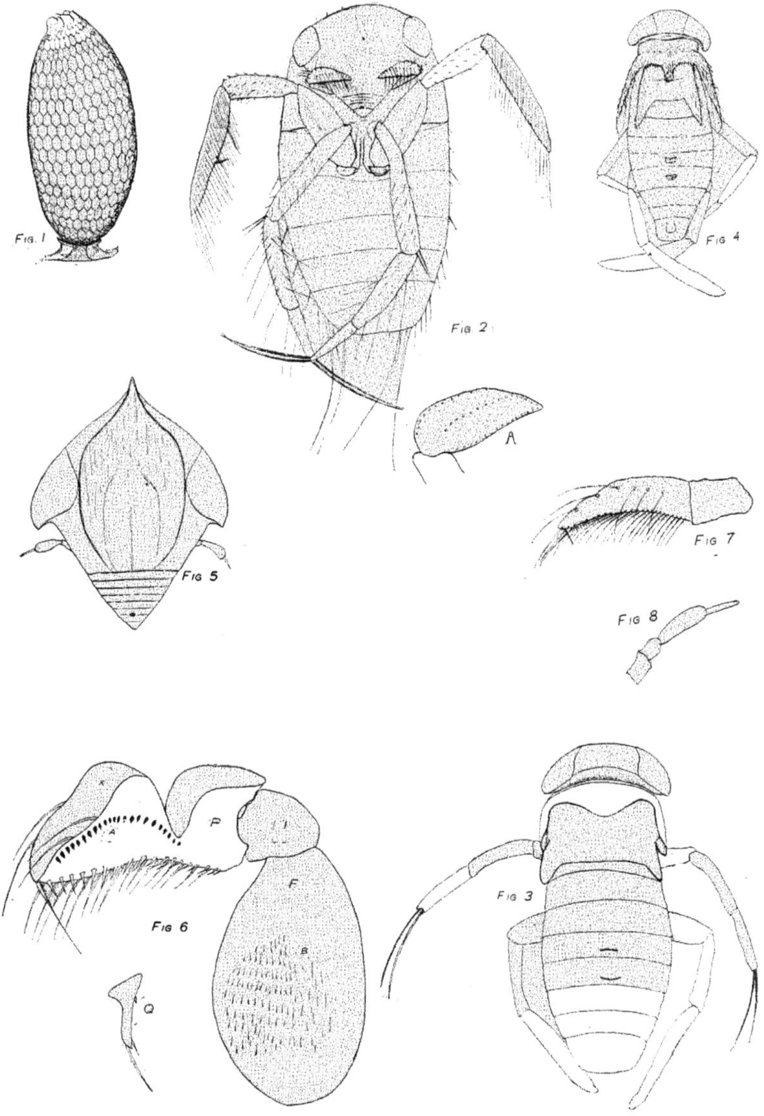

RAMPHOCORIXA BALANODIS, N. GEN. ET SP.

NEW SPECIES AND GENERA OF NORTH AMERICAN LEPIDOPTERA.

BY WM. BARNES, M.D., AND J. H. MCDUNNOUGH, PH.D., DECATUR, ILL.

(Continued from page 93.)

Subfamily *Hypeninæ.*

Epizeuxis terrebralis, sp. nov.

♂.—Antennæ rather lengthily ciliate; palpi upturned, 2nd joint smoothly scaled, attaining to front, 3rd joint long, narrow, pointed; thorax and primaries deep black-brown with indistinct maculation; t. a. line single, waved, black, inclined slightly outwardly with rather prominent outcurve in submedian fold; reniform an indistinct dark lunular mark, very faintly and only partially outlined in ochreous; from it a faint wavy shade proceeds to inner margin parallel to t. p. line; this latter most distinct of the maculation, slightly outcurved around cell and then parallel to outer margin, slightly wavy with faint inward angle on submedian fold; s. t. line only traceable in costal portion, faint, black; black terminal line and dusky fringes; secondaries dark smoky with broken terminal line. Beneath uniform dark smoky with darker terminal line and small discal spot on secondaries. Expanse, 23 mm.

Habitat: White Mts., Ariz., 1 ♂. Type, collection Barnes.

Smaller and darker than *suffusalis* Sm.; t. p. line less waved and without the ochreous blotch in reniform.

Bleptina flaviguttalis, sp. nov.

Palpi ochreous; head and thorax pale purplish gray sprinkled with black scales; primaries purple-gray, shaded outwardly with darker and with the remaining area thinly sprinkled with black scales; basal line an indistinct dark upright shade line across the whole area of wing; subbasal area very slightly shaded with ochreous; t. a. line single, black, upright, very slightly bent inwards at costa; orbicular a pale yellow dot, indistinctly outlined with brownish; reniform yellow, triangular, with apex produced towards costa, lower portion containing a black dot; a dark median shade line, indistinct in costal half, wavy below reniform and equidistant from t. a. and t. p. lines; t. p. line black, crenulate, concave below costa, again opposite reniform, outcurved between veins 3–5, incurved between veins 1–3, slight outward angle on vein 1; s. t. line pale yellow, even, outcurved between veins 3–5, otherwise straight, shaded anteriorly with dark brown ; a terminal row of black dots ; fringes dusky cut by a waved indistinct black line ; secondaries deep smoky with faint pale subterminal line only distinct near anal angle. Beneath pale smoky shading into dark brown outwardly ; a discal dot and faint dark median line on both wings and pale subterminal line, most distinct on secondaries ; fringes dusky. Expanse, 24 mm.

Habitat : Palmerlee, Ariz., 1 ♂. Type, collection Barnes.

A typical *Bleptina*, most closely related to a Texan species which we have identified as *inferior* Grt. In our species the t. a. line is straighter, the lower portion of the reniform larger and the subterminal line of secondaries is distinct at least in the anal half of wing.

Bleptina minimalis, sp. nov.

Palpi recurved, third joint long, very pointed, with slight tuft of scales on posterior side in both sexes ; antennæ finely ciliate ; general colour pale ochreous to dark gray, primaries often considerably shaded with black scales ; maculation varying in distinctness, often almost lacking, in other cases quite distinct ; when present, basal line a slight dark streak very close to base ; t. a. line single, dark, slightly waved, most prominent below orbicular ; this latter spot just beyond t. a. line, longitudinally oval, yellow, small, obscurely outlined in brownish ; reniform yellow, constricted medially, lower portion broader than upper and containing a blackish dot, the whole partially outlined in dusky ; at times the whole reniform is much reduced, becoming a mere yellow streak ; occasionally traces of a faint dusky median shade, usually however lacking ; t. p. line dark, crenulate, shaded outwardly with paler, excurved from just below costa around the reniform, with strong incurve in submedian fold ; s. t. line pale, shaded more or less strongly on both sides with dusky, slightly wavy, excurved in central portion ; a terminal series of dark dots ; secondaries smoky brown, occasionally with traces of a crenulate dark median line and pale subterminal one, mostly however immaculate. Beneath uniform smoky, lighter in basal area of secondaries in the paler forms, which also show traces of the lines of upper side. Expanse, 14 mm.

Habitat : Babaquivera Mts., Ariz.; Redington, Ariz., 2 ♂s, 6 ♀s. Types, collection Barnes.

The small size will readily distinguish this species. In general the ♀s are darker and more obscurely marked than the ♂s. Fresh specimens also appear considerably darker than those that have been in the cabinet some time.

Family *Notodontidæ.*

Hyperæschra stragula, var. *ochreata*, var. nov.

♀.—Primaries pale ochreous brown, maculation as in *stragula;* on costa between basal and t. a. lines a bar of blue-black extending downwards to first brown dash in cell and somewhat sprinkled with whitish; just beyond the discal lunule a large blue-black patch, rather irregularly circular, resting on costa, defined outwardly by the t. p. line, basally by vein 3; this contains two whitish patches, one between veins 4 and 5, the other smaller between veins 6 and 7, and opposite the discal lunule it is indented with white; costal area between it and t. a. line whitish, containing the reddish-ochre discal mark; terminal area blue-black; basal half of inner margin broadly blue-black crossed by the pale curved t. a. line. Secondaries white; fringes shaded with blue-black at inner angle. Expanse, 46 mm.

Habitat: Provo, Ut. (July 21st), 1 ♂. Type, collection Barnes.

The large circular black mark beyond cell is very characteristic, the ground colour is paler than in *stragula* and the s. t. line much less waved. May prove to be a distinct species on receipt of more material.

Family *Thyatiridæ.*

Habrosyne rectangula, var. *arizonensis*, var. nov.

Ground colour of primaries and thorax mouse-gray with none of the brownish shading characteristic of *rectangula;* white dash at base of wing reduced in size; pinkish-white costal and subterminal shading somewhat reduced; secondaries similar to ground-colour of primaries, not brownish as in the type form.

Habitat: White Mts., Ariz., 1 ♂. Type, collection Barnes.

Family *Geometridæ.*

Genus *Grossbeckia*, gen. nov. (type *G. semimaculata*, sp. nov.).

Antennæ of ♂ shortly bipectinate; palpi long, drooping, pointed at extremity, smoothly scaled; fore legs with the femur slightly hollowed out distally to receive the tibia which is ½ the length of the femur and possesses a slight hair pencil beneath, concealing the epiphysis; posterior tibiæ normal, spurred, outer spur about half as long as the inner; wings long, narrow, primaries with slightly convex costal margin, well rounded apex and rather oblique outer margin; inner margin straight; secondaries slightly crenulate with sharp angle at junction of inner and outer margins; primaries with vein R_1 from about middle of cell; areole long, narrow, exceeding the cell by about half its own length; R_2, R_3 and R_4 stalked from apex of areole; R_5 from a point with R_2; M_1 from lower angle of areole; M_2 from centre of discocellular; M_3 and Cu_1 from lower angle of cell; Cu_2 from near angle; anal vein parallel to inner margin; secondaries with vein M_2 absent, vein S. C. joined to the cell for about ⅔ of its length; veins R and M_1 stalked; M_3 and Cu_1 connate from lower angle of cell; Cu_2 from near angle; 2nd anal very close to inner margin; frenulum present.

The genus is remarkable in lacking vein M_2 of secondaries and having the subcostal vein distinctly united with the cell, which would place it in the family *Fernaldellinæ* of Hulst. The type species possesses very little resemblance to specimens of the genus *Fernaldella* and is in fact very distinctly *Hydriomenid* in general appearance. We take pleasure in naming the genus after our friend Mr. J. A. Grossbeck.

Grossbeckia semimaculata, sp. nov.

Palpi, thorax and abdomen dark gray, latter pale ochreous beneath; primaries with the costal half of wing alone showing maculation; below cubital vein, with the exception of the inner margin, the whole area is dull gray-brown; costal half of primaries brown, suffused with light and dark gray; a slight black streak at base of wing; from costa near base an obscure dark shade extends obliquely across wing to cubital vein, bordered outwardly slightly with gray; at apex of cell a large white quadrate patch; costal area above this patch largely dark smoky-brown with several pale gray streaks, beyond the patch the brown ground colour forms a small round blotch; from apex of wing inwards nearly to white patch a diffuse gray shade extends, bordered inferiorly by a slightly notched dark line ending in a dark shade, from which two curved dark lines, parallel to outer margin, arise, extending to vein Cu_2 and enclosing a dark gray blotch; terminal area shaded with dark gray; cubital and median veins marked with black; a black shade, broken twice by gray scaling, extends narrowly along the inner margin; secondaries rather pale smoky, fringes somewhat lighter. Beneath primaries uniform pale glossy brown ; secondaries whitish, shaded with brown along outer margin. Expanse, 31 mm.

Habitat : Palmerlee, Ariz., 1 ♂. Type, collection Barnes.

Bears considerable resemblance to *Cataclothis frondaria* Grt., but the pale immaculate area is much larger and the venation totally different. This pale area is due to the fact that when at rest the wings are so folded that only the costal half of wing is visible, the insect then having much the appearance of a Phycitid.

Diastictis (Cymataphora) pallipennata, sp. nov.

♂.—Palpi, head, thorax, and abdomen white ; ground colour of primaries white, marbled with gray, which shade is predominant beyond the t. p. line ; t. a. line single, rather broad, especially at costa, dark gray, angled outwardly below costa and with slight inward angle on submedian fold ; median space more or less shaded with gray ; a small oval gray mark at end of cell, slightly filled with white ; t. p. line dark gray-brown, curved outward from costa, inward opposite cell and with a rather sharp outward angle in submedian fold, in general parallel to t. a. line ; subterminal and terminal spaces almost entirely olive-gray with the exception of a white costal patch ; through this area a broad white subterminal line runs, rather irregular in course, angled slightly outwardly below apex, the angle preceded by a short dark dash in subterminal space, another slight angle or outcurve between veins 3 and 4, termination at inner angle ; fringes white, checkered with dark gray. Secondaries white, heavily mottled with dark gray, leaving only traces of the ground colour visible ; a faint discal dot and rather heavy dark postmedian line, parallel to outer margin, with slight outcurve before inner margin ; terminal area somewhat darker ; fringes white, faintly checkered with gray. Beneath white, heavily sprinkled with gray ; primaries with discal dot and straight postmedian line ; secondaries with dark postmedian and subterminal lines and discal dot.

Male paler than ♀ ; the subterminal and terminal spaces white with but little dark gray shading ; subterminal line wanting ; from costa near apex a dark dash inward to vein 7, semiparallel to t. p. line. Antennæ bipectinate, posterior tibiæ swollen but without hair pencil. Expanse, 19–22 mm.

Habitat : Redington, Ariz., 1 ♂, 4 ♀s. Types, collection Barnes.

The species has considerable resemblance with *Macaria s-signata* in general type of maculation. The extent of the dark olive-gray shading is variable, the median area is in some specimens pure white, in others quite smoky; the discal mark may be present or absent.

Phycitiinæ.

Euzophera strigalis, sp. nov.

Palpi upturned, 3rd joint moderate, pointed, smoothly scaled with whitish; pectus and legs pale gray; antennæ of ♂ lamellate and ciliate; front gray; thorax and primaries dark gray, sprinkled with lighter; primaries with the basal area and central area at end of cell slightly less sprinkled than remainder of wing; all maculation wanting, except that the veins are prominently outlined in black, giving a strigate appearance to the wing. Secondaries, hyaline, with white fringes, slightly smoky at apex of wing. Beneath primaries smoky; secondaries as above, costal margin sprinkled with dark scales. Expanse, 42 mm.

Habitat: Eureka, Ut., 1 ♂. Type, collection Barnes.

Related to *E. gigantella* Rag., but lacks all traces of the transverse lines.

BOOK NOTICE.

INSECTS OF FARM, GARDEN AND ORCHARD.

"INSECT PESTS OF FARM, GARDEN AND ORCHARD," by E. Dwight Sanderson. Publ. John Wyley & Sons, 43 East 19th Street, New York; also The Renouf Publishing Company, 25 McGill College Ave., Montreal. XII, 684 pp., 513 figs. $3.00.

The increasing number of workers in economic entomology and the consequent enormous output of literature embodying the results of their, or other people's, investigations is rendering it gradually more difficult for the student, farmer or fruit-grower to gain a knowledge of the life-histories of and means of controlling the insects with which they have to deal. Any means whereby this difficulty can be lessened is an addition to the insect-fighting organization as a whole, and its welcome is proportionate to its efficiency. We give a whole-hearted welcome to this last addition to our economic literature, and are glad that the author found the necessary leisure time to develop and complete a work which he is eminently fitted to carry out.

It is impossible in a review of this nature to give more than a superficial idea of the contents. In the author's words, it has been his effort "to discuss all of the important insects of farm, garden and orchard at sufficient length to give a clear idea of their life-histories and habits, and also the best means of control, so that the book may be used as a reference work both by the student of economic entomology and by the practical farmer, gardener or fruit-grower." His effort has certainly been successful, with the result that in addition to a well-balanced treatment of the insects affecting staple crops, such as his previous work, now out of print, gave, he has included insects affecting small bush and orchard fruits. The well-selected references which are given under each insect to the more important bulletins on that insect and its control, will prove of great value to the student or to the agriculturist with a thirst for more knowledge, and there are many such.

Like our injurious insects, the author recognizes no international boundary, in fact, his free annexation of our provinces is almost startling at times; Nova Scotia, however, is not "Southern Canada" (p. 619), but eastern.

While the author has succeeded to a remarkable degree in his use and choice of popular names, there are one or two instances where we believe that the name is too long to be suitable for popular use; nevertheless, we fully realize the difficulty of choosing a short name, which is at the same time distinctive. The author calls *Cephus pygmæus* Linn. the Wheat Saw-fly Borer; this name might not be recognized by persons accustomed to call the species the Wheat-stem Saw-fly or the Wheat-stem Borer. Such instances serve to indicate that we are still a long way from the solution of the question of popular nomenclature, and that there is much careful work yet to be done by not only our own Committee on Nomenclature, but by an International Committee.

The book is illustrated by over 500 figures, which, with a few exceptions, are excellent, and their clear reproduction is due to the fine quality of the paper used. With the exception of a certain number of typographical errors and one or two inverted figures, which are troubles from which all authors are compelled to suffer at the hands of their printers, there is no fault to find with the printer's share of the work. Its reasonable price should make it one of the chief books of reference, especially for those for whom the book is written, and we hope that the second edition will soon make its appearance. C. GORDON HEWITT.

Mailed April 9th, 1912.

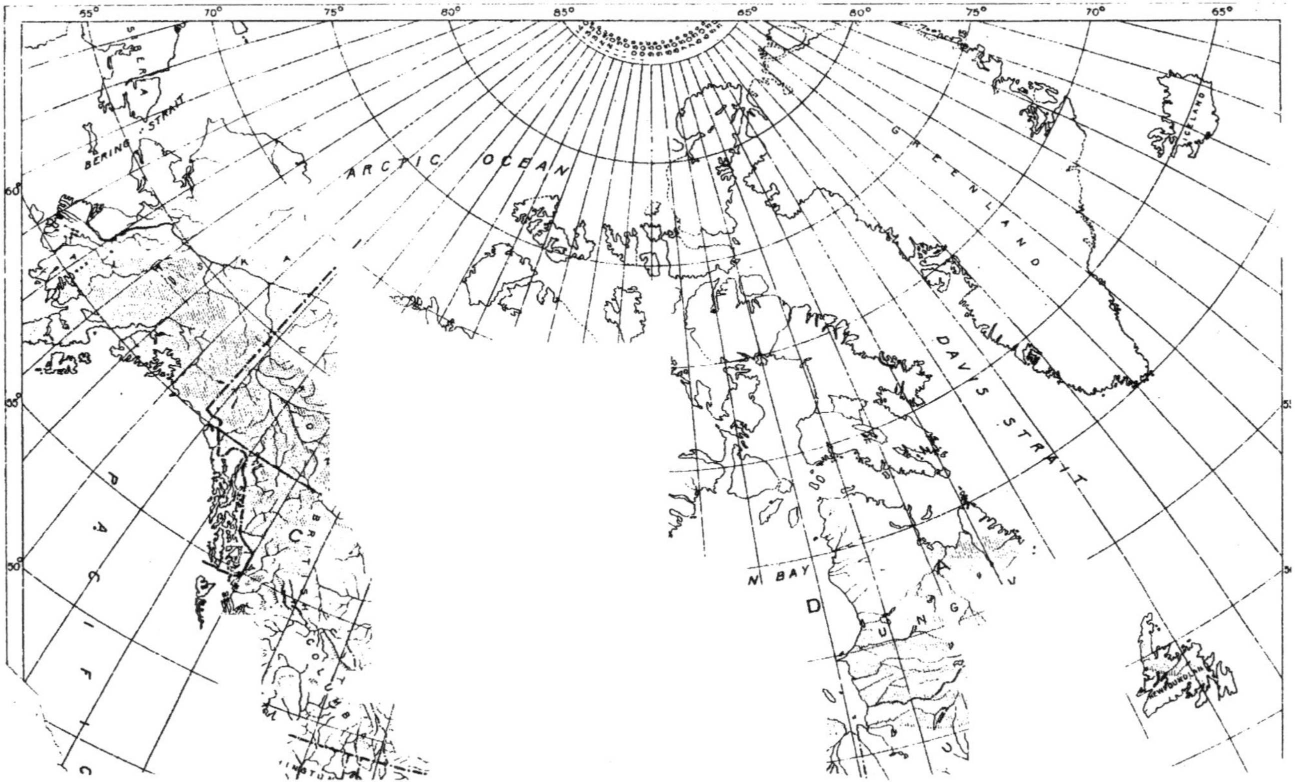

ARCTIC OCEAN
GREENLAND
ICELAND
DAVIS STRAIT
BERING STRAIT
SIBERIA
BRITISH COLUMBIA
NEWFOUNDLAND
N BAY
PACIFIC

The dotted parts of the Austral Zones east of the Great Plains indicate the extent of the humid divisions of these Zones, known respectively as the Alleghanian, Carolinian, and Austroriparian Faunas. The undotted parts of the same Zones are known as the Transition, Upper Sonoran, and Lower Sonoran.

The Canadian Entomologist.

VOL. XLIV. LONDON, MAY, 1912. No. 5

MAP ILLUSTRATING FAUNAL ZONES OF NORTH AMERICA.

Through the courtesy of Dr. H. W. Henshaw, Chief of the Biological Survey of the United States Department of Agriculture, we are enabled to publish the Fourth Provisional Zone Map of North America. This map has not yet been published by the Biological Survey, by whom it was prepared, to accompany a revised edition of their Bulletin No. 10, now in course of preparation, but has appeared in the American Ornithologists' Check List.

Our object in publishing this map is primarily to assist those engaged in the preparation of the Catalogue of the Insects of Canada and Newfoundland. (See pp. 273–275 of Vol. XLIII of this journal.) On page 274 it was stated that the geographical distribution of each species within Canada and Newfoundland will be given. "This will be indicated as a rule by Provinces, in order from east to west, e.g., N. S., Ont., B. C., etc. The characteristic faunal zones inhabited by the species will be indicated so far as it may be possible by abbreviations, thus: Ar.–Arctic, H.–Hudsonian, C.–Canadian, T.–Transitional" With the addition of Upper Austral, to be indicated by "U. A.", these are all the zones which are represented in Canada and Newfoundland, so far as we know at present. The entire map of North America has been published, as it is impossible to consider or discuss the faunal zones of Canada apart from those of the United States.

In stating the distribution of provinces, the recent extensions made to the boundaries of the Provinces of Manitoba, Ontario and Quebec should be noted. The northern boundary of Manitoba is marked by the 60th parallel, and the new north-eastern boundary is a line drawn from the north-eastern corner of the original boundaries to the shore of the Hudson Bay, where the latter is intersected by the 89th meridian. The Province of Ontario extends northward to the Hudson Bay, east of the eastern boundary of Manitoba. The Province of Quebec extends northwards, and includes the region of Ungava.

Mr. Edwin C. Van Dyke, of San Francisco, Cal., who has made a careful study of faunas of western North America, in a recent letter to me, proposes that the following should be included in a zone to be named *Vancouveran:* That portion on the southern side of the inner Aleutian Islands, South-eastern Alaska and the Islands of the Coast, Western British Columbia, including the islands, Western Washington, the western portion of Northern Oregon, and a strip along the coast of California, to a little south of San Francisco Bay. That this zone has not been included in our scheme is not evidence of its non-acceptance, for we believe that Mr. Van Dyke's proposal is supported by a number of facts. Pending further investigation, however, we have deemed it advisable to restrict ourselves to the zones already indicated.—[C. GORDON HEWITT.

REPORT ON THE CELEBRATION OF THE CENTENNARY OF THE FOUNDATION OF THE ACADEMY OF NATURAL SCIENCES OF PHILADELPHIA.

BY F. M. WEBSTER,

Delegate from the Entomological Society of Ontario.

This noted gathering was, thanks to the united efforts of the members, an entire success. The entire world of letters, personified by a company of more than one hundred distinguished men and women, representing the institutions of learnings and scientific societies of this country and Europe, participated in the ceremonies. The fine new lecture hall of the academy was given over to the carrying out of the set programme, made up of papers on purely scientific subjects, prepared and read by the distinguished delegates at the meeting.

On the platform were seated Mayor Blackenburg; Dr. Samuel G. Dixon, president of the academy; Sir James Grant, of the Royal Society of Canada; Dr. Edward J. Nolan and Dr. J. Percy Moore, the secretaries.

Among the smaller social events in connection with the centennial celebration was a dinner given by Dr. Henry Skinner, of the academy, at his home in Glenn road, Ardmore. There were present the following delegates, representing the entomologists, in which particular branch of natural science Dr. Skinner is especially interested: Dr. W. J. Holland, of the Carnegie Museum, Pittsburgh; Professor J. H. Comstock, of Cornell University; Professor C. W. Johnson, of Boston, representing the Boston Society of Natural History; E. T. Cresson, of Philadelphia, representing the American Entomological Society, of which he was founder; Dr. Philip P. Calvert, of the University of Pennsylvania, representing the Sociedad Aragonesa de Ciencias Naturales, and Professor F. M. Webster, Washington, D.C., representing the Entomological Society of Ontario.

The address of the president, Dr. Dixon, was exceedingly gratifying to entomologists, by reason of his laying some stress on the fact that among the original founders of the academy was Thomas Say, the father of American entomology. In his mention of the services many of the former members of the academy had made to science, he again spoke of Thomas Say, who went out with the Long Expedition to the Rocky Mountains, in 1819. This was followed by expressions of appreciation of the later works of Le Conte, Horn, Cresson and others. He told of the size and importance of some of the special collections, mentioning among others the collection of insects which now numbers 1,000,000 specimens and has world-wide renown.

Doctor Dixon showed the practical use of the work of the academy, and the real value to people and Government in the study of insect life; the now known cause of many preventable diseases, among them yellow fever, an outbreak of which was promptly suppressed in New Orleans, La., and malaria, which have been banished from Cuba and the Panama Canal section. He gave some figures showing the immense damage done to crops by insect life, and showed the money loss in this field, which economic entomology is trying to correct, to be more than $1,000,000,000 a year.

There were but two papers presented relating exclusively to insects. The first by Henry Skinner, M.D., D.Sc., on "Mimicry in Butterflies."

Dr. Skinner's long familiarity with these insects rendered his paper of unusual interest and value. First calling attention to the many cases of deceptive resemblance among butterflies, and the very much that had been written on protective mimicry, both in this and other countries, he called attention to the fact that actual observations on the feeding of birds on butterflies were almost entirely lacking here in America, so much so that at the present time protective mimicry among butterflies must be admitted to be far more fancied than real, and, that the proof justly demanded by science was here conspicuously lacking. The doctor rested his case on the scientific as well as legal objection of "not proven." The second paper, by Mr. Jas. A. G. Rehn, dealt with "The Orthopterological Inhabitants of the Sonoran Creosote Bush" throughout the country along the Mexican border so rich in new and unique species of insects. It was of much interest not only relative to Orthoptera but also from a faunal point of view.

Prof. Webster's Address.

Members of the Academy of Natural Sciences of Philadelphia:

A very pleasing duty has devolved upon me as an honourary member of the Entomological Society of Ontario, in having been delegated to represent that body, at this your 100th anniversary, and to convey, for the Society, its hearty congratulations and good wishes for your continued prosperity and success.

It will perhaps not be out of place for me to call attention at this time to the fact that this sister society, but four years the junior of the Entomological Society of Philadelphia, afterwards the American Entomological Society, expects, next year, to celebrate its 50th anniversary.

We, who have had the good fortune to attend the meetings of this Society across the border, cannot easily forget the cordial greeting and warm comradeship we have always enjoyed, and we all the more appreciate the hearty God-speed which I am expected to convey to you. Not only have our colleagues done a grand work in Canada, but the pages of the CANADIAN ENTOMOLOGIST have been as freely open to us as to their own numbers.

Insects know no national boundaries, therefore those who study them must be equally cosmopolitan in their investigations. So, also, science knows no race, nationality or creed, because it deals with the universal, and in recognition of this, my message becomes all the more appropriate.

F. M. WEBSTER.

ON THE LARVAL STAGES OF CERTAIN ARCTIAN SPECIES.

BY WM BARNES, M.D., AND J. MCDUNNOUGH, PH D., DECATUR, ILL.

A. phyllira Drury.

In a previous article (CAN. ENT., XLIII, 257), we described the final larval stage of this species. Since then we have been successful in breeding from the egg, and append our notes on the various stages. Packard has already described the larval history (Jour. N. Y. Ent. Soc., III, 178), but rather briefly, so that we feel justified in publishing our own account as a verification and amplification of Packard's. It has been suggested that *phyllira* is but a variety of *rectilinea* or vice versa. We would call attention to the fact that in *phyllira* larvæ the spiracles are orange, whilst in *rectilinea*, according to Gibson (CAN. ENT., XXXV, 117), they appear to be black; this would seem to suggest that we are dealing with distinct species. All our bred specimens showed (apart from slight increase or decrease in the heaviness of the white markings), very little tendency to variation, and in no case could we detect a specimen with traces of white markings on the veins in the outer portion of the wing; *phyllira* normally possesses a slight white dash on the subcostal vein and occasionally one on the cubitus near base of wing, as figured by Drury, but beyond this the veins are not outlined with white. The white markings on the veins of *rectilinea* on the other hand show apparently a marked tendency to obsolescence, and we possess several Illinois specimens in which we can detect but the merest trace of white on the veins; in fact, were it not for locality, they might easily be mistaken for *phyllira*. Continued breeding will be necessary to decide the question as to whether the above mentioned larval distinction holds good.

In the accompanying sketch we give a diagram of the position of the primary tubercles in the first stage of *phyllira;* this is apparently typical

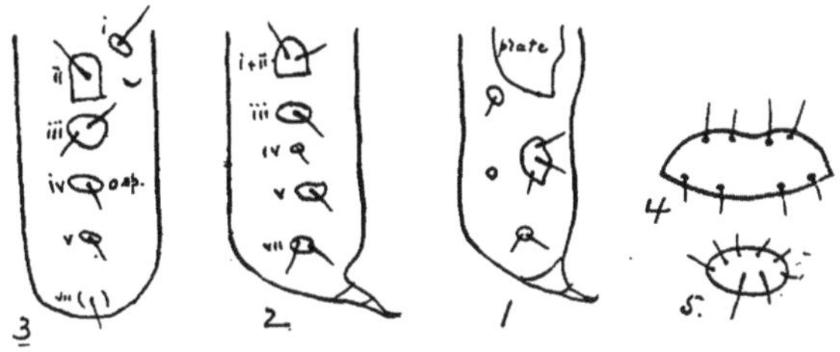

Fig. 7.

1. Prothoracic segment.
2. Mesothoracic segment.
3. Abdominal segment.
4. Prothoracic plate.
5. Small plate.

for the genus *Apantesis*, at least it holds good for all the species that are discussed in this present paper. On the meso- and metathoracic segments tubercles I and II coalesce forming a single wart with two setæ; on the abdominal segments, I and II are separate, the former being minute; III contains two setæ, a typical Arctian feature; IV is immediately behind the spiracle, V directly below IV, each one with seta; VI is absent and VII is represented only on those abdominal segments which bear no prolegs as a minute seta; on the thoracic segments VII is a more prominent wart with two setæ. On the prothorax and 9th and 10th abdominal segments considerable reduction of the tubercles takes place. In the following descriptions if no reference to the position of the tubercles is made, it may be taken for granted that they correspond with the above diagram.

Ovum.—Conical from a flat base; very slightly sculptured; pale yellow, shiny, with no colour change until just previous to emergence when it becomes blackish; deposited promiscuously on the ground.

Stage I.—Head black with sparse setæ; body pale greenish brown with blackish tubercles and large black thoracic plate, this latter containing 8 black setæ arranged in an anterior and posterior row of four. On the thoracic segments the setæ of tubercles I, II and III are black, the others being white; on the abdominal segments I–IX tubercle II contains a black seta and the dorso-anterior one of III is also black; the remainder are white. On the 9th abdominal segment the whole dorsal portion to each side of the central line is occupied by a large tubercular patch containing three black and one white setæ, the latter situated on the lower posterior portion; below this large patch is a small tubercle with a single white seta. The anal plate is small, oval, and dark in colour, containing two central setæ and a marginal row of six minute white ones on posterior side. Legs blackish; prolegs similar in colour to body. Length, 3 mm.

Stage II.—Head black with numerous short setæ; body greenish, strongly marbled with red-brown, the marbling so arranged as to leave a stripe of the ground colour extending along the body below tubercle II, giving the appearance of a subdorsal line. A series of white patches, more or less evident dorsally, situated between tubercles II. Thoracic plate tends to break up into two anterior mounds and a single transverse posterior one, the former each containing about 7 black setæ, inclined forward over the head, the latter with 4 long straight black setæ and two white ones, arranged three on each side of the central dorsal line. On the prothorax the small tubercle behind the plate contains two white setæ; the large tubercle anterior to the spiracle 6 black setæ and 1 white one, situated anteriorly; the lateral tubercle bears 3 white setæ. On the meso- and metathorax tubercle I and II contains 8–9 black setæ and 1 small anterior white seta inclined forward; tubercle III bears two long black setæ, a small white one on the anterior margin and a similar one on the posterior portion; tubercle IV is small with two white setæ; tubercle V bears 7 setæ, of which on the mesothorax 3 are black and 4 white, on the metathorax 2 black and 5 white, the posterior black one having changed colour; tubercle VII bears 6 white hairs. On the abdominal segments tubercle I is minute, bearing a short white seta, II has a small white seta on anterior margin inclined forward and a long central black seta surrounded by a ring of 5 others; there are usually 2 small white setæ on anterior margin of III and one on posterior portion, and further 5 long black setæ arising out of the central area; tubercles IV and V contain each 4–5 fine white setæ; tubercle VII is small, on leg-bearing segments with three small white hairs, on others with but one or two setæ, absent on 9th abdominal; the large tubercle of this segment bears about 12 black setæ, of which the central one is longest and points backward; further there are about 5 small white hairs around outer margin. The anal plate is small, heart-shaped with about 12 minute hairs, both black and white, situated mostly towards the posterior margin. All black setæ are strongly barbed. Legs black. Length, 6 mm.

Stage III.—Head black with short setæ as in previous stage; body red-brown with a pale yellow dorsal stripe; the mounds of the prothoracic plate are more distinctly separate than in previous stage, the anterior ones being kidney-shaped, each bearing 8–10 setæ, the posterior one narrowly oval with about 8 upright barbed bristles; almost all these setæ are black, an occasional short white one being intermingled. Tubercles shiny, black, prominently conical, tubercle I minute; tubercle II largest with about 12–15 spiculated black spines, those of the rear segment being longest and inclined backward; tubercle III with black setæ; tubercle IV with two central black and about 8 white setæ; other lateral tubercles entirely with white setæ. Legs black; prolegs colour of body.

In late stages the body colour appears greenish marbled strongly with red-brown and with central lateral portions of abdominal segments showing traces of orange. Length, 9 mm.

Stage IV.—Head as before; body dark gray; marbled with black, and with a pale creamy, usually continuous, dorsal stripe. All tubercles black, shiny, II very large; lateral tubercles, especially V and VII, shaded prominently around the base with orange. There is considerable increase in the number of setæ, which are mostly prominently barbed. Legs black; prolegs orange. Length, 12 mm.

Stage V.—Much as in previous stage. Pale yellow dorsal stripe shows a tendency to narrow or disappear intersegmentally; traces of a broken yellow subdorsal line, composed chiefly of a strip of colour between tubercles II and III. Tubercles I and II black, others either tipped with orange-brown or entirely suffused with this colour. Spiracle orange, prolegs pale reddish. Length, 23 mm.

Stage VI.—Similar to preceding. In freshly moulted specimens all lateral tubercles show very strongly orange, becoming however later more tinged with black; dorsal stripe prominent; subdorsal line almost obsolete. Length, 29 mm.

Stage VII.—We would refer to our previous article (CAN. ENT., XLIII, 259) for the description of this, the final stage. All larvæ examined agreed excellently with the description drawn up from the spring brood.

Apantesis placentia A. & S.

Ovum.—Practically identical with that of *A. phyllira*, deposited promiscuously.

Stage I.—When first emerged pale yellow with black head, turning later dirty brown. Tubercles large, blackish, with long setæ, arranged as in *phyllira;* setæ of I–III black, of lateral setæ white; prothoracic plate dark, large, with apparently 6 black setæ. Length, 3 mm.

Stage II.—Head black. Body red-brown with greenish intersegmental tinge; no trace of markings. Tubercles black, I minute with single white seta, II large, with about 5 black setæ, those of the posterior segments being longest and inclined backward, III with 3 or 4 black setæ, IV and V each with 3 small white outwardly inclined setæ. Thoracic plate semilunate, black, with a double row of four setæ. Length, 6 mm.

Stage III.—Head black. Ground colour greenish gray heavily mottled with dark brown and with lateral central portions of segments broadly reddish orange, giving the general appearance of a reddish-orange ground colour, the true ground colour being only apparent intersegmentally and dorsally. A dorsal series of orange diamond-shaped patches, more or less concealed by a thin line of the same colour. These patches are not prominent, being similar in colour to the lateral orange shading; they are most recognizable immediately following the moult, when the lateral colour is not so developed. Tubercles black with considerable increase in the number of setæ. Length, 9 mm.

Stage IV.—Head black. General appearance much darker than in preceding stage. Body gray-green with dark brown marbling; dorsal deep-orange stripe more prominent; lateral orange shading considerably reduced, being confined mostly to the base of the tubercles; these latter shiny black with numerous black setæ, except V and VII which still bear white ones. Spiracle black, prolegs reddish. In late stages the colour becomes paler and the lateral orange markings are again plain. Length, 12.5 mm.

Stage V.—Head and body jet black with large shiny black tubercles which show a great increase in setæ; these are barbed, but not nearly so prominently so as in *phyllira*. A broken dorsal reddish stripe is present. Prolegs reddish; stigma black. Length, 22 mm.

Stage VI.—Scarcely any change from previous stage; rather blacker, dorsal stripe often lacking, when present much broken into spots of reddish orange; tubercles very shiny and large. Length, 30 mm.

Stage VII.—This final stage has already been described by us. (Vide CAN. ENT., XLIII, 259.)

The resultant imagines showed but little variation; in one ♀ there was a slight indication of the W mark due to a few light dots in the subterminal area; in most specimens, however, the tendency was to a reduction rather than an increase of the light markings of primaries. The ♀s agreed well with the figure published with the above mentioned article.

(To be continued.)

ON NORTH AMERICAN PHLŒOTHRIPIDÆ (THYSANOPTERA), WITH DESCRIPTIONS OF TWO NEW SPECIES.

BY J. DOUGLAS HOOD, U. S. BIOLOGICAL SURVEY.

Trichothrips anomocerus, sp. nov.—(Plate VI, figs. 1–4.)

Female.—Forma brachyptera. Length about 1.5 mm. Colour clear brownish yellow, with conspicuous hypodermal pigmentation in head, thorax and abdomen, which is orange by reflected light and maroon-brown by transmitted light; tube heavily chitinized and darker at middle; segments 7 and 8 of antennæ blackish brown.

Head distinctly wider than long, blunt anteriorly, frons not at all produced between antennæ, dorsal and lateral surfaces with very minute spines; vertex flat, evenly declivous; genæ subparallel, rounded; postocular bristles pointed, moderately long. Eyes greatly reduced, only one facet visible on lateral profile. Ocelli wanting. Antennæ slightly more than twice as long as head, the last two segments compactly united, the separating suture scarcely visible; segment 3 subconical; 4–6 oval, pedicellate; 7+8 lanceolate, pedicellate; segments 1 and 2 exactly concolorous with body; 3–6 successively very slightly darker; 7+8 rather abruptly dark blackish brown; sense cones moderate in length, slender; formula: 3, 1–1; 4, 1–2; 5, $1–1^{+1}$; 6, $1–1^{+1}$; 7, 0–1; 8 with one at middle of dorsum. Mouth-cone not quite attaining base of prosternum; labium broadly rounded; labrum pointed, scarcely surpassing labium.

Prothorax large, massive, notum weakly chitinized; it is distinctly longer than head (about equal in length to width of head), and across the coxæ is just twice as wide as long; bristles long, pointed; anterior marginals wanting. Pterothorax greatly reduced, narrower and shorter than prothorax. Legs stout, concolorous with body; fore femora short, thick; fore tarsus armed with a strong, acute tooth.

Abdomen large, heavy, about one and one-fourth times as wide as pterothorax; all bristles pointed. Tube thickly chitinized, slightly shorter than head, about two and one-half times as wide at base as at apex; terminal bristles short, about half as long as tube.

Measurements: Length, 1.53 mm.; head, length .18 mm., width 20 mm.; prothorax, length .205 mm., width (inclusive of coxæ) .40 mm.; pterothorax, width .37 mm.; abdomen, width .47 mm.; tube, length .17 mm., width at base .101 mm., at apex .041 mm. Antennæ: Segment

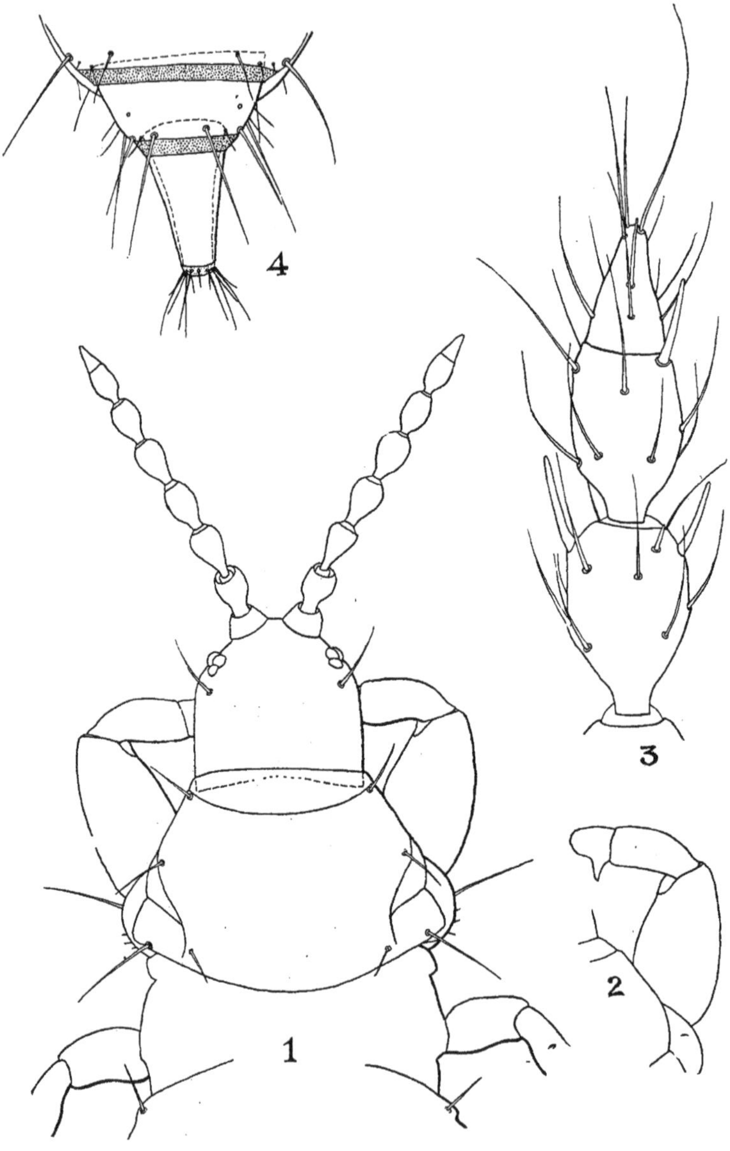

TRICHOTHRIPS ANOMOCERUS, SP. NOV.

1, 45μ; 2, 56μ; 3, 62μ; 4, 56μ; 5, 59μ; 6, 51μ; 7, 45μ; 8, 33μ; total, .41 mm.; width at segment 4, .039 mm.

Male.—Forma brachyptera. Slightly smaller than female. Length about 1.1 mm. Prothorax very slightly, if any, heavier than in female. Fore femora slightly more swollen; tarsal tooth a little stouter. Abdomen slender, tapering from near the base.

Measurements: Length, 1.09 mm.; head, length .17 mm., width .19 mm.; prothorax, length .192 mm., width (inclusive of coxæ) .37 mm.; pterothorax, width .32 mm.; abdomen, width .38 mm; tube, length .13 mm., width at base .083 mm., at apex .036 mm. Antennæ: Segment 1, 45μ; 2, 50μ; 3, 56μ; 4, 49μ; 5, 53μ; 6, 47μ; 7, 43μ; 8, 29μ; total, .37 mm.; width at segment 4, .034 mm.

Described from fifteen females and seven males, taken in February under sycamore bark at Plummers Island, Maryland (in the Potomac near Washington, D. C.), by Mr. W. L. McAtee.

The form of the apical antennal segments seems to ally this species quite closely to *T. ambitus* Hinds, from which, however, it is abundantly distinguished by the shorter tube, shorter and broader head, and the much heavier prothorax. The general facies of the species is thus that of *T. pedicularius* Haliday and *T. americanus* Hood.

Cryptothrips junctus, sp. nov.—(Pl. VII, fig. 1, *a*, *b*, *c*.)

Female.—Forma brachyptera. Length about 1.7 mm. Surface smooth, shining, anastomosing lines scarcely evident. Colour by reflected light bright crimson red; head and prothorax darkened with blackish brown; tube, legs and antennæ nearly black. Colour by transmitted light blackish brown; the head, prothorax and abdomen with a nearly continuous layer of bright crimson hypodermal pigment; antennæ dark blackish brown, segments 1 and 2 and pedicel of 3 slightly paler; legs slightly paler than antennæ, non-pigmented, tarsi pale yellow.

Head rectangular, about one and one-fifth times as wide as long; cheeks parallel, rounded very abruptly to eyes and slightly flaring at base; vertex rounded, slightly produced; postocular bristles long, explanate and divided at tip. Eyes small, flattened, protruding, anterior in position and directed forward. Ocelli small, subapproximate, anterior, the posterior far removed from the eyes. Antennæ seven-segmented, with an oblique suture at middle of ventral surface; spines and sense-cones long, slender; formula: 3, 1–2; 4, 2–2; 5, $1\text{–}1^{+1}$; 6, $1\text{–}1^{+1}$; 7, 0–1. Mouth-cone large, heavy, blunt; maxillary palpi more than half the length of pronotum.

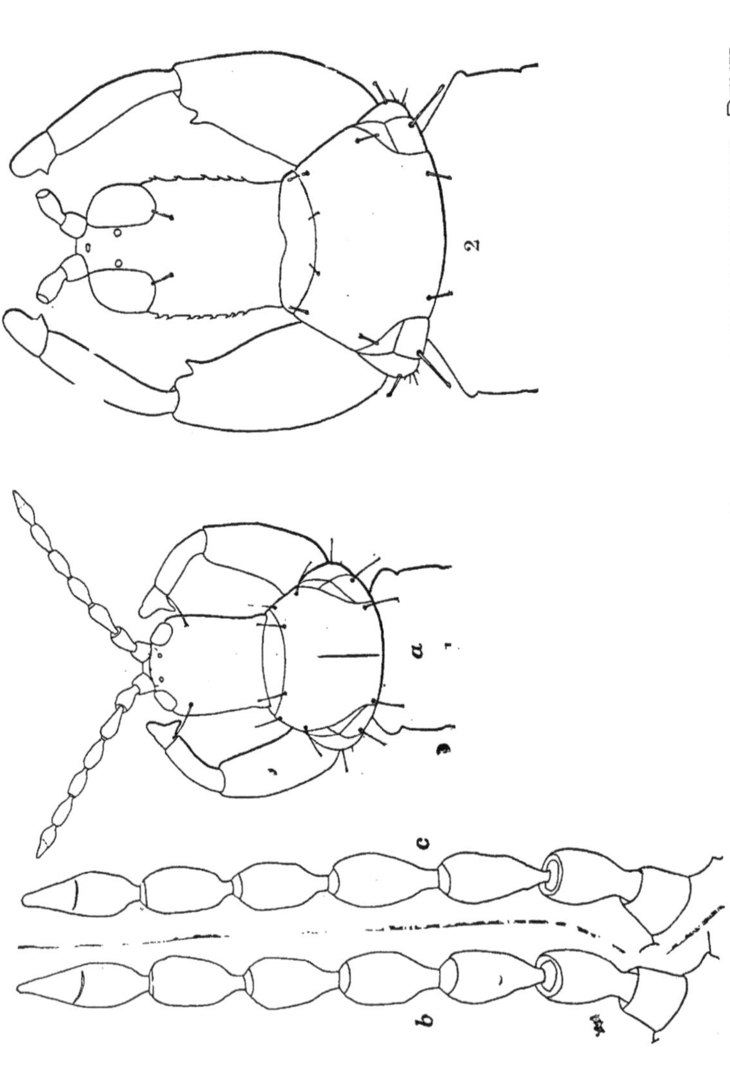

Cryptothrips junctus, sp. n., and Acanthothrips nodicornis Reut.

Prothorax three-fourths as long as width of head and (inclusive of coxæ) slightly more than twice as wide as long; usual bristles all present, long, dilated and divided at tip. Pterothorax much broader than long, sides subparallel. Legs short, rather slender; fore tarsi armed with a rather long, acute tooth.

Abdomen stout, about one and one-half times as broad as pterothorax; sides subparallel at base, converging roundly from segment 6 to tube. Tube about .6 as long as head, distinctly more than twice as wide at base as at apex, tapering evenly.

Measurements: Length, 1.75 mm.; head, length .30 mm., width .25 mm.; prothorax, length .18 mm., width (inclusive of coxæ) .40 mm.; pterothorax, width .39 mm.; abdomen, width .57 mm.; tube, length .17 mm., width at base .092 mm., at apex .039 mm. Antennal segments: 1, 48μ; 2, 66μ; 3, 66μ; 4, 68μ; 5, 64μ; 6, 64μ; 7, 90μ; total length of antenna, .47 mm.; width at segment 4, .037 mm.

Female.—Forma macroptera. Differs from the brachypterous form only in the presence of wings and the consequent increased development of the pterothorax.

Fore wings much broader than hind pair, sparsely fringed, and of equal width throughout; subapical fringe double for five or six hairs; the three subbasal spines knobbed; wings of both pairs uniformly brown in colour.

Male.—Forma brachyptera. Differs from the brachypterous female in the somewhat slenderer head with subconcave cheeks, as seen from above, larger prothorax with a thickened median line becoming obsolete before apex and base, stouter and slightly arcuate fore femora, longer and stouter tarsal tooth, and the slenderer abdomen.

Described from twenty females (two of which are macropterous) and eleven males from Baldwin, Michigan, and Mahomet and Murphysboro, Illinois. Specimens were taken April 17, August 16, September 4 and November 7, under bark on white oak, soft maple and sycamore, by Dr. H. E. Ewing, L. M. Smith and the writer.

Type locality: Baldwin, Michigan.

The seven-segmented antennæ, elongate maxillary palpi and the armed tarsus of the female distinguish this species at once from *C. rectangularis* Hood and *C. carbonarius* Hood, the only North American species properly referable to *Cryptothrips*.

During the latter part of August, 1908, I found pupæ of this species in abundance at Baldwin, Michigan, under the loose scales of the bark of some white oak trees (*Quercus alba*) which stood in a lowland sandy area between two small lakes; and with them was occasionally seen a wingless male or, more rarely, a wingless female. By August 31 adults were plentiful, always wingless, and the males greatly outnumbered the females. September 2 females were abundant, and one of those taken was macropterous. September 4 two males and a second winged female were found to have matured in a vial which contained pupæ taken September 2.

Acanthothrips nodicornis Reuter.—(Pl. VII, fig. 2.)

This species has long been known as *Acanthothrips nodicornis*, but Amyot and Serville's *Hoplothrips corticis*, dating from 1843, is probably identical with it. The only North American record of the species is that by Franklin (Psyche, Vol. X, p. 222, 1903), who found a single female under loose bark on a sycamore tree at Amherst, Massachusetts. My specimens, four females and six males, were taken in an open sandy forest about twelve miles from Baldwin, Michigan. One hot summer's day in August many were seen in copulation on the stump of a young poplar, which two weeks before had been cut for tent stakes; but when approached they scampered hastily away or dropped at once to the ground and secreted themselves among the fallen leaves. The few taken are all somewhat larger than European examples, averaging nearly one millimeter longer than several specimens (presumably cotypes) received from Prof. Reuter. The drawing and the following description, based on North American examples, may be of use to students of the group.

Female.—Length about 3 mm. Dorsal surface closely subreticulate; ventral surface smooth. Colour by reflected light nearly black; abdominal segments 3–8 marked at base with a pair of latero-dorsal white blotches, about equal in size to the second antennal segment. By transmitted light the colour is dark blackish brown with maroon pigmentation; antennal segments 1 and 2 concolorous with the body, 2 paler at apex; segments 3–5 with base and apex yellow, intermediate portion blackish brown; segments 6–8 slightly lighter than body, the base of segment 6 yellowish; legs concolorous with body, excepting tarsi and extremities of tibiæ, which are yellowish brown.

Head one and one-half times as long as wide; sides subparallel, converging slightly to eyes and to base, forming a slight neck-like constriction; dorsal and lateral surfaces sparsely spinose, the lateral spines arising from anterior surface of prominent tubercles, of which about eight are visible on each cheek; postocular bristles short, blunt, inconspicuous, one-third as long as eyes.* Eyes large, very finely faceted, one-third as long as head and about as wide as their interval. Ocelli moderate in size; anterior ocellus slightly overhanging the abruptly declivous vertex. Antennæ slender, about one and three-fourth times as long as head; segments 3–6 urn-shaped; 7 and 8 closely united, the latter conical; sense-cones long and slender, scarcely distinguishable from the antennal bristles; formula: 3, 1–2; 4, $1–2^{+1}$; 5, $1–1^{+1}$; 6, $1–1^{+1}$; 7 with one on dorsum near apex.† Mouth-cone pointed, attaining the mesosternum.

Prothorax about .6 as long as head and, inclusive of coxæ, about twice as wide as long; usual spines all present, expanded distally. Pterothorax slightly wider than prothorax; sides nearly straight, slightly converging posteriorly. Wings large, powerful, arcuate, of nearly equal width throughout; fore wings faintly washed at base with brown, and with the three subbasal spines nearly equal in length and blunt; apical fringe double for about thirty hairs; hind wings with a faint vein at costal third reaching about to middle. Fore femora large; subapical tooth acute and directed slightly anteriorly; fore tarsi armed with a broad acute tooth, the anterior margin of which is at right angles to the tarsus.

Abdomen large, broadly rounded at apex; marginal bristles dilated at tip. Tube about .8 as long as head, tapering evenly from base to apex; terminal bristles about as long as tube.

Measurements: Length, 3.2 mm.; head, length .43 mm., width .29 mm.; prothorax, length .27 mm., width (inclusive of coxæ) .56 mm.; pterothorax, width .65 mm.; abdomen, width .69 mm.; tube, length .34 mm., width at base .104 mm., at apex .052 mm. Antennal segments: 1, 48μ; 2, 73μ; 3, 148μ; 4, 129μ; 5, 120μ; 6, 87μ; 7, 75μ; 8, 44μ; total, .73 mm.; width, .042.

Male.—Shorter and slenderer than female. Length about 2.6 mm. Fore femora larger, stouter, nearly as wide as head; tarsal tooth larger, slightly curved. Abdomen tapering evenly from about segment 6 to base of tube.

*Moulton, in his Synopsis, Catalogue and Bibliography of North American Thysanoptera, Tech. Ser., 21, Bur. Ent., U. S. Dept. Agr., states in his key on page 19 that *H. magnafemoralis*, *nodicornis* and *doanei* have no postocular spines. This is incorrect as regards the first two species, at least.

†The formula for the antennal sense-cones is the same as this in both *H. magnafemoralis* Hinds and *H. albivittatus* Hood. In the original description of the latter species, however, their positions are not so described, the three rudimentary cones and the full-developed one on the outer surface of the third segment having been overlooked in the nearly opaque and otherwise unsatisfactory type specimen.

EXPLANATION OF PLATES VI AND VII.

Plate VI.

Fig. 1. *Trichothrips anomocerus*, sp. nov.—Female, × 117.

Fig. 2. *Trichothrips anomocerus*.—Apex of right antenna of female, × 514.

Fig. 3. *Trichothrips anomocerus*.—Tip of abdomen of female; membranous portions stippled; × 117.

Fig. 4. *Trichothrips anomocerus*.—Right fore leg of female, × 117.

Plate VII.

Fig. 1. *Cryptothrips junctus*, sp. nov.—*a*, head and prothorax of ♂, × 67; *b*, left antenna of ♀ from Michigan, × 199; *c*, left antenna of ♀ from Illinois, × 199.

Fig. 2. *Hoplothrips nodicornis*, Reuter; ♀, head and pronotum; × 67.

LASIOPTERYX MANIHOT, N. SP. (DIPTERA).

BY E. P. FELT, ALBANY, N. Y.

The small, yellowish midges were reared from *Cassava (Manihot utilissima)*, July 15, 1911, by Mr. W. H. Patterson, of the Agricultural School, St. Vincent, W.I. This species appears to be allied to *L. carpini* Felt, from which it is easily distinguished by the narrow wings. The longer, stouter antennæ in both sexes serves to separate it from a more closely allied undescribed form.

Male.—Length, 1 mm. Antennæ nearly as long as the body, thickly haired, fuscous; 13 segments, the fifth with a stem about ½ the length of the basal enlargement, which latter has a length ½ greater than its diameter and bears a thick whorl of long, stout setæ; terminal segment produced, with a length thrice its diameter and tapering to a narrowly rounded apex. Palpi yellowish. Mesonotum fuscous yellowish. Scutellum, postscutellum and abdomen yellowish, the latter sparsely haired. Wings subhyaline, broad, costa dark brown, the membrane rather thickly clothed with linear scales. Halteres yellowish. Coxæ and femora mostly yellowish, the tibiæ slightly darker, the tarsi fuscous yellowish; claws very long, slender, unidentate, the pulvilli rudimentary. Genitalia; basal clasp segment moderately stout; terminal clasp segment long, stout. Other organs indistinct.

Female.—Length, 1 mm. Antennæ extending to the base of the abdomen, rather thickly haired, fuscous yellowish; 13 subsessile segments, the fifth with a length about ½ greater than its diameter and with a thick whorl of long, stout setæ; terminal segment reduced, narrowly rounded apically. Palpi yellowish, the first segment subquadrate, the second narrowly oval, the third as long as the second, the fourth ½ longer than the third, somewhat dilated. Abdomen apparently lighter than in the male; ovipositor short, terminal lobes narrowly oval and sparsely setose. Other characters nearly as in the male.

NOTES ON CUBAN WHITE-FLIES WITH DESCRIPTION OF TWO NEW SPECIES.

BY E. A. BACK, VIRGINIA AGRICULTURAL EXPERIMENT STATION.

From an economic standpoint, there are probably no insects in Florida so detrimental to interests of citrus growers as the citrus white fly (*Aleyrodes citri* Riley and Howard), and the cloudy-winged white-fly (*Aleyrodes nubifera* Berger), which cause an annual estimated loss to the citrus industry of that state of over $1,125,000.*

For several years, the writer, with others, was engaged in an investigation of these insects and during that time many reports of white-fly infestations of *Citrus* in Cuba were brought to our attention. Considering the wide spread distribution of the citrus and cloudy-winged white-flies in Florida and the large amount of citrus nursery stock that had been shipped into Cuba from Florida nurseries, it was to be expected that these two species must necessarily have been introduced long ago. As a result of the demand by Florida citrus growers for an examination into the white-fly situation in foreign countries in hopes of discovering a parasite or other enemy that would be of assistance in controlling white-fly pests in Florida, the writer, while still in the employ of the Bureau of Entomology, U. S. Dept. Agric., made an investigation during October and November of 1910 in Cuba and Mexico.

During this search several species of white-flies were collected in Cuba. Heretofore only *Aleyrodes howardi* and *nubifera* have been correctly recorded from Cuba. A more extended collection will, beyond doubt, bring to light many species not listed here, but already recorded from other islands of the West Indies.

It is very generally believed throughout Cuba and Florida that the two great white-fly pests of Florida are present in abundance in Cuba. This is largely due to the fact that all aleyrodids whenever seen, no matter whether on guava or other vegetation, are thought to be the citrus white-fly. As a matter of fact, the citrus white-fly which causes the greatest loss of all white-flies now known has never been found in Cuba, and the cloudy-winged white-fly, next in injuriousness, only in slight numbers. While nine species are here recorded from Cuba, none are at present serious pests because of the work of parasites and fungus diseases.

The Citrus White-fly, *Aleyrodes citri* Riley and Howard.

There is no authentic record of this havoc-working species in Cuba.

***White-flies Affecting Citrus in Florida, Morrill & Back, Bulletin 92, B. E., U. S. Dept. Agric.**

Cook and Horne* refer to this species as being introduced into Cuba at Santiago de las Vegas on Florida nursery stock.

An examination of material in the Bureau of Entomology, Washington, by Prof. Quaintance, and of material collected by Cook and Horne at the Cuban Experiment Station by the writer, leaves no doubt that *A. nubifera* is the species regarding which they wrote. Frequent reports both in Florida and Cuba of injury caused by this insect to Cuban citrus groves are entirely groundless. In no citrus grove visited from Havana west beyond Bahia Honda, for over 100 miles east of Havana, or in the Isle of Pines in the general vicinity of Santa Fe, was any trace of this pest found. Over three thousand acres of citrus were examined. Mr. W. H. Hoard, of Victoria de las Tunis, who has been thoroughly familiar with this insect in Florida for many years, states that it does not occur, to his knowledge, in Central and Eastern Cuba. Cuban growers of citrus may well feel thankful that this pest has not yet secured a foothold on their island. As this species feeds on coffee almost as greedily as on *Citrus*, as evidenced by examinations made by Dr. E. W. Berger in Florida and by the writer in the Audubon Park greenhouses in New Orleans, the coffee industry of the island would be affected should this pest become abundant. Coffee plants examined by the writer at Seiba Mocha were free from white-fly.

The Cloudy-winged White-fly, *Aleyrodes nubifera* Berger.

This is the species referred to by Cook and Horne (l. c.) erroneously as *citri*. They stated in 1908 that since first seen this species had been decreasing so that at that time it was very difficult to find more than a few healthy specimens in one place. In their opinion the red fungus *(Aschersonia aleyrodis)* was responsible for this gradual decrease. While the writer examined many orange and grape-fruit trees of all ages in Cuba, even trees in the grove in which it was present in 1908, he was unable to find specimens, but Prof. P. Cardin found orange trees in the Vedado district of Havana badly infested during June, 1911, and sent specimens to the writer. As the red fungus is known in Florida to attack only sparingly this aleyrodid, it is more than probable that other causes have brought about this condition of scarcity, especially the wholesale mortality due to overcrowding as a result of the peculiar habit of the adults of this species to crowd small areas of the tenderest growth with eggs far beyond its capacity to furnish room for the development of the larvæ subsequently

*Bulletin 9, Estacion Central Agronomica de Cuba, 1608, page 30.

hatching therefrom and the deaths due to what is believed at present, in Florida, to be bacterial in origin.[1]

While the citrus white-fly feeds on a number of plants and trees, the cloudy-winged has been found only on citrus and the rubber tree, *Ficus nitida*. Its discovery on the latter food plant in New Orleans by the writer November, 1910, was such as to indicate its probable origin in India or China. While this species is doubtless still present in Cuba, it cannot be said to be of economic importance at this writing.

The Woolly White-fly, *Aleyrodes howardi* Quaintance.

This species was found quite generally distributed on orange trees wherever these grow in Cuba and Isle of Pines. It is this species which causes the blackening of foliage to which reference is frequently made. It is, however. not a serious pest and cannot be classed in destructiveness with *citri* and *nubifera*. Its spread is most rapid among old orange trees and during the drier seasons. Being a species possessing a thick pupa it is heavily parasitized. It is not only parasitized by the red fungus *Aschersonia aleyrodis*, as noted by Cook and Horne (l. c.) and by the writer at Seiba Mocha, Guinis, Santiago de las Vegas, and on the Isle of Pines, but is preyed upon by the larvæ of a Tortricid moth. Frequently colonies were found, each pupa of which showed the emergence hole of a hymenopterous parasite or devoured by Tortricid larvæ. The life history of the species and its occurrence in Florida has been treated by the writer.[2]

Besides occurring in Cuba and Isle of Pines, it has been found at Tampa, Ft. Myers and Miama, Florida. It occurs quite generally in the West Indies.

The Paw-paw White-fly, *Aleyrodes variabilis* Quaintance.

This white-fly has previously been reported only from Florida by Quaintance[3] and Back[4] and from Barbadoes by Gowdy.[5] Recently it has been found in abundance at Santiago de las Vegas on paw-paw (*Carica papaya*) by Prof. P. Cardin. It causes a severe blackening of the foliage at times.

1. Natural Control of White-flies Affecting Citrus in Florida, Morrill & Back, Bulletin 102, B. E., U. S. Dept. Agric.

2. The Woolly White fly: a New Enemy of the Florida Orange, Back, Bulletin 64, pt. 8, B. E., U. S. Dept. Agric.

3 Tech. Bull. No. 8, Div. Ent., U. S. Dept. Agric.

4. Florida Fruit & Produce News, 1910.

5. West Indian Bulletin, Vol. IX, No. 4.

Aleyrodes floridensis Quaintance.

This white-fly has been reported by Quaintance (l. c.) from Florida on guava *(Psidium)* and alligator pear *(Persea perseæ)*, and in Barbadoes on both these and smilax *(Theabroma cacao)* by Gowdy (l. c.) It was found by the writer in Cuba only on guava in the Botanical Gardens at Havana. While generally present in Florida wherever the alligator pear is grown, no evidence of injury has ever been known to follow even the heaviest infestations, and it will probably never be a pest in Cuba.

Aleyrodes mori Quaintance.

This aleyrodid with a black pupa case and white wax marginal fringe has been reported only from Florida, where it infests several plants, more especially the mulberry (Morus). Discovered by the writer on guava in the Botanical Gardens, Havana, but very scarce.

Paraleyrodes perseæ Quaintance.

This species, which has previously been reported as infesting *Persea carolinensis*, guava and citrus in Florida, was found on guava at Havana and Santiago de las Vegas in all stages ; and while it was not abundant, it was by no means rare.

Aleurodicus cardini, n. sp.

Egg.—About 0.26 mm. long, width about 0.076 mm. Elongate oval, uniformly pale yellowish, unmarked. Pedicle short ; egg lying prone on leaf, often entirely surrounded and concealed by fluffy waxen secretions of the adult. Eggs laid without regard to arrangement on leaf.

Larva.—Crawling first instar. (Fig. 1.) Length about 0.319 mm., width about 0.12 mm. Elongate oval, pale yellowish white in colour without darker markings or waxen secretions. Thirteen pairs of marginal spines, short, the posterior two pairs longer ; a fourteenth pair located on venter near margin on cephalic end of case. Spine on lower side of distal third of antennæ and terminal spine of antennæ proportionately longer and more distinct than in *A. citri* or *A. nubifera.*

Pupa Case.—(Fig. 2.) Length about 0.94 mm., width about 0.64 mm. Subelliptical, elevated on a vertical marginal waxen fringe. Colour yellowish to yellowish white, after emergence empty case whitish, semi-transparent ; parasitized specimens appear blackish either throughout or in spots. Margin entire without pattern of any sort ; near margin is a series of wax pores. On venter near margin are eighteen or twenty inconspicuous bristles seen only with high magnification ; of these, three pair, one cephalic, and two caudad, are more conspicuous. On dorsum nearer the margin than centre are five pairs of round well defined compound

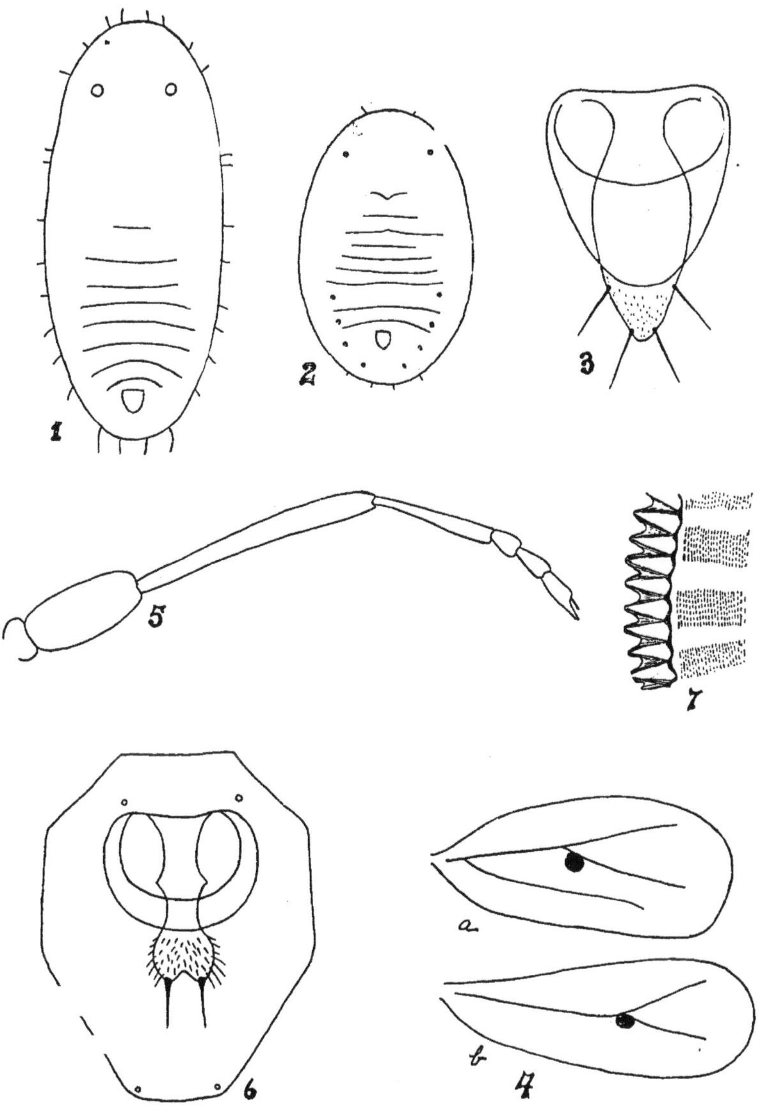

ALEURISCUS CARDINI, SP. N., AND ALEYRODES TRACHOIDES, SP. N.

pores ; four pairs on abdominal segments and one pair on cephalic region. Cephalad of vasiform orifice is a pair of minute bristles. Vasiform orifice elongate cordate (Fig. 3), about 0.09 mm. wide at base, and about 0.1 mm. long from base of operculum to tip of lingula ; cephalic margin straight, caudal and evenly rounded. Operculum subelliptical nearly one-half as long as orifice. Lingula broad, extending well beyond caudal end of orifice, on distal fourth which usually lies beyond caudal end of orifice with two pairs of comparatively long setæ. Rudimentary legs and antennæ as usual.

From wax pores on dorsum, there may be frequently seen protruding white glistening waxen rods which frequently break off and fall about the pupæ as in *P. perseæ*. The dorsal surface of case usually becomes, especially towards maturity, well dusted with a thin coating of white secretions, and at times a very narrow, downwardly directed marginal fringe may be seen outside the vertical fringe.

Adult.—Length, ♂, about 1.16 mm. Forewing, ♀, 1.39 mm. by 0.62 mm.; length *hind femur*, 0.26 mm. Length hind tibia, 0.35 mm.; length hind tarsi, 0.18 mm.; length claws, 0.08 mm.; ♀ proportionately larger. Yellow, covered with whitish waxen secretions; eyes red, not divided, but distinctly constricted. A line extending along side of head, interrupted by upper portion of compound eye, the lateral callosities of prothorax, indistinct traces along suture of proximal segments of abdomen, and portions of vasiform orifice, all blackish. Wings beautifully irridescent, with deep violescent reflections, a small prominent round brownish spot about 0.06 mm. in diameter (Fig. 4) on each fore and hind wing just behind the posterior distal branch of vein, usually enveloping vein but never filling the angle between veins as shown in *A. mimos* (Tech. Bull. 8, Div. Ent., Dept. Agric., Pl. VI, Fig. 6), wings otherwise unmarked. Antennæ (Fig. 5) seven jointed, the comparative lengths of the various segments as follows :

Segments $\frac{1}{25}$, $\frac{2}{6}$, $\frac{3}{137}$, $\frac{4}{8}$, $\frac{5}{1.5}$, $\frac{6}{2.5}$, $\frac{7}{24}$

Segments 3 to 7 show usual corrugations ; segment 7 with constriction on distal half at which point is borne a distinct bristle.

Habitat.—Type material collected at Havana and Santiago de las Vegas, Cuba, in November, 1910, by the writer.

Food Plant.—Guava, *Psidium guajava radii.*

Type.--Type material in collection of the U. S. D. A., Bureau of Entomology, and in that of the writer.

This species is really distinguished by its irridescent wing on which the spots described stand out prominently. It differs superficially from *iridescens* in having a spot on the hind wings and in colour of pupa case; from *minima* it differs in having no appreciable clouding of wings other than the spots described, in shape and location of the spots, and in the pupa case having but five instead of seven pairs of wax pores. The darkened portions of the vasiform orifice appear as a dark spot on the untreated adult. In crawling about the leaf, the female leaves behind a line of fine fluffy waxen secretion rubbed from a tuft of the same developing on the under side of her abdomen. Frequently her path can be distinctly followed by the aid of these lines of secretions. In mating, the sexes head in the opposite direction, and in this respect differ from those species of *Aleyrodes* that have come under the observation of the writer.

This species becomes quite abundant on the Guava at times, and when not parasitized becomes a nuisance. In November, 1910, it was causing noticeable blackening of the foliage at Santiago de las Vegas. The species is, however, heavily parisitized by a hymenopterous parasite and the red fungus (*Aschersonia aleyrodis*) which the writer found generally present on affected leaves. Prof. Patricio Cardin, for whom this species is named, sent the writer specimens in May, 1911, over 90% of which had been parasitized by a hymenopterous parasite. This is the species of White-fly figured by Cook and Horne as an undetermined aleyrodid on guava (Pl. XV, fig. 41, Bull. 9, Estacion Central Agronomica de Cuba), and beyond doubt is that referred to in the Primer Inforne Annual of the same station as "Guagua a mosca blanca de la guayabo." Cook and Horne (l.c., p 31), say that *Aleyrodes howardi* is the species referred to, but in this they are apparently mistaken, as the writer has not found *howardi* except very rarely on guava. While *howardi* was generally present on orange trees close by, this species was found only on guava.

Aleyrodes trachoides, n. sp.

Egg.—About 0.2 mm. long. Pale in colour, smooth, without reticulations or waxy secretions; curved with convex side approximating leaf, attached by short stalk arising from convex surface, about one-fourth distance from base to tip of egg. Eggs deposited promiscuously about lower surface of leaf.

Larva, crawling first instar.—Length about 0.27 mm., width about 0.14 mm.; elongate elliptical, yellowish white, with nine pairs of marginal bristles and one pair cephalad on venter near margin; the anterior and two posterior pairs of marginal bristles longest.

Pupa case.—Length about 0.83 mm., width about 0.5 mm. Sub-elliptical in shape, many specimens with more or less evident indentures on cephalo-lateral margin of case. Black, with whitish mealy waxen secretion on dorsum, not abundant enough to entirely obscure colour of case; case with conspicuous lateral waxen fringe of a cottony nature, averaging about one-half in length the width of the case. Along the dorsi-meso is a distinct elevation extending cephalad from and including the vasiform orifice to near the margin of the case, reduced to a mere ridge on thoracic region, but broader and evenly rounded on the abdominal region, and merging caudad into a more or less octagonal rim around the vasiform orifice. (Fig. 6.) Abdominal segments distinct, extending outward, but slightly beyond the rounded keel. Dorsum with three pairs of stout bristles; one on the mesothorax, one just cephalad of orifice and one at posterior end of keel. Two pairs of minute marginal bristles present; one cephalad, and one caudad. Marginal rim distinct, with colourless wax tubes distinct; wax tubes elongate, more or less rounded, some acute, incisions obtuse or acute. Between margin of case and line marking outer limits of pupa within is a series of dark elongate strictions parallel to case margin. (Fig. 7). Vasiform orifice (Fig. 0), sub-semi-circular, distinctly broader than long; operculum sub-semicircular, broader than long, reaching about 2/3 distance from base to tip of orifice. Lingula fully developed, reaching well beyond the caudal margin of orifice, consisting of a basal shaft and expanded tip; shaft acutely enlarged midway; tip nearly circular, with distinct constriction distally and armed distally with a pair of comparatively long weak bristles and numerous short hairs. Lingula seen with great difficulty, except in case pupal skins protruding from pupa case.

Adult.—Female dried specimen, about 0.8 mm. long, fore wing about 0.96 mm. long, posterior femur 0.2 mm. long, posterior tibia 0.3 mm. long, posterior tarsi 0.1 mm. long. Adults yellowish, without darker markings, after emergence becoming thoroughly coated with whitish secretions, eyes reddish, constricted at middle, but not divided; wings with typical *Aleyrodes* venation, whitish, without spots or clouds, slightly violescent. Males do not differ from females except in usual characteristics.

Habitat.—Type material collected at Santiago de las Vegas, Cuba.

Food Plant.—Indigenous solanaceous vine, *Solanum seaphorthianum* Andr. (Dr. Cauizares authority.)

Type.—Type material in collection of U. S. D. A., Bureau of Entomology, and in that of the writer.

This species is closely related to *Aleyrodes tracheifer* Quaintance, and runs to this species in the key (Technical Series, No. 8, Div. of Ent., U. S. Dept. Agric.). It differs, however, in that the wax marginal fringe is not as wide, that the marginal wax pores are more even, in the shape of the rim about the vasiform orifice, and in the development of the keel on the dorsum. In *tracheifer* this keel is narrow and of even width throughout, except for certain constrictions which produce an "arrow-shaped" effect anteriorally. In *trachoides* this keel is a mere ridge on the thoracic region, but very much broader on the abdominal regions. The lingula of *tracheifer* is very small and poorly developed as compared with that of *trachoides*.

Described from an abundance of material collected by Prof. P. Cardin, who states, that although extremely abundant (quite coating the under surfaces of leaves affected), no sooty mold (*Meliola*) follows its attack. Prof. Cardin is also authority for the statement that when abundant this species causes the foliage to fall.

EXPLANATION OF PLATE VIII.

Fig. 1. *Aleuriscus cardini.*—Crawling young, dorsal view.

Fig. 2. *A. cardini.*—Pupa case, dorsal view.

Fig. 3. *A. cardini.*—Vasiform orifice.

Fig. 4. *A. cardini.*—*a*, fore wing; *b*, hind wing.

Fig. 5. *A. cardini.*—Antenna.

Fig. 6. *Aleyrodes trachoides.*—Vasiform orifice and rim about same, circles showing location of 7 spines.

Fig. 7. *Aleyrodes trachoides.*—Margin of pupa case, enlarged.

APHID NOTES FROM OREGON.

BY H. F. WILSON, CORVALLIS, OR.

In a general study of the plant lice of Oregon, we have found abundant material in many old and some new species. We are making an effort to clear up the life history of a number of them and the present paper is the first of a series which we hope to get out, giving all stages of as many species as possible.

Illinoia osmaroniæ, n. sp.

This quite large aphis is found on the leaves of *Osmaronia cerasiformis* and is quite abundant about Corvallis, Oregon.

Stem-mother.—On March 18, 1911, the stem-mothers of this species, with newly-born young, were plentiful in the just opening leaf buds of the host plant.

General colour light green throughout with legs and antennæ slightly paler green. Body large and robust. Antennæ about two-thirds the length of body and quite slender. Nectaries quite slender and about one-fourth as long as the body. Cauda short and triangular in shape.

Measurements: Length of body, 3.33 mm.; width, 1.8 mm. Length of antennal segments, I, 0.135; II, 0.04; III, 0.7; IV, .33; V, .38; VI, .2; spur, .51 mm.; total length, 2.295 mm. Length of nectaries, .73 mm. Length of cauda, 0.22 mm.

Spring Migrant.—April 25th; what is probably the third generation of this species found abundantly on under side of leaves; numerous young present.

General colour light green; head and thorax orange; antennæ with six segments; spur of sixth as long or longer than third segment; sixth segment and spur dusky and basal two-thirds of third segment also dusky; legs light green except distal end of tibia and tarsi, which are nearly black. Wings hyaline and large. Nectaries long and cylindrical, slightly constricted at tip. Cauda large, slightly turned up and blunt at point. Third antennal segment with from 24 to 28 small, irregularly-placed sensoria, most of them scattering. Antennal tubercles large and distinct, and strongly gibbous; at upper inner edge two bristle-like hairs at highest part. First segment large and strongly gibbous; second segment small in comparison.

Measurements: Length of body, 2.88 mm.; width, 1.22 mm. Length of antennal segments, I, .176; II, .09; III, 1.11; IV, .84; V, .75; VI, .26; spur, 1.11 mm.; total length, 4.336 mm. Length of wing, 5.7 mm.; total expanse, 12.4 mm. Length of nectaries, 1 mm. Length of cauda, .44 mm.

Fall Migrant.—Oct. 11; winged individuals not very common and producing young on leaves of *Osmoronia* about Corvallis.

General colour green with head and thorax orange coloured. This form resembles the previous form entirely, except in the antennæ and size, the fall migrant being slightly smaller, and the third segment of the antennæ bears from 18 to 22 regularly placed round sensoria lying in a straight line along the outer edge.

Oviparous Female.—The egg-laying female is orange-green in colour, and is quite small in comparison with the stem-mother. The legs are a little lighter in colour than the rest of the body and the antennæ are slightly dusky at tip. Antennæ longer than the body and placed on large tubercles.

Measurements: Length of body, 2.33 mm.; width, 1 mm. Length of antennal segments, I, .11; II, .066; III, .4; IV, .33; V, .38; VI, .154; spur, .82 mm.; total length, 2.26 mm. Length of nectaries, .73 mm. Length of cauda, .33 mm.

Egg.—The eggs are deposited on the shoots at the base and on the under side of the buds. No measurements of the eggs were secured. They are very similar to other species in this group, although smaller than one would expect for the size of the insect. When first deposited they are light greenish yellow and later become deep shining black.

Alate Male.—Collected on underside of leaves November 3rd, 1911. General colour light yellow. Head, thorax, legs and antennæ dusky to black. Two basal segments of antennæ and basal half of femora yellow. Abdomen with six transversal bands broken in half, four in front of base of nectaries and two behind. Cauda of medium length, tapering and blunt at the tip. Antennal tubercles large and very distinct, being slightly gibbous on upper inner edge; first antennal segment large and strongly gibbous on inner side, second segment quite small in comparison. The third, fourth and fifth antennal segments with an irregular row of small widely separated circular sensoria on each segment. They vary from 18 to 26 on each segment.

Measurements: Length of body, 2.66 mm.; width, .9 mm. Length of antennal segments, I, .135; II, .066; III, 1.02; IV, .8; V, .8; VI, .198; spur, 1.29 mm.; total length, 4.299 mm. Length of wing, 5.95 mm. Total wing expanse, 12.85 mm. Length of nectaries, 7 mm. Length of cauda, 3 mm. The penis of this form is easily forced out by a little pressure on the abdomen.

Illinoia macrosiphum, n. sp.

First collected about Corvallis, Oregon, on *Amelanchier alnifolia*, July 4, 1911. Found in small colonies and not very plentiful. General colour whitish yellow. The specific name *macrosiphum* is applied on account of the extremely long nectaries. In the very young these are as long as the body, in the mature specimens they are ½ to ⅔ the length of the body. An effort was made to secure the winged forms of this species, but none were found excepting the alate male of what is supposed to be the same species collected on rose November 3, 1911, and on *Amelanchier* bushes the last of September. These last specimens measure in length, from forehead to tip of cauda, 2 mm., while the nectaries measure from base to tip 1.78 mm. Antennæ reaching to tip of nectaries. Comparative lengths of segments can be obtained from measurements. Antennæ except basal segment slender, basal segment very large in proportion to the others. Legs quite long, nectaries large at base and tapering, each long and ensiform.

Measurements: Length of body from forehead to base of cauda, 2 mm., width, .85 mm. Length of antennal segments, I, .176; II, .09; III, .82; IV, .622; V, .644; VI, .176; spur, 1.29 mm.; total length, 2.818 mm. Length of nectaries, 1.622 mm. Length of cauda, .33 mm.

Oviparous female resembles the viviparous female, except in the colour of the body, which is rosy red.

Alate Male.—What we supposed to be the males of this species were collected on wild rose bushes under *Amelanchier alnifolia.*

General colour green with rosy tint; five transverse bands may be found on the abdomen. These are broken so as to appear like ten spots. Head and thorax dusky. Antennæ except first two segments dusky. Legs with dusky joints and tarsi. Nectaries dusky, cauda rosy coloured. Third segment with numerous small sensoria; fourth with about thirty; fifth with about twenty. A very interesting character of this species is found in three small sensoria on the sixth segment besides those at the base of the spur. One of these may be found at each end of the segment and the third lies midway between.

Measurements: Length of body from forehead to base of cauda, 1.066 mm.; width, .52 mm. Length of wing, 2.71 mm.; width, 1 mm.; total wing expansion, 5.93 mm. Length of antennal segments, I, .11; II, .045; III, .6; IV, .49; V, .58; VI, .135; spur, 1.174 mm.; total length, 2.134 mm. Length of nectaries, .75 mm. Length of cauda, .198 mm.

Myzus rhamni Boyer.
Syn. *Macrosiphum rhamni* Clarke.

This species is very abundant about Corvallis on *Rhamnus purshiana.* The entire development is apparently passed on this plant as they were present throughout the year. I have not seen specimens of Clarke's *Macrosiphum rhamni* and repeated efforts to locate the types, if there are any, were unsuccessful. From the description I am led to believe that the California species is the same as the one found in Oregon, and there seems little doubt but that the Oregon species is the same as the one described by Boyer.

Stem-mother.—The first stem-mothers were collected on the 23rd of March and at that time they were about full grown.

General colour light green. Antennæ towards tips dusky; distal end of tibiæ and tarsi dusky. In this stage the characters resemble those of *Aphis* more than anything else. The antennæ are stout and measure less than one-half the length of the body. The legs are short and the antennæ and cauda are as in *Aphis*. Antennal tubercles distinct, but not long.

Measurements: Length of body from forehead to tip of cauda, 2.65 mm.; width, 1.30 mm. Length of antennal segments, I, .154; II, .066; III, .33; IV, .242; V, .27; VI, .11; spur, .44 mm.; total length, 1.612 mm. Length of nectaries, .5 mm. Length of cauda, .176 mm.

Viviparous apterous female of the summer generations collected June 4th, 1911, on underside of leaves of tree on college campus; pupa and alate forms also present.

General colour light green or lemon-yellow throughout; but the characters are like those of *Myzus*, and this form is quite distinct from the stem-mothers. The antennæ are quite long and slender and placed on prominent tubercles. First antennal segment strongly gibbous on inner side. Sixth antennal segment and spur almost setaceous in form. For comparative lengths see measurements. Legs rather stout, short and sparsely hairy. Nectaries thick at the base and slightly tapering with a slight inward curve. Cauda medium in length and blunt at the tip.

Measurements: Length of body, 2.5 mm.; width, 1 mm. Length of antennal segments, I, .135; II, .09; III, .778; IV, .55; V, .51; VI, .154; spur, .98 mm.; total length, 3.197 mm. Length of nectaries, .778 mm. Length of cauda, .154 mm.

Spring Migrant.—Collected on underside of leaves June 4th, 1911.

General colour lemon-yellow. Head and thoracic shield light orange. First two antennal segments green, the base of the third green, remainder of antennæ dusky to black. Basal half of nectaries green to dusky, outer half darker to black. Some specimens with an orange spot in the centre of the body just back of the thorax. Antennæ long and slender and placed on prominent tubercles. First segment large and strongly gibbous on the under side. Second segment small. Third segment with about twenty-five nearly circular sensoria of variable sizes and irregularly placed. Frontal tubercle of head quite prominent. Legs long and slender.

Nectaries long, slender and slightly curved in at middle. Cauda of medium length, bluntly pointed.

Measurements: Length of body, 2.48 mm.; width, .95 mm. Total wing expanse, 8 mm.; length of wing, 3.8 mm. Length of antennal segments, I, .15; II, .1; III, .85; IV, .56; V, .55; VI, .2; spur, 1.1 mm.; total length of antennæ, 2.51 mm. Nectaries, .78 mm., and cauda, .15 mm.

Fall Migrant.—This form so nearly resembles the above as to make a second description unnecessary.

Oviparous Females.—Taken on leaves and in the act of oviposition along young shoots, November 1, 1911; present until a late frost in November.

General colour green. First two antennal segments and basal half of third and legs, except tarsi, light green, remaining parts of antennæ and tarsi dusky to black. Other characters and measurements taken from specimens mounted in balsam. Antennal tubercles strong and prominent. First antennal segment large, remaining segments long and slender. Antennæ medium length and with decided *Myzus* characters, being slightly curved in and having the constricted tips. The portion of the abdomen back of the nectaries large and extending back nearly to the end of the nectaries. Cauda short and blunt.

Measurements: Length of body, 2.25 mm.; width, 1 mm. Length of antennal segments, I, .135; II, .066; III, .58; IV, .49; V, .4; VI, .11; spur, .55 mm.; total length, 2.331 mm. Length of nectaries, .71 mm.; length of cauda, .11 mm.

Alate Male.—Collected with oviparous females and alate viviparous females on underside of leaves, Nov. 1st and 3rd, 1911.

General colour green. Antennæ with first two segments and base of third light green; remaining segments, with distal half of femora and tarsi, dusky, other parts of less green. Nectaries light dusky at base, shading into black at tip. Antennæ on prominent tubercles and with first segment large and gibbous. Third and fourth segments with numerous slightly-raised sensoria, fifth with about eight slightly larger sensoria on outer edge and in a row. Wings large and venation regular. Legs long, femora stouter than in alate females. Nectaries shorter than in females, but with *Myzus* characters. Cauda of medium length, not tapering and quite blunt at the tip.

Measurements: Taken from specimens preserved in balsam. Length of body, 1.174 mm.; width, .4 mm. Length of wing, 2.15 mm.; total wing expansion, 4.60 mm. Length of antennal segments, I, .066; II, .047; III, .44; IV, .34; V, .33; VI, .09; spur, .622 mm.; total length, 1.935 mm. Length of nectaries, .242 mm. Cauda, .11 mm.

Eggs deposited on young twigs about base of buds.

(To be continued.)

NOTES ON SOME NORTH AMERICAN TINEINA.

BY ANNETTE F. BRAUN, UNIVERSITY OF CINCINNATI.

Argyresthia annettella Busck.

Argyresthia annettella Busck, Proc. U. S. Nat. Mus., XXXII, 12, 1907.

The larvæ of this species mine the leaves near the tips of the twigs of the Juniper (*Juniperus communis* L.). The leaf, except at its extreme tip, is reduced to a mere shell, containing a few scattered grains of excrement, as may be seen by holding the twig toward the light. In this manner each larva excavates about four leaves, passing from one to another through the stem. The mines are started in summer, and the larvæ winter within the mines, leaving them to pupate in May. The mined leaves later become discoloured, and ultimately the entire end of the twig dies. Where the miners are abundant, the numerous brownish dead ends of the twigs give evidence of their presence. The cocoon, which is an open meshwork of coarse silk, is attached to the upper side of a leaf near the mine. The imagoes appear during the early part of June.

Although the Juniper is widely distributed around Cincinnati, *A. annettella* seems to occur only in three or four isolated spots, where I have seen as many as 40 or 50 mines upon a single plant about five feet high.

Lithocolletis trinotella Braun.

Lithocolletis trinotella Braun, Ent. News, XIX, 99, 1908; Trans. Am. Ent. Soc., XXXIV, 279, 1908.

Since the original description of this species was published, I have been successful in rearing four specimens from small tent mines on the under side of leaves of Silver Maple (*Acer saccharinum* L.), collected in Clermont Co., Ohio. The mines are extremely small, about 8 mm. long, and much wrinkled at maturity. The pupa is enclosed in a loose web of silk.

The moths, while agreeing in all essential particulars with the types, are somewhat larger, and have a third costal white streak, which is often obscure and entirely unmargined.

Lithocolletis martiella Braun.

Lithocolletis martiella Braun, Trans. Am. Ent. Soc., XXXIV., 290, 1908.

A single specimen of this species, bred from *Betula lenta* L., at Balsam, N. C., July, 1911, confirms Dr. Dyar's somewhat doubtful record of its food-plant as birch, and gives two widely-separated localities for the species, the type locality being Kaslo, B. C.

The mine, which is placed on the lower surface of the leaf, is elongated, and the loosened epidermis is thrown into a series of fine ridges. The pupa is not enveloped in a cocoon, but the one-half of the mine containing the pupa is sparingly lined with silk.

Lithocolletis betulivora Walsingham.

Lithocolletis betulivora Walsingham, Ins. Life., III, 326, 1891; Braun, Trans. Am. Ent., XXXIV, 339, 1908; Dyar, List N. A. Lep., No. 6328, 1902.

A single specimen of this species was bred from *Betula lutea* Mich., at Balsam, N. C. The pale markings are suffused with yellowish to such an extent that they are scarcely differentiated from the ground colour of the wing, and dark scales are entirely lacking, except external to the pair of spots at the apical third and in the apex of the wing.

Coriscium cuculipennellum Hübner.

Coriscium cuculipennellum Hübner, Ges. eur. Schmett., VIII, Tin., VI, Al. B. f. 2, 1831; Fernald, CAN. ENT., XXV, 96, 1893; Dyar, List N. A. Lep., No. 6401, 1902.

I have found the mines of this species common in the vicinity of Oxford, Ohio, upon the leaves of Green Ash (*Fraxinus lanceolata* Borck.) and White Ash (*Fraxinus americana* L.). The mine, at first very narrow and shining white, begins on the upper side near the midrib, usually following the midrib downward more or less closely for a length of 3-4 cm., thence diverging and slanting outward to the margin of the leaf, where it is scarcely more than .5 mm. wide. Here it enlarges into an elongate white blotch 2-2.5 cm. long and 5 mm. wide. The epidermis in this blotch becomes so much wrinkled that the edge of the leaf is bent over, entirely concealing the mine, except at the extreme ends. The loosened epidermis is everywhere very thin.

The larva later feeds within conically-rolled leaves, and spins the characteristic suspended cocoon within the roll.

Lyonetia latistrigella Walsingham.

Lyonetia latistrigella Walsingham, Trans. Am. Ent. Soc., X, 203, 1882; Busck, Proc. Ent. Soc. Wash., V, 209, 1903; Dyar, List N. A. Lep., No. 6416, 1902.

Two specimens of this interesting species were bred at Balsam, N. C., from mines on *Rhododendron maximum* L. The mines were only observed upon the young tender leaves which had not yet attained their full size. The mine begins as a very fine black line, continuing thus for a length of about 3 cm., after which it becomes noticeably broader for about the same distance, but is still to be considered a linear mine. Beyond this point it rapidly enlarges to a brownish elongate blotch, 4 cm. in length or more, with an average width of 5 mm. The larva leaves the mine to pupate, suspending its naked chrysalis by means of a few silken threads stretched across a bent leaf.

The two imagoes agree closely with Walsingham's description, and exhibit no variation. They are easily distinguished from the allied species by the conspicuous ferruginous patch of scales in the apical fourth of the wing.

TWO NEW SPECIES OF COLEOPTERA FROM ILLINOIS.

BY A. B. WOLCOTT, CHICAGO, ILL.

The two apparently new species herewith described were collected by Prof. Arthur G. Vestal during the course of his biological studies in the Illinois sand region, and, as the result of his investigation will, no doubt, soon be published, it seems desirable that these nondescripts be made known prior to the appearance of his paper.

For the opportunity of describing these beetles I am indebted to Prof. Vestal, who, with rare generosity, likewise gave me the two unique representatives of the following species.

Saprinus illinoensis, sp. nov.

Broadly oblong-oval, strongly convex, shining black; the antennæ dull rufous, the basal joint and the legs rufo-piceous. Head impunctate, strongly margined at sides and apex; surface with a distinct and irregularly eroded chevron. Prothorax twice as wide as long; the sides rather strongly convergent and feebly rounded, more strongly rounded near apex; marginal groove distinct, deep throughout; disk feebly, rather densely rugulose, more feebly so toward middle and obsolete in small area at middle of base, coarsely and deeply but sparsely punctate along the base. Elytra rounded at the sides, three-fourths longer than the prothorax, and at basal third distinctly wider, finely and not very densely punctate at apex, punctate space extending narrowly along the suture to the middle and not entered by the dorsal striæ; each elytron with a vague impression at middle near sutural striæ; margined stria straight, deep, fine along the apex to suture; outer subhumeral fine, distinctly diverging from the marginal to the middle thence converging and extending very nearly to apex; inner subhumeral distinct from the middle to apical sixth, fragmentary and feeble before the middle; oblique humeral fine and feebly impressed, extending to basal third, not joining the internal subhumeral; dorsal striæ rather fine, broadly arcuate, the first extending to apical fourth, second and fourth to apical third, the third slightly shorter, one to three hooked at base, the fourth broadly arched at base; joining the entire sutural. Propygidium short, sub-impunctate in basal half, the punctures apically rather coarse and dense, but feeble, subcarinate at middle. Pygidium not densely but rather coarsely, feebly punctate. Prosternal striæ abbreviated at apical fourth, rapidly divergent posteriorly; lateral convergent carinæ very distinct; transverse suture punctate. Mesosternum feebly emarginate at apex, coarsely, remotely punctate. Metasternum with a distinctly limited transverse band of coarse, sparse punctures posteriorly. Anterior tibiæ with five subacute erect teeth, the outer three longer and broader.

Length, 4.5 mm.

One specimen, Havana, Ill. "Under a board at the Devil's Hole, July 29, 1910."

This species would by Dr. Horn's table fall with *sphæroides* J. E. Lec. In size and colour, however, it is nearest the recently described *lakensis* Blatch. It agrees with both these species in having the sutural striæ entire and the dorsals not entering the punctured space. *Illinoensis* may be distinguished by the very distinct chevron of the head, the irregular dorsal striæ, the manner and extent of punctuation of the prothorax and elytra and its somewhat larger size.

Bruchus arenarius, sp. nov.

Form very robust, black, densely evenly cinereo-pubescent.* Head subopaque, finely densely subrugosely punctate; front feebly subcarinate. Antennæ as long as half the body, not conspicuously incrassate externally; second joint slightly longer than wide; black, basal joint red beneath.

*Under a rather high power glass, sparse, evenly distributed, yellowish hairs are discernable; these are not numerous enough, however, to alter the general grayish tone of the pubescence.

Thorax more than twice as wide at base as long; sides strongly arcuate; disk moderately convex; basal lobe broadly rounded; finely densely feebly punctate. Scutellum small, broader than long, punctured and cinereo-pubescent. Elytra subquadrate, conjointly at middle as broad as long; sides distinctly arcuate; disk flattened, finely striate; striæ finely and feebly punctate; intervals broad, flat, finely rugosely punctate, each with a series of distant large punctures. Pygidium oblique basally, convex and vertical in apical half; the tip somewhat inflexed; rather coarsely, sparsely and feebly punctate, uniformly cinereo-pubescent. Hind femora mutic. Apical spur of hind tibiæ about one-third the length of the first tarsal joint.

Length, 2.25 mm.

One specimen, Havana, Ill. "On the sand, between tufts of bunch-grass at the Devil's Hole, April 9, 1911."

This species belongs to group IV of Prof. Fall's table, where it would seem to be placed best immediately after *leucosomus* Sharp. The small size of this species, in connection with its entirely black colour, uniform, not variegated pubescence and absence of spots of pygidium, renders it easily recognizable.

BASILARCHIA WEIDERMEYERII ANGUSTIFASCIA, A NEW GEOGRAPHICAL RACE.

BY WM. BARNES, M.D., AND J. MCDUNNOUGH, PH D., DECATUR, ILL.

A series of 2 ♂s and 5 ♀s, collected last summer in the White Mts., Arizona, differs from the typical form from Colorado and Utah, as depicted by Edwards (Vol. I, pl. 42), in that the median white band is much reduced in width, and the intersecting veins, especially on the primaries, are more broadly black. This difference is most noticeable in the ♀s, the band on the primaries being distinctly broken up into an irregular row of white semiquadrate spots, of which the third from the costa is greatly reduced in size; on the secondaries the spots are *not* broader than long. As this feature is remarkably constant in all the specimens before us, and as, furthermore, we have had for years a ♀ labelled Arizona in the collection which shows the same peculiarities, we consider a varietal name for the Arizona form warranted; the extreme form of this race, in which the white band has entirely disappeared, is the ab. *sinefascia* Edw., also from Arizona. The males are normal in size, having a wing expanse of 2½ in. (63 mm.); the females are somewhat larger than usual, all our specimens measuring 3 in. (60 mm.). The types are in coll. Barnes.

GEOMETRID NEWS—DESCRIPTIONS OF TWO NEW HYDRIOMENAS.

BY L. W. SWETT, BOSTON, MASS.

Hydriomena henshawi, nov. sp.

Palpi short; expanse of wings 35 mm.

Colour of fore wings light ash-gray, speckled with black atoms; the space between the basal line and body of the same colour. Basal line bent outwardly from body at vein Sc. (Smith's Glossary), then curved slightly inwardly toward body, the curve ending quite a distance out on inner margin with a black dash; mesial space gray, with black atoms; median band black and irregular; intradiscal line running from costa to inner margin almost diagonally, with irregular curves between the veins; mesial or discal space with a faint spot; extradiscal line black, starting in a dash at costa, then curved outward with irregular points, as in *autumnalis;* outer margin pale gray with black atoms, the usual watery black band curved more regularly than in *autumnalis*. Fringe long, pale gray, with double points at base of fringe.

Hind wings pale gray, with the usual two faint extradiscal bands.

Beneath, the discal points on the fore wings are represented by two pale dashes, the lines above showing through faintly. The dots on the hind wings beneath are round; beyond, the two pale gray lines show through from above. The fringe is long and pale ash-gray, as above.

Type, 1 ♂, Nevada, Museum of Comparative Zoology, Cambridge, Mass. I take pleasure in naming this species after my kind friend, Mr. Samuel Henshaw, who has assisted me much in my work on the Hydriomenas.

This species resembles slightly *H. quinquefasciata* Packard.

(To be continued.)

Extensive infection of the San José scale has been discovered on trees in the southern part of Wisconsin by Professor J. G. Sanders, of the University of Wisconsin. Professor Sanders, who is also State Nursery Inspector, reports that steps are being taken to control the pest and prevent its spreading beyond the area affected already.—[*Science*.

Mailed May 8th, 1912.

The Canadian Entomologist.

Vol. XLIV. LONDON, JUNE, 1912. No. 6

SOME PARASITIC BEES *(COELIOXYS)*.

BY T. D. A. COCKERELL, UNIVERSITY OF COLORADO, BOULDER.

Coelioxys moesta Cresson.—Peachland, B. C., Aug. 9, 1909 (J. B. Wallis, a53). ♀.

Coelioxys deplanata Cresson.—Wawawai, Wash., Aug. 30, 1908 (W. M. Mann). Both sexes.

Coelioxys rufitarsis Smith.—Four females, Wawawai, Wash., Aug. 30 and Sept. 6, 1908 (W. M. Mann).

Coelioxys immaculata, n. sp.—Male; Miners, Indiana, July ; collector unknown, but there is a label bearing the number 1525.

Length a little over 10 mm., robust, black, with rather dull white hair, faintly creamy on upper part of head ; eyes pale green, with abundant quite long hair ; antennæ and mandibles entirely black ; tegulæ bright apricot colour ; femora except the lower side, and tibiæ and tarsi entirely, bright ferruginous, as also are the tibial spurs ; hair on inner side of basitarsi creamy ; head and thorax with dense, large punctures, those of vertex larger than those on mesothorax ; lower part of cheeks with a broad bevelled space, which is shining and punctured ; thorax above without the usual white hair patches ; scutellum broadly rounded behind, without any median projection ; lateral teeth thick, not curved ; abdomen shining, but well punctured, the second and third segments with deep transverse constrictions ; fourth ventral segment with a weak emargination ; sides of fifth segment with very short spines ; sides of sixth with large thick spines; end of sixth with four teeth, the upper ones short, and directed obliquely upwards, the lower large and unusually broad. In Robertson's table (Trans. Amer. Ent. Soc., XXIX, p. 174), this runs out at 3, because of the red legs, punctured bevelled space, etc. Robertson says of male *octodentata*, "disc of abdomen opaque, densely punctured" ; *immaculata* has the abdomen very conspicuously shining, except the sublateral region of the second segment just beyond the sulcus, which is dull and very densely covered with minute punctures, in complete contrast with the corresponding areas on the first and third, and with the sparsely-punctured middle of the second.

Coelioxys grindeliæ denverensis, n. subsp.—Four males; Denver, Colorado, Aug. 6 to 25, 1908 (Mrs. C. Bennett). Eyes light red (green in *C. grindeliæ* Ckll,); fourth ventral segment strongly emarginate (entire in *grindeliæ*). Otherwise they seem about the same. Face densely covered with white hair; antennæ entirely black; bevelled space on cheeks rugose but shining; anterior coxæ with large flattened spines; tegulæ black, the margin sometimes dark reddish; legs black, including tarsi; spurs dark; second abdominal segment on each side sublaterally with a more or less evident but small shining raised area; teeth on each side of scutellum long; teeth at sides of sixth abdominal segment long; lower apical teeth of abdomen not broad. In Robertson's table this runs out at 3, although the first abdominal segment is very hairy at sides, and sublaterally has distinct indications of a basal band. The anterior part of the mesothorax is conspicuously but diffusedly hairy, instead of having well-defined spots as in *C. deplanata*.

Coelioxys angelica Cockerell.—The male, previously unknown, has been taken by Mr. F. Grinnell, jr., in Strawberry Valley, San Jacinto Mts., California, alt. 6,000 ft., July 18. By its small size and general appearance, it closely resembles *C. deani* Ckll, but the sulcus on the last abdominal segment is much broader. It agrees with the female *angelica* in having a series of large pits along the basal margin of the mesothorax. The anterior coxæ have short spines.

Coelioxys texana vegana, n. subsp.—Beulah, New Mexico, 8,000 ft., August, (Cockerell). I had erroneously placed this with *C. moesta*. It differs from *C. texana* as shown in the table; by the black legs, with red only at the apices of the joints, it resembles *C. alternata* Say. It differs from Say's description of *alternata* by the dark chestnut-red tegulæ, and the total absence of any white hair band bordering the mesothorax, though there is a little tuft of hair just before the axillæ. The abdomen is sparsely punctured, as in *texana;* the fourth ventral segment has slender apical spines.

Coelioxys erysimi, n. sp.—Male at flowers of *Erysimum parviflorum*; Rifle, Colorado, July 3 to 8 (S. A. Rohwer).

Length about 10 mm.; black, with white hair, abundant on head and thorax; tegulæ black; legs entirely black; hind spurs red; eyes pale green, with long hair; antennæ and mandibles black; cheeks hairy all over; vertex, mesothorax and scutellum with large, quite dense, punctures; scutellum rounded behind; axillar spines moderately long, obtuse; wings strongly dusky at apex; nervures dark; anterior tibiæ, and all the tarsi, with short fulvous hair on inner side; abdomen shining, strongly, not densely, punctured; apical hair-band on first segment dense and entire, the other apical bands successively thinner, except at sides, beyond the second segment hardly appreciable dorsally; transverse sulci on second segment oblique; a short white subbasal band at sides of second segment; on segments 3 to 5 very strong subbasal hair-bands, broadly interrupted in the middle; sixth segment deeply excavated in middle, the upper apical margin with seven short teeth, a broadly triangular median one, and three on each side; at the lower apical level are the usual two teeth, long and sharp, about one mm. apart; at the sides of the sixth segment the teeth are very long and sharp, but at the sides of the fifth are no teeth, although very minute tubercles can with difficulty be seen; fourth ventral bidentate.

Coelioxys quercina, n. sp.—Male; Oak Creek Cañon, Arizona, 6,000 ft., August (F. H. Snow, 1974).

Length, 11 mm. or rather over; black, with white hair; tegulæ clear red; mandibles black; antennæ black, the flagellum faintly brownish beneath; anterior femora above, in front and at apex, middle and hind femora at apex, tibiæ (the hind ones broadly suffused with blackish on outer side) and tarsi bright ferruginous; spurs red; eyes light green, with short hair (about half as long as in *C. erysimi*); thorax above with the usual large punctures; scutellum rounded behind; axillar spines long and straight; pits at base of metathorax minute and obscure; abdomen with a strong apical hair band on first segment, the others successively weaker, as in *C. erysimi;* first segment with a basal band; the others with interrupted basal or subbasal bands, becoming successively stronger, broader and less interrupted, that on the fifth almost entire; fifth segment not toothed at sides, sixth with well-developed sharp lateral teeth; apex formed as in *C. erysimi*, but the teeth are smaller; fourth ventral bidentate.

Coelioxys fragariæ, n. sp.—Male; Strawberry Valley, San Jacinto Mts., California, 6,000 ft., July 17 (F. Grinnell, jr.).

Length about 10½ mm. (abdomen extended); black, with white hair; tegulæ bright red, with a tuft of white hair in front; mandibles and antennæ black; legs black, the tarsi and spots at apices of femora and tibiæ rather dark red; eyes pale greenish-ochreous, the hair short, as in *C. quercina;* head and thorax above with the usual large punctures;

anterior border of mesothorax with the two hair patches distinct; scutellum not tuberculate in middle; axillar spines large, slightly curved; base of mesothorax without conspicuous pits; wings darkened apically; first r. n. meeting first t. c.; abdomen with hair bands as in *C. quercina;* apical structures of the same type as in *C. quercina*, but the median spine large; fourth ventral bidentate.

Coelioxys hirsutissima, n. sp.—Male; Kenworthy, San Jacinto Mts., Calif., 5,000 ft., June 8 (F. Grinnell, jr.).

Length about 8½ mm. (abdomen retracted); black, with white hair, abundant on head and thorax, and forming entire apical bands (but no subbasal ones) on all the abdominal segments; eyes light green, with long hair, as in *C. erysimi;* antennæ black; apical half of mandibles red; tegulæ red; legs red, with white hair; cheeks hairy; scutellum not tuberculate in middle; axillar teeth rather short; fifth abdominal segment without lateral spines, sixth with slender lateral spines; apex quadridentate, the two lower teeth broad, hardly so far apart as the length of one, slightly curved inwards; ventral hair bands very dense; ventral segments with numerous fine punctures, producing a rather rugose effect, wholly different from the smooth surface, with scattered strong punctures, of the venter of *C. erysimi*, *quercina* and *fragariæ*.

The following table compares the above-described species with various other male *Coelioxys:*

Abdominal bands bright orange-ferruginous, confined to the apices of the segments; no median tooth in apical emargination of abdomen (Assam) .... *turneri* Ckll.

Abdominal bands not orange or red .... 1.

1. Apex of abdomen red, each apical lobe strongly tridentate; anterior coxæ without spines (Willowmore, Cape Colony; *(Brauns)* .... *afra* Lepeletier.

Apex of abdomen not red .... 2.

2. Apex of abdomen multidentate, each lobe with more than two teeth . 3.

Apex of abdomen quadridentate, or quinquedentate by reason of a small median tooth .... 8.

3. Segments 2 to 5 with hair bands at apex only; mandibles red (Willowmore, Cape Colony; *Brauns*) .... *difformis* Friese.

Segments 2 to 5 with basal or subbasal hair bands, interrupted in middle; mandibles black; fourth ventral segment strongly bidentate apically; anterior coxæ with conical, stout, rather short spines; hind spurs red .... 4.

4. Tegulæ black; lower part of cheeks covered with hair; axillary spines rather long (Colorado) .... *erysimi* Ckll.
Tegulæ red .... 5.
5. Sides of middle of mesothorax very densely punctured; lower part of cheeks covered with hair .... 6.
Sides of middle of mesothorax with well-separated punctures; lower part of cheeks bare, bounded behind by a very strong keel; axillary spines very short .... 7.
6. Face broader; anterior tibiæ, all the tarsi, and other parts of legs bright ferruginous; median apical spine of abdomen very small (Arizona) .... *quercina* Ckll.
Face narrower; legs black, with the small joints of tarsi and apices of femora and tibiæ dark red; median apical spine of abdomen very long (California) .... *fragariæ* Ckll.
7. Fifth abdominal segment with small lateral spines; hair of face white (New Mexico) .... *texana*, subsp. *vegana* Ckll.
Fifth abdominal segment without lateral spines; hair of face with a yellowish tint (Washington Co., Wis.; *Graenicher*). *texana* Cresson.
8. Tegulæ entirely bright red; leg red .... 9.
Tegulæ black, or dull and dark red .... 12.
9. Lower apical teeth of abdomen long and slender .... 10.
Lower apical teeth of abdomen broadened; thorax above without hair spots .... 11.
10. Ventral surface of abdomen densely and very coarsely punctured (Boulder, Colorado) .... *edita* Cresson.
Ventral surface of abdomen shining, with widely-separated punctures (Falls Church, Va.; *Banks*) .... *sayi* Robertson.
11. Larger; lower apical teeth of abdomen more widely separated; second s. m. receiving recurrent nervures about equally distant from base and apex .... *immaculata* Ckll.
Smaller; lower apical teeth of abdomen less widely separated; first r. n. joining second s. m. very near base, very much nearer than second r. n. to apex (California) .... *hirsutissima* Ckll.
12. End of abdomen narrow and elongated, with a deep parallel-sided sulcus above; lower apical teeth very sharp; hind spurs red; very small species (Boulder, Colorado) .... *deani* Ckll.
End of abdomen broader, the median sulcus broadened .... 13.
13. Legs bright red; anterior coxæ strongly spined .... 14.
Legs black, or the tarsi red .... 15.

14. Sixth segment of abdomen, in lateral view, not much longer than high (Wawawai, Wash.) .... *deplanata* Cresson.
Sixth segment of abdomen, in lateral view, very much longer than high (Willowmore, Cape Colony; *Brauns*).... *penetatrix* Smith.

15. All the apical teeth of abdomen (including lateral ones) very short and blunt; spines of anterior coxæ strong, covered on outer side with snow-white hair; fourth abdominal segment with a subbasal hair band in the transverse sulcus (New Mexico).... *soledadensis* Ckll.
Apical teeth of abdomen at least partly elongated or sharp ...... 16.

16. Fourth ventral segment emarginate .... 17.
Fourth ventral segment entire .... 18.

17. Emargination of fourth ventral segment wide, the segment not produced in middle; hair on eyes short... *grindeliæ*, subsp. *denverensis* Ckll.
Emargination of fourth ventral small and narrow, in a produced median lobe; hair on eyes long (Beulah, New Mexico) .... *rufitarsis*, subsp. *rhois* Ckll.

18. Lower apical teeth of abdomen very sharp; very small species (California) .... *angelica* Ckll.
Lower apical teeth of abdomen obtuse .... 19.

19. Hair of eyes short; face narrower (Las Vegas, New Mexico) .... *grindeliæ* Ckll.
Hair of eyes very long (Olympia, Wash.).. *ribis* subsp. *kincaidi* Ckll.

A CORRECTION.

In the key to the species of *Metopia* given in my last paper on Tachinidæ (CAN. ENT., Vol. XLIII, Nos. 8 and 9), I have stated that in *Metopia lateralis* the third abdominal segment bears six or seven marginal macrochætæ, while in *Metopia leucocephala* it bears only a single pair. This distinction was based upon the study of a few specimens after I had left the National Museum, and a re-examination of a large series of specimens of both sexes shows that the character is a variable one. In both *lateralis* and *leucocephala* the number of marginal macrochætæ on the third abdominal segment varies from two to six or seven. The tendency to the development of a considerable number of strong setæ seems to be more marked in the males than in the females.

I am indebted to Mr. H. E. Smith, of the Gipsy Moth Laboratory, who called my attention to the inconstancy of the character.

W. R. THOMPSON, Naples, Italy.

THE BLATTIDÆ OF ONTARIO.

BY E. M. WALKER, TORONTO.

The Blattidæ, or cockroaches, are represented in Ontario by eleven species, only two of which, however, are natives, the others being, with two or perhaps three exceptions, merely accidental visitors from the south.

Ischnoptera pensylvanica (De Geer).—Generally distributed throughout Ontario as far north as the Temagami District, and locally common or even abundant. I have specimens from the following localities: Point Pelee; Toronto; De Grassi Point, Lake Simcoe; Stony Lake, Peterborough Co.; Lake Joseph, Muskoka District; Go Home Bay, Georgian Bay; Temagami Park.

This cockroach is very abundant on the rocky, sparsely-wooded country about Go Home Bay, where it occurs in rotten logs and under loose bark. It readily takes up its abode in the summer cottages, where it becomes as much at home in the kitchen and larder as its cosmopolitan relatives of the city, and is often regarded by the residents as a nuisance. I came across it also in considerable numbers on a rocky island in Stony Lake, Peterborough Co., while on a canoe trip. They were first seen at night running up and down a tree trunk in some numbers. Our provision bags became infested with them, and remained so during the rest of the trip.

More annoying still is their habit of eating the paste from book-bindings and nibbling the surfaces of the covers. On my first visit to the Georgian Bay Biological Station, being unacquainted with this habit, I left a water-colour drawing, which I had just made, upon a book-shelf in the laboratory. Next morning only a ghost of it was to be seen, so thoroughly had the cockroaches nibbled off the pigments from the surface of the paper.

The adults appear about the middle of June, remaining until some time in August. They are most abundant during July. The species hibernates in the nymph state. Full-grown nymphs are found in the latter part of May.

Ischnoptera borealis Rehn.—An adult male of this species, labelled

"Toronto," is in the collection of the Provincial Education Dept. I remember also seeing a similar pale *Ischnoptera* some years ago in the collection of the late Dr. Brodie, which I took for *I. uhleriana*, but as these two species had not been separated at that time, I am unable to say to which of the two it belonged.

I. uhleriana has also been reported by Caulfield from "Welland and westward" (Ann. Rep. Ent. Soc. Ont., 16, 1888, p. 71), but for the same reason, as pointed out by Rehn, this record may also belong to *borealis*.

Blattella germanica L.—The "Croton Bug" is probably common throughout the settled parts of the Province. I have specimens from Toronto, Hamilton, Goderich and De Grassi Point, Lake Simcoe.

Blatta orientalis L.—The "Black-beetle" is doubtless also common in every city and town in the Province, though I have specimens only from Toronto and Sarnia.

Periplaneta americana L.—I have never met with this cockroach in Canada, but it has been recorded from Essex County by Caulfield (loc. cit.).

Periplaneta australasiæ Fabr.—I have taken a single male adult, and Mr. C. W. Nash several nymphs of this insect from bunches of bananas at Toronto.

Nyctibora holosericea Burm.—Toronto. One nymph from a bunch of bananas.

Nyctibora sericea Burm.—Mr. Nash has an adult male which he took from a bunch of bananas at Toronto.

Leucophæa surinamensis L.—One specimen from bananas. Taken by Mr. Nash.

Pancheora virescens Thumb.—A single adult from bananas. Taken by Mr. Nash.

Pancheora acolhua Sauss. & Zehntn.?—Some years ago I sent a Panchlora for determination to Mr. A. N. Caudell, who labelled it somewhat doubtfully *P. acolhua* Sauss. & Zehntn. The specimen has since been destroyed by dermestids, so that the determination cannot be verified. It was taken at Toronto from a bunch of bananas.

NEW GENERA OF NORTH AMERICAN LITHOBIIDÆ

BY RALPH V. CHAMBERLIN, UNIVERSITY OF PENNSYLVANIA, PHILADELPHIA.

In a study of the North American species of the Lithobiidæ that fall into the old genus Bothropolys as originally defined by Wood—that is, all those species having the coxal pores, in several series the writer finds that they compose in reality two clearly separated groups of generic value. In addition a third genus, represented by a species here described for the first time, is found, which, while evidently close to the other two in some features, differs from them in having the coxal pores arranged in but a single series. Diagnosis of these genera are herewith given together with those of other genera.

Genus *Bothropolys* Wood (emended).

Head margined continuously from the caudal end cephalad to the eyes on each side, the lateral margin not broken. Prosternum with a well chitinized spine on or near the anterior margin at each ectal angle; prosternal teeth more or less uniformly spaced with no diastema separating them into two groups on each side. Gonopods of the male consisting of a single undivided article. Basal spines of the gonopods of the female 2 + 2. Anal legs with the tarsal claw single; penult legs with the tarsal claw armed at base with a single small or sometimes obsolete spine or accessory claw, or this sometimes quite absent. Coxal pores in several series. (Coxæ of last two pairs of legs armed each with a stout ventral spine.)

Type.—*B. multidentatus* (Newport).

In addition to the type, *B. hoples* Brolemann and *B. permundus* Chamberlin belong in this genus.

Genus *Ethopolys* gen. nov.

Lateral margination of head ending abruptly about one-third the distance forward from the caudal edge, each lateral margin being distinctly broken—that is, rectangularly bent in at this level. A wider interval of diastema separating an ectal group of from 1 to 4 prosternal teeth on each side from an inner larger group, a slender, often more or less bristle-tipped spine, occurring in the diastema, but none at the ectal angle. Gonopods of the male distinctly biarticulate. Basal spines of the gonopods of the female 3 + 3. Tarsal claw of anal legs with a very small spine or accessory claw at base; the claw of penult legs with two accessory claws. Coxal pores in several series. (Coxæ of last two pairs of legs each armed with a stout ventral spine).

Type.—E. xanti (Wood)

In addition to the type *sierravagus* Chamberlin, *E. pusio* Stuxberg, and *E. bipunctatus* (Wood) belong to this genus, as does probably also the doubtful species *E. monticola* of Stuxberg, which I formerly have regarded as probably the same as *sierravagus* but which agrees rather better with adult *pusio*, a species which Stuxberg based on a very young and immature specimen, though differing according to the published description from either.

Genus *Zinapolys* gen. nov.

Each lateral margin of the head distinctly broken a little in front of caudal third of its length as in the preceding genus. Prosternal spine immediately caudal of the ectal prosternal tooth on each side; prosternal teeth large and uniformly spaced, no diastema separating them on each side into two groups. Gonopods of male distinctly biarticulate. Basal spines of gonopods of female 6 + 6. Claws of legs, especially of the more caudal pairs, long and rather slender, all with two accessory claws. Coxal pores in a single series on each coxa of the last four pairs of legs. (Coxæ of the last two pairs of legs each armed with a stout ventral spine).

Type.—*Z. zipius* sp. nov.

The type is the only species of the genus thus far known.

Zinapolys zipius sp. nov.

Antennæ short, composed of 20 articles; the articles beyond the large second one of moderate length, becoming shorter proximad of the ultimate; hairs clothing the articles usually long, mostly oblique to surface; not very dense. Eyes composed of about twenty ocelli arranged in four series, eg. 1 + 5, 5, 4 (3). The single ocellus not very large. Organ of Tomosvary large, exceeding an ocellus, well removed from the eye-patch. Prosternal teeth 6 + 6–7 + 7, the most ectal on each side largest; the others decreasing from this one to the mesal incision. Spine moderately stout at base but apically long and bristle-like. Angles of none of the dorsal plates produced. Gonopods of male distinctly biarticulate, the distal article much narrower than the proximal; subconic, pale. Claw of gonopods of female entire, curved, deeply hollowed out on ventral side. Basal spines 6 + 6, mostly thickest at middle, being acuminate distad and somewhat narrower at base. Coxal pores circular to transversely subelliptic, moderately large and distinct; 4, 4 (5), 5, 5. Spining of legs: first to seventh, $\frac{0,0,3,2,2}{0,0,2,3,2}$; eighth to eleventh, $\frac{0,0,3,2,2}{0,0,3,3,2}$; twelfth and thirteenth, $\frac{1,0,3,2,2}{0,1,3,3,2}$; penult, $\frac{1,0,3,1,1}{1,1,3,3,2}$ and 1, $\frac{1,0,3,1,0}{1,1,3,3,1(2)}$. All legs terminating in three claws. Last two pairs of coxæ armed laterally as well as ventrally and dorsally.

Original coloration somewhat uncertain because of long preservation of specimens in too weak alcohol, but apparently ferrugino-testaceous with the legs and vender more yellowish, and the caudal ventral plates and legs and the prosternum and head darker; antennæ darker than legs but somewhat paler than head. Length 17–20 mm.

Locality.—Kooteno Co., Idaho.

Genus *Paitobius* gen. nov.

Head as in *Lithobius*, as also are the mouth-parts, nearly. Coxosternum of second maxillæ with narrow median membranous strip which is thin and bent dorsally. Prosternum bearing uniformly 2 + 2 teeth of which the inner one on each side is always borne conspicuously farther forward than the outer, the line tangent to apices of teeth curving cephalad from sides to middle, i.e., being procurved. Spine at ectal angle bristle-like apically. Anterior margin narrow, the lateral slope beginning almost directly from ectal tooth. Antennæ always short, consisting of from 27 to 35 articles. Coxal pores uniseriate, circular. (Last two pairs of coxæ laterally armed). Penult legs always armed with two claws; and legs also armed with two claws (excepting in *naiwatus*). Dorsal spines of anal legs always 1,0,3,1,0; of penult, 1,0,3,1,0 to 1,0,3,2,1. Anal and penult legs always short and distinctly furrowed longitudinally along dorsal surface of third to fifth articles; furrow more distinct on third article and especially in the male in which this article is wider or more crassate than in the female. Gonopods of male small, conical, directed caudo-ectad and nearly always wholly concealed by the sternite. Gonopods of female with the claw always distinctly partite, three lobes being typically present or rarely one of these almost obliterated. Basal spines rather slender and acuminate from base, distad. Body of adults always showing a deeper violaceous or purplish or reddish-purple pigment, modifying the coloration, more or less, of entire body; and in preserved specimens, at least, distinctly colouring especially the muscles. Anal legs always dark, proximally with the tibiæ and tarsi conspicuously paler, usually yellow. The head and dorsum smooth and shining, never rugose. (In all known species the 9th, 11th and 13th, of 6th, 7th, 9th, 11th and 13th, or of 7th, 9th, 11th and 13th dorsal plates with posterior angles produced.)

Type.—*P. carolinæ* Chamberlin.

Distribution.—The South-eastern States.

In addition to the type, the genus includes the following species: *naiwatus* Chamberlin, *tabius* Chamberlin, *juventus* Bollman, and *simitus* Chamberlin.

The genus, which is compact, can readily be detected by the character of the prosternal teeth.

Genus *Taiyubius* gen. nov.

This genus is very close to the preceding, but the species composing it may always be at once distinguished by the characters of the prosternum, the teeth of which are the same in number but differ in not having the inner teeth borne far forward, and in having the axis of each of the latter directed somewhat mesad of directly cephalad, with the line tangent to apices of teeth curving caudad from the sides mesad, i.e., this line clearly recurved. Antennæ short, or very short, consisting of from 26 to 39 articles. Posterior coxæ either entirely unarmed laterally or with each one of last pair or of last two pairs with a weakly developed spine which is often difficult to detect. Anal legs always with two claws; penult with two or three. Dorsal spines of anal legs, 1,0,3,1,0; of penult, 1,0,3,1,1. Anal and penult legs very nearly as in *Paitobius*. Gonopods nearly as in *Paitobius*, but basal spines characteristically much broader and wider near middle of length. Pigmentation much as in preceding genus. In known species posterior angles of 9th, 11th, and 13th dorsal plates produced.

Type.—*T. angelus* Chamberlin.

Distribution.—Western United States.

Other species belonging to this genus in addition to the type are *satanus* Chamberlin, *harrielæ* Chamberlin, and *purpureus* Chamberlin.

Genus *Sonibius* gen. nov.

Related to the two preceding genera which it replaces in the north central section of the country. The prosternal teeth small and subequal, 2 + 2 or 3 + 3 in number, with the line of their spaces recurved.

Readily distinguished from the preceding two genera in having the short antennæ composed normally of but twenty articles which are relatively long, whereas in *Paitobius* and *Taiyubius* they are mostly very short and crowded. Last two or three pairs of coxæ laterally armed, last four or five pairs dorsally armed. Anal legs armed with two or three claws as are also the penult, the number being mostly three. Dorsal spines of anal legs 1,0,3,1,0; of penult and 13th always 1,0,3,1,1. Gonopods of female with claw partite. Basal spines characteristically short and broad. Adults not showing the peculiar reddish-purple pigment in deeper tissues manifest in *Paitobius* and *Taiyubius*. Dorsum always smooth and shining. In known species posterior angles of 9th, 11th, and 13th dorsal plates produced.

Type.—*S. bius* Chamberlin.

Distribution.—North Central United States.

Besides the type, other species known to belong to this group are *politus* McNeil, *numius* Chamberlin , and *yanikans*, sp. nov.

Arenotini

Most of the species of the Lithobiidæ known from Central America and from Mexico compose a group which may be designated as the Arenobini. Among other features all have the gonopods of the male, although large and prominent, composed of but a single article and the claw of the gonopods of the female wholly undivided, with basal spines 2 + 2, large and stout, and the basal article with inner sides strongly chitinized and conspicuously excavated toward base. The dorsal spines of the anal legs from 1,0,3,2,1(0) to 1,0,3.2,2 ; of the penult always 1,0,3.2,2 in the female, and either the same or 1,0,3,2,1 in the male. Excepting the new genus *Sotimpius*, proposed for *Lithobius macroceros* and *L. decodontos* of Pocock, the prosternal teeth are 2 + 2 in number with the ectal spine in many species stout and tooth-like. In *Sotimpius*, which will not be further discussed here, the prosternal teeth number 5 + 5 or 6 + 6 and the ectal spines are bristle-like.

Genus *Arenobius* gen. nov.

Body conspicuously attenuated, cephalad with first dorsal plate narrower than the third. Dorsum smooth and shining, especially the first plate and the head. Prosternal teeth 2 + 2 with in most the spines stout and dentiform or more rarely these slender and bristle-like (only in Subgenus *Sititius*). Antennæ short to medium, occasionally equalling half the length of the body ; composed mostly of from 25 to 60 articles. Coxal pores uniseriate, circular or a little transversely elongate. With one exception the anal legs are armed with two or three claws. Dorsal spines of the anal legs 1,0,3,2,1 normally, occasionally varying in individual cases to 1,0,3,2,0 or 1,0,3,1,1 ; dorsal spines of penult legs in females always 1.0,3,2,2 ; in males nearly always 1,0,3.2,1, rarely the same as in female (frequently so in immature stages) With few exceptions both the anal and the penult legs are conspicuously modified in the male, the tibiæ of both bearing special lobes, furrows or bunches of hair ; more rarely with the penult legs normal and with the first tarsal joint specially modified. Gonopods of female with claw large, strongly curved and entire ; basal spines 2 + 2, stout; first joint with distal and inner edges strongly chitinized, excavated on mesal side toward base, leaving between the two a broadly triangular space with apex distad. Gonopods of male rather large and conspicuously exposed, but undivided.

Type.—*A. manegitus* Chamberlin.

Distribution.—From Colorado to the South-eastern States, and southward through Mexico to Central America.

In the United States, besides the type species, occur three other known species which differ from the type in having the ectal spines slender and apically bristle-like as well as in the character of the lobe on the penult legs of male. They may be placed in a separate sub-genus *Sibibius*. The species of this sub-genus are *coloradanus* sp. nov., *mississippiensis* sp. nov., and probably *œdipes* Bollman. Composing a new sub-genus *Kunobius* are the two species *pontifex* and *humberti* of Pocock, which differ in having the penult legs of male not at all modified while the first tarsal joint of the anal legs is very strongly enlarged, the tibiæ being specially modified as well. Both these species are from Province Guerrero, Mexico. The species *stolli* Pocock from Guatemala, differs from all other species of the genus in having all the dorsal plates with posterior margins straight, none of the angles being produced, and, according to the original description, in having the claw of the anal legs single. It may be placed in a subgenus *Sowubius*. Other Mexican and Central American species apparently belonging to *Arenobius* are the following: *godmani*, *salvini*, and *vulcani* of Pocock and *sontus* sp. nov., a rather aberrant species described below.

(To be continued.)

MEETINGS OF THE MONTREAL BRANCH.

Meeting Jan. 9th, 1912.—At the residence of Mr. H. H. Lyman, 11 members present. Mr. G. A. Southee in the chair. Mr. Lyman read a paper entitled "Further Notes on Types in the British Museum."

Mr. H. F. Wolley Dod, of Calgary, Alta., gave an interesting account of his visit to some of the U. S. collections of Lepidoptera. He referred particularly to those of Dr. J. B. Smith at New Brunswick, N. J.; the U. S. National Museum at Washington; the Strecker collection at Chicago, and that of Dr. Wm. Barnes at Decatur, Ill.

Discusssion followed on the various methods of collecting Noctuids.

Specimens of the genus *Xylina* were exhibited, representing about 25 North American species.

Meeting Feb. 10th.—At the residence of Mr. G. Chagnon; 6 members present. Mr. Southee in the chair.

A letter was read regarding supposed injury to greenhouse plants caused by a beetle, specimens of which were submitted. These were seen to be *Nacerdes melanura*, a European insect introduced through commerce. It is now very abundant in warehouses downtown, particularly near the wharves, and is often seen in numbers on sidewalks throughout the summer.

Mr. Winn read a short paper entitled "A Miniature Insectary," describing a space in his cellar boxed in around a south window, the inner of the double windows having been moved about three feet back, the outer window being left on during the winter and replaced by a wire screen in summer. The space only amounted to perhaps 40 cubic feet, but was sufficient to accommodate on shelves a number of breeding cages, jelly jars, boxes and tubes, and the conditions seemed to suit the insects, as there were practically no failures to get imagoes or parasites.

Mr. Moore read a paper on "Sexual Differences in Hemiptera," illustrated by specimens. Size and colour are the usual characters. Females of most species were much more seldom found in collecting than males, whether this was due to secretive habits of the females he did not know.

The Secretary followed with a paper on "The Determination of Sex in Lepidoptera. Several boxes of specimens were shown to illustrate the superficial points of distinction. The structure of the antennæ, the frenulum and some slides of genitalia were shown under the microscope.

Copies of Dr. Barnes' and Dr. McDunnough's "Contributions to N. A. Lepidoptera," Parts I–III, were shown.

Meeting March 9th.—At the residence of Mr. Lachlan Gibb; 10 members present. Mr. Chagnon, Vice-President, in the chair.

The question of finding a new place for the cabinet and book-cases was taken up, owing to Mr. Gibb's departure for England, and Mr. Lyman offered to look after it temporarily.

A paper on "Rye's Newest Moth," by Henry Bird, Rye, N. Y., was read by the Secretary. The paper dealt with the discovery of *Gortyna erepta* boring in the roots of a coarse grass on the shore of the Atlantic. The species had previously been taken only in Kansas.

The chairman then announced that he had a pleasing duty to perform, and handed to Mr. Lyman an illuminated address, signed by all the members of the Branch, expressing their good wishes on the occasion of his marriage. Mr. Lyman replied, thanking the members for their gift.

Copies of two parts of the "Genera Insectorum," dealing with the family Geometridæ, were shown by the Secretary, as well as some drawings of structure of the Brephinæ.—A. F. WINN, Secretary.

INSECTS BRED FROM COW MANURE.†

BY F. C PRATT,
Late Assistant Entomologist, Bureau of Entomology.

NOTE.—This paper has been compiled from numerous notes made by Mr. F. C. Pratt some time before his death. The investigation was prosecuted with skill and enthusiasm by Mr. Pratt, and the paper gives but a slight idea of the large amount of work on the subject which was done by this assiduous entomologist.—W. D. HUNTER.

INTRODUCTORY.

In 1907, at the suggestion of Mr. D. L. Van Dine, an attempt was made by the Bureau of Entomology to breed parasites of the horn-fly (*Hæmatobia serrata* Rob.-Dev.) to ship to Hawaii for experiments in the control of the pests. Although a number of predators were bred, no absolute records of parasitism were obtained. However, the work resulted in the rearing of a considerable number of species which have not been known to develop in cow manure. The following list may be considered supplementary to that of Dr. L. O. Howard (CAN. ENT, Feb., 1901, pp. 42–44), which dealt exclusively with Diptera. Dr. Howard's list included 25 species of Diptera, the present list contains 31 species of that order, 17 of Coleoptera and one of Lepidoptera. Of the species of Diptera included in Dr. Howard's list, 14 occur in this one and 20 are new.

Special notes were made on the occurrence of the various species in the stables and milking-houses on account of the possibility of the contamination of the milk by disease organisms or in other ways.

The records from Victoria, Texas, were obtained by Mr. J. D. Mitchell. The species previously bred from cow manure are preceded by an asterisk. The Diptera were determined by the late D. W. Coquillett, and the Coleoptera by Mr. E. A. Schwarz.

List of species bred from cow manure.

DIPTERA.

Family PSYCHODIDÆ.

**Psychoda minuta* Banks.—Victoria, June 9. Six specimens.

Family CHIRONOMIDÆ.

**Ceratopogon specularis* Coq.—Dallas, July 26, 29, 30, Aug. 31, Oct. 21, Nov. 4. Victoria, May 27, Dec. 1. Seventy specimens.

This species has also been bred from cow manure by Long See Biol. Bull., 1902, p. 7.

Family MYCETOPHILIDÆ.

*Sciara, sp.—Victoria, June 9. Nine specimens.

†Published by permission of the Chief of the Bureau of Entomology.

Family CECIDOMYIDÆ.

Lestremia leucophæa Meig.—Dallas, July. Two specimens.

Bred from droppings one day old in pasture.

*Diplosis sp.—Victoria, June 9. One specimen.

Cecidomyia sp.—Victoria, June 9. Dallas, May to December. Seventeen specimens.

The salmon-coloured larvæ are very conspicuous. The species apparently is perfectly at home in the manure.

Family BIBIONIDÆ.

Scatopse atrata Say.—Dallas, July 26, Sept. 18. Nineteen specimens.

This species was bred from fresh manure. The flies follow the cattle into the stables.

Family EMPIDIDÆ.

Tachydromia pusilla Loew.—Dallas, June 22. One specimen.

Family SARCOPHAGIDÆ.

**Sarcophaga (Helicobia) quadrisetosa* Coq.—Dallas, July 26, 29, 30, Sept. 10, 16, 18, 23, Oct. 8. Victoria, June 15, July 4, 23, Nov. 2, 24. One hundred and forty-two specimens.

This species is one of the most common in cow manure. From one dropping 78 specimens were bred. The developmental period varied from 7 to 14 days. The files frequently invade the stables, and are often seen on the manure immediately after it is voided.

Sarcophaga incerta Wills.—Dallas, Aug. 1, 16, Sept. 3, 10, 12, 22, Oct. 18, 21, 23. Fifty-nine specimens.

This species is exceedingly common in the pastures, but is seldom seen in the stables. It develops in from 14 to 18 days in the summer.

Sarcophaga assidua Wills.—Dallas, July 26, Aug. 3, Sept. 9, 12. Thirty-eight specimens.

This species has been bred from fresh manure in the milking-houses. The developmental period in August ranged from 17 to 20 days.

Sarcophaga helicis Towns.—Victoria, Aug. 7. One specimen.

On account of the published notes on the habits of this species it is likely that the occurrence in cow manure was accidental.

Sarcophaga varicauda Coq.—Dallas, July 30 Twenty-four specimens.

This species is found in and about the milking-houses more commonly than in the pastures. Development occupied from 10 to 14 days.

Family MUSCIDÆ.

Pseudopyrellia cornicina Fabr.—Dallas, July 29, Aug. 19, Sept. 10, 17, 22, Oct. 21, 23, 24, 26. One hundred and thirty-two specimens.

This species is frequently found in the stables. It places its eggs in clusters on fresh manure. In one case 285 puparia were found in one mass of manure. Development occupies from 9 to 20 days, depending upon the temperature.

Pyrellia cyanicolor Zett.—Victoria, Oct. 4. Twenty specimens.

**Morellia micans* Macq.—Dallas, 17, 22. Thirteen specimens.

**Musca domestica* L.—Dallas, Tex., June 17, 19, July 20, 31, Aug. 19. Fifty specimens.

This is one of the most common species in the stables. Fresh manure attracts it in great numbers.

Stomoxys calcitrans L.—Dallas, Aug. 3. One specimen.

This species is very abundant in and about the stables, but, judging by our records, does not breed commonly in cow manure.

**Hæmatobia serrata* Desv.—Dallas, June 22, July 26, Sept. 14, 18, 19, 21, 22, 28, Oct. 8, 14. Victoria, July 18, Oct. 4, Dec. 10. Fifty-three specimens.

This species is to be found in the milking-houses. It varies greatly in numbers with the weather. Dry weather prevents development, and a series of showers invariably brought about a sudden and conspicuous increase in numbers.

**Myiospila meditabunda* Fabr.—Dallas, July. Twenty-two specimens.

This species was not observed in the stables.

Family ANTHOMYIDÆ.

Limnophora discreta Stein.—Dallas, July 30, Aug. 31, Sept. 5, 12, 23. Victoria, May 3, 9, 25, Aug. 9, Sept. 10, Nov. 1. Ninety-eight specimens.

This is one of the most abundant species. It was not taken in the milking-houses, but was frequently bred from manure deposited in their immediate vicinity.

Limnophora debilis Will.—Dallas, July 26, Aug. 1, 24, Sept. 7, 16, 23, Oct. 7, 9, Nov. 4. Victoria, Aug. 9, Oct. 14, 20, Sept. 10, Nov. 4. Two hundred and thirty-four specimens.

This species and the preceding are apparently the most common flies breeding in cow manure in Texas, except the undetermined species of Limnosina.

Anthomyia albicincta Fall.—Victoria, June 27. Three specimens.

Apparently this species breeds in a great variety of substances.

Pegomyia fusciceps Zett.—Victoria, June 3. Fourteen specimens.

This record seems to be substantiated by the breeding from human excrement, as noted by Dr. L. O. Howard.

Family BORBORIDÆ.

Limnosina spp.—Dallas, Aug. 30. Three hundred and sixty specimens. *L. albipennis* Rond. has been recorded.

Apparently three species were bred from fresh as well as partially dried deposits.

Family SEPSIDÆ.

Sepsis pleuralis Coq.—Victoria, Oct. 10. One specimen.

**Sepsis violacea* Meigen.—Dallas, July 29, Sept. 12, 19, 20, 21, Oct. 26, Nov. 6. Victoria, Texas, July 9, Aug. 4, Oct. 10. Seventy-seven specimens.

This species was taken commonly in the stables. It breeds in fresh droppings.

Sepsis insularis Will.—Dallas, Aug. 22, Oct. 8, 18. Victoria, May 14, Sept. 10, Dec. I. Fifty specimens.

This species was taken in the stables repeatedly.

Family OSCINIDÆ,

Hippelates microcentrus Coq.—Dallas, Sept. 3. One specimen.

An allied species, *flavipes*, was bred from human excrement. See Dr. Howard's list.

Elachiptera costata Loew.—Dallas, Aug. 3. One specimen.

Family AGROMYZIDÆ.

Desmometopa m-nigrum Zett.—Dallas, Aug. 22. One specimen.

The single specimen bred was from fresh manure in a stable.

COLEOPTERA.

Family HYDROPHILIDÆ.

Cercyon nigriceps Marsh.—Dallas, Sept. 10, 14, Oct. 9, 17, 23, 26.

Two other species were bred from human excrement. See Dr. Howard's list.

Family STAPHYLINIDÆ.

Aleochara bimaculata Grav.—Dallas, Oct. 19.

This species is probably predaceous. It was taken in manure three days old.

Philonthus flavolimbatus Er.—Dallas, Oct. 19.

This species and the following were found in the breeding-cages, but there is no absolute proof that they were actually breeding in the manure.

Philonthus varians Payk.—Dallas, Aug. 6.

Philonthus longicornis Steph.—Dallas, Aug. 27.

Lithocharis ochracea Grav.—Dallas, Aug. 19, Sept. 10.

Cilea silphoides Er.—Dallas, Sept. 10.

Platystethus americanus Er.—Dallas, Sept. 23.

Platystethus spiculatus Er.—Dallas, July 30, Aug 3, 19.

Oxytelus sculptus Grav.—Dallas, Aug. 3, Oct. 28.

Family HISTERIDÆ.

Hister abbreviatus Fab.—Victoria, Oct. 14.

Hister cœnosus Er.—Victoria, April 15. Dallas, June, July.

This species is predaceous. It was found devouring the larvæ and puparia of *Pseudopyrellia cornicina*, and undoubtedly attacks other species, including the horn-fly.

Family SCARABÆIDÆ.

Canthon lævis Dury.—Victoria, April 15.

Aphodius fimetarius Linn.—Dallas, Oct. 19.

Aphodius lividus Oliv.—Dallas, Sept. 10, Oct. 18.

Aphodius vestiarius Horn.—Victoria, April 15, Oct. 20, Aug. 4.

Aphodius sp.—Victoria, April 15.

The various species of Aphodius are by far the most common beetles found ia cow manure.

LEPIDOPTERA.

Family TINEIDÆ

Setomorpha rutella Zeller (det. Aug. Busck)—Victoria, Nov. 23.

The occurrence of this species in cow manure may be accidental.

NEW SPECIES OF THE COLEOPTEROUS GENUS *COLLOPS* ER.

BY CHARLES SCHAEFFER,
Museum of the Brooklyn Institute, Brooklyn, N. Y.

Special help employed last year at our Museum to catalogue the collections made it necessary to rearrange certain boxes as well as to identify unnamed species. In the genus *Collops* several species collected on our museum trips, and from other sources, proved to be new, were given names and entered in our catalogue. It was my intention to revise the entire genus later on and the descriptions of the different forms drawn were kept back for this reason. However, as I have for some reason, to delay at present a revision of this genus, the descriptions of the new species are published in advance, in order that the names entered in the catalogue may stand.

The measurements of the species herein described are taken from specimens with the head deflexed.

Collops nigritus, new species.

Head black, densely punctate; clypeus reddish. Antennæ with first joint relatively strongly angulate at middle, red with a black spot; second joint reddish, on underside blackish; following joints black, rather feebly serrate. Prothorax red; surface densely punctate with the usual short, pale and erect black hairs. Elytra black, very densely punctate. Basal half of each ventral segment black, apex red. Legs black, apex of anterior coxæ reddish. Length 3.5 mm.

Arizona.

A single male found among the unmounted material of the Dietz collection.

From all species with unicolorous elytra, this species differs by the rather strongly angulated first antennal joint of the male, the densely punctate prothorax, and the black elytra.

Collops parvus, new species.

Head bluish-black; clypeus reddish. Antennæ black, first joint red, excavate on inner side. Prothorax red; surface shining, scarcely punctate on the disk. Elytra elongate-oval, blue, feebly shining; surface moderately densely punctate. Trochanters and femora black, tibiæ and tarsi red. Ventral segments red, black at sides. Length 3 mm.

A single male in the O. Dietz collection from Arizona labelled *punctatus*.

Collops eximius, var. *floridanus*, new var.

Like *eximius* Er., except thorax red, without large, black spot. Length 5 mm.

Florida, collected by R. F. Pearsall and received from A. Nicolay.

A large series of *eximius* Er. which I have seen, shows very little variation in the form of the black spot. The series in Mr. Nicolay's collection from Florida is also constant, except that in some specimens two faint, narrow, dark spots are visible on the thorax.

Collops aulicus Er.—Entomographien, p. 55.

I have taken a female specimen in the Huachuca Mts., Ariz., which, according to the figure in the Biol. Cent. Am. Col., Vol. III, pt. 2, pl. VI, fig. 21 and 22, seems to be that species. However, I have not seen the original description, but, as the species is also reported from Guanajuato, it is more than probable that my surmise is correct. A specimen in the Dietz collection from Arizona, which agrees well with the description of *marginicollis*, differs only from the above mentioned specimen by having the thorax black, with side margins near base, more or less narrowly pale. If the two should prove to be the same, *aulicus* Er. as the older name, has to be accepted for this species.

Collops argutus Fall—Occ. Pap. Cal. Acad., VIII, p. 242.

A few specimens from the Huachuca Mts., Arizona, agree closely with the description of this species, except that the abdomen is red, with last segment black. The abdomen in the description is said to be black in some specimens, and rufous at middle near base in others, which indicates that the colour of abdomen at least is variable.

Collops femoratus, new species.

In coloration, nearest to *pulchellus* Horn, but not quite as elongate; elytra more finely and closely punctate; basal antennal joint of male not excavate on the inner side; front and middle femora and hind femora at base red, tibiæ and tarsi black. Length 4 mm.

Huachuca Mts., Arizona.

The ventral segments are red in a few males; in one male and in the females they are spotted with black at sides. A female from Brownsville, Texas, with clear, red abdomen, does not seem to differ otherwise from the Arizona specimens.

Collops scutellatus, new species.

Red; metasternum and palpi black; head with central bluish spot and elytra with basal and elongate-oval, blue, apical spots; the two spots on each elytra narrowly divided. Thorax narrower than in *pulchellus*, *argutus* or *femoratus*; scutellum clear-red. Basal antennal joint rather slender, feebly enlarged towards apex, and not excavated on inner side; the outer joints feebly dilated. Length 3.25 mm.

One male, New Braunfels (O. Dietz.)

The coloration, the scarcely apically dilated basal joint of antennæ, the red scutellum and the more shining elytra will separate this from all previously described maculate species.

Collops tibialis, new species.

Head blue, clypeus and a large, oval, median spot from the clypeus to almost the middle of the head, reddish. First two antennal joints of male red, the outer blackish and feebly serrate. Prothorax shining, red, feebly punctate at middle, distinctly so at sides. Elytra with two large blue basal spots connected at suture, and oval sub-apical spot on each side, involving largely the lateral margin. Surface punctured similarly to *quadrimaculatus*. Femora black, tibiæ red, abdomen red. Length 3 mm.

Nogales, Arizona (F. W. Nunnenmacher.)

A small species with thorax less transverse than the other species.

Collops similis, new species.

Head bluish-black, inter-antennal space and clypeus red. Antennal joints red; first joint stout, the other serrate. Thorax red at middle with two, rather indistinct, darker lines. Elytra red, with basal and oval sub-apical spot blue; surface more coarsely punctate than in *quadrimaculatus* Fab. Front and middle femora red; hind femora, tibiæ and tarsi black. Abdomen red. Length 3.50 mm.

S. W. Utah, collected by J. Chr. Weidt, of whom I bought a few specimens some years ago. This type is in the Museum collection. This species has more coarsely punctate elytra than any other of the known maculate species.

Collops punctulatus, var *texanus*.

Like *punctulatus* Lec., except that the thorax is bright red. Length 2 mm.

Brownsville, Texas.

A few specimens have the blue elytral vittæ narrowly divided, forming two large spots on each elytron; in one specimen the spots are confluent as in typical *punctulatus*.

Collops punctulatus, var. *utahensis*, new var.

Differs from *punctulatus* in having the elytral vittæ broadly divided. The basal spot is small and more or less transverse. The apical spot oval. Length 2–2.25 mm.

Buckskin, Utah. (Doll & Engelhardt.)

Apparently a constant, local form of *punctulatus*.

Collops sublimbatus, new species.

Closely allied to *C. georgianus* Fall, from which it differs in having the head polished, the disk of prothorax shining, and scarcely punctate. Length 3.5–4 mm.

Clayton, Georgia.

Through the kindness of my friend, Mr. William T. Davis, I have seen a number of specimens which agree very closely with Fall's description of *georgianus*, except as above stated. From *limbellus*, it differs in having the outer antennal joints not serrate, and the second joint is much wider than long.

ON THE LARVAL STAGES OF CERTAIN ARCTIAN SPECIES.

BY WM. BARNES, M.D., AND J. H. MCDUNNOUGH, PH.D., DECATUR, ILL.

(Continued from page 136)

Apantesis incorrupta H. Edw.

We received a ♀ of this species about the middle of June from the neighbourhood of Redington, Ariz., which had deposited numerous ova en route. The young larvæ hatched within 1–3 days after receipt of eggs. In all probability therefore, the duration of ovum stage is about 5 days. Unfortunately, owing to our absence from home, the complete larval history could not be worked out. The early stages may however prove of value, especially when compared with those of *nevadensis*, of which Dyar lists *incorrupta* as a variety. We, ourselves, see no reason why it should not enjoy specific rank.

Ovum.—Very similar to that of other *Apantesis* species; rather conical, with flat base. Yellowish, turning black before emergence, laid promiscuously.

Stage I —Head and prothoracic shield blackish, latter with 4 anterior and 4 posterior setæ. Body pale reddish, with green of the food largely showing through the skin after eating. Tubercles blackish, with a similar arrangement to that of other *Apantesis* species. Tubercle I small, with minute, white seta. The seta of tubercle II and the upper one of III black on abdominal segments; all other setæ long, white, increasing in length on rear segments. On meso- and metathorax, tubercles I and II possess one white and one black seta. Length 3 mm.

Stage II.—Head, thoracic plate, and tubercles black; body purplish brown, shading into lighter ventrally, and tinged with orange at the base of the lateral tubercles III–V. A pale, dorsal line and a broken subdorsal one on a level with tubercle III. Tubercle I minute, with single short black seta; on thoracic and two posterior abdominal segments, tubercle II possesses a single long white seta, surrounded by 6 or 7 shorter black ones; on the remaining abdominal segments the white seta is lacking, and the black setæ are 5–6 in number. Tubercle III on thoracic segments with two long black setæ and several small basal ones; on abdominal segments with very long white central seta, a ring of about 4 shorter black ones, and a small cluster of minute basal white hairs. Tubercle IV similar in arrangement to III; ventro-lateral tubercle with short white setæ. Length 5 mm.

Stage III.—Very similar to preceding with an increase of secondary black setæ, single long white seta present as before; whitish dorsal line slightly enlarged in centre of each segment. Length 7 mm.

Stage IV.—Head black, body purplish brown mottled strongly with whitish; pale dorsal line very distinct; subdorsal line almost lost in the white marbling. Tubercles black, lateral ones distinctly orange at bases. Long white setæ of III and IV very prominent, especially on rear segments. Tubercle V of anterior segments also bears a white seta which is lacking in the posterior half; other setæ mostly black. Lateral abdominal tubercles with several pale setæ inclining to orange; spiracle black. Prothoracic plate split up into 4 chitinous mounds of which the two anterior are the larger, bearing numerous setæ projecting over the head; the posterior carry 4 6 setæ. Legs black. Length 13 mm.

Apantesis phalerata Harris.

The larval history of this moth has been already described by Gibson (CAN. ENT., XXXII, 369; id. XXXIV, 50). We venture to add, however, some more precise details on the earliest stages, as it is probably in these that we must look for points of distinction from closely allied species.

A ♀ of the form *radians* Wlk., i. e., with W mark absent and broad black border to red hind-wings, deposited ova freely in June; these were not dropped promiscuously as in the previous form but placed neatly in rows in irregular groups on the underside of any available object. The egg itself offers no points of distinction, being similar to those species already described. We were unfortunately not able to breed the species through.

Stage I.—Head, lobes blackish; clypeus and mouth-parts as well as suture between lobes, pale brown; body pale greenish red; tubercles dark; on abdominal segments I minute, with white seta; II and III as usual. Setæ black on all segments except on 9th abdominal, where 9th dorsal tubercle bears two long white and two black setæ; remaining tubercles bear white hairs. Prothoracic plate the colour of body, except for a dark mound on each side of the central line anteriorly; these mounds bear each 2 black setæ. There is further a posterior row of four setæ, the outer (lateral) one being white. Length 2.7 mm.

Stage II.—Head as before. Body greenish brown with pale dorsal stripes and orange tinges laterally at the base of the tubercles. Characteristic of this stage is the great increase of setæ on tubercle II which bears 10-12 short black hairs. On abdominal segments III has 2 or 3 black central setæ surrounded by a ring of 4-5 small white ones; on thoracic segments III bears merely 1-2 black setæ. The lateral tubercles bear one long central black seta, the remaining ones being smaller and white. The thoracic plate shows four slightly raised darker areas, two anterior and two posterior, each bearing the black setæ. All legs the colour of the body. Length 4.8 mm.

Euchætias spraguei Grt.

Ova sent by Mr. Kwiat of Chicago in June hatched during transmission, the young larvæ eating the egg shells. From a rather crushed unfertile egg following note was made.

Ovum.—Hemispherical, with flat base; pitted with numerous slight punctures; pale orange-yellow. Diameter .4 mm.

Stage I.—Head black, with sparse black setæ. Body pale red-brown with large black almond- or kidney-shaped tubercles; arrangement of setæ typically Arctian; tubercle I about half as large as II, with short black seta; seta of II black, longest on 1st and last two abdominal segments; III with black setæ; lateral setæ short and white; thoracic plate semilunate with 4 setæ on anterior margin. When at rest the first two abdominal segments appear slightly humped. Length 3 mm.

Stage II.—Head pale yellowish; body red-brown, becoming later yellowish; tubercles black, II and III being the largest; tubercle I with two long black setæ and a couple of short white ones inclined towards the head; II with about 8 black setæ; III with 6; IV small, situated immediately posterior to the spiracle and bearing two setæ; V with 5–6 setæ of which the central one is black, the remainder white; VII larger with 2 central black and 8–10 white basal setæ; spiracle pale with black rim. Length 6 mm.

Stage III.—Head and body pale orange yellow; tubercles blackish arising out of a pale base and with numerous long plumed gray hairs; traces of a pale subdorsal line and lighter shading laterally around tubercle V. Length 13 mm.

Stage IV.—Head, body and legs pale orange with traces of a pale yellow subdorsal and a similarly coloured subspiracular line: tubercles black, giving rise to numerous long, silky, plumed, dark gray hairs, which form a thick covering over the body. Length 20 mm.

Stage V.—Head pale reddish; body dorsally pale green, shading into pale orange laterally; traces of a dark dorsal line and the pale subspiracular line of the previous stage; tubercles pale; base of hair black; body thickly covered with long silky plumed pinkish gray hairs; the medio-dorsal hairs blackish, forming a dark dorsal line; prolegs reddish; spiracles creamy, rimmed with black. Length 28–30 mm.

Pupation in a coarse cocoon of mixed hairs and earth on or just beneath the surface of the ground.

Pupa of the usual type, with immoveable segments, dark brown. One pupa out of sixteen emerged during July; the remainder are hibernating.

Food plant.—Spurge (Euphorbia).

APHID NOTES FROM OREGON.

BY H. F. WILSON, CORVALLIS, OR.

(Continued from page 159.)

Stem-mother.—Collected on terminal shoots of *Pseudotsuga douglassii* about Corvallis, Oregon, March 15, 1911.

General colour light brown, with two rows of black spots extending midway along the dorsum to the middle of the abdomen. These spots sometimes join so as to give the appearance of two dark lines extending along the body. Body semi-shining and with faint traces of a light flaky powder on dorsum. Legs and antennæ dusky brown. After having been mounted on slides for some time this species turns red and a deep red colour is assumed by the balsam surrounding them. The abdomen is quite large in comparison with the head and is almost globular. Antennæ VI segmented, and about one-fifth the length of the body. The nectaries are but small round tubes slightly elevated; they are about as wide as long and are situated on the side of the abdomen about two-thirds of the way from the base of the thorax to the base of the cauda. Cauda broad and slightly angled, very short.

Measurements: Length of body, 3.8 mm.; width, 2.99 mm. Length of antennal segments, I, .09; II, .09; III, .3; IV, .135; V, .135; VI, .12; spur, .045 mm.; total length, .87 mm. Length of cauda, .3 mm.; nectaries, .022 mm.

Spring migrant.—Collected June 4, 1911, on terminal shoots of same host plant. General colour of head and thorax dark or dusky. Abdomen greenish brown, with colouring of white powder. Legs and antennæ, except tarsi and tips of third, fourth, fifth and six segments, light brown. Other parts dusky to black. Antennæ about one-fourth the length of the body. Head rounded in front and with a suture or line extending from back to front midway between antennæ. Wings hyaline. The first anal and cubital veins quite distinct while the median with its two branches, remains only as faint lines. The nectaries of this form are cone-shaped with a flanged mouth and are apparently not placed as far forward as in the earlier forms. Cauda short and broadly angular.

Measurements: Length of body, 2.84 mm.; width, 1.09 mm.; length of wing, 3.65 mm.; width, 1.1 mm; total wing expanse, 8.08 mm. Length of antennal segments, I, .66; II, .11; III, .44; IV, .154; V, .198; VI, .11; spur, .045 mm.; total length, 1.123 mm. Length of nectaries, .064 mm. Length of cauda, .22 mm.

The fall migrant was not secured.

Egg-laying female.—Collected on terminal shoots of above plant, Oct. 30, 1910, and Oct. 27, 1911, along with the alate males. General colour brownish with ash-grey powder on body, and with two more or less regular stripes down the back; and with a wide brown stripe extending across the body from one nectary to the other. At the base and above the cauda another transverse band is usually present. Antennæ and legs except tips, light brown; other parts dusky to black. Body robust and with large semi-conical nectaries which are brown in colour. Antennæ and legs hairy; antennæ one-third the length of the body.

Measurements: Length of body, 2.9 mm.; width, 1.7 mm. Length of antennal segments, I, .066; II, .09; III, .35; IV, .176; V, .176; VI, .11; total length, 1.013 mm. Length of nectaries, .06 mm., and cauda, .35 mm.

Alate male.—Collected on terminal shoots of host plant Oct. 30, 1910, and Oct. 27, 1911, about Corvallis, Oregon.

General colour: Head and thorax black with green abdomen. Abdomen with a series of black, transverse, more or less distinct, bands. Antennæ yellow at base, dusky at tip. Femora and tibiæ dusky at middle to black at ends; tarsi black. Wings hyaline but with costa dark brown, median vein and branches almost indistinct; other veins dusky. Nectaries slightly bell-shaped with a flanged opening. Third antennal segment about equal in length to fourth and fifth segments and with about 30 to 39 visible small circular sensoria. Fourth segment with 10 to 12 circular sensoria which appear slightly larger than those on the third segment. Fifth with about eight medium-sized, and one large, visible sensoria at the distal end. Sixth segment with one large and apparently six small sensoria at base of spur.

Measurements: Length of body, 2 mm.; width, .87 mm. Length of wing, 3.87 mm.; width, 1.52 mm.; total expanse, 8.61 mm. Length of antennal segments, I, .066; II, .11; III, .51; IV, .242; V, .3; VI, .154; spur, .066 mm.; total length, 1.448 mm. Length of nectaries, .045 mm.; cauda, .176 mm.

Females along tips of needles, depositing from 5 to 8 eggs.

Lachnus occidentalis Davidson.

Quite abundant in spring and fall on the terminal shoots, where it causes the undersides of the twigs to appear bluish or smoky.

Stem-mother.—Collected on *Abies grandis*, near Corvallis, Oregon, March 18, 1911. Gregarious individuals from just hatched to two-thirds grown. Colour dark brown, body pruinose. Head and legs black except tibiæ, which are dark brown. Body with four rows of spots on the abdomen, which resemble dried spots of some white crystalline substance. The more mature specimens taken later are darker and have more of the pruinose covering. The spots are more distinct, and the underside of the body and the sides are white with this covering.

Measurements: Adult females. Length of body, 3 mm; width, 2 mm. Length of antennal segments, I, .09; II, .09; III, .35; IV, .154; V, .198; spur, .198 mm. Spur tapers into joint so that it is not distinct from segment. Total length, 1.08 mm; length of cauda, .27 mm. Nectaries small, cone-shaped, with chitinous ring at opening. Body, antennæ and legs bearing long hairs.

Spring migrant—Found on twigs, on underside, and causing bluish colour intermixed with white flocculent castings of Aphids.

General colour; head black; rest of body dark green but whitish in appearance from flocculence which covers body and legs. Legs and antennæ light brown with dark joints, tarsi dark to black. Antennæ slender and apparently without sensoria except one large one at the distal end of the fifth segment. Third segment the longest and about equal to four and five in length. Nectaries cone-shaped, placed on the side of the body and much larger than those of the apterous forms. Fore wings with the median vein and branches indistinct, present as a mere outline. Hind wings with normal venation of two oblique veins.

Measurements: Length of body, 2 1 mm.; width, 0.9 mm. Length antennal segments, I, .066; II, .09; III, .44; IV, .176; V, .22; VI and spur, .22; total length, 1.172. Length of wing, 4 mm.; width, 1.5 mm.; total expansion, 8.75 mm. Length of cauda, .22 mm. Body, antennæ and legs bearing long hairs.

Fall migrants not obtained.

Oviparous female.—Collected at Corvallis, in colonies on under side of small twigs. Egg deposition takes place on the needles and they are laid in a row, about five being the most found on any one needle.

General colour brown, with white powdery wax on dorsal and ventral parts of body. That on the ventral part is thicker than above and extends half way up the sides. The characteristic rows of spots as are found on all the apterous forms of this species are found on the dorsum. This form is smaller than the viviparous forms.

Measurements: Length of body, 2.8 mm.; width, 1.7 mm. Length of antennal segments, I, .066; II, .09; III, .33; IV, .176; V, .198; VI and spur, .176 mm.; total length, 1.036 mm.

Alate male.—Collected near Corvallis, Nov. 3, 1911, on same tree that the spring forms were collected on. Only a very few specimens found and these were so active and so small as to be not readily located.

General colour of body, green. Head and thorax dusky. Antennæ and legs dusky. Body almost hid in white, fluffy threads of wax which also appears on the legs. Head almost as wide as thorax. Antennæ reaching to abdomen and with small circular, raised sensoria on the third, fourth and fifth segments. From top view there are shown about sixteen on the third segment, eight on each side; the fourth has about seven to nine, and the fifth has none in one specimen and apparently three in another. The wings are large in proportion to the body and the veins are but lines, the median veins almost obsolete. Nectaries cone-shaped with flanged opening and placed on the side of the body. Cauda more distinct than in other forms and broadly angular.

Measurements: Length of body, 2 mm.; width, .78 mm. Length of antennal segments, I, .066; II, .066; III, .33; IV, .154; V, .154; spur, .176 mm. Total length, .946 mm.

Egg.—Size not ascertained but they are covered with a powdery-like substance and resemble the eggs of *Longistigma caryæ* Harris.

Originally described by W. M. Davidson, Journal of Economic Entomology, Vol II, p. 300, 1909.

GEOMETRID NOTES—TWO NEW HYDRIOMENAS.

BY L. W. SWETT, BOSTON, MASS.

(Continued from page 164.)

Hydriomena transfigurata, nov. sp.

Palpi moderate; expanse of wings 29–30 mm.

Colour of fore wings greenish gray. in faded specimens yellow or reddish mixed in with the green. Possibly there are red and yellow varieties in series. The fore wings are quite pointed, much more so than in *H. autumnalis*, and the hind wings are quite dark. Basal band in fore wing black, running diagonally from costa to inner margin, with a slight projection at R. The basal band is more regular than in *autumnalis*, and the space between is greenish gray, where in *autumnalis* it is clear gray. The mesial band of *transfigurata* is broader than in most species of *Hydriomena*, and the usually characteristic watery, irregular central band is almost lacking or can be just faintly discerned. The general colour of the entire mesial band is green, with scarcely any central band, or, if present, represented by a round spot or series of spots. Outer or intradiscal band quite regular, running diagonally, almost directly from costa to inner margin, which it strikes farther out than in *antumnalis*. There is also a tendency in almost every specimen for the intradiscal line to unite with the extradiscal at C_1 and C_2. Discal space greenish gray with linear discal line. The extradiscal line is irregular, and is farther in from the apex of the wing than in *autumnalis*. It has also a tendency to unite with the intradiscal line near the inner margin. Outer marginal space greenish gray, with a faint trace of a black, irregular, watery line, which is narrower than in *autumnalis*, and appears as a mere trace, where in *autumnalis* it is very striking.

Hind wings quite dark and smoky gray, the usual two dark lines hardly visible.

Beneath, the markings of the fore wings shows through faintly; general colour ash-gray, speckled with black atoms. On the hind wings only the extradiscal lines show, and these but faintly.

This is a very peculiar species, quite distinct from *autumnalis*. It resembles *californiata* Pack. closely in some respects, but differs in the time of its appearance, as well as in the markings.

My attention was first attracted to this species by a unique specimen from Newfoundland, taken by Mr. Owen Bryant in early August, which I could not associate with any other. Later, my friend, Mr. William Reiff, turned up four specimens at Forest Hills, and Mr. Bryant seven males at Cohasset, Mass., and I also took a few myself.

This seems to be a very early species, occurring with *ruberata*, and seems distinct enough from *H. autumnalis* to deserve specific rank. The peculiarly shaped fore wings, the broad mesial band and the tendency for the extradiscal and intradiscal lines to unite near the inner margin will help to separate it from any other species.

Type ♂, Forest Hills, Mass., May 16, 1911 ; type ♀, Forest Hills, Mass., May 11, 1911. Paratypes, 8 ♂s, 3 ♀s, Newfoundland in August ; Forest Hills, Mass., May 5-11, 1911, and Cohasset, Mass., May 10, 1907, in coll. Boston Society of Natural History.

BOOK NOTICE.

MEMOIRS ON THE COLEOPTERA. By Col. Thos. L. Casey, Part III, 1912. The New Era Printing Co., Lancaster, Pa.

The third part of these Memoirs has been recently issued, the first and second parts having been published in 1910 and 1911 respectively.

This last one treats of three different families, viz: "A Descriptive Catalogue of the American Byrrhidæ," "A Revision of the American Genera of the Tenebrionid Tribe Asidini," and "Studies in the Longicornia of North America."

The first mentioned is a revision and synopsis of the family, occupying 67 pages, in which are fully described 6 new genera, 4 new subgenera, 48 new species and 10 new subspecies.

The second is a study in the tribe Asidini of the family Tenebrionidæ; to this is devoted 155 pages, in which are fully described 14 new genera, 129 new species and 28 new subspecies.

The third being studies commencing with the family Spondylidæ and embracing the Cerambycidæ as far as the genus Microclytus, 162 pages are devoted to this portion, in which are described 19 new genera, 4 new subgenera, 170 new species and 39 new subspecies.

In addition to the new genera and species, many of the older ones are described and very many useful notes are interspersed throughout ; the whole forming a valuable addition to the literature on these families.

J. D. E.

Mailed June 12th, 1912.

The Canadian Entomologist.

VOL. XLIV. LONDON, JULY, 1912. No. 7

NEW SCOLIOIDEA.

BY NATHAN BANKS, EAST FALLS CHURCH, VA.

The following notes and new species apply to that section of the fossorial Hymenoptera, in which the pronotum reaches back to the tegulæ. Ashmead transferred them to the Vespids, in a superfamily, but they have little affinity with them in most of their structure, and most authors still keep them in association with the other fossorial Hymenoptera.

PSAMMOCHARIDÆ

Priocnemis Schiödte.—This name should be used instead of *Cryptocheilus* in my table to genera (Journ. N. Y. Ent. Soc., 1911, p. 222). The type of *Cryptocheilus* is *C. annulatus* and in this species (which I have recently obtained) the hind tarsi have spines below on the last tarsal joint; thus *Cryptocheilus* will replace *Priocnemoides*, at least until this genus is divided.

Priocnemis semitincta, n. sp.—Similar to *P. arcuata*, but the wings are faintly uniformly tinged with yellowish, not darker at tip. The spurs are nearly as dark as the legs, and the second recurrent vein curved outward. The posterior margin of pronotum hardly angulate.

Length 8 mm.

From Las Vegas, N. Mex., 23rd July, 1902 (Oslar).

Priocnemis directa, n. sp.—Similar to *P. arcuata*, but the spurs are about as dark as the legs (pale in *P. arcuata*) and the hind margin of the pronotum is strongly angulate; the wings are coloured as in *P. arcuata* but the second recurrent vein is straight, not bent near tip as in *P. arcuata*.

Length 10 mm.

From Lee Co., Texas, Aug., Sept. (Birkman).

Priocnemis minorata, n. sp.—Similar to *P. conicus*, and runs to it in my table (Journ. N. Y. Ent. Soc., 1911, p.) but it has only about eight teeth on the hind tibia above (in *P. conicus* there are 10) and in this species the teeth are not nearly as long as the space between them, while in *P. conicus* they are as long. This species is also much smaller. The wings are dark, but darker on tip than elsewhere; the basal vein curved, and the clypeus truncate.

Length 10 mm.

From Great Falls and Falls Church, Va., in April.

Priocnemis relicta, n. sp.—Similar to *P. germanus* but much smaller, and the legs more brown; the hind tibia has only about six teeth above, and these very weak (in *P. germanus* stronger and about ten present); the veinlet between the second and third cells is nearly straight and the second recurrent vein arises but little beyond the middle of the anal cell; clypeus hardly truncate.

Length 6 mm.

From Sea Cliff, N. Y., in September.

Psammochares georgiana, n. sp—Black, except the abdomen which is entirely reddish; and the hind tibiæ and tarsi are also reddish, the latter with the tips of the points black. The third joint of the antennæ is not quite as long as the width of the vertex. The metanotum has a broad, deep, median furrow; the long spur of the hind tibia is about two-thirds the length of the metatarsus; and the third submarginal cell is no longer than the second, and with a very oblique vein on the outer side. The structure is very similar to *P. fuscipennis*, but the lateral ocelli are plainly nearer to the eyes than to each other; the abdomen has a compressed tip.

Length 12 to 14 mm.

From Bainbridge and Pomona, Ga., Sept. (Bradley), and Southern Pines, N. Car, June, Aug. (Manee).

Pseudagenia nanella, n. sp.—Bluish or purplish and black, clothed with whitish hair, dense on the clypeus; pronotum arcuate behind; metanotum grooved; abdomen polished; wings hyaline, venation black; third submarginal cell much longer than broad, its outer side very oblique; long spur on hind tibia nearly one-half the length of the metatarsus, latter not spined, with only very short, fine bristles. Similar to *P. architecta*, but uniformly much smaller, the face much more narrow, being as high as broad; antennæ less slender, the third joint but little longer than the fourth; and few if any long hairs above on the basal segment of the abdomen.

Length 6 to 7 mm.

From Sea Cliff and Ithaca, N. Y., Great Falls, Chain Bridge and Glencarlyn, Va., June to Sept.

SCOLIIDÆ

The true Scoliidæ are most easily recognized from all other families of this superfamily by the striated nature of the apical part of both wings. The strongly emarginated eyes also distinguish them, but some other forms have the eyes slightly emarginate.

SCOLIA

The species of *Discolia* have in the female a smooth macula on the sides of the second abdominal segment; it is less distinct in the male. Our species of this section known to me may be tabulated on colour marks as follows:

1. No pale marks on head or thorax; costal area, venation and the entire wing black..2.
   Pale marks on head or thorax................................6.
2. Abdominal segments fringed with black hair; abdomen black at tip..3.
   Abdominal segments fringed with fulvous hair; abdomen reddish at tip..4.
3. Broad, white bands on second and third segments, no spots on venter..*bicincta* Fabr.
   White spots or band on first segment; bands on second and third segments more or less broken into two spots; two spots on second ventral segment................................*undata* Klug.
   Widely separated white spots on sides of several segments, sometimes only on third; no ventral spots; abdomen more elongate................................*guttata* Burm.
4. No yellow spots on abdomen, more than apical half reddish................................*hæmatodes* Burm.
   Yellow spots on third, sometimes also on fourth segment...........5.
5. Black hair above on fourth and fifth abdominal segments; second segment usually black................................*dubia*.
   Only reddish or yellowish hair on fourth and fifth segments; second segment more or less reddish (from Palmerlee, Arizona)..*thalia*, n. sp.
6. Abdominal segments margined with dark or black hair; second and third segments with yellow spots; wings and venation all black................................*nobilitata* Fahr.
   Abdominal segments fringed with fulvous hair..................7.

7. Venation dark, costal area (except base) black....................8.
Venation largely yellow, costal area yellowish to the stigma.........9.

8. Large yellow spots on second segment, spots or bands on third and fourth.. *consors* Sauss.
No yellow spots on second segment, which is reddish ; spots or bands on third and fourth segments......................*amœna* Cress.

9. Abdomen reddish, with spots or bands of yellow on second, third, fourth and fifth segments ; metanotum and head mostly reddish..*ridingsi* Cress.
Abdomen with second and third segments black, with separated yellow spots ; metanotum and head mostly black..........*lecontei* Cress.

S. flavicostalis Cress. is probably the male of *S. lecontei; S. inconstans* Cress. runs to *S. lecontei* but has no spot on the second segment ; *S. amœna* may be only a variety of *S. consors*.

Trielis hermione, n. sp.—Black, densely clothed with long white hair; that on the posterior margin of segments three to five black, on sixth and seventh segments all black hair. Clypeus white, with a median black spot reaching to the front margin ; a small spot each side above clypeus near the eyes, and sometimes a median one below the antennæ, white ; pronotum with white band above reaching back to the tegulæ, but each side is excised below before tip ; band or spot on scutellum, and the post-scutellum whitish. Abdomen segments one to four with a white subapical band, narrowed in middle, the first sometimes interrupted ; on fifth segment a median transverse spot, and a dot on each side ; venter usually with narrow apical bands on third to fifth segments, and spots on the second, sometimes some or all bands absent. Femora black except white apical spots, hind tibiæ mostly blackish, and front tibiæ with black behind; rest of legs pale yellowish; all densely white haired. Wings nearly hyaline, stigma and venation brownish yellow ; third submarginal cell ends at about middle of lower edge of marginal cell. It is higher than broad and nearly twice as broad above as below.

Length 12 to 14 mm.

From Southern Pines, N. Car., June (Manee).

Dielis fulvopilosa, n. sp.—Black, densely clothed with long fulvous hair ; the ocellar region, posterior slope of metanotum, most of pleura, and inner posterior side of femora mostly free of hairs. Abdomen with fulvous hair at tip of segments, very dense near tip ; rest of surface with pale yellow hair. Venter with fulvous hair at tips of basal segments, the apical third almost covered with fulvous hair. Tarsi yellow; tibial spurs yellow; basal abdominal segment with two transverse yellow spots behind, not quite touching. Second and third segments each with two large yellow spots nearly reaching base, and more widely separated in front than behind, and each rather notched on inner side. Pygidial area covered with short scale-like tanny hairs. Head and thorax coarsely punctate, but the lower part of metanotal slope smooth; abdomen with scattered finer punctures; venter more coarsely punctate, second ventral with a transverse smooth area. Posterior side of hind femora smooth, except punctate near base and a streak of punctures reaching toward the tip; tibial spurs long, slightly spatulate at tip; antennæ short, curved and heavy, basal joints punctate. Wings slightly smoky, the costal area fulvous and with fulvous hair; beyond stigma is a dark cloud.

Length 16 mm.

From Palmerlee, Arizona. Similar to *Elis limosa* Sauss., but the abdomen marked differently, and that species has sharp pointed spurs and more white hair on femora and venter.

TIPHIIDÆ

Scoliphia, n. gen.—With the venation as in *Epomidiopteron* and *Paratiphia*; with marginal cells in female open, in male closed, first cubital cross-vein not reaching across; the first recurrent vein curved back above; stigma very small in both sexes. Mesonotum with a sulcus each side; tegulæ extremely long, twice as long as broad. Basal abdominal segment with a transverse carina, truncate in front; second ventral produced prominently in front. The large tegulæ and second ventral segment like *Epomidiopteron*; but the carina on basal segments like *Paratiphia*. In general appearance it is like *Epomidiopteron*.

Scoliphia spilota, n. sp.— ♀ black; coarsely punctate; a white spot on each side of clypeus, on each humerus of pronotum, on mesopleura, a median spot on scutellum and postscutellum; spot on each side of basal three segments of abdomen, that on the second largest; sixth segment white across base; wings black, violaceous. Clypeus elevated, rounded below, a faint transverse furrow above the ocelli, head sparsely gray-haired, pronotum with anterior carina black-haired above, mesonotum smooth in middle, likewise on scutellum and postscutellum; latter with a pit each side. Metanotum with large basal area coarsely confluently punctate and rugose; sides smooth except behind, posterior surface margined by a carina which is emarginate above, the enclosed area punctate only on the sides. Segments of abdomen with smooth posterior margins, first segment more coarsely punctate, apical segments with transverse median row of black hairs, black hairs on sides and on venter; second segment nearly as long as broad; ventral segments with apical margins also impunctate; legs with white hairs, tibial spurs two-thirds of metatarsus. Wings with second discoidal more than twice as long as broad, third cubital much longer than broad, receiving the second recurrent vein at the middle, second cubital longer than the first. Male similar to ♀; more slender and scutellar spots often absent; clypeus wholly white; basal area of metanotum with two submedian carinæ, the whole metanotum more hairy than in female. In the female the last dorsal segment is roundedly produced at tip, basal half before the row of bristles coarsely punctate, beyond extremely finely punctate. In the male the last dorsal segment is coarsely punctate, black-haired, a median carina on apical half; the curved apical spine lying in a groove of the tip and barely visible.

Length ♀ 20 mm., ♂ 16 mm.

From Palmerlee, Arizona (Biederman).

In markings this is close to *Epomidiopteron juli* of South America, but differs in markings from *E. elegantulum* as well as in sculpture and structure.

THYNNIDÆ

Glyptometopa eureka, n. sp.—Reddish yellow throughout; smooth, but with distinct, rather large, sparse punctures, and sparsely clothed with long white hair. Antennæ 12-jointed, no longer than width of head, first segment longest, punctate in front; practically no clypeus, the antennæ close down to the margin; mandibles long, broadest near tip where they are angulate on inner side and thence concave with a minute tooth to the acute tip. Head flat above, nearly twice as wide as long, a curved punctate groove on each temple, not fringed with hair; many punctures between eyes, a few on vertex behind. Pronotum three times as broad as long, but little punctured; the mesonotum shorter than pronotum, the pleura strongly produced in a vertical ridge fringed with erect white hairs. Metathorax about as long as the prothorax and mesothorax together, broadest behind where the sides are rounded, sparsely coarsely punctate all over, the posterior surface sloping. Abdomen as long as head and thorax together, slender, second and third segments about of equal width, each segment with a bowed finely punctate transverse line before tip; the first segment more densely punctate especially on the sides than the others and with a dark spot on the side at tip, on lower sides with three straight carinæ from base toward this dark spot, the upper carina well separated from the others, which are close together and parallel; last dorsal segment elongate, triangular, convex, with a median ridge on apical part and a few punctures each side near tip. A distinct bilobed process between the middle coxæ; legs short, middle and hind tibiæ spinulose on the outer sides, the spurs long, tarsi slender.

Length 12 mm.

From Palmerlee, Arizona, Sept. (Biederman).

Differs from type in larger size, in more punctate body, much broader head (compared with length), etc. This is the third Thynnid described from north of Mexico.

SAPYGIDÆ

Eusapyga carolina, n. sp.—♂ black; a transverse curved band on margin of clypeus, a large transverse spot above antennæ with a median upper projection, the blister each side by upper part of eye, and a stripe each side on the scape, yellow. Thorax with an interrupted band on front of pronotum, a band behind, two spots on the scutellum, two on the post-scutellum, a large spot on mesopleura and a rather smaller one on meta-pleura, yellow. Abdomen with a large spot each side on basal segment, two dots near middle of hind margin, second segment with broad, straight band over rather more than apical half, third with dot in middle near hind margin, fourth and fifth with narrow undulate bands, yellow. Venter black, with lateral yellow spots on second to fifth segments, growing smaller behind. Legs yellow, femora and coxæ mostly black, middle coxæ mostly yellow. Wings fumose, very dark in anterior part, stigma yellow. Body clothed with short, dense, mostly white hairs, head and thorax densely coarsely punctate. Clypeus with a minute tooth each side near middle on lower margin; the posterior pair of bosses on vertex are larger than the others; abdomen beyond first and second segment mostly smooth, the punctures on hind border of first segment rather large, this segment concave in front above.

Length 10 mm.

Southern Pines, N. Car., Aug. (Manee).

NEW GENERA OF NORTH AMERICAN LITHOBIIDÆ[1]

BY RALPH V. CHAMBERLIN, UNIVERSITY OF PENNSYLVANIA, PHILADELPHIA.

(Continued from p. 178)

Genus *Gosibius*, gen. nov.

Anterior margin of the prosternum wide. Teeth, 2 + 2; small, the line of apices clearly recurved; ectal spines springing from a rounded nodule, long and slender, being much more slender than the teeth, but stouter than the bristles.

Antennæ long, the articles all long and slenderly cylindric.

Coxal pores circular, uniseriate.

In the known species the last four pairs of coxæ are armed laterally. The penult legs bear two or three claws and the anal one or two. In both males and females the anal legs are armed dorsally with 1,0,3,2,2 spines, as are also the penult. Characteristic also of this genus is the presence on the tibiæ of most of the legs (e.g., the 5th to 12th pairs), of three ventral spines, as against but two in the related genera. The posterior legs are short and slender, with the prefemur and femur, and also sometimes the tibia, of anal pair more or less longitudinally furrowed above: in the male the penult legs always unmodified, and the anal legs also unmodified or with the femur alone modified, being then widened and complanate and more distinctly furrowed dorsally.

Gonopods of female nearly as in related genera, the claw being large and entire and the first article conspicuously excavated on mesal side of base, which side is also strongly chitinized. Basal spines, 2 + 2 or 3 + 3.

The species of this genus are less strongly narrowed cephalad than those of *Arenobius*, and the first plate is nearly as wide as or wider than the third.

Type.—*G. paucidens* (Wood).

Distribution.—Southern California, etc.

In addition to the type only one species, *G. monicus*, Chamb., is at present known, with certainty, though mutilated specimens from Los Angeles seem to represent a second.

1. Owing to a mistake of our own, the several new species of *Arenobius*, to have been described below, have been published elsewhere by Mr. Chamberlin. We regret the awkward division of the article which our error has necessit.ted.—[EDITOR.

A NEW PALÆARCTIC *GERANOMYIA* (TIPULIDÆ, DIPTERA).

BY C. P. ALEXANDER AND M. D. LEONARD, ITHACA, N. Y.[1]

The following species is described from material sent to the authors by Prof. Dr. M. Bezzi. It was received by him from a correspondent in Ile Djerba, off the northern coast of Africa. Our thanks are due to Dr. Bezzi, and we take pleasure in dedicating this interesting species to him.

Geranomyia bezzii, sp. n.

Male (alcoholic). Colour light yellow; proboscis with a brown subapical band; thoracic dorsum with four longitudinal brown stripes; pleuræ with a few dark brown spots. Wings hyaline with four rather indistinct spots.

Length, 7.2–7.5 mm.; wing, 6.3–6.4 mm.; head, total, 2.2 mm.; thorax, 1.7 mm; hind femora, 5 mm.

Head: Proboscis light yellow, with a conspicuous, brown, subapical band; palpi brown; antennæ yellow. Front, vertex and occiput light brownish yellow.

Thorax: Ground colour yellow; dorsum with two median and two lateral brown stripes. Mesothoracic præscutum pale yellow, with two brown longitudinal bands, a little wider than the dividing median line, these bands darker on the outer margin; they begin near the cephalic margin of the sclerite and continue caudad. fading out at about two-thirds the length of the sclerite. Just cephalad of the end of the median stripes begin the dark brown lateral stripes; on the præscutum they are arcuated, continuing back onto the scutum, where they are also broader; end of the scutellum darker brown on either side of the pale median line; caudal edge of the postnotum dark brown. Pleuræ concolorous with the dorsum, lateral margin of the mesothoracic præscutum dark brownish black, most intense on the margin of the sclerite; an intense brown semilunar mark on the pronotal pleuræ, midway between the anterior coxæ and the dark mark on the edge of the mesothoracic præscutum; an irregular, interrogation-like mark below the wing root; ventral portion of the mesothoracic episternum and sternum brown. Halteres light yellow, knob clear yellow. Legs light brownish yellow, tips of the segments not appreciably darker.

Abdomen yellow, with a brown mark on the ventral edge of the tergites, the first elongate, expanded over two segments; behind this there

1. Contribution from the Entomological Laboratory of Cornell University.

are five marks on successive segments. On the dorsal edge of the sternites are six corresponding marks, rather less distinct than the tergal marks. Hypopygium light yellow, fleshy apical appendages almost white.

Wings: Hyaline or nearly so; veins light brownish yellow; very pale brown clouds around the base of Rs, around cross-vein *r*, and in the middle of cell Sc. Venation: Sc ending about opposite the origin of Rs; Sc_2 slightly retracted proximad of the origin of Rs, about one-half the length

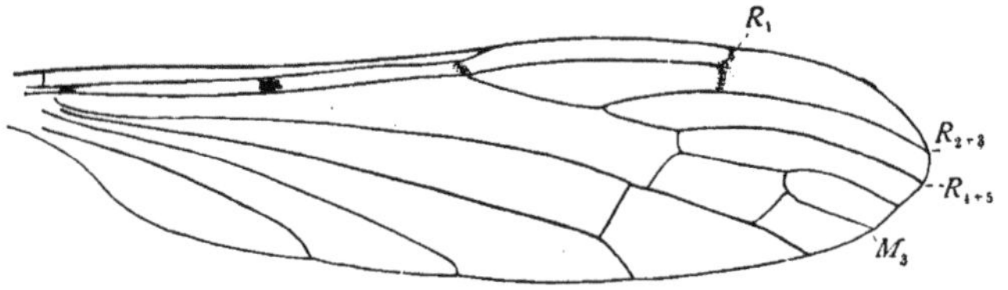

FIG 6.—Wing of *Geranomyia bezzii*, sp. n.

of Sc_1. Rs moderately long, about twice the length of the basal deflection of R_{4+5}; cross-vein *r* at the tip of R_1, which is abruptly upcurved beyond it, very indistinct; R_{2+3} and R_{4+5} arcuated and parallel; cross-vein *r-m* short, pale; basal deflection of Cu_1 about equal to Cu_2; Cu_1 fusing M distad of the fork M.

Holotype, ♂, Ile Djerba, Tunis. (Museo Torino.)

Paratype, 3 ♂s, Ile Djerba, Tunis. (One in Museo Torino, two in Cornell University.)

Remarks: Some venational variation occurs in the paratypes. In some Sc_2 is exactly opposite the origin of Rs, and Sc_2 is only a little shorter than Sc_1; basal deflection of Cu_1 at the fork of M, or even slightly proximad of it. (See figure.) The relative length of Cu_2 and the basal deflection of Cu_1 varies somewhat, Cu_2, however, being generally a little the shorter.

Key to the Palæarctic Geranomyiæ:

1. Wings unspotted........................(No Palæarctic species).
   Wings spotted..2.
2. Thoracic dorsum without distinct stripes........................3.
   Thoracic dorsum with distinct stripes..........................4.
3. Antennæ and palpi yellowish brown; femora and tibiæ black at tip....................................*atlantica* Woll.[1]

1. Wollaston—Ann. Mag. Nat. Hist., ser. 3, I, p. 115 (as *Limnobia*), (1858).

Antennæ and palpi black; femora and tibiæ not black at
tip.................................... *canariensis* Bergr.[2]
4. Thoracic dorsum with two dark stripes.......... *bivittata* Becker.[3]
Thoracic dorsum with more than two dark stripes..............5.
5. Costal margin of wings with six large equidistant brown
spots.......................................*caloptera* Mik.[4]
Costal margin of wings with four spots.......................6.
6. General colour yellowish brown; proboscis unicolorous..*unicolor* Hal.[5]
General colour light yellow; proboscis light yellow, with a dark subapical band.................................*bezzii*, sp. n.

This key is based entirely on the published description of the species hitherto proposed. Some of these descriptions are very insufficient, for example, those of *atlantica* Wollaston and *unicolor* Haliday. One, *maculipennis* Curtis,[6] is so brief and unsatisfactory that we have not attempted to include it in the above key. The complete description reads as follows: "Rather larger than *G. unicolor*, and is of a lurid ochre, the wings tinged with the same colour. It may be merely a variety, differing principally in colour, arising possibly from age.

Whether or not *Aporosa* Macq. (1838), in which Enderlein has placed *maculipennis* Macq. (=*canariensis* Berg.) and *vicina* Macq., is distinct from *Geranomyia* is uncertain. The character of a radial cross-vein should be sufficient to distinguish this group of species from the typical *Geranomyia* group. Enderlein[7] states that *vicina* has but one marginal cell; however, Macquart (Diptères Exotiques, V, 1, pt. 1, p. 70), states clearly that there are two marginal cells. It is doubtful whether *vicina* is a *Geranomyia;* the statement of "rostre un peu alongé" being quite insufficient to give it a position in the genus *Geranomyia.*

Acknowledgements are made to Mr. Frederick Knab for his kindness in supplying a reference not otherwise obtainable.

The drawing of the wing was made by means of the projection microscope in the Entomological Laboratory.

2. Macquart—Diptères Exotiques, Vol. I, pt. I, p. 63 (as *Aporosa maculipennis*) (1838); changed to *canariensis* by Bergroth, Wiener Entomol. Zeitung, Vol. 8, p. 118 (1889).

3. Becker—Berlin Mitt. Zoöl. Mus., Vol. 4, p. 187 (1908).

4. Mik—Verhandlungen Zoöl.-Bot. Gesellschaft Wien., Vol. 14, p. 791 (1864), as *maculipennis*, n. sp.; changed to *caloptera* Mik, Verh. Zool.-Bot. Gesellschaft Wien., Vol. 17, p. 423 (1867).

5. Haliday—Entomological Magazine, Vol. 1, p. 155 (1833); Curtis, Brit. Entomol., Vol. 12, p. 573 (excellen coloured figure); Macquart, Suit. à Buffon, Vol. 2, p. 652 (1835).

6. Curtis—Brit. Entomol., Vol. 12, p. 573 (1835).

7. Enderlein, G.—Zoologische Jahrbücher, Vol. 32. part 1, p. 79, 80 (1912).

DRAGON FLIES COLLECTED AT POINT PELEE AND PELEE ISLAND, ONTARIO, IN THE SUMMERS OF 1910 AND 1911.

BY F. M. ROOT, OBERLIN, OHIO.

Lestes unguiculatus Hagen.—Point Pelee. One specimen.

Lestes forcipatus Rambur.—Pelee Island. Very common.

Lestes vigilax Hagen.—Point Pelee. Common around ponds.

Enallagma carunculatum Morse —Pt. Pelee. Fairly common near ponds.

Enallagma pollutum Hagen.—Pt. Pelee. Fairly common near ponds.

Ischnura verticalis Say.—Pt. Pelee and Pelee Island. Common.

Gomphus vastus Walsh.—Pelee Island. Five specimens taken near woods.

Anax junius Drury.—Pt. Pelee and Pelee Island. Common. (See note at end.)

Æschna clepsydra Say.—Pelee Island. One specimen taken.

Æschna constricta Say.—Pt. Pelee and Pelee Island. Fairly common. (See note.)

Epicordulia princeps Hagen.—Pt. Pelee. Fairly common about large ponds.

Pantala hymenæa Say.—Pelee Island. One taken, others seen. (See note.)

Tramea carolina Linné.—Pt. Pelee. Rare. (See note.)

Tramea lacerata Hagen.—Pt. Pelee and Pelee Island. Common. (See note.)

Celithemis eponina Drury.—Pt. Pelee. Common near ponds.

Celithemis elisa Hagen.—Pt. Pelee. Rare.

Leucorrhinia intacta Hagen.—Pelee Island. Common at swamps.

Sympetrum rubicundulum Say.—Pelee Island and Pt. Pelee. Fairly common.

Sympetrum vicinum Hagen.—Pelee Island and Pt. Pelee. Very common. (See note.)

Sympetrum corruptum Hagen.—Pt. Pelee. Rare. (See note.)

Erythemis simplicollis Say.—Pt. Pelee and Pelee Island. Common near ponds. (See note.)

Pachydiplax longipennis Burm.—Pt. Pelee and Pelee Island. Common. (See note.)

Libellula basalis Say.—Pt. Pelee and Pelee Island. Fairly common near ponds.

Libellula incesta Hagen.—Pt. Pelee. Common at the ponds.

Libellula pulchella Drury.—Pt. Pelee and Pelee Island. Common. (See note.)

Plathemis lydia Drury.—Pt. Pelee. Rare, but seen regularly.

NOTE.—On Pelee Island in 1910, about the middle of August, or a little later, there were three days when dragon-flies of species hitherto not seen in large numbers swarmed around the end of the Point. Presumably they were migrating. The principal species concerned were *Anax junius*, *Æschna constricta*, *Tramea lacerata* and *Pantala hymenæa*.

On Point Pelee in 1911, about the middle of August, the deer-flies became suddenly much more numerous, and on August 17 great numbers of dragon-flies appeared (perhaps following the deer-flies). The great bulk of these were teneral *Anax junius* (with reddish-purple abdomens), and towards evening they clustered so thickly on the cedars near the end of the Point that eight or ten could be captured any time by a single sweep of the net. With them were large numbers of *Sympetrum vicinum* (which preferred the low junipers to the cedars) and smaller numbers of *Tramea lacerata* and *Æschna constricta*. There were also a few each of *Tramea carolina*, *Sympetrum corruptum*, *Erythemis simplicollis*, *Pachydiplax longipennis* and *Libellula pulchella* with the flocks. They remained until August 20.

THREE DAYS IN THE PINES OF YAPHANK. RECORDS OF CAPTURES OF HEMIPTERA HETEROPTERA.

BY J. R. DE LA TORRE BUENO, WHITE PLAINS, N. Y.

The name Yaphank (with the stress on the "*hank*") has a truly barbarous cadence. It is an interesting relic, one of the few remaining vestiges of the great Shinnecock tribe, once Lords of Long Island. The place that bears this cacophonous name is, indeed, one of the very few regions near New York and its teeming millions not utterly spoiled to the lover of nature by the "improvements" of modern progress as exemplified by its advance agents, the real estate dealers. Here and there in this land of sand and pines and scrub-oak, are still to be found ancient trees that stood when Hendrick Hudson first sailed into the Narrows. The present holders of the land are descendants of original Royal Patentees, and they own great stretches of wilderness. So it comes about that insect life is abundant in numbers and rich in species, not the least among them being the

Hemiptera. The chief collecting grounds are about two miles from the railroad station and the vegetation consists mainly of pine, scrub oak and along the roads, maple trees, and the weeds and shrubs common to this latitude.

Toward the end of September, 1911, I had the good fortune to spend three days there with Mr. G. B. Engelhardt, who was guide, philosopher and friend. We arrived about 11 a.m. the morning of the 23rd, and indulged in a little collecting before the noon-day meal, after which we went out and did some sweeping and beating with good results, one being the capture of a new *Corizus*, described elsewhere. In the evening, between 8 and 10, Engelhardt went sugaring, carrying a trap lantern, while I swept. The following day was rainy in the morning, but as soon as it cleared up sufficiently we took our way to the Carman River, a clear, shallow stream flowing over a bed of sand where a little dredging was done, which yielded among other things one specimen of *Belostoma lutarium* Stal. (taken by Engelhardt), which is the farthest Northern authentic record for the species known to me. In the afternoon sweeping and beating made up the programme, in the brush and trees about a cranberry bog and in the grasses growing in it. Night *sweeping* gave good results, no less than 16 species being taken in clearings in the woods, while *Ozophora picturata* Say flew to light, its great agility making it hard to catch. The morning of the 25th dawned grey and muggy, the day finally clearing in the late afternoon. Sunshine or rain being one to the waterbugs, Engelhardt and I betook ourselves to the lake, where wingless *Rheumatobates rileyi* Bergr. was far from uncommon, but only one *Trepobates pictus* H. S. was seen, although I was out in a boat looking for it. Here, in the floating duck-weed and algæ I secured what seems to be a new species of *Microvelia*, in goodly numbers. On the way to and from the lake sweeping and beating were done with good results, and this part of the programme repeated in the afternoon yielded among other things, no less than 11 specimens of the new *Corizus*, 2 being fully winged, the other brachypterous. In the evening our stay was wound up by Engelhardt visiting his sugared trees, while I watched the trap light and caught two *Ozophora*. Altogether, in the three days, in spite of unfavourable weather, we got between us some 300 specimens and 82 species of Hemiptera. The identified species are listed hereafter, with appropriate comment. Many of these are recorded from Long Island for the first time, and some of the other records are unusual or remarkable.

Apateticus (= *Podisus*) *cynicus* Say.—Was taken at sugar in the evening—a most unusual manner.

Apateticus maculiventris Say.

Apateticus serieventris Uhler.

Apateticus modestus Dallas.

Apateticus placidus Uhler.

Halcostethus (= *Peribalus* M. & R.) *limbolarius* Stal.

Trichopepla semivittata Say.

Euschistus euschistoides Voll. (=*fissilis* Uhler.)

Euschistus variolarius P. B.

Thyanta custator Fabr.

Nezara hilaris Say.—At sugar, taken by Mr. Engelhardt.

Dendrocoris humeralis Uhler.

Brochymena arborea Say.

Tetyra bipunctata Fabr.—Was taken at light.

Aradus shermani Heid.—This species was taken under bark of dead pine tree, a few adults and a number of nymphs in various stages. Apparently first notice other than the type loca.ity in Pennsylvania.

Aradus cinnamomeus Panz.

Mezira granulata Say.

Corynocoris typhaeus Fabricius.—Swept from weeds in a dry field. This appears to be the preferred habitat of this species.

Alydus eurinus Say.

Alydus pilosulus H. S.

Megalotomus 5-spinosus Say.—Common on false indigo (*Baptisia tinctoria*). Some specimens were also swept at night.

Harmostes reflexulus Say.

Corizus lateralis Say.

Corizus hirtus Bueno.—In a sandy spot, in short grasses, by sweeping.

Jalysus spinosus Say.

Lygaeus kalmii Fabr.

Nysius providus Uhler.—Swept and taken at light.

Nysius thymi Wolff.

Ischnorhynchus geminatus Say.

Geocoris piceus Say.

Phlegyas abbreviata Uhler.—One long-winged specimen was swept.

Crophius disconotus Say.—Beaten from oak.

Ligyrocòris diffusus Uhler.

Pamera basalis Dallas.

Antillocoris (= *Cligenes*) *pilosulus* Uhler.—Taken by sweeping grasses in dry cranberry bog.

Pseudocnemodus bruneri Barber.—Two long-winged specimens were swept, one by daylight, the other at night. This is a pretty common and widespread species.

Carpilis ferruginea Stal.—Two specimens taken by sweeping in a marsh. This species has apparently not been recognized since Stal described it in 1874, in En. IV, pp. 144, 153. This is a notable addition to our fauna, and serves to show how little is known of the Hemiptera of any given region.

Ozophora picturata Uhler.—A number of specimens were taken at light and one was beaten from oak. This is a most agile species.

Drymus unus Say.

Corythuca juglandis Fitch.—Taken by beating.

Corythuca crataegi Morrill.—Taken by beating.

Corythuca pergandei Heidemann.

Physatocheila plexa Say.—Beaten from oak.

Reduviolus sordidus Rent.

Reduviolus ferus Linné.

Mesovelia bisignata Uhler.

Rhagovelia obesa Uhler.

Microvelia americana Uhler.—There are also 3 seemingly undescribed *Microveliae*.

Gerris marginatus Say.

Gerris remigis Say.

Trepobates pictus H. S.

Rheumatobates rileyi Bergroth.—Abundant on the lake.

Næogeus (= *Hebrus* Curtis) *concinnus* Uhl.—Quite abundant on the damp edges of a cranberry bog.

Pygolampis, sp.—Nymph.

Pselliopus (= *Milyas*) *cinctus* Fab.—Beaten and swept. Found mating.

Zelus luridus Stal.—Nymphs.

Fitchia aptera Stal.—One large fully-winged female was swept in a little meadow.

Siena diadema Fab.

Sinea spinipes H. S.—Beaten from trees. Crandell states that this species ranges over the southern and central parts of the United States. It is now for the first time recorded from the northeastern part. It is a most interesting addition to the fauna of N. Y.

Ranatra americana Mont. (=*4 denta* Uhl., Bno. et auctt).

Triphleps insidiosus Say.

Piezostethus sp.—Beaten from pine, in company with *Aradus cinnamomeus*.

Cardiastethus sp.—One specimen only of this small species was beaten from pine.

Gelastocoris sp.—Nymphs in about 2nd instar were common, hiding in crevices in the debris at the shores of the cranberry bog. No adults were noted, whence it may be inferred they hibernate as nymphs.

Pelocoris femaratus P. B.

Belostoma flumineum Say.

Belostoma lutarium Stal.—The only specimen authentically northern of which I have any knowledge. This is a notable addition to the fauna of New York.

Corixa.—Two unidentified species.

Notonecta undulata Say.

Notonecta variabilis Filber.

Buenoa elegans Filb.

Plea striola Fiib.—As usual, in water weeds.

There are in addition six species of Capsids not identified.

LEPIDOPTERA FROM YUKON TERRITORY.

BY ALBERT F. WINN, WESTMOUNT, QUE

Through the kindness of Mr. Lachlan Gibb, I have had the opportunity of studying a collection of butterflies and moths taken during the summer of 1910 at, or near, Dawson City, Y. T. A smaller collection was made in 1911 but butterflies must have been scarce last year, the collector having been out on 17 days and capturing but 54 specimens in all. As but little has been published in this magazine on the insect fauna of this northern part of our country, the list which follows, covering both seasons' captures may be of interest to entomologists studying distribution and particularly to those engaged on the preparation of the list of Canadian Insects.

Papilio turnus Linn.—June 4th to 22nd, a number of specimens all small in size, average expanse 3¼ inches, but markings exactly the same as in those found about Montreal.

Papilio machaon var. *aliaska* Scud.—June 1912, one specimen.

Pontia napi var. *hulda* Edw.—May 15 to July 4, apparently the commonest butterfly about Dawson. Some of the females are darkly suffused and approach var. heyoniæ.

Pontia nelsoni Edw.—July 17th, one specimen only.

Euchlöe ausonides Bdv.—June 15th to July 2nd, several specimens.

Colias chippewa Edw.—June 20th to July 17th.

Colias skinneri Barnes.—July 14th to August 1st, three ♂, one ♀.

Argynnis electa Edw.—July 20th, one ♂ in battered condition.

Argynnis frigga var. *saga* Kaden.—June 15th to July 7th, several specimens. Apparently a common species. One specimen has, in the black portion at base of pormiaries, a circular fulvous spot with two black dots.

Phyciodes pratensis Behr.—July 4th to 15th, three specimens.

Vanessa milberti Godt.—Several specimens, May 20th to June 18th, hibernated, fresh specimens July 17th to August 8th; exactly similar to those found in Eastern Canada.

Vanessa antiopa Linn.—Many hibernated specimens May 15th to June 26th; those taken after June 15th are in fragmentary condition. Fresh specimens July 27th to August 8th. Most of the specimens have the dark portion of the wings beneath marked with yellowish transverse striæ, in some forming a central band nearly as wide and distinct as outer border.

Coenonympha kodiak var. *yukonensis* Holland.—Three specimens, June 25th to July 2nd.

Erebia discoidalis Kirby.—Three specimens, June 15th to 20th.

Erebia disa var. *mancinus* Doubl.-Hew.—One specimen, rather broken, June 27th.

Erebia epipsodea Butl.—One specimen, June 22nd.

Œneis jutta var. *alaskensis* Holland.—A number of specimens June 18th to July 2nd.

Chrysophanus helloides Bdv.—One female Aug. 5th, the latest date on any of the species of butterfly received. The specimen has a yellowish washed-out appearance.

Lycæna sæpiolus Bdv.—June 20th to 29th, several.

Lycæna lygdamus Doubl.—Several specimens more or less worn; 19th to 26th June.

Lycæna rustica Edw.—One ♂, July 22nd.

Lycæna lotis Lintn.—Several specimens, July 7th to 16th.

Lycæna pseudargiolus spring from *marginata* Edw.—One ♀ specimen much broken, June 19th.

Lycæna pseudargiolus, spring from *lucia*, Kirby.—The commonest form in the North, June 10th to 26th.

Carterocephalus palæmon Dallas—One specimen, June 25th.

Thanaos persius Scud.—One specimen, June 23rd.

Deilephila galii Rott.—One specimen, July.

Rhynchagrotis rufipectus Marr.—Two specimens, July 22nd.

Noctua baja Linn.—Two specimens, July 12th to 22nd.

Noctua clandestina Harr.—One specimen, July 8th.

Leucania yuconensis Hamps.—Eight specimens mostly rubbed, July 1st to 15th.

Leucania commoides Guen.—One specimen June 28th.

Falcaria bilineata Pack.—One specimen June 18th.

Euchœca albovittata Hübn.—One specimen June 23rd.

Eustroma testata Linn.—One, Aug. 7th.

Eustroma propulsata Walk.—Several July 8th to 20th.

Eustroma nubilata Pack.—Four specimens July 8th to 14th.

Eustroma triangulata Pack.—Aug 2nd to 11th, several.

Keolexia scylina Hulst.—Apparently the commonest moth, many specimens July 26th to Aug. 8th.

Rheumaptera hastata Linn.—June 20th to July 2nd. As usual with this species in the west, most of the specimens are much marked with white.

Rheumaptera sociata Bork.—Four specimens, June 28th to July 2nd. The white band on fore-wings is much more even and contrasting than in Eastern specimens.

Rheumaptera luctuata D. & S.—Several June 15th to 29th, very variable.

Mesoleuca silaceata Hübn.—Six specimens, July 16th to 28th.

Larentia multiferata Walk.—Several June 15th to 28th, all badly rubbed.

Hydriomena furcata var. 5-fasciata Pack.—Ten specimens, July 15th to 22nd.

Cœnocalpe magnoliata Guen.—One specimen, June 20th.

Gypsochroa designata Haw.—Numerous specimens, June 15th to 28th.

Xanthorhoe convallaria Guen.—Twenty specimens, July 2nd to 20th. A common species.

Xanthorhöe ferrugata Clerck.—Six specimens, July 19th to 26th.

Cosymbia lumenaria Hübn.—One specimen, June 20th.

Leptomeris frigidaria Moeschl.—Six specimens, June 29th to July 15th.

Eufidonia notataria Walk.—One specimen, June 23rd.

Sciagraphia granitata Guen.—One specimen, July 20th.

Sciagraphia denticulata Grote.—One specimen, June 19th.

Diastictis bicolorata Fabr.—Three specimens, July 15th to 22nd.

Diastictis inceptaria Walk.—Two specimens, July 12th; seem to agree with argillacearia Pack, which is considered a synonym of inceptaria Walk.

Diastictis subcessaria Walk.—One specimen, July 14th.

Gladela julia Hulst.—Three specimens, July 14th to 22nd.

Sicya macularia Harr.—One specimen, July 26th.

Metrocampa perlata Guen.—Eight specimens, June 30th to July 27th.

NEW NOCTUID SPECIES.

BY WM. BARNES, M.D., AND J. MCDUNNOUGH, PH.D., DECATUR, ILL.

Fotella olivia, sp. nov.

♀.—Head and thorax pale ochreous, more or less heavily sprinkled with dark gray, abdomen ochreous, untufted; primaries dark gray-brown, considerably sprinkled with pale ochreous scales, giving a general rough and mottled appearance; the ochreous scaling is often such that a basal dash of the ground colour is apparent, extending to below orbicular; t. a. and t. p. lines represented by a small dark patch on costa, the latter line at times being visible across wing as a row of dark dots bent inward in submedian fold; orbicular and reniform two small and rather diffuse whitish patches, not clearly defined, former with or without central dark dot; terminal area narrow, ochreous, defining, in contradistinction to dark subterminal space, the s. t. line, which is very irregular, angled outwardly below costa to nearly terminal border and slightly bent inward opposite cell and in submedian fold; terminal space broadest at costa; terminal row of dark dots; fringes dark, slightly dotted with ochreous. Secondaries dull white, slightly smoky outwardly. Beneath shiny white, with faint sprinkling of brown scales. Expanse 25 mm.

Habitat.—La Puerta Valley, San Diego Co., Calif. (G. H. Field, July.) Five ♀s. Type coll. Barnes. Cotype with Mr. Field.

With the exception of the depth of ground colour of primaries, all the specimens are constant in markings. The species possesses a small frontal navel-like tubeicle, and agrees generically with *fragosa* Grt., which we recently redescribed as *Hadenella cervoides*, differing from this species in the more mottled appearance, the pale terminal area and better defined orbicular. We place it in *Fotella*, as it seems to bear considerable resemblance to *notalis* Grt., the type of the genus; it is, however, to judge by the description, much smaller and has the orbicular present.

Phyllophila aleptivoides, sp. nov.

♀.—Thorax smoothly covered with flat gray and white scales; abdomen ochreous, primaries dark gray, mingled with white and ochreous; basal portion of costa broadly white; lower basal portion of wing between submedian fold and inner margin extending to t. a. line, ochreous; a broad wedge-shaped patch of dark gray extends obliquely inward from costa to submedian fold, broadest on costa and bordered outwardly by t. a. line; this latter very obscure, geminate; median area dark gray, sprinkled with white along costa and with ordinary spots prominent, white, ringed with black; orbicular round, with dark central dot; reniform constricted centrally, shaped like figure 8, with central dark dots in upper and lower halves; claviform oval, pure white, about same size as orbicular; between orbicular and reniform an ochreous patch; t. p. line geminate, indistinct at costa, inwardly oblique and semiparallel to outer margin; subterminal space ochreous, with the exception of a dark apical dash extending inward to t. p. line; terminal space narrow, dark gray, with terminal row of dots edged inwardly with white; fringes dusky. Secondaries smoky with whitish fringes.

♂.—Our single specimen differs considerably from the ♀, in that the whole median area is suffused with white, obscuring the three spots and leaving merely the two dark central dots of reniform visible; the dark costal patch near base of wing becomes on this account much more prominent; the secondaries are paler than in the ♀.

Beneath in both sexes dull white, immaculate. Expanse 19 mm.

Habitat.—La Puerta Valley, San Diego Co., Cal. (G. H. Field, July). One ♂, one ♀.

The species is distinctly quadrified in venation and would fall, according to Hampson (Cat. Lep. Het., Vol. X), close to the genus *Phyllophia* Gn., which as yet has no North American representatives. The front of our species has a strong rounded frontal protuberance, the surface of which is roughened by numerous minute conical tubercles; the antennæ in both sexes are almost simple, and there are apparently no tufts of hair on any of the abdominal segments; in this latter respect it differs from *Phyllophia*, which possesses a dorsal crest at base, but we hesitate to separate the species on such a minor point of difference.

The species, especially the ♀, has considerable superficial resemblance to *Aleptina inca* Dyar. The ♂ type is with Mr. G. H. Field, the ♀ type in coll. Barnes.

Eustrotia bifasciata, sp. nov.

Head, thorax and abdomen white; primaries white, with sub-basal and subterminal areas dark brown, giving the impression of two irregular bands crossing the wing; base of wing narrowly white; broad dark sub-basal band bordered outwardly by geminate t. a. line, which is filled with ochreous and somewhat irregular in outline; about centre of costa is a small dark patch, and a black dot at end of cell represents reniform; t. p. line geminate, partially filled with ochreous, irregular, bent strongly inward in submedian fold; s. t. line defined by difference between dark subterminal and pale terminal spaces, very irregular, bent inward and closely approaching t. p. line opposite cell, almost touching outer margin at vein M_2 and again incurved above anal angle; slight dark terminal line; fringes white, with faint dark checkerings. Secondaries, basal half white; outer portion smoky brown with pale fringes. Beneath white, with a broad dark border to both wings. Expanse 20.5 mm.

Habitat.—La Puerta Valley, San Diegc Co., Cal. (G. H. Fields, July.) Three ♂s.

Type coll. Barnes. Cotype with Mr. Field.

All three specimens are rather worn, and we have been unable to determine whether thoracic and abdominal tufts are present or not. As the venation is markedly quadrifid and the general appearance slight, we place the species for the present in *Eustrotia*. We know of no other species to which it bears much resemblance.

CORRECTION.

In our recent paper on North American Lepidoptera the description of *Diastictis pallipennata* (Vol. XLIV, p. 126), was drawn up from a ♀, and not a ♂, as stated.

J. M. McDunnough.

NOTES ON THE APHID GENUS, *ERIOSOMA* LEACH.

BY H. F. WILSON, OREGON AGRICULTURAL COLLEGE.

Eriosoma lanigera versus *Schizoneura lanigera.*

At various times since the description and naming of Housmann's *Aphis lanigera* different authors have erected generic names for this species.

There seems to be no doubt of the validity of the specific name for the species of Aphis originally described by Housmann as *Aphis lanigera*, but the generic names erected for this insect have been more or less in doubt.

The author of this paper has made a thorough investigation of all the known literature, and concludes that *Eriosoma* was erected and definitely placed with this species, and that *Schizoneura* and other later names are synonymic as far as this species is concerned. Samouelle is generally supposed to have originated the genus in his compendium of useful information, but such is not the case. In 1817 Sir Oswald Mosely gave a paper before the Horticultural Society of London, entitled, "*Aphis lanigera* or American Blight." At the end of this article a note is appended by Dr. William Elford Leach, in which he mentions *Aphis lanigera* of Housmann, and concluding that a new genus should be made for this species he proposes the name *Eriosoma.* The note appended to the original paper reads as follows: "Note on the Insect, by William Elford Leach, M.D., F.R.S., etc. The animal of which so accurate an account is given in the preceding paper is the *Aphis lanigera* of Housmann; it is described by the author in Illiger's Magazine for 1802, page 440, and is referable to Latreille's third division of the genus *Aphis*, but which division I consider to constitute a peculiar genus distinct from *Aphis*, and which I have named *Eriosoma.*"

Eriosoma has its body covered by woolly matter; its abdomen has neither horns nor tubercles, and its antennæ are short. The body of *Aphis* is naked, its antennæ are long and setaceous, and the abdomen is furnished with a tubercle or horn-like process on each side.

Although this paper was read in 1817, it evidently was not published until 1818, in the latter half of that year. The entire article is printed in Volume III, Trans. Hort. Society, London, 1820, pages 54 to 61. The preface to this volume is dated January, 1820, but reads, "When the Society completed the second volume of their Transactions in March, 1818, arrangements were made to insure, if possible, the publication of portions of the succeeding volumes, at less distant periods than had hitherto been done. They flattered themselves with the hope of being able to complete a volume every second year, by the publication of one-fourth part at intervals of six months, and it is very satisfactory that, so far, that hope has been realized."

The author takes this as sufficient proof to definitely establish this generic name.

ENTOMOLOGICAL SOCIETY OF ONTARIO.

Meetings of the Entomological Society of Ontario were held during the winter months of 1911 and 1912 in the Biological Lecture Room of the Ontario Agricultural College. Before Christmas the meetings were held on alternate Thursday afternoons and after New Years joint meetings were held with the Wellington Field Naturalists Club, weekly. The meetings were well attended by the staff and students of the Ontario Agricultural College and interested citizens of Guelph. The first meeting was devoted to observations by the various members, and during the rest of the season the following papers were read in order.

Observations in Algonquin Park.—Prof. J. E. Howitt.
Foul Brood of Bees.—Mr. G. L. Jarvis.
Ants.—Mr. W. H. Wright.
The Nursery Question.—Mr. L. Caesar.
Mosquitoes.—Mr. C. A. Good.
The Economic Importance of Calosoma sycophanta.—Mr. J. Noble.
Insect Intruders in Indian Homes.—Mr. G. J. Spencer.
Birds in Relation to Insects.—Mr. E. N. Calvert.
Fall Collecting of Coleoptera.—Mr. A. W. Baker.
Insectivorous Birds.—Prof. T. D. Jarvis.

NOTE ON GEOPHAGUS.

Geophagus as a name for a genus of the Geophiloidea (Attems, 1897), is preoccupied by Geophagus in Pisces (Hæckel, 1840), and must accordingly be replaced. *Sogophagus* nom. nov., may be substituted.

R. V. CHAMBERLIN.

During July and August communications for the Editor may be addressed to the Biological Station, Go Home Bay, via Penetang, Ont.

NOTES ON THE LIFE HISTORY OF *ESTIGMENE PRIMA* SLOSSON.

BY ALBERT F. WINN, WESTMOUNT, Q.

My acquaintance with this "many-spotted ermine moth" was first made on June 12, 1897, when Mr. Dwight Brainerd and I visited the entomologically famous Gomin Swamp near Quebec city, under the guidance of Rev. Dr. Fyles, in search of *Œneis jutta* and other Lepidoptera. In one particularly moist spot my eye was attracted downwards and observed a pair of these moths in coitu. My first impression was that they must be a northern variety of *Spilosoma congrua (antigone)* with the black spottings exaggerated. The female was kept alive and laid a good supply of eggs which duly hatched, but through illness I was unable to attend to their needs. Mr. Lyman afterwards took the moths to Washington where Dr. Dyar determined them as *E. prima* Slosson. Figures of this species are given in this magazine, Vol. XXXII, pl. 4, figs. 9 & 10.

No further specimens came my way till June 4th, 1910, when at Shawbridge, Que., in the Laurentian Mts. about 40 miles north of Montreal I captured a battered specimen flying, or rather driven by the cold high wind. It was found to be a female and was therefore boxed for eggs. Three were laid almost immediately. Next day, June 5th, a batch of 45 was deposited; on the 6th, 27; on the 7th, 11; and on the 8th, 36. Total 122 eggs, all laid in daytime and arranged in irregular masses. The moth was then killed to preserve what little was left of it.

In order to have a better chance of breeding the larvæ I asked my friend Mr. Arthur Gibson of Ottawa to take half of the eggs, which he kindly consented to do. For some reason his little larvæ refused both plantain and dandelion, and of other foods offered they selected apple, but unfortunately soon died. I at once re-divided mine which were thriving on plantain and also gave about a dozen to Mr. Lyman who was just leaving on a trip to Europe. We all succeeded in rearing the caterpillars to full growth and into pupa, Mr. Lyman having considerable difficulty in obtaining a supply of plantain leaves in the beautifully kept lawns of England, but mine alone produced moths.

For various causes none of us kept a complete record of all the stages but for the following imperfect record I am indebted to both Mr. Gibson and Mr. Lyman for the notes they made which have been included with my own.

Egg.—Similar to other eggs of this genus in colour and shape, being dome-shaped; .815 mm. wide, .65 mm. high; appearing smooth to naked eye, but under microscope distinctly pitted all over with depressions of irregular shape. Colour pale honey-yellow when laid, turning buff colour after two days and gradually becoming darker through orange shades till a day before hatching when they appear nearly black. Laid in clusters side by side, on under side of a leaf in cofinement. First larva hatched June 14, leaving little of egg shell except the base; duration of egg stage 10 days.

Stage I.—Length 2.25 mm nearly cylindrical, head slightly larger. Head black, shining, cordate; mouth-parts yellowish. Body whitish before feeding turning greenish, with black tubercles from which arise black and gray hairs. Legs black; prolegs grayish. Fed readily on common plantain *(Plantago major)*.

Stage II.—Length 4.5 mm. Not much change noted in appearance but colour darker with faint dorsal stripe.

Stage III.—Length 7 mm. Head black, body cylindrical, blackish with tufts of black and grayish hairs from the conspicuous black tubercles. Segments 2, 3 and 4 pale orange with black spots, 11 and 12 similar. A pale yellowish dorsal band from segment 5 to 10.

Stage IV.—Length 11 mm. Head black, mouth parts yellowish; body black with longitudinal stripes yellow, hairs black; legs and prolegs black.

Stage V.—Length 20 mm. No further notes taken.

Stage VI.—Mature larva (description by Mr. Arthur Gibson) "Length 38 mm. at rest, 45 mm. extended. Head 2.8 mm. wide, rounded, somewhat quadrate, conspicuously depressed at vertex, flattened in front, median suture pale, setæ dark brown, long and slender. Skin of body streaked and blotched with dark reddish brown. Dorsal stripe chrome-yellow, conspicuous, distinct on all segments, wider on abdominal ones. Tubercles jet black, each with a bunch of radiating stiff, barbed bristles mostly of uniform length. Tubercle I about one-half size of II; II larger than III; IV, V, VI, elongated. Bristles from tubercle I mostly silvery with a few black ones intermingled; from II, III and IV silvery and black in about equal numbers; from V and VI mostly silvery. Tubercle VII larger than VIII, bristles from these mostly black and short. An indefinite, broken yellow lateral line is also present, most apparent along upper edge of tubercle III. Spiracles black, close in front of tubercle IV. Thoracic feet black, shiny, bearing short dark bristles; prolegs also black, claspers reddish, bristles dark."

The first of my larvæ began to make its cocoon on July 16th. Mr. Gibson's spun up on July 10th, July 11th and July 14th. Larval period in confinement 32 days; out of doors would probably be about 6 weeks.

Cocoon.—Oval, about 22 mm. long, 10 mm. wide; thinly made of brownish silk, in which the hairs from body are woven. The ends of many hairs project, giving the cocoon a spiny appearance. Some larvæ spun up between leaves, but the majority in corners of the boxes.

Pupa.—Length 18 mm., width 7 mm.; nearly black in colour inclined to dark crimson particularly on wing cases. Body closely punctured, thorax creased, spiracles pearly glistening. Cremaster short consisting of about 8 short stiff reddish capillate bristles.

The first moths emerged (in a cool cellar) on April 21st. The moth is doubtless single-brooded throughout its habitat, which seems to be limited to the northern part of Quebec and Ontario, westward to Winnipeg, Man.; and to the White Mts. of New Hampshire and the Adirondacks and Catskills of New York.

The larvæ are voracious feeders and never seem to stop eating day or night. When disturbed in their repast, or put upon a table for examination they are most active creatures. If a large plantain leaf or an arch made of a sheet of note paper be provided they lose no time in scurrying along to take refuge beneath it.

BOOK NOTICE

BUTTERFLY-HUNTING IN MANY LANDS. Notes of a Field Naturalist. By George B. Longstaff, M. A., M. D., Oxon.; F. R. C. P., F. S.A., F. G. S. Longmans, Green and Co., London, New York, Bombay and Calcutta. Price 21 s.

The writer of this volume has attempted a very difficult task—that of incorporating into a readable form the entomological diaries kept by him during many years of butterfly collecting in many lands. We think that, considering the difficulties presented by such an undertaking, he has been remarkably successful in carrying out his object, and we attribute this success largely to a marked literary sense and gift of narrative, of which he is the happy possessor. We fear, however, that there are very few, even among entomologists, who will read the book from cover to cover; only such lepidopterists as are more or less familiar with the butterfly fauna of the entire world will find all of its chapters readable.

Dr. Longstaff's travels have taken him to India, Ceylon, China, Japan, Algeria, Egypt and the Soudan, South Africa, the West Indies, South America, Canada, Australia and New Zealand; so that he has enjoyed the somewhat unusual experience of having collected butterflies in every continent of the globe. His sojourn in Canada was limited to a rapid journey across the continent in 1904 on his return to England from the Orient and very little opportunity for collecting was had on the way, but in tropical countries Dr. Longstaff's experience has been wide and varied and he shows himself to be thoroughly familiar with butterfly life everywhere.

Many amusing anecdotes and interesting impressions of the various countries visited by the author, and of the customs of their inhabitants, are scattered through the volume, greatly helping to enliven it; while all that is of real scientific value is encompassed in the last chapter, entitled "Bionomic Notes". This chapter contains many interesting notes under the following headings: "The scents of butterflies"; "The coloured juice exuded by certain Lepidoptera"; "The tenacity of life of protected species"; "Butterflies bearing marks of the attacks of foes"; "Experimental evidence as to the palatability of butterflies"; "Mimics in the field deceiving man"; "Notes on the flight of sundry butterflies"; "Heliotropism"; "List and shadow"; "The inverted rest attitudes of Lycænids and some other butterflies"; "General remarks on rest attitude of butterflies"; "Cosmopolitan Lepidoptera"; "Seasonal dimorphism"; "The selection as resting-places of yellow leaves by yellow butterflies".

As a supplement translations by Ernest A. Elliott, F. Z. S., F. E. S., of a series of important papers by the late Fritz Müller on the scent-organs of Lepidoptera have been appended to the book, together with an introductory note by Prof. E. B. Poulton, by whose suggestion they were included.

The book, including the appendix and the very full index occupies 728 pages. It is illustrated by six good coloured plates, upon which many other insects besides butterflies are depicted, and 19 text figures. The appendix is also illustrated by nine lithographic plates.

Mailed July 15th, 1912.

The Canadian Entomologist.

VOL. XLIV. LONDON, AUGUST, 1912. No. 8

GEOMETRID NOTES.

REVISION OF THE GENUS HYDRIOMENA HUB. GROUP WITH MODERATE PALPI.

BY L. W. SWETT, BOSTON, MASS.

8. *Hydriomena autumnalis* Ström (Det. Kgl. Danske Vid. Skrift. Selsk., p. 85, 1783). Palpi moderate.

We have generally regarded as typical *autumnalis* a grayish form with clear mesial space and slate-coloured bands, bordered with olive. It occurs in June and July in both Europe and America, but there is a May brood which we do not commonly get here. We imagine our species to be the same as the one that is found in Europe, but do we really know the form that Strom described as *autumnalis?* It seems strange that a geometrid occurring only in the early summer should be called *autumnalis*. This same doubt seemed to arise in the mind of a German specialist, for Hoyningen-Heune, in the Berl. Ent. Zeit., Vol. 51, p. 254, 1906, discusses this question, and pleads for the name *trifasciata* Bockhausen for this form. He claims it answers Bockhausen's description better than Ström's, which seemed to him to apply to some unknown or unrecognizable species, possibly *Larentia autumnalis* Bockhausen. Perhaps he is correct, but it would be hard to prove, as Ström might have called it *autumnalis* from his having found the larvæ in the fall or for various other reasons. Since nearly every collector recognizes it by the older name, *autumnalis,* and it is difficult to prove otherwise, I think it better to retain this name. The form found in America is very close to the European in colour and markings, but there are slight differences which, should the genitalia prove different, would refer it to *pluviata* Guenèe (Phal., II, p. 378, No. 1505, 1857). Guenèe states that the American form has more pointed fore wings, the lines are not so close together, and those bordering the mesial band are more oblique and that the median space is larger. He mentions having a specimen with the mesial space shaded with pink, and to this form I have given the varietal name *perfracta* (CAN. ENT., XLII, p.279, 1910). The differences that Guenèe points out seem pretty constant, but the enlarged central band does not hold, as he has every variation of it among European specimens. The pointed wings and green or gray ground colour with slate-coloured bands seem to be rather distinctive of the American form; also the basal band is apt to be thicker and the first mesial more pronounced, but I think a careful study of the genitalia with extensive breeding will be necessary before we can be sure. However, I feel that the American form will at least prove to be a good variety of the European, and it might be well later to adopt the name *pluviata*. The uncus in *Hydriomena autumnalis* is very peculiar, being forked while it is spatulate with a tendency to fork in *furcata* Thmb.

Hydriomena autumnalis, trifasciata or *pluviata* Gn., if our form proves distinct from the European, is more common in the Atlantic States than the Pacific. In Europe it seems to vary less in colour than here, though the bands are variable in width and the specimens in size. There is more tendency to melanism in the European specimens, but I have some from Saskatchewan, from Mr. A. J. Croker, that are as dark as those from Norway.

H. autumnalis Ström appears to be more common in Maine than Massachusetts, as my friend Mr. Frost took a very large number one day in a cedar swamp at Monmouth, and Mr. Emerton even took a specimen on the summit of Mt. Kataadn. In Europe it is said to feed on alder and willow, but the American food plant has not been recorded to my knowledge. The palpi of the European and American forms are all moderately long and do not vary. My kind friend, Mr. Chagnon, of Montreal, is working on the genitalia, and I hope to publish some notes on his results later. There is an excellent figure of *pluviata* or *autumnalis* in Packard's Monograph, Pl. VIII, fig. 29, which is typical of eastern specimens and of certain of the European specimens in markings, but not wholly in colour. The venation is shown in the Monograph, Pl. I, fig. 6.

Taking *autumnalis* as a whole, it is far less variable in markings than *furcata* or *quinquefasciata*, but is about the same in regard to colour variation. It is very easy to separate the species with moderate palpi, as there are fewer closely-allied forms, the only difficult one being *californiata* Pack., which is liable to be confounded with var. *perfracta* of *autumnalis*. Nearly all the collections I have seen contain *autumnalis*, so I do not think it can be rare in any particular locality in the North Atlantic States.

Var. (a) *perfracta* Swett (CAN. ENT., XLII, p. 279, 1910).

This looks like the normal *autumnalis*, only the mesial space is suffused with a deep pink and the course of the basal line is different. This variety was referred to without name by Gueree) Phal., Vol. 2, p. 398, 1505, 1857) under *pluviata*. It approaches superficially *H. californiata*, but I have pointed out the differences in the description. The variety seems to be very rare. It was taken in the Catskills by Mr. R. F. Pearsall. There are no other records of it to my knowledge.

Var. (b) *crokeri* Swett (CAN. ENT., XLII, p. 278, 1910).

This variety occurs in the Northwest, most commonly around Victoria. The green shadings of the typical form are replaced by yellow in the variety, and it has an intensely black median irregular band, instead of being of the usual slate colour. It is a very striking and distinct form; there is nothing approaching it among the European varieties.

Var. (c) *columbiata* Taylor (CAN. ENT., XXXVIII, p. 399, 1906).

I have a photograph of the type, kindly sent me by the Rev. G. W. Taylor for comparison, and it approaches the European var. *constricta* Strand very closely, but it lacks the cinereous ground colour, and I think it will remain in good standing. It may be known by the narrow mesial area, shaded with dark where the typical form species is white. The intradiscal band near the inner margin lacks the long-toothed projection of *crokeri*, and the hind wings are light ash with two bands beyond the discal spot. It is also distinguished by the larger size and greenish ground colour.

Var. (d) *constricta* Strand (Ent. Zeit., Gub. XIV, p. 61, 1906) also Berl. Ent. Zeit., Vol. 51, p. 254-257, 1906, and Ach. Naturv. Christiania, XXII, No. 5).

This is a peculiar cinereous form, in which the mesial white space is suffused with smutty striations, giving the whole a rather smoky appearance, approaching melanism. The blue irregular median band is very faint and the lines are all hardly discernible. The variety can best be separated by the dark narrow central space and dark hind wings with prominent discal spots and two extradiscal bands. I have examples from Norway, Germany and North America, the latter being represented by a ♂ and a ♀ from Forest Hills, Mass. (June 20, 1911), and Monmouth, Maine (June 25, 1903), which agree with the European specimens. I have also seen it from Redvers, Sask., through Mr. Croker. It is not so rare in Europe as in North America, where I have only seen four examples. I have a specimen from England, through Mr. L. B. Prout, which approaches the variety, but it is more suffused and is almost unicolorous.

Var. (e) *nigrescens* Hoyningen-Heune (Berl. Ent. Zeit., 51, p. 254, 1906).

This variety is almost unicolorous, and has the white mesial space entirely suffused with cinereous, giving the wings a dusty appearance. The markings are very indistinct, and the whole insect has a smoky aspect. I have specimens from Berlin, Germany; Southport, England, and have seen a specimen from Redvers, Sask., through Mr. Croker.

These are all the varieties, so far, that have turned up, and are easily distinguished from the normal form by their colour. Walker described in 1860 (Cat. Brit. Mus., XXI, 489) *Boarmia divisaria*, which has been stated to be a synonym of *autumnalis* in Dyar's List, but the description does not seem to apply, and I doubt the reference. The type is in the D'Urban collection. Walker also described *renunciata* (Cat. Brit. Mus., XXIV, 1187, 1862), from Hudson Bay, and *frigidata* (Cat. Brit. Mus., XXVI, 1729, 1862), from Nova Scotia, which are synonyms of *autumnalis* in all probability, as Packard states in his notes on the North American Moths of the family Phalænidæ in the British Museum (5th Rep. Peabody Acad., p. 88, 1873), that they are our common *pluviata*.

9. *Hydriomena transfigurata* Swett (CAN. ENT., XLIV, p. 195, 1912).

This is a pointed-wing species and closely resembles *irata* Swett in markings, though the antennæ of the latter will quickly separate it. It does not resemble *autumnalis* greatly, but I have generally found it mixed with the latter in collections. *H. transfigurata* can be readily separated by the time of appearance (early May), the tendency of intra- and extra-discal lines to unite near the inner margin, the dark hind wings and the distance of the extradiscal line from the outer margin in fore wings; also the faint median and marginal bands which are prominent in all other species. I should say this species is confined to the Atlantic States, and has occurred more commonly in the last two years than ever before. There is no form like it in Europe so far as I have seen, and it is not very common here, the only localities being Forest Hills and Cohasset, Mass., and New Brighton, Pa.

10. *Hydriomena californiata* Pack. (Proc. Bost. Soc. Nat. Hist., XIII, 396, 1871).

This was a very puzzling species to me for some time, until I received a specimen from Rev. G. W. Taylor like the one I had determined as *californiata*. We had both come to the same conclusion independently, and were the first, I believe, to determine *californiata* correctly. Packard did not understand *californiata* clearly, as he merged all sorts of forms under that name later, but the figure (pl. VIII, fig. 30) is excellent, and leaves no doubt as to the species. The type is not in the collection at Cambridge, and must have been returned to the sender by Packard; it was probably from Jas. Behrens. This species occurs in late July, and bears a slight resemblance to var. *perfracta* Swett, but is only found on the Pacific Coast to my knowledge.

11.—*Hydriomena lanavahrata* Strecker (Lep. Rhop. Het. Suppl., 2, 11, 1899).

According to Dr. Dyar, this is a variety of *californiata*, but I do not feel sure he knows *californiata* correctly, and until I can verify his assertion I think it better to let matters stand as they are. The locality would seem good for varieties, for the vicinity of Berkeley seems to be their Mecca. I have nothing in my collection that answers to the description, though Strecker does not give the essential points for differentiating Hydriomenas.

12. *Hydriomena glenwoodata* Swett (CAN. ENT., XLI, p. 231, 1909).

This is a peculiar, small, slender species with the slate-gray irregular line, *s* curved. The mesial space is somewhat suffused with dark atoms, and the whole insect is somewhat suffused with cinereous. It resembles slightly var. *nigrescens* of *autumnalis*. The palpi are very slender and a little longer than normal. I have never seen it from anywhere but the Rocky Mts., my type being from Pike's Peak, 8,000 to 10,000 ft. elevation.

13. *Hydriomena magnificata* Taylor (Ent. News, XVII, No. 6, 1906).

This resembles *speciosata* somewhat, differing in the moderate palpi. The description is clear, and the species cannot be confounded with any other.

This completes all the species and varieties known to date of the group with moderate palpi.

GROUP WITH MODERATE PALPI.

8. *Hyd. autumnalis* Ström or *pluviata* Gn.	White mesial space, olive shadings, slate-coloured bands.
Var. (a) *perfracta* Swett.........	(a) White mesial space suffused with pink.
Var. (b) *crokeri* Swett..........	(b) Green ground colour replaced with yellow, black bands.
Var. (c) *columbiata* Taylor......	(c) As typical, mesial space suffused with cinereous and narrow, very large size.
Var. (d) *constricta* Strand......	(d) Mesial lines smutty, lines indistinct.
Var. (e) *nigrescens* Hoyn.-Heune..	(e) Unicolorous; smutty suffused.
9. *Hyd. transfigurata* Swett.....	Green ground colour, tendency of extra- and intradiscal lines to join near inner margin.
10. *Hyd. californiata* Pack......	Red shadings to lines and mesial space.
11. *Hyd. lanavahrata* Streck.....	Said to be var. of *californiata*.
12. *Hyd. glenwoodata* Swett...	Long-winged slender species with s-shaped intradiscal line.
13. *Hyd. magnificata* Taylor.....	Black and green.

I wish to thank Mr. A. F. Winn and Mr. G. Chagnon for the loan of specimens, and the latter for his help on the genitalia, on which we hope to publish something later. I append a few of the more important references to *Hyd. autumnalis* and its varieties :

1783. *autumnalis* Ström, Kgl. dansk. Vid Selsk. Skr., p. 85.

1776. " Dennis & Schiff, Syst. Verz. Wien, 109, 5 (not described).

1786. *autumnalis* Sepp., II, pl. 5, figs. 1–8.

1794. *trifasciata* Bork., Eur. Schmett., V, p. 308, No. 141.

1797. *impluviata* Hüb., Eur. Schmett, 223 (post 1797).

1810. " Haw., Lepid. Brit, part 2, p. 321.

1822. " Hüb., Verz. Schmett., 322, 3, p. 106.

1828. " Treits., Schmett. Europa, VI, part 2, p. 21.

1829. " Stephens, Nomenc. Brit. Ins., p. 44.

1830 " Dupon., Lep. France, V, p. 424, pl. 200, fig. 3.

1831. " Stephens, Ill. Brit. Haust., III, p. 254.

1839. *impluviata* Wood, Lep. Ins. of Great Brit., p. 610.

1840. " Boisd., Index Method, 1767, p. 214.

1847. " H.-S., Syst. Bearb. Schmett. Eur, III, p. 168, pl. 31, fig. 193.

1850. *impluviata* Stephens, Cat. Brit. Lep, p. 195.

1853. " La Harpe, Schmett., p. 295.

1857. " Verz. Wien., p. 109, K. 5.

1857. " Gn., Phal., II, p. 377, 1504.

1857. *pluviata* Gn, Phal., II, p. 378, 1505.

1860. *(divisaria) ?* Walk., List Lep. Brit. Mus., XXI, p. 489.

1862. *renunciata* Walk., List Lep. Brit. Mus., XXIV, p. 1187.

1862. *impluviata* Walk., List Lep. Brit. Mus., XXIV, p. 1267.

1862. *pluviata* Walk., List Lep. Brit. Mus., XXIV, p. 1268.

1862. *frigidata* Walk., List Lep. Brit. Mus., XXVI, p. 1729.

1876. *trifasciata* Pack., Monog. Geom., p. 91, pl. VIII, fig. 29.

1889. *trifasciata* Zieg., B. E. Z, XXXIII, p. 7, S. B.

1895. *autumnalis* Hulst, Ent. News, XI, p. 43.

1904. *autumnalis* Dyar, List Koot. Dist., Proc. U. S. Nat. Mus., XXVII, p. 899.

1904. *autumnalis* Petersen, Mem. Ac. St. Petersb., XVI, p. 73.

1906. *constricta* Strand, Ach. Naturv. Christ., XXII, No. 5.

1906. *nigrescens* Hoyn.-Heune, Berl. Ent. Zeit, LI, p. 254.

1906. *constricta* Strand, Ent. Zeit., Gub. XIV, p. 61.

1906. *columbiata* Taylor, CAN. ENT., XXXVIII, p. 399.

1906. *autumnalis* Hoyn.-Heune, Berl. Ent. Zeit., LI, p. 254.

1907. " Gatnar, Wien. Jahr. Berl. Ent. Ver., 18, p. 37-42.

1910. " Dyar, Harriman Alaska Exped., p. 222.

1910. *perfracta* Swett, CAN. ENT., XLII, No. 8, p. 279.

1910. *crokeri* Swett, CAN. ENT., XLII, No. 8, p. 278.

BATS VS. MOSQUITOES.

We have received the following interesting letter from Dr L. O. Howard, Chief of the U. S. Bureau of Entomology, who has kindly given us permission to publish it :

STATE BOARD OF HEALTH OF FLORIDA

Jacksonville, Fla., June 26, 1912.

Dr. L. O. Howard, Chief of Bureau of Entomology, Washington, D. C.

Dear Doctor,—I thank you very much for your favour of the 24th. I had looked askant at the idea of bats reducing the number of mosquitoes appreciably. Some twenty years ago, perhaps longer, at Tavares, Fla., a development company undertook to build a winter resort. Tavares was at the time a small municipality with perhaps two or three hundred inhabitants located among the lakes in the southern part of the State.

Among the earlier efforts at developments an opera house was constructed, but owing to the freeze of 1895 it was never completed. The municipality never grew to amount to anything; in fact, I think the number of inhabitants now is what it was about then. The doors and windows of the lower floor of this opera house were securely fastened up to keep intruders out, but the upper windows were only closed by loose boards, which soon dropped out, making it easily accessible to bats. They took advantage of it and in the course of a few years were there in countless thousands. I know of no way of estimating the number, but you may get some idea of it from the fact that the only time I was ever there at the right hour was on a trip to Eustis. The train stopped at Tavares one half-hour before sunset, and remained there something like forty-five minutes. I took advantage of the occasion to see the bats emerge from the building. I had only been watching a few minutes when they began, first a single one, then two or three together, and as if the rustle started them, then they began seriously flying out of the window with incredible swiftness. There must have been at least half a hundred a second. I watched this stream of bats pouring out for half an hour or so, and was told by some of the residents of Tavares that it would continue until something like half an hour after dark, making probably two hours altogether.

It was on this trip, now seven years ago, that I was making some mosquito observations, and I have to confess that I have never seen more mosquitoes in the interior of the State, than I saw at that time.

Some two years ago the opera house in question was cleaned out and converted into a packing house. I have since made inquiry of the citizens in the vicinity of Tavares and Eustis, as to whether they have experienced any appreciable difference in the number of mosquitoes now, and when the bat roost was at its height, and am convinced that the difference, to say the least, is not such as to cause one to notice it.

Again thanking you for your information, I am, very truly and cordially yours,

(Signed) HIRAM BYRD.

SOME NEW SPECIES OF DELPHACIDÆ.*

BY C. S. SPOONER, GEORGIA STATE BOARD OF ENTOMOLOGY.

The following species were, with one exception, taken by the author during the past five years. In the genus *Pissonotus* they form a very considerable addition to the list of our species.

My thanks are due to Professor A. D. MacGillivray of the University of Illinois for going over the manuscript and for valuable help and suggestions. I also wish to thank Professor J. Chester Bradley of Cornell University for the loan and exchange of specimens and Mr. C. P. Alexander also of Cornell University, for the gift of specimens of several desirable species.

The author has planned an extensive study of the Delphacidæ and would be grateful for the loan or exchange of material. He will gladly name specimens for the privilege of retaining desirable duplicates.

Pissonotus guttatus, n. sp.

Brachypterous ♀.—Eyes oval, deeply indented below to receive the antennæ, colour grey; vertex considerably longer than wide, slightly rounded in front, projecting slightly before the eyes; carinæ of the vertex all present, rather indistinct; vertex uniformly black except the caudo-lateral angles which are slightly yellowish. Carinæ meeting on the front just below the curve of the vertex, extending distinctly from this point throughout the length of the front, indistinct on the vertex; front considerably constricted between the eyes, deep pitchy black above, becoming gradually lighter until it is white at the base; clypeo-frontal suture curved, the clypeus deep uniform black with the median carina distinct; the black colour of the clypeus extends as a band across the anterior coxæ as is characteristic of the genus.

The second segment of the antennæ about one-fourth as long again as the first, with a few protuberances; antennæ uniform pale honey-yellow.

The length of the prothorax from the anterior to the caudal margins about equal to that of the vertex; caudal margin slightly concave, carinæ distinct; anterior portion deep, shiny black; narrow band on posterior margin dirty white.

Legs normal, pale honey-yellow except tips of the tarsi which are black and two slender brown lines on the outer sides of the tibiæ.

*Published with the permission of E. L. Worsham, State Entomologist of Georgia.

Scutellum triangular, sides very slightly arcuate ; median carina distinct ; lateral carina inconspicuous, reaching the posterior margin ; colour uniform honey-yellow.

Elytra short, extending to the middle of the first abdominal segment, coriaceous, highly polished, veins almost obliterated ; colour pale honey-yellow with an oval blotch of white in the centre of the apical margin ; abdomen honey-yellow but slightly darker than the scutellum and elytra.

Genitalia uniform honey-yellow, paler than the tergum ; pygofers tapering to a rounded point ; plates extending about one-third the length of the pygofers ; anal style white. Length 3 5 mm.

Described from a single female taken at Ithaca, N. Y., Aug. 1st, 1896. Type in collection of Cornell University.

Very close to *P. delicatus* Van Duzee but easily separable from it by the black front, vertex and prothorax. The prothorax lacks the foveæ so conspicuous in *delicatus* and the shape of the front is quite different.

Pissonotus foveatus n. sp.

Brachypterous ♀.—A fairly large species, form long, oval. Vertex slightly rounded in front ; eyes oval, deeply indented below to receive the antennæ ; colour gray, irregularly mottled with black ; vertex slightly longer than wide, slightly wider behind the eyes than before ; fully carinated ; carinæ sharp, except the posterior median, which tends to fade out posteriorly ; the anterior median carina forked just below the apex of the head , colour of the vertex yellowish white, with a pair of dark brown foveæ just posterior to the forking of the anterior median carina, one on either side of the carina, and another pair similarly situated, but slightly farther caudad ; there is a yellowish fovea located in each posterior lateral angle of the vertex.

Front with sides slightly convex ; median carina sharp throughout ; colour yellowish white marked with brown. These markings vary somewhat, but may in general be described as follows : According to colour, the front may be divided into four regions ; first, just under the vertex a darker area in which may be found two deep brown lines extending across the median carina ; these lines extend to the edge of the front ; the second area is light, and the only markings on it are a pair of brown dots on each side at the outer margin; this pale area is about equal in width to the first area. Covering about two-thirds of the remainder of the front is the third area, dark and marked like the first area, except that it is darker and less distinctly marked ; in some specimens it is scarcely more than a dark brown band. The remainder of the front, comprising the fourth area, is yellowish white devoid of markings.

Clypeus uniformly black, with the median carina distinct; the black colour extending in a band across the anterior coxæ as is typical of the genus.

Second segment of the antennæ about one-fourth as long again as the basal segment; basal segment and basal half of the second segment yellowish white; distal half of the second segment deep brown.

Prothorax shorter than head, caudal edge concave; the carinæ distinct; a fovea on each side of the median carina, midway to the lateral carinæ and slightly nearer the cephalic than the caudal margin; colour yellowish white, with varied dark brown spots and blotches on the caudal margin; legs yellowish, with a brown band across the proximal portion of the tibiæ.

Scutellum triangular, sides very distinctly arcuate, carinæ distinct, the lateral carinæ extending to the caudal margin; a brownish fovea on each side of each lateral carina, about midway between the cephalic and caudal margins; colour yellowish brown.

Elytra short, extending slightly beyond the middle of the second abdominal segment, coriaceous, polished, veins indistinct; colour greyish white, irregularly spotted with brown.

First, second and third abdominal segments with the dorsum yellowish brown, with a few brown dots; fourth, fifth and sixth segments with their median portions yellowish brown and their lateral portions deep brown.

Genitalia reddish-brown, pygofers tapering to a blunt point, plates extending about one-third the length of the pygofers, light yellow, tip of the ovipositor much lighter brown, style yellow. Length, 3.25 mm.

Brachypterous ♂.—Smaller; form and markings about as in female, with not quite so much brown.

Genitalia, opening of the pygofers rather narrow, oval; superior wall of the anal tube prolonged into two incurving tusks, the points of which rest upon two large pointed projections extending inward from the inferior wall of the pygofers; colour yellow, except the ends of the tusks and the points of the projections, which are reddish brown. Length, 2.5 mm.

This species was taken quite abundantly on a species of *Compositæ* by the author at Corpus Christi, Texas. Types, taken May 19, 1907, in the author's collection.

Pissonotus variegatus, n. sp.

Macropterous ♀.—Form and size of *P. pallipes*. Head rounded in front; eyes large and oval, slightly indented below to receive the antennæ; colour yellowish around the edge, black in the centre; vertex about one-third again as long as broad, projecting slightly beyond the eyes, quite strongly rounded in front, in which it differs from *pallipes* where it is nearly straight; the posterior carina wanting, other carinæ sharp and distinct; colour light yellow, except outside the anterior median carinæ, where it is marked by two brown dots on each side.

Front slightly wider below the eyes than between them; median carina distinct throughout the length of the front, forked just below the vertex; colour, anterior two-thirds yellow, much mottled with brown, posterior third pure light yellow; clypeus uniformly black, with the median carina distinct, the black band continuing across the anterior coxæ.

First and second antennal segments subequal, proximal segment and proximal half of the second segment yellow, distal half of the second segment reddish brown, roughened by numerous protuberences.

Prothorax equal in depth to the head, posterior edge concave, almost angled at the centre; carinæ all distinct, lateral ones reaching caudal margin of the prothorax, quite widely divergent; colour brown, marked with yellow, especially in the centre; legs yellow, with a brown band around the distal end of the femora and proximal end of the tibiæ; tips of the tarsi black.

Scutellum triangular with the sides very much arcuate, apex rounded, scutellum about twice as deep as the prothorax; carinæ distinct, reaching to the caudal margin; colour brown with yellow markings. One-fourth of the elytra extending beyond the tip of the abdomen; milky in colour with brown dots along the veins. Genitalia dark brown; pygofers bluntly pointed, plates dark yellow, extending over one-third the length of the pygofers, very much curved; style white. Length, including elytra, 3.5 mm.

Described from a single female taken by the author at Corpus Christi, Texas, June 19, 1907. Type in the author's collection.

Pissonotus divaricatus, n. sp.

Macropterous ♀.—Form and general appearance of *P. basalis*, although not quite so large and heavy an insect; eyes oval, not deeply emarginate below to receive the antennæ.

Vertex a little longer than wide, sides nearly straight, very slightly wider in front of the eyes than behind them; all the carinæ distinct, posterior foveæ fairly deep ; colour reddish brown with the carinæ and margins light yellowish brown.

Front about twice as long as wide, sides very slightly arcuate, widest a little below the eyes ; clypeo-frontal suture slightly curved, median carina distinct, forked just below the apex of the head ; colour uniform reddish brown except a narrow strip along the clypeo-frontal suture, which is light yellow.

Clypeus uniformally black with the median carina fairly distinct.

Second segment of the antennæ a little more than twice as long as the first segment, second segment roughned by tubercles, proximal segment black, second segment yellowish brown.

Prothorax not quite so deep as the vertex, carinæ distinct, the lateral carinæ quite widely divergent, fading out just before the posterior margin, posterior margin distinctly concave ; colour reddish brown, except the narrow posterior edge, which is almost white.

Coxæ light yellow ; femora and most of the tibiæ reddish brown, tips of the tibiæ and the first two tarsal segments very light yellow, last tarsal segment dark brown, spur almost white with large, prominent, black serrations.

Scutellum almost twice as deep as the prothorax, sides very decidedly arcuate, tip a rounded point, carinæ all distinct, lateral carinæ reaching the posterior margin of the scutellum and curving outward; colour dark brown, except the tip, which is light yellow, almost white.

Elytra extending one-third their length beyond the tip of the abdomen, veins brown, membrane smoky ; abdomen uniformly dark brown.

Genitalia : Pygofers ending in a blunt point, reddish brown, ovipositor reddish brown at the tip, lighter at the base, plates dark within, lighter on the free edge, apex very gradually curved, extending about one-third the length of the pygofers, style light yellow. Length, including elytra, 3 mm.

Macropterous ♂.—Same form, size and general characters as the female. Genitalia : Opening of the pygofers broad oval, superior wall of anal tube prolonged into long, incurving, tusk-like horns, the points resting in indentations of the inferior wall of pygofers, styles large hook like organs with the hooks pointing outward ; colour dark reddish brown.

Described from a pair taken by the author at Middletown, N. Y., July 12, 1910. Two females of this species were taken at the same locality on July 11 and 18, 1910. Types in the author's collection.

Pissonotus piceus, n. sp.

Brachypterous ♀.—A small species slightly smaller than *P. brunneus* and not so stout. Head short, very slightly curved in front. Eyes slightly indented below to receive the antennæ; colour light gray around margins, black in the centre; vertex about as deep as wide in front, sides curving between the eyes, narrower behind the eyes than before; all the carinæ present, but all rather weak, foveæ not very deep; colour deep shiny black.

Front rather wide and short, sides nearly straight. median carina very faint; anterior three-fourths deep shiny black, posterior one-fourth pure white.

Clypeus uniformly black; median carina but a mere suggestion; the black band carried across the anterior coxæ as usual; posterior of this a band of white and towards the tip of the abdomen, black.

Basal segment of the antennæ about one-third as long as the second segment; basal segment brown, second segment honey-yellow spotted with white, tuberculate.

Prothorax about as deep as the head, posterior margin very slightly concave, median carina practically indistinguishable, lateral carinæ distinct for about two-thirds of their length and then fading out; colour pure white.

First and second pairs of legs with the coxæ light brown, shading through black on the femora and tibiæ to white on the first two tarsal segments, the last tarsal segment black, tibiæ of the first and second pairs of legs foliaceous; third pair of legs shading from dark brown at the base of the femora to honey-yellow on the tibiæ to white on the tarsi, tips of the tarsi black.

Scutellum triangular, sides straight, not visible for its entire breadth, covered by the prothorax on its outer edges; median carina indistinct, lateral carinæ short, curved outward, poorly defined; colour uniform deep shiny black.

Elytra short, not quite covering the first abdominal segment, coriaceous, polished, veins indistinct; colour, basal three-fourths deep shiny black, posterior one-fourth pure white.

Abdomen uniform shiny black; genitalia deep reddish brown, almost black; pygofers tapering to a blunt point, plates very short, extending only one-fourth of the length of the pygofers, only a small edge visible; style white. Length, 2.5 mm.

A very pretty and delicate insect. The foliaceous tibiæ recall *Phyllodinus*, but the carinæ of the prothorax are straight and the other characters agree with *Pissonotus;* it may deserve to be placed in a new genus, but for the present I prefer to place it in *Pissonotus*. The species is very easily identified by the white thorax and white margin of the elytra and by the deep shiny black of the rest of the body. Described from a specimen taken by the author at Middletown, N. Y., July 11, 1910. Two other specimens were taken at the same locality July 18 and 21, 1910. Type in the author's collection.

Pissonotus binotatus, n. sp.

Brachypterous ♀.—Form and general appearance of *P. marginatus*, but considerably smaller.

Eyes oval, deeply indented below to receive the antennæ; colour dark gray, almost black, with a yellow margin; vertex about as long as wide, very slightly produced before the eyes, anterior margin slightly curved; carinæ all present, very pronounced, posterior foveæ very deep; colour uniform dark reddish brown.

Front about one and one-third times as long as broad, widest below the eyes, sides slightly arcuate; clypeo-frontal suture straight, median and lateral carinæ quite prominent, median carina forked just below the apex of the head; colour uniform reddish brown, except a very narrow band along the clypeo-frontal suture, which is light yellow.

Clypeus of the form of a truncated triangle; median carina fairly prominent; colour black, the black band extending across the anterior coxæ.

Basal segment of the antennæ a little less than one-half the length of the second; the second segment lacks the protuberences so often found; basal segment reddish brown, second segment light yellow.

Prothorax a little deeper than the head, caudal margin almost straight, very slightly emarginate on the sides and a suggestion of an emargination in the centre; median carina very prominent, lateral carinæ strong on proximal two-thirds of the prothorax, fading out before reaching the posterior margin; colour uniform reddish brown.

Anterior legs yellow, lineated with brown; second and third pairs of legs with yellow coxæ, femora and proximal half of tibiæ brown, the tibiæ becoming gradually lighter in colour toward the distal end, the distal end of the tibiæ and first two tarsal segments light yellow, almost white; last tarsal segment dark brown; tarsal spur rather small, light yellow, almost white.

Scutellum triangular, about one and one-third times as deep as the prothorax, sides straight, median and lateral carinæ prominent, the latter attaining the posterior margin ; colour uniform reddish brown.

Elytra short, practically covering the first abdominal segment, highly polished, veins indistinct ; colour reddish brown, with two yellowish white dots on the apical margin of each elytron.

Abdomen uniformly reddish brown, a prominent carina extending along the middle of each tergum ; genitalia reddish brown, pygofers ending in rather a sharp point; plates short, extending only one-third the length of the pygofers, a little lighter in colour; style white. Length, 2 mm.

Brachypterous ♂.—Form and markings the same as that of the female, considerably smaller in size. Genitalia, aperture of the pygofers quite long and narrow, superior wall of the anal tube produced in long outcurving horns, these rest on projecting points of the ventral margins of the pygofers ; styles small ; anal style light yellow, rest of the genitalia dark reddish brown. Length, 1.5 mm.

This species resembles *P. marginatus* quite closely. It is a much smaller species, there are distinct differences in the proportions of the front and antennæ, the spur is smaller proportionally. There are also differences in the genitalia and some noticeable colour differences. *P. binotatus* lacks the white on the prothorax, the front is much darker, and has two white spots on the apical margin of the elytra instead of a full white band as in *marginatus*.

Type of the male and female taken at De Witt, Mitchell Co., Ga., April 6, 1912, by the author. Types in the author's collection.

Liburnia dolera, n. sp.

Macropterous ♂.—A medium-sized form for this genus. Eyes oval, deeply and narrowly indented below to receive the antennæ ; colour gray, darker in the centre. Vertex slightly longer than wide, projecting a little beyond the eyes, carinæ distinct, except the posterior median, which is quite faint ; foveæ deep ; colour dark reddish brown.

Front widest a little below the eyes, constricted considerably between the eyes, sides curved, the median and lateral carinæ sharp and prominent, the former forked at the vertex ; clypeo-frontal suture slightly curved ; colour reddish brown, with the ventral portions of the lateral carinæ dark yellow. Clypeus a lighter brown than the front, the carina distinct.

Basal segment of the antennæ one-third the length of the second segment, dark brown ; second segment rather thick, tubercled ; light yellow in colour.

Prothorax a very little deeper than the vertex ; hind margin concave, almost angled at the centre ; carinæ very distinct ; colour uniform shiny reddish brown.

Legs brown, becoming lighter toward the distal end of the tibiæ ; the tarsi yellow ; spur wide at base and very finely serrate.

Scutellum triangular, about twice as deep as the prothorax, sides arcuate, apex a rather sharp point, carinæ distinct, the lateral carinæ attaining the posterior margin, the median carina is obsolete toward the apex ; colour shiny, reddish brown, with the margins and apex yellow.

Elytra nearly twice as long as the abdomen ; colour smoky brown, veins darker brown with dark dots along them.

Tergum of the first two abdominal segments yellow, remainder of the abdomen reddish brown. Genitalia, aperture of the pygofers large, oval, wider than long ; ventral edge of the pygofers deeply notched, styles large, broad, divergent from their base, following the curve of the pygofers to the anal tube, a few long setæ at their apical end. Length, including elytra, 3.5 mm.

Taken on reeds by the author in Renwick Swamp, Ithaca, N. Y., July 20, 1908. Five specimens are before me, one of these shows much more yellow on the vertex and front. Type in the author's collection.

This species suggests the macropterous form of *L. lineatipes* Van Duzec, but the colouring is different and, besides other minor differences, the genitalia are decidedly unlike that species.

Achorotile foveata, n. sp.

Macropterous ♀.—Eyes oval deeply emarginate below to receive the antennæ ; colour gray. Vertex as wide as long, rounded before, extending a little beyond the eyes ; carinæ distinct, except the posterior median, which is weak ; posterior foveæ deep ; colour yellowish brown, lighter posteriorly.

Front twice as long as wide, widest near the middle, sides gently curving ; clypeo-frontal suture straight ; the median carinæ curved, following the curve of the sides to the front ; just outside of each median carina there is a row of pustules, six on each side, three near the vertex and three near the clypeus ; between these two groups are two pustules on each side along the outer margin of the front ; colour deep reddish brown.

Clypeus shiny black with an indistinct carina. Second segment of the antennæ about one and one-third times as long as the first, covered with pustules ; basal segment reddish brown, second segment yellowish brown.

Prothorax two-thirds as deep as the vertex; lateral carinæ distinct, following the curve of the eye; behind each carina a row of seven pustules; median carina faint, a puncture on each side of it about the middle of the prothorax; posterior margin quite deeply concave; colour deep reddish brown.

Legs yellow lineated with brown; tarsal claws black; spur triangular, finely toothed. Scutellum a little more than twice as deep as the prothorax; triangular, sides strongly arcuate, terminating in a rather sharp point; two pustules on each side near the middle of the lateral margin; median carina distinct, lateral carinæ rather faint, divergent; colour polished black with yellowish tip.

Elytra extending one-third of their length beyond the abdomen; veins brown, membrane slightly smoky.

Abdomen black, except the dorsum of the first two segments, which is yellowish; along the lateral margins of the dorsum of each segment is a transverse row of four pustules.

Genitalia: pygofers tapering to a blunt point, dark reddish brown; plates about one third as long as the pygofers, only a narrow edge showing, light brown in colour; ovipositor and anal tube dark honey-yellow.

Length, including elytra, 3.5 mm.

Described from a female taken by Professor J. Chester Bradley at Felton, Santa Cruz Mts., California, May 17, 1907. Type in the collection of Cornell University.

This species may be readily told from *A. albosignata* by the deeper vertex, the different coloration, and by the presence of four instead of two pustules on each side of the abdominal segments.

THE NORTH AMERICAN ÆSHNID DRAGON-FLIES.

At the present time, when the air is full of nomenclatural discussion, when there are many entomologists who are devoting themselves almost exclusively to naming and classifying insects from dried skins, "systematists" they are called, and we often seek in vain for the system, it is as refreshing as a woodland brook to a tired traveller to read a monograph of the nature of Dr. Walker's "North American Dragon-flies of the genus Æshna."* Here we have a systematic study of a group in which the

*University of Toronto Studies, Biological Series, No. 11, VIII, 213 pp., 28 plates (6 coloured). Publ. by the Librarian, University of Toronto Library, 1912. $2.00.

life-histories, ecology and seasonal and geographical distribution have been given the attention which they merit and which is necessary for a sound basis of classification. A classification which is not based on morphological characters considered in relation to and together with biological data must of necessity be incomplete. One thing is certain, that only further study of the bionomics of insects will settle the disputes of the "lumpers" and "splitters," to use colloquial but expressive definitions. The present monograph is an admirable illustration of this fact, and were this the only outstanding feature of this most thorough piece of work, the author would deserve the thanks of his entomological confreres. But his complete treatment of what he rightly characterizes as a "neglected group" of insects renders the volume additionally welcome both to entomologists and to those interested in zoögeography,

The monograph may be roughly divided into three sections, namely, taxonomic, bionomic and systematic. Perhaps the most important feature of the section on the taxonomy of the group is the fact that the author calls attention to the necessity of a study not only of a large series but of the colour pattern. The exclusive reliance upon structural features and the neglect to take into consideration the colour pattern has resulted in a "lumping" of species which a study of the natural colours does not support. A very careful study of colour patterns has therefore been made, and the six excellent coloured plates illustrating the same make this section of the work invaluable to the Odonatist.

Perhaps the most interesting, and, to the writer's mind, certainly the most important part of the bionomic section, is that dealing with variation and geographical segregation. If more than a brief reference were attempted here this review would exceed the appointed limits. In this section the author has, as it were, struck a rich metalliferous vein, and we are eager to follow it; it is too rich and promising to be left, and we hope it wi.l be followed up by further investigation. It is found that there occur in the females varieties in colour, in the length of the apparently functionless abdominal appendages and in the depth of the third abdominal segment and further, that there is a distinct correlation between the variations of the last two structures. These variations are dependent to a large extent upon locality, and hence, possibly upon climatic conditions. Here then is an unrivalled field awaiting the attention of the biometrician. Important observations have been made by the author upon the life-history, and his work is made increasingly valuable by the excellent illustrations, especially of the acts of oviposition, copulation and emergence of the adult.

The North American species of Æshna appear to resolve themselves into six groups, and about two-thirds of the monograph is devoted to very full and orderly descriptions of the twenty species and geographical subspecies found in North America. As an example of monographic treatment, this volume would be difficult to surpass, both in its broad and thorough character and in the unusual excellence of the author's numerous illustrations.

In congratulating the Editor of our journal on the production of so useful and excellent a monograph, which will bring great credit to Canadian entomology, we should also like to express our great indebtedness to the author's father, Sir Edmund Walker, for his generosity in rendering possible the illustration of the monograph by so excellent and large a series of plates, and which has enabled justice to be done to the exceptionally well-drawn figures. C. Gordon Hewitt.

A NEW PAPILIO FROM CENTRAL AMERICA.

BY GEORGE A. EHRMANN, PITTSBURG, PA.

Papilio chromealus, sp. n.

Closely allied to *P. copanæ* Reakert. Upper side of all the wings with a golden sheen (bluish-green in *P. copanæ*); submarginal arrow-shaped spots of the fore-wings orange instead of yellow. On the upper side of the hind wings the submarginal row of spots is identical with that of *P. copanæ* except in colour, the costal spot between nervules 1 and 2 being pale buff, the other four spots between nervules 2 and 5b deep orange chrome. The spot between nervules 5b and 5c is pure white. The spot between nervules 5c and 6 is the last spot and is also deep orange chrome.

The colour and all the markings of the underside are the same as in *P. copanæ*.

Habitat.—Honduras, Central America.

This fine Papilio was collected near the base of the Congrehoy Peak in the Province of Yoro by the late Dr. Carl Thime and sent to me with several thousand other Lepidoptera from various localities in Honduras.

P. chromealus is a very conspicuous Papilio and if not a distinct species it is assuredly a beautiful variation of *P. copanæ* Reakert, from Guatemala.

NEW SPECIES OF COLEOPTERA OF THE GENUS AGRILUS.

BY C. A. FROST, SOUTH FRAMINGHAM, MASS.

For several years considerable study has been given the specimens of this genus and some of the more important results are here presented:

Agrilus champlaini, new species. *Holotype* a male. Form of *muticus*, robust, colour black with a purplish tint, subopaque. *Antennæ* reaching the middle of the prothorax, serrate from the fourth joint, bronzed; *head* densely coarsely punctured, strigate above and pubescent below the middle, occiput somewhat concaved, the median impressed line extending to the middle of the front, eyes dark. *Prothorax* one-fourth wider than long, widest at the middle, one-fourth wider at the base than at the apex, sides regularly arcuate, hind angles obtuse with a rather strong carina, lateral margin sinuous, two discal foveæ in front of the middle, basal depression distinct, lateral faint; surface transversely rugose, becoming confused and finer at the sides and anterior angles. *Scutellum* not carinate, notched. *Elytra* nearly parallel to the apical seventh, tips separately and broadly rounded with coarse unequal serrations, disc slightly flattened with a vague costa, suture elevated behind the middle, surface rather coarsely irregularly granulate imbricate, finer in the basal depressions which are moderately deep. *Body* beneath more shining, bronzed, with purplish reflections especially along the sides and apex of the abdomen; *prosternal* lobe sinuate-truncate, intercoxal process broad, acute at tip, surface roughly densely punctate, propleuræ granulate; *metasternum* rather closely granulate becoming subasperate posteriorly like the coxal plates; *abdomen* finely punctate, sparsely along the middle, more closely along the sides, resembling imbricated scales on the first segment and with undulating lines connecting the punctures on the others; *pygidium* coarsely punctate with a strong projecting carina. First abdominal segment flat or slightly concave, second with smooth groove extending nearly to the posterior suture, last ventral eroded-truncate at the tip, vertical portion of abdomen granulate, inferior margin serrate, apex smooth and truncate. Front and middle tibiæ mucronate; *claws* similar on all feet, cleft with the inner parts broad and curving inward, the apices nearly touching.

Length 8 mm. Width at base of elytra 2.3 mm.

The females at hand have the front less pubescent and more concave, granulations of the eye more distinct, sides of the prothorax more arcuate, being widest in front of the middle, discal depressions more pronounced and with the intervals between the strigæ more distinctly punctate, surface of the elytra more finely and regularly imbricate; the abdomen is smoother and more shining bronze, with the first ventral slightly flattened; the claws are alike on all the feet, the inner portion quite broad and curving inward slightly, leaving a much wider space between the apices than in the male.

The pubescence of this species is not at all evident except on the post-clypeal area, the anterior angles of the prothorax, and the apices of the elytra. The sexes do not vary much in this respect, but I suspect that my males do not show the normal elytral pubescence; possibly it is discoloured. On the under side of the body the vestiture is short and sparse, giving a silver tint to the abdomen; it becomes more dense on the prosternum of the males.

This species would naturally be placed next to *angelicus* in Dr Horn's table (Trans. Am. Ent. Soc., XVIII, p. 283). Through the kindness of Mr. H. C. Fall, of Pasadena, Cal., I have been enabled to examine a specimen of *angelicus*, and can say that it does not resemble the present species in colour, form or punctuation. *Champlaini* might be confused with *anxius* and allies, but the darker colour, more robust and shorter form, sculpture, and structure of the claws should at once distinguish it.

This species is represented by three specimens (emerged May 29, 1911), two females and one male, all bred from the twigs of the hornbeam, *Ostrya virginica*, by Mr. Alfred B. Champlain, at New Haven, Connecticut. The specimens and two of the galls were sent to me by Dr. W. E. Britton from the State Agricultural Experimental Station in that city. It is through the kindness of these two gentlemen that the above description has been made. The galls were collected at Lyme, Conn., April 30, 1911.

The gall is in each case about one inch in length, fusiform, expanding to a diameter of 12 millimeters in the middle, and on branches of about six millimeters in diameter. One of the galls was split open and the bark removed from one side, so that the course of the larva was shown to be a spiral from the point of entrance toward the end of the twig. It circled the twig in four distinct courses, each one increasing in diameter about one-half the previous one and leaving a ridge between them. They are tightly packed with debris and are wholly in the wood, leaving the bark intact. On the fourth spiral the gallery leads directly to the heart of the branch, from whence it is hollowed out in an arcuate course downward until it intersects the bark in an oblong exit very near the point of entrance. The exit in the two specimens at hand measures three by two millimeters, the long diameter being transverse to the twig.

Since the above was written a male specimen has been received from Mr. H. B. Kirk, which was taken at Harrisburg, Pa., June 16, 1911. It differs slightly from the Connecticut male in the following particular: The front is more concave, the two discal foveæ of the prothorax are very indistinct, the prosternal lobe is distinctly emarginate in front, the scutellum is not notched, and the form is slightly more cuneate; with the elytral tips normally serrate.

The *holotype* and *allotype* are in my collection, and the remaining female *paratype* is in the collection of the Connecticut Agricultural Experimental Station.

A. cratægi, new species.—*Holotype* a male. Form of *obsoletoguttatus*, elongate. *Colour* olive-æneous, suffused with cupreous on apical third, varying to entirely cupreo-æneous, shining. *Antennæ* reaching the middle of the prothorax, serrate from the fourth joint, æneous. *Head* slightly flattened, greenish, granulate-punctate, more closely in a post-clypeal pubescent area, smooth spaces near the eyes finely alutaceous; occiput impressed, punctures tending to form strigæ posteriorly; median line extending to the middle of the front, where it ends in a slight depression. *Prothorax* one-fourth wider than long, varying in the specimens at hand from sides nearly parallel to regularly arcuate, hind angles rectangular, with a strong, nearly straight carina extending almost half the length of the prothorax, lateral margin slightly sinuate; disk with an anterior circular depression and a posterior oblong one, lateral oblique depressions moderate; surface transversely strigate with punctures between the strigæ. *Scutellum* transversely carinate. *Elytra* slightly sinuate behind the humeri and faintly dilated behind the middle, apices separately rounded and serrulate, disk slightly flattened, with the faintest indication of a costa, basal depressions rather slight, surface imbricate in very regular transverse series, gradually becoming finer toward the apex, pubescence very indistinct, visible only in the basal depression of the elytra and the anterior angles of the prothorax. *Body* beneath shining æneous with a faint cupreous tinge; prosternal lobe distinctly emarginate, granulate-punctate, with a thick mat of erect brownish hair extending from the anterior margin down the intercoxal process to the acute tip, and covering a small patch between the meso-coxæ; propleuræ and sides of metasternum granulate-imbricate, sparsely pubescent; metasternum smoother at middle. *Abdomen* imbricate on the first ventral, finely imbricate and sparsely punctate on the others; first ventral with a small patch of brownish hair on the intercoxal process, last ventral oval and eroded at tip; inferior margin of the vertical portion of the abdomen serrulate. *Pygidium* with an evident carina, not projecting, coarsely sparsely punctured. *Claws* similar on all the feet in both sexes, nearly bifid. Length 6 to 8 mm. Width 1.5 to 2.3 mm.

The females at hand differ as follows: Head cupreous or æneus, more concaved along the median line, more coarsely sparsely sculptured, an opaque depressed area above the clypeal carina; sides of prothorax slightly more arcuate; beneath sparsely pubescent, no erect brownish hair.

The two extremes in size, as given in the above description, are both females; the other specimens average 6.5 mm. in length.

A pair of this species first came to me from Mr. A. B. Champlain, with the label, "Chinchilla, Pa., VII, 2." The past year Mr. H. B. Kirk sent me four male and two females bred from the dead fallen trunk of Cratægus, which was also infested with *Xylotrechus colonus* and *Neoclytus luscus*. The material was collected by Mr. W. S. Fisher, of Highspire, Pa., and Mr. Kirk, at Harrisburg, Pa., March 11, 1911, and the specimens are labelled "Emerged IV, 6–11."

This species should be placed between *politus* and *fallax* in Horn's table, but it resembles neither of them so much as *obsoletoguttatus* in size and shape. From the two latter it can at once be distinguished by the lack of pubescent spots on the elytra, and the structure of the claws. It resembles the narrower forms of *politus* somewhat, but the hairy prosternum will at once separate the males, and the claws, form of prothorax, broader head, and the more parallel form will be sufficient to distinguish either sex in a series of *politus*.

The distribution of the types is as follows: *Holotype*, *allotype* and two *paratypes* in my collection, a male and female *paratype* in the collection of Mr. H. B. Kirk, a male *paratype* in the collection of Mr. W. S. Fisher, and a male *paratype* in the collection of Mr. Chas. Liebeck, to whom I am much indebted for comparing many specimens with the Horn types.

A. cephalicus Lec.—Original description (Trans. Am. Phil. Soc., XI, p. 249). Obscurus, ænescens, capite cupreo, haud pubescente fortiter haud confluenter punctato, sat profunde canaliculato, thorace latitudine haud breviore, dorso canaliculato et biimpresso, lateribus subrectis fortiter impressis, basi bifoveato, angulis posticis oblique carinatis, elytris sat fortiter dense granulatis, subunicostatis, apice subserratis rotundatis. Long, .18–.25. Locality: "Middle States and Lake Superior."

The above species was suppressed as a synonym of *egenus* by Dr. Horn in his monograph without explanation or remarks, and it seems to be entirely unwarranted. In making an examination of the *egenus* series in the Le Conte collection at Cambridge, I was much surprised to find that the specimen bearing the label "*A. cephalicus* Lec." (and also "*egenus* No. 11") belonged to a different group, having the antennæ serrate from the fourth joint. Numbers 3, 5 and 16 were also this species, and there were seven specimens having the fourth joint serrate and the inner lobe of the claws incurved; these are probably *otiosus*. The type of *puncticeps* Lec., which has the fourth joint serrate, is placed as No. 13 in the *egenus* series. This was also made a synonym of *egenus* by Dr. Horn. The exact standing of *puncticeps* is at present doubtful; I was inclined to place it in the *otiosus* group, although I was unable to see the claws of the middle tarsi, which were the only ones intact; but since reading Le Conte's synopsis, in which he places it in the group with the inner lobe of the claws contiguous, I consider it to be a valid species. If it should prove to be identical with *cephalicus*, the name *puncticeps* will have priority.

The Le Conte specimen bearing the label *cephalicus* is a female, and I have prepared the following re-description from five males from Highspire, Pa., June 12 to 20, 1909, and June 14, 1910, all collected by Mr. W. S. Fisher, of that place. He also sent me two males from Jeannette, Pa.

A. cephalicus Lec. Re-description: Form of *otiosus;* colour æneous-olive. *Antennæ* moderate, bluish æneous, serrate from the fourth joint, second and third joints with rather long pubescence on the under side. *Head* convex, bluish, median line varying from distinctly to faintly impressed, and extending to a post-clypeal pubescent area, sparsely punctate, finely alutaceous, strigate on the occiput. *Prothorax* a little wider than long, narrowed at the base, sides feebly arcuate, lateral margin nearly straight, hind angles with a well defined carina, disk convex, subequally bi-impressed on the median line, an oblique lateral depression; surface transversely strigate, strigæ becoming confused anteriorly. *Scutellum* transversely carinate. *Elytra* subparallel, narrowed at the apical third to the rounded serrulate apices; disk flattened, faintly costate; surface densely imbricate, basal impressions moderate. *Body* beneath bluish varying to greenish aeneous. *Prosternal* lobe distinctly emarginate and covered with an erect grayish pubescence, or hair, that extends to near the middle of the first ventral segment; it is shorter and less noticeable on the metasternum; tip of the intercoxal process of prosternum acute; propleuræ and coxal plates subgranulate, sparsely pubescent; metasternum scabrous-imbricate; ventrals imbricate, becoming finer toward the apex and sparsely punctate along the middle, tip of last ventral oval, granulate. *Pygidium* coarsely irregularly punctured without distinct carina. *Tibiæ* mucronate on all the legs. *Claws* with a broad obtuse tooth at the base, similar on all the feet. Length 4.5 to 6 mm. Width 1.2 to 1.4 mm.

A male from Lyme, Conn. collected by Kirk and Champlain July 4, 1911, seems to belong to this species but does not quite accord in some particulars. A male from Vicksburg, Miss., sent me by Col. T. L. Casey, agrees very well with the Pa. specimens, but it is somewhat smaller and of a lighter bronze-aeneous colour.

A female from Highspire, Pa., which agrees well with the type, shows the following sexual differences: form slightly more robust, colour bronze-aeneous, head larger and broader between the eyes, more densely punctate, less pubescent, aeneous in colour; prothorax less narrowed behind; elytra slightly dilated behind the middle; the median area of denser pubescence of the male is here lacking. Two females from Jeannette, Pa., are somewhat smaller and resemble the males more nearly in form. I have also seen a male from Lafayette, Ind., June 21, 1907; collection of Mr. A. B. Wolcott, Chicago, Ill.

In the type there is an apparent carina of the pygidium which is caused by a median smooth space sharply limited by the coarse and irregularly confluent punctuation on each side of it. The punctures are very sparsely placed toward the margin. The pygidium of a single male that I have dismembered shows similar characteristics.

Agrilus auricomus, new species. *Holotype* a male. Form elongate, depressed, broadest at the base of the elytra, black or olivaceus black, shining; thorax shining aeneous with a path of golden pubescence in the lateral depression extending to the anterior angles. *Antennæ* reaching the middle of the prothorax, aeneous, serrate from the fourth joint. *Head* densely coarsely punctate, becoming transversely strigate above the middle; occiput feebly impressed; the median impressed line extending from the back of the head nearly to the clypeus; a triangular patch of golden pubescence above the clypeus, the suture of which is indicated by a fine carina; granulations of the eye unusually fine or indistinct. *Prothorax* one-third wider than long, base very slightly wider than the apex, at the middle equaling the elytra at the base, sides regularly arcuate, lateral margin sinuate; disk depressed and with a shallow median impression; two small circular foveæ each side in front of the middle, vaguely defined by surrounding smooth space; lateral depressions deep, causing sides to appear explanate; surface transversely strigate; hind angles with a faint carina. *Scutellum* carinate at the sides, interrupted at the middle. *Elytra* slightly sinuate behind the prominent humeri, feebly broadened behind the middle, basal depressions deep with two very faint impressions behind them ; the junction of the discal flattened portion with the convex sides of the elytra has the appearance of obtuse costæ curving inward from the humeri and vanishing near the middle of the elytra ; apices separately rounded, subacute and serrulate ; surface coarsely imbricate-granulate. *Body* beneath shining aeneous with golden pubescence which is sparse on the middle of the abdomen and becomes very dense on the episterna, sides of the metasternum, outer half of the coxal plates, vertical portion of the abdomen, and a triangular patch on each side of all the abdominal segments. *Prosternum* thickened, swollen behind the lobe which is slightly sinuate-truncate in front and with a distinct marginal bead ; inter-coxal process slightly concave, longitudinally with an acute and depressed tip ; surface densely punctate. *Metasternum* rather densely punctate, becoming rather strigate at the sides. *Abdomen* with the first segment flattened and rather coarsely densely punctate at the middle, becoming strigate at the sides ; second segment with a deep smooth groove with sharply defined edges narrowing and vanishing at the posterior third, sparsely coarsely punctate ; last three segments finely and less sparsely punctate ; apex of the last segment subtruncate with a slight tendency to emargination; median carina of the pygidium strong and projecting. *Claws* cleft alike on the middle and hind tarsi ; nearly bifid on the front pair with the inner lobes less incurved ; front and middle tibiæ distinctly mucronate. Length 10 mm. Width 2.4 mm. at the base of the elytra.

Three males of this species were taken at Framingham, Mass., May 31, 1909. One of these is now in the collection of Prof. H. C. Fall, Pasadena, Cal., and to him I am indebted for an examination of this specimen with reference to the above description. There is very little variation in the two specimens in my collection, the type being slightly more cuneate in form.

A female collection by Mr. A. B. Champlain at Lime, Conn., now in the Experiment Station collection at New Haven, May 29th, 1910, is referred to this species. It differs in being more black in colour (but not the opaque black of *bilineatus)* and more robust in every way while being only slightly longer —10.8 mm. The occiput is more impressed, and the granulations of the eyes normal. The carina of the hind angles of the thorax is more distinct ; the prosternal lobe shows a slight emargination in front and is less swollen behind it. The punctuation is more dense beneath and the golden pubescence covers less area. The claws are cleft alike on all the feet, the inner lobe being nearly as long as the outer and somewhat incurved. The last abdominal segment is slightly eroded-granulate and faintly truncate. The thorax at the middle, the elytra at the base and at the enlargement behind the middle measure 3 mm.

Another female from New Haven, Conn., June 12, 1911, collected by Mr. B. H. Walden is similar to the preceding except in the following details : occiput less impressed ; thorax with a larger and more elongate basal depression; lacking the two anterior foveæ; lateral margin more sinuate; at middle slightly broader than the elytra at the base; carina of the hind angles obsolete; tips of the elytra more acute and prolonged and having a sutural angle equal to one-seventh the length of the elytra; between the first and second segments of the abdomen there is a distinct suture extending half way to the middle of the body; last segment rounded with the edge eroded-granulate; the pubescence is a yellowish white. Length 11.8 mm., width 3 mm. at base of elytra and 3.5 mm. behind the middle at the enlargement. The colour of this specimen is slightly olivaceous as in the type.

This species is closely related to those specimens that have been referred to by Dr. Horn (Trans. Am. Ent. Soc, Vol. XVIII., page 308) as the olivaceous variety of *acutipennis* but the form and the golden pubescence should at once separate it from that variety.

It appears to me that the term "last abdominal segment serrate" has not been hitherto clearly defined or the serrations have escaped notice in many species. In the present species the lower edge of the vertical portion of the abdomen is strongly serrate. The serrations begin near the middle of the last segment where the overhang of the superior part commences to be prominent, and, increasing in coarseness, extends to the smooth apical area where the two edges of the superior portion merge directly beneath the pygidial carina. The inferior portion of the last segment at the tip, which is the part referred to in the previous description, is granulate near the edge. In the females the first four abdominal in the larger specimen and the first three in the other are visible when the specimens are viewed from directly above. In the males only the first two segments are so visible.

The short grayish pubescence that covers the elytra and thorax in specimens of *anxius* and related species is here almost invisible except on the apices of the elytra and for a short distance along the suture. This pubescence arises from slight depressions in the furrows between the rugæ and is seen to be arcuately decumbent in a lateral view across the elytra toward the light. Under a high-power hand lens it appears as minute silvery points on the elytral disk of this species. By placing specimens with the head toward the light and the body inclined backward toward the observer pubescent spots and apical vittæ can be seen on many species that have been described as being without elytral pubescence.

In conclusion it may be said that the studies in the *otiosus* and *anxius* groups have been, so far, rather disappointing, due to the difficulty of getting series of both sexes. The only species at all abundant in this locality is *otiosus*, taken on oak leaves. The olivaceous variety of *acutipennis* has been encountered quite often on oak, and *bilineatus* occurs in favourable places on oak sprouts; but in general the species turn up singly or in pairs, with aggravating slowness. Several very interesting problems are suggested by the material at hand, and more specimens from widely separated localities may present a solution.

Mailed August 6th.

The Canadian Entomologist.

Vol. XLIV. LONDON, SEPTEMBER, 1912. No. 9

THE ODONATA OF THE PRAIRIE PROVINCES OF CANADA.

BY E. M. WALKER, TORONTO.

With the exception of the short lists of captures in the Entomological Record, published in the Annual Reports of the Entomological Society of Ontario, and a few other isolated records, no information appears to be extant on the Odonata of the vast territory between Ontario and British Columbia. Before the section on the Odonata of the new Catalogue of Canadian Insects is issued, it seems, therefore, desirable to place on record in detailed form all the information we have been able to obtain on the distribution of the dragonflies of this region.

The source of this information is mainly to be found in the collections made by Messrs. J. B. Wallis, N. Criddle, T. N. Willing and N. B. Sanson, and to these gentlemen the writer wishes to express his sincere thanks for the privilege he has enjoyed of retaining specimens for study for an indefinite length of time, or permanently for his collection. The list is of a preliminary nature, and no doubt many species will be added to it in the future.

In looking over almost any collection of dragonflies from the prairie country one is apt to be struck with the large preponderance in individuals of the genera *Lestes*, *Sympetrum*, *Enallagma* and *Æshna*. These genera are also best represented in number of species, *Leucorrhinia* coming fifth. The latter genus is probably nowhere better developed in North America than here. There are doubtless also more species of *Somatochlora* from this region than appear in the present list, particularly in the less explored northern parts. Apart from this genus, the *Corduliinæ* are apparently poorly developed. The absence of *Agrioninæ* (*Calopteryginæ* Auctt.) and *Cordulegasterinæ* is probably also due to insufficient exploration. The occurrence of two species of the genus *Coenagrion* (*Agrion* Auctt.) is of much interest, one of the species being almost identical with the Palæarctic *C. lunulatum*. Finally, attention may be drawn to the fact that if we include *Æshna cærulea septentrionalis*, which has been taken at Fort Resolution, Great S'ave Lake, and doubtless occurs also in Northern Saskatchewan and Albeita, the list includes all the species of dragonflies that are common to the Palæarctic and Nearctic Regions, except the essentially tropical *Pantala flavescens.* These species are *Enallagma cyathigerum*, *Anax junius*, *Æshna cærulea*, *Æshna iuncea*, *Æshna palmata*, *Sympetrum scoticum* and *Libellula quadrimaculata*. The only genera not represented in the Palæarctic region are *Argia*, *Amphiagrion* and *Tetragoneuria*.

In the following list the names of the collectors, Messrs. Wallis, Criddle, Willing, Sanson and the late Dr. Fletcher, are abbreviated: Ws, C, Wg, S and F, respectively.

LIST.

1. *Lestes congener* Hagen.

MANITOBA.—Aweme, Aug. 29, 1907, 1 ♂; July 10, 1909, 1 ♀ (C) Westbourne, July 27, 1908, 3 ♀s; Aug. 1, 1908, 1 ♂; Aug. 20, 1908, 2 ♂s, 1 ♀; Aug. 26, 1908, 2 ♂s, 2 ♀s (Ws).

This species ranges across the continent, and is apparently most abundant in the Canadian Zone.

2. *Lestes unguiculatus* Hagen.

MANITOBA.—Aweme, Aug. 5, 6, 1907, 2 ♂s; July 10, 1909, 1 ♀ (C). Westbourne, July 27, 1908, 2 ♂s, one teneral; July 29, 2 ♂s, 1 ♀, incl. pair in cop.; Aug. 10, 14, 2 ♂s (Ws). Winnipeg, July 9, 1908, 2 ♂s, 1 ♀ (Ws).

SASKATCHEWAN.—Regina, 3 ♀s (Wg); Aug. 7, 1903, 2 ♂s (F). Goose Lake, July 20, 1907, 1 ♂, 2 ♀s, teneral (Wg). Davidson, Aug. 21, 1907, 1 ♀ (Wg). Radisson, July 29, 1907, 2 ♂s, 2 ♀s (Wg, F). Lumsden, Sept. 10, 1906, 1 ♂ (W. J. Alexander).

ALBERTA.—Near Waterton Lake, Aug. 5, 1908, 1 ♂; Aug. 10, 1908, 1 ♂, 2 ♀s (E. V. Cowdry).

A transcontinental form, inhabiting chiefly the Transition and Upper Austral Zones.

3. *Lestes uncatus* Kirby.

MANITOBA.—Aweme, Aug. 18, 30, 1907, 2 ♀s (C). Westbourne, July 27, 1908, 2 ♂s, 2 ♀s, incl. pair in cop.; July 29, 1908, 1 ♂; Aug. 10, 1908, 1 ♂ (Ws). Winnipeg, July 24, 1908, 1 ♂, 1 ♀ (Ws)

SASKATCHEWAN.—Regina, July 17, 1907, 1 ♂, 1 ♀ (F); June 19, 1908, 1 ♀ (Wg). Goose Lake, July 19, 20, 1907, 3 ♂s, 5 ♀s (F).

Another transcontinental species. Very common on the Canadian prairies.

4. *Lestes disjunctus* Selys.

MANITOBA.—Aweme, Aug. 5, 1905, 1 ♂ (C). Westbourne, July 27, 1908, 6 ♂s; July 29, 1908, 1 ♂, 2 ♀s; Aug. 10, 1908, 1 ♂; Aug. 29, 1908, 1 ♀ (Ws). Winnipeg, July 4, 1908, 4 ♂s, 2 ♀s (Ws).

SASKATCHEWAN.—Regina, July 17, 1907, 1 ♀ (F), 1 ♀ (Wg). Duck Lake, July 22, 1907, 10 ♂s, 9 ♀s (F, Wg).

ALBERTA.—Banff, July 11, 18, 1908, 3 ♀s (S).

This is probably the commonest Canadian *Lestes*, and like the other species listed here, is widely distributed, occurring from Nova Scotia to British Columbia.

5. *Argia vivida* Hagen.

ALBERTA.—Banff, swamp off Hot Springs Road, June 21, 1908, 1 ♂, teneral (S).

This species has already been reported from this locality and from Glacier, B. C., by Osburn (Ent. News, XVI, 1905, p. 187). It probably does not belong to the prairie fauna.

6. *Nehalennia irene* Hagen.

MANITOBA.—Aweme, July 25, 1908, 2 ♀s; July 4, 1909, 1 ♀ (C). Westbourne, July 27, 29, 1908, 1 ♂, 2 ♀s (Ws). Winnipeg, July 7, 1908, 1 ♂, 1 ♀ (Ws). Winnipeg Beach, Lake Winnipeg, June 19, 1909, 3 ♂s, 4 ♀s (Ws).

These are the most westerly records for this species in Canada.

7. *Amphiagrion saucium* Burm.

MANITOBA.—Aweme, June, 1911, 1 ♀, teneral (E. Criddle).

This species is known also from Quebec, Ontario and British Columbia, but appears to be very local in Canada.

8. *Coenagrion resolutum* Hagen. (Pl. IX, figs. 1, 1a.)

Though the males of this species are readily distinguished by the peculiar form of the abdominal appendages, it may be worth while to record a description of the colour-pattern of both sexes, as I have before me some excellently preserved alcoholic specimens, received from Mr. T. N. Willing, of Regina, Sask.

Male: Head bronze-black above, postocular spots blue, posterior margin of occiput yellowish green. Eyes pale green, dark olivaceous above. Face, including a broad front margin of the frons, pale green or greenish yellow, except the nasus, which is bronze-black. Pronotum bronze-black, the anterior and lateral lobes, a marginal line along the sides of the posterior lobe and a spot on each side mesad of the lateral lobes, black. Thorax bronze-black, the humeral bands pale green to bluish green, slightly curved, rounded at both ends, widest in front, more or less constricted towards the posterior end. Pleura pale bluish to yellowish green, becoming more yellowish beneath. Abdomen pale blue above, yellowish green beneath, marked with bronze-black as follows: Segs. 1–3 as in fig. 1; slightly more than apical half of 4 and 5; 6 and 7, except a very narrow interrupted basal line; 10 dorsally, except a greenish median spot at the posterior margin. The superior appendages black, their slender inferior processes and the inferior appendages black-tipped.

Female: Colour variable, the pale markings being sometimes blue above, as in the male, but varying to wholly greenish yellow. Markings of head and thorax similar to those of the male, but the postocular spots are larger, and the posterior pale marginal line of the pronotum is entire or barely interrupted. Abdominal segments marked above with dark bronze as follows: Segs. 1–3 as in figure 1a; 4–6 except a basal interrupted line; 7 except a basal interrupted line and a bluish apical line; 8 and 9 except a bluish apical band; 10 with a subtriangular dorsal spot.

MANITOBA.—Winnipeg, July 7, 1908, 1 ♂ (Ws). Winnipeg Beach, Lake Winnipeg, June 19, 1909, 12 ♂s, 1 ♀ (Ws).

SASKATCHEWAN.—(Locality not given.) June 20, 1908, 5 ♂s, 3 ♀s.

A widely-distributed boreal species, occurring locally also in the Transition Zone.

9. *Coenagrion angulatum*, sp. nov. (Pl. IX, figs. 2, 2a, 2b, 2c.)

Closely allied to *C. lunulatum*, from which it differs somewhat in the form of the abdominal appendages of the male.

The pale terminal tubercle of the superior appendages is shorter and more broadly rounded, and the angle between it and the inferior process is shallower; the apices of the inferior appendage are much smaller and do not project beyond the latter process, as in *lunulatum;* while the inferior process of the inferior appendage is shorter, broader and blunter than in *lunulatum*.

Male: Azure blue above, greenish yellow beneath. Head black above, postocular spots blue, rather large; eyes pale green, dark olivaceous dorsally. Face pale green; nasus and a line between rhinarium and labrum black, middle lobe of labrum pale bluish. Pronotum black, anterior lobe blue, lateral lobes pale yellowish green. Thorax bronze-black, the blue humeral bands about as broad as the black bands laterad of them, straight or but feebly curved, the margins subparallel; pleura blue, fading beneath into pale yellowish. Legs pale yellow, outer surfaces of femora and inner surfaces of tibiæ and whole of tarsi black. Abdomen blue above, yellowish beneath, marked above with bronze-black as follows: Seg. 1 with a transverse basal spot; 2 with a narrow transverse angular spot and an apical transverse band; 3 except the basal two sevenths; 4 except the basal fifth; 5 except a pair of spots on the basal sixth; 6 and 7 except a narrow interrupted basal line; dorsum of 10, or a laterally constricted spot upon it. Superior appendages black, with a pale terminal tubercle, inferior appendages pale, the sides and apices black.

The female resembles that of *C. resolutum*, differing as follows: The posterior margin of the pronotum is slightly trilobate, the middle portion arcuate as seen from behind (not at all trilobate in *resolutum*), the pale posterior margin is narrower and sometimes confined to this middle lobe; the thoracic bands, as in the male, are straighter, somewhat broader and the sides more nearly parallel. The abdomen is marked similarly to that of *C. resolutum*, but the dark areas are somewhat more extensive on segs. 1–3, and on 7 there is a basal interrupted pale band, which is absent in the latter species. The dorsum of 10 is entirely dark, except a narrow posterior marginal line.

The two alcoholic specimens which I have are pale yellowish, with a reddish tinge on the thorax, the transverse bands on 7 and 8 faintly bluish.

Length of body, ♂, 29–31 mm., ♀, 28–30 mm.; abdomen, ♂, 22–23.5 mm., ♀, 22–23.5 mm.; hind wing, ♂, 16–17 mm., ♀, 18–18.7 mm.

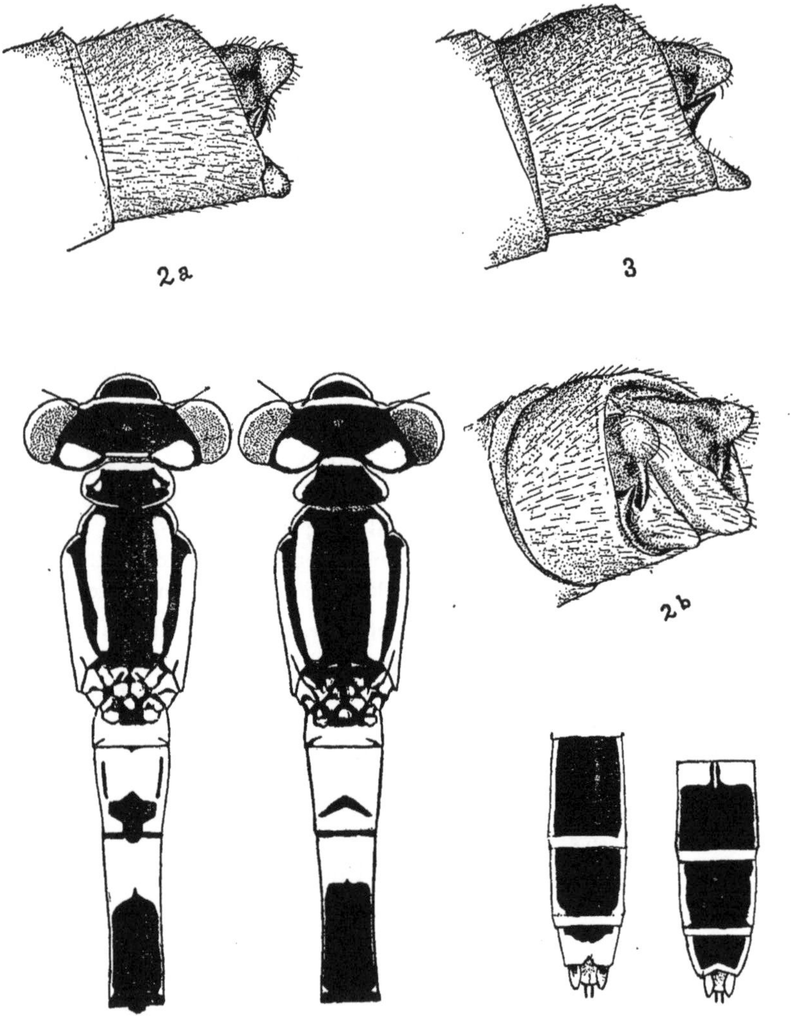

THE ODONATA OF WESTERN CANADA

Types.—♂, Carnduff, Sask., July 16, 1900 (Wg).

Cotypes: MANITOBA.—Aweme, July 4, 1905, 1 ♀ (C). Winnipeg Beach, Lake Winnipeg, June 19, 1909, 4 ♂s (Ws).

SASKATCHEWAN.—Prince Albert, June 18, 1905, 1 ♀. Also 1 ♂, 3 ♀s from Saskatchewan without further data (Wg).

To Mr. Kenneth J. Morton is due the credit of first recognizing the close relationship between this species and *C. lunulatum*.

Can this be the *Agrion interrogatum* Selys, described from the female only, from Saskatchewan? (Bull. Acad. Belg. (2) 41, p. 1254, 1876).

10. *Enallagma cyathigerum* Charpentier.

SASKATCHEWAN.—Prince Albert, June 18, 1905, 1 ♂ (Wg). Kinistino, July 22, 1 ♂ (F). Duck Lake, July 22, 1907, 6 ♂s (Wg, F).

ALBERTA.—Lethbridge, July 5, 9, 1907, 2 ♀s (Ws). Calgary, July 10, 1903 (Wg). Near Waterton Lake, Aug. 10, 1908, 1 ♂ (Cowdry).

I am unable to distinguish the females of this species from the following, and therefore have not included them in the above list Females probably of both species have been received from the following localities: Aweme, Man., July 1, 1909, 1 ♀ (C). Winnipeg Beach, Man., June 19, 1909, 5 ♀s (Ws). Abernethy, Sask., June 27, 1903, 1 ♀ (Wg). Duck Lake, July 22, 1907, 18 ♀s (Wg, F). Lethbridge, Alta., July 5, 9, 2 ♀s (Ws). Banff, Alta., June 17, 1908, 2 ♀s (S). Near Waterton Lake, Aug. 5, 1908, 1 ♀ (Cowdry).

This circumpolar species doubtless occurs also in Manitoba, as I have taken it in Northwestern Ontario (Nipigon).

11. *Enallagma calverti* Morse.

MANITOBA.—Aweme, June 24, 1909, 1 ♂; July 4, 1909, 2 ♂s (C). Winnipeg Beach, Lake Winnipeg, June 19, 1909, 2 ♂s (Ws).

SASKATCHEWAN.—Prince Albert, June 19, 1905, 1 ♂ (F). Duck Lake, July 22, 1907; 2 ♂s (Wg, F).

ALBERTA.—Medicine Hat, June 29, 1904, 1 ♂ (Wg). Lethbridge, July 5, 9, 2 ♂s (Ws). Banff, June 17, 1908, 1 ♂ (S). Laggan, 1 ♂ (J. E. Bean).

The females, as stated above, are listed with the preceding species, from which they are apparently inseparable.

These two closely-allied boreal species seem to be the commonest Enallagmas of the prairies. I believe that they are one and the same species, as I have seen males which could be placed about equally well in either species.

12. *Enallagma hageni* Walsh.

MANITOBA.—Westbourne, July 27, 1908, 2 ♂s, 1 ♀ (Ws).

SASKATCHEWAN.—Regina, July 17, 1907, 1 ♀? (F).

Apparently rarer than in Ontario, where it is by far the commonest Enallagma, except, perhaps, in the far north.

The record from Regina is quite doubtful, as the females of this species are difficult to separate with certainty from certain allied species.

13. *Enallagma ebrium* Hagen.

MANITOBA.—Westbourne, July 27, 29, 1908, 3 ♂s, 5 ♀s (Ws). Winnipeg, July 7, 1908, 1 ♀; July 24, 1908, 2 ♂s (Ws).

SASKATCHEWAN.—Carnduff, July 16, 1900, 2 ♂s (Wg).

This species is not known in Canada west of Saskatchewan.

14. *Enallagma civile* Hagen.

MANITOBA.—Winnipeg, July 9, 1908, 1 ♂; July 28, 1908, 1 ♀ (Ws).

This is the northern limit of this species as far as known.

15. *Ophiogomphus rupinsulensis* Walsh.

MANITOBA.—Aweme, June 30, 1907, 2 ♂s; July 19, 1910, 2 ♀s (C).

SASKATCHEWAN.—Saskatoon, July, 1907, 1 ♂ (Wg).

The dark markings of the thorax are less distinct than in specimens from Algonquin Park, Ont.

The females were quoted doubtfully in the Entomological Record for 1911 as *O. severus* Hag.

16. *Ophiogomphus severus* Hagen.

ALBERTA.—Lethbridge, July 8, 1909, 1 ♂ (Ws).

This specimen was compared with specimens of *O. severus* in the Hagen collection (Museum of Comparative Zoology, Cambridge, Mass.).

17. *Gomphus externus* Hagen.

MANITOBA.—Aweme, June 30, 1907, 1 ♀; July 22, 1909, 1 ♂; July 9, 1910, 1 ♂ (C, Ws). Winnipeg, June 25, 1910, 1 ♂, 1 1 ♀ (C).

18. *Gomphus notatus* Rambur.

MANITOBA.—Aweme, July 19, 1910, 4 ♂s, 3 ♀s (C, Ws).

This species was determined by Mr. E. B. Williamson, as I had never met with it before. It has also been recorded from the Province of Quebec.

19. *Æshna sitchensis* Hagen.

MANITOBA.—Winnipeg Beach, Lake Winnipeg, Sept. 6, 1909, 1 ♂ (Ws). Westbourne, Aug. 19, 1908, 1 ♂ (Ws).

SASKATCHEWAN.—2 ♂s, 1 ♀. (Exact locality not given. Scudder, Museum of Comparative Zoology.)

This boreal species ranges from Newfoundland to Alaska.

20. *Æshna juncea* Linné.

ALBERTA.—Banff, July 17, 1902, 2 ♀s (R. C. Osburn, S).

Circumpolar and common in the Boreal Region of North America.

21. *Æshna subarctica* E. M. Walker.

MANITOBA.—Winnipeg, Sept. 9, 1909, 1 ♀ (Ws).

This is the western limit of this species so far as known.

22. *Æshna interrupta lineata* E. M. Walker.

MANITOBA.—Aweme, July 20, 1906, Aug. 10, 1907, Aug. 16, 1908, 2 ♂s, 2 ♀s (C). Winnipeg, July 6–24, Sept. 7, 1908, 5 ♂s, 5 ♀s (Ws). Winnipeg Beach, Lake Winnipeg, Aug.–Sept. 6, 1909, 2 ♂s, 3 ♀s (Ws). Westbourne, July, 1907, Aug. 16, 1908, 2 ♂s, 2 ♀s (C). Swan River, Sept. 8, 1906, 1 ♀ (W. J. Alexander).

SASKATCHEWAN.—Meota, July 8, 1907, 1 ♂, 1 ♀ (Wg). Carlton, July 28, 1900, July 22, 1907, 3 ♂s (F. Wg). Duck Lake, July 22, 1907, 2 ♂s, teneral (F, Wg). Goose Lake, July 21, 1907, 1 ♂ (Wg). Parkside, July 24, 1907, 1 ♂, teneral (Wg). Regina, July 18, 1905, 4 ♂s, 2 ♀s (Wg). Moose Jaw, Aug. 24, 1908, 1 ♀ (Caudell).

ALBERTA.—Banff, Aug. 4, 1906, Aug. 16, 1908, 1 ♂, 1 ♀ (Currie, S). Waterton Lake, Aug. 7–10, 1908, 2 ♂s, 3 ♀s (E. V. Cowdry).

I have also 1 ♂, 2 ♀s from Banff, taken Sept. 6, 1906, and Sept. 1, 10, 1908 (S), which approach the race *interna* Walk. Similar intermediates are found in British Columbia. In the prairies country only the pure *lineata* occurs.

This is the most characteristic dragonfly of the Great Plains in Canada.

23. *Æshna eremita* Scudder.

MANITOBA.—Husavick, July 8, 1910, 1 ♂ (Ws).

SASKATCHEWAN.—(Without definite locality.) 6 ♂s, 3 ♀s (Kennicott).

ALBERTA.—Banff, 1 ♂ (S).

This is the most generally distributed *Æshna* of the Boreal Region, and is very common in the wooded parts of the north.

24. *Æshna canadensis* E. M. Walker.

MANITOBA.—Westbourne, Aug. 24, 1908, 1 ♂ (Ws).

Abundant in the Canadian Zone in the Eastern Provinces, and occurring also in British Columbia and Washington.

25. *Æshna palmata* Hagen.

ALBERTA.—Near Waterton Lake, 4,100 ft., 3 ♂s (Cowdry). Banff, July 10, Aug. 6, 1908, 2 ♂s, 1 ♀ (S). Laggan, July 22, 1901, 1 ♂ (Osburn).

These are the extreme eastern limits of this species, which is abundant on the Pacific Coast.

26. *Æshna umbrosa* E. M. Walker.

MANITOBA.—Winnipeg Beach, Lake Winnipeg, Sept. 6, 1909, 1 ♂ (Ws). Hilton and Treesbank, July 28, 1910, 2 ♂s (Ws).

These specimens belong very decidedly to the eastern race *umbrosa*. The western form, *occidentalis*, occurs in British Columbia, and will very likely turn up in the Rockies of Alberta.

27. *Æshna constricta* Say.

MANITOBA.—Westbourne, Aug. 26, 29, 1908, 2 ♂s, 1 ♀ (Ws).

This eastern species is not likely to be found in Saskatchewan or Alberta. There is, however, a single record from British Columbia, which needs confirmation.

28. *Anax junius* Drury.

MANITOBA.—Aweme, Sept. 9, 1906, 1 ♀ (C).

29. *Tetragoneuria spinigera* Selys.

MANITOBA.—Winnipeg, June 17, 19, 1910, 3 ♀s (Ws, C). Aweme, June 11, 1905, 1 ♂; June 30, 1907, 1 ♂ (C).

This species occurs commonly also in British Columbia and Ontario. It is the most northern species of the genus.

30\. *Dorocordulia lintneri* Hagen.

MANITOBA.—Lake Winnipeg (Hagen).

31\. *Somatochlora semicircularis* Selys.

ALBERTA.—Banff, July 7, 15, 18, 1908, 1 ♂, 2 ♀s (S).

Also recorded from Laggan by Osburn. It occurs in abundance in British Columbia, and although recorded from Maine, I consider it very doubtful whether the latter record belongs to the same species.

32\. *Somatochlora macrotona* Williamson.

MANITOBA.—Winnipeg Beach, June 19, 1909, 3 ♀s (Ws). Winnipeg, June 16, 19, 1910, 2 ♂s, 4 ♀s (Ws). Husavick, Aug. 17, 1910, 1 ♀ (Ws).

In one of the males of this beautiful insect the wings are slightly tinged with yellow, in the other there is a clear yellow patch on the basal half of the wings, most distinct on the hind pair. Of the females, the wings vary considerably in the amount of flavescence. In one of the Winnipeg specimens, e.g., the basal half of both pairs is only slightly tinged with yellowish, the rest clear ; in others the basal half to three-fifths is yellowish brown, while in the specimen from Husavick the wings are wholly suffused with a rather dark yellowish brown.

A male of this species was sent to Mr. Williamson for determination. It was hitherto known only from Duluth, Minn.

33\. *Somatochlora albicincta* Burmeister.

MANITOBA.—Aweme, June 11, 1909, 1 ♀ (C).

ALBERTA.—Banff, July 7, 1908, 1 ♀ (S).

A species of the north, probably distributed through almost the entire Boreal Region, except the Arctic Zone.

34\. *Somatochlora* sp.

I have also examined a female *Somatochlora* from Banff, Alta. (S), which I was unable to determine.

35\. *Libellula quadrimaculata* Linné.

MANITOBA.—Aweme, July 12, 1906, 1 ♂; July 9, 1907, 1 ♂ Winnipeg, July 8, 9, 1908, 2 ♂s (Ws).

SASKATCHEWAN.—Duck Lake, July 22, 1907, 1 ♀ (F).

Circumpolar and very generally distributed. It is common in the Eastern Provinces and in British Columbia.

36. *Libellula pulchella* Drury.

MANITOBA.—Aweme, July 9, 13, 1910, 2 ♂s (C). Husavick, July 11, 1910, 1 ♂, 1 ♀ (Ws).

These are the most northwesterly records for this common eastern species.

37. *Sympetrum scoticum* Donovan.

MANITOBA.—Aweme, Sept. 3. 1907, 2 ♀s; Aug. 10, 23, 1908, 4 ♂s, 1 ♀ (C); Aug. 28, 1908, 1 ♂ (E. Criddle). Winnipeg, Sept. 7, 1908, 1 ♂, 1 ♀ (Ws). Westbourne, July 27, 29, 1908, 4 ♂s, 6 ♀s; Aug 1, 1908, 1 ♂, 1 ♀ (Ws); Aug. 20, 29, 1908, 14 ♂s, 12 ♀s (Ws). Grandview, Sept. 18, 1906, 2 ♂s, 1 ♀ (W. J. Alexander).

SASKATCHEWAN.—Regina, Sept., 1908, 1 ♂ (Wg). Moosomin, Sept. 13, 1906, 1 ♂ (Alexander).

ALBERTA—Banff, July 24, 30, 1908, 1 ♂, 2 ♀s; Aug. 28, 1908, 1 ♂ (S). Beaver Lake, 1907 (A. Halkett).

Circumpolar and generally common in the Canadian Zone.

38. *Sympetrum costiferum* Hagen.

MANITOBA.—Aweme, Sept. 7, 1907, 1 ♀; Sept. 1, 1908, 1 ♂, 1 ♀ (C). Westbourne, July 24, 29, 1908, 2 ♂s, 4 ♀s; Aug. 5, 1908 (gravel pit), 3 ♂s, 5 ♀s; Aug. 14, 24, 1908, 6 ♂s, 5 ♀s (Ws). Winnipeg, Sept. 7, 1908, 1 ♂, 1 ♀ (Ws). Carberry, Sept. 6, 1906, 2 small ♀s (Alexander).

SASKATCHEWAN.—Regina, Aug. 8, 10, 1886, 1 ♂, 2 ♀s (F); Sept., 1908, 1 ♀ (Wg). Moosomin, Sept. 13, 1906, 7 ♂s, rather large (Alexander).

39. *Sympetrum madidum* Hagen

SASKATCHEWAN.—Battleford, July 2, 1907, 1 ♂ (Wg).

The two rows of cells between Rs and Rspl are imperfectly developed in this specimen.

S. madidum is also known from British Columbia.

40. *Sympetrum rubicundulum decisum* Hagen.

MANITOBA.—Aweme, Aug. 5, 1907, 1 ♂, 1 ♀, teneral; Aug. 19, 1908, 1 ♀ (C). Treesbank, July 20, 1908, 1 ♂ (Ws). Winnipeg, July 6, 1908, 1 ♂ (Ws). Westbourne, July 27, 31, 1908, 3 ♂s, 7 ♂s; Aug. 10, 20, 1908, 1 ♂, 3 ♀s (Ws). Deloraine, July 6 (F). Grandview, Sept. 18, 1906, 1 ♀ (Alexander).

SASKATCHEWAN.—Indian Head, July 22, 1903, 1 ♀ (Wg). Regina, Aug. 10, 1886, 2 ♂s, 4 ♀s (F); Aug. 7, 1903, 1 ♂, 1 ♀ (Wg). Lumsden, Sept 10, 1906, 1 worn ♂ (Alexander). Davidson, Aug. 21, 1907, 1 ♀ (Wg) Goose Lake, July 19, 20, 1907, 2 ♂s, 2 ♀s (F, Wg). Saskatoon, July 16, 1906, 1 ♂ (Wg) Duck Lake, July 22, 1907, 1 ♂, 2 ♀s (F). Kinistino, July 25, 1907, 4 ♂, 2 ♀s (F, Wg). Birch Hills, July 25 1907, 3 ♀s (Wg). Carlton, 1 ♀ (F).

ALBERTA.—Cardston, July 23, 1900, 1 ♀ (Wg). Beaver Lake, 1907, 4 ♀s (O. Halkett). Banff, July 21, 1908, 3 ♂s; Aug. 3, 1908, 1 ♂; Aug 10, 1908 (summit of Sulphur Mt.), 1 ♀; Aug. 24, 1908, 1 ♀ (S). Near Waterton Lake, Aug. 10, 24, 1908, 4 ♂s, 4 ♀s (Cowdry).

This is perhaps the commonest dragonfly of the prairie region.

41. *Sympetrum obtrusum obtrusum* Hagen.

MANITOBA.—Westbourne, July 27, 1908, 1 ♂; Aug 10, 26 4 ♂s, 1 ♀ (Ws). Swan River, Sept. 8, 1906, 3 ♂s, 1 ♀ (Alexander).

42. *Sympetrum obtrusum morrisoni* Ris.

ALBERTA.—Near Waterton Lake, Aug. 5, 1908, 1 ♀ (Cowdry).

A pale specimen, the femora pale in the basal half.

43. *Sympetrum corruptum* Hagen.

MANITOBA.—Aweme, Sept. 15, 1907, 1 ♂; Sept. 8, 1908, 2 ♀s (Ws). Winnipeg, July 8, 1908, 1 ♀; Sept. 7, 1908, 2 ♀s, teneral (Ws)

SASKATCHEWAN.—Regina, Aug. 10, 1886, 1 ♀ (F).

ALBERTA.—Lethbridge, July 7 1909, 1 ♀ (Ws). Banff, Sept. 13, 1897, 1 ♂, 1 ♀; July 15, 1908, 1 ♂; July 27, 1908 (trail above Hot Springs), 1 ♀ (S). Laggan, 1 ♀ (J. E. Bean).

This species appears to be very much commoner in the Western Provinces than in Ontario, where it is very local.

44. *Leucorrhinia borealis* Hagen,

MANITOBA.—Bird's Hill, June 5, 1909, 3 ♂s, 2 ♀s (Ws). Aweme, July 15, 1907, 1 ♀, somewhat teneral; June 4, 9, 1909, 2 ♀s (C).

ALBERTA.—Banff, June 17, 1908, 1 ♂ (S).

Except the orginal record from Hudson's Bay, these are the only localities for this species known to me.

45. *Leucorrhinia proxima* Calvert.

MANITOBA.—Winnipeg Beach, Lake Winnipeg, June 19, 1909, 1 ♂ (Ws).

Occurs in the Canadian Zone across the continent.

46. *Leucorrhinia glacialis* Hagen.

MANITOBA.—Aweme, June 11, 1911, 1 ♀ (C).

47. *Leucorrhinia hudsonica* Selys.

MANITOBA.—Winnipeg, July 27, 1908, 1 ♀ (Ws). Winnipeg Beach, June 19, 1909, 1 ♂, 2 ♀s (Ws).

ALBERTA.—Laggan, 1 ♂ (J. E. Bean).

A transcontinental species of the Boreal Region.

48. *Leucorrhinia intacta* Hagen.

MANITOBA.—Aweme, June 21, 24, 1911, 2 ♀s, teneral; July 9, 1910, 1 ♀ (Criddle). Winnipeg, July 9, 1908, 1 ♂, 2 ♀s (Ws). Winnipeg Beach, June 19, 1909, 2 ♂s (Ws).

This last record is the most northerly for this species.

49 *Pantala hymenæa* Say.

MANITOBA.—Husavick, July 8, 1910, 1 ♂ (Ws).

The capture of this southern species in Manitoba was wholly unexpected. It has since been taken on Pelee Island, Ont., by F. M. Root.

EXPLANATION OF PLATE IX.

Fig. 1.—*Coenagrion resolutum* (♂), dorsal view of head, thorax and abd. segs. 1–3; 1a (♀), dorsal view of segs. 8–10.

Fig. 2.—*Coenagrion angulatum* (♂), dorsal view of head, thorax and abd. segs. 1–3; 2a (♂), lateral view of seg. 10 and appendages; 2b, same, viewed obliquely from behind; 2c (♀), dorsal view of segs. 8–10.

Fig. 3—*Coenogrion lunulatum* (♂), lateral view of seg. 10 and appendages. (Specimen from Jena.)

Dr. C. Gordon Hewitt, Dominion Entomologist, left for England on July 26th to represent Canada at the International Congress of Entomology, which was held at Oxford from August 5th to 10th.

Subsequently he attended a conference which the Secretary of State for the Colonies arranged at the Colonial Office for the purpose of working out a scheme for Imperial co-ordination in the prevention of the spread of insect pests and the more extended investigation of the noxious insects which occur in the different parts of the Empire.

PROVINCIAL ENTOMOLOGIST FOR ONTARIO.

The Ontario Department of Agriculture has appointed Mr. Lawson Cæsar, B. A. and B. S. A (University of Toronto), Provincial Entomologist for Ontario. Mr Cæsar will have charge of all the Inspection work throughout the Province and will carry out investigations concerning the depredations of insects and injuries caused by plant diseases. He will continue his connection as Lecturer in the Department of Entomology at the Ontario Agricultural College, during the winter months. This is the first appointment of the kind to be made in Canada and will it is hoped be followed by some of the other Provinces before very long. Mr. Cæsar is well known as an economic entomologist and as a Lecturer to Farmer's Institutes and Meetings of Fruit Growers. He is also the writer of Bulletins on the Lime-Sulphur Wash, The Codling Moth, Little Peach Disease and a Spraying Calendar.

THE DIVISION OF ENTOMOLOGY, OTTAWA.

It has been suggested that an account of the recent work and developments of the Division of Entomology of the Dominion Department of Agriculture would be of interest to the readers of The Canadian Entomologist. Believing this to be the case, the following notes have been written for our entomological friends.

The most important development of the activities of the Division is the extension of the field work by means of field laboratories or stations. It is obvious that work on the more serious insect pests occurring in different parts of Canada could only be carried out in the regions where such pests occur. The method of carrying on all investigations at the Central Experimental Farm, Ottawa, where the Division of Entomology is located, not only had serious limitations but the results might not be applicable to local conditions. The Brown-tail Moth and its parasites must be studied in the regions where the insect occurs; so also in the case of such pests as the Apple Maggot, the Bud-moth and other pests.

During the present season six field stations have been established. In three cases the laboratory consists of a two-roomed portable cottage, in one case the Ontario Department of Agriculture has given us the use of a room in the Jordan Harbour Experiment Station and in the other two cases temporary quarters at farm houses are being used. The stations and work are distributed as follows:

In Nova Scotia, Mr. G. E. Sanders occupies an entomological laboratory located at Bridgetown, N.S., from which he is directing the work of San José Scale inspection pending the appointment of a provincial entomologist. The San José Scale was dicovered in Nova Scotia by Mr. Sanders during the Brown-tail Moth work. Investigations are also being made on the Bud-moths (*Tmetocera ocellana*), Green Apple-Worm (*Xylina* sp.) and the Brown-tail Moth, special attention being devoted to parasitic work. For experimental work in the control of the Bud-moth a ten-acre orchard has been placed at the disposal of the Division of Entomology by Mr. R. S. Eaton at Kentville, N. S. where spraying experiments are being carried on.

An entomological field station is located at Fredericton, N.B., in the grounds of the University of New Brunswick. Mr. J. D. Tothill, who is in charge of the Brown-tail Moth work in New Brunswick is in charge and is devoting his attention chiefly to the breeding of the predaceous enemy of the Brown-tail and Gipsy Moths, the *Calosoma* Beetles, and their colonisation, and the study of Tachinid parasites. Through the kind co-operation of Dr. L. O. Howard and Mr. A. F. Burgess, of the United States Bureau of Entomology, we have been able to procure a supply of the European *Calosomas* and Tachinids (*Compsilura concinnata*) colonized in Massachusetts and we are now endeavouring to establish these enemies of the Brown-tail and Gypsey Moths in New Brunswick where the Brown-tail Moth was found during the last winter's scouting work, to have spread over a very large area and the intensity of the infestation will undoubtedly increase. Mr. Tothill recently visited Massachusetts for the purpose of collecting parasitised material. Studies are also being made on other insects as opportunities occur.

Mr. C. E. Petch has been recently appointed a field officer of the Division and placed in charge of the field station at Covey Hill, Quebec, where he will study chiefly insects affecting the apple, namely, Apple Maggot, Apple Curculios and the Capsid or other bugs injuring fruit. The unusual prevalence of these pests in the locality where the entomological station has been placed will afford splendid opportunities for useful work.

From the office of the Division at Jordan Harbour, Ontario, Mr. W. A. Ross is carrying on a very thorough study of the Apple Maggot or Railroad Worm which he began to investigate last year when employed by the Ontario Department of Agriculture. At St. Ives, in Middlesex County,

Ontario, Mr. H. F. Hudson is conducting an investigation on an outbreak of the Chinch Bug which proved very destructive to grass land last year. He is also making observations on Wireworm and White Grub.

In British Columbia Mr. R. C. Treherne, in addition to studying certain more serious apple pests, is investigating the Strawberry Weevils, chiefly *Otiorynchus sulcatus*, which are responsible for serious losses to strawberry growers not only in British Columbia but also in other regions. With a view to studying similar conditions elsewhere he is visiting the States of Oregon and Washington. For the present season the field laboratory is located at Hatzic, B. C.

Although no field work is being carried out in the Western or Prairie Provinces it is hoped that work will be commenced there in the near future, as there are several insect pests affecting cereals and field crops demanding attention.

Since his appointment as Assistant Entomologist, Mr. J. M. Swaine has made an excellent start in the Division's work on forest insects. In May he visited the Riding Mountain Forest Reserve in Manitoba for the purpose of liberating the parasites of the Larch Sawfly, which were imported from England, and also to study the native Bark-beetles. Other parts of Eastern Canada are being visited during the present season.

Mr. Gibson has been continuing his work on Cutworms, studying especially species from Alberta where Cutworms have been responsible for very considerable losses in certain sections of Southern Alberta where growing grain was completely destroyed. He also superintended the work on Root-Maggot control, this being the third season during which these experiments have been carried on.

The distribution of Ticks in Western Canada, with especial reference to the Rocky Mountain Spotted Fever Tick, *Dermacentor venustus* is being studied with interesting results. In the campaign against the House-fly which is now in full swing in various cities, especially in Eastern Canada, the Division is continuing to provide much powder and shot.

It is, we hope, hardly necessary to add that any assistance which local entomologists are able to render us in the various branches of work outlined above will be most welcome. This is especially the case where we are working out the distribution of certain insects, The territory which we have to cover is so great in extent that we wish to enlist all the co-operation possible. C. G. H.

NOTES ON TAYLOR'S TYPES OF GEOMETRIDÆ.

BY WM. BARNES, M.D., AND J. H. MCDUNNOUGH, PH.D., DECATUR, ILL.

Recently Dr. Barnes has acquired by purchase the whole of the Taylor collection of Geometridæ including the types and cotypes of about 65 species described as new. Most of these types have the type-label attached but in some few instances the author, probably due to the poor condition of his health, has neglected to affix more than a small written label with the name of the species on one of the specimens, even when the original description states that several specimens were present. As however in the original descriptions he has been very careful to give full detail regarding localities and dates we have in nearly every case been able to recognize these other specimens and have considered ourselves justified in affixing "type" or "cotype" labels as the case might be. In order to account for the fact that these labels are not in Taylor's handwriting, and to give workers on the *Geometridæ* accurate knowledge concerning these types, we publish in the present article a full list of the Taylor types contained in the Barnes collection together with such notes on the same as are necessary.

Rachela pulchraria.—One ♂ with type label in Taylor's handwriting attached, dated Oct. 11th, '06, but without locality label. According to the original description the locality is Tacku River, B. C.

Eupithecia insignificata.—One ♂, type label attached, labelled "W. 15.4.04". A second specimen from Regina was apparently included under this name but to us appears to be different.

Eupithecia sublineata.—One specimen with hand-printed type label attached, labelled "W. 18.4.04". Four other specimens from Wellington, B. C. (2); Vancouver, B. C.; and Field, B. C. (Dod), are contained in the series but are not cotypes.

Eupithecia obumbrata.—Two ♀ types with printed labels "Type" and "Type 2", from Gold Stream, B. C. "10.5.03." Four other specimens taken early in May and labelled "Vancouver Is., G. W. Taylor" are also present in the series.

Eupithecia modesta.—The original description calls for both ♂ and ♀ types; the ♂ however only bears the label "*T. modesta*" in Taylor's handwriting in addition to the locality label "Vanc., 6.6.05". We have affixed to this a type label. The ♀ type with printed label is almost totally destroyed, portions of the wings being pinned under the label.

Eupithecia minorata.—Type label in Taylor's handwriting attached; there is no locality label but the round date label "22 4.05"; characteristic of material from Kaslo, B C. sent out by Mr. Cockle, is affixed to the specimen.

Eupithecia packardata.—Type label in Taylor's handwriting on specimen labelled "Ottawa, Can., 7, VIII, 1906., C. H. Young". The series contains 5 other specimens; 2 from Montreal (A. F. Winn); 1 from Trenton (Evans); 1 from Catskill Mts.; and 1 without locality label.

Eupithecia grata.—Type label in Taylor's hand on specimen labelled "Ottawa, Can., 5 VI. 1906, C. H. Young".

Eupithecia lagganata.—1 ♂ in fair condition only, labelled type in Taylor's hand; locality "Laggan, Alta. (Dod)".

Eupithecia compactata.—The single ♀ type with label in Taylor's hand; locality "Windermere, Upper Columbia River, B. C. (Dod) 13 VII. 07".

Eupithecia spaldingi.—The specimen labelled type in Taylor's hand is very poor and almost unrecognizable. We doubt if the median band mentioned in the original description will appear so prominently in fresh specimens. The locality label is "Stockton, Ut. (Spalding), IX., 2, 03".

Eupithecia dyarata.—A hand printed type label was present but not attached to any specimen. The original description specifies as type a ♀ from Kaslo (Cockle) dated "24 IV. 06." A specimen without abdomen but apparently a ♀ is contained in the series and labelled "J. W. Cockle, 24 IV. 02." This agrees well with the original description and also with topotypes received from Mr. Cockle; we imagine that the "06" is therefore either a printer's error or an oversight and have placed the label on this specimen.

Eupithecia scelestata.—This species was described from four specimens labelled Kaslo, B. C., and dated respectively 21st April, 2nd and 3rd May, 1903 and 4th May 1905. All these specimens are in the series but only the specimen dated 2nd May, 1903, which, by the way, is the poorest specimen of the lot, bears a type label. We have affixed cotype labels to the other three specimens.

Eupithecia winnata.—A ♀ in rather poor condition has a type label in Taylor's handwriting. There should be two cotypes in Mr. Winn's collection.

Eupithecia alberta.—Both specimens mentioned in the description are present with hand printed type labels "Type 1" and "Type 2".

Eupithecia regina.—Type and 3 cotypes with hand printed labels attached; the fourth cotype mentioned dated "Aug. 2nd, 1903" is missing.

Eupithecia youngata.—Type and 2 cotypes with printed labels. The type specimen from Ottawa has only two wings left but the cotypes from Catskill Mts. are in good condition.

Eupithecia dodata.—Type specimens labelled "Type 1" and "Type 2" present.

Eupithecia adornata.—Type specimen with printed label "From Calgary, Alta., 10.VI. 95". Four cotypes are also labelled, the fifth one (of the date June 14th) being missing.

Eupithecia olivacea.—Both specimens which served for the original description are present, correctly labelled.

Eupithecia terminata.—Two specimens present both labelled "co-type" in Taylor's handwriting.

Eupithecia helena.—A single somewhat faded specimen with printed type label attached.

Eupithecia perbrunneata.—There are no specimens labelled type; one specimen from Kaslo "31.V. 07" is labelled in Taylor's hand "E. perbrunneata Tayl. = laricata Dyar, not Freyer, three types". Two of the specimens mentioned in the original description we have found in the series, viz. those labelled Kaslo, May 23rd, and June 2nd; these we have made type and cotype respectively. The Victoria specimen mentioned is not to be found.

Eupithecia placidata—A single ♀ specimen labelled "cotype" in Taylor's hand is present.

Eupithecia slocanata.—The ♂ and ♀ mentioned in the original description are present, both labelled "cotype".

Eupithecia fletcherata.—One type, "Ottawa, Can., 3. VIII. 06"; type label in Taylor's handwriting.

Eupithecia bryanti.—The type ♀ alone has printed label attached; of the six cotypes mentioned in the description we have discovered four, all in poor condition, viz. those dated 27th (2), 28th and 29th July; to these we have affixed cotype labels.

Eupithecia harveyata.—Both specimens present with printed labels "Type" and "Type 2".

Eupithecia hanhami.—One of the two specimens mentioned by Taylor in the description, one labelled "Victoria, B. C., 5. VI. 03" has a type label affixed; the second specimen dated 25th June, 02 (not 05 as stated). simply bears a label in Taylor's hand "*Eupithecia hanhami*".

Eupithecia indistincta.—Single type with two written type labels affixed.

Eupithecia gibsonata.—The specimen with type label is dated "9 4. 03, No. 94" not "9 VI." as stated in description. A second specimen "4 4.03" is also present.

Eupithecia fasciata.—Two types with labels; type 1 from Ottawa, type 2 with no locality label; a very poor specimen, presumably the one mentioned as having been received from Mr. Kearfott.

Eupithecia quebecata.—Two specimens with written type labels; "Type 1" from Kamouraski 26.8. 98; type 2 from Biddeford, 23. 7.99.

Eupithecia fumata.—A single type specimen labelled in Taylor's handwriting; locality as stated in description.

Eucymatoge rectilineata.—A written type label attached to a very worn specimen; locality as stated in description.

Eucymatoge vancouverata.—The single type is a ♀ with handprinted type label attached; the collection contains a splendid series of both sexes.

Entephria takuata.—Of the four specimens which served for the original description only three were in the collection and none marked as types; one however is labelled "Mesoleuca takuata T." in Taylor's hand and this we have made the type; it is ♀, dated 4th Aug., 06 and in good condition; the other two specimens which bear date labels agreeing with the dates of original description, we have made cotypes.

Entephria lagganata.—No type labels affixed but all four type specimens can be identified by the dates. A specimen labelled "Laggan, Alta,, 9.VIII. 00, 5700 ft." and marked in Taylor's handwriting "Mesoleuca lagganata T." we have made type, the others becoming cotypes.

Eustroma harveyata.—Single type specimen as stated in description type label attached.

Zenophleps victoria.—The single ♀ type is labelled; one wing is damaged. A second ♀, also from Victoria, B.C., is present.

Mesoleuca hulstata.—No type labels attached. Five specimens labelled "Claremont, Cal. Baker" as stated in description were present; one of those bore the label "Mesoleuca hulstata Taylor" in Taylor's hand; this we have made the type, the others cotypes. All are in rather poor condition.

Mesoleuca occidentata.—Both types present with labels in Taylor's handwriting.

Mesoleuca occidentata, var. *mutata.*—No type labels; the three specimens mentioned in the description are present, one labelled "var. *mutata*" in Taylor's hand. On this we have placed a type label, on the other two cotype labels.

Mesoleuca hersiliata var. *mirandata.*—Type with label attached.

Mesoleuca boreata.—Two specimens with written type labels, both in very rubbed condition. They appear to us nothing but an extreme form of var. *mutata* and it is even possible that the lack of orange scales is due to the poor condition of the specimens. We have several specimens from Victoria and Duncans, B.C., which show only traces of orange scaling.

Mesoleuca casloata.—Of the three specimens mentioned in the description only two are present, one with label "M. casloata Taylor". This we have made the type, the other cotype.

Mesoleuca decorata.—Three types present as stated.

Hydriomena multipunctata.—The single type with written type label; the species is very close to *indefinita* Grossbeck.

Hydriomena manzanita.—Type with handprinted label attached.

Hydriomena autumnalis, var. *columbiata.*—Of the four specimens mentioned in the description only three are present, one bearing a handprinted type label; the other two dated respectively May 23rd and 27th we have made cotypes.

Hydriomena magnificata.—One ♀ type with written type label.

Petrophora fossaria.—The specimen from Mt. Cheam, B.C., bears a type label; the other specimens mentioned from Laggan, B.C., we have also found and to these affixed cotype labels.

Petrophora pontiaria.—One ♀ type as stated in description; the other three specimens mentioned are present and have received cotype labels.

Petrophora circumvallaria.—No type labels; the four specimens mentioned as types are present and we have labelled them so. The species cannot be distinguished from the European *turbata.*

Petrophora planata.—A pair from New Brighton, Pa., have type labels attached; the name falls before *iduata* Gn. as pointed out by Dr. Dyar.

Leptomeris subfuscata.—1 ♂ from Goldstream, B.C., only bears a type label; the ♀ type mentioned is present and to this we have affixed a type label. Of the other species referred to in the description we have found the three from Vernon, B C., and 1 ♂ from Goldstream, B.C., and on these have affixed cotype labels. Concerning the three specimens from Victoria, B.C., there are two labelled "Victoria" without date, and 1 specimen with date "30 V" but no locality; we have refrained from making these cotypes as they cannot be definitely identified.

Aplodes unilinearia.—One ♂ type with written label.

Aplodes hudsonaria.—Two types, one with handprinted, the other with written label.

Delinea bryantaria.—One ♂ type with printed label in rather poor condition.

Sciagraphia purcellata.—Type and cotype with written labels; type in good condition, other specimen poor.

Diastictis hulstiaria.—A type label unaffixed is present; there is evidently no type specimen, the name being merely *nom. nov.*

Macaria quadrifasciaria.—The one damaged type specimen as stated in description.

Enemera simularia.—♂ and ♀ types with written labels, both specimens in wretched condition.

Enypia packardata.—A single ♂ bears a printed type label; the other five specimens mentioned are not clearly enough indicated to warrant type labels being affixed.

Gabriola dyari.—Taylor states that four specimens, all dated "August 1903" served for the description. The type label however is on a specimen dated "28.VII.03". The three other specimens dated "19th and 21st Aug." are present and to these we have affixed cotype labels.

Sabulodes costinotata—The two cotypes present are from Prescott, Ariz., not Phoenix as stated in the description.

Sabulodes arizonata.—Two cotypes with written labels are in the collection.

Besides the above the following types were contained in the collection:—*Eupithecia latipennis* Hulst.; *Eupithecia frostiata* Swett; *Eupithecia kootenaiata* Dyar (cotype); *Eupithecia niphadophilata* Dyar (cotype); *Anthelia taylorata* Hulst. (one side broken off); *Phigalia denticulata* Hulst.; *Euclæna abnormalis* Hulst. (without much doubt a suffused aberration of one of the other species, possibly *pectinaria*.)

SOME CANADIAN SAWFLIES COLLECTED BY FREDERICK KNAB.

BY S. A. ROHWER, WASHINGTON, D. C.

In June, 1907, Mr. Frederick Knab spent some time in the Oxbow region at Saskatchewan, where he collected a number of insects. The sawflies which were collected have been given to the United States National Museum, and have been determined and placed in the collection. As some of the species are new to Canada, and as nearly all the records will add to the distribution of the species, a list of them is now presented, as it may be useful to those who are editing "Insects of Canada."

The following were collected at Oxbow, Saskatchewan, which is a prairie region with a number of small ponds, around which various willows grow.

Pamphilius (*Pamphilius*) *nigritibialis* Rohwer.

Arge spp.

Cimbex americana, var. *decemmaculata* Urban.—Female.

Zaræa inflata Norton.—June 15. Female.

Dolerus agristus MacGillivray.—June 11 to 18, 1907. Three typically-coloured females. Four females which have the venter and the terminal dorsal segments blackish.

Macrophya succincta Cresson.—Female, June 17, 1907.

Tenthredella erythromera (Provancher).—Female, June 18, 1907.

Empria maculata (Norton).—Female, June 19, 1907. A small specimen with a narrow sheath.

Hoplocampa xantha Rohwer.—Three females, June 15, 1907.

Lycaota spissipes (Cresson).—Females, June 17 and 21 ; male, June 15.

Paracharatus rudis (Norton).—Female, June 15, 1907.

Paracharatus nigrisoma Rohwer.—June 21, 1907.

Monophadnus truncatus Rohwer.—June 1, 1907.

Amauronematus lincolnensis Rohwer.—Two females, June 1, 1907, not typical.

Amauronematus semirufus (Kirby).—May 30, 1907. Differs in a few minor points.

Amauronematus knabi Rohwer.—June 15 and 19, 1907.

The following species were collected at White River, Ontario, which is in the forest region to the north of Lake Superior :

Dolerus dysporus MacGillivray.—June 24, 1907.

Nematus erichsoni (Hartig).—June 24, 1907.

Monsoma inferntia (Norton).—Female, June 25, 1907.

COLLECTING BEES AT GUALAN, GUATEMALA.

BY WILMATTE P. COCKERELL, BOULDER, COLORADO.

In going from Quirigua to Guatemala City, we passed through a desert region—a place of curious forms of cacti, but especially interesting because cf the trees and shrubs, at that time of the year, late February, covered with splendid blossoms, and usually without leaves. One tree (*Gliricidia maculata* H.B.K.) was very common and with its delicate pink flowers reminded one of the peach of the temperate zone, but inspection showed it to have a papilionaceous flower. This, I thought, would be a wonderland for bees, since bees are peculiarly adapted to desert areas.

When we returned to Quirigua, I determined to spend two or three days at Gualan, and I anxiously inquired of every one whether there was some one in the village who would befriend me, a missionary perhaps, a priest, an American who owned a coffee finca or a hotel-keeper who spoke English; and at last I found a young man who sometimes went to Gualan to buy cattle for the commissary of the United Fruit Company, and he said there was a hotel and that the negro-French proprietor did speak English, but that the place was usually full of drunken natives and was absolutely impossible for an American lady. That settled the hotel question, but I could at least go up between trains, though even for so short a time it was not considered wise for me to go alone, and Mr. Earl Morris was detailed to go with me. There was much joking about the biological altar needing a sacrifice, for my friends at Quirigua were archæologists and were uncovering one of the wonderful old Maya temple cities, and bees looked very small to eyes focused for forty feet doorways. But Mr. Morris was a splendid assistant, and helped in every way, even if in his heart he was sighing for sculptured walls and ornate pottery. The train left Quirigua at ten o'clock and arrived at Gualan at eleven-thirty, the down train picked us up at two-thirty. It was a wonderful three hours ! The lovely pink and white blossoms of *Gliricidia maculata* were visited by great Carpenter bees (*Xylocopa*), but unfortunately the flowers were so high, and the bees flew so swiftly that I secured only a few specimens.

The best catch of that day was a very small bee belonging to the genus *Perdita*, and if you saw it I am afraid you would agree with the Indians who said, "So small bugs can be of no use." The Perditas are among the smallest of bees, and yet the finding of one on *Cordia alba*, a yellow flowered tree, at Gualan, was a distinctly dramatic and interesting thing to me. Years ago my husband described seventy of these small bees which he had collected in New

Mexico, and half a dozen have been found in Northern Mexico, and I had often wondered whether the little bees were in Central America if some one who was interested in finding them would only look. And there it was, a new species that extended the distribution of the genus Perdita a thousand and more miles, and I had added a tiny fact to the everlasting why of the universe.

A few hundred yards below the village there was a number of trees covered with cardinal flowers, and I was especially anxious to collect from them, but we were beguiled into chasing butterflies, and the yellow-flowered trees had other bees than the *Perdita tropicalis*, so that it was time for the train, and we had seen only the glow of the cardinal tree from a distance.

Another trip was imperative, and on that day I had an amusing experience. The conductor of the train, a rather interesting Guatemaltecan brought me a Ladies' Home Journal and a little note which said that if I were English he would lend me the magazine; what did I do with my veil (net) and did I, like other strangers, think them savages to be conquered ?

Judging that he wrote English better than he spoke it, and read better than he understood, I wrote that I was grateful for the magazine (and really, even the Ladies' Home Journal looked good to me; that I had the net to catch bees, because my husband studied bees from all parts of the world. He answered : "Thanks. Good for him and the world. Hope that he finds the Bee that carries the strength of life — like they do honey. So that the wise live long to be learned, and the fools long enough to learn."

There was more correspondence about the duty of one nation to another, the books that would give a Guatemaltecan an idea of the United States, all of which is too lengthy to record here, but just before we reached Gualan he wrote : "Guess I tire you, I like to write English to get acquainted with. Excuse me—My wish that the bees won't bite you while searching for flowers. That they sometimes on the mountains sing you a chorus. Remembering you of God, the father of all peoples."

When we reached Gualan, we went at once to the Cardinal Tree, and found it even more wonderful than we had thought. Imagine a great tree, fifty to seventy-five feet high, with branches literally covered with fragrant cardinal flowers, and the flowers swarming with wasps and bees, and on the branches great gaily colored birds assembled to eat the insects. I too wanted to collect insects, but the lowest branches were just out of reach. Mr. Morris offered to climb up and collect for me. Many Indians gathered in the path just below us, and called out to Mr. Morris that the tree was full of 'sarpents,' and that the branches would break and dash him to destruction, but he climbed on. Soon he began to beat himself, and I knew that the ants, the little guardians of the tree, were after him. Then, too, he had been obliged to crawl over some of the curious flat cactus that grows along the trunks of trees in that country, and when I added my voice to the Indians cry, "Come down," Mr. Morris said that he thought he would. We made a pile of stones and boxes, and so were able to get a few wasps and bees, but I shall never cease to envy the birds so gracefully collecting from the beautiful Guacamaya. With wings I might have secured a dozen forms new to science. I carried home a flower covered branch, and later Mr. Morris secured leaves and pods from the same tree, and great was my surprise and delight when Captain John Donnell Smith, of Baltimore, said that the tree itself was new !*

The excavations at the ruins became daily more interesting, and I could not ask Mr. Morris to spend more time with me, but most fortunately I learned that the station agent's wife spoke English, and she generously kept me at her house one night, thus, giving me the better part of two days for collecting.

I found the walls around one of the patios, here a place for chickens and turkey-buzzards instead of ferns and orchids as in Guatemala City, alive with red woolly *Centris* (*C. tarsata* Smith) nesting—there were literally thousands of them, and I spent the most of one afternoon getting specimens of these bees—and the bees (*Mesocheira bicolor* Fabr.) that were parasitic in their nests.

Then, too, there were some Megachiles(*M. gualanensis* n. sp.), leaf-cutting bees, nesting in the same wall, and they had interesting parasites (*Coelioxys sanguinosus* n. sp.). Dozens of small Indian boys watched me, and occasionally begged to be allowed to use the net. Some native teachers came out to drive the boys into school, but stayed to watch the strange 'Inglese' catching 'musca.'

"For what does she want the little bugs, "they inquired of my hostess. "Does she make medicine of them?" Not such a strange supposition, since they grind up all sorts of insects and use them as medicine.

"The Senora does not gather them for medicine," they were told, but the fame of the medicine-maker spread, and a woman brought a little child with a terrible sore on his neck, and begged me to give her the fly that could cure her baby. It was pitiful !

***Phyllocarpus* n. sp.; the genus previously known only from a single species occurring in Brazil.

A more amusing incident followed. A larger boy asked which made the best medicine, and I begged Senora Caldero to explain that the bees were for study. "How can you explain that to such ignorance," she asked, but I begged her to try, and the boy said that he understood; but a few minutes later he was telling a young girl that the little black bees were for pains in the stomach—the red ones for pains in the legs. When reproached, he excused himself by saying : "The other is much too difficult for a girl to know." The inferiority of woman serves its purpose the world over.

My adventures did not end with the day, for in the middle of the night I was awakened by a great ringing of bells, and the light from a burning house lighted my room. "Get up ! Get up !" my hostess called, " there is a terrible fire. Do not try to save anything but come quickly." Fortunately I had lain down with my clothes on, so that I was ready in a minute, carrying with me my precious box of bees. I found my hostess and her children wrapped in blankets, and we all hurried out into the street. The fire was only a few doors from our house, and with a brisk wind blowing it looked as though nothing could save any house in the village. Some way in the crowd I was separated from Senora Caldero and her family, and I found myself in the middle of the road surrounded by people wailing and crying to the saints. It was a weird moment ! The men had formed a chain from the fountain and passed water in every sort of jar and pan, but they worked effectively, and I soon saw that the fire would be conquered. I thought I would be safer in the house, for I did not like being in the midst of that excited crowd, so I crept back into the dark house, still holding jealously my little box of bees,

It was not long before my host came up from the office where he slept, and the family was brought home. There was much embracing and much excited talk, and more wine and whiskey offered to everyone in the good Latin-American fashion, and the daylight was almost upon us before the village became quiet again.

The next morning a horse and a moso were ordered for seven o'clock, and came at eight, the usual custom of the country. Until two o'clock I rode along the river collecting here and there, and enjoying the bright-hued birds, and the beautiful plants. Two plants stand out in the memory of that forenoon; *Antigonon guatemalense* Meissn., a vine with great racemes of most exquisite pink flowers; the other (*Adenocalymna macrocarpum* Donn. Smith?)* a bush with great violet-purple bells, like a glorified pentstemon,

*Capt. Donnell Smith wrote that he was not quite sure of the species of *Adenocalymna*. More material is needed. The plant belongs to the *Bignoniaceæ*.

but with a dreadful odor. The little moso who carried my press could hardly be induced to carry a piece. He made me understand that it would give me diseases unnumbered, but I insisted, and so far not a single disease has resulted.

The plants collected were all kindly identified by Capt. Donnell Smith. The following list of Gualan bees has been prepared by my husband. The new species are in course of publication in the Annals and Magazine of Natural History.

BEES OF GUALAN.

(1.) *Prosopis quadratifera* n. sp. At flowers of *Iresine paniculata* (L.).

(2.) *Prosopis gualanica* n. sp. One male.

(3.) *Halictus hesperus* Smith. 27 females. One at flowers of *Cordia alba*; five at flowers of *Phyllocarpus* n. sp.; the rest at *Vernonia aschenborniana* Schauer, collectimg the white pollen.

(4.) *Halictus townsendi* Ckll. One female, Feb. 23, at flowers of *Tithonia diversifolia* A. Gray.

(5.) *Augochlora binghami* Ckll. One female.

(6.) *Augochlora* sp. 1 female.

(7.) *Angochlora cordiæfloris* Ckll. One female, Feb. 23, at flowers of *Calopogonium cœruleum* Desv.

(8.) *Agapostemon nasutus* Smith. Seven males, seven females. Six of the males and six females at *Vernonia aschenborniana;* one male at *Calopogonium cœruleum;* one female at *Tithonia diversifolia.*

(9.) *Agapostemon nasutus gualanensis* n. var. Four males.

(10.) *Perdita tropicalis* n. sp. At *Cordia alba.*

(11.) *Centris totonaca* Cresson. One female, "at flowers of yellow vine."

(12). *Centris tarsata* Smith. Eleven males. One from flowers of *Iresine paniculata.*

(13.) *Centris inermis gualanensis* n. subsp. At flowers of *Calopogonium cœruleum.* Also at Quirigua.

(14.) *Leptergatis toluca* (Cresson). One male at flowers of *Cordia alba.*

(15.) *Mesoplia azurea guatemalensis* n. subsp. At flowers of *Calopogonium cœruleum.*

(16.) *Mesocheira bicolor* (Fabr.). Two females.

(17.) *Exomalopsis callura* n. sp. At flowers of *Vernonia aschenborniana.*

(18.) *Exomalopsis similis* Cresson. One female at flowers of *Cordia alba.*

(19.) *Xylocopa wilmattæ gualanensis* n. subsp.
(20.) *Xylocopa fimbriata motaguensis* n. var.
(21.) *Xylocopa barbata* (Fabr.). At flowers of *Calopogonium cæruleum*.
(22.) *Ceratina nautlana* Ckll. One female, at flowers of *Vernonia aschenborniana*.
(23.) *Ceratina virescens* Friese. One male.
(24.) *Ceratina regalis* n. sp.
(25.) *Ceratina xanthostoma* n. sp.
(26.) *Ceratina xanthostoma rufipennis* n. var.
(27.) *Coelioxys sanguinosus* n. sp.
(28.) *Megachile gualanensis*. Both sexes.
(29.) *Dianthidium gualanense* n. sp.
(30.) *Euglossa cordata* (L.). One male at flowers of *Arthrostemma fragile* Lindl.
(31.) *Melipona fulvipes* Guér. One male.
(32.) *Trigona zexmeniæ* n. sp. At flowers of *Vernonia aschenborniana*. Also found at Quirigua.
(33.) *Trigona mellaria* Smith. One at flowers of *Calopogonium cæruleum*.
(34.) *Trigona cupira* Smith. Twelve workers, eleven at *Vernonia aschenborniana*.
(35.) *Trigona amalthea* Oliv. Two workers at *Calopogonium cæruleum*.

All these bees are new to the fauna of Guatemala. The bees recorded from Guatemala up to the beginning of 1912 are : *Halictus providens* Smith, *Augochlora chryseis* Smith, *A. radians* Vachal, *A. nigromarginata* Spinola, *Agapostemon brachycerus* Vachal, *Emphoropsis fulva* (Smith), *Centris clypeata* Friese, *C. labrosa* Friese, *Tetrapedia moesta* Cresson, *T. maura* Cresson, *Bombus lateralis* Smith, *B. unifasciatus* Smith (*mexicanus* Cresson) *Melipona nigripes* Friese, *Trigona fuscipennis* Friese, *T. schulthessi* Friese, *T. flaveola* Friese.

BOOK NOTICE.

WOODLAND IDYLS. By W. S. Blatchley. The Nature Publishing Company, Indianapolis. Price $1.25, post-paid.

In this little book are recorded the observations and reflections of one who pitched his tent, and spent his summer vacation, apart from the haunts of men, living, in gipsy style, upon squirrels, berries, and other woodland supplies.

The author has contrived, by an unusual construction of his sentences, to give an air of quaintness to his work—as in:—

"The prunella, favourite of my summer blossoms, did I find on yesterday," page 86.

"Tiny the stream, yet this broad valley has it carved," p. 87.

"The writing off my mind, squirrels and marmots do I seek," p. 167.

One passage, at least, in *Woodland Idyls*, will be of interest to entomologists. It is that in which the author tells that he saw an ichneumon light upon a spider, *which a wasp was carrying off*, and deposit an egg in it (pp. 206—9). Does not this afford us a glimpse into the life-histories of such insects as Zabriskie's *prædator*, in Ashmead's genus, *Sphecophagus*?

A few brief quotations from the book under consideration will set the author's style and trend of ideas fairly before the readers of the Canadian Entomologist.

The author's descriptive powers:—

"I saw a skeedoodlum of a wren, his feathers half gone from moulting, his body not bigger than thirty seconds, yet with his head in air he was rolling forth sound enough for a cardinal or other bird ten times his size. 'Cher-whitty—cher-whitty'." * * * *"A cheery little cuss is he, who would sing were his tail on fire." (p. 42).

"Fuzzy gnats dance in rhythmic mazes before my eyes, while their cousin, a slender reddish-gray mosquito, probes my flesh, I do not feel him until his body is red and gored with my blood. After swatting him the itch begins. Niches they fill in the great scheme of nature. Organs they have for performing all the duties of life. Those duties are but few—to eat, grow and reproduce their kind. Lowly creatures we call them, yet "lowly" only because we esteem ourselves "high." (p. 79).

The author as a botanist:—

"The densely flowered spikes of the vervain before me, some of them two feet in length, have but an inch or two in blossom at a time. The seed pods or fruit of the past are below, the unopened buds of the future above. The flowers are now close to the top, the fruiting portion long, the budding part short, for its season is near the close. Life, present work, is now in the flowering part; duty performed, finished work, in the seed part; promises or hopes for the future in the buds. Only the present blooming part, that which is active, is beautiful. That is the part attractive to the human eye, in the plant as well as in the human. What are you doing? Be up and at work. Live not upon a past reputation. Chance not your happiness upon the budding unlived future, which may be seared by a night's hoar frost into something dull and dead." (pp. 46-7).

The author as a Darwinian:—

(The Red-headed Woodpecker). "In a century from now the bills of his descendants will be broader, their eyes keener, their throats wider, and they will be part swallow, part woodpecker, creatures better adapted to the life they have adopted. For he is slowly changing from a simon-pure woodpecker, where the struggle for life grows ever more bitter, as the forests grow fewer, into a cleaver of the air, a swallower on the wing, a contortionist who can rise and fall, twist and turn in rapid flight after his oft-times elusive prey." (p. 203).

The author's philosophy:—

"Long may, and doubtless long will, the world wag on without me. My turn at the wheel has ended. Content am I to sit in the shade and practice shooting at a marmot's head." (p. 171).

The author's religious opinions:—

"Great oaks like these were most worthy to be the Gods of the Druids. As much right to worship them had they as I the sun. I revere or worship only that which I *know* exists—that which is the highest, most powerful of all things known to me. Back of or above the sun there may be somewhere—but where we know not, nor shall we ever know—a power higher than the sun, master of him, and of all other suns—the Overlord of all. Until I know, which I shall never do, that there is such an Overlord, until then I worship, if you may call it worship, that highest power, that ruler which my senses ken." * * * * "Then let the oak tree my Sabbath temple be, let the sun be the God unto whom this morn my reverence is due, and this spot of mother earth the altar at which I kneel to do homage unto him." (pp. 228 and 229).

The "God-gifted organ-voice of England," telling of other devotions, breathes a different spirit from that expressed in the last quotation . It says:—

"Thou sun, of this great world both eye and soul,
Acknowledge HIM thy greater; sound His praise
In thy eternal course, both when thou climb'st
And when high noon hast gain'd, and when thou fall'st."
—Adam's Prayer in *Paradise Lost.*

The writer of this article ventures to express an earnest hope that the author of *Woodland Idyls* may attain unto the higher knowledge—the knowledge spoken of by the "MASTER," in His address to His Father Almighty:—"This is life eternal that they might *know* thee the only true God, and Jesus Christ whom thou hast sent."—St. John XVII.: 3. T. W. F.

Mailed September 18, 1912.

REV. G. W. TAYLOR, F.R.S.C.

The Canadian Entomologist.

VOL. XLIV. LONDON, OCTOBER, 1912. No. 10

THE REV. GEORGE W. TAYLOR, F. R. S. C., F. Z. S.

The subject of this memoir was born in Derby, England, in 1851, and came to Canada when he was twenty-five years of age. He settled in Vancouver Island and studied for the Ministry under the Rt. Rev. George Hills, D. D., Bishop of Columbia. He was made a deacon in 1884 and ordained to the priesthood in 1886. His first clerical charge was Cedar Hill. He had already given attention to the attractive and but little known fauna of the Pacific Coast, for in the preface to the Toronto Check List of Insects, which was published in 1883, Messrs. Brodie and White speak of him as a collector to whom their thanks were due, and at the annual meeting of the Entomological Society of Ontario, held at London, Ont., October 15, 1884, Mr. James Fletcher presented, on behalf of Mr. Taylor, a collection of Diurnal Lepidoptera to the society. On this occasion Mr. Fletcher said that although Mr. Taylor was but a new member "he had already done good work." In the report of this meeting the first contributon, by the Rev. G. W. Taylor, to the annals of the Society, appears. It is entitled "Notes on the Entomology of Vancouver Island." In the CANADIAN ENTOMOLOGIST for the same year (Vol. XVI) other papers written by him will be found.

In 1887, Mr. Taylor was appointed Honorary Provincial Entomologist of British Columbia. In the Annual Report of the Entomological Society for that year he published a very interesting account of a series of expeditions made by himself, Mr. Fletcher, Professor Macoun, Mr. Tolmie and others, to the summit of Mount Finlayson, in search of *Chionobas gigas* Butler.

After some years' active service during which he had built a church, he resigned his charge in Columbia Diocese, and moved to Ottawa. There he was favourably received by the Ecclesiastical authorities; and there he founded the church of St. Barnabas. But after some years, for the benefit of his health, he returned to British Columbia and became rector of the church at Wellington, near Nanaimo. He retained this charge until five years ago when he was appointed by the Federal Government Curator of the Biological Station at Departure Bay.

It was during the period of his second residence in Vancouver Island that Mr. Taylor became a constant contributor to the pages of the CANADIAN ENTOMOLOGIST. In the volumes of that magazine numbered from XXXVI to XLII inclusive no less than eighteen papers from his hand appear. His last contribution, entitled "On some New Species of Mesoleuca," is given in the number for March, 1910.

Of late years Mr. Taylor gave much attention to the Geometridæ, especially those belonging to the genera *Eupithecia* Curtis and *Mesoleuca* Hübner. Of these he described and named many new species. The whole of his collection of Geometridæ has been recently purchased by Dr. Wm. Barnes, of Decatur, who, without doubt, will make excellent use of it; but we cannot but regret that so much of the fruit of our late friend's research and ability should have passed from the Dominion.

In 1881 Mr. Taylor was made a Fellow of the Royal Society of Canada. He had been for many years a Fellow of the Zoological and Entomological Societies of England, and fifteen years ago he was elected a Corresponding Member of the Ottawa Field Naturalists' Club. All these societies have been benefitted by his labours.

In the Thirty-fourth Annual Report of our own society appears a highly-appreciative and eulogistic account of Mr. Taylor from the pen of the late Dr. Fletcher. From it we learn that many naturalists have given honour to Mr. Taylor by naming after him new species of various kinds, as, for example: *Melitæa taylori* W. H. Edwards, *Mediolaria taylori* Dall, *Leucandra taylori* Lambe.

Undoubtedly Mr. Taylor's chief scientific work was done in connection with Marine Zoology, and in recognition of this the Federal Government, in 1905, appointed him a member of the Dominion Fisheries Commission for British Columbia. In the report of that Commission, Mr. Taylor described as many as thirty kinds of edible shell-fish.

"There is in course of publication by the Dominion Government at the present time a very long and valuable report on the crabs, shrimps, and other crustacea of British Columbia."—(*Ottawa Evening Journal*, Aug. 24th, 1912.)

The following words, written by Dr. Fletcher in the lifetime of Mr. Taylor, and in the paper above referred to, convey much in few words, and were justly due to the deceased: "Mr. Taylor is an indefatigable collector and a generous correspondent, who considers no trouble too much to make observations or secure specimens when specially desired. In his parish work he is painstaking, gentle and self-denying—always ready to help. A clear and forcible preacher and an earnest liver, who shows in his works that religion is not an accessory of every-day life, but an integral part of it."

Mr. Taylor died of paralysis, on August the 22nd last, and was buried in the cemetery at Nanaimo. He leaves to mourn his loss a married daughter and two sons. The funeral service was read by a dear friend of the deceased, the Venerable Archdeacon Scriven.

It is to be hoped that measures will be taken to secure for the benefit of posterity the very valuable conchological and (remaining) entomological collections left by Mr. Taylor. T. W. F.

ON THE DIPTERA OF BAJA CALIFORNIA, INCLUDING SOME SPECIES FROM ADJACENT REGIONS.—II.*

BY C. H. T. TOWNSEND, LIMA, PERU.

This paper embodies a report on a lot of flies sent me for determination some years ago by the California Academy of Sciences. They were secured on a later expedition than those mentioned in the first paper.† Unless otherwise stated, they were collected jointly by Dr. Gustav Eisen and Mr. Frank H. Vaslit, who, together, visited and collected in San José del Cabo in September, and Tepic in October and November, 1894. Species already listed in the first section appear here with their original numbers.

*The present paper has been in manuscript for nearly ten years, but with many others was never reached by the Publication Committee of the Cal fornia Academy of Sciences, owing to lack of funds. It was returned to me many years ago, has since that time been overlooked, and is now offered on account of the fact that the results it contains appear to have lost none of their interest during the lapse of time.

†Section I appeared in Proc. Cal. Acad. Sci., Ser. 2, Vol. IV, pp. 593–620.

BIBIONIDÆ.

2. *Dilophus stygius* Say.

Tepic.—Twelve ♂s and forty-three ♀s, Nov. One pair *in coitu*, which verifies my conclusion that the two sexes associated together in my former determination (see No. 2 of Section I) are the same species. The females vary considerably in size, some being as small as the larger males. The small linear blackish stigma in the whitish wings of the ♂ is often nearly or quite obsolete.

It is worthy of note that, among the specimens sent me of the females of this species, there was inadvertently included a specimen of a black sawfly, which would easily pass for a ♀ *Dilophus stygius* if not looked at a second time. This sawfly is of the same uniform deep shining black as the ♀ *stygius*, is of the same size, and has the same black wings of corresponding shade. I can hardly resist the conclusion that the sawfly mimics the ♀ *stygius*, though for what reason cannot at present be said. The sawfly is a remarkable form, in that it possesses long-branched antennæ. Each antenna is split nearly to base into two branches, the stalk or pedicel being short and bare, and the branches hairy. The abdomen of the sawfly is more shining than that of the ♀ *stygius*, but this does not show save on close inspection, while its general form closely approaches that of the ♀ *stygius* abdomen. Of course, the head of the sawfly is totally different from that of the ♀ *stygius*, but this is not conspicuous on first sight, the effect being lost in the uniform colour resemblance and otherwise close similarity.

D. stygius is an abundant Mexican species. The length of the body in the ♀ does not average over 6 mm. in the present specimens. I believe that the ♀ *Dilophus* identified by Bellardi as *orbatus* Say (Saggio I, p. 19) was not that species, but *stygius* Say. My reasons for this opinion are as follows :

D. stygius was described by Say from Mexico. *D. orbatus* was described by Say from Pennsylvania, and Osten Sacken has identified as *orbatus* two sexes of a species collected in Florida by himself. It is very common for a Middle Atlantic Coast species to extend into Florida, but rarely does a northern species extend so far southwest as Southern Mexico. Bellardi's specimens were from Orizaba. I have myself taken in numbers in the outskirts of Orizaba what I believe to be *stygius*. The males from Florida, which Osten Sacken identifies as *orbatus*, and which I consider to be that species in all probability, are described as having the wings yellowish, whereas all the Mexican males that I have seen have the wings distinctly whitish with no yellowish tinge. Bellardi gives the length of his specimen (♀) as 8 mm., which is much longer than ♀ *orbatus* as given by Say ($^1/_5$ inch = less than 5 mm.), and Wiedemann (2 lines = about 4 mm.). The median cross-vein of the wing is always present in both sexes; it is often situated in both sexes exactly at the furcation of the vein, at other times being just a little distance before the furcation.

62. *Plecia bellardii* Towns., n. nom.

Sym. *vittata* Bell. (nec. Wied.) preocc.

Tepic.—Eight ♂s and three ♀s, Nov. Aside from the genital characters the females may be known by the eyes not being contiguous but well separated, the front being fully as wide as the eyes. Bellardi says that the ♀s are larger than the ♂s, and have the wings longer and wider. In my specimens there is hardly any difference in size of body or wings, except that the abdomen of the ♀ may be slightly larger. Length of body about 7 mm., of wing about 9 mm.

I identify these specimens with Bellardi's (not Wiedemann's) *Plecia vittata*, which Schiner (Nov. Reise Dipt., p. 22), makes a synonym of *plagiata*. I believe that this synonymy is incorrect. If *vittata* Bell. is distinct, as I believe, it must be called by another name, as *vittata* is preoccupied by Wiedemann. I have therefore proposed the name *bellardii*.

There is no brownish tinge to the wings, which vary from a dense to a dilute black, with an iridescent greenish to violet reflection in oblique lights. Wiedemann describes the darker parts of the wings of *plagiata* as blackish brown, Schiner gives no reason whatever for placing *vittata* Bell. as a synonym of *plagiata*.

TABANIDÆ.

63. *Pangonia tepicana* Towns., n. nom.

Syn. *P. basilaris* Wd., Aus. Zweifl., II, 621 (preocc.).

Tepic.—One ♀, Oct. I believe this to be *P. basilaris* Wd., Aus. Zweifl., II, p. 621 (not *basilaris* Wd., Aus. Zweifl., I, pp. 554–5, and not *wiedemanni* Bell., Saggio Ditt. Mess., I, p. 48). Von Röder has pointed out (Dipt. gesam. Süd-Amerika von A. Stübel, p. 7) the differences in the wing coloration of *basilaris* (Wd., Aus. Zw., I, pp. 554–5) and *wiedemanni* Bellardi. In the latter the black of wings is confined to the extreme base, and extends only as far as the cross-veins at base of basal cells. In the former it extends to the cross-veins, closing the basal cells, and takes up the whole basal third of the wing. Röder's specimen of *basilaris* was from the Rio del Cinto (Ecuador), about 5,000 ft.

My specimen agrees perfectly with Wiedemann's description (Aus. Zw., II, 621). It differs markedly, as does also Wiedemann's description, from *P. wiedemanni*, as described by Bellardi. I therefore believe that Bellardi was in error in identifying his species with Wiedemann's. *P. tepicana* differs as follows from Bellardi's description of *wiedemanni:*

♀.—Length, 12½ mm.; proboscis, hardly 3½ mm. Front brownish-yellow pollinose; first two antennal joints light brownish yellow, third wholly reddish yellow, apex not fuscous. Apical annulus elongate, narrow and pointed, hardly half length of rest of third joint. Third joint is swollen at base, but flattened, and the annuli are strongly contiguous. Face brownish-yellow pollinose. Palpi not unusually elongate, last joint about as long as third antennal joint, flattened and curved, but pointed at tip. The palpi and the six lancet-like organs are clear reddish yellow. There are four faint lines apparent on thorax, distinguished from the fuscous-yellow pollinose surface by being more thickly pollinose. Abdomen brownish yellow, first segment black under scutellum, from which a black median vitta extends back to fifth segment (subobsolete for a short distance on third segment in my specimen). Third, and especially fourth and fifth segments tinged with brownish, due to age of specimen no doubt. First two segments with yellowish hair only; third with black hair on anterior two-thirds and yellow hair on posterior one-third; fourth and fifth (these segments are short) with black hair anteriorly, and yellow hair behind, giving the hind border of abdomen a good fringe of yellow hair. Femora blackish, rest of legs wholly orange-yellow, with front femora distally tinged with same colour. Wings tinged with fuscous-yellow, the extreme base blackish brown. All else as in Bellardi's description.

This species will be distinguished at once by the smaller size, shorter proboscis, black femora, and the median abdominal vitta and black hair of third to fifth abdominal segments.

15. *Tabanus punctifer* O. S.

Mesa Verde, L. Cal. One ♀, Oct., 1893 (Eisen).

ASILIDÆ.

21. *Proctacanthus arno* Towns.

San José del Cabo. Four ♀s and seven ♂s, Sept. One of the ♂s measures 33 mm.

22. *Eccritosia amphinome* Walk.

Syn. *Proctacanthus zamon* Towns. (Section I, No. 22).

San José del Cabo. Twenty ♂s and eleven ♀s, Sept. Four of the

♀s and three of ♂s measure 29 to 31 mm. I observed this species at Hermosillo, Sonora, in Sept. 1894, on the sand of the dry bed of the Rio Sonora.

64. *Doryclus distendens* Wd., var. *varipennis* Walk.

San José del Cabo. One ♀, Sept.

Dr. Williston places Walker's species as synonymous with *distendens*, but it may be considered a good variety on the strength of the two brown cross-bands of the wings. This is the first exact record of the species from north of Guatemala.

The present specimen is a strongly aberrant one, with body almost wholly brownish red, front tibiæ and metatarsi not at all blackish; and the fourth posterior cell wide open, being as wide on margin of wing as the first posterior cell. The lateral thoracic vittæ are grayish pollinose, but the two middle vittæ are tawny grayish. The middle vittæ are not elongate cuneiform from a hind view (see O. S., Biol. C. A. Dipt., I, 182), but are distinctly equilateral, well separated and parallel. From a front view they do appear elongate cuneiform. Abdomen is almost wholly brownish red, with only flakes of blackish in places, especially on underside.

Since Jænnicke's figure represents *Doryclus distendens* with the fourth posterior cell completely closed, I infer that this is the normal venation of the genus. Whether the present form should be separated on account of this cell being wide open I cannot now decide. As it otherwise agrees so closely with *Doryclus* in the more important characters, I refer it here.

APIOCERIDÆ.

26. *Rhaphiomidas xanthos* Towns.

San José del Cabo. One ♂, apparently not maturely coloured, seems to be this species. Sept.

Length, nearly 25 mm. This is the only specimen of *Rhaphiomidas* in the lot, which seems strange since so many occurred in the previous sendings. The wings do not quite reach the tip of the abdomen. Segments 6, 7 and 8 of abdomen together about as long as 5, which is but a little shorter than 3. It seems that in the previous description a segment was missed, which is revealed in this less matured specimen.

SYRPHIDÆ.

65. *Chrysogaster bellula* Will.

Tepic.—One ♀, Nov.

Length, 4½ mm. Resembles *intida* in antennæ, which are much longer than face, and with second and third joints nearly equal in length. The face is not more than three-fifths the length of antennæ. It agrees perfectly in the wings with Williston's description of *bellula*, and not at all with *intida*; therefore I place it here. The antennæ are brown, with first two joints tinged with yellowish. The disk of abdomen is pronouncedly opaque blackish, but with some cupreous and green. The face is quite rugose and the epistoma is hardly produced downwards. I am unable to restore the markings of the eyes, doubtless because the specimen was originally an alcoholic one, and therefore cannot say toward which species it inclines in the pattern of the eye-picture. (See Williston, Biol. C. A. Dipt., III, p. 7.)

66. *Volucella obesa* Fab.

Tepic.—Two specimens, ♂ ♀, Oct.

Length, 10 to 11 mm. Metallic green. The third antennal joint is only moderately short in the ♀, and hardly shorter in the ♂.

67. *Volucella dichroica* G.-Tos.

Tepic.—One ♀ I consider as this species. Oct.

The face is strongly conically projected below, ending in two teeth formed by a median longitudinal notch in the apex of the cone, and I should hardly call it obtuse. The scutellum is not reddish-coppery *(rosso-rame)*, but of the same greenish-violaceous colour as the thorax and abdomen. The metatarsi and next two joints, especially in the hind legs, are pale brownish-yellowish, as are also bases of antennæ. Otherwise it agrees well with Giglio-Tos' description. The brownish spot at distal end of submarginal cell is subobsolete, and a similar cloud is apparent on last section of fourth vein at distal end of apical cell, and along last section of third vein. There are bristles on the edge of the scutellum, and the eyes are hairy, both of which characters are unmentioned by Giglio-Tos.

The specimen agrees well with the more important characters in Williston's description of *V. viridis*, from Chapada, Brazil, except that the ♀ front is not of equal width, but is very noticeably widened anteriorly. While the marginal cell is short petiolate, the legs are more luteous than in Giglio-Tos' specimen, yet their prevailing colour is black. In the colour of the scutellum it agrees better with *viridis*, and it possesses the ciliate-like pile of femora and tibiæ. These two species must be very closely allied.

48. *Eristalis tricolor* Jænn.

Tepic.—One ♀, Oct.

Length, 9½ mm. Has much more black on the abdomen than Lower California specimens. The black triangle of second segment expands on each side along posterior margin, widening at posterior corners of segment into a spot. Third segment is black, with a yellow spot on each side, which reaches anterior border only. Narrow hind margin of second to fourth segments light yellow. Tibiæ quite yellowish, even hind pair.

San José del Cabo. One ♀ and three ♂s, Sept. These have more yellow on the abdomen than the above specimen. The ♀ has even the fourth segment yellowish (brownish yellow), with black spot in middle. Second and third segments same. The three ♂s are the same, except that the fourth segment is wholly black in two, and with only the anterior lateral angles yellow in the other.

TWO BEES NEW TO CANADA.

BY T. D. A. COCKERELL, BOULDER, COLORADO.

Chelynia ricardonis, n. sp. (? *rubi*, subsp.).

♀.—Length, 9 mm.; similar to *C. rubi* (Ckll.), but sides of head above, and sides and anterior part of mesothorax, with conspicuous white hair; tubercles densely fringed with dull white hair; abdominal markings bright lemon yellow (cream-coloured in *rubi*), the band on first segment broad and entirely curved at sides; that on second interrupted sublaterally, the lateral pieces of it pyriform; third segment with a rather short median stripe and small lateral spots; fourth with a median butterfly-shaped yellow mark; hind basitarsi long, subclavate, with reddish hair on inner side.

Hab.—Vernon, British Columbia, June 19, 1902 (Miss Ricardo). British Museum. This has the structure of *C. rubi* (*betheli* Ashm.), but differs in the colour of the markings and pubescence. It is probably a valid species, but it may prove to represent only a local race or subspecies of *C. rubi*. The latter occurs at Seattle and Olympia, in the State of Washington. Although the two species are not very far apart geographically, Vernon is an inland locality, with doubtless a very different fauna from that of the coast.

Anthidium porteræ Ckll.

Calgary, one male (Miss Ricardo). British Museum. Also from Calgary, from some collector, is a male *A. tenuifloræ* Ckll., a form with the scape of the antennæ entirely black.

NEW GENERA AND SPECIES OF XYELIDÆ AND LYDIDÆ.*

BY ALEX D. MACGILLIVRAY, UNIVERSITY OF ILLINOIS, URBANA, ILL.

The most of the following descriptions have been in manuscript for many months. The names in this paper and some others to be published later are to be used in another place, and they are offered for publication at this time for that reason.

Paraxyela, n. gen.—Front wings with the free part of M arising distinctly before the point of separation of R and Sc_2, the free part of R_5 distinctly shorter than $R + Sc_2$, frequently less than one-half the length of $R + Sc_2$; the hind wings with the free part of R_5 present; clypeus triangular in outline, the median portion two or three times as long as the lateral portions; the antennæ with the third segment longer than all the following segments together; the claws cleft, the two parts of the cleft parallel. Type, *Xyela tricolor* Nort.

Macroxyela bicolor, n. sp.—Male: head with a flat depressed area in front of the median ocellus, never crossed by the median fovea; median fovea represented by a linear smooth spot only slightly if at all depressed below the surface of the front; the area of the head between the antennal sockets and the ocellar furrow blackish and coarsely punctured, the remainder of the head and the notum finely shagrined; antennæ with the third segment five times as long as all the following segments together; the fourth and fifth segments subequal, each longer than any of the following segments; the body black with the clypeus, the labrum, the malar space, the supraclypeal area, the basal plates at sides above the abdomen, and the legs, rufous. Length, 8 mm.

Habitat.—Columbus, Ohio. Professor J. S. Hine, collector.

Differentiated from the males of all other species of the genus known to me by the greater abundance of rufous.

Macroxyela obsoleta, n. sp.—Female: head with a flat depressed area in front of the median ocellus, never crossed by the median fovea; median fovea a broad, flat, indistinct, depressed area, more distinct near the median ocellus; antennæ with the third segment many times longer than all the following segments together, with a black ring at base; the fourth, fifth and sixth segments subequal in length, the following segments shorter; the head below the ocelli sparsely, coarsely punctured; front wings with

*Contribution from the Entomological Laboratory of the University of Illinois, No. 32.

the radial cross-vein much nearer the point of separation of R_2 than the radio-medial cross-vein; the saw-guides strongly convex above on the basal half and straight or slightly convex below; the body rufous with a spot about the ocelli, a spot near the base of each wing, and the base of the abdomen more or less black. Length 8 mm.

Habitat.—Ithaca, N. Y. J. O. Martin, collector.

Similar in appearance to *infuscata* Norton, but readily separated by the sculpture of the head.

Macroxyela distincta, n. sp.—Female: head with a flat depressed area in front of the median ocellus, never crossed by the median fovea; median fovea a distinct, narrow, elongate, diamond-shaped depression, flat on the bottom; antennæ with the third segment many times longer than all the following segments together; the fourth, fifth and sixth segments subequal in length, the following segments shorter; the head below the ocelli roughened by elongate punctures; the radial cross-vein nearer the point of separation of R_2 than the radio-medial cross-vein; the saw-guides convex above on the basal half and straight below; the body rufous with two spots on the lateral lobes of the mesonotum, and the postscutellum black. Length 8 mm.

Habitat.—Ithaca, N. Y. J. O. Martin, collector.

The male is black with the clypeus, labrum, legs and venter, except at base, apex, and lateral margin of abdomen for the most part, yellowish-rufous.

Separated from *infuscata* Norton and *distincta* Mack by the form of the median fovea.

Protoxyela, n. gen.—Front wings with the free part of M arising distinctly before the point of separation of R and Sc_2, the free part of R_5 distinctly shorter than $R + Sc_2$, frequently less than one-half the length of this vein; the free part of Sc_2 almost twice as long as the free part of Sc_1; Sc_1 much more oblique than Sc_2; the cell R_3 usually divided by a supernumerary cross-vein; the hind wings with free part of R_5 present; the clypeus not triangular in outline, the median portion but little if any longer than the lateral portions; the antennæ with the third segment as long as all the following segments together; the claws with an erect tooth at middle. Type, *Xyela ænia* Nort.

Itycorsia angulata, n. sp.—Female: body olivaceous with the basal segments of the antennæ, the clypeal suture, the furrows of the head, the postocellar area in great part, two irregular spots on the posterior orbits, a crescent-shaped mark on each side on the vertex between the postocellar area and the orbital spot, an irregular band between the dorsal margins of the compound eyes, including the ocelli, the pronotum except the lateral and caudal margin, a spot on the cephalic half of the median lobe of the mesonotum, a spot on each lateral lobe, a round spot on the disk of the mesonotum, the dorsum of the metathorax in great part, the basal plates, the pleural and sternal sutures, and the caudal surfaces of the femora, for the most part, black; the median fovea a pit nearer the ocelli than the antennæ, with a tubercle at its ventral end ; antennæ with about thirty-five segments, the third segment as long as the next two ; the postocellar area broadly convex, higher than the ocelli ; mesal eye-margin distinctly angulate ; the head sparsely, punctately roughened except the declivous area, which is polished ; front wings with the free part of R_5 and the radial cross-vein interstitial. Length 14 mm.

Habitat.—Axton, N. Y. (C. O. Houghton and the author, collectors) ; Manchester, Conn. (A. B. Chamberlin, collector); Wallingford, Conn. (J. K. Lewis, collector).

This species is closely allied with *luteomaculata* Cress.

Cephaleia distincta, n. sp.—Male : Body black with the clypeus, the supraclypeal area, the head between the compound eyes and the antennal sockets, a faint spot on each vertical furrow, the posterior orbits, the tegulæ, a band on the mesopleuræ, the prosternum, the legs beyond the coxæ and the lateral margin of the abdomen, yellow ; antennæ with about twenty-five segments, the third segment longer than four and five together; the median fovea extending to the median ocellus ; the clypeus slightly carinated ; the head sparsely punctured, punctures confluent in the region above the antennal sockets ; the mesonotum sparsely punctured ; the scutellum almost smooth. Length 10 mm.

Habitat.—Mt. Washington, N. H. ; Mrs. Annie Trumbull Slosson, collector.

This species would fall in a table near *mathematicus* Kirby, from which it can be differentiated by the black head.

Cephaleia criddlei, n. sp.—Female : body black with the clypeus, a spot on the inner margin of the compound eye, a broad spot on each vertical furrow, the posterior orbits, a long spot on the lateral lobes of the mesonotum including the scutellum and the dorsum of the abdomen, and extending as an angulated band along the lateral margin, rufous ; the antennæ beyond the pedicel, and the legs beyond the tip of the femora, yellowish white; the head deeply, sparsely punctured; the pleuræ and notum deeply, closely punctured; the median fovea indistinct, not connected with the median ocellus; the wings with a dusky band in the region of the stigma. Length 14 mm.

Habitat.—Aweme, Manitoba; Norman Criddle, collector.

This species is similar to *fascipennis* Cresson. The densely banded wings will differentiate it.

Cephaleia jenseni, n. sp.—Female: body rufous with dusky spots on the antennal sockets; median fovea, posterior orbits, postocellar area, the meson of the prothorax and line at sides of the prothorax, a line on the median lobe of the mesonotum, the coxæ and the cephalic and caudal margins of the femora, black; the posterior orbits and the antennæ, white, somewhat rufous at base; median fovea a rounded pit; median ocellus in a rounded depression; the head sparsely punctured; the median lobe of the mesonotum, the shoulders of the lateral lobes and the scutellum, polished; the remainder of the notum sparsely punctured; the third segment of the antennæ longer than the fourth and fifth together; the wing-veins brownish, slightly infuscated along the veins. Length 11 mm.

Habitat.—Eagle Bend, Minnesota. J. P. Jensen, collector.

This species is similar to *criddlei* Mack. It lacks the fuscous banded wings and the form of the median fovea is different.

Pamphilius transversa, n. sp.—Female: body black, with the clypeus, the face, the first segment of the antennæ beneath, the antennæ on its apical third, the cheeks, the posterior orbits, the tegulæ, the scutellum, the legs except the posterior tibiæ, and the abdominal segments three and four and part of five, varying from whitish to yellowish and rufous; the wings hyaline; the veins brownish; the stigma dark; the head finely sparsely punctured; the notum almost smooth, sparsely punctured on the posterior angles; the scutellum roughened; the antennæ with the third segment wider and slightly longer than the fourth; antennæ with about twenty-eight segments; the mesopleuræ finely roughend and setaceous. Length 12 mm.

Habitat.—Franconia, New Hampshire. Mrs. Annie Trumbull Slosson, collector.

This species is similar to *perplexa* Cresson.

Pamphilius dentatus, n. sp.—Body black with the terminal half of the antennæ, the clypeus, a dentate spot on the inner orbits extended as a parenthesis-shaped mark to the caudal aspect of the head; an emarginate spot in front of the median ocellus and an angular line behind it, a parenthesis-shaped mark at the lateral margin of the vertical furrow, the margin of the pronotum, the tegulæ, the V-spot, the scutellum, the postscutellum, a spot on humeral angle beneath the wings, the front and middle legs and the hind legs except the tibiæ, white; the abdomen rufous beyond the basal plates; antennæ with about twenty-seven segments; supraclypeal area carinated; head depressed about the median ocellus, sparsely punctured; declivous area smooth; median lobe of mesothorax smooth, lateral lobes densely punctured and scutellum sparsely punctured: wings hyaline; veins brownish. Length 8-10 mm.

The male differs in having the entire declivous area yellow and the notum, except the scutellum and the postscutellum, black.

Habitat.—Wilbraham, Mass.—J. O. Martin, collector. Hamden, New Haven, and Wallingford, Connecticut—B. H. Walden, collector.

This species is near *rubi* Rohwer.

Pamphilius fletcheri, n. sp.—Male: body black with the front and clypeus below the transverse ridge, the proximal segment of the antennæ beneath, the apical half of the antennæ, the inner and posterior orbits, a line on each side of the caudal margin of the head, a narrow line on the collar, the tegulæ, the scutellum, the postscutellum, and the legs except the posterior tibiæ, the tarsi becoming more or less rufous, white; abdominal segments three to five rufous; antennæ with about twenty-six segments, the second and third segments equal in length; the declivous part of the head roughened; the median ocellus in a heart-shaped depression, the apex being behind the ocellus, the median fovea a pit below this depressed area; the head strongly elevated and roughened on each side between the lateral ocelli and the compound eyes; the mesonotum polished; the scutellum sparsely punctured; wings hyaline. Length 8 mm.

Female.—Body black, with an anchor-shaped area on the head, the front margin of the clypeus, the mandibles, the distal half of the antennæ, a line on the posterior orbits, the cheeks, the inner orbits, bifurcating near the middle of the compound eyes, one part extending obliquely toward and almost to the lateral ocelli, the other extending along the margin of the compound eyes, swollen at their upper inner margin, narrowed again on the posterior orbits, triangularly expanded at the caudal margin of the head and extending along its caudal margin on each side, two spots in front of the median ocellus, a line on the collar, the tegulæ, two converging bars on the median lobe of the mesonotum, the scutellum, the postscutellum, and the legs below the middle of the coxæ, except the distal five-sixths of the posterior tibiæ, white; the abdomen beyond the first segment rufous; the head and mesonotum sparsely punctured; the third segment of the antennæ distinctly longer than the fourth; antennæ with about twenty-four segments; the median foveæ wanting; the median ocellus located in a heart-shaped depression; the frontal declivity broadly and deeply broken by the antennal furrows; wings hyaline, the veins and stigma brown. Length 9 mm.

Habitat.—St. John, New Brunswick.

Described from two males and a female received from Dr. C. Gordon Hewitt, Division of Entomology, Ottawa, Canada, where the type is deposited. These specimens were reared from larvæ received from St. John, New Brunswick. The larvæ feed on the leaves of raspberry (see Annual Report of Experimental Farms for year 1899 (1900), pp. 180-181). The The species is named for the late Dr. James Fletcher.

This species is near *rubi* Rohwer and *dentatus*.

SMERINTHUS CERISYI KIRBY AND *SMERINTHUS OPHTHALMICUS* BDV.

BY F. H. WOLLEY DOD, MILLARVILLE, ALTA.

It is not very often that I take notes on Sphingidæ, or take much notice of them at all outside my own district; but recent observations casually brought to my notice the fact that two good species were probably involved under the above two names, though I had long ago taken it for granted that such was not the case, and I became immediately interested, and followed the matter up. My first observation in the matter was made while I was somewhat hastily glancing through this family in Mr. Winn's collection at Montreal last January. Thereon I wrote: "Under *cerisyi*, two specimens, Biddeford, Me. and Montreal; have much crenate s. t. lines and apical mark almost lunate as in *geminatus*," and "two from B. C. under *opthalmicus* have lines fairly even, wavy, and apical marks not lunate." Shortly afterwards, whilst in England, I compared this note with Kirby's figure and concluded that it must really represent the form I have so long known as *cerisyi* at Calgary which is the one Mr. Winn has taken on the east coast. Kirby's figure is probably somewhat exaggerated, and has the apical mark almost as lunate, well defined and contracting as *geminatus*. The dark marks near the anal angle of primaries are also more like those in *geminatus*. These exaggerated characters may of course have been the result of figuring from a specimen with worn margins.

In the British Museum I found a Calgary *cerisyi*, another from Vernon, B. C, and a third from Ashnola, taken by Mrs. Nicholl. Under "Subspecies of *ophthalmicus* Bdv.", I found Butler's type of *vancouverensis* from Vancouver Island, and other specimens from there, Frazer Pines and California, which appeared to agree with it. This is the form which Mr. Winn had as *ophthalmicus*, and is that of which Holland figures a female on Plate VII, Fig. 3, as *cerisyi*. Without having seen Boisduval's type, which, if it still exists, is probably somewhere in France, I must assume that it is the form subsequently described by Butler. All the Calgary specimens at present in my collection are *cerisyi*, and had I even taken *ophthalmicus* here I should probably have noticed the difference. I have a series of the latter from Vancouver and the Island, but no *cerisyi* from outside Alberta, though it evidently occurs right across the continent. Besides the differences mentioned, *ophthalmicus* has the terminal dark shade wider centrally. The two have exactly similar antennal structure, and the only structural difference I can find elsewhere is in the outer margin which has fewer dentations and more acute apices in *ophthalmicus*. I may be in error about their distinctness, and the point requires working out carefully with far more material than I have been able to examine ; and, know, for all I may have been so worked out. Holland for instance states that "they run into each other to such an extent as to make it often impossible to distinguish them" and treats them as do most others, as subspecies. Crenations rather than undulations is the rule throughout in *cerisyi*, in lines, apical marks and outer margins, though I feel bound to admit that the variation in my two series is such as to suggest that a large increase of material might result in increased difficulty in separating them. But with closely allied species such is often the case.

European *ocellatus*, of which I have four specimens, resembles *ophthalmicus* rather than *cerisyi*, though the top one of three figures given in Mr. Richard South's "Moths of the British Isles" has the cranate apical marks exactly as in *cerisyi*. The outer margin is more entire than in *ophthalmicus*, and antennal structure is similar to both.

SOME COCCIDÆ FROM THE GRAND CANON, ARIZONA.

BY T. D. A. COCKERELL, BOULDER, COLORADO.

Mr. E. Bethel, when recently visiting the Grand Cañon, was so good as to collect some Coccidæ for me, and the species prove so interesting that they are herewith recorded:

1. *Ceroplastes irregularis* Ckll.—In quantity on *Atriplex*, July 22

2. *Orthezia garryæ* Ckll.—On *Fendlera*, July 21. Previously known only on *Garrya*, from a single locality in New Mexico. The following notes are based on the Arizona material:

Pale pea-green; last antennal joint very slender, dark; first joint narrow, bent; legs very long; skin densely spiny as usual.

Measurements in microns: Middle leg, femur and trochanter, 800; tibia, 832; tarsus (without claw), 384.

Antennal joints: (1) 176, (2) 128, (3) 192, (4) 192, (5) 160, (6) 120, (7) 120, (8) 192–208, or counting apical spine, 208–224.

3. *Phenacoccus betheli*, n. sp. (possibly subsp. of *P. cockerelli* King).—On *Amelanchier*, July 21. Adult females solitary on twigs; hemispherical or nearly; about 4 mm. long, 2¾ broad, a little over 2 high; dark raspberry-red, covered dorsally with white mealy secretion in small tufts, like a deposit of alkali on the soil, the surface more or less visible between; short, thick irregular marginal tufts. On boiling in KHO, turns the liquid red. Legs and antennæ extremely small, as also are the mouth-parts; legs slender, claws with a very distinct inner tooth.

Measurements in microns: Middle legs, femur and trochanter, 165; tibia, 118; tarsus (without claw), 63.

The same measurements for hind leg are 175, 125, 70.

Antennal joints: (1) 33, (2) 48, (3) 45, (4) 18, (5) 38, (6) 33, (7) 33, (8) 35, (9) 53.

Larva pale yellow, elongate oval, 560 μ long and 240 wide, of the ordinary Pseudococcine type; antennal joints in μ (1) 20, (2) 23, (3) 20, (4) 18, (5) 18, (6) 60.

Related to *P. cockerelli* King, but peculiar for the very short fourth antennal joint, and in spite of the rather large size of the insect, the very small legs. It may be only subspecifically distinct, but no intermediates are known, and it has the aspect of a distinct species. The first three antennal joints are like those of *P. rubivorus*, but not so the others.

A NEW APHID FROM OREGON.*

BY H. F. WILSON, CORVALLIS, OREGON.

Lachnus pseudotsugæ, n. sp.

Stem-mother.—Collected on terminal shoots of *Pseudotsuga douglassii* about Corvallis, Oregon, March 15th, 1911.

General colour light brown with two rows of black spots extending midway along the dorsum to the middle of the abdomen. These spots sometimes join so as to give the appearance of two dark lines extending along the body. Body semi-shining and with faint traces of a light flaky powder on dorsum. Legs and antennæ dusky brown. After having been mounted on slides for some time this species turns red and a deep red colour is assumed by the balsam surrounding them. The abdomen is quite large in comparison with the head and is almost globular. Antennæ VI segmented, and about one-fifth the length of the body. The nectaries are but small round tubes slightly elevated; they are about as wide as long and are situated on the side of the abdomen about two-thirds of the way from the base of the thorax to the base of the cauda. Cauda broad and slightly angled, very short.

Measurements: Length of body, 3.8 mm; width 2.99 mm.. Length of antennal segments, I, .09; II, .09; III, .3; IV, .135; V, .135; VI, .12; spur, .045 mm.; total length, .87 mm. Length of cauda, .3 mm.; nectaries, .022 mm.

Spring migrant.—Collected June 4, 1911, on terminal shoots of same host plant. General colour of head and thorax dark or dusky. Abdomen greenish brown, with colouring of white powder. Legs and antennæ, except tarsi and tips of third, fourth, fifth and sixth segments, light brown. Other parts dusky to black. Antennæ about one-fourth the length of the body. Head rounded in front and with a suture or line extending from back to front midway between antennæ. Wings hyaline. The first anal and cubital veins quite distinct while the median with its two branches, remains only as faint lines. The nectaries of this form are cone-shaped with a flanged mouth and are apparently not placed as far forward as in the earlier forms. Cauda short and broadly angular.

*We regret that, owing to a printer's error, which escaped us, the present description of *Lachnus pseudotsugae* was published in the June number of the Canadian Entomologist (pp. 192) without the name. We therefore republish the description in full with the name added.

Measurements: Length of body, 2.84 mm.; width, 1.09 mm.; length of wing, 3.65 mm.; width, 1.1 mm; total wing expanse, 8.08 mm. Length of antennal segments, I, .066; II, .11; III, .44; IV, .154; V, .198; VI, .11; spur, .045 mm.; total length, 1.123 mm. Length of nectaries, .064 mm. Length of cauda, .22 mm.

The fall migrant was not secured.

Egg-laying female.—Collected on terminal shoots of above plant, Oct. 30, 1910, and Oct. 27, 1911, along with the alate males. General colour brownish with ash-grey powder on body, and with two more or less regular stripes down the back; and with a wide brown stripe extending across the body from one nectary to the other. At the base and above the cauda another transverse band is usually present. Antennæ and legs, except tips, light brown; other parts dusky to black. Body robust and with large semi-conical nectaries which are brown in colour. Antennæ and legs hairy; antennæ one-third the length of the body.

Measurements: Length of body, 2.9 mm.; width, 1.7 mm. Length of antennal segments, I, .066; II, .09; III, .35; IV, .176; V, .176; VI, .11; spur, .045 mm.; total length, 1.013 mm. Length of nectaries, .06 mm., and cauda, .35 mm.

Alate male.—Collected on terminal shoots of host plant Oct. 30, 1910, and Oct. 27, 1911, about Corvallis, Oregon.

General colour: Head and thorax black with green abdomen. Abdomen with a series of black, transverse, more or less distinct, bands. Antennæ yellow at base, dusky at tip. Femora and tibiæ dusky at middle to black at ends; tarsi black. Wings hyaline but with costa dark brown, median vein and branches almost indistinct; other veins dusky. Nectaries slightly bell-shaped with a flanged opening. Third antennal segment about equal in length to fourth and fifth segments and with about 30 to 39 visible small circular sensoria. Fourth segment with 10 to 12 circular sensoria which appear slightly larger than those on the third segment. Fifth with about eight medium-sized, and one large, visible sensoria at the distal end. Sixth segment with one large and apparently six small sensoria at base of spur.

Measurements: Length of body, 2 mm; width, .87 mm. Length of wing, 3.87 mm.; width, 1.52 mm.; total expanse, 8.61 mm. Length of antennal segments, I, .066; II, .11; III, .51; IV, .242; V, .3; VI, .154; spur, .066 mm.; total length, 1.448 mm. Length of nectaries, .045 mm.; cauda, .176 mm.

Females along tips of needles, depositing from 5 to 8 eggs.

COLLECTING COLEOPTERA IN A MAINE SAWMILL YARD.

BY C. A. FROST, FRAMINGHAM, MASS.

One of the most prolific and interesting collecting places that I have ever found, is the yard of an old sawmill situated on the banks of the Cochnewagin Stream, below the village of Monmouth, Maine. This mill was a picturesque and weatherbeaten structure as long ago as I can remember, and has been built and in constant operation for at least a hundred years. The logs are hauled into the yard during the winter months and remain there under natural conditions of moisture until they are converted into lumber; thus they do not come in contact with the water for a long period as is usually the case. I have not been able to find many specimens in the yard of a steam sawmill near the lake where the logs remain in the water all the spring.

For several years past, while on my vacations, I have spent many hours collecting on the logs, and the slab and board piles, to the neglect of other localities. During the first hot period of 1909 (June 20 to 26) I collected nearly eight hours each day for three days in this mill-yard and secured over five-hundred specimens, exclusive of a hundred *Monohammus scutellatus* which swarmed in such numbers that I think three hundred more could have been taken.

From June 20 to 25, 1910, fairly hot weather prevailed, but, on account of the previous cool weather and the greatly diminished amount of lumber, the collecting was not as good as in former years. The slab piles below the mill, however, yielded many good things as did the alder bushes and dying trees along the brook.

The following notes, concerning species that have been taken in or near this mill-yard, may be of interest.

The Carabidæ were limited to specimens of *Tachys nanus* and *flavicauda* under bark, and a few specimens of *Pterostichus adoxus* which were hiding under bits of wood near the slab piles beside the brook. *Pterostichus rostratus*, *Chlænius sericeus*, and several species of *Bembidium* and *Platynus* have also been taken in this locality if not actually in the yard.

The *Staphylinidæ* have been represented thus far by a few bark-haunting species, and one or two others on the flowers of *Viburnum* growing near the logs. On these flowers were taken *Cercus abdominalis* and *Epuræa* sp. representing the *Nitidulidæ*, with *Cryptorhopalum hæmorrhoidale* of the *Dermestidæ*.

Silvanus bidentatus, *Læmophlæus biguttatus* and *adustus* were found under the bark of the slabs, while a fungus growing on an elm log yielded *Cucujus clavipes* and *Tritoma thoracica*.

Hister lecontei, and *coarctatus*, *Plegaderus transversus* and *sayi*, *Teretrius americanus*, and two specimens of *Hister carolinus*, were all the *Histeridæ* discovered; all were taken under bark or crawling over the logs.

Of the *Elateridæ*, one or two *Alaus oculatus* were found resting on the side of a log as if they had just alighted from a flight; *Adelocera brevicornis* and *obtecta* were taken, the latter always on the board piles. Several *Elater apicatus* were taken on the trunk of a partly dead elm in the daytime, while a single specimen of this species was found on an elm log at the same place by use of a dark lantern. This was the only specimen of Coleoptera (excepting two or three *Magdalis armicollis* which were evidently hiding in crevices in elm bark) that was taken by this method of collecting. Several specimens of *Elater sanguinipennis* were beaten from *Alnus incana* which grows abundantly in the pasture near the mill.

Chalcophora brevicollis Casey was taken quite commonly on the slab piles, while *liberta* was rarely seen. *Dicerca divaricata* was common on the maple cordwood, and *caudata*, a very distinct species, was beaten in numbers from *Alnus incana* sprouts. *D. tuberculata* L. & G. and *chrysea* Mels. (commonly confused with *tenebrosa*) were taken on the slab piles. Fifteen specimens of the latter were found on the trunks of a few fire-injured fir-trees (*Abies balsamea*) from June 20 to 25, 1910. From my observations it seems probable that it breeds in this conifer. It may be noted here that *tuberculata* was taken at Wales, Me., ovipositing in the bark of a healthy twelve-inch hemlock, two or three feet above the ground.

Buprestis sulcicollis was taken once on the logs and once flying near a steam saw-mill about half a mile away. (This species was also taken at South Paris, Me., on slab piles, June 14, 1910.) In previous years single specimens of *B. maculiventris*, *consularis*, and *impedita* Say (commonly called *striata*) have been taken on the logs. It may be recorded that this latter species was taken ovipositing on the stump of a large white pine that had been cut the previous winter. The beetle was hidden by the scarf of the cutting and was laying the eggs in the cut surface of the stump; the date was June 23, 1910. *Poecilonota erecta* L. & G., formerly called *cyanipes* Say, has occurred at Monmouth once or twice. Large numbers of *Melanophila fulvoguttata*, and a few *acuminata* were taken, the former always on hemlock logs. It was also very abundant on hemlock bark in a clearing at Wales, Me. A single specimen of *Anthaxia viridicornis* has been taken.

Chrysobothris dentipes and *scabripennis* were running over the white pine logs in numbers, and there were also a few *femorata* about these logs, but the majority of this species showed a preference for the beech and maple cordwood. *C. trinervia* was taken several times at South Paris, Me., and *azures* has been taken twice on the dead twigs of the beech and willow at Wales, but neither species has yet been seen at Monmouth. *C. sexsignata* has been taken but once, and that emerald gem, *harrisii*, has evaded capture on several occasions.

It was noticed that the species of *Dicerca* and *Chalcophora* could be picked up in the fingers or caught in the hand as they dropped, while the species of *Chrysobothris* were exceedingly lively and it required active use of the net to secure them. The species of *Buprestis* were generally easy to get by brushing into the net, but those of the genus *Melanophila* were active or sluggish according to the temperature. *M. fulvoguttata* has a habit of running around the log and slipping out of reach between it and an adjacent one.

Near the mill-yard *Eupristocerus cogitans* was beaten from *Alnus incanus* sprouts in numbers, on June 22. This species rests on the upper side of the leaf near the centre and slides off over the edge when disturbed; it was unusual to find more than a pair on a single bunch of the bushes.

Large numbers of *Enoclerus quadriguttatus*, together with the variety *rufiventris* Spin., were running over the logs in company with lesser numbers of *Thanasimus dubius*. The former species and variety were seen feeding on adult *Scolytidæ*.

Some, but not all, of the following *Ptinidæ* have been taken in the mill-yard; *Ernobius mollis* and *luteipennis*, *Hadrobregmus carinatus*, *Microbregma emarginatum* Duft., *Trypopitys sericeus*, *Xyletinus fucatus*(?) and *Ptilinus ruficornis*.

The species of *Cerambycidæ* have been nearly as well represented as those of the *Buprestidæ*. Three specimens of *Physocnemum brevilineum* were taken running up the trunk of a decaying elm. A large number of *Orthosoma brunneum* were found hiding under the bark of pine stumps. A dozen or two *Phymatodes dimidiatus* were taken on some spruce logs, which are rarely seen in the yard.

Hiding in the inequalities of the bark of the logs or slabs, *Asemum moestum* was often found. *Callidium antennatum* and *janthinum*, much more abundant, were often seen perched on the slab piles; this latter species and *Monohammus scutellatus* were much sought after by birds *Acmæops proteus*, varying from black to nearly all testaceous, were common on the piles of newly sawn lumber. *Xylotrechus fuscus* occurred on the logs and board piles at rare intervals, and a specimen was taken on the trunk of a partly dead fir high up among the branches. This specimen has the pubescence of the prothorax and elytra in excellent preservation and clearly shows that *fuscus* is entitled to specific rank. It is undoubtedly, as noted by Col. T. L. Casey (Mem. on the Col. III, p. 359), more nearly allied to *nauticus* than to *undulatus*. *X. colonus* was taken but once.

Neoclytus erythrocephalus was often seen running over ash logs, and a few specimens of *N. muricatulus*, so much resembling ants that they may have been unnoticed many times, were taken. The variation in size of these species was strikingly great; specimens at hand measure from five to twelve millimeters in length. It may also be noted here that specimens of *Monohammus scutellatus*, selected from the very large amount of material available at that time, show a variation in each sex of from 13 to 25 mm.

Acanthocinus obsoletus was abundant and almost invisible against the bark of the white pine logs on which they rested. *Ecyrus dasycerus* was beaten from the branches of a dead poplar near the logs, and a single *Purpuricenus humeralis* was once swept from a low bunch of *Salix* near the road.

I have at hand a single example of *Saperda calcarata* of uncertain date but which was undoubtedly bred in the poplars that fringe the high bank between the yard and the pond.

Even now in fancy, I can see the old sawyer as he stood with hand on the lever that controlled the log carriage and watched the saw tear through the huge pine logs. One day he called me, then a small boy, to see "this funny looking bug" pinned on the beam behind him. It was *Saperda calcarata* and the specimen formed the nucleus of a very heterogeneous collection which has, like the old sawyer, long since crumbled into dust.

Saperda obliqua, and the variety of *lateralis* having the post-median cross bar on the elytra, have been beaten from *Alnus incana* on two occasions. A variety of *Oberea tripunctata* Swed. (possibly *amabilis*) has occurred in some numbers on this plant, both at Monmouth and Wales. Variations of *affinis* and *mandarina* have been beaten and swept from bushes. One specimen of *S. vestita* was taken on a board pile in the yard.

One of the rarest and most interesting of the Cerambycidæ was taken wandering over the pine needles beneath a huge white pine near the yard. It was *Pachyta rugipennis* and nearly escaped me by its superficial resemblance to *Rhagium lineatum* which is not rare on the logs.

The Chrysomelidæ were not represented except for a number of *Calligrapha scalaris* and larvæ which were swept from the low *Alnus incana* in the pasture.

Tenebrionidæ.—*Alobates pennsylvanica, Iphthimus opacus, Tenebrio tenebrioides, Hypophloeus parallelus*, and *Xylopinus saperdioides* were taken commonly under bark or on the slab piles. *Upis ceramboides* was common, flying and on slabs. One specimen of *Platydema americanum* was taken. *Arthromacra ænea* was beaten from *Alnus* in large numbers.

Of the Melandryidæ, *Penthe obliquata, Synchroa punctata, Melandrya striata, Phloetrya liturata, Eustrophus bifasciatus, Orchesia castanea* were taken on the logs or slabs. *Enchodes sericea* was beaten from the dead branches of a large rock maple, and also found on a log of this tree at Wales, Maine. *Salpingus virescens* and *Pytho planus* were both taken on one occasion under bark or slabs. *Ditylus cœruleus* was captured a few times on the logs and was also seen flying in the daytime. *Dendroides canadensis* and *concolor* were both taken; the former under bark and by beating, the latter by beating maple sprouts in a wood clearing at Wales.

Tomoxia lineella was seen on the trunk of a decayed elm, and a single specimen was captured while unguardedly trying to rid itself of a large mite. Several specimens of this species were taken several years before on a dead tree in tne woods.

Hylobius pales was sometimes present in large numbers especially under bits of slabs that had fallen to the ground. *Pachylobius picivorus* was less common, in fact rare. *Pissodes affinis, strobi*, and a single *dubius* were taken from the piles of lumber; *affinis* was the most abundant. *Mononychus vulpeculus* was swept from the flowers of the blue iris near the brook.

Large numbers of Scolytidæ were taken flying in the late afternoon. The spruce logs were very badly riddled just beneath the bark by some species of this family. Among those that have been identified are *Dendroctonus valens, Hylurgops pinifex, Hylesinus aculeatus, Xyloterus bivittatus* and *Hylastes cavernosus*.

Taking into account the undetermined species and a few specimens that may be in alcoholic material not yet examined, the number of species taken in and about the mill-yard will not fall much short of 125; many of them which are rare under ordinary conditions are here abundant, while excessively rare species appear with gratifying frequency. If it were possible to collect in this yard frequently from June 1st to August 15th, I feel sure that the results would be surprising. Any collector who can visit a place where lumbering operations are carried on, even on a moderate scale, will be amply repaid.

OBSERVATIONS ON THE LIGHT-EMISSION OF AMERICAN LAMPYRIDÆ.—FOURTH PAPER.

BY F. ALEX. MCDERMOTT, PITTSBURG, PA.

In continuation of his former observations on the light-emission of American Lampyridæ and its relation to the sexual life of the insects (CAN. ENT., 1910, Vol. 42, pp. 357-363; 1911, Vol. 43, pp. 399-406; 1912, Vol. 44, p. 73), the writer has made during 1912 the observations recorded below on species of Lampyridæ not heretofore encountered by him, which support his former observations and show specific distinctions which are of interest. About the time of the publication of the second paper of this series, Dr. S. O. Mast, of Johns Hopkins University, read the a paper embracing very similar observations on (probably) *Photinus ardens* Lec. and emphasizing the bearing of the behaviour of these insects on the theory of phototactic orientation (Abstract in Science, 1912, Vol. 35, p. 460.)

Photinus marginellus Lec.—This species was first observed in Ashland, Ohio, during the latter part of June. In the late afternoon, several hours before sunset, both sexes were found in flight and resting on the leaves of low plants. As mentioned by Leconte, the male greatly resembles *P. scintillans* while the female, instead of being apterous as in the latter species, has wings and elytra as fully developed as those of the male. The flash of the male is a single, short, sharp one, and in colour appears to the eye more yellowish than that of *scintillans*, though resembling that of the male of the latter species in intensity and manner of delivery. The flash of the female *marginellus*, however, differs distinctly from that of the female *scintillans;* instead of being a single flash, somewhat slower than that of the male, as in *scintillans*, the flash of the female *marginellus* consists of two coruscations, the first being brighter and of shorter duration than the second, which follows the first immediately. The flash of the female is delivered with only a very short interval after the flash of the male she is answering. It will be noticed that this double flash of the female *marginellus* differs decidedly from the double flash of the male of *P. consanguineus* previously described.

*Photinus castus** Lec.—This species was found during June and July in open places, particularly in Schenley Park in Pittsburg, Pa.

*See following footnote.

As with *scintillans* and *marginellus*, the flash of the male is a single, short, bright scintillation. The males of *castus* and *marginellus* were frequently found flying together over the same plot of ground, and it proved quite easy to distinguish them by the characteristics of their light-emission. The flash of the male *marginellus* is decidedly shorter and more sudden than that of the male *castus*. *P. scintillans* was not observed at this time, and hence could not be compared.

The female *castus* has the wings and elytra fully developed, but like *marginellus* female, flies but little. The flash of the female *castus* is very much like that of the female *scintillans*—a single, short scintillation, slightly more prolonged than that of the male, somewhat less intense, and with no indication of doubling (difference from *P. marginellus* ♂). It is delivered immediately after the flash of the male answered, without any distinct pause.

The mating process in both *P. marginellus* and *P. castus* is exactly the same as described for *P. pyralis* and *P. scintillans*, and needs no further comment. The females of *marginellus* and *castus* will not answer with any certainty the flare of a match; in fact, in a large number of trials, only one distinct answer, from *P. castus* ♂ was observed. As in the other species of *Photinini* heretofore observed, the females of these two species are much less numerous than the males, and with *marginellus* the writer observed for the first time among our Lampyrids the attempted coupling between males, reported by Oliver (Ier. Cong. Internat. d'Entomol., Brux., 1910, pp. 143-144). While the females of both species fly readily they are comparatively rarely found in flight, preferring to creep to the tips of blades of grass, upper edges of leaves, etc., where they remain until mated. As in the species of *Photinus* previously described, the luminous apparatus of the males of both *castus* and *marginellus* covers the entire ventral surfaces of the 5th and 6th abdominal segments, and a good portion of the 4th, while the organ of the female consists of a small rectangular spot on the 5th abdominal segment; in both species the eyes and antennæ of the male are somewhat larger than those of the female.

Where the males of both species were seen flying over the same area, careful watching showed the presence of the females of both species in the vegetation. Although especially watched for, no case of interbreeding was encountered, and indeed no case of approach between the sexes of the different species. In both species, flying males have been seen to respond (apparently) to males of the same species in the vegetation. In both of these species one will frequently find a pair in copula, surrounded by several more males; at first this suggests that a specific odor may also play a part in the attraction, but the observations of Mast (ante) are opposed to this view, as are also those of Emery (see the writer's second paper). The presence of these extra males is probably accounted for by the attraction of several males before actual mating, as females of *Photinus* do not appear to flash while coupled, unless disturbed.

P. marginellus and *P. castus* also differ in their conduct in the late evening; *marginellus* ♂ continues to flit around the vegetation for a long time, thus resembling *P. scintillans*, while the *castus* ♂ flies aimlessly high above the vegetation after the time of maximum activity, thus more closely resembling the habit of *P. pyralis*.

*Rileya (Lucidota)** *atra* Oliv.—The first specimen of this species taken this year was a larva, found in a decayed stump at Niagara Falls, N. Y., on May 6th. This glow-worm was kept alive and subsequently developed, the imago proving to be this species. The adult was but faintly luminous, and for only a very little while after emergence, and no excitation produced light-emission during the remainder of the time (about a week) that it was kept alive. Subsequently several adults of this species were taken in flight in the daytime in July, near Sharpsburg, Pa., and in no instance was there any indication of luminosity. All specimens taken were apparantly males. The luminous apparatus of the adult is represented by two small brownish scales on the last segment of the abdomen; the larva was quite as luminous as that of *Photinus pyralis*, which it much resembled. The reared adult, with its larval and pupal skins, is deposited in the U. S. National Museum.

Rileya (Lucidota) punctata Lec.—This species closely resembles the foregoing, but is only about 3/5 as long. Both sexes were taken in flight in the daytime, in woods near Sharpsburg, in July, and neither showed any indication of luminosity, although the brownish scales representing the luminous apparatus were present, as in *R. atra*. In this species the eyes of the male are larger and the antennæ longer than in the female.

Lecontea angulata Say.—Numerous males of this species were observed at Canajoharie, N. Y., on June 28th, flitting about over the flats on the south bank of the Mohawk River, in a manner very similar to that described last year for *L. lucifera*. As the time available for observation was very brief, no females were found.

*Ern. Olivier (Revue scient. du Bourb. et du cent. de la France, 1911-12) has recently segregated these two species, giving them the generic name *Rileya*.

A NOTE ON *PHOTINUS CASTUS* LEC.

BY F. ALEX. MCDERMOTT, PITTSBURG, PA.

During the course of the observations recorded in the preceding paper, it became evident to the author that the insects usually classified as *Photinus marginellus* and *Photinus marginellus* var. *castus*, are actually distinct species. As brought out in the foregoing paper, the manner of the light-emission of the males of the two differ somewhat, and that of the females very distinctly. Such distinctions among the Lampyridæ cannot be other than specific, especially in view of the very close resemblance which many of the known distinct species bear to one another, and while a definite boundary between species and varieties, which is satisfactory to everyone, has not been established, and possibly never will be, it still seems proper to consider that when a Lampyrid shows such differences from other forms in its manner of light-emission as to almost preclude the possibility of interbreeding, it is due the position of a distinct species.

LeConte established the species "*Photinus casta*" for an insect from Georgia, U. S. A., in the same paper in which he established "*P. marginella*" for specimens from Missouri and elsewhere. (Proc. Acad. Nat. Sciences, Phila., 1851, p. 335.) Subsequently he appears to have abandoned this arrangement, and grouped the "*casta*" as a variety of "*marginella*," and in his List of the Coleoptera of North America (Miscellaneous Collections of the Smithsonian Institution, No. 140, p. 51, 1866), he gives *Photinus marginellus* var. *castus*; this is the earliest reference to the change in the classification of this insect which I have been able to locate. The insect is thus listed as a variety of *marginellus* in the Genminger-Harold Catalogue (Vol. 6, p. 1,643), and in Ern. Olivier's recent lists (Wytsman's Genera Insectorum, Fasc. 43; Schenkling-Junk Catalogus Coleopterorum, Pars 9).

In view of the specific differences in light-emission, above referred to, it has seemed best to re-establish the species under LeConte's (corrected) name. The species differs from the other species heretofore described in these papers, most particularly in the pale gray colour of the elytra; by transmitted light the latter appear to be almost pigmentless, except at the margins; the central black spot on the disc of the thorax is small and more frequently wanting than in the true *marginellus*; the insects average somewhat larger than either *P. marginellus* or *P. scintillans*, but are distinctly smaller than *P. consanguineus*.

The writer is indebted to Dr. Samuel Henshaw for his kindness in comparing specimens sent him with LeConte's types in the Museum of Comparative Zoology.

BOOK NOTICES.

ELEMENTARY ENTOMOLOGY, by E. Dwight Sanderson and C. F. Jackson, Ginn and Co., Boston. Price $2.00.

The appearance of this work so soon after the senior author's excellent book on "Insect Pests of Farm, Garden and Orchard" comes as a surprise but a very welcome surprise to students of entomology. The book is intended primarily as a text book for short courses in entomology, but covers the systematic side of the work so well that it will probably be used by many teachers for all but their most advanced classes.

The book contains three hundred and seventy-two pages with four hundred and ninety-six illustrations of a very superior character. It is divided into three main divisions. Part I, consisting of sixty-six pages, deals with the structure and growth of insects. A few pages of this section are given to a description of the differences between insects and closely allied invertebrates; the remainder is devoted to a concise and clear account of the external and internal anatomy of insects, and to their growth and transformations, the latter being illustrated by the life history of a few common species. The section throughout shows abundant evidence of skilful handling of a somewhat difficult subject. The only criticism that suggests itself is that instead of taking two examples of complete metamorphosis from Lepidoptera, it might have been better to have chosen one of these from some other order. However, this is a minor point.

Part II, which contains two hundred and two pages, deals with the classes of insects. For convenience these have been divided into nine groups: Aptera, Orthoptera, Neuroptera and Pseudoneuroptera, Platyptera, Hemiptera, Coleoptera, Lepidoptera, Diptera, and Hymenoptera. The various orders which have been put into one group, such as, for instance, the Neuroptera and Pseudoneuroptera, are mentioned and their characteristics briefly given. In each order a large percentarge of the families are described, those of economic importance being given the preference. In bringing out the characteristics of the different families, copious illustrations have been used. One can scarcely give this side of the work too much credit as the photographs and drawings are not only excellent works of art in themselves but in a very large proportion of cases show the different stages in the life history of the insect described and thus enable the student to understand and remember them much better.

By conciseness of language and the use of illustrations the senior author, who is responsible for parts I and II, has succeeded remarkably well in giving a good general view of all the orders. About three hundred different species of insects have heen described and illustrated in this section.

Part III, consisting of almost one hundred pages, is devoted to laboratory exercises and outlines a fairly comprehensive course of study of the external and internal anatomy of types of the more common and important orders, giving special attention to a comparative study of the mouth-parts. Hints are also given on the proper methods of studying life histories. One of the most useful chapters in this division of the book consists of a series of keys to the various orders and families. Most of the keys are simple and easy to use, being based on characteristics that can readily be seen with a hand lens. In the case of the Lepidoptera and Diptera it was of course found necessary to use the wing venation in constructing a satisfactory key. In doing so the author has inserted diagrams of wings of most of the families included in the keys to these two orders. It is doubtful, however, whether it would not have been an improvement to have dovoted a short chapter of four or five pages to a study of wing venation and the method of clearing wings of the Lepidoptera.

The remainder of part III outlines methods of collecting, preserving and studying insects, and many suggestions are given that will be helpful to all but the most experienced entomologist. This part does much to remedy a long-felt defect in entomological text books and will help greatly to give the book a wide circulation among students and teachers of entomology.

As is usual in a work of this character, there are a few errors of minor importance, chiefly of a typographical character. Promethia (page 216), Velidæ (page 309) and Physopodæ (page 308) are clearly cases of this nature. In part II (page 116) Negro Bugs are classed as Corimelænidæ and in the key as Thyreocoridæ. *Œcanthus niveus* (page 87) should clearly be *O. nigricornis*. On page 161 it is stated that "the lady bird beetles form the only family of the Trimera. This is rather misleading, as is also the statement on page 75 that springtails "are never injurious." These, however, are insignificant mistakes and do almost nothing to lessen the value of the book.

L. CÆSAR, O. A. C., Guelph, Ont.

THE FUNGUS GNATS OF NORTH AMERICA. By Oskar A. Johannsen, Ph.D. Parts I-IV, from Bulletins 172, 180, 196 and 200. Maine Agricultural Experiment Station, Orono, Me.

The concluding part of this admirable work was issued in June of the present year, the first part having appeared in Dec., 1909, and the second and third parts in June, 1910, and Dec., 1911, respectively. It is a work of 306 pages, whose aim is to "present a synopsis of the fungus gnats, or Mycetophilidæ, of North America, giving descriptions of and tables to all the genera and species and life-histories when known." On account of their small size and unattractive appearance, these flies have hitherto received little attention from systematic entomologists, so that the preparation of the keys and descriptions of the 82 genera and 428 species must have been a task of unusual difficulty, demanding an exceptional amount of patience and industry on the part of the author, who is therefore the more to be congratulated upon his having so successfully accomplished it.

In the first part the general characters of the family in their various stages are discussed, special attention being given to the venation and the form of the hypopygium of the male. There is also a short discussion of the habits and economic relations of the groups as a whole and an analytical key to the eight subfamilies. The remainder of Part I is devoted to the systematic treatment of the first five subfamilies (Bolitophilinæ, Mycetobiinæ, Diadocidiinæ, Ceroplatinæ and Macrocerinæ), all of which are comparatively small groups. Fifteen genera and 71 species, of which 11 are new, are described in this part.

Part II deals with the Sciophilinæ, and includes a short account of their habits and earlier stages, in addition to the tables and descriptions. Twenty-nine of the 69 species described in this part are new, most of these belonging to the two largest genera, Sciophila and Mycomya.

Parts III and IV treat of the Mycetophilinæ and Sciarinæ, the two groups of the most economic importance. The former is the largest of the subfamilies, embracing 48 genera and 110 species, of which 54 are described as new. In the Sciarinæ 9 genera and 56 species are described, 46 of the latter belong to Sciara, including 22 new species.

In all the tables the venational characters take the most prominent place, but many others are also employed, such as the position of the ocelli, structure of the palpi and antennæ, and particularly that of the hypopygium of the male. There is considerable variation in the length of the descriptions, some of the diagnoses of new species being only four lines in length, while others occupy 20 lines or more. The descriptions are supplemented by a series of tables giving the relative measurements of the various joints of the legs. There is also an index to the genera and a series of well executed plates, mostly in half-tone, illustrating structural features, chiefly the wings and the hypopygia of the males. It is not quite clear why the plates have not been numbered.

THE Canadian national collections in the Victoria Memorial Museum at Ottawa should be the results of efforts to increase knowledge and the exhibit should be made in a manner to diffuse knowledge. They should not be collections of curios. All Canadians, especially educators, will be interested to make use of the collections often, and should direct others to do so as well as students. It is hoped that time may dispel rapidly the idea, which unfortunately too many people have, that the place is a storehouse for curiosities or abnormal and monstrous things rather than that it is an institution of learning. Some of the staff will always be glad to meet classes or visitors and to give them such assistance as is possible. Pending the completion of the lecture hall, informal talks in the laboratories or offices may occasionally be arranged, especially if a few days' notice is given. As time goes on, the institution will probably be able to loan pictures, lantern slides, maps, labels, casts, and even specimens for educational purposes, HARLAN I. SMITH.

CHANGE OF NAME.—Dr. Brölemann has called my attention to the fact that my genus *Poabius* is preoccupied by Poabius of C. L. Koch. It may be replaced by *Pokabius*.—R. V. CHAMBERLIN.

CORRECTIONS.—In my last paper on the genus Hydriomena (CAN. ENT., XLIV, Aug., 1912) the following corrections should be made:—P. 225, lines 16, 17 and 19, for "*Bockhausen*" read "*Borkhausen.*" P. 226, line 12, for *Thmb.* read *Thunb.* P. 229, No. 11, and p. 230, No. 11, for "*lanavahrata*" read *banavahrata.* P. 230, line 11, for "(d) *Mesial lines smutty*" read "(d) *Mesial space smutty.*" L. W. SWETT.

Mailed October 12th, 1912.

The Canadian Entomologist.

Vol. XLIV. LONDON, NOVEMBER, 1912. No. 11

SYNONYMY OF THE PROVANCHER COLLECTION OF HEMIPTERA.

BY E. P. VAN DUZEE, BUFFALO, N. Y.

Through the kindness of Rev. A. Huard, of Quebec, I recently had an opportunity to examine the Provancher collection of Hemiptera now deposited in the Museum of Public Instruction in the Parliament Buildings in that city. This collection has been well cared for and is in excellent state of preservation. The main part of the collection seems to represent the exact material used by the Abbé in the preparation of the Hemiptera volume of his Petite Faune Entomologique du Canada, practically all the species included in that volume being in the collection in the same order as in the book; the few additional species being in most cases placed between the regular rows of the arrangement. Usually there is but one or at most two specimens of each species and the labels seem to be in Provancher's own handwriting. There are no "types" so indicated nor could I find any trace of the types of his species published in 1872 in Vol. IV of the NATURALISTE CANADIEN, and I am convinced that he incorporated this material with his general collection at the time he published the Petite Faune, or so much of it as he then possessed, and consequently that it will be impossible definitely to locate all of his earlier species. The Petite Faune collection however contains Provancher's determination of most of his 1872 species and so far as these specimens agree with his first descriptions they must be taken as representing the nearest approximation to types of his earlier species now in existence.

When starting for Quebec I took with me a good series from my own collection for comparison, covering all the species of which I felt in doubt, and by this means I was enabled to locate nearly all of the species in the Petite Faune and most of those of 1872.

In the following notes I have thought it best to give my determination of each of the Petite Faune species, indicating all uncertain forms where I had no material with me for direct comparison and so was obliged to depend upon my memory for the determination. Under each species I give first the page in the Petite Faune, followed by the name as there printed. Where the determination is correct this word follows the name and after it is the name now used for the species where it differs from that employed by Provancher.

20. *Thyreocoris unicolor* P. B., correct.

21. *Thyreocoris pulicarius* Germ., correct.

21. *Homœmus æneifrons* Say, correct.

22. *Eurygaster alternatus* Say, correct.

27. *Canthophorus cinctus* P. B., correct. *Sehirus cinctus* P. B.

28. *Pangæus bilineatus* Say, correct.

29. *Podisus cynicus* Say, correct. *Apateticus cynicus* Say.

30. *Podisus modestus* Dall. Under this name is one *Podisus serei-ventris* Uhl. pinned to the label and one *modestus* at the side.

31. *Podisus spinosus* Dall., correct. *Podisus maculiventris* Say.

32. *Perillus circumcinctus* Stal, correct. *Perilloides circumcinctus* Stal.

33. *Perillus exaptus* Say, correct. *Perilloides exaptus* Say.

34. *Rhacognathus americanus* Stal., not in the collection.

35. *Brochymena annulata* Fabr. is *4-pustulata* Fabr. (Under the name *4-pustulata* Fabr. is one example of *myops* Stal.)

36. *Euschistus fissilis* Uhler, correct.

36. *Euschistus tristigmus* Say, correct.

38. *Aelia americana* Dall., is *Neottiglossa undata* Say; a dark specimen but not as dark as the western *trilineata* Kirby.

39. *Neottiglossa undata* Say, correct; a pale example. (In the collection is a western specimen of *Thyanta antiguensis* Westw., labelled *Neottiglossa sulcifrons*.)

40. *Hymenarcys nervosa* Say, correct.

40 *Cœnus delius* Say, correct.

41. *Lioderma ligata* Say is *Pentatoma persimilis* Horvath.

42. *Thyana custator* Fabr., correct.

43. *Mormidea lugens* Fabr., correct.

44. *Cosmopepla carnifex* Fabr., correct.

46. *Banasa calva* Say; under this name is a pale example of *dimidiata* Say.

46. *Banasa dimidiata* Say, correct.

46. *Banasa euchlora* Stal, not in the collection.

48. *Acanthosoma cruciata* Say. On this label is an example of *Elasmostethus atricornis* Van D., and by it one of *E. cruciata* Say.

In this collection are the following erroneous determinations: *Thyanta custator* labelled *Trichopepla atricornis* Stal. ; *Euschistus servus* Say labelled *E. impictiventris* Stal. ; *E. fissilis* Uhler labelled *E. variolarius* P. B. ; *Apateticus bracteatus* Fh. labelled *Podisus grandis* Dallas ; and *Perilloides exaptus* Say labelled *Perillus splendens* Uhler.

53. *Anasa tristis* De G., correct.

55. *Chelinidea vittigera* Uhler. Under this name is one example of *vittigera* Uhler and one of *tabulata* Burm. Judging from the description the former must have been the one stated to have been taken in Quebec.

55. *Alydus eurinus* Say, correct.

56. *Alydus 5-spinosus* Say, correct.

56. *Alydus pluto* Uhler. Under this name are two females of *eurinus* Say.

57. *Tetrarhinus quebecensis* Prov., is *Protenor belfragei* Hagl. In the collection it stands under the correct name showing that Provancher must have corrected his own determination later.

58. *Capys muticus* Say, correct. *Neides muticus* Say.

58. *Jalysus spinosus* Say, correct.

60. *Corizus punctiventris* Dall, correct. *Stictopleurus crassicornis* Linn.

60. *Corizus lateralis* Say, is *nigristernum* Sign. as usually determined.

61. *Harmostes fraterculus* Say. On this label is an example of *Ortholomus longiceps* Stal.

The following are incorrectly determined; *Ceraleptus* sp. determined as *Orsillus scolopax* Say; *Metapodius terminalis* Dall. as *Anisoscelis corculus* Say; *Metapodius femorata* Fabr. as *Anisoscelis declivis* Say; and *Harmostes reflexulus*, pink form, as *H. serratus* Fabr.

69. *Lygæus bistriangularis* Say, correct.

70. *Lygæus turcicus* Fabr. is *L. kalmii* Stal.

70. *Nysius grœnlandicus* Zett, is *N. thymii* Zett.

71. *Helonotus abbreviatus* Uhl., correct. *Phlegyas abbreviatus* Uhl.

72. *Cymus tabidus* Stal is *Cymus discors* Horv.

73. *Œdancala crassimana* Fabr. is *O. dorsalis* Say.

74. *Ischnorhynchus didymus* Zett., correct. *I. resedæ* Panz.

75. *Oxycarenus disconotus* Say, correct. *Crophius disconotus* Say.

75. *Ischnodemus falicus* Say is *Peritrechus fraternus* Uhler.

76. *Blissus leucopterus* Say, correct.

77. *Emblethis arenarius* Linn. is *E. vicarius* Horv.

77. *Plociomerus nodosus* Say is *Scolopostethus* sp., probably *diffidens* Horv.

78. *Carpilis ferruginea* Stal, correct.

79. *Ligyrocoris constrictus* Say, correct. *Perigenes constrictus* Say.

80. *Heræus insignis* Uhl. is not in the collection.

81. *Eremocoris ferus* Say, correct.

82. *Trapezonotus nebulosus* Fall., correct.

82. *Pamera bilobata* Say is *Ligyrocoris contractus* Say.

84. *Pterometus canadensis* n. sp. This is the species lately described as *Pseudocnemodus brunneri* by Mr. Barber and must be known as *Pseudocnemodus canadensis* Prov. The following are incorrectly determined: *Oncopeltus fasciatus* Dallas determined as *Lygæus gutta* H. S.; and *Dysdercus mimus* Say as *Lygæus pulchellus* H. S. *Geocoris limbatus* Stal is correctly named. Under the name *Cnemodus mavortius* is an example of the larger dark form which I now consider to be distinct.

85. *Geocoris bullatus* Say, correct.

89. *Anthocoris musculus* Say. This is *A. borealis* Dallas which is probably a synonym of *musculus* Say.

90. *Tetraphleps canadensis* n. sp., correct. *Lyctocoris canadensis* Prov.

91. *Triphleps insidiosus* Say, correct.

102. *Collaria meilleuri* Prov., correct.

103. *Collaria oculata* Reut., correct.

103. *Miris instabilis* Uhler, correct. *Stenodema instabilis* Uhler.

104. *Trigonotylus ruficornis* Fall., correct.

104. *Trigonotylus pulcher* Reut., correct.

104. *Leptopterna dolobrata* Linn., correct. *Miris dolobrata* Linn.

106. *Resthenia insignis* Say. Under this name is the black form with the pronotal collar only fulvous. It pertains to Reuter's genus *Platytylellus*.

NOTE —Under the name *Resthenia nigricollis* is a large black *Lopidea*, and under the name *Resthenia maculicollis* stands *Lopidea confluens* Say. There is also an *Orthotylus congrex* Uhler under the name *Lomatopleura caesar* Reut., but this placing must have been an accident.

106. *Lopidea confluens* Say, correct.

108. *Phytocoris scrupeus* Say is *P. lasiomerus* Reut.

108. *Phytocoris pallicornis* Reut. is *P. tibialis* Reut.

109. *Phytocoris eximius* Reut., correct.

110. *Phytocoris inops* Uhler, correct.

111. *Neurocolpus nubilus* Say, correct.

112. *Compsocerocoris annulicornis* Reut. This is not Reuter's species but a large dark coloured *Phytocoris* of the *eximius* group, perhaps still

undescribed. I have taken the same form about Buffalo and Mr. Moore has sent me specimens taken by him at St. Hilaire, Que.

113. *Calocoris rapidus* Say, correct. *Adelphocoris rapidus* Say.

114. *Calocoris bipunctatus* Fabr., correct.

114. *Pycnopterna amœna* n. sp. This is the *Closterocoris ornata* Uhler and must hereafter be known as *Closterocoris amœna* Prov. Its occurrence at Ottawa, if really taken there, was probably accidental. Its range seems to be restricted to the Pacific region.

116. *Camptobrochis grandis* Uhler, correct.

116. *Camptobrochis nebulosus* Uhler, correct.

118. *Coccobaphes sanguinarius* Uhler, correct.

119. *Lygus pratensis* Linn. is *L. convexicollis* Reut.

120. *Lygus flavonotatus* Prov. is *L. pratensis* Linn., var. *lineolaris* P.B.

120. *Lygus belfragei* Reut. is the red variety of *L. pratensis* Linn.

121. *Lygus invitus* Say. Pinned on this label is a *Lygus tenellus* Van D., and next to it is a *L. invitus* Say.

121. *Lygus contaminatus* Fall. is *L. pabulinus* Linn.

122. *Pœcilocapsus lineatus* Fabr., correct.

123. *Pœcilocapsus affinis* Reut., correct. *Horcias dislocatus affinis* Reut.

123. *Pœcilocapsus marginalis* Reut. I did not find this in the collection, but the determination is undoubtedly correct.

123. *Pœcilocapsus goniphorus* Say, correct. *Horcias dislocatus goniphorus* Say. With this specimen is pinned an example of var. *nigrita* Reut. of the same species.

124. *Orthops scutellatus* Uhler is *Tropidosteptes amœnus* Reut., var. *palmeri* Reut.

125. *Systratiotus venaticus* Uhler, correct. *Pœciloscytus venaticus* Uhler.

127. *Pamerocoris brunneus* Prov. On this label is pinned an example of *Plagiognathus politus* Uhler, but as it does not agree at all well with Provancher's description of 1872 I am inclined to think that the type specimen was lost and the present one substituted by error.

127. *Pœciloscytus sericeus* Uhler. On this label is an *Orthotylus flavosparsus* Fall.

128. *Pœciloscytus basalis* Reut. On this label is a *Sthenarops malinus*. Neither this nor the preceding specimens agree with the descriptions in the Petite Faune and may have been placed there by mistake.

129. *Capsus ater* Fieb., correct. Authority should have been *Linnæus*

130. *Monolocoris filicis* Linn., correct.

131. *Pilophorus bifasciatus* Fabr. is *P. clavatus* Linn.

132. *Stiphrosoma stygica* Say, correct.

133. *Trichia punctulata* Reut. This specimen is in poor condition but is undoubtedly a *Tropidosteptes*, perhaps *pettiti* or *palmeri*. With it stands a pale example of *Lygus pratensis* Linn.

134. *Stenarops chloris* Uhler is *Tropidosteptes commissuralis* Reut.

134. *Stenarops malinus* Uhler is a large pale *Lygus pratensis* Linn.

135. *Labops hesperius* Uhler, correct.

136. *Orthocephalus saltator* Hahn. A Capsid new to me but certainly not the European *saltator* Hahn.

137. *Chlamydatus luctuosus* n. sp. On this label is a broken specimen of *Dicyphus agilis*, but it does not agree with the description entirely and may be an error.

138. *Orthotylus dorsalis* Prov. is *O. congrex* Uhler. This specimen agrees in all respects with Provancher's description of 1872 and the name must take precedence over Uhler's published in 1887.

140. *Dicyphus californicus* Stal, correct.

141. *Idolocoris famelicus* Uhler is *Macrolophus separatus* Uhler.

141. *Idolocoris agilis* Uhler is correct.

143. *Hyaliodes vitripennis* Say, correct.

144. *Malacocoris provancheri* Burque is a good species of *Diaphnidia* near *pellucida* Uhler.

146. *Parthenicus psalliodes* Reut. On this label is a very poor specimen of *Ilnacora stalii* Reut.

147. *Globiceps flavomaculatus* Fabr. is *Mimoceps gracilis* Uhler.

148. *Oncotylus decolor* Fall., correct. *Lopus decolor* Fall.

148. *Oncotylus pulchellus* Reut. is *Orthotylus flavosparsus* Fall.

149. *Oncotylus punctipes* Reut. Probably correctly determined. Our American species differs from Reuter's description only in wanting the black pubescence on the antennæ and venter and in having the areoles scarcely darker than the rest of the membrane.

150. *Macrocoleus coagulatus* Uhler, probably correct.

150. *Amblytylus 6-guttatus*, n. sp. A distinct and beautiful species belonging to genus *Macrotylus* I took it at Ottawa.

152. *Psallus delicatus* Uhler is a form of *Plagiognathus obscurus* Uhler.

153. *Plagiognathus fuscosus* Prov. Under this name is placed the ordinary form of *P. obscurus* Uhler, but it is not the *fuscosus* nor the *dorsalis* of the NAT. CAN, 1872, as quoted by Provancher. The former is very close to if not identical with *P. politus* Uhler.

154. *Plagiognathus rubricans*, n. sp. A good species pertaining to genus *Rhinocapsus* Uhler. It differs from *vanduzeei* Uhler in being larger and in having the second antennal joint entirely black. Mrs. Slosson has recently taken this species at Lake Toxaway, N. C.

155. *Agilliastes associatus* Uhler, correct. NOTE.—Among the Capsids in this collection is a *Garganus fusiformis* Say named *Megocœlum signatum* Dist., and a *Ceratocapsus pumilus* determined as *Ceratocapsus lutescens* Reut

155. *Agalliastes verbasci* H. S., correct. *Chlamydatus* is now used for this genus.

158. *Corythuca ciliata* Say is a variety of *arcuata* Say.

158. *Corythuca juglandis* Fitch is a typical *arcuata* Say.

159. *Leptostyla oblonga* Say. This seems to be a *Leptobyrsa*, probably *explanata*, but unfortunately I had no specimen for comparison.

159. *Gargaphia tiliæ* Walsh, correct.

160. *Pysatochila plexa* Say, correct.

160. *Leptophya mutica* Say, correct.

162. *Phymata wolfii* Stal is *erosa pennsylvanica* Handl.

165. *Aradus robustus* Uhler is *4 lineatus* Say.

165. *Aradus aequalis* Say is *robustus* Uhler.

166. *Aradus acutus* Say. On this label was a species still undetermined in my collection but quite distinct from *acutus* Say.

166. *Aradus 4-lineatus* Say. The species under this name was new to me and was quite distinct from either *4-lineatus* or *robustus*.

167. *Aradus similis* Say. This seemed to be *tuberculifer* Kirby as nearly as I can tell without comparing specimens directly.

167. *Aradus rectus* Say, correct. *A. lugubris* Fallen.

167. *Brachyrhynchus granulatus* Say. New to me but not *granulatus* as determined in my collection.

168. *Brachyrhynchus lobates* Say is the *granulatus* of my collection.

169. *Aneurus politus* Say is *septentrionalis* Walker.

169. *Aneurus inconstans* Uhler, correct.

170. *Cimex lectularius* Linn., correct.

175. *Coriscus subcoleoptratus* Kirby, correct.

175. *Coriscus propinquus* Reut. is the young of the preceding species.

175. *Coriscus vicarius* Reut. is the larval form of *Alydus eurinus* Say.

176. *Coriscus inscriptus* Kirby is *C. rufusculus* Reut. The name *Reduviolus* is now used for this genus. NOTE.—In the collection is an *Acholla mutispinosa* De G. labelled *Sinea coronata* Stal, and a *Diplocodus exsanguis* Stal. labelled *Acholla tabida* Stal.

176. *Coriscus ferus* Linn., correct.

180. *Sinea diadema* Fabr., correct.

181. *Diplodus luridus* Stal. is *Diplocodus luridus* Stal., female.

181. *Darbanus georgiæ* Prov. is a worn female specimen of *Diplocodus luridus* Stal.

182. *Darbanus palliatus*, n. sp. is the male of *Diplocodus luridus* Stal.

182. *Evagoras marginata*, n. sp., is *Zelus cervicalis* Stal.

183. *Melanolestes picipes*, H. S., correct.

183. *Melanolestes abdominalis*, H. S., correct. Leconte is authority for this species.

184. *Opsicœtus personatus* Linn., correct. An immature specimen is his *Reduvius albosignatus* as suggested by him. The name *Reduvius* is now used for *Opsicœtus*.

186. *Emesa longipes* De Geer. Under this name is a *Ploiariola*, probably *errabunda* Say.

186. *Cerascopus errabundus* Say. The insect on this label seems to be a *Barce* but I could not locate the species without material for comparison.

189. *Salda ligata* Say, probably correct.

190. *Salda obscura* Prov. is *littoralis* Linn.

190. *Salda major* Prov. is *deplanata* Uhler which name it must supercede as it has priority by one year.

191. *Salda littoralis* Linn. is *interstitialis* Say.

191. *Salda lugubris* Say. Apparently *repleta* Uhler but I could not be certain without specimens for comparison.

192. *Sciodopterus bouchervillei* Prov. is *coriacea* Uhler which name it must supercede having priority.

193. *Limnobates lineata* Say, correct.

195. *Gerris rufoscutellatus* Latr., correct.

195. *Gerris remigis* Say, correct.

195. *Gerris marginatus* Say, correct.

196. *Gerris canaliculatus* Say. This may be *buenoi* Kirk. It is smaller than *marginatus* and stouter than *canaliculatus* usually is but I

could not be certain of the determination without specimens for direct comparison.

197. *Belostoma grisea* Say, correct. *Benacus grisea* Say.

198. *Zaitha fluminea* Say, correct. *Belostoma fluminea* Say.

199. *Ranatra fusca*, P. B. is *R. americana* Montd.

200. *Notonecta irrorata* Uhler, correct.

201. *Notonecta undulata* Say, correct.

202. *Corisa*, spp. My own material in this genus is still unworked and I did not attempt to locate the Provancher species.

204. *Prionosoma villosum*, n. sp. does not differ in any respect from *podopoides* Uhler.

204. *Euschistus jugalis*, n. sp. I could not find this in the collection but from the description I am now strongly inclined to consider it the form of *servus* with acute *humeri* although it might be *conspersus* Uhler.

205. *Platygaster pacificus*, n. sp., correct.

211. *Cicada pruinosa* Say, correct.

212. *Cicada septendecim* Linn., correct. *Tibicina septendecim* Linn.

213. *Cicada canadensis*, n. sp. This is *rimosa* Say more strongly marked with orange on the base of the elytra and wings than usual. I have an exactly similar specimen which I cannot distinguish from *rimosa*. It belongs to genus *Okanagama* Dist. and not to *Tibicen*, and is not *noveboracensis* Emmons as I had conjectured.

214. *Cicada rimosa* Say. The ordinary dark form of this species.

217. *Amphiscepa coqueberti* Kirby is *Otiocerus degeeri* Kirby.

218. *Hysteropterum semivitreum*, n. sp. This species was a surprise to me and I had taken nothing at all allied to it for comparison. It seems to belong to the Californian group of *Issids* and is very close to *Dictyobia permutata* Uhler. It may be an accidental introduction from the west.

219. *Scolops sulcipes* Say, correct.

220. *Helicoptera septentrionalis*, n. sp. This is the *Elidiptera* I have been determining as *septentrionalis* Prov.

221. *Helicoptera vestita*, n. sp., is *Elidiptera opaca* Say.

222. *Cixius stigmatus* Say. This is the *C. stigmatus* of my table published in CAN. ENT., XXXVIII, p. 408, Dec., 1906.

223. *Oliarus quinquelineatus* Say, correct.

223. *Oliarus cinnamomeus*, n. sp., correct.

224. *Delphax unipunctata* Prov. is *Stenocranus dorsalis* Fitch.

225. *Delphax furcata* Prov. This seems to be a good species of *Liburnia*.

229. *Enchenopa binotata* Say, correct.

229. *Enchenopa latipes* Say, correct. *Campylenchia latipes* Say.

230. *Archasia canadensis*, n. sp., is *A. belfragei* Stal.

231. *Janthe expansa* Germ, correct. *Antianthe expansa* Germ.

232. *Entylia sinuata* Germ. is male of *E. bactriana* Germ.

232. *Entylia carinata* Germ. is female of *E. bactriana* Germ,

233. *Entylia concava* Germ. is *E. concisa* Walk.

234. *Ceresa diceros* Say, correct.

235. *Ceresa bubalus* Fabr. On the label is *C. taurina*, Fh. and by it a male of *C. bubalus* Fabr.

235. *Ceresa brevicornis* Fitch. On this label is a female *C. basalis* Walk., and next it is a female *bubalus* Fabr.

235. *Ceresa semicrema* Say. Under this name is a dark male and female of *Ceresa basalis* Walk.

237. *Stictocephala inermis* Fabr., correct.

237. *Stictocephala festina* Say is *lutea* Walk.

238. *Cyrtosia vau* Say is probably correct. This specimen has no cloud at apex of the elytra and is larger and darker than usual.

239. *Crytosia trilineata* Say, correct.

239. *Cyrtosia fenestrata* Fitch. Under this name is a dark male of *vau* Say.

240. *Cyrtosia ornata*, n. sp., is the male of *C. cinereus* Emmons.

241. *Thelia univittata* Harr. is *godingi* Van D.

242. *Thelia bimaculata* Fabr., correct.

243. *Telamona scalaris* Fairm., correct. *Heliria scalaris* Fairm.

243. *Telamona tristis* Fitch, correct.

244. *Telamona unicolor* Fitch, correct.

144. *Telamona fasciata* Fitch, correct. Male of *unicolor* Fh.

244. *Telamona reclivata* Fitch, correct.

245. *Publilia concava* Say, correct.

246. *Carynota mera* Say, correct.

246. *Carynota picta*, n. sp., is *C. porphyrea* Fairm.

247. *Ophiderma marmorata* Say is *Carynota stupida* Walk. (*muskokensis* Godg.).

248. *Ophiderma inornata* Say is *flava* Godg., a little clouded with reddish.

248. *Tragopa brunnea* Prov. is *Acutalis semicrema* Say.

250. *Platycotis 4-vittata* Say, correct.

251. *Platycotis nigromaculata*, n. sp., is *P. sagittata* Germ.

[Under the name *Telamona querci* Fh. is a species I cannot distinguish from *obsoleta* Ball (from memory only) but it is certainly not *querci*.]

253. *Bruchomorpha oculata* Newm., correct.

254. *Embolonia tricarinata*, n. gen. et. sp., is the *macropterous* form of *Bruchomorpha oculata* Newm.

255. *Aphrophora parallela* Say, correct.

256. *Aphrophora 4-notata* Say, correct.

256. *Aphrophora quadrangularis* Say, correct. *Lepyronia id.*

257. *Philænus spumarius* Linn., correct.

258. *Philænus lineatus* Linn., correct.

258. *Philænus albiceps* Prov., is *spumarius* var. *leucocephala* Linn.

259. *Clastoptera obtusa* Say, correct.

260. *Clastoptera proteus* Fitch is *obtusa* var. *tristis* Van D.

260. *Clastoptera saint cyri* Prov. is the variety of *proteus* later named subspecies *flava* by Ball.

263. *Tettigonia viridis* Fabr. This is the European *viridis* Linn. As there seems to be no reason to question Provancher's statement that this specimen was taken in Quebec we must add the species to our list of North American *Hemiptera*.

263. *Tettigonia tripunctata* Sign is *Kolla tripunctata* Fitch.

265. *Proconia undata* Fabr., correct.

265. *Proconia costalis* Fabr., correct.

266. *Diedrocephala mollipes* Say is *Dræculacephala noveboracensis* Fitch.

267. *Diedrocephala coccinea* Forst., correct.

267. *Diedrocephala hieroglyphica* Say is *Tettigoniella gothica* Sign.

268. *Helochara communis* Fitch, correct.

268. *Acopsis viridis* Linn. is *Dræculacephala mollipes* Say.

269. *Gypona quebecensis* Prov. Under this name is straight *cana* Burm. It is not the species formerly sent to me by Provancher as *quebecensis* which was smaller and of a deeper green color.

269. *Gypona hullensis*, n. sp., is *pectoralis* Spangb.

270. *Eucanthus orbitalis* Fitch. Dr. Ball now places this as a synonym of *E. acuminatus* Fabr.

270. *Penthimia picta* Prov. The specimen is missing from this label but there can be no question that it is the male of *americana* Fitch.

275. *Platymetopius acutus* Say, correct.

· 275. *Platymetopius magdalensis*, n. sp. This is the species later described by Prof. Osborn as *obscurus*.

276. *Scaphoideus immistus* Say, correct.

277. *Scaphoideus auronitens*, n. sp., correct.

278. *Deltocephalus curtisii* Fh., correct. *Athysanus curtisii* Fh.

278. *Deltocephalus inimicus* Say, correct.

279. *Deltocephalus citronellus* Prov. The insect on this label is a *Thamnotettix* probably still undescribed. It is not the form described in the NAT. CAN., p. 378.

279. *Deltocephalus minkii* Fieb., correct.

280. *Deltocephalus sayi* Fitch, correct.

280. *Selenocephalus placidus*, n. sp. This is an *Acucephalus* new to me. It has a dark shade under the sharp lunately rounded anterior margin of the vertex and the apex of the elytra are coarsely alternated with fuscous points which are not properly indicated in Provancher's description. It may be one of the European species already recognized from this country.

281. *Athysanus obsoletus* Kirsch. is now known as *relativus* Gill. & Baker.

282. *Athysanus plutonius* Uhler. This has more recently been separated out as a distinct species under the name *uhleri* Ball.

282. *Acocephalus circumflexus*, n. sp., is the male of *albifrons* Linn.

283. *Thamnotettix citronellus* Prov. Under this name is a very pale specimen of *eburata* Van D., but it does not answer to the description in the NAT. CAN., p. 378, and cannot be that insect.

284. *Thamnotettix clitellarius* Say, correct.

284. *Thamnotettix subcupræus* Prov., correct.

284. *Thamnotettix melanogaster* Prov., correct.

285. *Thamnotettix decipiens*, n. sp. The only specimen on this label is much paler than this species is generally found in the east.

285. *Thamnotettix seminudus* Say, correct. *Eutettix seminuda* Say.

286. *Jassus unicolor* Fh., correct. *Chlorotettix unicolor* Fh.

286. *Allygus irroratus* Say, correct. *Phlepsius irroratus* Say.

287. *Cicadula 6-notata* Fall., correct.

288. *Bythoscopus clitellarius* Fitch is *Idiocerus provancheri* Van D.

289. *Bythoscopus fenestratus* Fh. is a pale form of *Oncopsis nigrinasi* Fh.

289. *Bythoscopus variegatus* Fh., correct. *Oncopsis variegatus* Fh.

290. *Bythoscopus pruni* n. sp., correct. *Oncopsis pruni* Prov.

291. *Idiocerus pallidus* Fitch, correct.

292. *Idiocerus verticis* Say. I was not able to locate this nearer than to place it in the *alternatus* group. It is not the western *verticis*.

292. *Idiocerus duzeei*, n. sp., correct.

292. *Idiocerus subbifasciatus* Say is *lachrymalis* Fitch.

293. *Idiocerus alternatus* Fitch, correct.

293. *Idiocerus novellus* Say, correct. *Agallia novella* Say.

294. *Pediopsis viridis* Fitch, correct.

295. *Pediopsis basalis* Van D., correct.

295. *Pediopsis insignis* Van D., correct. Now known as *trimaculata* Fitch. This genus must be known as *Macropsis*.

295. *Pediopsis flavescens* Prov. A small female of *Oncopsis nigrinasi* Fh.

296. *Agallia sanguinolenta* Prov., correct.

296. *Agallia 4-punctata* Prov., correct.

298. *Erythroneura mali*, n. sp., is *Dicraneura communis* and must be known as *Dicraneura mali* Prov.

298. *Erythroneura vitis* Harris, correct.

299. *Erythroneura vitifex* Fitch, correct.

299. *Erythroneura vulnerata* Fitch, correct.

299. *Erythroneura rosæ* Linn., correct. These are now placed in *Typhlocyba*.

300. *Typhlocyba jocosa* Prov. A reddish form of *Balclutha punctata* Thunb.

301. *Typhlocyba punctata* Thunb. is the common green form of that species.

335. *Cymus angustatus* Stal., correct.

336. *Aradus abbas* Bergr., correct.

336. *Coriscus flavo-marginatus* Scholz., correct.

337. *Delphax bifasciatus*, n. sp., is *Stobaera tricarinata* Say, a little faded.

338. *Ceresa subulata* Say is *constans* Walk.

338. *Helochara bifida* Say, correct. *Kolla bifida* Say.

339. *Deltocephalus chlamydatus*, n. sp. is an *Athysanus* later described as *infuscata* by Gillette and Baker.

339. *Deltocephalus superbus*, n. sp., is a *Xestocephalus* later described by me as *fulvocapitatus*.

340. *Erythroneura obliqua* Say, correct. *Typhlocyba obliqua* Say.

340. *Typhlocyba unica*, n. sp., is an *Empoasca* later described as *splendida* by Gillette.

ON SOME UNDESCRIBED FORMS OF FLORIDA COLEOPTERA.

BY W. S. BLATCHLEY, INDIANAPOLIS, INDIANA.

In 1911 I spent the time from January 8th to April 17th in Central and Southern Florida, and while there collected insects in a number of localities, notably near Sanford, St. Petersburg, Sarasota, Ft. Myers, Little River, and Ormond. The time of year was not the best for the most successful collecting, as the insects of Florida hibernate in much the same manner as they do farther north, many of them being represented there, as here, during the winter months in the egg, larval and pupal stages. However, about 500 species of Coleoptera were taken, and also many Orthoptera, Hemiptera and butterflies. It is at present my intention to again visit Florida in January, 1913, and, perhaps, stay later in the spring, making collections in the same orders, and then publish notes on the "catch" of the two seasons. Meanwhile, a few forms of Coleoptera, which have apparently hitherto escaped observation, are herewith described.

Cœlambus marginipennis, sp. nov.

Short, rounded, oval, subdepressed above, moderately convex beneath. Head, thorax, under surface, femora and tibiæ reddish brown; elytra piceous-black, shining, with narrow side margins, broadening into a rounded lobe at middle, reddish brown, tarsi and apical fourth of antennæ dusky. Clypeus broadly rounded, distinctly margined. Head and thorax finely, evenly but not densely punctate; the elytra more coarsely, densely and rather shallowly punctate. Meso- and meta-sterna coarsely, rather sparsely and deeply punctate, the punctures of abdomen finer and more shallow. Length 2.5—2.8 mm.

Frequent in shallow brackish ponds, one to two miles inland, near Sarasota, Florida. March 1—3. This beetle has the form of *C. acaroides* Lec., but the elytra are differently coloured, and without the carinæ of that species. It is a little larger, more rounded, and much more coarsely punctate than *C. farctus* Lec. In a few specimens the elytra are mostly wholly piceous, but in the great majority the paler side margins broaden at middle to form a rounded lobe.

Aphodius campestris, sp. nov.

Elongate-oblong, convex. Head and thorax reddish or pale chestnut brown, shining, the latter with front margin darker; elytra, under surface and legs brownish yellow. Head very finely and sparsely punctate, not tuberculate; clypeus broadly and shallowly emarginate at middle, the sides curved. Thorax not narrowed in front, sides broadly curved, hind angles obtusely rounded; base very distinctly margined, disk finely and sparsely punctate. Elytra equal in width to thorax, finely striate, the striæ finely and indistinctly punctate; intervals feebly convex, smooth. Front tibiæ stout, broad, distinctly punctate in front, strongly 3-toothed. Hind tibiæ rather slender; first joint of hind tarsi as long as the next three together. Length 3 mm.

Two specimens from beneath dry cow-dung in company with *A. vestiarius* Horn, near Sarasota, February 17th. Closely related to *A. rubeolus* Beauv., but smaller, more slender, with paler elytra, longer basal joint of hind tarsus, and with base of thorax distinctly margined. One of the types is in the collection of F. Blanchard, Tyngsboro, Massachusetts.

Hymenorus granulatus, sp. nov.

Oblong-parallel, subdepressed, sparsely pubescent with fine recumbent grayish hairs. Black, shining; palpi and mandibles reddish brown, tarsi piceous. Eyes large, separated by about one-half their own diameters; antennæ stout, half the length of the body, the joints flattened, triangular, the third twice as long as second, half the length of fourth. Thorax at base one-third wider than long, sides broadly rounded into the front margin, base slightly sinuate each side near middle; disk strongly declivent in the region near the front angles, obsoletely foveate each side of middle at base, very densely and coarsely punctate, the punctures feebly separated or in part confluent. Elytra slightly wider than base of thorax, their sides parallel to apical fourth; disk striate, the intervals convex; densely granulate-punctate. Basal joint of hind tarsi slightly curved, one-half longer than the remaining joints together. Length 7—7 5 mm.

Described from four specimens beaten from scrub-oak foliage near Sanford, March 28—29 The dense punctuation of thorax and elytra, taken in connection with the uniform shining black color, readily distinguishes this from all other described species of *Hymenorus*.

Andrimus confusus, sp. nov.

Elongate-oval, sparsely clothed with short, suberect yellowish hairs. Head, thorax, under surface and legs reddish brown; elytra and antennæ dark chestnut brown. Head tansversely sulcate in front of eyes, finely and rather closely punctate, alutaceous between the punctures; eyes large, separated by nearly their own width; antennæ slender, one-half the length of body, the joints obconical, the third more than twice the length of second, one-half the length of fourth. Thorax three-fourths as long as wide, sides parallel from base two-thirds their length, then broadly rounded into the front margin; disk evenly convex, very finely alutaceous, finely and sparsely punctate, without trace of median line or basal foveæ. Elytra one-fourth wider than thorax, their sides parallel to apical fourth then rounded to a blunt apex; disk striate, the striæ with rows of close-set punctures; intervals feebly convex, each with two rows of minute punctures. Abdomen smooth. Basal joint of hind tarsi equal in length to the other three combined. Length 9 mm. Two species beaten from live oak near Sanford. March 29.

Diaperis maculata floridana, var. nov.

This variety differs from typical *maculata* (*hydni* Fab.) in the colour of the elytra, the elongate submarginal dark spot near the humeral angle of *maculata* being absent and the large irregular black spot on apical third here uniting with the sutural black stripe to form a common cross-bar.

Frequent beneath bark of fungus-covered oak log near Sarasota. Feb. 28. Horn in his remarks on *D. maculata** states, "This species is remarkably uniform in its system of elytral coloration.

Mr. F. W. L. Sladen, F. E. S., has been appointed Assistant Entomologist for Apiculture in the Division of Entomology, Ottawa. Up to the time of his appointment, Mr. Sladen devoted his whole time to Apiculture in England, where he possessed a large apiary and made a special study of queen-rearing according to scientific methods. His writings on the subject include "Queen-rearing in England", "Breeding the British Golden Bee", and several articles on the collection of pollen, etc. His studies of the *Bombi* are recorded in "The Humble Bee", reviewed in the present number of this Journal. As he has travelled in Europe, India and North America his knowledge of Apiculture and native bees is unusually wide and his appointment will prove an additional source of strength to the Division of Entomology, where he will have charge of the apicultural work. He will also study the Canadian *Bombi* and native bees.

C. G. H.

*Trans. Amer. Philosophical Soc., XIV, 379.

NEW SPECIES OF *FURCOMYIA* (*TIPULIDÆ*).

BY CHAS. P. ALEXANDER, ITHACA, N. Y.[1]

The crane-flies herein characterized as new are, with one exception, Neotropical forms. There have been described by previous writers 15 species of South American *Limnobini* that I have no hesitation in referring to the genus *Furcomyia* (= *Dicranomyia* of authors). With the single exception of *F. muscosa* End. (Ecuador), the forms are Chilian or Patagonian, and are species named by Macquart,[2] Blanchard,[3] Philippi[4] and Bigot.[5] No species have been mentioned from the various countries of Middle America, and it is probably because of this fact that so many of the forms received proved to be novelties.

The material included is the property of Eastern Museums, as follows: U. S. National Museum, received through Mr. Frederick Knab, and the American Museum of Natural History, received through Mr. J. A. Grossbeck. I express my sincere gratitude to both of these gentlemen for their kind help in this respect.

A Key to the spotted-winged Furcomyiæ.

(South America (northern portion), Central America and the Antilles.)

1. Sc short, ending before, or opposite, or only slightly beyond, the origin of Rs.......................................2.
   Sc long, ending far beyond the origin of Rs. ....5.
2. Wing-marking abundant, forming a network.......................3.
   Wing-marking scanty, confined to the neighbourhood of veins......4.
3. Legs with the femora uniform brown apically; wing pattern regular.................................*reticulata*, sp. n. (Cuba)
   Legs with the femora yellowish apically with a broad gray subapical ring; wing pattern irregular....................*muscosa* Enderl. (Ecuador)
4. Legs black, a reddish annulus far before the tip of the femur; no supernumerary cross-vein in cell R_3; seam on cord of wing, dark brown, narrow; antennæ dark except at base.....*osterhouti*, sp. n. (Panama)

1. Contribution from the Entomological Laboratory, Cornell University.
2. Macquart, Pierre Justin; Dipt. Exot., Vol. I, pt. 1, p. 72 (1838).
3. Blanchard, Emile; in Historia fisiça y politica de Chili Zoologia, Tome 7, pp. 337-344, esp. pp. 340-343 (1852).
4. Philippi, Rodolfo; Verhand. zool-bot. gesells. Wien., Vol. 15, pp. 597, 598, 602-617, 780, 781; esp. 612-614 (1865).
5. Bigot, Jacques; Mission Scientifique du Cap Horn, 1882-1883; Tome 6, 2nd part, pp. 5-10; esp. pp. 8, 9, pl. 2, fig. 2 (1888).

Legs with the femora dark brown at the tip with an indistinct subapical ring; a cross-vein in cell R_3; seam on cord pale brown, broad; antennæ pale........................*translucida*, sp. n. (Panama)

5. Wings with an abundant pattern in the cells..........*gloriosa*, sp. n. (Guatemala)

Wings with the markings scanty and more or less confined to the neighbourhood of veins....................................6.

6. Wing hyaline, with the markings brown; pleuræ with a brown band; tibiæ and tarsi uniform dark......................*eiseni*, sp. n. (Guatemala)

Wing dusky, with the markings dark brown; no pleural band; tibiæ at tip, and tarsi, orange brown......................*lutzi*, sp. n. (Brit. Guiana)

Furcomyia reticulata, sp. n.

Antennæ brown; thorax yellow, with an irregular brown median stripe; legs yellow, darkening to brown apically; wings hyaline, reticulated with brown marks.

♀.—Length, 4.5–6 mm; wing, 5.3–5.4 mm.

♀.—Head: rostrum yellowish brown; palpi dark brownish black. Antennæ, basal segments pale, whitish; flagellum light brown, the segments rounded, becoming oval and then elongated toward the tip of the antennæ. Front, vertex and occiput dull yellow, the vertex and occiput prolonged caudad, with two brown stripes above and brown on the sides.

Thorax: pronotum brown, thickly yellow pollinose; a small brown median spot at the caudal margin of the scutum. Mesonotum, præscutum dull yellow sericeous, a broad, light brown median stripe, overlain by a dark brown stripe, whose margins are very irregular; two interrupted brown stripes on either side of the median mark, the outermost very pale on the margin of the sclerite; scutum dull brown, with four brown stripes, continuations of the lateral præscutal vittæ; the two stripes on each side unite at the caudal margin of the sclerite and run half across the scutellum; scutellum very pale, whitish yellow, sending a median prolongation cephalad onto the scutum; postnotum brown. Pleuræ light brown, thickly pale yellowish pollinose. Halteres very pale yellow, the knob brown. Legs: coxæ, trochanters and femora dull yellow, the femora darkening to brown apically; extreme base of the tibiæ whitish, rest of tibiæ and the tarsi dark brown. Wings, veins brown, except costa, which is light yellow and black alternated; membrane hyaline, costal cell with small, equally-spaced brown marks ; from the base to the tip of R_1 about 19, these marks a trifle narrower than the hyaline interspaces ; five large brown blotches along the radial cells, the first at the base of vein M ; second in middle of cell R ; third just before the origin of Rs ; fourth over the fork of Rs, and the last at the tip of R_{2+3}, irregular ; all the cells with narrow brown marks across them producing a net-work. Venation (see fig. p) : Sc short, Sc_1 ending before the origin of Rs, Sc_2 about opposite it ; Sc_2 longer than Sc_1 ; Rs angular at base ; basal deflection of M_{1+2} long, so that the inner end of cell 1st M_2 is almost on a level with cell R_3 ; basal deflection of Cu_1 before fork of M, sometimes far before ; cross-vein *m* far out, so that the deflection of M_3 is much longer than *m*.

Abdomen, tergum, segments brown, darkest on caudal margin, paler on the sides ; sternum dull yellow ; a dark brown median spot on caudal margin of each sclerite.

Holotype, ♀.—Pinar del Rio, Cuba ; 1900 (Palmer and Riley).

Paratype, ♀.—Type locality, March 27, 1900 (Palmer and Riley).

Types in U. S. Nat. Mus. coll. (No. 15,133).

Furcomyia osterhouti, sp. n.

Whitish ; mesothoracic præscutum with a broad median stripe and two short lateral ones ; femora black, with a postmedian reddish annulus ; wings with brown spots, bands and seams.

♀.—Length, 6.5 mm (about) ; wing, 5.7 mm.

♀.—Head : rostrum and palpi dark brownish black. Antennæ, basal segments yellowish brown, flagellum very dark brown, almost black. Front, vertex and occiput pale, whitish, tinged with brown.

Thorax : pronotum dark brown above, abruptly pale, yellowish white on the sides. Mesonotum pale yellowish white, the median stripe broad, dark brown ; the lateral stripes appear on the hind margin of the præscutum and run back across the scutum and scutellum ; at the caudal end of the latter sclerite they unite and form a very broad median band, which occupies the dorsum of the postnotum. Pleuræ pale, whitish. Halteres, knob and most of the stem dark brown. Legs : coxæ and trochanters yellowish brown ; femora black, with a distinct reddish annulus at about three-fourths the length ; tibiæ reddish at base, rest of tibiæ and tarsi shiny black. Wing with a slight yellowish tinge, especially in the cephalic cells ; a very narrow brown mark from *h* caudad ; a brown mark from the tip of Sc_1 down beyond Rs ; a brown mark at tip of R_1 and on *r* ; a narrow seam along the cord ; outer end of cell 1st M_2 seamed with brown ; most of the veins seamed with brown; apical portions of the radial cells suffused with brown. Venation: (See fig. q.) Sc ends beyond origin of Rs, Sc_2 at its tip; cross-vein *r* at tip of R_1; Rs arcuated at origin; basal deflection of Cu_1 before the fork of M.

Abdomen, tergum yellowish, the apex of each sclerite brown, with a narrow brown median band; sternum, markings less clearly defined.

Holotype, ♀.—Bocas d'Toro, Panama; Sept. 28, 1903. (P. Osterhout, coll.)

Type in U. S. Nat. Mus. coll. (No. 15,130.)

Furcomyia translucida, sp. n.

Whitish; mesothoracic præscutum with a narrow median brown stripe; femora darkened at the tip, pale subapically; wings with brown spots and bands; a supernumerary cross-vein in cell R_3.

♂.—Length, 5.8 mm.; wing, 6.9 mm.; middle leg, femur, 5.7 mm.; tibia, 5.2 mm.

♂.—Head: rostrum and palpi dark brown. Antennæ, basal segments brown, flagellum yellowish, the terminal three or four segments brown; segments of the flagellum short, globular, the apical segments more elongated. Front, vertex and occiput light yellow, the vertex with a large brown spot in the centre.

Thorax: pronotum dark brown, becoming paler, yellowish white on the sides; mesonotum, præscutum very pale, almost white, with a clearly-defined dark brown median stripe, rather narrow, ending at the suture; scutum and scutellum pale, whitish, with a dark brown stripe on each lobe, running backward and meeting on the caudal margin of the scutellum; postnotum with a very broad brown median mark resulting from the confluence of the scutellar stripes. Pleuræ very pale, whitish; a brownish mark on the propleuræ above the fore coxa. Halteres pale, knob dark brown. Legs: coxæ and trochanters whitish; femora yellowish brown; a clearer yellow subapical ring, tip broadly brown, the extreme apex again rather lightened; tibiæ and tarsi brown, gradually increasing to dark brown. Wings: subhyaline or very faintly yellowish; a brown mark at the humeral cross-vein extending down across the arculus; a second mark at tip of Sc_1 and down across Rs almost to M; a third, extending into a cross-band, from the stigma, where it is darkest, unbroken across the cord; a brown seam on the supernumerary cross-vein in cell R_3; outer end of cell 1st M_2 seamed with brown. Venation: (See fig. r.) Sc short, ending just beyond the origin of Rs; Sc_2 just opposite origin of Rs; R_1 extending beyond cross-vein *r-m*, *r* at its tip. Rs square at its origin and spurred, in a line with R_{2+3}; a strong cross-vein in cell R_3 at about two-thirds of the length of the cell; cell 1st M_2 rather elongate; basal deflection of Cu_1 at the fork of M.

Abdomen: tergum pale yellowish white, apical fourth dark brown; apex sternum similar, but the dark apex not so clearly defined.

Holotype.—♂. Bocas d'Toro, Panama; Sept. 28, 1903. (P. Osterhout, coll.)

Type in U. S. Nat. Mus. coll. (No. 15,129)

Furcomyia gloriosa sp. n.

Antennæ brown; thorax gray, dorsum striped with darker; legs, femora dark on apical half, with a subterminal yellow ring; wing spotted and suffused with brown.

♀.—Length about 6.5 mm.; wing, 8.4 mm.

♀.—Head: rostrum and palpi dark brown. Antennæ, basal segments very dark brown; basal five flagellar segments lighter brown, apical segments dark brown. Front, vertex and occiput dull gray, with a black mark on vertex along inner margin of the eye.

Thorax: pronotum dull greenish gray pollinose, with a broad black stripe on the side of the scutum. Mesonotum, præscutum dark brown, thickly grayish pollinose, with a black stripe on either side of the narrow median gray line, running from the anterior margin of the sclerite almost to the suture. Lateral stripes short, broad, beginning behind the pseudosutural fovea, running across the suture and covering most of the scutum; scutum in middle and along the caudal margin dark brown; scutellum and postnotum dark brown. Pleuræ black, greenish gray pollinose. Halteres, stem pale yellowish brown, knob dark brown. Legs, coxæ and trochanters dark brown, the former gray pollinose; femora light yellow, the apical quarter dark brown, with a subapical yellow ring. Wings hyaline or nearly so; costal cell with four brown marks, the last at Sc, the 3rd over the origin of Rs; a large square mark at the tip of R_1 (stigmal) extending down over the fork of Rs; cells 2nd R_1 and R_3 with large brown spots filling most of the cells; cells R_5 to Cu_1 suffused with lighter grayish brown and with hyaline spots; basal and anal cells with smaller brown spots; a series of about four in cell 1st A. Ends of veins Cu_2, 1st and 2nd A, with broad, grayish brown suffusions. Veins brown; Sc and R yellow, except where located in the brown markings, where they are black. Venation: (See fig. j.) Sc long, ending far beyond the origin of Rs, but slightly before its middle; Rs long; basal deflection of Cu_1 far before the fork of M.

Abdomen, tergum dark brown; sternum lighter brown, extreme caudal margins of the sclerites light yellow.

Holotype.—♀. Totonicipan, Guatemala, Cent. Am., 1902. (Dr. G. Eisen.)

Type in U. S. Nat. Mus. coll. (No. 15,132.)

This insect agrees superficially with *muscosa* End.* of Ecuador, but has *Sc* much longer, legs very different in colour, and is a much smaller species. *Muscosa* has a supernumerary cross-vein in cell R_3, but this may not be normal, as it is not mentioned in the specific description.

Furcomyia eiseni, sp. n.

Antennæ black throughout; body yellow; legs, femora yellow, passing into brown on the tibiæ and tarsi; wings hyaline, with six brown spots along costa, the second, largest, at origin of Rs.

♂.—Length, from 4.5–5 mm.; wing, 6.3–7.5 mm.

♀.—Length, from 4.5–6 mm.; wing, 5.7–7 mm.

Head: rostrum and palpi black. Antennæ black throughout in the ♂, with conspicuous long hairs, not so noticeable in the ♀. Front, vertex and occiput blackish, grayish pollinose in front.

Thorax: pronotum dull yellow; mesonotum dull reddish yellow, with a very indistinct darker median stripe and darker lateral stripes which are brownish, these continued back on the scutum, where they cover the lobes; scutellum and postnotum brownish. Pleuræ yellow, with a more or less conspicuous dark brown stripe running from the cervical sclerites to the postnotum. Halteres yellow at base; apical half of the stem and the knob brown. Legs: coxæ and trochanters light yellow; femora yellow at base, passing into brown; tibiæ and tarsi darker brown. Wings hyaline; cells *C* and *Sc* slightly yellowish; six brown marks along the costal margin on the cross-veins, as follows: A brown mark at the wing base; a large brown rectangular mark at the origin of Rs; a third at the tip of Sc, where it is continued down over the fork of Rs, here meeting the fourth blotch, located at the tip of R_1; the marks continuing across the cord; wing subapically largely dark; outer end of cell 1st M_2 seamed with brown; a brown mark in the end of cell 2nd R_1 and cell R_3; ends of veins Cu_1, Cu_2 and 1st A, with small brown clouds; a large spot at end of 2nd A. Venation: (See fig. s.) Sc long, ending just before the fork of Rs, Sc_2 at its tip; Rs square at its origin; base of cell 1st M_2 arcuated, nearly on a level with the inner end of cell R_3 (as in *stulta* O. S.);

*1912. Zoöl. Jahrbuch.; pt. 1, pp. 75, 76; fig. W[1]. *(Dicranomyia.)*

basal deflection of Cu_1 just beyond the fork of M ; Cu_2 generally shorter than the deflection of Cu_1.

Abdomen, tergum dark brown, the bases of the sclerites somewhat paler ; sternum light yellow, the caudal and lateral margins conspicuously dark brown.

Holotype.—♂. Aguna, Guatemala, Cent. Am. (2,000 ft.) ; Sept., 1902. (Dr. G. Eisen, coll.)

Allotype.—♀. With the type.

Paratypes.—5, ♂ ♀. With the type.

Types in U S. Nat. Mus. coll. (No. 15,131), except one paratype in author's collection.

Furcomyia lutzi, sp. n.

Antennæ black ; body orange ; abdomen brown ; legs black, tip of tibiæ and the tarsi pale, orange yellow ; wings dusky, with brown marks.

♀.—Length about 6 mm.; wing, 7.3 mm.; middle leg, femur, 5.4 mm.; tibia, 5.8 mm.

Head : rostrum and palpi dark brownish black. Antennæ dark brownish black. Front thickly gray pollinose ; vertex and occiput dark orange brown, brighter orange on the occiput.

Thorax : pronotum and mesonotal præscutum and scutum deep orange ; scutellum and postnotum much lighter coloured, yellowish orange. Pleuræ orange yellow, lighter coloured ventrally. Halteres, stem yellowish basally, darkening to the blackish knob. Legs : coxæ and trochanters orange yellow, extreme base of femora yellow ; remainder of femora and most of the tibiæ dark brownish black ; tibiæ with the apical eight pale orange brown ; tarsi orange brown. Wings suffused with dark brown, costal and subcostal cells and the radial cells very dark ; dark brown spots arranged as follows : a rounded mark at the origin of Rs ; one at fork of Rs, continued down the cord as a broken seam ; a round spot at end of R_1 ; outer end of cell 1st M_2 seamed with dark brown. Venation : Sc long, ending nearer to the fork of Rs than to the origin, Sc_2 at tip of Sc_1. Cross-vein *r* at the tip of Rs ; deflection of R_{4+5} long ; basal deflection of Cu_1 far before the fork of M.

Abdomen, tergum, segments dark brown ; sternum light yellow.

Holotype.—♀. Tukeit, British Guiana ; July 19, 1911. (F. E. Lutz, coll.)

Type in American Museum of Natural History.

Furcomyia omissa, sp. n.

Small; dark brown; wings dark, stigma present; Sc_1 short, Sc_2 apparently lacking.

♀.—Length, 3.7–4 mm.; wing, 4–4.2 mm.

♀.—Head: rostrum and palpi dark brownish black. Antennæ brownish black. Front, vertex and occiput brown.

Thorax: mesonotum, præscutum with a thick brownish pollen, becoming grayish on the sides of the sclerite; a dark brown median stripe beginning near the anterior end of the sclerite, becoming narrower and finally obsolete before the suture; scutum, scutellum and postnotum dark brown. Pleuræ dark brown, with a sparse gray bloom on the middle of the thorax. Halteres dark brown; remainder of femora, tibiæ and tarsi dark brown. Wings somewhat suffused with darker; a small oval brown stigma. Venation: (See fig. o.) Sc short, ending far before the origin of Rs, Sc_2 not evident. Rs rather short, about one and one-half times the length of the deflection of R_{4+5}; cross-vein *m* present in the type, absent in the paratype.

Abdomen dark brown.

Holotype.—♀. Aguna, Guatemala, Cent. Am. (Dr. G. Eisen.)

Paratype.—♀. Same as the type.

Types in U. S. Nat. Mus. coll. (No. 15,139.)

Furcomyia knabi sp. n.

Like *liberta* O. S., but ventral lobe of ♂ hypopygium produced entrad in a long slender arm.

♂.—Length, 6.5–7 mm.; wing, 8.8–9 8 mm.

♀.—Length, 7 mm.; wing, 9 mm.

Head: rostrum and palpi dark brownish black; antennæ black. Front, vertex and occiput clear gray.

Thorax: dorsum of the mesonotal præscutum suffused with brown, general colour brownish gray, much browner than the clear gray of the head; stripes on thoracic dorsum ill-defined; scutum dull gray, the scutellum very light gray; postnotum gray. Pleuræ grayish. Halteres yellow, knob brown. Legs: coxæ and trochanters brown; femora, tibiæ and tarsi dark brown. Wings almost as in *liberta* O. S., not pallid at base; a faint stigma at the tip of R_1. Venation: (See fig. m.)

Abdomen gray. Hypopygium: (See fig. w.) Dorsal aspect, 9th sternite very convex, ending in a small knob deeply bifid; pleuræ long, cylindrical, bearing two apical lobes; the dorsal lobe slender, chitinized, ending in an acute point; ventral lobe yellow, produced entad into a long arm chitinized, its apex blunt but slightly notched. Ventral aspect, 9th tergite almost straight on caudal margin ; pleuræ short, the inner caudal angle produced into a long appendage, which is tufted with yellow hairs at its tip ; guard of the penis long, enlarged basally, projecting slightly beyond the apices of the pleural appendage ; ventrad of the pleural arm is a slender acicular appendage.

Holotype.—♂. Totonicipan, Guatemala, 1902. (Dr. G. Eisen.)

Allotype.—♀. Antigua, Guatemala. (Dr. G. Eisen.)

Paratypes.—♂♂. Totonicipan, Guatemala. (Dr. G. Eisen.)

Types in U. S. Nat. Mus. coll, (No. 15,135). One paratype in author's collection.

Like *liberta* O. S. (Proc. Acad. Nat. Sci. Phil., 1859, p. 209 ; Monograph Dipt. N. Am., Vol. 4, p. 69), of the Eastern U. S., but larger, the mesothoracic præscutum browner and the stripes indistinct. In *liberta* the ♂ genitalia (fig. u) consists of short pleuræ, the swollen ventral lobes produced entad in a blunt knob, which bears two conspicuous caudad-projecting spines at its tip, the ventral one very stout, spine-like, the dorsal one more slender. In *knabi* the pleuræ are longer, the lobes short, the ventral one produced into a long arm, which is slightly notched apically.

Explanation of Plate XI.

Fig. h. Wing of *Furcomyia andicola*, sp. n.
" i. " *F. insignifica*, sp. n.
" j. " *F. gloriosa*, sp. n.
" k. " *F. argentina*, sp. n.
" l. " *F. liberoides*, sp. n.
" m. " *F. knabi*, sp. n.
" n. " *F. simillima*, sp. n.
" o. " *F. omissa*, sp. n.
" p. " *F. reticulata*, sp. n.
" q. " *F. osterhouti*, sp. n.
" r. " *F. translucida*, sp. n.
" s. " *F. eiseni*, sp. n.
" t. " ? *F. fumosa*, sp. n.
" u. Hypopygium of *F. liberta* Osten Sacken.
v = ventral apical appendage.
d = dorsal apical appendage.
" v. Hypopygium of *F. liberoides*, sp. n.
" w. " *F. knabi*, sp. n.

(To be continued.)

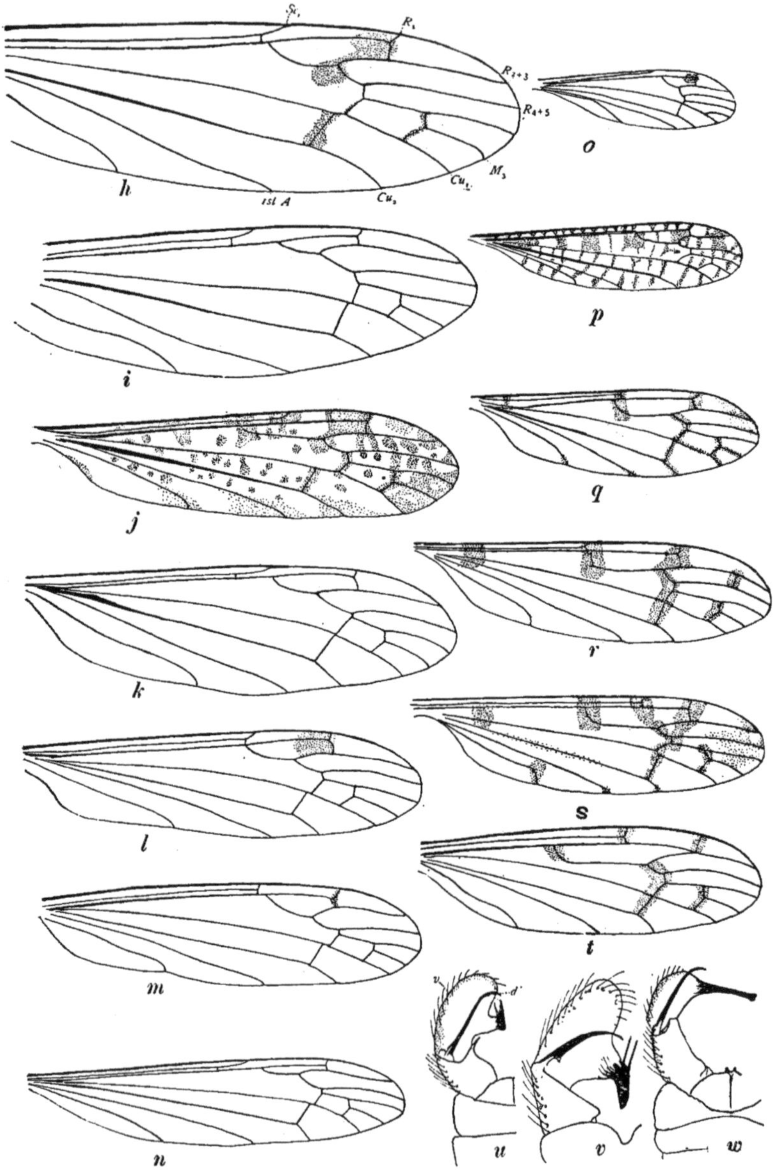

FURCOMYIA (TIPULIDAE, DIPT.).

THE INTERNATIONAL CONGRESS OF ENTOMOLOGY.

The Second International Congress of Entomology was held at Oxford (England), from August 5th to 15th, the first Congress having been held at Brussels in 1910. It was attended by representative entomologists from Australia, Belgium, Canada, Borneo, British East Africa, Chili, Egypt, France, Germany, Holland, Hungary, Luxenbourg, Sandwich Islands, Spain, Sweden, Switzerland, Turkey and the United States, besides a large number from Great Britian and Ireland.

As representative of the Canadian Government and a delegate from the Entomological Society of Ontario, I sailed from Quebec on July 26th, but an unfortunate collision at sea necessitated my return and re-embarkation from New York, and on this account I missed the proceedings of the first day, during which the President, Prof. E. B. Poulton, F.R.S., delivered his presidential address.

In welcoming the entomologists of all nations, the President alluded to the suitability of Oxford as the meeting place of such a gathering, and referred to the celebrated meeting of the British Association in 1860 in the same place, when Huxley made his celebrated and crushing retort to Wilberforce's attempt to throw ridicule on the evolutionary doctrines recently set forth by Darwin and valiantly championed by Huxley. Prof. Poulton traced the history of the Hope Department of Entomology at Oxford, of which he has charge, and referred to the great work of Prof. Westwood, his predecessor and former teacher. He described a splendid exhibit of the polymorphic African *Papilio dardanus*. Tracing its geographical variations and illustrating the gradual development of mimicry by the female, the polymorphism of the same sex and the proportions of the different mimetic forms hatching out from the eggs of a single female.

The meetings of the Congress were general and sectional. At the various sectional meetings, which were usually held at the same time, economic and medical entomology, evolution and bionomics, mimicry and distribution, systematic entomology and nomenclature and morphology were discussed. It was naturally impossible for one to attend all the sections or to hear all the papers which one would have wished to hear. On this account, therefore, I shall refer only to certain of the papers which I was able to hear. In any case, space would forbid the writing of a more lengthy account, which will be given in the official reports of the Congress.

Mr. G T. Bethune-Baker and Rev. G. Wheeler brought forward and discussed a proposal from the Entomological Society of London for the formation of the International and National Committees to deal with the vexed and complicated question of nomenclature. The matter was wisely referred to the Executive Committee of the Congress for consideration and report. As a result of the Committee's report, the Congress decided upon the formation of an International Committee on nomenclature and of National Committees to be elected by the entomological societies. It was also resolved to request better representation for entomology on the International Committee on Zoological Nomenclature. In reply to a question as to what would be the result of a disagreement of the part of the newly-established International Committee on Entomological Nomenclature with the International Committee on Zoological Nomenclature, I was pleased to receive from Dr. Jordan, the General Secretary, the assurance that the finding of the Entomological Committee would prevail and would be accepted.

In the case of such a meeting, presided over by Prof. Poulton, and held in the Hope Department, which might well be called the home of the study of mimicry, it was natural that in the section on evolution and bionomics there should be presented some most interesting and valuable papers on mimetic resemblances, their evolution and distribution. Prof. Poulton described the researches of Mr. C. A. Wiggins and Dr. C. H. Carpenter on the forest inhabiting Pseudacræas of Uganda. The polymorphic character of the mimetic species has led to the creation of a number of species. Breeding experiments are throwing considerable light as to the relationships of the different forms. On the same subject Prof. Punnett read a very suggestive paper by Mr. J. C. F. Fryer, who is attempting to work out the polymorphism of *Papilio polytes* on Mendelian lines. Dr. R. C. L. Perkins discussed the colour groups of the Hawaiian wasps, in which the influence of a well-protected intruder upon the superficial aspect of the members of a native fauna is shown.

In the morphological section, Dr. F. A. Dixey read a paper on "The scent organs in the Lepidoptera." The specialized scales which serve to distribute scent in many species may be either generally scattered over the wing surface or collected in patches. In the latter case there is a special supply of air tubes to the sockets of the scales. Prof. G. H. Carpenter described the prescence of maxillulæ, small-paired appendages connected with the hypopharynx in certain beetle larvæ. Papers by Dr. G. Horvath (Budapest), Padre L. Navas (Barcelona), and Prof. J. Van Bemmelen (Gröningen), dealt with the morphology and phylogeny of insect wings. Mr. L. Doncaster (Cambridge), gave an account of his investigations on the question of sex-limited inheritance of characters in insects.

Papers on the geographical distribution of insects were read by Dr. Anton Handlirsch (Vienna), who ranged over the whole field of fossil insects, and Dr. P. Speiser (Labes), Baron von Rosen (München), discussed the forest Termites, while Prof. Calvert gave an exceedingly interesting review of the advance which has been made in the knowledge of the dragon-flies since 1895.

Among the papers on insect bionomics which were read, three may be specially mentioned, on account of their exceptional interest. Dr. W. M. Wheeler (Harvard), gave an account of his recent investigations in Central America in ten Acacia-inhabiting ants. His results do not confirm the popular idea as to the adaptation of the acacias for the purpose of encouraging ants with a view to protecting themselves against the leaf-cutting ants. The ants merely frequent the convenient hollow places in the acacias because it suits their convenience so to do. Dr. A. Seitz (Darmstadt), described the results of an interesting experiment which he devised to test the sense of sight of insects, in this case butterflies. The character of this sense may be judged from the fact that the male butterflies of the species observed were found to be so short-sighted that they attempted to copulâte with paper-coloured replicas of the females mounted on pins. A masterly and extremely suggestive paper on the Mallophaga was read by Prof. Vernon L. Kellogg. It was demonstrated that the association between these parasitic insects and their avian hosts was of a most remarkable character, tending to show that the parasites had become associated with their present hosts before the latter had become divided into separate species. They showed to an extraordinary degree the effect of isolation brought about by specific association.

Before the sections for economic and medical entomology many valuable and important papers were read. Sir Daniel Morris read a paper by Mr. W. A. Ballou (Government Entomologist for the West Indies), on the more important insects of the West Indies and the methods adopted for their control. Prof. J. Jablonowski (Budapest), contributed two papers, one of which on the methods of fighting the locust, *Stauronotus maroccanus*, in Hungary was of unusual interest and importance from the Canadian point of view. Mr. A. G. L. Rogers (Board of Agriculture and Fisheries, England), contributed a paper on the necessary investigation with relation to insect pests preliminary to legislation. While the ideas set forth by the author were in the main sound, he made many assertions which were not borne out by facts. This was shown in the subsequent discussion, which proved so interesting that it was postponed until the following day, when a resolution was passed, and subsequently submitted to and passed unanimously by the Congress, supporting the proposed formation of an International Commission by the International Institute of Agriculture at Rome to deal with the problems connected with the spread of insect pests. Prof. F. V. Theobald gave an account of his investigations in the Aphid genus *Macrosiphum*, a most valuable piece of work both to the systematist and to the economic worker. In the medical section, Prof. S. A. Forbes (Illinois), read a paper on *Simulium* and Pellagra in Illinois. The results of his enquiry do not, so far, lend support to the theory that this disease is transmitted by *Simulium* flies. A paper on the methods of combating *Musca domestica* led to a discussion on the subject, but no new facts were brought forward.

The social side of the Congress was not the least enjoyable feature of the meeting nor the least important. In the garden of Wadham College a private café was installed, where luncheon, tea and light refreshments were served. Here delegates from all lands were able to meet. The members of the Congress were entertained most hospitably at Nunham by the Rt. Hon. L V. Harcourt, M. P., Secretary of State for the Colonies, and by the President and Fellows of St. John's College at Bagleywood, on August 7th. A banquet was held in the hall of Wadham College on August 8th, at which many but short felicitous speeches were made On August 15th the members made a visit to Tring Park, where they were entertained by the Hon. Walter Rothschild and shown over his celebrated zoological museum and entomological collections.

The next Congress will be held in Vienna, in 1915, under the presidency of Dr. Anton Handlirsch, who will undoubtedly prove a most genial host.

Mr. Henry H. Lyman, of Montreal, also represented the Entomological Society of Ontario. Dr. E. M. Walker was elected an additional member of the Permanent Committee of the Congress.

C. GORDON HEWITT.

Dr. R. Matheson has been appointed Provincial Entomologist for Nova Scotia The recent appointment of Mr. L. Caesar as Provincial Entomologist for Ontario and the appointment of Mr. W. H. Britton as Plant Pathologist and Entomologist for British Columbia are pleasing evidences of the fact that the Provincial Governments are realising the importance of applied entomology.

Dr. Matheson is a native of Nova Scotia and after studying at the

Provincial Agricultural College at Truro, N. S., he graduated at Cornell University, Ithaca, N. Y. Later he was appointed State Entomologist and Professor of Entomology at the Agricultural College for South Dakota. Two years ago he returned to Cornell University where he assisted on the staff of the entomological department, taking his doctor's degree last year. Dr. Matheson's training makes him well qualified for the position he now holds, and with the recent introduction of the San José scale, the presence of the Brown-tail moth and the occurrence of several other serious insect pests in Nova Scotia, he will find problems of importance and interest awaiting him.

C. G. H.

BOOK NOTICE

THE HUMBLE-BEE.—Its life-history and how to domesticate it, with descriptions of all the British species of *Bombus* and *Psithyrus*. By F. W. L. Sladen. 13–283 pp., 34 figs., 5 coloured plates (Macmillan). $2.50.

"Everybody knows the burly, good-natured bumble-bee," the author states in his opening sentence, and while this is true, the author has shown, in giving us the results up to date of what has been a life-study, how little even the entomologist knows of these people of the hedgerows, whose homes he no doubt laid waste when a boy.

Roughly speaking, the book can be divided into three parts. In the first part the life-histories and habits of *Bombus* and of the parasitic usurper *Psithyrus* are described in full and in a fascinating manner, a manner which makes the general reader feel the intense interest of the real naturalist. We see the queen in her solitude anxiously choosing the site of the future nest and brooding over her eggs and young; then the gradual development of the little community. Some of the author's descriptions are the best we have read in entomological literature; one of these is the description of the death of the queen: "In the case of *B. pratorum*, and probably of the other species whose colonies end their existence in the height of summer, the aged queen often spends the evening of her life very pleasantly with her little band of worn-out workers. They sit together on two or three cells on the top of the ruined edifice, and make no attempt to rear any more brood. The exhausting work of bearing done, the queen's body shrinks to its original size, and she becomes quite active and youthful-looking again. This well earned rest lasts for about a week, and death, when at last it comes, brings with it no discomfort. One night, a little cooler than usual, finding her food supply exhausted, the queen grows torpid, as she has done many a time before in the early part of her career ; but on this occasion, her life-work finished, there is no awakening."

The interesting and important discoveries which the author made as to the parasitic nature of the *Psithyrus* form a valuable portion of the first part of the work. We are told how the *Psithyrus* queen, protected by her coat of mail, impervious to the attacks of the *Bombus* queen, enters the home of the latter, and, after treacherously killing her, makes slaves of the workers, as she herself neither produces workers nor is provided with the pollen-collecting baskets in her hind legs.

The second part of the book describes the author's experiments in domesticating the *Bombi*, in which many types of domiciles were used. This section will prove of great value to future workers on the same lines. We should point out, in passing, that there is undoubtedly an important field of investigation in the encouragement of the *Bombi*. The economic significance of the presence of these insects where clovers are grown is now more generally appreciated, but we are not aware of any efforts having been made for the encouragement of these fertilizing agents. It is proposed to carry out in Canada investigations of the nature indicated.

In the third section of the book all the British species of *Bombus* and *Psithyrus* are described. Not only is a clear description of the queen, worker and male of each species given, and a brief description of their habits as observed by the author, but coloured illustrations render the identification of the species possible to anyone. The coloured figures, and there are five plates of them, are photographed direct from the specimens, and are undoubtedly the finest specimens of this kind of work which we have seen. The work is further enriched by the author's own drawings and photographs. The author has demonstrated, we believe for the first time, the importance of the structure of the male genitalia in separating the species and groups of species. The great variations in colour render such a method of separation of greater importance and significance. Illustrations are given of the male genitalia of the British species.

Although the author has confined himself to specific descriptions of the British forms, the book is none the less valuable to entomologists on this side of the Atlantic. From a monographic standpoint alone it is a work which should find a place on the bookshelf of every entomologist whose desires are not confined to the killing bottle and cabinet, but sit, like the fairies, astride the velvet-backed bumble bee and sail along the hedgerow, over field and forest and into every nook where insect creeps.

G. GORDON HEWITT.

Mailed November 8th, 1812.

The Canadian Entomologist.

Vol. XLIV. LONDON, DECEMBER, 1912. No. 12

NEW SPECIES OF THE FAMILY IPIDÆ (COLEOPTERA).

BY J. M. SWAINE,

Assistant Entomologist for Forest Insects, Division of Entomology, Ottawa.*

An undescribed species of the genus Trypophlœus was recently received from Weymouth, N.S. Mr. G. E. Sanders, who collected the material, has found the species in the vicinity of Weymouth only, in dying stems and twigs of Alnus.

This species is closely related to *T. alni* Lind., of Russia, which breeds in the bark of *Alnus incana* but is distinguished by its shorter wing covers, coarsely punctured but not granulate hinder half of the pronotum, and unimpressed elytral striæ.

Trypophloeus nitidus, n. sp.—Black (when matured); length, 2 mm.; width, ¾ mm.; clothed with short, inconspicuous, grey hairs of two lengths; pronotum small, from above subtriangular; elytra with rows of punctures, interspaces finely, confusedly punctured; the whole body shining.

The *head* is subglobular, punctured rather variably with coarse, very shallow punctures and faintly aciculate behind the eyes; the whole head often reticulate from very minute, dense, shallow punctures; the front more coarsely, closely, and rather roughly punctured. A median, longitudinal impression extends down the front and ends in a V-shaped impression at the base of the epistoma. The front is rather sparsely clothed with short, gray hairs. The epistoma, which bears a few long, yellowish hairs, is widely margined, shining and produced at the median line into a broad lobe, the upper surface of which is distinctly concave. The eyes are wide, coarsely granulated, slightly emarginate in front. The antennal fossa is small, rounded, and lies in the very short space between the eye and the base of the mandible. The antennal scape is strongly curved and clavate; the first segment of the funicle is large and pedunculate, the remaining four segments saucer-shaped and rapidly widened. The club is elongate, narrowed distally, and truncate, with three transverse sutures, of which the third is indistinct. The sutures are more densely hairy on the outer side, on the inner side the first suture is sometimes incomplete. The outer surface of the truncate tip bears a large, stout seta at each end.

*Contributions from the Division of Entomology, Ottawa.

The *pronotum* is subtriangular in outline; as wide as the elytra; with the base broadly rounded and finely margined; the hind angles broadly rounded; the sides sinuate, swollen behind, strongly convergent cephalad, and the apex not very narrowly rounded. The apical margin bears two larger recurved points on the median line, with two or three smaller ones on each side. The asperations of the front half of the pronotum are strong, wider and more strongly compressed towards the centre and concentrically arranged. The caudal half of the pronotum is coarsely, densely punctured, with a few very fine punctures intermixed, but is not granulate. The hairs of the pronotum are short, fine, and point towards the summit.

The *scutellum* is triangular, distinct, not depressed, closely and coarsely punctate, and surrounded by a narrow, transversely rugose area.

The *elytra* are slightly over twice the length of the pronotum, 16:7, with the sides parallel as far as the level of the top of the declivity, then gradually rounded and narrowed to the narrowly rounded tip. The elytra are punctured in rows, the striæ hardly impressed, with the interspaces confusedly punctured with very fine punctures and with a row of widely separated, slightly larger punctures which bear long, stout bristles. The punctures of the strial rows are round, small at the bottom, large at the surface, deeply impressed individually, so that the surface is somewhat transversely wrinkled. The pubescence is short on the disc, longer and more conspicuous on the sides and declivity. The interspaces bear each a row of longer, stout bristles, with minute, slender setae irregularly placed. The declivity is convex, with the striæ distinctly impressed.

The front coxæ are prominent and contiguous. The prosternum is short, with a sharp, slender, intercoxal process. The metasternum is longitudinally sulcate on the median line. The hind coxæ are very elongate and sharp-pointed distally. The foretibiæ are strongly widened distally, the inner margin strongly sinuate, the outer margin straight on the distal half and finely serrate.

Can. Div. Ent. Col., No. 2087; Weymouth, N.S.; Alnus incana.

Dryocoetes pubescens, n. sp.—This species is represented in our collection from Colorado. It is allied to *affaber* Mannh., but is distinctly more elongate, with the elytra more densely and coarsely punctured on the declivity.

The front of the female is densely clothed with long yellow hairs, shorter at the centre; with a smooth median space extending over the vertex. The front in the male is rather roughly granulate-punctate, thinly clothed with long hairs, with a distinct, smooth median space extending from a moderate pit in the centre of the front caudad to the vertex, very narrow at first and wider behind.

The *pronotum* is much as in *affaber* Mannh., widest behind and narrower towards the front; the sides usually curved, but sometimes nearly straight for a short distance. The pronotum of both sexes is more distinctly granulate than in the male of *affaber*; the whole upper surface is strongly granulate. The prothorax is margined behind.

The *elytra* are very closely, deeply and coarsely punctured in rows. The striæ are not impressed, and the punctures of the interspaces are as large and about as numerous as those of the striæ. The punctures of the first interspaces are confused behind. The declivity has the first two striæ impressed, as usual, but the suture is not raised, so that the declivity appears from above as quite distinctly flattened. The punctures of the declivity are coarse and very numerous. The densely and coarsely punctured declivity distinguishes this species from others described from North America.

The pronotum and elytra are rather densely clothed with long, erect, yellow hairs.

The type bears the labels; Col., Cornell U., lot 302, sub. 37. 189, type ♀.

Dryocoetes confusus, n. sp.—Length, 3½ mm. Dark red to nearly black, front densely hairy, and elytral interspaces confusedly punctured; pubescence long, erect, straight, and rather dense on the pronotum and on the elytra.

The front of the female is almost entirely covered with a very dense, circular brush of short, yellow hairs, with the marginal hairs longer and thicker; a fine median carina is visible. The frontal hairs are very much denser than in *eichoffi* Hopk. The front of the male is densely, coarsely, roughly punctate and sparsely clothed with long hairs, with a shallow impression below and a fine medium carina above. The eyes are emarginate and the antennal club obliquely truncate as usual.

The *pronotum* is slightly longer than wide, widest behind the middle, about as wide as the elytra, gradually narrowed cephalad of the posterior third, broadly rounded in front and very broadly rounded behind. The entire surface is roughened, but the asperations are finer and closer behind the middle. The smooth median line is nearly obsolete. The pronotum is sparsely pubescent, with long hair on the sides and in front, and the disc nearly glabrous.

The *elytra* have the sides nearly parallel, slightly wider behind, with the declivity somewhat flattened from the depth of the first two striæ. The striæ of the disc and sides are hardly impressed except the sutural striæ which are rather distinctly impressed, and convergent towards the base. The stria punctures are moderate to rather coarse, close, and at times somewhat irregular towards the declivity. The interspaces are wide, with the punctures nearly as large as those of the striæ, and irregular, except that the first three are uniseriately punctured towards the base. The interspaces are granulate on the declivity. The sutural interspaces are convex throughout. The elytra are rather densely pubescent, more noticeably so on the declivity.

Cornell University Collection; Colorado. The type bears the labels; Cornell U., no. 302, sub. 35, Col., 18, ♀.

Dryocoetes minutus, n. sp.—Length, 1¾–2¼mm.; width, ⅔–¾ mm.; a small slender species, nearly black, with legs and antennæ lighter.

The front is densely granulate-punctuate, clothed with long, rather dense, yellow hair, less dense than in the female of *eichoffi* Hopk. Probably one sex only is represented. There is a faint, traverse, linear impression across the middle of the front at the level of the upper part of the eyes, and a small, central, frontal tubercle. The eyes are slightly emarginate. The first segment of the antennal funicle is larger than usual, and the club truncate and strongly compressed.

The *pronotum* is distinctly longer than wide, with the hind margin very broadly rounded; the hind angles distinct; the sides nearly parallel from the base to beyond the middle, then regularly rounded in front; cephalic half punctured and rather coarsely asperate, caudal half coarsely punctured on the disc, punctures nearly as large as those of the elytral striæ, and with minute asperations on the sides; rather densely clothed with short yellow hair, longer in front and very short on the disc.

The *elytra* are slender, much longer than the pronotum, with istinctly impressed striæ of medium, close, rounded punctures; the sutural striæ more strongly impressed, parallel, more closely punctured; the interspaces wider than the striæ, flattened, rather sparsely, uniseriately punctured and pubescent, with the punctures smaller than those of the striæ on the disc, but on the sides as large as those of the striæ, and granulate towards the declivity. The declivity is convex, rapidly narrowed, compressed towards the apex, with the sutural striæ deeply impressed and the sutural interspaces granulate; the other striæ not impressed, and the strial and interstrial punctures equal, confused, and granulate. The pubescence is much denser on the declivity.

The fore tibiæ are strongly widened distally, with four very long teeth on the distal half of the other margin.

Type from Colorado, in the Cornell University collection; lot 302, sub. 94, 130.

Ips pilifrons, n. sp—Length, 4½–5mm.; width, 1⅞mm. Larger and stouter than *pini*, with the sutures more strongly angled, the elytral striæ impressed, the elytral interspaces punctured, the front with a dense mass of short hairs, and the declivital armature of the *pini* type. Color, dark reddish to nearly black.

The front of the female is convex, granulate above and in front of the eyes, punctured on the sides, with a swollen area in front presenting a flat, oblique, anterior surface, which is covered with a circular, dense mass of short, yellow or brownish hairs. The front of the male has the pubescent area of the female replaced by a convex densely granulated area, moderately pubescent, with long yellowish hairs. The antennal club has the first suture bisinuate, the second sharply angled in front, not prolonged, the third suture angled but often indistinct, and the sutures strongly recurved at the sides.

The *pronotum* is shorter than the elytra, 2: 2½; longer than wide, 2: 1⅔; broadly rounded behind; slightly rounded on the sides, and gradually narrowed cephalad or subparallel for over three-fourths the length, then rapidly narrowed and rounded in front; with the disc rather coarsely roughened in front; coarsely and deeply punctured behind, but not very densely except on the sides, and clothed with light slender hairs on the sides and in front.

The scutellum is very small and distinctly channelled. The *elytra* are punctate-striate, with the striæ distinctly impressed and wider on the disc; the punctures of the discal striæ large, deep, subquadrate, and usually closely placed; the punctures of the lateral striæ usually distinctly smaller than those of the disc, and near the lateral margin sometimes easily confused with those of the interspaces, which are there small, numerous, and irregular; the sutural striæ deep, variably widened towards the declivity; the interspaces convex, with setigerous punctures, smaller than those of the striæ, usually extending from the base to the declivity; the punctures of the first two interspaces rather closely placed; those of the third, fourth and fifth more distant, except near the declivity; the first two interspaces with granules which become much larger near the declivity, with smaller granules intermixed; the remaining interspaces from the sixth outward confusedly punctured and granulate at the declivital margin. The *declivity* is deeply excavated, coarsely and confusedly punctured, not pubescent, with the sutural interspaces raised and the elytra dehiscent at the tip. The declivital teeth are coarser than in *pini*, and the acute apical margin is usually more strongly produced. The elytra are clothed with light, soft hairs, rather dense along the sides, around the margin of the declivity, along the base and along the suture, but sparse on the central areas of the elytra.

The type is from the Cornell University Collection. Colorado; ♀.

CANADIAN BEES IN THE BRITISH MUSEUM

BY T. D. A. COCKERELL, BOULDER, COLORADO.

The bees in the British Museum are now being rearranged by Mr. G. Meade-Waldo, who has sent me for determination a number of species, some of them Canadian. In recording them, I give the accession numbers, which show when they were received at the museum. Thus, 99–303 means accession 303 of the year 1899. It will be seen that the three species of *Osmia* here introduced as new were received at the museum in 1844, more than 20 years before the birth of their describer. Other species were received at the museum long before they were described in this country.

Megachile femorata Smith.—♂, Canada, pres. by Mrs. Farren White, 99–303. ♂, Canada, 59–130. Smith's *femorata* is usually regarded as a synonym of *M. latimanus* Say, but Titus has treated it as a distinct species. If it is to be separated, the form with hardly any dark color on the anterior tibiæ, and the coxal spires stout, must be referred to *femorata*, while *latimanus* male has approximately the basal half of anterior tibiæ on outer side black and the coxal spines more slender. According to this separation, the usual Rocky Mountain insect is *latimanus*, but I have a male *femorata* from as far south as Las Vegas, New Mexico (at flowers of *Asclepias verticillata*; W. Porter). It seems probable that the two insects do not represent distinct species.

Megachile latimanus Say.—♂, British Columbia (Miss Ricardo) 1903–134. ♂, Calgary, Canada (Miss Ricardo), 1902–55. These females differ from the ordinary form by the distinctly longer black hair on the dorsal surface of the abdomen. They look a little like *M. vidua*, but are readily separated by the densely punctured mesothorax and the light hair of last dorsal abdominal segment.

Megachile wootoni Ckll.—♀, Calgary (Miss Ricardo), 1902–55. ♂, Calgary, with same data. ♂, Arctic America, 55–42.

Megachile melanophaea Smith.—♂, Hudson's Bay, 44–17.

Megachile relativa Cresson.—♂, Chulukwayuk trail, British Columbia, Aug. 1859.

Megachile vernonensis, n. sp.—♀, Length, about 11 mm.; black, with long dull white hair; antennæ not enlarged at apex; eyes green; anterior coxæ with short but well-formed spines, largely hidden by hair; anterior femora broad, smooth, concave and ferruginous beneath, above with a rather obscure red patch; hair on inner side of tarsi pale orange; sides of vertex with black hair, but none on thorax above; apical carina of sixth abdominal segment with a large rounded (semicircular) emargination, the margin on each side of it jagged with short irregular teeth;

morphological apex of sixth segment with four short dentiform projections, the middle ones not quite so near to one another as to the lateral, the margin between the middle ones convex. Almost exactly like the male of *M. cleomis* Ckll., but differing in the apex of sixth segment (*cleomis* has the middle teeth considerably nearer to one another than to the lateral, and the margin between them concave), and in having the densely granular concave upper surface of sixth segment so feebly white-tomentose that the tomentum is only visible in lateral view (*cleomis* has this part densely tomentose); the hair of the face has a creamy tint, instead of being clear white as in *cleomis*. The lateral ocellus is a trifle nearer to edge of vertex than to nearest eye.

♀.—Length, 11 mm.; mandibles 4-dentate, reddish apically; eyes light green, narrow; clypeus shining, closely punctured, its lower margin straight, a transverse depression above the margin; vertex with brown hair; abdomen with white hair-bands; sixth segment sloping (not concave) in profile, with coarse black hair, its apical third with very fine white tomentum; ventral scopa white, entirely black on last two segments. Very like a small *M. cleomis*, but distinguished by the wholly black hair on last two ventral segments, the narrower eyes, and the last dorsal abdominal segment as described. Also near to *M. generosa* Cress., but considerably smaller, and with the same distinctive characters as those separating it from *cleomis*. *M. anograe* Ckll., another similar species, is at once separated by its brilliantly shining sixth abdominal segment, with coarse black hair to the apex.

In Friese's table (Das Tierreich) the female runs nearest to *M. addenda*, but Robertson describes *addenda* as having the margin of clypeus denticulate, while only the last ventral segment of abdomen has black hair. The male runs best to *M. texana*, i.e., Cresson's male *texana* which appears to be *cleomis*.

Hab.—Vernon, British Columbia (Miss Ricardo). The type (male) taken July 7, 1902; the female, Aug. 18, 1902. This is possibly to be considered a subspecies of *M. generosa*, but with the evidence available it seems a distinct species.

Megachile montivaga Cresson. — ♂, N. Ontario, Canada (H. Edwards), 89–113.

Megachile vidua Smith.—♂, British Columbia, 60–112. The specimen is unusually large.

Dianthidium pudens (Cresson).—♀, British Columbia, 60–112. Described from Nevada.

Osmia novaescotiae, n. sp.—♀, Length, about 9 mm.; head rather large, dark steel-blue, densely, punctured; mesothorax and scutellum more tinged with greenish but pleuræ and metathorax dark blue; abdomen short, broad-oval, shining steel-blue; hair of head greyish-white, pale fuscous on middle of face; hair of thorax white, with a creamy tint above, and no dark hair intermixed; tegulæ piceous, with a greenish spot in front; wings dusky hyaline, reddish, distance from base of first s.m. to insertion of first r.n. as great as length of first t.c.; b.n. going just basad of t.m.; legs reddish black, not at all metallic, with pale pubescence, reddish on inner side of tarsi; abdomen closely but rather shallowly punctured, the punctures going nearly to the margins of the segments; sublateral region with quite long black hair; ventral scopa black. The clypeal margin is entire, and the mandibles are 3-dentate; the area of metathorax is densely granular basally, but more shining apically.

Hab.—Nova Scotia (Ent. Club), 44–12. I have been much perplexed to decide whether this could be the female belonging to the male from Nova Scotia described as *O. simillima* by Smith. This may indeed be the case, but the type of *simillima* must be considered to be the female, which may not be conspecific with the male. Smith says that the female *simillima* is so like the European *O. caerulescens* that it is difficult to distinguish; but *novaescotiae* is easily separated from *caerulescens* by the broader, less deeply punctured abdomen, without white marginal fringes or bands. In our fauna it is *O. purpurea* Cresson, which closely resembles *caerulescens.** In my brief notes on Smith's types, I observed that according to Robertson's tables the female type of *simillima* was an *Osmias.* str., while the male was a *Monilosmia.* Dr. Graenicher has, however, obtained at Milwaukee, Wisconsin, what he regards as true *simillima*, and has both sexes from the nest. The male of this species is a *Monilosmia*, but the female h as a black ventral scopa and clypeus with entire margin, quite contrary to Robertson's definition of *Monilosmia.* The Milwaukee females are larger than *novaescotiae*, with a dark greenish abdomen, and the hind margins of the segments more broadly smooth. They are very unlike *O. caerulescens.* They have the hair on inner side of middle tibiæ black; in *novaescotiae* it is pale, with a reddish tint. The b.n. goes more broad of the t.m. than in *novaescotiae.* There is no doubt, I think, that the Milwaukee "*simillima*" is distinct from *novaescotiae*, but I find that except for the smaller amount of dark hair on the head (a variable character) it is scarcely or not to be separated from the western *O. densa* Cresson. This probably explains why we have never been able to find a male for *densa*;

*Can it be that *O. purpurea* is *cærulescens?* From the British Museum I have a female marked North America, 40, 4-2, 484, and it is quite impossible to distinguish it from European *cærulescens*, while, at the same time, it agrees with Cresson's description of *purpurea.* It has the shiny metathoracic area of *cærulescens*, which Smith expressly says is wanting in *simillima.*

it is doubtless of the *Monilosmia* type. *Osmia chlorops* Ckll. and Titus, which occurs in the same localities as *densa* (e.g., at Florissant), and like it visits *Pentstemon*, is doubtless the male of *densa*. It is hardly different from the Milwaukee "*simillima*."

O. novaescotiae, compared with a number of species which have white hair on the pleura in the female, differs (1) from *albolateralis* by the deep blue (not green) abdomen, total absence of black hair on vertex, etc.; (2) from *coloradella* by the non-metallic legs, abdomen seen from above showing black hair projecting at sides, etc.; (3) from *densa* by absence of black hair on front, blue abdomen, etc.; (4) from *dubia* by the shining middle of mesothorax, largely pale hair on legs, and blue abdomen; (5) from *faceta* by the simple clypeus, etc.; (6) from *felti* by the much smaller size, dorsal abdominal segments sculptured almost to apex; (7) from *melanotricha* by the blue, more densely sculptured abdomen, and absence of black hair on vertex; (8) from *pentstemonis* by absence of coarse black hair on head and vertex; (9) from *phaceliae* by absence of black hair on vertex and scutellum, and blue abdomen.

Osmia subarctica n. sp.— ♀, Length, nearly 7½ mm.; dark steel-blue, the femora and tibiæ metallic, the tarsi piceous; head rather large, densely punctured, clypeus and sides of face a fine dark blue; clypeal margin entire; mandibles tridentate; flagellum obscure ferruginous beneath; hair of clypeus black, but of sides of face white; hair of vertex pale, but front with a slight intermixture of fuscous hairs; hair of thorax entirely rather dull white; mesothorax and scutellum closely punctured, but shining; area of metathorax dull and granular; tegulæ dark rufopiceous, blue in front; wings dusky hyaline; b.n. exactly meeting t.c.; hair of legs largely black, brown-black on inner side of tarsi; abdomen moderately shining, with shallow sculpture, dorsally with extremely short and scanty hair, white at sides; no hair-bands; ventral scopa black, with coppery tints.

Hab.—Hudson's Bay, 44-17. Closely related to *O. pentstemonis* Ckll., but separated by the absence of dark hairs on thorax above, and the less shining abdomen. It is perhaps not more than a subspecies of *pentstemonis*, but I have a series of the latter, all looking different from *subarctica*. *O. subarctica*, compared with other species having white hair on the pleura of the female, differs thus: (1) From *albolateralis* by the metallic legs, much smaller size, etc ; (2) from *coloradella* by dark hair of middle of face, etc.; (3) from *densa* by the metallic legs and much smaller size ; (4) from *dubia* by the metallic legs, etc.; (5) from *faceta* by the much smaller size, simple clypeus, etc.; (6) from *felti* by the much smaller size and metallic legs; (7) from *melanotricha* by the metallic legs, absence of long black hair in sublateral region of abdomen, and narrower head ;

(8) from *phaceliæ* by the metallic legs, absence of black hair on scutellum, and narrower head.

Osmia atriventris Cresson.—♂. Ent. Club, 44-12. The accession number is the same as that of *O. novæscotiæ*, but there is no locality label.

Osmia tersula, n. sp.—♂, Length about 8½ mm.; head and thorax densely granular-punctate, very dark green, with abundant long, wholly pale hair, slightly creamy-tinted on thorax above and head; mandibles stout, strongly curved, bidentate, the teeth subequal; face narrowing below; antennæ entirely dark, moderately long, not moniliform; area of metathorax blue-black, dull, granular; tegulæ piceous, punctured; wings dusky hyaline, b. n. meeting t.m.; second s.m. unusually long and low; legs brown-black, the femora and tibiæ not metallic, their hair long and pale, yellowish on inner side of tarsi; middle tarsi simple; hind basitarsi toothed; abdomen shining, very dark blue-green, the hind margins of the segments obscurely reddish; basal segment with long, pale hair; apical segments with conspicuous, long hair, wholly pale; middle segments almost hairless; no hair bands; venter with pale hair; margin of sixth segment with a shallow notch; seventh emarginate, hardly bidentate; first ventral entire; third ventral with an orange tuft or pencil of hair on each side of emargination.

Hab.—Hudson's Bay, 44-17. Distinguished especially by its dark color and toothed hind basitarsus. In Schmiedeknecht's table of European species it runs close to *O. panzeri*, but differs entirely in the antennæ and the pubescence. It may also be compared with *O. angustula*, which is smaller, with quite different pubescence. In our fauna there is closer resemblance to several species of the Rocky Mountains. The following table separates it from some of these:

Hair of vertex partly dark.............................. 1
Hair of vertex wholly pale.............................. 2
1. Scutellum with a median smooth line; teeth of mandibles subequal............................ *pulsatillæ* Ckll.
   Scutellum without a smooth line; apical tooth of mandibles very long .... *vallicola* Ckll.
2. Abdomen shining blue; seventh segment very strongly bidentate............................ *wheeleri* Ckll.
   Abdomen dark greenish; seventh segment emarginate............................ *tersula* Ckll.

From *O. tersula*, *O. amala* Ckll. differs at once by the bright blue abdomen and much larger antennæ; *O. enena* Ckll. by the blue abdomen, strongly bidentate seventh abdominal segment, and broader face; *O. seneciophila*, by the abundant black hair on apical part of abdomen.

NOTES ON SOME CANADIAN BEES

BY J. C. CRAWFORD, WASHINGTON, D. C.

The specimens here recorded are part of a collection made at Medicine Hat, Alberta, by Mr. J. R. Malloch, between September 1st and October 15th, 1911. Other species of bees, which have not been studied, were also collected.

Agapostemon viridulus Fabr.—2♂.
Bombus huntii Greene.—1♂, 1♀.
Calliopsis coloradensis Cress.—6♂, 1♀.
Diadasia diminuta Cress.—1♂.
Dialictus anomalus Robt.—3♂, 1♀.
Halictoides marginatus Cress.—2♂, 2♀.
Halictus aberrans Cwfd.—1♂, 2♀.
Halictus lerouxii Lep.—8♂, 4♀.
Halictus provancheri D.T.—3♂, 1♀.
Halictus pruinosiformis Cwfd.—5♂, 4♀.
Halictus pruinosus Robt.—3♂.
Neopasites illinoiensis Robt.—1♂, 2♀.
Panurginus innuptus Ckll.—1♂, 1♀.
Perdita cockerelli Cwfd.—2♂, 1♀.

Perdita citrinella Grænicher—1♂, 2♀. Both the females have the hind tibiæ darkened. In one female the first recurrent vein is interstitial; in the other it is received by the first submarginal cell as in a paratype of the species.

Sphecodes minor Robt.—2♀.

Phileremulus mallochi, new species.—Male: Length, 3 mm; head and thorax black, closely punctured (but the sculpture concealed by the pubescence), closely covered with white appressed pubescence, that on the dorsum of the thorax slightly tinged with yellowish; abdomen red, disks of segments 4–6 more or less suffused with dusky; apical margins of segments 1–6 with bands of white appressed pubescence, disk of first segment with similar pubescence; labrum and mandibles, except the reddish tips, testaceous; scape and pedicel black, rest of the antennæ reddish; axillæ produced, tooth-like; tegulæ dark, the outer edge at middle obscurely reddish; scutellum bilobate at apex; metanotum medially strongly produced into a bilobate process; propodeum with a roughened basal triangular area without pubescence and with a strong median longitudinal carina; wings hyaline, marginal cell squarely truncate at apex; submarginal cell appendiculate at apex, receiving the recurrent vein slightly apicad of the middle; femora dark, tibiae and tarsi yellowish more or less suffused with dusky and the apical joints of the tarsi dark; abdomen rugoso-punctate.

Habitat.—Medicine Hat, Alberta, Canada. One specimen collected by Mr. J. R. Malloch. Type Cat. No. 15212, U.S.N.M.

This species in general appearance, very closely resembles *Neolarra pruinosa*, but in addition to the generic differences, differs also in the dark tegulæ, the carina on the propodeum more elevated, the appressed pubescence not covering the abdomen, etc. *P. vigilans* and *P. nanus* are both smaller, with light coloured tegulæ and with the appressed pubescence covering abdomen; *P. vigilans* also has the carina on the propodeum indicated at base only, the process on metanotum only indicated, etc.

Perdita canadensis, new species.—Female. Length about 9 mm. Head blue-green, thoracic notum green, pleuræ blue-green; clypeus and labrum black, the former with purplish tinges, smooth, with a few fine punctures and produced anteriad of a line connecting the lower ends of eyes fully one-third the length of the distance between eyes at lower ends; face without markings; antennæ dark, scape with a narrow yellow line, apical joints of flagellum reddish beneath; pubescence of head and thorax long, erect, strongly tinged with ochraceous; collar with two small yellow spots; tubercles dark; wings milky white; veins hyaline; the stigma and costal vein light brown; first recurrent vein interstitial or almost so; legs dark, anterior knees and a short narrow stripe on anterior tibiæ yellow; pubescence on outer side of legs greyish, on inner side ochraceous, on tarsi reddish; first abdominal segment with a small yellow spot on each side; segments 2-5 with yellow bands, the ends of which are turned caudad on segments 3-5; band on second segment dilated laterally so that the posterior margin is diagonal; bands on segments 2-3 notched medially on posterior margin; pygidium broad at apex with the apical martin emarginate.

Habitat.—Medicine Hat, Alberta, Canada. Two females collected by Mr. J. R. Malloch. Type Cat. No. 15213, U.S.N.M.

This species belongs to the group with *albipennis* Cresson, *lacteipennis* Swenk and Cockerell, and *pallidipennis* Grænicher; it differs from all of them in the dark face and other markings; *albipennis* and *pallidipennis* have the first recurrent vein received by the second submarginal cell and the pygidium rounded apically; *albipennis* has the clypeus produced hardly half as much as *canadensis; pallidipennis* has it produced about as much; *lacteipennis* has the clypeus produced about as in *albipennis* the first recurrent vein received by the second submarginal cell and the pygidium emarginate at apex. Of the value of this last character I am at present doubtful.

NEW SPECIES OF *FURCOMYIA (TIPULIDÆ)*.

BY CHAS. P. ALEXANDER, ITHACA, N. Y.

(Continued from page 341.)

Furcomyia libertoides, sp. n.

Closest allied to *liberta* O. S. of the Eastern U. S., but differs as follows: The præscutal stripes are not clearly defined, the middle of the dorsum being suffused with bright brown; tergum of abdomen brownish, not clear gray; wings with the stigma conspicuous, rectangular, not a narrow seam to cross-vein *r*. Hypopygium from above—see fig. 5.). The pleural piece triangular, the ventral apical appendage fleshy, its inner margin produced into a point which is directed cephalad; two short spines about equal in size, projecting caudad on the middle of this appendage; dorsal arm, or apical appendage, rather short, gently curved. Venation, fig. 1.

Length about 6.5–7.5 mm.; wing, 8.7–8.8 mm.

Holotype.—♂. Marin Co., Cal.; March 23, 1897.

Paratypes.—♂s 5. With the type.

The material is part of the Wheeler collection; one paratype in author's collection.

Furcomyia simillima, sp. n.

Yellowish thorax, with a dark brown median stripe; halteres very long.

♂.—Length about 5.5 mm.; wing, average, 6.8 mm.

♀.—Length about 5.8 mm.; wing, average, 7.4 mm.

Head: rostrum and palpi dark brown. Antennæ, first segment dark brown, thickly gray pruinose, remaining segments dark brownish black. Front, vertex and occiput brown, thickly gray pruinose, producing a gray effect.

Thorax: cervical sclerites dark, almost black; pronotum light dull yellow, dark brown along the dorsal median line. Mesonotum bright brownish yellow, becoming grayish on the sides; a broad dark brown median stripe continued from the pronotum, ending just before the suture; lateral stripes indistinct, grayish brown, beginning behind the pseudo-suture, continued across the suture and suffusing the lobes of the scutum; median line of the scutum and the scutellum paler yellowish white; postnotum brown; metanotum light yellow. Pleuræ light yellow, becoming grayish toward the metapleuræ. Halteres very long, extreme base yellowish, rest dark brown. Legs: coxæ and trochanters yellowish; femora yellow becoming somewhat darker apically; tibiæ and tarsi yellowish brown. Wings subhyaline; no stigmal spot; veins yellowish brown. Venation: (See fig. n.) Sc ending before origin of Rs, Sc_2 far before tip so that Sc_1 is long, somewhat shorter than Rs; basal deflection of Cu_1 before the fork of M.

Abdomen: tergum yellowish brown, apices of the sclerites narrowly paler; sternum light yellow.

Holotype.—♂. Totonicipan, Guatemala. (Dr. G. Eisen.)

Allotype.—♀. Antigua, Guatemala. (Dr. G. Eisen.)

Paratypes.—11 ♂s, 8 ♀s. Quichi (July, 1902); Antigua and Totonicipan (July, 1902); Guatemala.

Types in U. S. Nat. Mus. coll. (No. 15,134.) Paratypes in author's collection.

Resembles *particeps* Doane (Ent. News, Jan., '08, p. 7), from north-western U. S, but head is more gray, abdomen much lighter coloured and the thoracic stripes different.

Furcomyia andicola, sp. n.

Head gray; thorax brownish yellow; wings with scanty brown marks.

♀.—Length, 8.1 mm.; wing, 11.2 mm.

♀.—Head: rostrum and palpi dark brown. Antennæ, basal segments brown, flagellar segments very dark brown. Front, vertex and occiput gray.

Thorax: pronotum dull yellow, the dorsum indistinctly suffused with brown. Mesonotum dull brownish yellow, a broad brown median stripe and shorter, less distinct lateral ones; scutum reddish brown, suffused with darker brown; scutellum and postnotum brown, with a grayish brown bloom. Pleuræ dark brown. Halteres, stem greenish at base, darkening to brown at the tip. Legs: coxæ greenish, femora brownish yellow, the tip clearer yellow; tibiæ light brown, darkened at tip; tarsi brown. Wings subhyaline, veins brown, C, Sc and R, more yellowish; a large, rectangular brown stigma, which is continued back over the fork of Rs as a rounded spot; narrow brown seams on the cord and outer end of cell 1st M_2. Venation (see fig. h.): Sc ending just beyond origin of Rs; Sc_2 removed from the tip so that Sc_1 is rather more than half as long as Rs; Rs about one and one-half the length of the deflection of R_{4+5}; basal deflection of Cu_1 before the fork of M.

Abdomen: tergum and sternum brown, the apices of the sclerites yellowish. It is probable that, in life, the insect is quite greenish.

Holotype.—♀. San Antonio, Bolivia. (Received from Staudinger-Bang-Haas.)

Type in author's collection.

Agrees most closely with *phatta* Phil., which has the thorax gray and the wing-pattern very different, three black spots in cells 1st R_1 and 2nd R_1

Furcomyia insignifica, sp. n.

Head brownish gray; thorax reddish brown, darker medially.

♀.—Length, 8.5 mm.; wing, 9.6 mm.; fore leg, femur, 5 9 mm.; tibia, 7 3 mm.

♀.—Head: rostrum, palpi and antennæ dark brown. Front, vertex and occiput brownish gray.

Thorax: pronotum yellowish brown. Mesonotum, præscutum reddish brown, darkest brown medially on præscutum; paler, yellowish, on the humeral angles; pleuræ brownish yellow, brightening to yellow on the sternum. Halteres long, slender, brown, brighter at the base. Legs long, slender; coxæ and trochanters yellowish; femora yellowish brown; tibiæ and tarsi brown. Wings hyaline, veins light brown; stigma barely indicated, rectangular, very pale. Venation (see fig. i.): Sc short, Sc_2 quite removed from the tip of Sc_1; Rs short, not much longer than the deflection of R_{4+5}; basal deflection of Cu_1 far before the fork of M.

Abdomen: tergum dark brown on the basal segments, lighter brown on the apical segments; sternum light brown.

Holotype.—♀. Iquico, Peru. (Received from Staudinger-Bang-Haas.)

Type in author's collection.

This species cannot be referred to *pallida* Macq, which has a triangular cell 1st M_2 which bears a spur, this character of an appendiculate cell also separating *elquiiensis* Blanch. The other species with unspotted wings, *flavida* Phil. and *chlorotica* Phil., are quite different insects, specimens of which are before me, and will be redescribed in a later paper.

Furcomyia argentina, sp. n.

Head gray; thorax gray, darker on dorso-median line.

♀.—Length, 8 mm.; wing, 8.9 mm.; fore leg, femur, 6 mm.; tibia, 7 mm.; hind leg, femur, 7.1 mm.; tibia, 7.7 mm.

♀.—Head: rostrum and palpi dark brown. Antennæ dark brown, grayish pollinose; segments submoniliform. Front, vertex and occiput gray.

Thorax: pronotum brownish gray, the gray being pollen. Mesonotum, præscutum gray, with an indistinct, broad, brown, median stripe; scutum, scutellum and postnotum pale, with a gray pollen. Pleuræ pale gray pollinose. Halteres short, stem dull yellow, knob brown. Legs: coxæ and trochanters dull yellow; femora similar, rather darkened toward the tip; tibiæ and tarsi light brown. Wings hyaline, veins dark brown, conspicuous; stigma indistinct, brownish. Venation (see fig. k.): Sc ends opposite the origin of Rs; Sc_2 far retracted so that Sc_1 is almost as long as the stigma; Rs only a little longer than the deflection of R_{4+5}; basal deflection of Cu_1 at the fork of M.

Abdomen: tergum dull brown; sternum yellowish brown.

Holotype.—♀. Neuquen, Argentina, 1907. (Dr. Adolf Lenol.)

Type in author's collection.

Differs from the hitherto described species by the characters given in under *insignifica*. From *insignifica* it differs in its wing venation, colour of veins, and body tone.

? *Furcomyia fumosa*, sp. n.

Wings infumed, with darker clouds.

♀.—Length about 5.5 mm; wing, 6.3 mm.

♀.—Head: rostrum and palpi dark brown. Antennæ dark brownish black. Front, vertex and occiput brownish, with a grayish pubescence.

Thorax: pronotum dark brown. Mesonotum light brown, the postnotum darker. Pleuræ dark brown. Halteres dark brown, base of the stem light coloured. Legs: coxæ and trochanters dark brown, rest of legs broken. Wings infumed with brown, darker brown clouds arranged as follows: At origin of Rs, at tip of Sc, at tip of R_1, along cord; most of veins and tip of wing clouded with dark brown. Venation (see fig. t.): Sc long, Sc_1 ending slightly before the fork of Rs, Sc_2 at its tip; R_1 bends down near its end and touches R_{2+3}, obliterating the cross-vein *r*; basal deflection of Cu_1 beyond the fork of M.

Holotype.—♀. Amatuk, British Guiana; July 14, 1911. (F. E. Lutz.)

Type in American Museum of Natural History.

This insect is closely allied to *Limnobia insularis* Will. (Dipt. St. Vincent, Trans. Ent. Soc. Lond, 1896, p. 287, pl. 10, fig. 58), but the wing has quite a different pattern, cell 1st M_2 less elongated, basal deflection of Cu_1 farther distad, etc The two species are certainly as close to *Furcomyia* as they are to *Limnobia*, but seem to represent a peculiar group which needs further study with more material.

MR. EDWARD P. VAN DUZEE, of Buffalo, leaves early in December for a four months' vacation in California. His temporary address will be San Diego, Calif.

SYNONYMICAL NOTES ON ŒDIONYCHIS

BY F. C. BOWDITCH, BROOKLINE, MASS.

In the rearrangement of my Oedionychis material I note the following:

In the Biologia, p. 418, speaking of *oculata* Fabr., Mr. Jacoby mentioned a figure of a species taken by himself. This figure is now before me, and agrees exactly with a specimen I have from Cayenne (typical locality). I think additional material will prove that the Central American form is distinct.

The name *illigeri* (Jac.) Proc. Zool. Soc.,1905, p. 441, for a Trinidad species was previously used in the Biologia, p. 421, for a Panama insect, so I would suggest for the later form the name *trinidadensis*.

The name *inconspicua* (Jac.) l.c., p. 424, for an Amazon form, was previously used in the Biologia, p. 417, for a Mexican species, so I would suggest for the later form the name *amazona*.

The name *colombiana* (Jac.) l.c., p. 445, was previously used in the same paper, p. 427, evidently some uncorrected error, for the species described on p. 445. I suggest the name *confusa*.

The name *rustica* (Jac.) l.c., p. 433, for an Argentine form, was peviously used by Von Harold, Deut, Ent. Zeit. XXI, p. 434, for a species from Bahia. For the Jacoby species I suggest the name *similis*.

The name *intersignata* (Jac.) l.c., p. 433, for a species from Espirito Santo, Brazil, was previously used (P. Z. S., 1894, p. 617), for a form from Surinam. I suggest the name *santoensis* for the Brazilian form.

Asphaera femorata (Jac.) seems to me the same as *chontalensis* (Jac.).

In explanation I would add that the late Mr. Jacoby in his working current catalogue (since 1885), for some reason or other, had not entered such of the "Biologia species" as were published after the date of Duvivier's list, so that all the names he used in the Biologia were not before him when he wrote in 1905.

NOTES ON *SYNTOMASPIS DRUPARUM* BOH. AND *ICHNEUMON NIGRICORNIS* BERGER

BY C. R. CROSBY, ITHACA, N.Y.

In Bulletin 265 of the Cornell Agricultural Experiment Station, April,1909, I gave an account of the habits of the Apple-seed Chalcis, (*Syntomaspis druparum* Boh.), and a resumé of the literature known to me at that time. Since then three more important papers have come to my notice.

In 1888 Dr. D. H. R. von Schlechtendal (Zeits, f. Naturwiss., Halle, ser. 4, VII., (LXI.), p. 416) records having reared this insect from the seeds of Cratægus. He states that the insect usually spends two or three winters in the larval state, only rarely emerging the first spring. He observed oviposition and found that the egg is deposited in the kernel. The ovipositor is inserted through the micropyle, the seed coat being very hard and thick.

In my former article I stated that the first account of this insect was given by Guérin-Méneville in 1865. This is an error, for over sixty years before Francais Berger of Geneva, Switzerland, published (Bull. Sci. Soc. Philomatique Paris, An. XII, 1803, p. 141 —wrong pagination for 241) a brief account of its habits and gave excellent figures of the larva, pupa and adult. This article has been overlooked so long because the insect was identified as *Ichneumon nigricornis* Fab. It is catalogued by Dalla Torre as *Ichneumon nigricornis* Berger although Berger stated that Jurine believed it should go in the genus Chalcis. The *Torymus nigricornis* of Boheman (Svensk. Vet.—Akad. Handl. p. 355, 1833) to which *Ichneumon nigricornis* Fabr. has been referred by Dalla Torre is an entirely different insect.

Soon after the publication of Bulletin 265 I obtained a copy of a paper entitled, "Commentatio de Torymidis, quarum larvæ in seminibus pomacearum vitam agunt," by W. N. Rodzianko, 1908, in which he gives an excellent review of the literature relating to *Syntomaspis druparum* and gives an extensive account of careful rearing experiments.

THE MIGRATION OF *ANOSIA PLEXIPPUS* FAB.

BY F. M. WEBSTER, WASHINGTON, D.C.

Regarding a phenomenon that has attracted so much attention, as has the migrations of the milkweed butterfly, among scientific men both at home and abroad, more especially of entomologists, we seem to possess a surprisingly limited amount of definite information. These migrations have been frequently reported in the newspapers and they are often observed by entomologists, as they appear to take the form of scattered bands, but where the members of these bands originate no one seems to know. Not all of the butterflies in a locality join the migration, as, after the bands have appeared from out of the north and passed onward toward the south, there are many others left behind. At least, this is true in the United States, and the writer has observed three of these migratory bands in the last twenty years.

September 21, 1892, in the clear, calm afternoon, there were swarms of these butterflies flying about in the city of Cleveland,

Ohio, on the south shore of Lake Erie. Whether the members of this band were migrants from the shores of Hudson Bay and Lake Athabasca, far away to the northwest, or whether they had gathered there from the east or west it was of course impossible to say.

The next band to be observed was at Urbana, Illinois, September 12, 1902, also in the afternoon, but at a temperature of 55° Fahr., with a brisk northwest wind and clear sky. Either this or another band of butterflies of this species was reported at Milledgeville, Illinois, about 160 miles to the northwest of Urbana, three days prior, while evidently still another was reported at Hoopstown, Illinois, some 35 to 40 miles north-east, a few days later. Whether or not these all belonged to the same band of migrants, from whence they came, or how the members came to be associated together, is still an unsolved problem. At Urbana, the company moved away on the morning of the 13th, but the usual number were observed wandering about, in a perfectly natural way, during the remainder of the month.

The third migration, observed by the writer, took place on September 12, about 3 p.m., on the Mall in Washington, D.C. The weather was cold, with light n.w. wind, but the sky was unclouded. This last, however, was not further investigated.

The daily press of Chicago, Illinois, September 13, one day prior to the occurrence in Washington, called attention to swarms of this butterfly observed congregating in the parks and gardens of the city and starting southward on their journey.

While it is true that this insect is of no economic importance, and of far too common occurrence to interest the collector, yet it seems to me that studies of the migrations of this species are well worth while, and the results would, beyond a doubt, prove of material aid in studying a similar habit in much more important species. The migration of insects is of itself an interesting problem, and a little care in observing and recording the appearance of these migrations and under what conditions these took place, would surely repay the many entomologists, amateurs and professionals scattered over Canada and the United States.

ON THE STATUS OF SOME SPECIES OF THE GENUS PANURGINUS.

BY J. C. CRAWFORD, WASHINGTON, D. C.

In a paper on the bees of Nebraska,* Messrs Swenk and Cockerell say that a comparison of cotypes of *Panurginus nebrascensis* with specimens of *P. ornatipes* shows that the two are synonyms and that *P. boylei* is a subspecies. The types of all of the involved

*Ent. News, XVIII., 183, 1'07.

species being in the collections of the U.S. National Museum has led to a re-examination of them and the characters given show them to be abundantly distinct. In view of these characters, what Messrs. Swenk and Cockerell had under the name *ornatipes* is somewhat of a mystery.

Panurginus ornatipes Cresson.—Male type: Process of labrum emarginate; punctures covering clypeus; punctures of mesoscutum small, sparse, at median anterior margin the punctures more sparse than at sides; a yellow stripe exteriorly on middle tibiæ (hind tibæ missing, but in a specimen from Paris, Texas, which is certainly conspecific with the type, the hind tibiæ have a similar stripe); wings yellowish and slightly dusky.

Panurginus nebrascensis Crawford.—Male type: Process of labrum rounded apically, punctures covering clypeus; punctures of mesocutum, large, close, at anterior ends of parapsidal furrows separated from each other by about the diameter of a puncture; punctures at median anterior margin of mesoscutum finer and crowded; middle and hind tibiæ completely annulate with black; wings dusky, more so apically.

Panurginus boylei Cockerell.—Male type: Process of labrum emarginate apically; clypeus with a median impunctured space which has a median depressed line; punctures of mesoscutum as large as in *nebrascensis* but not crowded along anterior median margin; middle and hind tibiæ completely annulate with black; wings slightly yellowish.

NOTE ON VANESSA CALIFORNICA AT PEACHLAND, B. C. IN 1912.

BY J. B. WALLIS, WINNIPEG, MAN.

A somewhat remarkable visitation of *Vanessa californica* came to my notice when in Peachland, B. C., during July, 1912.

Almost immediately on my arrival I was questioned concerning a caterpillar (descriptions decidedly remarkable!) which had occurred in such numbers as to defoliate its food-plant, and had been compelled to migrate by thousands. I was also told of the appearance, in very large numbers, of a brown butterfly which was believed to be connected with the "worms."

In neither of my two previous visits (1907-9) had *californica* been seen, so I was quite at a loss to place a caterpillar whose food-plant was *Ceanothus sp.*

Next day the problem was solved. On going a mile or two into the hills, *californica* was found in very great numbers. There must hsve been many thousands of them, and in favored spots they almost filled the air. Being in a wagon, I made little effort to secure specimens, although five were taken at one almost aimless sweep of the net.

Four days later I made a special trip after *californica*, but, with the exception of three deformed specimens, not one was seen, and during the remainder of my five-week stay not more than a dozen were noticed.

It would be interesting to know if a large influx of this beautiful butterfly was noted in any locality.

Practically every plant of *Ceanothus* was entirely defoliated, and the pupa cases were hanging everywhere. Nine were counted on a twig four inches long, eight on another five inches long, and so on; while some young pine trees about seven feet high looked to be well laden with strange fruit.

The percentage of parasitism appeared to be very small. I did no actual counting of large numbers, but estimated it was no greater than one per cent.

ON THE LARVA OF *PLEUROPRUCHA* (*DEPTALIA*) *INSULSARIA* GUEN.

BY LOUIS B. PROUT, LONDON, ENGLAND.

My esteemed correspondent, Dr. Eugenio Giacomelli, of La Rioja, Argentine Republic, recently sent me the description of the larva and pupa of a small Geometrid moth unknown to him, together with imago bred therefrom. Knowing how extremely little had yet been done with the early stagesof the Neotropical *Geometridæ*, he naturally hoped that his discovery might prove entirely new. This is not actually the case, for the moth turns out to be the very widely distributed *Pleuroprucha insulsaria* Guen. (var. ? *asthenaria*, Walk.; compare my memoir on the Argentine *Geometridæ*, Trans. Ent. Soc., Lond., 1910, 215.) But as the larva is evidently very variable, and it seems likely that the Southern form constitutes a local race, it is well worth while to give a translation of Dr. Giacomelli's note on his larva. His account of the pupa, both as to its structure and its activity, agrees very exactly with Hulst's (Ent. Amer., 3, 175, 1887, erroneously as *Acidalia* "*insularia*").

"Ground color delicate green, more intense dorsally, ventral region glaucous green; above on the central segments three small, crescent-shaped spots, yellow, paler than the ground; mediodorsal and lateral lines also paler. Setæ simple, not numerous, short, inconspicuous. Later it changes color as follows: The delicate green becomes glaucous, the longitudinal lines a dull vinous red, laterally and dorsally, between them some round dots of the same colour [the tubercles], bearing the short, simple hairs.

"The larva lives on *Prosopis* (Mimosæ) and *Acacia ripari* (Mimosæ). It pupated five days after I took it, so that it would appear that thechange of colouring indicates that the transformation from caterpillar to chrysalis is near at hand."

I am indebted to Mr. Grossbeck for calling my attention to Hulst's descriptions, as well as a note by Bruce (Ent. Amer., 3, 48), to the effect that he bred the species from the egg of *Galium*. Packard gives *Celastrus scandens*, and it is evidently not very particular about its food.

THE SECOND INTERNATIONAL CONGRESS OF ENTOMOLOGY.

BY HENRY H. LYMAN, MONTREAL.

I confess that it is with considerable diffidence that I approach the above subject. Reports of the meeting have already been published by Mr. H. Rowland-Brown in "The Entomologist," and by the Canadian Government representative, Dr. C. Gordon Hewitt, in this journal, but there are certain aspects of the subject which these gentlemen have not dwelt upon that appear to me, at least, to be of considerable importance.

I hope I am not wrong in assuming that the *raison d'etre* of an international scientific congress is primarily to study the subject in its international aspect, and to secure, as far as possible, co-operation among the scientists of all the countries represented, and that this aspect should never be lost sight of. Yet, it appears to me, as one who has attended both congresses so far held, that this aspect was less in evidence at the Oxford meeting than at the one held two years previously at Brussels; while the social aspect, which was almost absent in Brussels, was very strongly developed. I fully admit the agreeable nature and also the important character of the social aspect, but I think there is a danger of overdoing it, and that it should never be allowed to obscure the more serious business of the gathering.

These congresses being from now on held only every three years, and, considering the very considerable expense incurred by governments and institutions in sending representatives to them, is it not of the highest importance that they should not be merely very pleasant reunions where highly interesting papers are read by eminent scientists, and where afterwards the pipes of social peace are smoked around the social board, but that the many pressing questions of international importance should be given first place and some attempt made to solve them, instead of referring them to committees from one congress to another, while every year confusion, at least in nomenclature, is becoming worse confounded? It is quite true that some attempt was made by some authors to deal with matters of international concern, but such attempts were few, and, unfortunately, some of the ideas were crude.

The programme, including the President's annual address,

comprised about fifty-five papers, but of these not more than ten were of international or semi-international interest.

Mr. Charles Oberthur in his paper advocated the adoption of a rule that no description should be accepted as valid unless accompanied by a good figure. Such an idea could not, of course, be entertained, for, apart from the difficulty of determining what is a good figure, it would, in the absence of highly-endowed journals or expensive government publications (which would only be open to official entomologists), throw the work of describing new species into the hands of wealthy entomologists who could afford to furnish the illustrations.

The suggestion of Mr. Ernest Olivier, that the Latin language should be used in all entomological descriptions is equally impracticable; and, even if it could be adopted, would certainly not mend matters, judging from the extremely meagre and inadequate Latin descriptions of the past.

The centralization of diagnostic descriptions, advocated by Mr. E. E. Green, while a consummation devoutly to be wished, seems impossible of attainment, but certainly a great improvement over the present chaotic condition could be made by a little co-operation between the entomologists of each country.

Of the other papers of international import, the only ones which led to any action were those by Mr. A. G. L. Rogers and Messrs. Wheeler and Bethune-Baker, the latter being accompanied by a communication on nomenclature from the Entomological Society of London, which led to important action being taken, as detailed by Dr. Hewitt.

There is another point which certainly merits consideration, and that is the serious disproportion among the representatives of the different nations, the English members of the congress equalling, if not outnumbering, the representatives of all the other countries combined. This was referred to by one of the German entomologists to whom I spoke while waiting on the Tring platform for the London train, who pointed to the whole page of names of English representatives, and said there were too many.

Disproportionate representation is, of course, inevitable, as there will always be a fuller representation of the entomologists of the country in which the congress is held, but if it should ever be desired to settle any disputed point by majority vote, some scheme of proportional voting power would probably have to be adopted.

If the congress could be brought to seriously consider and decide such questions as to whether or not the law of priority should be rigidly enforced in all cases, irrespective of consequences, whether the Tentamen of Hubner should or should not be recognized, and similar troublesome questions, it would do more to justify its existence than it has yet done.

BOOK NOTICE.

HOUSE FLIES AND HOW THEY SPREAD DISEASE. By C. G. Hewitt, D. Sc. (The Cambridge Manuals of Science and Literature, Cambridge; the University Press. One shilling.)

Although in the last few years the public has at last become more or less awake to the fact that the house fly is not merely a troublesome nuisance but a serious enemy of mankind, few even among the educated realize the many ways in which this ubiquitous insect can make itself a source of danger to public health. The present little book is just what has been needed to bring this important matter home to the public mind and we believe it will have a far-reaching influence in increasing such efforts as are being made to keep this pernicious insect under control.

The book is not a record of new observations, the subject matter having already been set forth at much greater length in Dr. Hewitt's earlier work—"The House-fly: A Study of its Structure, Development, Bionomics and Economy."* It is written for the benefit of the layman, and it has been the author's endeavour "to avoid as far as possible the use of technical terms unfamiliar to the lay mind and the inclusion of matter which is of interest chiefly to the specialist." In this aim we think he has been eminently successful, the book being written in a clear, simple style, as easy to read as a novel.

Of the two parts into which the book has been divided, the first deals with the natural history of the fly; the second with its relationship to disease. In Part I. the author has wisely restricted the description of the fly's structure and metamorphosis to such features as are necessary to a proper understanding of the general facts of its life history. In this part are also given short accounts of other species of flies commonly found in houses and of the parasites and natural enemies of the house-fly. In Part II. the relationship of flies to a number of diseases is discussed, a special chapter being devoted to the important *role* which they play in the dissemination of typhoid fever and the summer diarrhœa of infants. In the last chapter is a full description of the methods of control which a thorough study of the fly's life history have proved to be most efficient in destroying it.

The little volume, which is small enough to be carried in one's pocket, has a most attractive appearance, the paper and print being of excellent quality and the illustrations, with few exceptions, are reproduced from the beautiful plates which accompanied the author s earlier work on the house-fly (l.c.). The quaint design on the title page is a reproduction of one used by the earliest known Cambridge printer, John Siberch, 1521.

*Manchester University Press, Biological Series, No. 1, 1910.

Mailed December 31st, 1912.

INDEX TO VOLUME XLIV.

ENTOMOLOGICAL SOCIETY OF ONTARIO, 49th ANNUAL MEETING (P. 15).

The
Canadian Entomologist

VOLUME XLV.
1913.

EDITED BY

DR. E. M. WALKER,
Biological Department,
UNIVERSITY OF TORONTO, TORONTO

Editor Emeritus: REV. C. J. S. BETHUNE.
ONTARIO AGRICULTURAL COLLEGE, GUELPH, ONT.

London, Ontario:
The London Printing and Lithographing Company Limited.
1913.

LIST OF CONTRIBUTORS TO VOLUME XLV.

ABBOTT, J. F., Washington, University	ST. LOUIS, MO.
ALEXANDER, CHAS. P., Cornell University	ITHACA, N.Y.
BARBER, H. G., Bureau of Entomology	WASHINGTON, D.C.
BARNES, DR. WM.	DECATUR, ILL.
BERGROTH, E.	TURTOLA, FINLAND.
BETHUNE, REV. C. J. S., Ont. Agri. College	GUELPH, ONT.
BEUTENMULLER, WM., Am. Museum of Natural History	NEW YORK, N.Y.
BIRD, HENRY	RYE, N.Y.
BLATCHLEY, W. S.	INDIANAPOLIS, IND.
BUENO, J. R. DE LA T.	WHITE PLAINS, N.Y.
CAESAR, LAWSON, Ont. Agri. College	GUELPH, ONT.
CAUDELL, A. N., Bureau of Entomology	WASHINGTON, D.C.
CHAGNON, G.	MOTREAL, QUE.
CHAMBERLIN, R. V.	CAMBRIDGE, MASS.
CLEMENS, W. A., Cornell University	ITHACA, N.Y.
COAD, B. R.	WASHINGTON, D.C.
COCKERELL, PROF. T. D. A., University of Colorado	BOULDER, COL.
COCKLE, J. W.	KASLO, B.C.
COSENS, DR. A., Parkdale Collegiate Institute	TORONTO, ONT.
CRAWFORD, J. C., Bureau of Entomology	WASHINGTON, D.C.
DOD, F. H. WOLLEY	MIDNAPORE, ALTA.
DODD, A. P.	NELSON, Q., AUSTRALI
FELT, E. P., State Entomologist	ALBANY, N.Y.
FYLES, REV. THOS. W.	OTTAWA, ONT.
GAHAN, A. B., Maryland Agricultural College	COLLEGE PARK, MD.
GIBSON, ARTHUR, Central Experimental Farm	OTTAWA, ONT.
GIRAULT, A. A.	NELSON, Q., AUSTRALI
HEWITT, DR. C. GORDON, Central Experimental Farm	OTTAWA, ONT.
HOOD, J. D., U. S. Biological Survey	WASHINGTON, D.C.
JOHNSON, C. W., Boston Society of Natural History	BOSTON, MASS.
KIRKALDY, THE LATE G. W.	
McCULLOCH, J. W., Kansas Agricultural College	MANHATTAN, KAS.
McDUNNOUGH, DR. J.	DECATUR, ILL.
MACGILLVRAY, PROF. A. D., University of Illinois	URBANA, ILL.
McGLASHAN, MISS XIMENA.	TRUCKEE, CAL.
MALLOCH, J. R., Bureau of Entomology	WASHINGTON, D.C.
MOORE, G. A.	MONTREAL, QUE.
MORRIS, F. J. A., Trinity College School	PORT HOPE, ONT.
MUIR, F., H. S. P. A. Station	HONOLULU, HAWAI.
NAKAHARA, WARO	HONGOKU, TOKIO, JAPA
RICHARDSON, C. H.	FOREST HILLS, MASS.
SKINNER, DR. HENRY	PHILADELPHIA, PA.
SLADEN, F. W. L., Central Experimental Farm	OTTAWA, ONT.
STRICKLAND, E. H., Central Experimental Farm	OTTAWA, ONT.
SWETT, LOUIS W.	BOSTON, MASS.
TOTHILL, J. D., Central Experimental Farm	OTTAWA, ONT.
TOWNSEND, C. H. T.	LIMA, PERU.
VAN DUZEE, E. P.	SAN DIEGO, CAL.
VENABLES, E. P.	VERNON, B.C.
WALKER, PROF. E. M., University of Toronto	TORONTO.
WALLIS, J. B., Machray Public School	WINNIPEG, MAN.
WEBSTER, F. M., Bureau of Entomology	WASHINGTON, D.C.
WEISS, H. B.	NEW BRUNSWICK, N.J.
WHEELER, PROF. W. M., Bussey Institution	FOREST HILLS, MASS.
WINN, A. F.	WESTMOUNT, QUE.
WOODRUFF, L. B.	NEW YORK, N.Y.
YOTHERS, M. A., Washington State College	PULLMAN, WASH.
ZETEK, JAMES	ANCON, CANAL ZONE.

The Canadian Entomologist.

Vol. XLV. LONDON, JANUARY, 1913 No. 1

NOTES ON AMERICAN HEMIPTERA.

BY DR. E. BERGROTH, TURTOLA, FINLAND.

II.

(Continued from Vol. XXXVIII, 1906, p. 202).

Aradidæ

1. *Aradus aequalis* Say.—Mr. Heidemann has sent me a specimen of *A. duryi* Osb., communicated to him by Mr. Dury. I quite agree with Heidemann in considering this species a synonym of *aequalis*.

2. *Aradus montanus*, n. sp.—Ovate (♀), finely granulated, uniformly dark brownish black including legs and antennæ. Head somewhat longer than broad and as long as the pronotum in the middle, with two parallel longitudinal impressions, the tubercle near the anterior angle of the eyes low and obtuse, antenniferous spines a little divergent, not quite reaching the middle of the first antennal joint and with a small tooth on their outer margin, rostrum reaching the anterior coxæ, antennæ moderately robust, a little more than half as long again as the head, second joint 2½ times longer than first and a little longer than half the breadth of the head (including eyes), a little thicker at apex than at base, third and fourth joints taken together scarcely longer than second, third joint a little thicker than second, fourth conspicuously narrower and shorter than third. Pronotum kidney-shaped but with a short lobelet anteriorly on each side near the neck of the head, a little narrower than the hemelytra between their dilated subbasal part and a little more than twice broader than long in the middle, its greatest width in the middle where the lateral margins are rather broadly rounded and from where they are strongly convergent toward the apex, much less so toward the base, the whole lateral margins distinctly crenate, the four median discal ridges subequally distant at their base, the inner pair reaching the anterior margin, thicker and more approximated before the middle, the outer pair not reaching the apical margin. Scutellum a little longer than the pronotum in the middle, with a blunt median longitudinal keel in the basal half, the lateral margins convergent from the base to beyond the middle, then slightly rounded to the apex. Hemelytra (♀) slightly passing the base of the dorsal genital segment, exocorium moderately dilated and reflected in

its basal part, mesocorium with a single oblique transverse ridge behind the middle, endocorium without distinct transverse ridges, membrane scarcely reticulate. Abdomen (♀) one-third broader than the pronotum, with the apical angles of the segments very slightly obtusely prominent, those of the fifth segment more distinctly so and those of the sixth segment strongly prominent owing to the lateral margin being deeply arcuately indented between the apex of this segment and the base of the genital lobes; sixth ventral segment in the middle scarcely longer than the fifth and than the two genital segments combined, its apical angles reaching the transverse level of the apex of the second genital segment which is half the length of the first, apical margin of dorsal genital segment notched in the middle, genital lobes convergent, approximated interiorly, their inner margin strongly rounded, the outer margin slightly rounded with a short tooth-like process in the basal half. Length, ♀ 8 mm.

Colorado (Leadville, 10,000—11,000 ft.: H.F. Wickham).—Coll. Schouteden. A plain-looking species, but not closely allied to any described North American form.

3. *Aradus curticollis*, n. sp.—Broadly ovate (♂), finely and thickly granulated, jet-black, membrane brown, third antennal joint of dirty whitish color, anterior and intermediate tibiæ paler in the middle. Head a little longer than broad, vertex with two longitudinal impressions which are slightly divergent forwards, eyes very prominent, substylated and directed a little upwards, intra-ocular tubercle scarcely perceptible, antenniferous spines slightly divergent, reaching beyond the middle of the first antennal joint, without a tooth on their outer margin, rostrum reaching fore coxæ, antennæ almost fusiform, incrassated in the middle and equally tapering toward base and apex, as long as head and pronotum together, first joint rather narrower, second joint as long as the breadth of the head (with the eyes) and considerably longer than the last two joints combined, cylindrically incrassate from apex to beyond middle, then moderately narrowed toward base, third joint cylindrical, narrower than apex of second and not much more than one-third its length, suddenly narrowed at base, fourth joint narrower and conspicuously shorter than third. Pronotum with entire, not crenulated lateral margins, distinctly shorter than head and three times broader than long, a little narrower than the hemelytra between their dilated subbasal part, its greatest width before the middle from which point the lateral margins are very strongly convergent to the apical angles and moderately roundedly convergent toward the base, the two median ridges parallel, reaching apical margin, the following pair at the base more distant from the median ridges than these from each other, a little convergent anteriorly, not reaching anterior margin,

ending in a flattened tubercle, the outermost (humeral) ridges very short, nodiform. Scutellum three-fourths longer than pronotum in the middle, with a transverse tubercle before the middle, lateral margins broadly and slightly rounded. Hemelytra (♂) reaching the apical lobes of the abdomen, roundedly dilated and reflected near base, exocorium and endocorium with some transverse ridges, mesocorium with a single oblique transverse ridge behind the middle. Abdomen three times broader than the membrane, apical angles of fifth segment very slightly obtusely prominent, male genital lobes obliquely slightly rounded at apex, meeting interiorly. Length, ♂ 5.8 mm.

North Carolina (Southern Pines: A. H. Manse).—Coll. de la Torre Bueno. A very distinct species, somewhat allied to the quite differently coloured *A. behrensi* Bergr., but more broadly ovate with longer and less incrassate antennæ and much shorter pronotum having the two median keels much more approximate at base. I have not seen the female, but the abdomen is probably not or not much broader in this sex. *A. cincticornis* Bergr. and *curticollis* Bergr. belong to the very few *Aradus* species having the abdomen broadly ovate also in the male.

4. *Aradus cincticornis* Bergr.—This species stands in some collections under the unpublished name *A. nasutus* Uhl.

5. *Aradus tuberculifer* Kirby.—Black, sometimes tinged with greyish brown, apical margin of connexival segments yellowish, corium with a dark luteous costal patch before the middle, this patch being sometimes diffused over a large part of the corium, legs fuscous black. Head distinctly longer than broad with a U-shaped impression above, intraocular tubercle distinct, antenniferous spines a little divergent with a small tooth, sometimes indistinct or wanting, on the outer margin, rostrum reaching or slightly passing the anterior margin of the mesosternum, second joint of antennæ a little shorter than the head, almost linear from the base to the middle, then strongly and rather suddenly clavately incrassate, third joint a little shorter than half the length of the second joint, incrassate, even thicker than the apex of the second joint, parallel-sided except at the constricted base, fourth joint distinctly shorter and a little thinner than the third. Pronotum two and one-half times broader than long in the middle, lateral margins very finely crenulate or almost smooth, antero-lateral margins slightly sinuate, the four median discal keels parallel, the inner ones approximated in their anterior half, the outer ones abbreviated before the middle; the greatest width of the pronotum is immediately behind the middle, from which point the lateral margins are very distinctly convergent towards the base. Scutellum subtriangular, a little longer than the pronotum in the middle, with a blunt median tubercle. Hemelytra (♀) passing the base of

the dorsal genital segment, costal margin of corium moderately ampliated toward the base, veins of membrane narrowly bordered with white. Female dorsal genital segment triangularly narrowing toward the apex, leaving the apical lobes free and separated from them by a fine suture. Discal lobes of the sixth female ventral segment distinctly longer than those of the fifth segment, either lobe considerably longer than broad, rounded at apex, taken together broader than long, first ventral genital segment about half the length of the sixth ventral segment, the apical lobes convergent, rounded at the external margin and at the apex, almost twice longer than broad. Length, ♀ 7.3 mm.

A. tuberculifer Kirby in Richardson, Fauna Bor. Amer. IV., 278, pl. VI., fig. 5 (1837).

A. caliginosus Walk., Cat. Hem. Het. Brit. Mus. VII., 36 (1873).

This is a Boreal species which has hitherto been recorded only from Canada and even there it seems to be rare. From the United States I have seen but one specimen, taken in Colorado (probably in the high mountains) by Morrison, and the species has apparently remained unknown to Uhler and Heidemann. It is closely allied to but specifically distinct from the Palearctic *A. crenaticollis* F. Sahlb.

6. *Aradus funestus*, n. sp —Black, apical margin or at least apical angle of connexival segments yellowish, corium with a more or less distinct yellow costal spot before the middle. Antennæ thin, second and third joints slightly and gradually incrassated from the base to the apex, third joint about one-third the length of the second, fourth joint equal to third in length and thickness. Pronotum twice broader than long in the middle, the lateral margins parallel from the middle to the base. Scutellum pentagonal, as long as the pronotum in the middle, the lateral margins parallel from the base to the middle. Hemelytra reaching the apex of the dorsal genital segment, costal margin of corium scarcely (♂) or slightly (♀) ampliated toward the base. Female dorsal genital segment dilated toward the very broad apex, laterally covering the basal part of the apical lobes, reaching almost to their outer margin, its apical margin broadly rounded. Apical lobes of the genital segment seen from the ventral side at least twice longer than broad. Other characters as in *A. tuberculifer*. Length, ♂ 57 mm., ♀ 7- 7.5 mm.

This species is common in Canada as well as in the Northern U. S. from the Atlantic to the Pacific ocean, and I have also seen a specimen from Colorado. In the various writings of Prof. Uhler it is recorded under the name *tuberculifer* Kirby; and under this name it stands in most if not all American collections. Kirby's description fits both these species equally well. Fortunately he had figured his species, clearly showing the structure of the antennæ. There can thus be no doubt as to which of the two species he had before him.

7. *Aradus lugubris* Fall.—In his Catalogue of the Heteroptera of the British Museum, Walker described as new an *Aradus fenestratus*, founding the species on many specimens from St. Martin's Falls, Albany River, Hudson's Bay, and from Nova Scotia, and on two specimens from the Rocky Mountains. In his Revision of Walker's Aradidæ Distant marked *fenestratus* as a good species, and in arranging the Aradids of that Museum Distant has apparently left the types of this species in the same state as Walker, as I stated when I examined them a year ago. The first specimen bears a round label with the word "type" upon it and belongs to *lugubris* to which also several other specimens appertain, but intermixed with them are a few specimens of *Aradus abbas* Bergr., easily recognized by the very slender antennæ narrowly biannulated with white. Walker's description exclusively refers to *lugubris*, of which *fenestratus* should be cited as a synonym.

Gen. *Calisius* Stal.

To the characters of this genius should be added: Metanotum et segmentum primun (verum) dorsale abdominis ad latera corporis visibilia. Orificia distincta, punctiformia, mox ante coxas posticas sita.

In all species of this genus the scutellum is constricted in the middle, but the margins appear to be straight owing to the linear corium being so closely attached to the scutellum that it seemingly forms a part of it. The connexivum in this genus is split from the lateral margin, being, as Champion correctly observed, "divided into two parts, a dorsal and ventral," but these are not always similarly armed, as will be seen from the descriptions given below. To get a correct view of the manner in which the connexivum is armed it is necessary to examine the upper lateral margin at a right angle to the margin (thus more or less horizontally, as the margin is more or less reflected) and the lower lateral margin obliquely from above, lest the tubercles of the ventral lateral margin will make the impression of being situated on the dorsal lateral margin.

8. *Calisius elegantulus*, n. sp.—Subelongately ovate (♀), light brownish testaceous, last antennal joint fuscous, scutellum with two transverse oblique black spots immediately behind the basal callosity at the median ridge and a cretaceous streak on each side between the black subbasal spot and the lateral sinuosity, the first connexival segment whitish testaceous, the three following segments infuscated, the three last segments with a whitish bloom and a small rectangular denudated fuscous spot before the middle. Head considerably longer than broad and longer than the

pronotum, granulated, with short longitudinal fuscous impression on either side within the eyes, antennæ much shorter than the head, the first three joints subequal in length, third reaching apex of head, narrower than second, fourth joint incrassated, longer and thicker than second. Pronotum on the anterior lobe with two strongly convergent waved keels, each bearing a black tubercle, on the posterior lobe with four short ridges, each bearing a brown tubercle. Scutellum rather strongly and densely punctured with a series of three or four black granules immediately within the lateral sinuosity, apex unarmed, the basal elevation broadly crescent-shaped, three times broader than long, granulated and with two small broadly separated black tubercles at the very base, partly overlapping the pronotal basal margin; a short faint oblique ridge between each side of the basal elevation and the scutellar margin; the median scutellar keel slightly granulated, narrowing toward apex. Abdomen with the first connexival segment but slightly narrower exteriorly than interiorly, the posterior margin of this segment scarcely oblique, the upper lateral margin of the three following segments with two short apically subtruncate lobes, anterior lobe blackish, posterior brown, upper lateral margin of all subsequent segments with two tubercles, anterior tubercle black, posterior pale, the lower lateral margin of all segments (first excepted) with three tubercles, the middle one of which is black, the others being pale. Legs pale testaceous, a submedian ring to femora and tibia, apex of tibiæ, and tarsi fuscous. Length, ♀ 3.7 mm.

Guadeloupe Island, West Indies; communicated by Mons. A. Montandon.

Allied to *C. pallidipes* Stal, but with differently constructed antennæ and differently coloured scutellum and legs.

9. *Calisius contubernalis* n. sp.—Oblong (♀), fuscous-ochraceous, sometimes darker, last antennal joint fuscous, first connexival segment whitish ochraceous. Head a little longer than broad, granulated, the apical process with a lateral impression, antennæ as long as the head, first joint slightly passing apex of antenniferous spines, second joint as long as first and almost reaching apex of head, third joint as thick as and a little longer than second, fourth joint thicker and somewhat longer than third, rostrum almost reaching base of head. Pronotum slightly shorter than the head, the posterior lobe finely and rather thickly punctured, with four keels, the two median ones parallel, the outer pair strongly convergent and continued over the anterior lobe where they are granulated, reaching the apical margin. Scutellum rather more strongly punctured than the posterior pronotal lobe, with a series of five black granules immediately within the lateral sinuosity and a transverse series of 8 or 9 such granules close to the apical margin, the median carina granulated, attenuated toward apex, at the base

forming a triangular elevation bearing three small tubercles (placed in a triangle) on each side and emitting a short oblique granulated ridge to the lateral margins. Sternum scarcely granulated. Abdomen with the first connexival segment forming a transverse triangle the apex of which reaches the lateral margin and the posterior margin of which is oblique; the upper lateral margin of all the following segments with two very short lobelets, anterior lobule brown, posterior pale, the lower lateral margin of all segments (first excepted) with three small tubercles, the median one blackish, the others pale; last dorsal segment in the male subquadrately elevated and granulated in the middle, dorsal male genital segment very short, transversely sublinear, venter scarcely granulated, fifth male segment arcuately sinuated in the middle almost to the base, sixth segment also deeply arcuately sinuate in the middle, yet scarcely shorter there than at the sides, first genital segment as long as sixth ventral segment, broadly sinuate at apex, the apical angles with a short upturned cylindrical process not reaching the apical angles of the last ventral segment, second genital segment shortly protruding beyond the dorsal genital segment, with two small teeth or granules at apex, on the underside divided into three lobes by two longitudinal impressions, the median lobe narrower than the somewhat tumid lateral lobes. Legs ochraceous, in dark specimens somewhat infuscated. Length, ♂ 3.7–3.8 mm.

St. George Island, Florida; Guadeloupe Island, W. T.

At once distinguished from *C. elegantulus*, apart from colour, by the structure of the antennæ, the sculpture of the scutellum, the form of the first connexival segment, etc. The structure of the antennæ and scutellum also separates it from *C. pallidipes*.

The male type from Florida, taken by Mr. Pergande, is in the Washington Museum; the female cotypes from Gualeloupe have been communicated by Dr. H. Schouteden.

10. *Calisius anaemus*, n. sp.—Closely allied to the preceding species, but entirely very pale ochraceous without darker markings and with all granules and tubercles as pale as the ground-colour. The very short first joint of antennæ reaching apex of antenniferous spines (remaining joints wanting). The two convergent keels of the anterior pronotal lobe connected at apex by a short transverse ridge. Scutellum close to apical margin without the transverse series of small tubercles, but the apical margin itself distinctly crenulated. Underside of body very finely and thickly granulated. First male genital segment a little shorter than sixth ventral segment, the apical margin a little sinuate in the middle. Second male genital segment with the median lobe very narrowly triangular. Length, ♂ 3.8 mm.

Biscayne, Florida.

This species was determined as *C. pallidipes* Stal by Uhler and was recorded under this name by Heidemann in Proc. Ent. Soc. Wash. VI., p. 229. The specimen is a ♂, not a ♀, as Heidemann says. Stal's species (from Rio Janeiro) is a darker, brown-speckled insect with differently sculptured pronotum and scutellum.

11. *Calisius major*, n. sp.—Oblong-ovate (♂), dark testaceous, scutellum (except basal elevation and median keel) whitish cinereous with a large median area and the apex sprinkled with fuscous and with an oblique black fascia on each side immediately behind the basal elevation, last dorsal segment and upper side of the protruding male apical genital segment blackish. Head slightly longer than broad, antennæ scarcely longer than the head, rather stout, first two joints short, second a little thicker and more oval than the first, third joint much the longest, thicker and more than three times longer than the second, attenuated at the base, fourth joint as thick as the third and more than twice longer than the second Pronotum a little shorter than head, anterior lobe remotely granulated, posterior lobe with four keels, the two median keels convergent from base to apex, the outer keels almost parallel, slightly convergent apically, the lateral margins of the lobe also somewhat elevated. Scutellum superficially and concolorously punctate, the transversely triangular basal elevation at its base on each side with two short keels, the outer one of which is obliquely continued to the scutellar lateral margin, the median scutellar ridge scarcely granulated, attenuated toward apex, the lateral margins immediately within the sinuosity with a series of three black granules, the apical margin neither granulated nor crenulated. Abdomen with the lateral margins of all the seven connexival segments provided with two tubercles, the anterior black, the posterior pale and sometimes notched, last male dorsal segment transversely convex, its apical margin sinuate, the apical male genital segment protruding considerably beyond the extremely short dorsal genital segment. Legs pale testaceous. Length, ♂ 4 mm.

Venezuela (La Guayra); in my collection.

Readily distinguished from all other species by the structure of the antennæ and other characters. The specimen being strongly carded I am unable to describe the ventral genital segments.

12. *Proxius gypsatus* Bergr.—Of this species, described from Venezuela and also found in Panama and Guatemala, I have seen two specimens from Florida; one is without precise locality, the other from St. Augustine and bears the label *Syrtidea diffracta* Uhl., apparently an unpublished name. Two species of *Proxius* are now known from Florida.

A NEW SPECIES OF CALISIUS.

BY DR. E. BERGROTH, TURTOLA, FINLAND.

Calisius annulicomis, n. sp.

Ovatus (♀), fuscus, pronoto fusco-nigro, scutello subtestaceo, medio vittis duabus albis antrorsum leviter convergentibus, antice arcuato-conjunctis, postice extrorsum curvatis et latera attingentibus signato, inter has vittas et elevationem basalem nigram fusco-conspurcato, carina media nigra, medio late albo-interrupta, abdomine magna parte obscure rufescente, subtus latera versus parce albo-granulato, antennis fuscis, articulo tertio (ima basi excepta) flavo, articulo quarto nigro, pedibus sordide flavidis, femoribus (ima basi et summo apice exceptis) fuscis. Caput pronoto distincte brevius, antennis capiti subaeque longis, articulo secundo primo crassiore et paullo longiore, tertio secundo fere dimidio longiore, quarto tertio sat multo longiore et crassiore. Pronotum lateribus rectis, irregulariter nigro-spinulosis, medio vix sinuatis insigne, lobo postico carinis sex instructo, duabus mediis antrorsum leviter convergentibus, usque in lobum anticum extensis, carinis subsequentibus in parte basali levius, deinde fortiter convergentibus et usque ad apicem carinarum mediarum extensis, cum his angulum acutum formantibus, carinis extimis prope marginem lateralem sitis. Carina media scutelli granulata. Margo lateralis superior segmentorum connexivi granulis tribus perminutis, margo lateralis inferior granis tribus majoribus albis instructi. Long. ♀ 4 mm.

Tasmania (Launcestown, J. J. Walker). Mus. Brit.

This remarkable species is by many characters very distinct from *C. interveniens* Bergr., the only Australian species hitherto known.

THE NORTH AMERICAN SPECIES OF THE GENERA ARTHROPEAS AND ARTHROCERAS.

BY CHARLES W. JOHNSON, BOSTON, MASS.

The species *Arthropeas leptis* Osten Sacken seems to be the cause of some confusion in these two genera. This is probably due to the comparative scarcity of material, to an oversight in Aldrich's Catalogue, and to the fact that Osten Sacken in describing the species and referring to a number of minor characters wherein it differs from the typical *Arthropeas* failed to mention the most important feature—the absence of spurs on the anterior tibiæ. This character, however, he mentions in 1882 (Berl. Ent. Zeits., XXVI, 365), as follows: "In the notes to my Catal. N. Am. Dipt., 1878 (p. 223), an insect is described which I referred provisionally to the genus Arthropeas. It has the body of a Leptid (Symphoromyia), with the antennæ of a Cœnomyia. It will probably form a new genus because, besides the differences in the venation noticed by me in the description, it has no spurs on the front tibiæ, while such spurs are distinct in *Arthropeas siberica*."

In 1886 Dr. Williston erected the genus *Arthroceras* (Ent. Americana, II, 107), based chiefly on the character above mentioned with *A. pollinosum* (a new species), and *A. leptis* O. S. as the types. That *A. leptis* belonged to the genus *Arthroceras* was recognized by Coquillett in determining the species for Mrs. Slosson's list of Mt. Washington insects (Ent. News. VI, 6, 1895).

The two species may be separated by the following table:

Thorax unicolor, yellowish pollinose; halteres yellow. Colorado, Washington................................*pollinosum* Will.
Thorax blackish, with two yellowish pollinose stripes; halteres brown. White Mts., N. H........................*leptis* O. S.

Both species seem to be confined to the Canadian zone. The former I have received from Clear Creek, Col., May 20, 1891 (Oslar); Happy Hollow and Little Beaver, Col., July 14 and 19 (Gillett) The latter has only been taken in New Hampshire, White Mts., "woods and alpine" (E. P. Austin); "Alpine region of Mt. Washington, at or above 5,500 ft." (Mrs. Slosson); "near summit," Mt. Washington, July 25, 1875 (Dr. Geo. Dimmock); Mt. Washington, July 7, 1909 (F. A. Sherriff); Base Station, Mt. Washington, July 30, 1912 (F. W. Dodge).

The species of the genus *Arthropeas* are likewise comparatively rare, and also seem to be confined to the Canadian zone. The species may be tabulated as follows:

Anal cell closed; wings distinctly banded; length, 8-9 mm..................................*americana* Loew.
Anal cell narrowly open; wings not banded; length, 12-14 mm....................................*magna*, n. sp.

Arthropeas americana Loew.

The following brief description is given chiefly as a comparative one to Say's *Xylophagus fasciatus:*

Thorax black, covered with a yellowish pollen, leaving three wide black vittæ; scutellum and metanotum black; abdomen yellow, basal half of the first to fourth segments black, the remaining segments yellow; halteres entirely yellow; legs yellow, outer half of the tarsi brown; apical third of the wing smoky black, base of the submarginal and first posterior, tip of the first basal, the entire discal and fourth posterior and outer portion of the fifth posterior cells whitish and forming a wide band across the centre of the wing; the greater portion of the first and second basal cells, base of the fifth posterior and tip of the anal cell and the anal angle smoky black; base of the wing and greater portion of the anal cell whitish. Length, 8.5 mm.

The following records constitute our present knowledge of its distribution: N. Wisconsin (Loew); Mass. (O. Sacken); Cheshire Harbor, near Mt. Graylock, Mass., June 30 (I. W. Beecroft); Lake Ganoga, North Mt., Pa., 2,300 ft., Aug. 29, 1897 (C. W. Johnson).

Xylophagus fasciatus Say.

"Wing dusky, fasciated; abdomen fasciated. Inhabits Indiana.

"Body dusky; thorax, *posterior portion honey-yellow; poisers blackish at tip;* wings dusky, a more distinct band on the middle and at the tip; feet honey-yellow; *hind tibiæ blackish;* tergum yellow, basal half of the four basal segments black; *remaining segments nearly all black.* Length over two-fifths of an inch.

"By an accident the head and anterior part of the thorax of this fine specimen were destroyed, but the above description will sufficiently indicate the species. The wing nervures resemble those of the *maculatus* Fabr."

In the above description by Say, based on an imperfect specimen, I have italicized the parts showing discrepancies to Loew's species. The differences are too great to consider them the same; the description of the bands on the wings, "on the middle and at the tip," also does not agree with Say's usual accuracy. The locality, "Indiana," which is entirely in the upper Austral, would also indicate a different species. Say's reference to *maculatus*, which is a *Xylomyia* (= *Solva* Walk.), would indicate a closed fourth posterior cell.

Arthropeas magna, n. sp.

Arthropeas, n. sp.? Townsend.—Trans. Amer. Ent. Soc., XXII, 61, 1895.

♂.—Face blackish, covered with a dull yellowish pollen and pile, beard whitish, face with a deep Λ-shaped groove bordering the oral cavity, from which extends a deep groove between the antennæ to the frontal triangle, ocelligerous tubercle black, palpi, proboscis and antennæ yellow. Thorax black, thinly covered with hair (blackish on the dorsum and yellowish on the sides), through which show four dull yellow pollinose stripes, the lateral stripes broad, the middle one narrow, but expanding at the ends and connected at the humeri and post-alar callosities with the lateral stripes, the black areas between the stripes shining behind the suture; pleuræ black, brownish pollinose; scutellum black. Abdomen black, middle and sides shining, first segment with a wide yellow, pollinose, posterior band, almost interrupted in the middle and expanding until it attains the full width of the segment at the lateral margins; second, third and fourth segments posteriorly margined with a yellow pollinose band, contracted in the middle and at the ends; on the second and third segments the bands are brown in the middle and at the ends, the remaining segments yellowish pollinose; venter entirely yellow. Legs yellow, coxæ blackish, halteres yellow. Wings brownish, slightly darker in the middle and along the fifth longitudinal vein; veins and costal cells yellow, basal half of the marginal cell white, the greater portion of the first and second basal cells noticeably lighter than the rest of the wing. Length, 12 mm.

♀.—Face, front and occiput covered with a dense brown pollen, the front about one-fourth the total width of the head, with five grooves above the base of the antennæ, the four outer ones slightly diverging below, above fusing and deflecting towards the ocelli, the middle one obsoletely divided into three smaller ones below the ocelli. The thoracic stripes are more prominent and a brighter yellow than in the male; scutellum velvety-brown, with three transverse ridges. The abdomen is shining and brownish black, with the posterior pollinose bands on the first, second and third segments, broadly interrupted. Length, 14 mm.

Three specimens, Beulah, Manitoba, received from Mr. C. T. Brues. Holotype and allotype in the author's collection. Paratype in the Museum Comparative Zoology, Cambridge, Mass.; "Hill City, So. Dakota" (Townsend).

This interesting species has the thick heavy form of *Cænomyia*, but the generic characters are those of *Arthropeas*, except that the anal cell is narrowly open. It seems to more clearly show the relationship of the two genera than the other species.

TWO NEW CANADIAN BEES.

BY T. D. A. COCKERELL, BOULDER, COLORADO.

Sphecodes hudsoni, n. sp.

♀. Length about 7 mm.; head and thorax black, legs dark rufo-fuscous, abdomen entirely clear yellowish-ferruginous; head broader than long, face very broad, thinly covered (including the clypeus) with fine pale hair; mandibles bidentate, the apical half dark chestnut-red, the inner tooth short and rounded, about 208μ from apex of mandible; process of labrum very broad, shallowly depressed or subemarginate in middle; only the first three points of the flagellum remain in the types, but they are dull ferruginous beneath; clypeus strongly punctured; front extremely, densely and minutely punctured in middle, not quite so densely at sides, the punctures are so small as to be hard to see with a hand lens; mesothorax brilliantly shining, with scattered punctures, the median sulcus well marked; pleura, beneath the wings, with a large shining raised area, the pleura below this with fine close rugæ; area of metathorax large, fully 320μ. long, with about 20 coarse rugæ, the lateral ones parallel, radiating, the middle ones irregular, some branching, Y-like in form; regular dark rufous; wings dusky hyaline, distinctly reddish, stigma and nervures red-brown; second submarginal cell broad, receiving first recurrent nervure just beyond the beginning of its last third; legs thinly clothed with pale hair; abdomen almost entirely impunctate, quite broad; apical plate about 170μ broad.

Hab.—Hudson Bay. British Museum (44. 17). In Robertson's tables of *Sphecodes* this runs nearest to *S. minor*, which is a larger and evidently different species. In the table of Maine species it runs to the group of *S. dichrous*, to which it is not closely allied. In my table of allies of *dichrous* it runs to the very much larger *arroyanus*. Superficially it is much like *S. washingtoni* Ckll., but aside from other differences, the metathoracic area is much larger than in *washingtoni*. It is a much larger species than *S. cressoni*, and has a broader head. Among the species of the northwest, it falls nearest to *S. patruelis* Ckll. (formerly recorded in error as *minor*), but *patruelis* has the front more coarsely punctured, and area of metathorax with stronger, irregular (not radiating) rugæ. It is quite different from *S. sulcatulus* by the densely punctured front, etc. The specimen has been in the British Museum for 67 years.

I take this opportunity to record two other interesting specimens of *Sphecodes* belongiing to the British Museum.

(1.) *Sphecodes falcifer* Patton. Colorado (*Cockerell*). A common species of the Eastern United States, but new to Colorado. comes from my old collection of 1887–1890. The material which went to the British Museum was mostly in papers, and nearly all came from Wet Mountain Valley. A statement of the exact locality was furnished for each lot, either in a letter or on the box, but unfortunately the data were only preserved when they accompanied the specimen itself, and all the rest were simply labelled "Colorado (Cockerell)". It is nearly certain that all the specimens labelled in this way were from Wet Mountain Valley.

(2.) *Sphecodes persimilis* Lovell & Cockerell. Trenton Falls, New York; from F. Smith's collection. The specimen (♀) has the junction of the first and second dorsal abdominal segments rather evidently depressed, to this extent slightly approaching *S. pecosensis*. F. Smith, who owned the specimen, died in 1879, but the species was not described until 1907.

Anthidium wallisi, n. sp.

♀. Length about 10 mm.; black with chrome yellow markings, those on face, consisting only of an oval spot on each side touching upper part of clypeus, paler yellow; a large yellow spot above each eye; mandibles, tegulæ and thorax wholly without yellow; antennæ black; pubescence dull white, on vertex shining and yellowish; ventral scope shining cream-colour; wings strongly brownish; femora and tibiæ black; front tibiæ with a yellow subapical more or less cuneiform mark; middle tibiæ with a yellow mark extending from before middle to apex; hind tibiæ with a yellow band, interrupted not far from base; tarsi ferruginous, more or less blackened at base, their hair mainly ferruginous; hind basitarsi with a broad yellow band; no pulvilli; first abdominal segment with a diamond-shaped yellow mark at each extreme side; second segment with a larger mark on each side, deeply notched inwardly, and a pair of transverse, discal stripes; third segment with an interrupted band, broad at sides, broadly and deeply notched in front sublaterally; fourth like third; fifth with the notch less developed, and the interruption narrower; sixth with two large yellow patches.

Hab.—Peachland, British Columbia, August 9, 1909 (*J. B. Wallis*, a 64.)

This has nearly the face-markings of *A. porteræ personulatum* Ckll., but *personulatum* is considerably larger, the spots at side of face are lower down, the abdominal markings are much paler, and the abdomen is not so densely punctured. I asked myself whether *A. wallisi* could possibly be a colour-variety of *A. tenuifloræ* Ckll., but it differs as follows, aside from the colour-markings: eyes paler and lighter green; teeth at lower corners of clypeus larger, nearly equal (the outer one much smaller in *tenuifloræ*); lateral tooth-like angles of sixth abdominal segment very prominent; broad depressed apical margins of abdominal segments excessively, minutely and densely punctured, not shining (shining and less densely punctured in *tenuifloræ*).

PHENACOCCUS BETHELI AGAIN.

BY T. D. A. COCKERELL, BOULDER, COLORADO.

When recently describing *P. betheli* in THE CANADIAN ENTOMOLOGIST, I remarked that it was possibly a subspecies of *P. cockerelli* King. I was surprised, a few days ago, to receive from Mr. E. Bethel a quantity of *P. betheli* on branches of *Amelanchier*, collected by Mr. L. J. Hersey at Steamboat Springs, Colorado. This looked suspicious, as Steamboat Springs is the type locality of *P. cockerelli*. However, the new material is twice the size of *cockerelli*, and yet the legs are not merely relatively, but actually smaller, and the fourth antennal joint is very short as in the Grand Canon insect. The insects, on being boiled in caustic potash, stain it a deep wine red. The larva is light orange.

Although I transmitted the original *cockerelli* material to Mr. King, I did not study it. I have, however, studied abundant material, agreeing with King's description, found by Mr. L. C.

Bragg on wild plums at Boulder, Colorado. Without feeling altogether confident that *betheli* is a distinct species, it seems that it must be so regarded unless proof can be brought to the contrary.

A word may be added concerning the value of antennal measurements in the determination of Coccidæ. Those who have used the antennal formulæ have found them very unreliable, and I have long ago given them up. It does not follow, as some uncritically assume that antennal measurements are therefore useless. The best way is to measure the joints of several antennæ, and from the measurements plot a "curve." I do this by using semitransparent typewriter paper, through which I can see my lined standard sheet. The curves so made will, of course, vary, hardly any two antennæ being exactly alike; but except for abnormalities (pathological specimens), nearly every species gives quite different curves, while two species, very different in other respects, will give nearly the same curve. Some of the widely distributed species give curves almost too variable to be of much service, but in these cases it is possible that the material contains more than one thing. It is also probable that widely distributed forms, living on various plants, are strongly heterozygous, while native species, with uniform environment and more or less restricted distribution, are prevailingly homozygous. It would be worth while for someone to carefully investigate a number of species with this point in mind.

The names of the members of the Society in the group on Plate I. are as follows:

First row, reading from right to left.—J. D. Evans, Prof. W. Lochhead, Rev. T. W. Fyles, H. H. Lyman, Dr. E. M. Walker, Dr. C. Gordon Hewitt, G. Beaulieu, A. F. Winn, Rev. L. Marcotte.

Second row, reading from left to right.—L. Cæser, Dr. E. H. Blackader, Rev. J. B. Mignault, Rev. Brother Germain, H. F. Hudson, Arthur Gibson, J. M. Swaine, G. E. Sanders, Dr. R. Matheson.

Top row, reading from right to left.—A. G. Turney, J. D. Tothill, Prof. J. E. Howitt, A. W. Baker, J. A. Guignard, J. I. Beaulne.

CORRECTIONS.

Page 214, line 7, for *heyoniæ* read *bryoniæ*.

Page 214, line 15, for pormiaries read primaries.

Page 215, line 24, for *Keolexia* read *Neolexia*, and for *scylina* read *xylina*.

AN EARLY REFERENCE TO THE OCCURRENCE OF THE ARMY WORM IN PENNSYLVANIA, NEW YORK AND CANADA.

BY F. M. WEBSTER, WASHINGTON, D. C.

The year 1743 seems to have been the first of which we have what is generally accepted as undoubtable evidence of the occurrence of this pest in the United States in destructive numbers. This information has always been based solely upon a statement made by Chas. L. Flint in a report on the Climatology of New England,* and is as follows: In 1743 there were "millions of devouring worms in armies, threatening to cut off every green thing. Hay very scarce, £7 and £8 a load."

There, however, is another bit of evidence of this outbreak of the Army worm in the year 1743 that appears to have been entirely overlooked. This is contained in a small but somewhat rare volume, by John Bartram, printed in London, England, in 1751.†

Mr. Bartram, as he states, "set out from his house on Skuylkil River the 3rd day of July, 1743." Under date of July 16th, near the Indian town of Tohicon, situated between the east branch of the Susquehanna and the main river, he says: "Here I observed for the first time in this journey that the worms which had done much mischief in the several parts of our Province by destroying the grass and even corn for two summers, had done the same thing here, and had eaten off the blades of their maize and long white grass, so that the stems of both stood naked four-foot high; I saw some of the naked dark-coloured grubs half an inch long, the most of them were gone, yet I could perceive they were the same that had visited us two months before; they clear all the grass in their way in any meadow they get into, and seem to be periodical as the locusts and caterpillar, the latter of which I am afraid will do us a great deal of mischief next summer."

Under date of 28th of the same month, having reached Oswego, New York, Mr. Bartram makes this entry in his record: "This was a rainy, thundering warm day, and two deputies arrived from the Oneidas. News came that the worms had destroyed abundance of corn and grass in Canada."

*Second Annual Report of the Secretary of the Massachusetts Board of Agriculture, 1854 (printed in 1855), p. 36.

† Observations on the Inhabitants, Climate, Soil, Rivers, Productions, Animals, and other matters worthy of notice. Made by Mr. John Bartram in his travels from Pennsylvania to Onondago, Oswego and the Lake Ontario, in Canada, to which is annexed a curious account of the Cataracts at Niagara. By Mr. Peter Kalm, a Swedish gentleman who travelled there. London: Printed for J. Whiston and B. White, in Fleet Street, 1751.

ENTOMOLOGICAL SOCIETY OF ONTARIO.—ANNUAL MEETING.

The Forty-ninth Annual Meeting of the Entomological Society of Ontario was held on Tuesday and Wednesday, Nov., 19th. and 20th. During the day meetings, which were held at the Carnegie Library, the chair was occupied by the president, Dr. E. M. Walker, while the evening meeting, held at the Normal School, was opened by the Hon. Martin Burrell, Minister of Agriculture.

Among those present were the Rev. T. W. Fyles, Dr. C. G. Hewitt, Messrs. J. H. Grisdale, W. H. Harrington, A. Gibson, J. M. Swaine, F. W. L. Sladen, J. A. Guignard, J. I. Beaulne and Rev. Bro. Germain, Ottawa; Messrs. H. H. Lyman and A. F. Winn, Montreal; Prof. W. Lochhead, Macdonald College, Que.; Mr. J. D. Evans, Trenton; Prof. J. E. Howitt, Messrs. L. Cæsar and A. W. Baker, Guelph; Dr. R. Matheson, Truro, N. S.; Mr. A. G. Turney, Fredericton, N. B.; Rev. Father Marcotte, Sherbrooke, Que.; Rev. J. B. Mignault, St. Therese, Que., and Messrs. J. B. Tothill, G. Beaulieu, G. E. Sanders, W. A. Ross, H. F. Hudson, C. E. Petch, Field Officers of the Division of Entomology.

On Tuesday morning the members met at the Experimental Farm, where a pleasant hour was spent looking over the specimens exhibited by those present and in examining the fine collections belonging to the Division. A meeting of the Council took place at eleven o'clock at which the report of the proceedings of the Society during the past year was drawn up and various questions of interest to the Society were discussed. A committee was appointed to consider certain changes in the constitution of the Society, which were proposed at a recent meeting at Guelph. In view of the fact that next year will mark the event of the Society's fiftieth annual meeting, it was decided that a Jubilee meeting be held in honour of the occasion, to which delegates from other Societies be invited and that this meeting be held at Guelph, about the beginning of September, the exact date to be decided upon later. A special committee was appointed to take charge of the arrangements in connection with the meeting.

In the afternoon the Society met at the Carnegie Library, the proceedings commencing with the reading of the reports of the various officers of the society, including those of three of the directors on the insects of the year in their respective districts, viz., Messrs A. Gibson, Ottawa; A. Cosens, Toronto; and W. A. Ross, Jordan Harbour. These were followed by the reports of the Montreal, Toronto and British Columbia branches.

The Annual Address was then delivered by the president, Dr. Walker, the subject being "Faunal Zones of Canada."

Dr. Hewitt then gave an interesting "Review of Canadian Entomology for 1912" in which he outlined the work of the division for the year, illustrating the valuable results that have already followed the establishment of field stations in various parts of the country. Prof. Lochhead next addressed the meeting on "The Teaching of Entomology in the Agricultural Colleges", a subject which evoked much interesting discussion. A particularly enjoyable feature of the meeting was the next paper, "The Rise in Public Estimation of the Science of Entomology", by the Rev. Dr. Fyles, whose charming style and dramatic delivery were once again the delight of all the members present.

At the evening meeting, which was held in the Auditorium of the Normal School, the chair was occupied by the Hon. Martin Burrell, Minister af Argiculture, who in a highly entertaining address, introduced the lecturer, Mr. F. W. L. Sladen, of the Division of Entomology. Mr. Sladen, who is a leading authority on apiculture, gave a very interesting and instructive lecture on "Bumble-bees and their Ways," which was illustrated by a number of beautiful lantern slides.

A special feature of the Wednesday meeting was an entertaining address by Mr. J. H. Grisdale, Director of the Dominion Experimental Farms, in which a keen appreciation was shown of the work that is now being done in Canada in economic entomology.

During the meeting the following papers were read; "The Chinch Bug in Ontario," by Mr. H. F. Hudson; "The Importation and Establishment of Predaceous Enemies of the Brown-tail Moth in New Brunswick," by Mr. J. D. Tothill; "The Discovery of the San José Scale in Nova Scotia," by Mr. G. E. Sanders; "Observations on the Effect of Climatic Conditions on the Brown-tail Moth in Canada," by Messrs Tothill and Sanders; "Observations on the Apple Maggot in Ontario in 1912," by Mr. W. A. Ross; "Notes on Injurious Orchard Insects in Quebec in 1912," by Mr. C. E. Petch; "Insects of the Season in Ontario," by Mr. L. Cæsar; "Injurious Insects in Quebec for the year 1912," by Prof., W. Lochhead; "Forest Insects in Canada in 1912," by Mr. J. M. Swaine; "The Elater Beetles," by Mr. G. Beaulieu; "Aquatic Insects," by Dr. R. Matheson; "The Entomological Record for 1912," and "Flea Beetles and their Control," by Mr. A. Gibson; "Insect Pests of Southern Manitoba during 1912," by Mr. Norman Criddle; "Some New and Unrecorded Ontario Fruit Pests" and "Arsenite of Zinc as a Substitute for Arsenate of Lead," by Mr. L. Cæsar.

The election of officers for the ensuing year resulted as follows:

President:—Rev. C. J. S. Bethune, M. A., D. C. L., F. R. S. C., Professor of Entomology and Zoology, O. A. Callege, Guelph.

Vice-president:—Dr. C. Gordon Hewitt, Dominion Entomologist, Central Experimental Farm, Ottawa.

Secretary-Treasurer:—Mr. A. W. Baker, B. S. A., Demonstrator in Entomology, O. A. College, Guelph.

Curator:—Mr G. J. Spencer, Assistant in Entomology, O. A. College.

Librarian:—Rev. C. J. S. Bethune.

Directors:—Division No. 1, Mr. Arthur Gibson, Div. of Entomology, Central Experimental Farm, Ottawa; Division No. 2, Mr. C. E. Grant, Orillia: Division No. 3, Mr. A. Cosens, Parkdale Collegiate Institute, Toronto; Division No. 4, Mr. C. W. Nash, East Toronto; Division No. 5, Mr. F. J. A. Morris, Port Hope; Division No. 6, Mr. R. S. Hamilton, Collegiate Institute, Galt; Division No. 7, Mr. W. A. Ross, Jordan Harbour.

Delegate to Royal Society:—Mr. A. F. Winn, Montreal.

Auditors:—Messrs J. E. Howitt and L. Caesar, O. A. College.

DESCRIPTION OF TWO NEW SPECIES OF ORTHOPTERA FROM PERU.

BY A. N. CAUDELL.

Bureau of Entomology, U .S. Dept. Agriculture, Washington, D. C.

Among Orthoptera recently received from C. H. T. Townsend, Piura, Peru, for determination were the following two species which seem to be undescribed.

Plectoptera huascaray, n. sp.

Description.—♂ (the ♀ unknown): Most closely allied to the *P. micans* of Bolivar from the West Indies but is decidedly larger. Is also allied to *P. picta* S. and Z. but has darker elytra and wings and the pronotal disk is not margined in front.

General colour black variegated with brown. Head black with a small ashy variegation and transverse stripe about the insertion of the antennæ; antennæ black, the first few segments lighter. Pronotal disk broadly elliptical, black in colour with the lateral margins broadly and the posterior margin very narrowly and interruptedly bordered with yellow. Elytra brown and black, the humeral area and a large subquadrate spot at about the apical third of the posterior margin black, the rest yellowish brown flecked with black, the black flecks assuming a definite elongate shape and regular arrangement along the posterior half of the costal margin. Wings large, smoky brown, the apical area nearly black, the costal margin almost entirely so; the apical area is very large, being nearly as long as the rest of the wing, and the base is straight, not at all angulate. Abdomen black; supra-anal plate twice as broad as long, mesially produced apically and narrowly rounded; subgenital plate asymmetrical, diagonally incised apically and furnished with short style-like organs; cerci short and stout, widest beyond the middle. The legs are black with a pre-apical yellowish band on the tibiæ and the base of the basal segment of the tarsi also lighter.

Measurements. Total length from front of pronotum to the tip of the elytra, 7 mm.; of elytra 5.5 mm.; of wing, 10 mm.

Type a single ♂, Huascaray, Peru, September 21, 1911, altitude 6500 feet. C. H. T. Townsend collector. Catalogue No 15321 U. S. Nat. Museum.

Cocconotus charape, n. sp.

Description.—♂ (the ♀ unknown): Allied to *C. pulcher* Brunn. and runs to that species in the tables in Brunner's Monograph of the Pseudophilinæ. It differs, however, very distinctly from that species.

In size and general coloration agreeing fairly well with *pulcher*. Head black and yellowish, the occiput blackish shading into yellowish on the cheeks and continuing yellowish down to the mouth-parts; mandibles, labrum and base of the clypeus and the sides of the face piceous, front of the face dark reddish brown with an apically pitted tubercle in the centre; on each side and just above the ends of the clypeal suture the face bears a large erect piceous pointed tubercle about as long as the clypeus; antennæ piceous basally, shading gradually to reddish brown. Pronotum without carinæ, the shoulders only slightly squared; disk slightly rugose, truncate behind, gently rounded before, the main transverse sulcus profound and situated distinctly behind the middle; prosternal spines long, sharp and piceous, the rest of the lower surface of the thorax light yellowish; the disk and the lateral lobes of the pronotum margined with piceous and the central portion of the disk, especially anterior of the principal sulcus, light yellowish brown, which colour continues down diagonally forwards entirely across the lateral lobes. Legs stout and yellowish, the coxæ, the geniculations and the dorsal surface of the anterior tibiæ more or less infuscated; fore tibiæ furnished with conchate foramina and armed above on the inner margin with four tubercular swellings and armed beneath with a double row of spines; fore femora less than one and one half times as long as the pronotum, smooth above but armed beneath on the front margin with three short black spines; middle legs similar to the front ones but the tibiæ have three distinct spinules above; hind femora very stout and short, the greatest width about three and one half times the length, smooth above, beneath armed on the outer side with seven or eight stout spines and on the inner margin with a smaller number of smaller spines, all the spines piceous to the base; hind tibiæ slightly curved, armed above and beneath on both margins with piceous spines, those beneath smaller and placed more remote from each other. Elytra fully developed, surpassing the tip of the abdomen, the anterior half greenish, the posterior half brownish; tympanum small, that of the left elytron the smaller and margined with piceous; wings about as broad as long and very gently infumate, when folded just reaching the tip of the elytra. Abdomen moderately plump, dark brownish, apically growing lighter; supra-anal plate small, vertical apically, obtusangularly rounded, entire; subgenital plate moderately elongate, truncate apically and furnished with a pair of elongate club-shaped apical styles, black in colour; cerci short, stout and apically cut squarely off, the tip slightly excavate and armed dorsally with a subapical tubercle.

Measurements. Entire length of body from the front of the head to the tips of the subgenital stylets, 33 mm.; pronotum, 7 mm.; elytra, 26 mm.; wings, 23 mm.; fore femora, 10 mm.; hind femora, 20 mm.; width of hind femora at the widest part, 6 mm.; of elytra at widest point, 7 mm.; three millimeters from the tip, 3 mm.; of wings at widest point, 21 mm.

Type a single ♂. Rio Charape, Peru, September 17, 1911. C. H. T. Townsend, collector. Catalogue No. 15320 U. S. Nat. Museum.

ON SOME APPARENTLY NEW COLEOPTERA FROM INDIANA AND FLORIDA.

BY W. S. BLATCHLEY, INDIANAPOLIS, INDIANA.

On of the most common of the Chrysomelid beetles taken in Florida in February and March was *Lema brunnicollis Lac.*, which was abundant on the flowers and foliage of the thistle *Carduus horridulus* Push. The first blossom of this thistle opened near Sarasota on February 6th, and the first *Lema* was taken on the 8th. They were found mating on February 16th. and again at Sanford on March 28th.

A careful comparison of these Florida specimens with those from Indiana discribed under the name *brunnicollis* Lac. in my "Coleoptera of Indiana", p. 1111, shows that the two are very distinct, the Florida example being much larger, with less convex elytra and having the frontal tubercles less prominent, the thorax less constricted at base, with two rows of coarse punctures along the median line and with numerous similar punctures scattered over the apical half. In colour the Florida specimens are darker, the elytra being blackish blue and the thorax in most specimens having the apical half clouded with greenish fuscous. These differences were pointed out to the late Frederick Blanchard, of Tyngsboro, Maassachusetts, and to Frederic Knab, of Washington, D. C., both of whom agreed with me that the southern species was undoubtedly the one described as *brunnicollis* by Lacordaire, although Mr. Blanchard wrote that examples of the form discribed from Indiana had been in his collection for many years under that name. The northern form is apparently unnamed and is herewith described more in detail as follows:

Lema palustris sp. nov.

Elongate-oblong. Head, thorax, scutellum and under surface, except abdomen, dull red; antennæ, legs and abdomen black; elytra bright greenish blue. Head very finely and sparsely punctate, the front with a strong bilobed tubercle. Antennæ with joints 1 to 4 subequal, the others longer and gradually stouter. Thorax as long as wide, finely and very sparsely punctate, with a single row of 5 or 6 coarser punctures along the median line; sides constricted behind the middle. Elytra impressed on the inner side of humeral angles, each with 10 rows of rather coarse, scarcely impressed punctures; intervals wholly smooth; abdomen distinctly but rather sparsely punctate. Length 4-4.5 mm.

In Indiana the species here discribed has been taken by sweeping herbage only in the tamarack swamps of the northern third of the State, hence the specific name given. It is probably a member of the Alleghanian fauna. The principal differences between it and the southern form, believed to be the true *brunnicollis*, have been given above. The length of the latter is 5-5.5 mm., and the body is proportionally much stouter. From the description of *L. coloradensis* Linell, *palustris* differs in having the antennæ and legs wholly black and in the abdomen being distinctly punctate.

Chlamys nodulosa, sp. nov.

Subquadrate, robust. Uniform dark brownish bronze. Antennæ paler at base, serrate from the fifth joint, the third and fourth joints subequal. Eyes large, reniform, deeply emarginate on the inner side, separated by an interval less than their longer diameter. Front with a number of fine scattered punctures. Thorax without trace of strigæ, the central gibbosity large, its crest with a pair of tubercles, its anterior face with four interrupted carinæ, each pair confluent at apex; a prominent tubercle each side one-third from apex and near the outer of these carinæ, and another, semi-obsolete, midway between this and the side of thorax, the intervals between the carinæ and tubercles deeply, coarsely but not densely punctate. Elytra each with about 9 prominent tubercles, the intervals between these with coarse punctures. Pygidium coarsely and sparsely punctate and with three short carinæ extending from a median gibbosity nearly to the posterior border. Under surface, especially the meso- and meta-episterna, very coarsely, densely and shallowly punctate. Length 4-4.5 mm.

Described from 10 specimens beaten from scrub-oak near Arch Creek, Sanford and Ormond, Florida. March 12th.–April 3rd. A pair of the cotypes are in the collection of Fredric Knab, and another in that of the late Mr. Blanchard.

This is a smaller species than *C. plicata.* very different in the sculpture of thorax and elytra and in the narrower separation of the eyes. It is more subquadrate and robust than *Exema gibber* Oliv. and has also a wholly different sculpture from that species, the tubercle being more pointed and prominent and the punctures more rounded, distinct and deeper. The character usually given as separating the genera *Chlamys* and *Exema* is very slight and more or less variable, and it is my opinion that the latter genus should be abandoned, *Chlamys* having the priority.

Cryptocephalus sanfordi, sp. nov.

Short, robust, subcylindrical. Head, thorax, scutellum, legs and under surface reddish yellow; elytra straw-yellow, the basal fourth of second interval, the entire fourth interval except a small oval spot at apical fourth, and three oblong spots on sixth interval shining black; joints 6–11 of antennæ fuscous. Front of head with a few minute scattered punctures. Thorax wholly without punctures. Elytra with six entire punctured dorsal striæ, the sutural stria represented by only 3 to 4 punctures, the first dorsal forking at the middle and therefore double on basal half, the fourth and fifth striæ sinuous and approaching in the black spaces; alternate intervals wider and wholly pale. Abdomen minutely and sparsely punctate, each puncture bearing a fine prostrate hair; fifth ventral deeply concave at middle. Length 3.5–4mm.

Described from 2 specimens beaten from willow near Sanford, Florida, March 25th.—27th., 1901.

Brachys cuprascens, sp. nov.

Ovate, shorter and stouter than *B. ovata* Web. Dark bronze, thickly clothed above with short coppery-red and whitish hairs, those on elytra arranged in three irregular very sinuous cross-bands composed mainly of the reddish hairs, but bordered anteriorly with the whitish ones. Head and thorax as in *B. ovata*, the median groove of the former narrower and less prominent. Rows of elytral punctures much coarser and more distinct and regular than in *ovata*, those of the interval next to the marginal carina so arranged as to give the appearance of ribs or plicæ beneath the vestiture. Shallow punctures of the under surface much less evident than in *ovata*. Last ventral of female more deeply emarginate, or impressed, and with the fimbriate hairs much more dense than in *ovata:* last ventral of male small, its hind border in both sexes finely pectinate. Length 4.5–5.2 mm.

Nine specimes beaten from the flowers of the farkle-berry, *Vaccinium arboreum* Marsh, near Sanford and Ormond, March 29th.—April 6th. Easily told at a glance from *ovata* by the much more dense and coppery vestiture. Mr. Blanchard wrote me that he had had it separated but not named in his collection for more than 40 years.

Hallomenus fuscosuturalis, sp. nov.

Elongate-oblong. Dull brownish yellow, sparsely clothed with fine prostrate yellowish hairs; elytra with a common sutural fuscopiceous stripe which is widest in the region of the scutellum, the margins also often darker than the disk. Head finely and evenly punctate; eyes small, deeply emarginate on the inner side; antennæ with second joint one-half the length of third, joints 3–11 subequal and one-half longer than wide instead of subquadrate as in the other species of the genus except *serricornis*. Thorax at base one-half wider than long, sides gradually rounded to apex which is one-third narrower than base; disk finely and densely punctate, the basal impressions feeble. Elytra as wide at base as thorax, sides parallel for three-fifths their length, thence gradually converging to the rounded apex; their surface, as well as that of abdomen, very finely and much less closely punctate than thorax. Length 3 mm.

Six specimens beaten from scrub-oak and willow near Sanford. March 28—29, 1911.

NEWFOUNDLAND LEPIDOPTERA.—In a little box of insects collected at St. Anthony's during the summer of 1910, were specimens of *Argynnis freija* Thunb.; *A. myrina* Cran.; *Colias pelidne* Bdv.; *Coenonympha inornata* Edw.; *Apantesis virguncula* Kirby; *Aplectoides livalis* Smith; *Anarta cocklei* Dyar; *Mamestra sutrina* Grote; *Autographa alias* Oltol.; *Epirrita dilutata* D. &. S.; *Epelis truncataria* Walk.; *Pyrausta insequalis* Guen.; *Crambus unistriatellus* Pack. A. F. WINN, Westmount.

Lycaena comyntas Godt.—While collecting Geometridæ after sundown at Valcour, N. Y., July 25th., I found a male *L. comyntas* asleep on a blade of grass. Like many other "Blues," it rests for the night head downwards, the tails of the hind-wings and the black spot strongly resembling a pair of antennæ and an eye at the wrong end. A second specimen was found in the same attitude after a few minutes' search. A. F. WINN, Westmount.

GEOMETRID NOTES—DESCRIPTION OF A NEW EOIS.

BY L. M. SWETT. BOSTON, MASS.

Eois brauneata, n. sp.

Expanse 13 mm; palpi quite long, blackish; head light ash gray at base of antennæ, darker beyond. Thorax and abdomen light ash gray. Fore wings long lanceolate, light gray shaded with brown. Five black dots on costa, from the first basally runs a black hair-line irregularly across the wing showing strongly on the veins as dots. Above the black discal dot there is a spot on the costa; from the discal dot to inner margin there runs a black line. There is a narrow pointed brown band which runs just outside of discal dot and extends toward tip of forewing as it runs beyond discal spot. Beyond this narrow band which resembles slightly the band of *Leptomeris occidentaria* Pack., only much narrower, there is a wide light ash space. Outer margin with a broad brown band crossing to inner margin, through the middle of which runs a zig-zag white line. Venular dots at base of long brown fringe. Hind wings grayish brown with an irregular, broad fuscous band crossing as two zig-zag lines on each side of the discal spot. This band appears to be projected outwardly on the veins. There is a marginal irregular black line. The fringe is long and brownish gray in colour. Beneath fore wings reddish brown, especially near apex of wing. There is a trace of the fuscous band crossing wing at discal spot, which is apparently in the middle of it. The costa has dots of black and yellow at intervals from base to tip. Hind wings lighter coloured than fore wings. Fuscous band at discal spot showing through faintly and bent toward opposite spot; rest of wing beyond light ash.

This delicate little Geometer may be known by the peculiar band across the fore and hind wings and by the ash colour shaded with brown. I take pleasure in naming this species after Miss Braun.

Type.—1 ♀, May 20, 1906, Cincinnati, Ohio; from Miss A. F. Braun; in my collection.

GEOMETRID NOTES.—A NEW DIASTICTIS HÜB.

BY L. W. SWETT, BOSTON, MASS.

Diastictis anataria, n. sp.

Expanse, 24 to 30 mm; palpi, 1 mm, grayish; head light gray at base of antennæ; thorax and abdomen light ashen gray. Fore wings light ash with a bluish tinge; there are four distinct costal patches of rusty brown, the outer being the largest. From each of the costal patches pale rust-coloured bands run sinuately to inner margin and at the outer margin is a broad reddish brown patch extending almost to inner margin. The first costal patch is linear and points towards first venular dots of outer margin; a pale rust-red line runs from this patch almost to inner margin. The second costal patch is not as strongly angulated on costa as the first. A line runs from it to vein 3 in a straight line just inside the black linear discal dot, and then curves backward towards body to inner margin. The third costal patch is smaller than the others but has a pale line running from it to inner margin, bending outwards in a curve opposite discal line to the middle of vein 2, then going straight to inner margin. The fourth costal patch is very large, 1½ mm wide and 2 mm long; it seems to have a notch near the bottom on the outer side and from the base of this runs a pale rust-red line to inner margin straighter than the other lines. The apex of the fore wing is light bluish ash, and near the base of the fourth costal spot is the broad reddish-brown band, widening as it approaches inner margin, being more distinct in some specimens than others, as are also the costal lines, which hardly show in one specimen. The fore wings are finely powdered with red-brown strigæ and have black venular dots on outer margin. The hind wings have traces of two pale brown lines beyond discal spot, but this may be an arrangement of strigæ as it shows in only one specimen. The general colour is lighter ashen with a yellowish tinge and the venular dots are the same as on fore wings. Beneath the wings are densely strigate, the lines of fore wings showing through very faintly from the second and third costal patches on each side of discal spot. Beneath third and fourth patches of fore wings the costa is bright orange, and the apex has brown cloudings. The hind wings are densely strigated also and there are two pale red brown lines crossing wing beyond discal spot. The veins are ochreous and the dots are between, at base of fringe.

Type,—1 ♂, August 8, 1909. Half Way House, Mt. Washington, N. H; taken at light by myself.

Cotypes—2 ♂s, July 27, August 11, 1909, N. E. Harbor, Maine; taken by Dr. Charles S. Minot, in Boston Society of Natural History collection. This species resembles *M. praeatomata* very slightly.

We owe our readers an explanation of the extremely late appearance of our December number. Part of the proofs went astray in the mails and the discovery was not made until after a considerable loss of time. This is the more regrettable as it has resulted in the delay of the January number also.

BOOK NOTICES.

The Large Larch Sawfly; with an account of its parasites, other natural enemies and means of control. By C. Gordon Hewitt. D. Sc. (Bulletin No. 10.—Second Series, Entomological Bulletin No. 5. Division of Entomology, Dept. Agriculture, Ottawa.)

It was a fortunate circumstance that when Dr. Hewitt came to Canada three years ago, he was already intimately acquainted with our most injurious Canadian forest insect, the Larch Sawfly (*Nematus erichsonii* Hartig), this species being apparently a native of Europe and more or less destructive there also. Dr. Hewitt had already spent three years in the investigation of the life history and economics of this insect in England and having thereby determined the means by which its ravages can be checked in its native country he was particularly well fitted to grapple with the more difficult problem of its control in the vast larch or tamarack districts of North America.

Since coming to Canada Dr. Hewitt's studies of the Larch Sawfly have been continued and the results of these and the earlier investigations are embodied in the present report, in which a detailed account is given of the life-history, parasites and other natural enemies of this insect in both Europe and North America and the means by which it can be controlled.

The artificial means of control which have proved useful in the English larch plantations are, of course, impracticable in the vast forests of Canada and we must therefore rely altogether upon the parasites and other enemies. These are, however, not potent enough in North America to check the extensive outbreaks of the sawfly, which have several times occurred in this country, until most of the trees of the affected region have been killed by repeated defoliation. Dr. Hewitt has accordingly been engaged in the importation of sawfly cocoons from England, where this species is largely controlled by an ichneumon fly, *Mesoleius tenthredinis* Morley, and has succeeded in rearing from the cocoons a considerable number of these useful parasites and liberating them in various parts of Canada where the Larch Sawfly is prevalent. There is thus much reason to hope that the *Mesoleius* will become established here and in time increase in numbers to such an extent as to materially aid the other natural enemies of the saw-fly, and perhaps entirely prevent the occurrence of such serious outbreaks as that which we have been experiencing in Canada of late years.

Among the noteworthy facts which have been brought to light in this study are the following: There are four or five ecdyses during the larval feeding period and another ecdysis within the cocoon, whereas according to Packard, whose statements have been followed in all subsequent accounts, there are only there moults. The period from the time of hatching to the spinning of the cocoon is about 16 to 21 days in Canada, which is about twice as long as given by Packard.

The habits of the Field Vole (*Microtis agrestis*), in England of extracting the larvae from the cocoons and feeding on them is parallelled in North America, as observed by Dr. A. N. Fisher of the United States Biological Survey, by the Deer-mouse (*Peromyscus maniculatus artemisiae*). Both of these rodents are normally phytophagous. Insectivorous birds are also an important aid in the control of the Larch Sawfly and their protection and encouragement is strongly recommended.

The bulletin is illustrated by an excellent coloured plate showing the adult and larva of *N. erichsonii*, the effect of its oviposition in the terminal shoot of the larch, and two of its most important parasites, the ichneumon fly, *M. tenthredinis*, and the fungus, *Isospora farinosa*, which attacks the larva within the cocoon. There are also a number of excellent drawings and half-tones from photographs.

Copies of this bulletin may be obtained from the Division of Entomology, Central Experimental Farm, Ottawa.

A Preliminary List of the Insects of the Province of Quebec. Part I, Lepidoptera, by A. F. Winn.

This is a most important contribution to our knowledge of the distribution of Canadian insects. The list embraces nearly 1,300 species and is modelled upon the last edition (1909) of Smith's Insects of New Jersey. It is published as an appendix to the Annual Report of the Quebec Society for the Protection of Plants.

Brief diagnoses of each family are given and under each species is a full list of localities, dates of capture and names of collectors. There are also annotated lists of the collectors whose records have been included and of the localities referred to. A few of the commoner species are illustrated.

Mailed January 22nd, 1913.

The Canadian Entomologist.

VOL. XLV. LONDON, FEBRUARY, 1913 No. 2

FURTHER NOTES ON ALBERTA LEPIDOPTERA, WITH DESCRIPTION OF A NEW SPECIES.

BY F. H. WOLLEY DOD, MIDNAPORE, ALTA.

(Continued from Vol. XLIV., page 39.)

299. *Mamestra mutata*, sp. nov.—Closely allied to *trifolii* Rott., by comparison with which it is best described. Ground colour paler than in *trifolii*, with less irroration; orbicular elongate, oblique, sometimes produced to a point anteriorly, outlined in blackish, with pale annulus and dark centre. In *trifolii* it is round or nearly so, with the pale annulus less contrasting. Reniform much narrower in upper half than lower, the upper extremity the shape of an inverted V with the apex curved slightly outwards. A pale annulus is traceable round the reniform, but is conspicuous only as the strokes, particularly the outer stroke, of this V. In *trifolii* the reniform is kidney-shaped, symmetrical, and the annulus less contrasting superiorly. The subterminal line is paler and more contrasting than in *trifolii*, and the W is rather deeper, and usually preceded by black dentate marks. Fringes more contrastingly cut with pale than in *trifolii*. On the underside, on both primaries and secondaries, there is a smoky discal dot at the end of the cell in both species. In *trifolii* these dots are centred by a fine whitish line on the cross-vein. In *mutata* this line is absent. Size of *trifolii*, but apices rather more acute.

Described from 6 ♂s and 10 ♀s. Calgary, Alta. (4 pair, by the author, June 22nd–Aug. 9th); Miniota, Cartwright, Winnipeg, Man. (1♂, 3 ♀ s, Dennis, Heath, and Hanham, Aug. 3rd–Sept. 20th); Stockton and Provo, Utah, (2 ♀s, Spalding, Aug 5th and 27th); and Prescott, Ariz. (1 pair, Kunze, Sept. 8th and 10th).

Type, ♂, Calgary, in collection of the author.

This is the *albifusa* of Smith (Ent. News, XXI, 360, Oct. 1910,) in part, but is not the *albifusa* of Walker. The character of the orbicular and reniform, and of the discal spots beneath will best serve to distinguish *mutata* from *trifolii* and its var. *albifusa*, to which latter my description might otherwise almost apply. I have held the species under the manuscript name for some time, thinking that I might ultimately find it to be a mere variation, which I am now satisfied that it is not. It is the *trifolii* of my former list which I cited as common some seasons, but I do not seem to have met with the species here for a good many years, my latest specimen being dated 1898. I have no Calgary *trifolii* in my collection at all, and if any one has a species from here under that name it will probably prove to be *mutata*. I have, however, *trifolii* from all the other localities mentioned for the new species except Miniota, and have 54 North American specimens now under examination, including one from Montreal, which I have compared with Walker's type of *albifusa* from Nova Scotia, in the British Museum. I have forms similar to *albifusa* from Ontario and several points in Manitoba, the latter showing a gradation from typical *trifolii*. Mr. J. B. Wallis of Winnipeg kindly lent me a splendid series to select from. One from Treesbank, which I returned him, appeared to be *mutata*, but none of the others. I have seen a specimen taken at Peachland, B. C. by Mr. Wallis. In Smith's paper above referred to, in designating this form as *albifusa*, he mentioned that I had labelled a Maine specimen for him as typical *albifusa*, but adds that he considered that specimen the only doubtful one of the series. I remain under the impression that my labelled specimen was correct. From his description, the bulk of his series were obviously *mutata*. *Albifusa* is a pale, strongly marked form of *trifolii* with contrasting shades and often sienna brown tints.

I have ten British examples of *trifolii* and have examined a long European series in the British Museum. They do not differ essentially from our North American forms, nor have I noticed any specimens, or any figured by Barrett or South, as referred to by Tutt, suggesting my new species. I am aware that there remain two names standing in our lists as synonyms of *trifolii* that remain to be identified,viz., *glaucovaria* Walker, and *major* Speyer. The type of the former, if still in existence, should be in the collection of the Entomological Society of Ontario. That of *major* I cannot locate. But the new species requires a name and I think it best to give it one, in view of the projected Canadian list, at the small risk of creating a synonym, supposing the other names to be now recognizable.

229, a. *M. morana* Smith (Ent. News, XXI, 361, Oct. 1910). —*oregonica* Grt., in part. The form I had listed as "var. *oregonica* Grt." Smith subsequently described as *morana*, and the species is certainly not a variety of *trifolii*.

I am under the impression that Grote described a mixture of two species as *oregonica*, and attached a type label to one of each. In the British Museum is a female type *oregonica* and those other specimens from Oregon which seem to me certainly distinct from *trifolii*, though Hampson makes them "Ab. 2 greyer, fore wing more thickly irrorated with pale brown." In the New York Museum are five Colorado specimens which I took to be the same species. This form, besides being more thickly irrorated and greyer, differs from *trifolii* in having less of a W in st. line, and the terminal space not darker than subterminal, or scarcely so. I saw the form in Smith's collection, and it is probably the one he refers to as *oregonica* in his paper above mentioned. Together we agre d that it fitted Grote's description better than did *morana*. In the Brooklyn Museum I found a male type from Mt. Hood, Oregon, which struck me at once as the "var. *oregonica*" of my Calgary list. It is larger than the British Museum type, and browner, with a deeper W, and impressed me as distinct therefrom, especially as Mr. Doll showed me a long ser es like it from the Yellowstone. I have a Yellowstone female which I compared with it, though mine is distinctly ochreous throughout. By the descr.ption this is evidently *morana*. I have taken no more than one specimen at Calgary, but have one from Laggan (July 17th) and it occurs at Kaslo and elsewhere in B. C. I have no specimens quite like the British Museum type in my collection, and am not positive that Grote's name really involves two species, but if it does, then by the strict law of priority, as the ma'e sex in such cases should hold the name, *oregonica*, male type at Brooklyn, would have preference over *morana*. That law having been voted down, it remains to be decided whether *morana* shall stand.

300. *M obesula* Smith.—High River (Baird) and Red Deer River.

303. *M. picta.* Harris.—High River, May 31st, 1910 (Baird), Red Deer River, July 7th, 1905. Apparently rare in Alberta.

313. *M. ectrapela* Smith.—Two specimens at timber line on Mt. St Piron, Laggan, on July 17th and 18th, 1907, about 7,000 feet.

315. *M. lucina* Smith.—In Prof. Smith's collection I found a figure of the type of *vau-media* from Colorado. The description is made from a single specimen collected by David Bruce, and is stated in Smith's Catalogue to be in the collection of Mr. Jacob Doll. A Calgary specimen in Smith's collection was almost exactly like the figure. The t. a. and t. p. lines are direct, and meet about the middle of the inner margin, forming a V, giving the name to the form which has a striking appearance. I do not imagine it to be anything but an aberration of *lucina-olivacea*, but the resemblance of the Calgary specimen to the type is rather peculiar. Without seeing the type I have no wish to condemn the name, which Hampson lists as a species "incog." but I have no intention of recording *vau-media* as a species from Calgary.

318. *M. larissa* Smith.—I agree with Sir George Hampson in making this a synonym of *anguina* Grt.

319. *M. vicina* Grt.—Since publishing my notes I have studied a good deal of material under the names *pensilis* and *vicina*, including both types, with the result that I have found that Calgary specimens are really most typical of the former. The type of *vicina* is from the Eastern States, that of *pensilis* is from Vancouver Island. The latter has the subterminal line less distinct, more direct, and has less prominent preceding dashes. I have a good series of this from the type locality, and a Kaslo series is only rather more strongly marked. All material from Alberta to the Atlantic coast, and from Utah, I have arranged under *vicina*, but do not believe that there is really any specific distinctness, and specimens from Manitoba and Saskatchewan would fit either series equally well. Sir George Hampson treats them as two species but I have failed to apply the separation given in his tables. "*Vicina*; fore wing moderately b oad, reniform extending well below cell," and, "*pensilis*, forewing narrow, reniform extending slightly below cell." These characters seem very variable.

322. *Scotogramma luteola* Smith= *phoca* Möschl.—The reference is Sir George Hampson's, and with the evidence at my disposal I

prefer to accept it. *Phoca* was described from Labrador and there is a specimen from there in the British Museum from the Standinger collection agreeing with Laggan specimens, though there are none there from Calgary as stated in the Catalogue. The *phoca* of Prof. Smith's collection was *Anarta impingens* Walker, which he also had elsewhere under its correct name. Möschler's figure certainly did suggest *impingens* rather strongly at first sight, but on closer inspection I agreed with Sir George Hampson that it really represented Smith's *luteola*.

323. *S. uniformis* Smith.—I have seen the type of this species in the Washington Museum and have a female in my collection taken by Mrs. Nicholl on Mt. Saskatchewan in the Rockies of Northern Alberta, on July 27th, 1907. Other specimens taken by Mrs. Nicholl are in the British Museum, some of them apparently mixed with *phoca*, which it resembles most nearly, but from which it is probably distinct. It is a large species, and generally more uniform in colour, as figured in Holland's Moth Book, Pl. XXIV, fig. 26, under the erroneous name of *inconcinna*. Hampson's figure of a Colorado specimen is not good, and is not certainly this species. Other records which I have of this species from Alberta are, Mt. Athabasca, 7,500 ft., July 27; Sheep Mountain, July 30th; and Broboktan Creek, Aug. 12th, 1907. Mr. Sanson has taken what I believe to be the species at Banff, July 21st, below 5,000 feet. Some specimens resemble the following.

324. *S. infuscata* Smith.—This is the species I had listed as "*phoca* Moeschl.?" which is probably prior to *luteola* Smith. Hampson makes *promulsa* prior to *infuscata*, though Smith objected to the synonym, stating that Hampson's figure of a Colorado specimen was *infuscata*, and not *promulsa* (Journ. N. Y. Ent. Soc., XV, 151, Sept. 1907). I must leave *promulsa* out of consideration for the present, as I have no means of identifying it, but my No. 324 is less brown than Hampson's figure, though not ochreous enough for true *infuscata*, of which I have seen the types from Park Co., Colo., 10,000 ft., and Gibeon Mt., Colo., 12,500 ft.

325. *S. perplexa* Smith.—This I had listed as *inconcinna* on Smith's own authority, on the strength of which also I permitted Sir George Hampson to figure one of my specimens under the name. The specimen figured is in my collection, though the

figure is not very good of it. The species, however, does not resemble *inconcinna* in the very least The type of that species, a female from Colorado, is in the Washington collection, and I associate it closely with *Mamestra oregonica* and *M. morana*. The description says; "It agrees with *submarina* in the peculiar modification of the last ventral segment, which is carinate at middle and foveate at each side." This seems as applicable to *morana* Smith as it is to *submarina*. Under *perplexa* I have in my collection specimens from Calgary and Laggan, Alta., Kaslo nd Nelson, B. C., and Provo, Utah. Those from the latter locality are the palest of the series, and are evidently the same species as that figured by Barnes and McDunnough from Stockton under this name. The series shows considerable variation in the distribution of the shades, and the paler specimens are nearer *sedilis*, which seems only a variety. Dr. Dyar records it as *sedilis* in the Kootenai list, and the *sedilis* of Sir George Hampson does not differ. Mr. Sanson has taken the species at Banff, July 15th to 27th. *Subfuscula* Grote is doubtfully distinct.

(To be continued.)

THE BEE GENUS HOPLITELLA.

In Canadian Entomologist, 1910, I described a genus of bees from California as *Hoplitella*. I now find that the same name was applied by Davidson in 1909 to a genus of Bryozoa. I propose to change the name of the bee to *Hoplitina;* type *Hoplitina pentamera* (Ckll.) = *Hoplitella pentamera* Ckll. 1910.

T. D. A. Cockerell.

Hepialus auratus Grote.—I am glad to be able to report having captured a specimen of this beautiful moth at St. Therese Island, about 3 miles from St. Johns, Que., on July 10, 1912. This is the second specimen recorded from Canada, the other having been taken by Dr. Fyles in Brome Co., Que., in July, 1865.

G. Chagnon, Montreal.

THE BEE-GENUS THRINCHOSTOMA IN ASIA

BY T. D. A. COCKERELL, BOULDER, COLORADO.

In 1891 Saussure described *Thrinchostoma*, a very remarkable genus of Halictine bees, from Madagascar. Since that time several species of the same genus have been found to occur in Africa, and we have come to look upon *Thrinchostoma* as one of the most characteristic members of the purely Ethiopian bee-fauna. Yesterday I received a box of bees from Mr. F. W. L. Sladen, and in it were two specimens marked "genus?", collected by him in the Khasia Hills, India, in 1895. To my utter astonishment, I recognized a perfectly typical member of *Thrinchostoma*, even to the unique patches of hair on the wings of the male! Thus a genus of bees is added to the fauna of Asia, and we are warned once again of the probable errors arriving from imperfect data on insect distribution. The study of fossils has indicated that the several groups of insects were formerly more widely distributed than at present, and so explains the occurrence of species stranded as it were, in remote regions, far from their nearest relatives.

Thrinchostoma sladeni n. sp.

♂.—Length about 12 mm. (head extended), expanse nearly 19; head and thorax black, with the usual short white hair; inner orbits concave; clypeus greatly extended as usual in the genus, its broad apical margin and the labrum cream-colour, but the sharp simple mandibles rufopiceous; molar space about as broad as long; clypeus shining, distinctly but not densely punctured; upper part of front shining and finely punctured, but its lower two-thirds dull and opaque; scape wholly dark; middle of mesothorax and scutellum brilliantly shining, with scattered minute punctures, but margins, especially broad anterior corners of mesothorax, duller and minutely rugulosopunctate; area of metathorax triangular, finely rugosopunctate; tegulæ light testaceous wings hyaline, slightly brownish, especially on apical margin; nervures and stigma dark rufous; b. n. falling a considerable distance short of t. m.; submarginal cells subequal, the second very broad; first r. n. joining second s. m. almost at end; second t. c. running through a patch of black hairs; legs red-brown, the basitarsi (except more or less at apex, and the hind ones on inner side) creamy white; anterior tibiæ clear red in front; hind femora incrassate, arched above, flattened and concave beneath; hind tibiæ incrassate, whitish above near apex, and below produced into a large flattened white apical lobe, which carries on its surface the widely separated spurs; abdomen claviform, narrowed basally; th.'first segment (except a dusky apical cloud), and the second except a transverse band (narrower in middle) clear ferruginous; rest of the abdomen black, with the hind margins of the segments broadly colourless hyaline; venter light red beneath as far as the fourth segment, which is broadly emarginate; fifth segment dull black emarginate.

♀.—More robust, the produced clypeus very broad, clear ferruginous (as also part of supraclypeal area), flattened and impunctate in middle, strongly lobed at sides, the shining sparsely punctured sides of face forming an acute angle on each side between the clypeus and its lobe; labrum and greater part of the broad bidentate mandibles clear red; sides of face and lower part of front with short golden tomentum; scape reddened apically; apical half or more of flagellum obscurely reddish beneath; hair of thorax (dense on prothorax above) pale fulvous; disc of mesothorax more strongly and closely punctured; area of mesothorax with small basal plicæ; first r. n. entering basal corner of third s. m.; third s. m. broader above; legs with golden hair; anterior tibiæ and tarsi, and middle tibiæ in front, clear red; only the first abdominal segment red, with a pair of subapical brown spots; second segment with the broad apical margin orange; the shining short hairs of the apical margin are golden on the second segment, but white on the others.

Hab.—Khasia Hills; the male is the type. The female is dated June. The sexes differ sufficiently to suggest that they may represent two species, but they are probably identical. The male is quite similar to the African *T. orchidarum* Ckll., differing principally by the claviform abdomen with red base, and the much less broadened hind tibiæ. The fifth ventral segment of *T. orchidarum* carries a broad dense brush of hair, wanting in *T. sladeni*.

It is perhaps possible that the Indian *Halictus wroughtoni* Cameron is a *Thrinchostoma*, although Bingham's figure of the male shows ordinary hind legs and gives no indication of hair-patches on the wings. It is in any event distinct from *T. sladeni*.

INQUIRY INTO THE RELATIONSHIPS AND TAXONOMY OF THE MUSCOID FLIES.

BY C. H. T. TOWNSEND, LIMA, PERU.

Dissections of the female reproductive system and studies of the eggs, first-stage maggots and reproductive habits of these flies, carried on for the past five years, have proved a golden key for unlocking many of the secrets connected with their relationships. Throughout the work, however, the problem of harmonizing these characters with those of the external adult anatomy has been a difficult one. At first sight the results seemed to indicate that the family groups heretofore recognized do not exist in the commonly accepted sense. The ordinary divisions seemed almost untenable, being often at variance with the results of the dissections or with external adult characters of well known utility.

It was soon evident that no satisfactory classification could be built up on the reproductive system characters alone. As examples of the disagreement between reproductive and external adult characters, the *Phasiidae* show in part flat-ovate macrotype eggs without uterus, in part elongate eggs deposited subcutaneously, also without uterus; and, if the Rutiliine and related flies are included in the family, in part elongate subcylindrical eggs hatching in an elongate uterus. The *Exoristidae*, after being restricted greatly from their former limits, are still more markedly differentiated in type of reproductive system and egg, showing not only the three Phasiid types but a half dozen or more additional ones as well.

It is now quite apparent that the external adult characters can not be subordinated to the reproductive characters in quite a good many cases, though they can so be in other cases. It seems practically certain, for example, that parallel specializations of the reproductive system have arisen quite independently in these flies, and that marked and parallel differentiations of the facial plate have so arisen with far less frequency. Facial plate differentiation is largely dependent on a greater or less lapse of oral and antennal functions, and such lapse is not of frequent occurrence. Reproductive system and egg modifications manifestly play an extensive part in the economies of these flies, wherefrom we may conclude that the reproductive system is plastic in a greater degree than is the facial plate. Subtribal to subfamily should be attributed to the characters of the reproductive system, eggs and first-stage maggots, with family values under certain circumstances. When the true or incubating uterus is present its main type is a character of high value. The structure of the egg-chorion and certain structural details of the first-stage maggot are characters of still higher value. When such characters as these are supported by others they may well be used in family definitions. Practically all the early-stage, egg and reproductive - system characters are especially important and serviceable to us as indicating positively the natural limits of taxonomic groups, whereby we can with certainty draw a fixed line between those groups whose individual forms often can not be separated on the external adult characters.

A classification of the Muscoidea into family divisions founded on the general character of the egg, whether elongate-subcylindrical or flattened-ovate, is quite out of the question considering the external characters of the flies themselves; one founded on elongate uteri, or the absence of uterus, or on maggots developing in the uterus, would result similarly in an artificial and unnatural grouping. This may be realized by studying the tabular summary at the end of this paper. But there are certain other characters exhibited here that will apply to family divisions. For example the old family *Sarcophagidae* may well be restored in a new sense on the characters of the cordate and V-shaped uterus, both types being a double-sac specialization of the uterovagina quite distinct in character from all the other forms of uterine specialization in the Muscoidea. This division is strengthened by the generalized character of the cephalopharyngeal skeleton in the first-stage maggot of most of the forms, and by the deeply-sunken anal-stigmatic cavity of all the maggot stages. Employing the uterine character it becomes now for the first time possible to define positively and accurately the limits of this family.

It now seems equally desirable to restore the old family *Dexiidae*, but in a new sense, on a combination of facial plate and accessory supporting characters, definitely limited by the reproductive and especially by the first stage cephalopharyngeal characters. This is a natural group intermediate in facial-plate evolution between the Muscid-Masiceratid stocks on the one hand and the Megaprosopid-Cuterebrid stocks on the other. We return at this point to an approximation of the group concepts of Schiner, who had an excellent eye for main natural distinctions in the Diptera.

The Megaprosopid type is clearly, though not closely, allied with the Dexiid, but its facial plate structure is closer to the Oestrid type than to the Dexiid, while the uterus and first-stage maggot as well as the cephalopharyngeal skeleton of latter differ markedly from those of the Dexiidae; hence it is advisable to maintain the group separately for the sake of uniformity. There has been a differentiation of the facial plate in the *Sarcophagidae*; the Paramacronychiine, Miltogrammine and Macronychiine types exhibiting a successive specialization in the direction of the Dexiid and Megaprosopid-Oestrid types. This is a case in which the facial plate characters are subordinated to the reproductive. It must be noted that the family group *Dexiidae* as restored does not include the many forms of the Pseudodexiine and Pyrrhosiine types, all of which have the Exoristid facial plate, though many of them possess pubescent and even densely plumose arista.

There are two large and taxonomically very practicable groups heretofore left in the *Exoristidae* that may most advantageously be accorded family rank at the present time. These are the microtype-egg forms with leaf-oviposition habit so far as known, which constitute the *Masiceratidae*; and the minute-platelet, coloured-maggot forms with foliage-larviposition habit so far as known, which constitute the *Hystriciidae*. There are three main categories of the former differing in the shape of the maggots and eggs; besides which there are numerous types differing in the structure of the chorion, which quite certainly indicates much diversity of origin. Yet they form a group easily defined on dissection of the females, and taxonomically quite as tenable as the *Oestridae* and several other long-accepted families.

The group of which *Phasiopteryx* is the type merits family rank on the remarkable and, so far as yet known, unique change of the eggs in the uterus from microtype ovate to macrotype subcylindrical, indicating wide separation from other stocks; not to mention the very exceptional structure of the first-stage maggot, which is no doubt largely adaptive.

The Cuterebrine flies are likewise too aberrant a type to be longer included in the same family with any of the other groups. They seem to have sprung from some old Mesembrinine stock, but are to-day well removed from their nearest living relatives.

It is now 23 years since Brauer and von Bergenstamm used the names *Masiceratidae* and *Hystriciidae*, but in different senses from those here employed. The family names must be accredited to them, since they employed them for the family types. The *Masiceratidae* as here revised includes but a fragment of the group to which they gave the name, only two of their genera so far as we yet know falling in it, these being *Masicera* and *Ceromasia*. But it takes in many of their *Phocoreratidae* and *Blepharipoda*, all of their *Willistoniidae* and *Goniidae*, their section *Myxexorista* (1893) and some at least of their *Baumhaueriidae* and *Germariidae*, The *Hystriciidae* as here revised includes all of their *Hystriciidae* except *Tropidopsis* which belongs in the Pyrrhosiine subfamily (*Hexamera* is not known to me), all of their *Tachinidae*, *Tachinoidae*, *Micropalpidae* (*Homoeonychia* unknown to me) including their section *Erigone* (1893), and a very few of their *Pyrrhosiidae*. It is profitable to note these comparisons as showing how nearly these authors in certain cases approached and how widely in others they deviated from proper definition of the groups on a study of the external adult characters alone.

If the peculiar reproductive and early-stage characters of *Phasiopteryx* are found to exist in *Oestrophasia*, the family will take the name *Oestrophasiidae* B. B. (1889). The name *Cuterebridae* was used in the present sense by Brauer and von Bergenstamm in 1889, but the family was ranked as an "Unter-Gruppe."

The *Sarcophagidae* of the present paper includes a large part of the *Sarcophagidae* B. B., a part at least of their *Rhinophoridae*, probably a part of their *Phytoidae*, probably all of their *Miltogrammidae* and *Paramacronychiidae*, and *Macronychia* alone of their *Macronychiidae*. In 1893 they referred *Melanophrys* to their *Paramacronychiidae*, but this genus belongs to the *Hystriciidae* of the present paper. The *Dexiidae* as here revised includes practically all of the *Dexiidae* B. B., and nearly all of their *Paradexiidae*.

From various comparisons we are able to judge with considerable certainty that the characters of the less adaptive structures of the egg and first-stage maggot are, on the whole, of prime taxonomic rank in the Muscoidae. They are therefore available for family definition in the case of large groups or pronounced types where other characters fail us. We may also justly conclude that the reproductive system and general egg and maggot structures furnish characters of inferior rank but of great service in the definition of such taxonomic categories as genera, group-units, subtribes, tribes and subfamilies, and even at times of families if they are supported by other important characters.

A comparative study of plant and animal taxonomy suggests (1) that the eggs, embryos, early and adolescent stages of animals will always furnish us the main key to their affinities whether such is present or lacking in the adult; (2) that the characters of the reproductive system, while of less rank, will enable us to fix definitely the limits of the lower taxonomic categories when their definition is obscured in the adult; and finally (3) that the more a structure becomes specialized, the more the taxonomic value of its characters contracts. The first point justifies the erection of the eleven families outlined and recognized in this paper. The last point emphasizes again the extreme taxonomic difficulties that exist in the muscoid flies, which are undoubtedly not only among highly specialized but also among the most recently specialized of all arthropods and hence the most difficult to classify in a convenient system. However much the values of certain characters may contract, in other words however obscured may become the group relationships in the structures exhibiting these characters, we are nevertheless often compelled, in the absence of others more distinctive, to use them if we wish to define certain of the higher taxonomic categories. Thus, in order to attain the greatest degree of clearness and practicability, we should in actual practice limit our main group-definitions to the fundamental group-categories or lowest groups of genera in these flies, which have been called group-units. Each group-unit consists of the *typic genus* together with those *atypic genera* which are found to be more closely related to it than to any other typic genus. For definition of typic and atypic genera, see Tax. Musc. Flies, p. II.; and for many pertinent considerations, pp. 7–13. As an example, *Exorista* may be taken as a typic genus, and *Euphorocera* as an atypic genus belonging with it; these together form a basis for the Group-unit *Exoristiæ*.

The eleven family types indicated in the diagram farther on show pronounced adult characteristics which cause them to stand forth prominently, as in bas-relief, from the mass of the Muscoidea. These characteristics are reinforced by valuable characters drawn from the eggs, early stages and reproductive system. But the limits of the groups which these eleven family types represent are often greatly obscured in the external anatomy of the adult, and it is the function of the egg, early-stage and reproductive characters to clear up this obscurity in all cases. This is the first time in the history of muscoid taxonomy that we have had the means of definitely segregating these various families and accurately determining their limits, and they may well be maintained now on the sum-totals of their respective characters. But in synoptic treatment, as will appear later, these family groups are unwieldy and do not aid us as such, though their divisions may be employed as leading directly to the group-limits.

The characters of the facial plate apparently continue to hold better in the main for the indication of family types than do those of any other single external adult structure. They become subordinated to the characters of the female reproductive system and early stages occasionally, as in the case of the Sarcophagidae and Dexiidae, but this is in accordance with the well-known law of contraction of values, and the consequent fluctuation of charactacters, which cannot hold for all groups. They are reinforced by various other external adult characters in the several groups. Where they fail from lack of differentiation to mark off otherwise prominent groups, the characters of the first and second categories are always available. The following scheme of derivations illustrates well, reading the group from left to right, the successive retrograde modifications of the facial plate that appear to have taken place in these flies.

(1) Phasid stem (Facial plate remains wide and elongate).

(2) Muscid (3) Exoristid (4) Hystriciid (5) Masiceratid stem (Facial plate shortens but remains wide).

(6) Phasiopterygid (7) Dexiid stem (Facial plate shortens further and becomes constricted below).

(8) Sarcophagid stem (Facial plate in the typical stock very

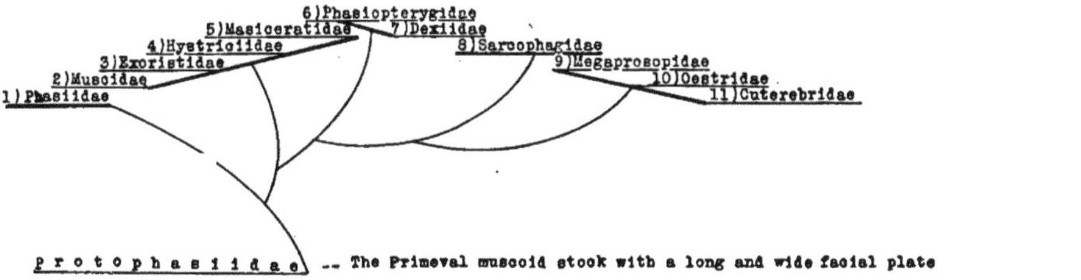

similar to that of the Muscid stem, but some stocks show a shorttening of clypeus and inferior constriction of the facial plate, foreshadowing and even approximating those of the extreme types).

(9) Megaprosopid (10) Oestrid (11) Cuterebrid stem (Facial plate reaches extreme of clypeal shortening and epistomal constriction).

The lines of descent have not been simple, but on the contrary quite complex, and the plan merely indicates the general trend in facial plate modification. From the Phasiid to the Oestrid extremes the successively increasing differentiation may be traced in successive types of ever greater clypeal shortening and epistomal constriction. It seems almost certain that the facial plate has specialized according to the retrogressive evolution here indicated. There are several facts that appear to confirm this view quite conclusively. Australia possesses no endemic oestrid nor cuterebrid stock, but it has been the focus of a considerable number of forms which must be considered as survivors of primitive phasiid stock. These are *Rutilia*, *Amphibolia*, *Microtropeza*, *Paramphibolia*, *Amenia*, *Senostoma* and *Chrysopasta*. Certain relatives of these, also evidently to be classed as survivors of the same stock, occur in the Australasian or Austromalaysian regions and strengthen the case in hand. These are *Paramenia* of New Zealand, *Pseudoformosia* of New Guinea, *Stilbomyia* of Java, and others. It is to be noted that none of these, however, reaches either South America or South Africa. Both of these continents were apparently cut off from the Australian-Antarctic landmass at a time antedating the greater or main dispersals of that branch of the primitive phasiid stock which gave rise to these forms. These facts indicate that the phasiid stocks are much older than the oestrid and cuterebrid stocks.

Other facts point to the same conclusion. The specialization toward partial and complete atrophy of the mouthparts in the oestrid stocks, toward partial atrophy of the same in the cuterebrid and megaprosopid stocks and toward antennal reduction in these and kindred stocks indicates that the extreme shortening and constriction of the facial plate are connected with a more or less complete loss of mouth and antennal functions. Certainly this is comparatively recent specialization, for the primitive stocks must have had highly functional mouthparts as well as high antennal development.

Facial plate reduction has probably followed antennal and mouth reduction. In other words it is a consequence of loss of nutritive and olfactory functions in the fly, and thus marks an extreme stage of parasitism and host-adaptation, particularly one in which the sexes may easily find each other, in which the female may easily find the host, and in which the maggots may easily store a large food-supply. The two muscoid stocks which are apparently of most recent evolution, the Masiceratid and Hystriciid, in which the mouthparts and antennæ are both still highly functional and the facial plate in consequence still retains its full development, have much less perfect host-relation, sex-relations and food-supply conditions. They must search assiduously for their hosts; the large fecundity which is necessary to their peculiar host-relations demands extensive feeding in the adult female, especially as she has not an unlimited food-supply during her larval life; and the necessity for feeding and host-searching makes the female a wanderer, whose discovery by the male calls for well-developed olfactory organs.

The comparison of *Cobboldia* with other types shows conclusively that pharyngeal atrophy (atrophy of pharynx and rostrum of proboscis, and not necessarily of haustellum or palpi, with more or less complete closure of pharyngeal cavity) is directly correlated with the evenly receding and gently-convex profile of the facial plate and peristomalia, and the consequent more or less complete recession of the epistoma; further that the great shortening of the clypeus is primarily dependent on and thus directly correlated with antennal atrophy (atrophy practically only of the third joint, and consequent loss of olfactory function).

As correlated in importance with facial plate specialization among external adult characters but of less value, it is interesting to note that excessive macrochaetal development has taken place in several stocks and probably by parallelism. The following groups, arranged by families, exhibit spinose-macrochaetal specialization:—*Hystriciidae*, the climax of all (*Dejeanini*, *Saundersiini* pt., *Hystriciini*, and *Larvaevorini* pt.); *Masiceratidae* (*Blepharipezini*, *Belvosiini* pt.); *Exoristidae* (*Pyrrhosiinae* pt.—*Tropidopsis* and *Paragymnomma*); *Dexiidae* (G.-U. s. *Echinodexiiæ*, *Tropidodexiiæ*); *Megaprosopidæ* (G.-U. *Megraprosopiæ*); *Phasiidae* (G.-U. *Amphiboliiæ*).

The wisdom of separating the *Megaprosopidæ* from the *Dexiidæ* and of maintaining them on a par with and more allied to the *Oestridae* may be questioned. It may be argued that the presence of macrochaetae allies them more with the *Dexiidae*. We know, however, that their maggots are of peculiar structure, that of *Microphthalma* at least being quite thickly clothed with long bristly hairs and representing the extreme development of bristly vestiture in the first-stage maggots so far as known, while its cephalopharyngeal skeleton is of a distinct type from the dexiid. Their uteri are of markedly different type from the form typical of the *Dexiidæ*, being known to be very long and irregularly coiled in both *Microphthalma* and *Megaprosopus*. Their segregation is thereby demanded since these characters strongly reinforce those of the facial plate. The absence of macrochaetae in the oestrids is due to their aerial life-habit, which is not shared by the Megaprosopids.

It is possible, notwithstanding the facial and oral characters, that the *Trixodini* may be found on investigation of their reproductive system and first-stage maggots to belong with the *Dexiidæ* rather than with the *Megaprosopidae*. They almost certainly have a uterus of the continuous-canal type and it is quite possibly of the fat and shortened dexiid type, but the final test of family position here will lie in the type of pharyngeal sclerite possessed by the first-stage maggot. These flies are very rare, at least in collections. The only known specimens are two collected by myself on tree-trunks in the mountains of the Rio Gila headwaters in

New Mexico and the Sierra Madre of Chihuahua. During a trip across the Sierra Madre of Chihuahua and Sinaloa in August and September, 1909, I made especial search for these forms but found none. A similar search earlier in the season would probably have been successful. These flies are of unusual interest as exhibiting facial and oral characters intermediate between those of the *Megaprosopidae* and those of the *Oestridae*, while their weak macrochaetae show a further trend toward the latter family. It is probable that they parasitize wood-boring larvæ.

If, as seems very certain, *Rutilia* and *Amphibolia* represent an old stock, then uterine development must be of very long standing. Both forms have coiled uterus in which the elongate eggs hatch. Certainly a type without incubating uterus would seem to be the original, and elongate subcylindrical eggs should be the more primitive form. If this is true, we must go well back into the past for the beginnings of the remarkable specialization in reproductive system, eggs and maggots of these flies. These specializations have quite certainly been largely adaptive, and thus we are better prepared to accept their independent origin in several stocks. Ovate, flattened eggs are an adaptation for attachment to surfaces, the larger or macrotype forms being designed for fastening externally to host and the small or microtype forms for fastening to leaf-surfaces to be swallowed by host. Here is extensive adaptation even in size —a specialization to a microscopic egg that can be swallowed by leaf-feeding insects without injury to the contained maggot. This last specialization seems to have arisen independently in several stocks, since these eggs exhibit a wholly unexpected variety of structure, the choria of some being reticulate after a honeycomb pattern, those of others having a pattern of raised arcs or wrinkles, while some have a perfectly smooth and unreticulate chorion, and still others have the chorion finely or coarsely punctured or finely or coarsely set with raised points.

Pediceled eggs are for attachment to hosts in place of flattened eggs. If neither pedicel nor flatness can be secured, nor viscid secretion for gluing the eggs nor structures for depositing them subcutaneously, then in order to meet the requirements of parasitism the eggs must be held in the uterus until the maggots are fully developed and have become highly active. Hence the need for special uterine development. So far as yet known no elongate unpediceled eggs are ever deposited on hosts except by the Gastrophiline. Cobboldiine and Cuterebrine flies, whose eggs are provided with a profuse viscid coating for attachment by their lateral or latero-anal surfaces to the hairs of the host. In this connection it also becomes evident that forms affecting a host to which the fly can not gain access must possess a uterus in which to develop active maggots that can search for and penetrate to such hosts.

Some maggot-depositing flies, on the other hand, which have what would seem the most perfect access to the host are most careful to keep at a certain distance from the latter. Such are the *Hystriciidae* or leaf-larvipositing forms which are greatly specialized in their coloured maggots, long coiled strap-like uterus, consolidated cephalopharyngeal skeleton and excessive macrochaetal development. Their very divergent host-relations may be in part due to certain of their hosts living in webbed nests and being in the habit of spinning sundry silken threads both for enlarging and changing their habitations and for marking their feeding trails whereby they may retrace their way to the nest. Silken webs are especially dangerous to forms of excessive macrochaetal development; and it may be that there is some connection between this and the origin of this remarkable host-habit, with the consequent coloration of the maggot.

Even more consolidated than in the *Hystriciidæ* is the cephalopharyngeal skeleton of the first-stage maggot in the *Masiceratidæ* or leaf-ovipositing forms, in which his structure has reached the extreme of reduction and consolidation. This argues for a high degree of specialization here, of longer standing than that of the leaf-larviposition forms. The conclusion is borne out by the elongate intestiniform uterus, microscopic size of the egg, and the remarkably divergent host relations whereby it becomes necessary fully to develop the maggot within the chorion without allowing it to escape therefrom until it shall have arrived in the alimentary canal of the host, notwithstanding that it may remain for a considerable time deposited and unswallowed. Such provisions mark an extreme specialization of very long standing. How these microscopic eggs could have originally arisen from a larger type, as we must needs conclude they have done, in sufficient mode to become established, necessitating corresponding marked changes in the oviposition habits and thus in the instincts of the female, is a fascinating problem for solution. It seems certain, moreover, that such eggs have arisen independently in several different stocks, but probably largely through parallelism due to evolution trend.

The flies with subcutaneous host-larviposition habits and those with subcutaneous host-oviposition habits are likewise much specialized. The remarkably specialized piercers, larvipositors, ovipositors, combinations of these, and accessory structures such as the ventral carina and its spinules denote high specialization. During copulation the piercing structures have evidently to be extended or thrown far backward, in the *Compsiluriæ* at least, for effecting the union of the vaginal orifice with the male.

While *Phasiopteryx* appears to be a waning survivor of an old stock with dexiid affinities, it exhibits a large amount of specialization in its very long and slender uterus and especially in its very differentiated isopodiform maggot with chitinized segmental lateral and dorsal plates. But what holds the utmost attention and interest in this form is the wholly unique character of the ova accomplishing in the upper part of the uterus their final growth or increase in size to the fully formed macrotype egg, which should by analogy with other macrotype-egg forms have been completed in the ovarioles. This seems difficult of explanation and, of course, at once suggests some connection with the microtype-egg or leaf-ovipositing forms. But *Phasiopteryx* is to all appearances of external structure far removed from the microtype-egg stocks. Its uterus is very similar in general form to that of the *Phasiatactiæ* and *Cnephalomyiiæ*, both of which have an elongate and more or less pointed microtype egg that is flattened ventrally. Its facial plate is not so divergent in type as to preclude a common origin with the masiceratid stocks. It seems probable that we must look on *Phasiopteryx* as a remnant of an offshoot from some ancient microtype-egg stock. If this view is correct, we may expect important light on phylogeny of the microtype-egg stocks from a thorough study of this genus and its allies. It appears quite certain from the facts in the case that the ancestors of *Phasiopteryx* possessed a microtype egg.

It is a general rule throughout the Muscoidea that those groups with greatest fecundity comprise parasitic forms whose host-habits afford their maggots the least favourable opportunity for encountering the host. Conversely the opposite is the case. The fecundity runs highest in the *Masiceratidæ* and *Hystriciidæ*, leaf-ovipositing and leaf-larvipositing parasitic forms, the latter exhibiting the extreme Thus we may conclude that in these groups there occurs the highest maggot mortality. Those forms which are parasitic in white grubs, wood-boring grubs, and hosts in general which the maggot must seek out for itself with limited chance of finding them also have a high fecundity. The *Myiophasiiæ*, which are weevil-grub parasites, have a much lower fecundity, and it is evident that their maggots usually reach the host. Forms which deposit eggs or maggots on the host also have a comparatively low fecundity, and those which inject the maggots or eggs subcutaneously have a still lower fecundity. The typical Sarcophagine flies, which are non-parsaitic in the strict sense, show on the whole the lowest fecundity of all, due to the nature of their larval food-substances on which the highly active maggots are deposited and which is ordinarily bountiful for their needs.

The Sarcophagine flies have perhaps developed maggots *in utero* on account of the generally perishable nature of their larval food-substances, combined with a fairly long incubation period necessary to the development of the maggot. On the other hand the muscine and calliphorine flies have not done so, on account of a marked difference in the nature of their food-substances which are in general less perishable, combined with an incubation period sufficiently short to meet the conditions and requirements of oviposition. It may be here pointed out that the most generalized type of cephalopharyngeal skeleton so far known in the Muscoidea is that exhibited in the first-stage maggots of the Sarcophagine flies and their allies. Evidently the sclerites have here remained almost unspecialized, being unreduced and freely articulated, as best fitted for their larval life-habit.

Returning again to taxonomic considerations, it is necessary to point out more fully that however well the family types already outlined may stand forth on general characters, it is nevertheless true that the recognition of the family groups which they typify does not facilitate synoptic treatment. Their employment does not elucidate the subject, but rather obscures it. They are often incapable of concise limitation and hence of compact synoptic definition on either external, reproductive or early-stage characters. Thus it is necessary to sidetrack them in actual synoptic practice and drop to lower categories. The group-unit is the category that here lends itself most conveniently to taxonomic manipulation. The reason for this lies in the fact that the characters of the reproductive system, egg and early stages, which can not always be conveniently interpreted as of family value, and often of subfamily or even of subtribal value, are much more pronounced and readily apparent, therefore more comprehensive, than those characters of the external anatomy of the fly which largely mark convenient family to subtribal divisions. While the characters of the facial plate and various supporting characters of family to subtribal importance exhibited by other external adult structures are often by themselves almost impossible of correct interpretation, so much so that hardly any two persons can be expected to read them alike, those of the reproductive system, egg and early stages are unmistakable and impossible of confusion.

Such external adult characters as the more or less ciliate facialia, degree of hairiness of eyes, apical cell ending at or near wing-tip, presence of true macrochaetae, hind tibiæ ciliate or pectinate, relative length of aristal and antennal joints, relative development of antennæ, mouth-parts and palpi, and especially exact plan of facial plate specialization including degree of constriction by vibrissal angles and their comparative degree of removal from the oral margin with the conformation of latter, in fact the majority of the external adult characters in these flies, are very difficult to describe accurately and few persons will be able correctly to interpret the descriptions in any event. Moreover these characters indicate close relationships only in certain cases, while in others they are the result of evolutional trend in stocks considerably removed from each other. For this and other reasons their value runs out at times. In certain groups some of these characters become highly untrustworthy, though they may hold good throughout other groups. It is often quite impossible to decide their values correctly without the aid of the reproductive and early-stage characters to guide us in the matter of close relationships. An intricately interrelated system of specialization in external adult anatomy has resulted in producing in distinct stock forms which closely approach each other in external characters. This was not realized until the investigation of the reproductive and early-stage characters had considerably progressed.

Until this work was well under way no one could interpret the genera as they actually exist, and all generic work was largely guess-work. Even at the present time muscoid genera as they commonly appear in the literature are in numerous cases complexes of widely different stocks. Forms belonging to distinct families have for a century been classed as congeneric, and the external differences between them are sometimes so inconspicuous that careless workers have even pronounced them conspecific. These facts serve to emphasize the invaluable aid to be derived from the reproductive and early-stage characters, and the necessity for taxonomic manipulation of the superfamily by means of smaller groups than families, subfamilies, tribes, and subtribes.

The following tabular summary will be useful. It shows the known main differentiations of the female reproductive system, eggs and first-stage maggots in the eleven muscoid families here recognized, exemplified by group-units. The group-unit consists, as already stated, of the typic genus plus the atypic genera which belong with it, and is a division of the subtribe.

Its ending is *iæ*, which is added to the root of the name of its typic genus.

I. PHASIIDÆ	1. Elongate macrotype eggs deposited subcutaneously without incubation, no uterus—Phasiiæ. 2. Flattened subovate macrotype eggs deposited supracutaneously without incubation, no uterus—Ectophasiiæ, Trichopodiæ, Xanthome-

I. PHASIIDÆ
(Continued from page 51.)

lanodiæ, Cistogasteriæ, Rhodogyniæ.

3. Elongate subcylindrical macrotype eggs hatched in coiled uterus—Rutiliiæ, Amphiboliiæ.

II. MUSCIDÆ

1. Elongate subcylindrical macrotype eggs, no uterus—Musciæ, Stomoxydiæ, Calliphoriæ.
2. Elongate macrotype eggs incubated in uterus, deposited as free maggots or maggots in choria—Mesembriniæ, Hypodermodiæ, Eumusciæ.
3. Elongate macrotype eggs hatched in uterus and maggot carried to or through its third stage therein—Dasyphoriæ, Glossiniæ.

III. EXORISTIDÆ

1. Flattened ovate macrotype eggs deposited supracutaneously without incubation, no uterus—Exoristiæ, Plagiopiæ, Winthemiiæ, Neophoroceratiæ, Chaetotachiniæ.
2. Flattened subovate macrotype eggs incubated but not hatched in coiled uterus, deposited supracutaneously—Meigeniiæ, Vivianiiæ, Cyrptomeigeniiæ, Thrixioniæ.
3. Elongate subcylindrical pediceled macrotype eggs incubated but not hatched in coiled uterus, deposited supracutaneously—Carceliiæ.
4. Elongate macrotype eggs deposited subcutaneously without incubation, no uterus—Phaniiæ, Hemydiæ, Leucostomiæ, Dionaeiæ.
5. Elongate macrotype eggs hatching

III. EXORISTIDÆ
(Continued from page 52.)

to white maggots, slender coiled uterus, the maggots deposited subcutaneously—Compsiluriæ, Celatoriiæ, Oxynopiæ, Weberiiæ.

6. Elongate macrotype eggs hatching to white maggots in slender coiled uterus like preceding, but the maggots deposited supracutaneously—Pseudomyothyriiæ, Hyalomyodiæ, Thryptoceratiæ.

7. Elongate subcylindrical macrotype eggs hatching to coloured maggots in coiled very long and slender to fat gut-like uterus, the maggots deposited near host—Eugymnochaetiæ, Bigonichaetiæ, Glaucophaniæ, Eriothryginæ, Macquartiiæ, Ophirioniæ, Steinielliæ.

8. Elongate macrotype eggs hatching to white maggots in short, fat gut-like coiled uterus, the maggots deposited near host—Eumyobiiæ, Pyrrhosiiæ, Ophirodexiiæ, Atrophopodiæ, Thelairiæ.

9. Elongate subcylindrical macrotype eggs hatching to white maggots in coiled strap-like uterus, the maggots deposited supracutaneously—Zygosturmiiæ, Azygo bothriiæ, Voriiæ, Siphosturmiiæ, Eryciiæ.

IV. HYSTRICIIDÆ

1. Elongate subcylindrical macrotype eggs hatching to coloured maggots in long coiled strap-like uterus, the maggots deposited on foliage near hosts—Melanophryoniæ, Ernestiiæ, Micropalpiæ, Copecryptiæ, Servilliiæ, Larvævoriæ, Hystriciiæ, Saundersiiæ, Dejeaniiæ.

V. MASICERATIDÆ

1. Microtype flattened—subovate eggs in coiled subtubular slender to fat uterus, incubated to full development of the shortened subovate maggot but not hatched therein, deposited on foliage near hosts to be swallowed by latter in feeding (chorion varying from gray to yellow and black in colour and exhibiting a great variety of minute structure) — Ceromasiopiæ, Epidexiiæ, Phasmophagiæ, Baumhaueriiæ, Ophirosturmiiæ, Eusisyropiæ Ommasiceratiæ, Dimasiceratiæ, Metopiopiæ, Euceromasiiæ, Euexoristiæ, Eumasiceratiæ, Masiceratiæ, Brachymasiceratiæ, Sturmiiæ Otomasiceratiæ, Chætophorocératiæ, Gaediiæ, Germariiæ, Atactiæ, Triachoriæ, Belvosiiæ, Blepharipeziæ G.-U. s.
2. Microtype flattened elongate pointed or oval eggs, incubated and deposited same as preceding but maggot elongate, uterus very long and slender (chorion black)—Cnephalomyiiæ, Phasiatactiæ, Salmaciiæ.
3. Microtype slightly flattened, elongate-subcylindrical eggs, incubated and deposited same as preceding, uterus very long and slender (chorion smoky-yellowish) — Cylindromasiceratiæ.

VI. PHASIOPTERYGIDÆ

1. Microtype slightly flattened ovate-rounded eggs, growing in the upper part of uterus to macrotype elongate subcylindrical eggs which hatch in lower part of uterus

VI. PHASIOPTERYGIDÆ (Continued from page 54.)

to maggots with chitinized dorsal and lateral segmental plates, the maggots deposited where they must seek the host for themselves, the uterus extremely long and slender—Phasiopterygiæ.

VII. DEXIIDÆ

1. Elongate subcylindrical slender macrotype eggs sharply pointed at anal end, hatching to white maggots with anal setæ borne at ends of anal stigmatic processes, the maggots deposited in choria on soil which they enter in search of white grubs, uterus fat and gut-like—Billæiæ, Microchaetiniæ, Mochlosomiæ, Dexiiæ (I am not certain that the last two groups possess the anal setæ of maggot, nor that the second group is parasitic in white grubs, but both are indicated by my studies).
2. Elongate subcylindrical macrotype slender eggs, hatching to white maggots with anal stigmatic processes but lacking the anal setæ, the maggots deposited at entrances of galleries of woodboring grubs to which they penetrate, uterus fat and gut-like—Sardioceratiæ, Eutheresiiæ, Paratheresiiæ .
3. Elongate subcylindrical slender macrotype eggs, hatching to white maggots in the very fat gut-like uterus, the maggots lacking both anal processes and setæ and evidently deposited near the hosts—Tropidodexiiæ. (In this group

VII. DEXIIDÆ
(Continued from page 55.)

the abdomen is rather densely set with subspinose macrochaetæ.)

4. Elongate subcylindrical macrotype eggs, hatching to coloured maggots in the fat gut-like uterus, maggots deposited near host—Myoceropiæ.
5. Very elongate subfiliform small macrotype eggs, hatching in the fat gut-like uterus to very slender subfiliform white maggots, which are deposited at weevil oviposition-punctures in various green fruits and buds, the maggots making their way to the weevil grubs inside—Myiophasiiæ.

VIII. SARCOPHAGIDÆ

1. Elongate subcylindrical macrotype eggs, hatching to white or yellowish-white maggots in a cordate double-sac uterus, the maggots deposited on the food-substance—Sarcophagiæ G. U.
2. Elongate subcylindrical macrotype eggs, hatching to white maggots in a V-shaped double-sac uterus, the maggots deposited in the nests of various wasps and bees where they feed on stored insect food when such is present and on the early stages of the host—Metopiiæ Eumacronychiiæ, Paramacronychiiæ, Miltogrammiæ, Macronychiiæ.

IX. MEGAPROSOPIDÆ

1. Elongate subcylindrical macrotype eggs, hatching in a very long irregular coiled uterus, the maggot clothed with long bristles—Megaprosopiæ.

X. OESTRIDÆ	1. Subcylindrical macrotype eggs rapidly tapered at anal end and with operculum at the obliquely-truncate cephalic end, incubated in uterus and attached by lateroanal surface to hairs of host by means of a viscid fluid—Gastrophiliæ. 2. Elongate subcylindrical pediceled macrotype eggs without operculum incubated in uterus and attached by the broad claspers of the pedicel to hairs of host by means of a viscid fluid, the chorion cleaving longitudinally for the escape of the maggot—Hypodermiæ. 3. Elongate subcylindrical macrotype eggs, hatching in uterus, the whitish maggots deposited free or in choria in the nostrils of host—Oestriæ.
XI. CUTEREBRIDÆ	1. Elongate subcylindrical large macrotype eggs with heavy chorion and operculum at cephalic end, incubated in uterus and attached by lateral surface to hair or skin of host by means of a profuse viscid fluid—Cuterebriæ, Dermatobiiæ.

SOME HETEROPTEROUS HEMIPTERA FROM SOUTHERN PINES, N. C.

BY J. R. DE LA TORRE BUENO, WHITE PLAINS, N. Y.

The Heteroptera listed below were all collected by Mr. A. H. Manee, of Southern Pines, N. C., whose labours have made possible the preparation of this paper. It is interesting not only on account of the records of distribution, data of great value in themselves, but also because it represents the fauna of a restricted area. It is hoped that it will be of value as a contribution to faunistics.

Euthyrhynchus floridanus Linné.
Mineus strigipes H. S.
Apateticus serieventris Uhler.
Stiretrus anchorago Fabr. var. *fimbriata* Say.
Mormidea lugens Fabr.
Solubea pugnax Fabr.
Euschistus servus Say.
Euschistus tristigmus Say var. *pyrrhocerus* H .S.

There is also one specimen intermediate between the var. and the typical form.

Euschistus crassus Dallas.
Neottiglossa undata Say.
Neottiglossa sulcifrons Stal.
Thyanta custator Fabr.
Nezara pennsylvanica P. B.
Nezara hilaris Say.
Banasa euchlora Stal.
Banasa dimidiata Say.
Brochymena 4-pustulata Fabr.
Brochymena annulata Fabr.
Stethaulax marmoratus Say.
Diolcus chrysorrhoeus Fabr.
Chelysoma guttata H. S.
Tetyra bipunctata H. S.
Cyrtomenus mirabilis Perty.
Amnestus pusillus Uhler.
Thyreocoris unicolor P. B.
Thyreocoris lateralis Fabr.
Thyreocoris pulicarius Germ.
Aradus falleni Stal.
Aradus curticollis Bergr.
Acanthocerus galeator Fabr.
Acanthocephala terminalis Dall.
Acanthocephala femorata Fabr.
Leptoglossus oppositus Say.
Leptoglossus phyllopus Linné.
Leptoglossus corculus Say.
Spartocera diffusa Say.

Chariesterus antennator Fabr.
Chelinidea vittigera Uhler.
Anasa tristis Deg.
Anasa armigera Say.
Alydus eurinus Say.
Alydus pilosulus H. S.
Megalotomus 5-spinosus Say.
Stachyocnemus apicalis Dallas.
Harmostes reflexulus Say.
Harmostes fraterculus Say.
Corizus lateralis Say.
Jalysus spinosus Say.
Largus succinctus Linné.
Arhaphe carolina H. S.
Arhaphe cicindeloides Walk.
Oncopeltus fasciatus Dallas.
Lygaeus facetus Say.
Lygaeus bicrucis Say.
Lygaeus Kalmii Stal.
Lygaeus turcicus Fabr.
Nysius californicus Stal.
Geocoris punctipes Say.
Phlegyas annulicrus Stal.
Oedancala dorsilinea A. & S.
Paromius longulus Dallas.
Perigenes constrictus Say.
Myodocha serripes Oliv.
Heraeus plebejus Stal.
Pamera bilobata Say.
Pamera basalis Dallas.
Antillocoris pallidus Uhler.
Cnemodus mavortius Say.
Ozophora picturata Uhler.
Cryphula parallelograma Stal.
Corythuca ciliata Say.
Corythuca arcuata Say.
Gargaphia angulata Heid.
Teleonemia belfragei Stal.

Reduviolus subcoleoptratus Kirby.
Reduviolus annulatus Reut.
Microvelia americana Uhler.
Gerris marginatus Say.
Barce uhleri Banks.
Barce fraterna Say.
Ploiaria carolina H. S.
Ploiariopsis hirticorins Banks.
Pygolampis pectoralis Say.
Narvesus carolinensis Stal.
Conorhinus sanguisugus Lec.
Arilus cristatus Linne.
Sinea diadema Fabr.
Melanolestes picipes H. S.
Rasahus biguttatus Say.
Sirthenea carinata Fabr.
Hammatocerus purcis Drury.
Apiomerus crassipes Fabr.
Apiomerus spissipes Say.
Pselliopus cinctus Fabr.
Zelus (Diplodus) luridus Stal.
Zelus (Diplodus) cervicalis Stal.
Zelus (Pindus) socius Uhler.
Fitchia aptera Stal.
Phymata fasciata Gray.
Phymata vicina Handl.
Macrocephalus prehensilis Fabr.
Lyctocoris campestris Fabr.
Triphleps insidiosus Say.
Acanthia ligata Say.
Gelastocoris oculatus Fabr.

Mailed February 10th, 1913.

The Canadian Entomologist.

VOL. XLV. LONDON, MARCH, 1913 No. 3

FURTHER NOTES ON ALBERTA LEPIDOPTERA, WITH DESCRIPTION OF A NEW SPECIES.

BY F. H. WOLLEY DOD, MIDNAPORE, ALTA.

(Continued from page 34.)

326. *Anarta cordigera* Thunb.—Didsbury, June 11th, 1906; C. G. Garrett.

327. *A. melanopa* Thunb.—Common in the mountains, from just below imber line upwards. Middle July and Aug.

328. *A.*, sp.—This cannot be *quadrilunata* of which I have seen the type from Colorado in the British Museum, and have a similar specimen from the same state. The only other specimen of No. 328 that I have seen is the ♀ before mentioned in Prof. Smith's collection, labelled July 25th, 1889.

329. *A. zetterstedtii* Stand.—Sir George Hampson in CAN. ENT., XL., 104, refers this form to *Sympistis zetterstedtii* Staud., recording a male and female taken by Mrs. Nicholl, on Mt. Athabasca, Alta., and Kicking Horse Pass, B. C. The species occurs in Northern Europe. The form is extremely near *lapponica*, and there are Labrador specimens under both names in the British Museum.

330. *A. zetterstedtii*, var. *labradoris* Stand.—This form so stands in the British Museum, but under *Sympistis*. I suspect it of being distinct from No. 329, but have only one specimen of the latter.

331. *Nephelodes tertialis* Smith,=*emmedonia* Cram.—In Ent. News, XXII., 397–401, I published some notes on this genus, and expressed the opinion that the Pacific coast *pectinatus* was possibly distinct, and at any rate recognizable as a variety. On the other hand *tertialis* does not seem in any way separable from eastern *minians*, to which *emmedonia* is merely a prior name.

333. *Leucania minorata* Smith,=*luteopallens* Smith,=*pallens* Linn.—*Minorata* was described in 1894 from California and Oregon, and compared with *oxygala* Grt.; "Smaller throughout, the ground

colour reddish, the secondaries darker." I have seen the type of *oxygala* Grt. (not *oxygale*) from Colorado, in the British Museum. It is a very smoky thing, with secondaries wholly dark, in fact darker than the types of *minorata*, or anything of this series that I have elsewhere seen. Sir George Hampson's figure of it is poor and misleading. The note I took the first time I saw the specimen was, "Suggests a melanic *minorata*. Suspiciously like European *fuliginosa* (*impura*)." I noted on my next visit, however, that it was not the same as *fuliginosa*. Having seen nothing else quite like it, I must for the present let it stand as possibly a good species, but feel quite satisfied that nearly all the references of Smith and Dyar to *oxygala* really refer to *minorata*. I have only two specimens from California, which agree with Smith's figure and description of the latter, except that they cannot be called reddish. They certainly might easily be confused with *fuliginosa*, but are not as dark as type *oxygala*, though more uniform smoky than the rest of my series.

In his "Revision of Leucania" Smith claims that the eastern North American form previously known as *pallens* is distinct from that European species, and describes it as *luteopallens* on somewhat indefinite characters, emphasizing, however, a difference in the genitalia. He there includes the Alberta and B. C. races under *minorata*, and says; "It stands between *oxygala* and the European *pallens*, being really the American representative of the latter species." Hampson refers *luteopallens* as a synonym of *pallens*, placing it in a group in the tables, "Fore wing without fuscous shade below median nervure," and holds *minorata* as distinct, and having "fore wing with fuscous shade below median nervure." I could not see that this character held good in the British Museum series, and it seems to be a variable feature all over this continent. The most justifiable separation in the group would seem to be between the European *pallens* and our new world form. In most European specimens the longitudinal strigation is similar, and shading evenly distributed all over the primaries. In none of my specimens i there an obviously darker shade below he median vein, though such a variation is mentioned by Tutt in "British Noctuæ and their Varieties," Vol. I., p. 42, under the name *suffusa*. In American specimens the region of the cell and of the submedian interspace is frequently somewhat paler than the rest of the wing, and a smoky streak is usual above vein 5. Alberta and California specimens are the darkest in my series, especially as to secondaries, but the variation overlaps, and my most European-like examples are from Vancouver Island. The secondaries vary similarly on both continents, and Mr. G. Chagnon, of Montreal, has exactly duplicated genitalia from both sides of the Atlantic. A pink variation is locally common in England, and it is probable that *rubripallens* Smith will prove to be the corresponding variety with us, but I am not yet sufficiently familiar with this to be able to form a definite opinion.

334, 335. *L. albilinea* Hübn,=*diffusa* Walker.—I have taken specimens here which connect the two series I had previously separated, and agree with Hampson in uniting the names. Walker's type is a female from Nova Scotia. Hampson also includes *obscurior*, *tetera* and *neptis* as synonyms, with which I agree, and would add *limitata* Smith.

336. *L. dia* Grt.=*heterodoxa* Smith.

336a. *L. dia* Grt. var. *megadia* Smith.—I have examined the type of *dia* Grt. in the British Museum, which, according to the catalogue comes from California, and some Calgary specimens are exactly like it. The male and female type *heterodoxa* are from the Sierra Nevada. *Megadia* will stand for that variation with a black basal streak, merely an evanescent character. A Calgary cotype of *megadia* is in the British Museum, and is correctly referred as a synonym of *dia* by Hampson. His reference of *heterodoxa* to *insueta* is based on a Minnesota specimen sent him by Smith. Whether this is the Minnesota example mentioned in Smith's description, of course I cannot be sure. Sir George Hampson's reference of the specimen to *insueta* appears to me correct, though it is unusually pale, and certainly very like some western *dia*. My knowledge of *insueta* is at present rather limited, but those I have from eastern localities suggest dark streaky *dia* with a rufous tinge, and not always a very pronounced one either. I quite expect that *insueta* will ultimately prove to be the same.

337. *L. multilinea* Walker.—I consider this form correctly named. I have a series from Vancouver Island. Besides the Calgary cotype of *anteroclara* Smith, previously referred to, a female

cotype of that from Vancouver, in the Rutgers College collection, is also *multilinea.*

{339. *L. ant roclara* Smith.

{340. *L. an'eroclara,* var. *calgariana* Smith.—I am convinced of the distinctness of *anteroclara* from *phragmitidicola,* though confusion with that species is certainly easy. *Calgariana* is pretty obviously a reddish variation of *anteroclara,* and bears the same relation to it as *roseola* does to *farcta.* But whether *anteroclara* is really distinct from *farcta* is another matter. *Farcta* was described from California, and I have a good series from Oakland. It is paler and more even, with median vein less contrastingly whitish, and has pure white secondaries. As a rule they may be separated also by the presence o' a dark shading below the median vein in *anteroclara,* but this does not always exist. I strongly suspect *anteroclara* of being a dark race of *farcta,* but so closely do species of Leucania sometimes resemble one another that I dare not risk the reference at present. I have very rarely seen true *anter clara* from west of the Rockies, but have compared and so named a single Kaslo specimen for Mr. Cockle.

Roseola was described from a single specimen from B. C., as a variety of *farcta,* but was subsequently treated by its own and all other authors as a species. It is common on Vancouver Island, and also at Kaslo, and occasional specimens, generally females, have dusky shading on secondaries. But without the pink coloration they are *farcta* exactly, and I see no reason for separating them. I have Kaslo specimens, and have compared others, so dark and streaky as to make separation from Calgary specimens of *calgariana* almost impossible, and have so named one for Mr. Cockle, but must for the present allow the names the benefit of the doubt.

341. *Himella contrahens* Walker,= *quadristigma* Smith,= *infidelis* Dyar.—I have six specimens from the Red Deer River, one from Lethbridge, Alta., and others from Regina, Sask., and Cartwright, Man. These show exactly similar variation to a Kaslo series, which are typical *infidelis.* A long series from Stockton and Provo, Utah, are similar, but run to a darker and more suffused form, one of which I have compared with Grote's type of *contrahens* from Nova Scotia, in the British Museum, and believe it to be the same.

Grote mentions after his description that he had found a specimen in the collection of the Canadian Entomological Society labelled "*Celæna contrahens* by Walker. This was presumably Walker's type. I have seen a male and female type of Morrison's *thecata*, from New Hampshire, in the Strecker collection, and they are the same species, as already referred by Smith and others. I do not feel quite sure that *conar* is the same species. It was described from "New Mexico, near the borders of Chihuahua". My notes on the type say that it is "almost flesh-coloured, faintly pink, and not reddish or brown." I have seen nothing else quite like it, and must for the present leave it alone. Hampson's figure under the name *conar* is *contrahens*, or more exactly the paler *infidelis* from Nebraska. The types of *quadristigma*, from Bluff, Utah, and Santa Rita Mts., Ariz., are paler still, and have less of the black suffusion usually found in more northern specimens. I might add that Strecker's description of *conar* says the colour is "very light silky grey, or ashen." Though this could scarcely be translated into "pinkish," as the specimen looked to me, still it is not the way I should describe any *contrahens* in my collection

343. *Tæniocampa malora* Smith,= *hibisci* Guen.—In Vol. XLII., p. 190, June, 1910, I published a note on *hibisci*, pointing out that *alia* was prior to *suffusca*, and citing the B. C. form, previously known as *pacifica*, as a local race of *hibisci* under the new name *latirena*, of which I called *quinquefasciata* a variation. On page 317, (October), Smith admitted the distinctness of *pacifica*, eliminated the name *latirena* as valueless, and made *hibisci* Guen.= *confluens* Morr., and a variety of *instabilis* Fitch. He also reinstated his *quinquefasciata* as a species, and created six more to keep it company, figuring genitalia. Dr. Dyar replied to him on page 399. I have to admit that I erred in producing the name *latirena* rather vaguely, though I thought I made it clear that it was applied to all B. C. forms of *hibisci* hitherto erroneously called *pacifica*. Smith was near the mark in saying that it could only be applied as a synonym of the entire *pacifica* Smith series, with the exception, of course, of *pacifica* Harvey. As Smith then described two variations of the B. C. forms, both of which I consider variations of *hibisci*, to avoid future confusion I refer the first of those names, *inflava*, to *latirena* Dod. His other name, *inherita*, applied to B. C.

specimens signifies a more strigate and irrorate form of the same. What he calls *instabilis* Fitch is, of course, a citation of Fitch of European *instabilis* Schiff,= *incerta* Hübn., which however is not very obviously distinct. *Malora* Smith, described from three males and two females from Calgary, is applied to a dull smoky-grey form of the same species. The variation seems to be very wide wherever the species occurs, though it may be said that, in general, Pacific coast specimens are richer in colour than those from the east, and Alberta specimens intermediate. I can, however, match Calgary and Vancouver specimens almost exactly, and also some from Calgary, Chicago and Montreal. *Quinquefasciata*, as the name implies, stands for a form with five distinct transverse lines. *Brucei* Smith from Denver, and Garfield Co., Colorado, and *proba* Smith from Alameda Co., Calif., I cannot believe to be distinct from *hibisci*, but *nubilata* Smith may prove distinct. I have a note that I found specimens from that region in Smith's collection suggesting a new species, but failed to make a satisfactory separation. Hampson makes *insciens* Walker, and *confluens* Morr. synonyms of *hibisci*, but calls *hibisci* "ab.I." with the spots joined. This aberration, according to Smith, is also Morrison's *confluens*. Walker's type of *insciens* is a female labelled "U. S. A., (Doubleday)" and has the subterminal line, and annuli to the spots, particularly the orbicular, unusually pale and wide. I have a note to the effect that when I was at the British Museum last March, a specimen labelled *confluens* Morr., the type of *insciens* Walker, and three pale, even, Calgary specimens stood separated in the collection. I fail to see that any such separation is warranted. I also found *latirena* and *pacifica* in the same series under *pacifica*, though I feel satisfied that the latter is distinct.

345. *Cleoceris populi* Strk.—The type of *populi* is from Loveland, Colo., and is a pale, slightly marked thing, and not unlike the form figured in Holland. I have seen Colorado and Wyoming specimens in other collections, but all were paler and less maculate than my Calgary series.

347. *Xylina amanda* Smith.—There are male and female types in the Washington collection, the former from Pullman, Washington, and the latter from Calgary. The Pullman specimen is much paler in colour than the other, and my notes say that they may

possibly be two species. I have specimens from Miniota and Aweme, Man., and Vineyard, Utah, which resemble the Calgary form, and two from Wellington, Vanc. I., are paler, more luteous and less maculate, probably like the Pullman form. The closest ally of this secies is *petulca* Grt., which occurs on Vancouver Island also. *Amanda* is a narrow-winged species, with a rather conspicuous pale yellowish patch in the cell, obscuring the upper portion of the reniform and reaching to the t. p. line. In *petulca*, though the spots themselves are yellowish filled, there is no such patch. Another conspicuous character in *amanda* is that the lower edge of the reniform is, in all my series, filled with dark fulvous. It is probable that I may sometimes, in naming offhand without comparison, have given the name *amanda* to pale specimens of *petulca*. In fact I have suspected them of being variations of one species, but am convinced of their distinctness. I have often seen them mixed in collections. *Amanda* narrowly escaped redescription by its author about two years ago.

348. *X. fagina* Morr.—I have not seen the type of this species, but the Alberta form is the same as the *fagina* of eastern collections. It is very rare here.

349–350. *X. georgii* Grt.—I have taken no more specimens of this species than those I originally listed as *oregonensis* Harvey, and *ancilla* Smith, but after studying material from all over the continent for some years I have long ago come to the conclusion that *oregonensis* and *ancilla* of my list are the same species. I have a specimen from Miniota, Man., compared with the type of *georgii* from Orillia, Ont., in the British Museum, and this scarcely differs from my cotype of *ancilla*. The species is one of the most variable of our Xylinas, the variation consisting in differences in shade of the ground colour, distinctness of maculation, and size and shape of the discoidal spots. Some specimens have slightly brown, almost reddish scales in the reniform, though this is rather unusual. I offer a list of what I consider synonyms of this species, with the type localities of each:

Oregonensis Harvey, Oregon.

Holocinerea Smith, Winnipeg, Man.; Vancouver, and N.W. British Columbia; Pullman, Washington; Sierra Nevada, Cal.

Fletcheri Smith, Ottawa.

Ancilla Smith, Calgary; Cartwright, Man; Wellington, B. C. (The male type is from Cartwright, and is practically a dead mate for the male type of *fletcheri*).

Vertina Smith, Corvallis, Oregon; B. C.

Var. *emarginata* Smith, Colorado Springs and Glenwood Springs, Colo. A pale, slightly marked form.

The only type of *oregonensis* that I have seen is an Oregon male in the Henry Edwards' collection, and my notes say that it is a pale *holocinerea*. I am in doubt as to the identity of Fig. 26., Plate IV, of Smith's Monograph, and it struck me that the material in his collection under *oregonensis* probably included two species. The *oregonensis* of Hampson's Catalogue, Plate CIII., Fig. 7, is a Californian specimen, and is certainly not *georgii*. I have in my collection a male and two females of a species from Glenwood Springs, Colorado, from Dr. Barnes. One of them is labelled "*oregonensis* Harvey, identified by Smith," and the other is labelled "*torrida* Smith" by Dr. Barnes. In my opinion these specimens are undoubtedly rather poorly marked *antennata*, and agree well with my eastern series of that. Under the description of *torrida*, Smith says that "the more obscure examples remind one of the *antennata* type" and it is possible that the latter species was included in the type material. It will be necessary to re-examine the types to decide, as I have previously mixed the forms myself, but the species I have at present under *torrida* is a brightly marked thing from Vancouver Island, and is that figured by Smith in his monograph under the name, on plate V., fig. 31.

(To be continued.)

DURING the latter half of January Mr. F. W. L. Sladen, Assistant Entomologist for Apiculture in the Division of Entomology, Ottawa, has been travelling in Nova Scotia and New Brunswick. A short course in Apiculture was given at the Agricultural College, Truro, and subsequently Mr. Sladen investigated apicultural conditions and possibilities and addressed meetings in the two provinces.

CORRECTION.—P. 367, line 19, after September 12 add 1911, December Number, CANADIAN ENTOMOLOGIST. F. M. WEBSTER.

TACHINIDÆ AND SOME CANADIAN HOSTS.

BY J. D. TOTHILL, DIVISION OF ENTOMOLOGY, OTTAWA, ONT.

In working over the Tachinidæ in the collection of the Division of Entomology, at Ottawa, a number of breeding records were encountered. Thirty-nine of these are new, in so far as the writer is aware, and these form the basis of the present list. In addition to these thirty-nine, there are seven that have been published since the appearance of Coquillett's "Revision of the Tachinidæ of America, north of Mexico," and Aldrich's "Catalogue of North America Diptera." These seven are included in the present list; they are indicated by an asterisk (*), and reference is made in each case to the published record.

The majority of these records were obtained by Dr. James Fletcher and Mr. Arthur Gibson. To the latter colleague, whose kindly assistance has made possible the compilation of the present list, the writer is under numerous obligations. The letters J. F. or A. G. placed in brackets after the species indicate the person who was responsible for the rearing. A few records were obtained by others than the above; in these cases the names or initials of the persons responsible for the rearing are given.

No doubtful records are included in the list. The arrangement follows that of Mr. D. W. Coquillett in the excellent list of tachinid flies and their hosts contained in his "Revision" (loc. cit., p. 9).

PARASITES.	HOST INSECTS.
Blepharipeza adusta Loew........	*Halisidota caryæ* Harris.—Bred from cocoons of host collected at Ottawa; 11 specimens, issued June 16–July 4. (A. G.)
	Halisidota maculata Harris.—Bred from cocoons of host collected at Ottawa; 1 specimen. (A. G.)
	Malacosoma disstria Hubn.—Bred at Fredericton, N. B., from larvæ and pupæ collected locally and at Ottawa; numer-

ous specimens, issued spring. (J. D. T.)

Blepharipeza leucophrys Wied....*Sphinx chersis* Hbn.—Bred from specimen of pupa collected at Ottawa; 1 specimen, issued May 30. (A. G.)

Euphorocera claripennis Macq....*Heliophila unipuncta* Haw.—Bred from larva collected at Ottawa; 1 specimen. (J. F.)

Malacosoma sp.—Bred at Ottawa from larva; place of collection not known. (J. F.)

Exorista affinis Fall.............*Phragmatobia fuliginosa* Linn.—Bred from cocoons collected at Ottawa; 2 specimens, issued April 13 and 27. (A. G.)

Exorista cheloniæ Rond..........*Apantesis ornata* Pack., var. *achaia* G. & R.—Bred at Ottawa from larvæ collected at Kaslo, B. C.; 3 specimens, issued June 6. (A. G.)

Malacosoma disstria Hbn.—Bred at Fredericton, N. B., from larvæ collected locally and at Ottawa; numerous specimens, issued spring. (J. D. T.)

**Phragmatobia assimilans* Walk., var. *franconia* Slosson.—Bred at Ottawa from larvæ collected at Hymers, Ont,; 1 specimen, issued May 7; c.f. A. Gibson, Ent. Record, 1910, p. 18, and A. Gibson, CAN. ENT., Vol. XLIII., p. 127.

Exorista eudryæ Town...........*Euthisanotia grata* Fab.—Bred from larva collected at Ottawa; 1 specimen, issued Aug. 27. (J. F.)

Exorista futilis O. S.	*Isia isabella* S. & A.—Bred from larva collected at Ottawa; 1 specimen, issued May 21. (J. F.)
	Hyppa xylinoides Gn. —Bred from larva collected at Ottawa; 1 specimen, issued May 23. (C. H. Young.)
Exorista helvina Coq	*Lycia cognataria* Gn.—Bred at Ottawa from larvæ collected at Coldstream, B. C.; 4 specimens, issued May 29 and June 12. (A. G.)
Exorista nigripalpis Town	*Tortrix fumiferana* Clem.—Bred by G. E. Sanders at Ottawa from larvæ collected at Chicoutimi, St. Sylvestre and Maniwaki, P. Q., and Duncans, B. C.; 68 specimens, issued June 18–July 10.
Exorista pyste Walk	*Tortrix fumiferana* Clem.—Bred at Ottawa by G. E. Sanders from larvæ collected at Chicoutimi, P. Q.; 2 specimens, issued July 3.
Exorista vulgaris Fall	*Tortrix fumiferana* Clem.—Bred at Ottawa by G. E. Sanders from larvæ collected at Chicoutimi, St. Sylvestre, St. Gabriel de Brandon and Montcalm, P. Q.; 16 specimens, issued June 18–July 10.
Frontina frenchii Will	*Papilio daunus* Bdv.—Bred at Ottawa from pupæ collected at Kelowna, B. C.; 9 specimens, issued Aug. 23. (A. G.)
	Papilio eurymedon Bdv.—Bred at Ottawa from pupæ collected

	at Nanaimo, B. C.; 10 specimens, issued April 12, 16, 17 and 19. (A. G.)
	Samia columbia Smith.—Bred at Ottawa from pupæ collected at Cartwright, Man.; 7 specimens. (J. F.)
Frontina tenthredinidarum Town. .	*Cladius pectinicornis* Fourc.—Bred from larva collected at Ottawa; 1 specimen, issued Aug. 29. (J. F.)
	Emphytus canadensis Kirby.—Bred from larva collected at Ottawa; 1 specimen, issued Sept. 1. (J. F.)
	**Nematus erichsonii* Hartig.—Bred from cocoons collected at St. John, N. B.; c.f. C. G. Hewitt, "The Larch Sawfly," Bull. No. 5, Div. of Ent., Ottawa.
Gonia capitata DeG.	*Paragrotis ochrogaster* Gn.—Bred at Ottawa from pupæ collected at Cow Bay, C. B.; 1 specimen. (J. F.)
	Agrotid sp.—Bred from pupa collected at Ottawa; 1 specimen, issued spring. (J. F.)
Linnæmyia anthracina Thompson.	**Hyphoraia parthenos* Harris.—Bred at Ottawa from larvæ collected at Hymers, Ont.; 2 specimens, issued May 16; c.f. W. R. Thompson, CAN. ENT., Vol. XLIII., No. 8, p. 266.
Masicera eufitchiæ Town.	*Halisidota tessellaris* S. &. A.—Bred from pupæ collected at Ottawa; 2 specimens, issued Aug. 8. (A. G.)

Masicera myoidea Desv. . . . . **Papaipema appassionata* Harvey.—Bred at Ottawa from larva collected by C. H. Young, at Meach Lake, P. Q.; 1 specimen, issued Aug. 13; c.f. James Fletcher, Ent. Rec. 1906, p. 102.

Papaipema purpurifascia G. & R.—Bred from larvæ collected at Ottawa; 9 specimens, issued Aug. 24–28. (A. G.)

Masicera rutila Meig. . . . . *Tortrix fumiferana* Clem.—Bred at Ottawa by G. E. Sanders from pupæ collected at Duncans, B. C.; 2 specimens, issued July 10 and 19.

Phorichæta sequax Will. . . . . *Heliophila commoides* Gn.—Bred from specimens of host collected at Ottawa; 7 specimens, issued June 10. (A. G.)

Phorocera leucaniæ Coq. . . . . **Euproctis chrysorrhœa* Linn.—Bred at Fredericton, N. B., from larva collected by P. N. Vroom at Chamcook, N. B.; 1 specimen, issued spring. A valuable parasite of the same host in Massachusetts; c.f. Fiske & Howard, Bull. 91, Bureau of Ent., U. S. Dept. of Agr., Washington, D. C. (J. D. T.)

Phorocera saundersii Will. . . . . *Ennomos magnarius* Gn.—Bred at Ottawa from pupa collected at Mt. Hebron, N. B.; 1 specimen, issued in office, Dec. 31. (J. F.)

Euvanessa antiopa Linn.—Bred

from pupa collected at Ottawa; 1 specimen. (J. F.)

Plagia americana V. d W. . . . . *Plusia æreoides* Grt.—Bred from larvæ collected at Ottawa; 2 specimens, issued June 10. (A. G.)

Sturmia albifrons Walk. . . . . *Heliophila unipuncta* Haw.—Bred from larvæ collected at Ottawa; 4 specimens, issued July 8. (J. F.)

Sturmia inquinata V. d W. . . . . . . *Sphinx chersis* Hbn.—Bred from larvæ collected at Ottawa; 20 specimens, issued May 22. (A. G.)

Sturmia normula V. d W. . . . . *Lemonias taylori* Edw.—Bred at Ottawa from larva collected on Vancouver Island; 1 specimen. (J. F.)

Sturmia phyciodis Coq. . . . . *Phyciodes tharos* Drury.—Bred from specimen of host collected at Ottawa; 1 specimen, issued June 23. (J. F.)

Tachina mella Walk. . . . . *Datana ministra* Drury.—Bred at Ottawa from larvæ collected at Armstrong, B. C.; 2 specimens, issued in office, Oct. 29. (A. G.)

Notolophus antiqua Linn.—Bred at Ottawa from larvæ collected at Rudolph, N. S.; 5 specimens, issued Sept. 3. (A. G.)

Tachina simulans Meig. . . . . *Harpiphorus tarsatus* Say.—Bred from specimen of host collected at Ottawa; 1 specimen, issued June 30. (J. F.)

Winthemia fumiferanæ Tothill. . . . **Tortrix fumiferana* Clem. — Bred at Ottawa by G. E.

Sanders, from Maniwaki, Que., and Duncans, B. C.; 26 specimens; c.f. J. D. Tothill, CAN. ENT., Vol. XLIV., No. 1, p. 2.

Winthemia quadripustulata Fab. . . *Cucullia convexipennis* G. & R. —Bred from larva collected at Ottawa; 1 specimen. (J. F.)

**Marumba modesta* Harris.— Bred from larva; c.f. James Fletcher, Ent. Record, 1903, p. 99.

Pholus achemon Drury.—Bred from pupa collected at Ottawa; 1 specimen. (A. G.)

The specimens reared from the above three hosts issued May 11, 16 and Sept. 21.

GEOMETRID NOTES—NEW VARIETIES

BY L. W. SWETT, BOSTON, MASS.

Cleora pampinaria var. *nubiferaria*, n. var.

Expanse 29 mm.; palpi very short. Fore wings smoky black with line running from inner margin up to vein a 1, then along the vein for about 3 mm., where it stops. In the centre of the fore wings on the median vein there is a dark line, especially broad at the vein, where it curves upward at right angles to the costa. On the outer fourth there is another band parallel to this, which runs from the median vein to costa just beyond the faint black discal spot. These lines are practically the same as in normal *pampinaria* but the black colourings of the wings render them indistinct. There is a very distinct white zig-zag line running parallel to the outer margin from costa to inner margin with a long white projection near M_2. The fringe is quite long and black with points at base. The hind wings are of the same smoky colour as the fore wings, with a faint extra-discal black line, which shows as black points on the veins. The discal spot appears as a ring and touches the extra-discal line. There is a trace of an irregular white line near outer margin. The outer edge of the wing is slightly scalloped and the fringe is rather long. Beneath the fore wings are smoky black with discal spot showing through; the apices of the wings are tipped with white. Hind wings of the same colour as fore wings with black discal spot instead of ring as above.

Type 1 ♂, Cincinnati, Ohio; through the kindness of Miss A. F. Braun.

This is no doubt a case of melanism and was so identified for me by Mr. Grossbeck. Melanism seems to be rare in this country but is common in Europe where it seems to represent a more recent type.

Ania limbaria var. *chagnoni*, n. var.

Expanse 22 mm.; palpi very short. Fore wings bluish yellow with chocolate border, basal space and mesial space of the same colour up to the chocolate-coloured margin. The basal line about 4 mm. out from body runs at right angles from costa to median vein, then almost straight to inner margin. There are traces of a large lunule near where the discal spot would be and expanding to the extra-discal line. Beyond the extra-discal line the entire margin is chocolate-coloured. Hind wings bluish yellow to the extra-discal line, beyond the margin chocolate as in fore wings. There is a large lunule in the discal space. Fore wings of the same colour as above, the chocolate margin showing through. The hind wings are the same as above, bluish yellow with a chocolate margin. This seeems to be a case of melanism but the markings are not identical with *limbaria*. Possibly this is a northern species. It is so different in appearance from *limbaria* that one would hardly recognize it, or where it belonged, were it not for the peculiar spur of the hind tibiae.

Type 1 ♂, St. Therese Isle, St. Johns Co., Que., VII.–9–1912; through the kindness of Mr. G. Chagnon, after whom I take pleasure in naming this unique variety.

Mr. Frederick Knab, of the U. S. Bureau of Entomology, has been recently appointed Honorary Custodian of the Diptera in the U. S. National Museum, to succeeed the late Mr. D. W. Coquillett.

THE SPRING GRAIN APHIS OR "GREEN BUG."

This aphis *Toxoptera graminum* Rond., must not be confused with the wide-spread grain aphis *Macrosiphum granaria* Buckton, formerly known as *Siphonophora avenæ* Fab., which is destructive from time to time in Canada. *Toxoptera graminum* has been found in western Canada; but it has not as yet inflicted depredations of so serious a character as have been recorded from time to time since 1890 in the United States. The very destructive nature of the "Green Bug," as it is popularly termed, in the United States in 1907, in which year it was also recorded in Manitoba and Saskatchewan, led the United States Congress to make a special appropriation for its investigation. These investigations have been continued up to the end of 1911 and the Bureau of Entomology of the United States Department of Agriculture have now published a record of the entire investigation by F. M. Webster and his assistant, W. T. Phillips (Bull. No. 110, Bur. Ent., U. S. Dept. Agric., Washington, 153 pp., 48 figs., 9 pls., 1912).

It is not possible within the compass of a short article to refer in more than a brief manner to the varied and valuable results of this study. The study is of unusual interest in that it affords results of value not only to the economic worker but also to the embryologist and to the student of insect bionomics, all of which results are necessary to a complete interpretation of this remarkable insect's habits and depredations.

South of the 35th parallel it appears to be permanently viviparous and to breed without the appearance of the sexes. It is unable to survive hot and dry conditions in the Southern States. In Indiana the overwintered eggs hatch from the end of March to about April 10th giving rise to wingless stem-mothers which pass through five instars. These stem-mothers reproduce viviparously producing wingless viviparous females and winged viviparous females.

The great fecundity of the aphids, due to their viviparous habits, is well known and the results of the authors' study of the progeny of single lines are of great interest in this connection. In Indiana the eggs hatched on March 27th, and the first-born aphids produced twenty-one generations before the adult ovipositing females appeared in November; the last born females produced ten generations. In Texas there were twenty-five generations between March 31st. and November 3rd. The age at which the females begin to reproduce varies according to the season; early in the season it is from twenty to twenty-seven days, from May to September it is from about six to sixteen days and later in the season from twelve to fifty-three days, the average for the three seasons of the year being early spring twenty-two days, summer nine days, and early fall nineteen days. In Texas the time is shorter, the shortest time being six days. The reproductive period is longer in the average in the spring and fall than in the summer; in the spring the average is eighteen days, in the summer twenty-six days and in the fall, forty-five days. The longest likewise is greater in the spring and fall than in the summer, the average is thirty-five days and the longest is seventy-eight days. The rapidity of production it very great: in Indiana the greatest number of young produced by one female in twenty-four hours was eight, in Texas ten. The greatest number of young produced by one individual was ninety-three. The average number of young for the entire viviparous breeding season, over a period of three years (1907-9), was 28.2; the average number of young produced in a day is greatest in the Spring.

The sexual forms, male and female, appear in Indiana at the end of September and adults may be found from October until the cold kills them off in December.* The oviparous females become adult in 11 to 41 days according to weather conditions, and if males are present they oviposit in from three to nine days. The males live from 8 to 10 days after reaching maturity, the females from 31 to 68 days if males are present; if males are not present they can live 88 days after reaching maturity. Aberrant individuals were found containing both living embryos and true eggs.

Throughout the Northern United States and no doubt in

* On page 77 the author states "one agamic female may reproduce all agamic individuals, a combination of agamic males and oviparous females or only true females and males." What are "agamic males"? An agamic female we know is a female which produces young in a parthenogenetic manner, that is, without fertilisation by the male. Huxley was the first, I beleive, to use the term "agamic" for this form of reproduction in the Aphids. As the male aphid, fortunately, cannot give birth to young either sexually or asexually is it not misusing the word to apply it to the male?

Canada also, Blue grass (*Poa pratensis*) is the most common host of Toxoptera.

The diffusion or natural spread of the Green Bug is dependent upon a number of factors both meteorological and biological. For example the influence of wind in dispersion depends upon whether the insect is in a winged or apterous condition and this is, of course, dependent upon those factors producing these conditions such as the curtailing of food supply, etc. The most favourable conditions for natural diffusion appear to be a decreasing food supply with a fairly high temperature and a not excessive parasitism.

The effects of temperature varied according to the locality whether in a north or southern region. In the north, where the effects of the temperature concern us most, the insect winters in the egg state. Here warm winters are of less importance and cool weather during spring and early summer exert a far greater influence on the numerical abundance of the insect.

As the early developmental stages in the winter eggs are effected by the temperature a complete study of the biology of *Toxoptera* necessitated the study of the embryology. The results of this study and the figures of the embryonic stages which are given are a welcome addition to our knowledge of insect embryology, the observations on that peculiar embryonic structure which the present authors have termed the "polar organ" being of special interest. The general results of their study, however, does not materially affect the early observations of Witlaczil, Will and others and the more recent work of Tannreuther.

The study of the natural enemies of the Green Bug naturally forms one of the most important sections of the work. The efforts made in certain quarters in the direction of distributing the chief parasite *Aphidius testaceipes* Cresson (known also by a host of other synonyms under the genus *Lysiphlebus*) and the reported success of these efforts made it extremely desirable that the biology and distribution of this parasite should be carefully studied and this fact is especially borne out by the results of the present thorough study of *Aphidius*, its biology and its relation to meteorological conditions in Kansas and other States. It was found that not only did this parasite occur over almost the entire United

States but that it would breed interchangeably from *Toxoptera* into other species of Aphids and in addition was reared from a large number of common and widespread species of Apids. Taking these facts into consideration it is very easy to see, as the authors rightly point out, "that it would be only in rare instances and under peculiar conditions that a locality would be found where *Aphidius testaceipes* would not be lurking, waiting for favourable weather conditions and abundant supplies of its host aphids to make its appearance in greater or less numbers." The effectiveness of this parasite will be appreciated when it is realized that a single female *Aphidius* may parasitize no less than 301 *Toxoptera*. No wonder their natural control is, at times, so sweepingly effective! Regarding the artificial distribution of the parasites, these investigations naturally point to the "futility of attempting materially to increase its numbers or efficiency by artificial introduction into grain fields" and further, I would add, they point to the necessity of making as careful studies as possible of the parasites before adopting any extensive system of artificial distribution. The account of the remedial and preventive measures is prefaced by the statement that with "an outbreak of this pest fully established and the winged adults being carried by the wind and scattered over the fields there to settle down and reproduce, the difficulties in the way of control are quite insurmountable." Bush-drag experiments, and spraying did not give satisfactory results or were impracticable. Cultural methods of prevention are the most important and the chief of these is the destruction of volunteer grain. In this connection I would venture to suggest, would it not be well to leave the volunteer growth as a trap crop, then seed later or sow spring oats? In the north the close grazing of waste lands is recommended; this would result in the destruction of a considerable proportion of the eggs laid on the Blue grass (*Poa pratensis*) which appears to be the normal host of the Green Bug in northern localities.

Great credit is due to Mr. F. M. Webster and his very able assistants, particularly Mr. Phillips, for the thorough character of this investigation, the results of which will be of great assistance to others working in the same field and confronted with similar problems.

C. GORDON HEWITT.

GENERIC TABLES FOR THE CIMICID SUBFAMILIES PHYLLOCEPHALINÆ, PHLŒINÆ AND DINIDORINÆ.

*BY THE LATE GEORGE W. KIRKALDY.

TABLE OF GENERA OF PHYLLOCEPHALINÆ.

1 (34) Pronotum rounded laterally, or if produced, then the produced part does not extend apically as far as the eyes.

2 (19) Lateral angles of pronotum obtuse or rounded, or if acute, then scarcely prominent.
(For *Delocephalus* No. 19.)

3 (18) Interoanterior angles of pronotum not produced.

4 (17) Hind angles of pronotum near the scutellum, not angulate.

5 (16) Lateral margins of pronotum not prominent anteriorly.

6 (9) Head short, scarcely longer than its breadth between the eyes, if at all.

7 (8) Costal margin of corium not lævigate, unless anteriorly, or rather sparsely punctured......5. *Metonymia* Kirk.

8 (7) Costal margin of corium entirely pale, lævigate, sometimes marked with spots or transverse impressions or black points, in remote transverse series..........................6. *Dalsira*† A. & S.

9 (6) Head distinctly longer than its breadth between the eyes.

10 (15) Antennæ extending apically as far as the apex of the head.

11 (12) Lateral margins of head laterally not or scarcely converging towards the apex, till close to it......................16. *Phyllocephala* Laporte

12 (11) Lateral margins of head distinctly converging to the apex.

13 (14) Lateral lobes of head plane or somewhat concave........................10. *Schyzops* Spinola.

14 (13) Lateral margins of head convex....14. *Dichelorhinus* Stal.

15 (10) Antennæ not extending as far as apex of head........................15. *Randolotus* Distant.

*These three tables are a beginning of the numerous uncompleted papers left by my late friend, which I purpose to publish from time to time, as I am able to edit them and tie up loose ends, if such there be. Fragments though these be, they will, nevertheless, prove highly useful in the absence of any late general work on these subfamilies.—J. R. T. B.

† 22. *Frisimelica* Distant seems to be near here, but the description is defective.

16 (5) Lateral margins of pronotum prominent anteriorly..................24. *Delocephalus* Distant.

17 (4) Hind angles of pronotum acutangular near the scutellum..................23. *Megarrhamphus* Bergr.

18 (3) Interoanterior angles of pronotum produced, or at least strongly dentate..............3. *Lobopeltista* Schout.

19 (2) Lateral angles of pronotum strongly acute or very prominent.

20 (27) Lateral angles of pronotum not turned forward.

21 (24) Lateral lobes of head contiguous in front of the median lobe.

22 (23) Anterolateral margins of pronotum straight.......................7. *Mercatus* Distant.

23 (22) Anterolateral margins of pronotum sinuate.......................9. *Schismatops* Dallas.

24 (21) Lateral lobes not contiguous, unless at the base.

25 (26) Interolateral margins of lateral lobes as long as from apex of median lobe to base; lateral angles of pronotum acuminate...................18. *Diplorhinus* A. & S.

26 (25) Interolateral margin very short in front of median lobe; lateral angles of pronotum acute. 4. *Storthogaster* Karsch.

27 (20) Lateral angles of pronotum turned more or less forward.

28 (29) Lateral angles of pronotum prominent, but apically blunt.........................8. *Sandehana* Distant.

29 (28) Lateral angles of pronotum acute.

30 (33) Anterolateral margins of pronotum slightly sinuate.

31 (32) Antennæ longer, second joint reaching or scarcely exceeding apex of head; lateral angles of pronotum produced.....................12. *Gonopsis* A. & S.

32 (31) Antennæ shorter, second joint scarcely reaching apex of head; lateral angles of pronotum rounded, scarcely prominent......................13. *Kaffraria* Kirk.

33 (30) Anterolateral margins of pronotum deeply emarginate.....................19. *Macrina* A. & S.

34 (1) Pronotum with lateral angles prominently extending forwards very distinctly beyond the eyes.

35 (44) First segment of antennæ not reaching to apex of head.

36 (41) Lateral margins of pronotum not extending as far as apex of median lobe of head.

37 (38) Lateral lobes of head contiguous.... 11. *Salvianus* Distant.

38 (37) Lateral lobes of head not contiguous, except at base.

39 (40) Anterolateral angles of pronotum acutely prominent.................... 17. *Roeburnea* Schout.

40 (39) Lateral angles of pronotum acuminately prominent................ 22. *Melampodius* Schout.

41 (36) Lateral angles of pronotum extending as far as the apex of the median lobe of the head.

42 (39) Lateral lobes of head acutely produced. 20. *Tetroda* A. & S.

43 (42) Lateral lobes of head exteriorly rounded... 21. *Gellia* Stal.

44 (35) First segment of antennæ extending beyond apex of head............................ 1. *Cressona* Stal.

TABLE OF GENERA OF PHLŒINÆ.

1 (4) Antennæ inserted close to the eyes (*Neogeic*).

2 (3) Scutellum as long as the space from its hind angles to the base of the laminate portion of the abdominal apex, distinctly shorter than the corium; lateral lobes of the head contiguous or overlapping... 1. *Phlœa* Lep. & Serv.

3 (2) Scutellum more than twice as long as the space between the hind angles and the basal part of the laminate apex of the abdomen, very slightly shorter than the corium; lateral lobes of the head contiguous only basally in the middle.................... 2. *Phlœophana* Kirkaldy.

4 (1) Antennæ inserted above one-third from apex of the head (*Palæogeic*)..................... 3. *Serbana* Distant.

The fossil *Palæophlœa* is not included.

TABLE OF GENERA OF THE SUBFAMILY DINIDORINÆ.

1 (16) Lateral margins of abdomen not tuberculately dentate.

2 (15) Pronotum anteriorly not wider than the head with eyes; pronotum laterally sometimes marginate, never laminate.

3 (12) Tarsi 3-segmentate..

4 (9) Antennæ 4-segmentate.

5 (8) Lateral lobes of head either not longer than the median, or, if so, then contiguous, at least partly.
6 (7) Labium scarcely extending to middle coxæ; hind femora not widened at base. . . . . 1. *Cyclopelta* A. & S.
7 (6) Labium extending to hind coxæ; hind femora (at least in ♀) strongly widened at base. . 2. *Patanocnema* Karsch.
8 (5) Lateral lobes of head much longer than the median, and not at all contiguous . . . . 3. *Dinidor* Latr.
9 (4) Antennæ 5-segmentate.
10 (11) Head subæquilateral, or scarcely transverse, lateral margins straight or slightly sinuate; eyes sessile; ♂ pygophor not emarginate, apically rounded, rarely with an obsolescent sinuation in the middle. . 4. *Aspongopus* Laporte.
11 (10) Head transverse, deeply sinuate in front of the stylate eyes; fore femora distinctly spinose towards the apex; ♂ pygophor distinctly sinuate apically . 5. *Colpoproctus* Stal.
12 (3) Tarsi 2-segmentate.
13 (14) Lateral margins of head not contiguous, head laterally with a spine in front of the eyes. . . . 6. *Thalma* Walker.
14 (13) Lateral lobes of head contiguous, head spineless in front of the eyes. . . . . 7. *Urusa* Walker.
15 (2) Pronotum anteriorly much wider than the head with eyes; pronotum laterally distinctly laminate. 8. *Sagriva* Spinola.
16 (1) Lateral margins of abdomen tuberculately dentate.
17 (18) Lateral margins of pronotum obliquely straight . . . . 9. *Byrsodepsus* Stal.
18 (17) Lateral margins of pronotum angularly sinuate. . . . . 10. *Megymenum* Laporte.

THE Province of Quebec is to be congratulated on its decision to appoint a Provincial Entomologist. The Rev. Abbe V. A. Huard, the Conservator of the Provincial Museum at Quebec, has been appointed to the office of Entomologist. As the editor of "Le Naturaliste Canadien" and successor to Provancher, he is well known to entomologists in Canada, and we wish him all success in his new duties in a field which offers unparalleled opportunities for entomological work and assistance to those whose livelihood depends on successful husbandry in the farm, field and forest.

SOME TROPIC REACTIONS OF *MEGILLA MACULATA* DE G. AND NOTES ON THE HYDROTROPISM OF CERTAIN MOSQUITOES.

BY HARRY B. WEISS, NEW BRUNSWICK, N. J.

This ladybird, which is the only species in New Jersey hibernating in sufficient numbers to be considered a colony, lends itself readily to experimentation, and the colonies containing as a rule about a thousand individuals may be found in different localities usually under a piece of bark or a mass of dried leaves.

This colonial hibernation is the result of various reactions to tropic stimuli. First the question arises as to just why they congregate in large numbers and this may be explained by chemotropism. All Coccinellidæ emit peculiar odors and as the colony increases, so does the odor, thereby making the chemotropic stimuli stronger and more effective. Mr. Edward K. Carnes in bulletin No. 5, Vol. I, of the California State Commission of Horticulture, writes that he has located colonies of *Hippodamia convergens* in that state simply by the odor alone. Here, however, the individuals in a colony number two and a half millions or more.

A lowering of the temperature as winter approaches with a corresponding decrease in the food supply undoubtedly renders them exceedingly susceptible to chemotropic stimuli. With *Megilla maculata*, there is no evidence at present that anemotropism plays any part in the selection of the hibernating quarter. Once in their place of hibernation, they become positively thigmotropic and negatively phototropic. Two hundred individuals were removed from a colony and placed in a glass breeding cage, one end of which was constructed so that they could if they desired act positively photo- and thigmotropic and the other end so that they could act only negatively phototropic and positively thigmotropic. Every one selected the dark end. This happened on both sunshiny and cloudy days. During all operations the temperature of the entire cage was uniform as indicated by thermometric tests. During the above experiment the temperature was gradually lowered in eight hours from 70° F. to 36° F.

At a temperature of 54° F. they remained as before. At a temperature of 64° F. about one third became positively phototropic and negatively geotropic, and their activity undoubtedly made them susceptible to chemotropic stimuli from a food viewpoint.

At a temperature of 70° F, about one-half were active and at 75° F. all were active. When the temperature was suddenly lowered as from 75° to 36° F. all became dormant at once and exhibited no tropic reactions. By at once I mean within ten or twelve minutes. Without doubt thermotropism plays an important if not the most important part in deciding just what reactions are to occur. A gradual lowering of the temperature such as would naturally result in the beetles acting phototropically and thigmotropically while a sudden drop resulted in what might be called immediate partial hibernation. Of course with a soft bodied insect this would have resulted in death.• When the temperature of the air was 42° F., that of their natural hibernation place was 54° F. which indicates an effort to secure optimum conditions.

After emerging from winter quarters, the females of *Culex pipiens* are at first positively chemotrophic. After having fed they become positively hydrotropic and deposit their eggs on the surface of water. While in hibernation during which time they may be fairly active, depending on the temperature of their hibernation quarters, they are strongly negatively hydrotropic. Food and water placed within easy reach of hibernating specimens were always avoided, even when the temperature of their surroundings was 75° or 80° F.

Aedes sollicitans and *Aedes cantator* are also positively hydrotropic but not to the extent of most other mosquitoes. With these species eggs are deposited in damp depressions and not on the surface of the water. Sterile females of both of these species are strongly negatively hydrotropic and fly long distances away from salt marshes where they breed. However this migratory habit, or at least the direction they take, is undoubtedly influenced by anemotropism inasmuch as they allow themselves to be carried by strong breezes and will fly inward against light breezes. Sterile females of *Aedes tæniorhynchus*, which has a similar life history to *sollicitans* are to a certain extent negatively hydrotropic.

Aedes salinarius, another salt marsh form is as strongly positively hydrotropic as *Culex pipiens*, in fact its hydrotropic reactions are similar to those of *pipiens*, as is its life history.

At different periods during a mosquito's life, its hydrotropic reactions are overshadowed by responses to chemotropic and phototropic stimuli and in some cases, negative hydrotropism might be mistaken for positive chemotropism. In the cases of the sterile females of *Aedes sollicitans*, chemotropism plays very little if any part in explaining their migratory habit. If it did the migrations would not be so extensive or cover the long distances they do.

Negative hydrotropism seems to be more prevalent among the salt marsh than other forms, in fact other species are negatively hydrotropic only for short periods and the females responding to such stimuli are not barren. For some reason the sterility of *sollicitans* seems to render it exceedingly susceptible to negative hydrotropic stimuli.

ANNUAL MEETING OF THE BRITISH COLUMBIA BRANCH.

The annual meeting of the British Columbia Entomological Society took place on January 9th, 1913, in Victoria. A morning, afternoon and evening programme was arranged. From 18 to 27 members were present during the day. A varied programme was rendered which included several reports from districts in the Province, viz., the Victoria District, the Lower Mainland, the Okanagan and the Kootenay.

An interesting lecture was given on the use of Carbon Bisulphide as a fumigant under coastal conditions by Mr. W. H. Lyne, Assistant Inspector of Fruit Pests. Mr. W. H. Brittain followed with a paper prepared on the important subject of Beneficial Insects, bringing the notice of the members forcibly to the fact that applied parasitic entomology was well to the forefront of present day economic entomology. He gave a number of interesting records which had taken place during the past few years in this especial connection.

Mr. G. O. Day, F. E. S., Duncans, presented a paper on *Xanthia pulchella* Smith, and offered a few systematic notes on its life history. Mr. R. C. Treherne gave the members present a brief outline of the life history of the Strawberry Root Weevil (*Otiorhynchus ovatus*), illustrating his points by means of diagrammatic charts.

Mr. Thomas Cunningham gave a long and very interesting paper on the strides that had taken place in the United States and in the world in general in regard to the placing of quarantine measures against injurious insects liable to importation through the medium of trade. His memorandum was listened to with great interest as it contained a summary of all the acts and regulations that had been passed during the past few years, and the reasons for the consideration of these acts and regulations and followed his paper with an outline of the insects at present in B. C., and drew attention to the ones liable to importation.

Mr. Tom Wilson, President, 1912-1913, offered his Presidential Address to the members at the evening session. He drew the attention of the members to the establishment of an investigational station under the Dominion Division of Entomology, a fact that will in all probability be accomplished by the spring. He also desired to welcome Mr. Brittain, the recently appointed Entomologist and Plant Pathologist to the Province under the auspices of the Provincial Department of Agriculture. He added his own sorrow to the resolution of commiseration at the recent death of their late President and father of their Society, the Rev. G. W. Taylor of Departure Bay, Vancouver Island. He closed with a feeling of congratulation at the successful resuscitation of the Society and hoped it would continue as successful as this meeting promised in the future,

Mr. Arthur H. Bush followed with an account of the Flora and Fauna that was common to meet with in the mountains at high and arctic elevations. He closed with a wish that the Society not forget the systematic side of entomology in its endeavours to become a force in the Province. Dr. Seymour Hadwen closed the evening session with a lantern slide lecture on Blood-sucking Flies. He was able to establish the fact of the existence at Agassiz of the English Warble fly (*Hypoderma bovis*,) which previously had not been recorded as existing on the North American continent.

Various resolutions were passed, chief among which was a tribute to the life and work of the late Dr. Fletcher. It was decided also to hold a semi-annual meeting of the Society at Vernon during June.

The following officers were appointed for the year 1913;—Hon. President, E. Baynes Reed; President, G. O. Day, F. E. S.; Vice-president, R. S. Sherman ; Secretary, R. C. Treherne; Asst. Secretary, W. H. Brittain. Advisory Board.—G. O. Day, R. S. Sherman, R. C. Treherne, W. H. Brittain, W. H. Lyne, A. H. Bush, Tom Wilson.

The proceedings of the Annual meetings are duly being printed at the cost of 25 cents each copy and can be had on application to the Secretary, Mr. R. C. Treherne, 1625 Nelson Street, Vancouver, B. C. R. C. TREHERNE, SEC.-TREAS.

ENTOMOLOGICAL SOCIETY OF AMERICA.

The seventh annual meeting of the Entomogical Society of America was held in the Normal School, Cleveland, Ohio, December 31, 1912, January 1, 1913. The meetings were all large and enthusiastically attended.

The following list of papers was presented:

C. Betten.—An interesting feature in the venation of *Helicopsyche*, the Mollannidæ, and the Leptoceridæ,

T. J. Headles.—Some facts regarding the influence of temperature and moisture changes on the rate of insect metabolism.

Lucy Wright Smith.—Mating and egg-laying habits of *Perla immarginata* Say,

Alvah Peterson.—Head and mouth-parts of *Cephalothrips yuccæ.*

J. E. Wodsedalek.—Life history and habits of *Trogoderma tarsale*, a museum pest.

Leonard Haseman.—Life cycle and development of the Tarnished Plant-bug, *Lygus pratensis* Linn.

J. F. Abbot.—The strigil in Corixidæ and its probable function.

Victor E. Shelford.—The ontogeny of elytral pigmentation in *Cicindela.*

N. L. Partridge.—The tracheation of the pupal wings of some saturnians.

L. B. Walton.—Studies on the mouth-parts of *Rhyparobia maderiæ* (Blattidæ), with a consideration of the homologies existing between the appendages of the Hexapoda.

James Zetek.—Determining the flight of mosquitoes.

William A. Riley.—Some sources of laboratory material for work on the relation of insects to disease.

Y. H. Tsou and S. B. Fracker.—The homology of the body setas of lepidopterous larvæ.

Anna H. Morgan.—Eggs and egg-laying in may-flies.

Herbert Osborn.—Remarks on the Cicadidæ with special reference to the Ohio Species. 2. Notes on insects of a lake beach.

Edna Mosher.—The anatomy of some lepidopterous pupæ.

Frank E. Lutz.—On the biology of *Drosophila ampelophila*.

E. P. Felt.—Observations on the biology of a blow-fly and a flesh-fly.

C. K. Brain.—Some anatomical studies of *Stomoxys calcitrans* Linn.

Edith M. Patch and William C. Woods.—A study in antennal variation.

Alex. D. MacGillivray.—Propharynx and hypopharynx.

F. L. Washburn.—A few experiments in photographing living insects.

The following officers were elected for 1913.

President.—Charles J. S. Bethune, Ontario Agricultural College, Guelph, Ontario.

First Vice-President.—Philip P. Calvert, University of Pennsylvania, Philadelphia, Pennsylvania.

Second Vice-President.—William M. Marshall, University of Wisconsin, Madison, Wisconsin.

Secretary-Treasurer.—Alex. D. MacGillivray, University of Illinois, Urbana, Illinois.

Additional members of Executive Committee.–Herbert Osborn, Ohio State University, Columbus, Ohio; C. P. Gillette, Colorado Agriculture Experiment Station, Fort Collins, Colorado; Vernon

L. Kellogg, Leland Stanford Jr. University, Stanford University, California; James G. Needham, Cornell University, Ithaca, New York; C. T. Brues, Harvard University, Cambridge, Massachusetts. Nathan Banks, U. S. National Museum, Washington, D. C.

Member of Committee on Nomenclature.—E. P. Felt, New York State Entomologist, Albany, New York.

The Society will hold its next meeting with the American Association for the Advancement of Science at Atlanta, Georgia.

ALEXANDER D. MACGILLIVRAY, Secretary,

BOOK NOTICE.

THE SPIDER BOOK.—A manual for the study of the Spiders and their allies, the Scorpions, Pseudo-scorpions, Whip-scorpions, Harvestmen, and other members of the Class Arachnida, found in America, north of Mexico, with analytical keys for their classification and popular accounts of their habits. By John Henry Comstock. Doubleday, Page & Co., New York.

Spiders have received relatively little attention on this continent from systematic zoologists, considering the large size of the order, the abundance of many of the species in every locality, their exceedingly interesting and varied habits and the important *role* that they play in the economy of nature. The same statement might, indeed, be made to include the whole of the Class Arachnida, but, whereas the other order of the class are less obviously attractive, it is difficult to understand why the spiders have never been favourites.

The "Spider Book," which is an excellent introduction to the study of the Arachnida, and the spiders in particular, is therefore to be welcomed as a most important addition to American arachnological literature, particularly as it is not only adapted to the needs of the beginner, but will doubtless also form a useful book of reference for teachers and entomologists generally.

In the first chapter the general characteristics of the Arachnida and their relationships to other classes of Arthropods are discussed. The characteristics of the various orders are also given, with tables for the separation of the families and genera, and in some cases the species. On account of its great size, the order Acarina (mites and ticks) is necessarily dealt with more briefly than the other groups, only the superfamilies being defined. Less space, for example, is given to this group than to the *Phalangida* (Harvestmen), a much smaller order.

In chapters II. and III. the external and internal anatomy respectively, of spiders, are discussed in considerable detail. A special section of the former is given to the description of the different types of male pedipalps, whose highly-complex structure is of great taxonomic importance, and has been a subject of special investigation by the author. Following the description of the different kinds of spinning glands at the close of chapter III. is a table, giving the names of these glands, with their number, the position of their spinning-tubes, their distribution among the various families and their functions.

Chapter IV. is an account of the life of spiders, and deals with this subject under a number of headings. Much attention is given to the description of the different kinds of silk and their functions, the types of webs, and to the structure and building of the orb type of web. The account of the development is rather brief, the embryological part being omitted altogether. This is, of course, to be expected in a popular work, but the "Spider Book" is more elaborate than popular works usually are, and we therefore think that a brief outline of the early stages of development would not have been out of place, considering the important bearing which the development of some of the organs, such as the book-lungs, tracheæ and spinnerets have upon the phylogeny of the group.

The systematic part of the book, comprising chapters V.-VII., is enlivened by interesting notes on the habits peculiar to the various families and genera, and by the numerous illustrations. Brief descriptions of many of the commoner species are given, as well as keys to all the families and genera inhabiting North America.

The copious illustrations, which are largely photographic reproductions of living or recently killed specimens and their webs and nests, are scattered throughout the text, and give the book a very attractive appearance by reason of their unusual excellence.

Mailed March 13th, 1913.

The Canadian Entomologist.

Vol. XLV. LONDON, APRIL, 1913 No. 4

FURTHER NOTES ON ALBERTA LEPIDOPTERA.

BY F. H. WOLLEY DOD, MIDNAPORE, ALTA.

(Continued from page 68.)

356. *Cucullia montanæ* Grt.—I have no note of having seen the type of this species, which, according to Smith's Catalogue, is in the Neumœgen collection at Brooklyn, nor have I seen Grote's description. The Calgary form, however, agrees with the description of *montanæ* in Smith's Monograph, and is also the *montanæ* of the British Museum collection, with the exception of the actual specimen figured by Hampson, which happens to be a Denver, Colo., specimen of *asteroides*, of which the type is correctly figured on the next plate. It had not, until recently, occurred to me that there was any likelihood of confusing the two, but I must admit that I have examined, and now possess, specimens which I have had considerable trouble in determining. Generally speaking, whilst the arrangement of colour in the two is about the same, the shades in *montanæ* are more intense, that is, the pale shades are paler, and the dark shades darker. But the colour varies somewhat in different localities, and more reliable points of distinction are as follows. In *montanæ*, the basal area, as far as the t. a. line, is very pale fulvous. The t. a. line is double, with the included space of the same pale gray colour as the central and outer middle portion of the wing below the spots. In *asteroides* the basal space is unicolorous with the central and outer middle area, and the t. a. line is single, though traces of an inner portion are sometimes discernible. In *asteroides* the tegulæ have a black line near the base, which seems to be lacking in my *montanæ*, though Hampson gives it as present in both. Slightly worn or poorly-marked specimens are occasionally extremely difficult to place. I have not both species from one locality. *Montanæ* is recorded from Colorado, and I think I saw it from there in the British Museum, but a Colorado specimen in my collection, sent as *montanæ*, appears to me to be *asteroides*. I am certainly strongly under the impression that the two are distinct.

358. *C. florea* Guen., =*obscurior* Smith, =*indicta* Smith.—I have specimens compared by myself with all these types. That of *florea* is a female in the British Museum, from Trenton Falls, N. Y. *Obscurior* was described from two females taken by Bruce in Colorado, and a type is at Washington. My specimen compared with this type is from Glenwood Springs, and a Calgary specimen compared with types *florea* and *indicta* is exactly like it. I have three specimens from Kaslo. The "*florea*" of my original list (No. 360) was wrongly identified, and the Calgary specimen figured by Sir George Hampson as *florea* is, in my opinion, a strongly-marked form of *postera*. The two are more nearly allied than I at first thought, as my male type of *indicta* happens to be an unusually pale gray, even specimen. I have two Calgary specimens which puzzled me for a long time, and seemed almost to connect them. Generally speaking, *postera* is better marked, and has more obvious reddish brown shades on costal region of primaries. In *florea* such shades are absent, or nearly so, as in the type, and never conspicuous. What appears to me a more reliable character exists in the dark cloud or shade preceding the crescent-shaped mark formed by the t. p. line below vein 2. In *postera* this shade is itself somewhat crescent-shaped, and about concentric with the t. p. line crescent. In *florea* it is direct, oblique, and if produced would meet the inner margin below the orbicular, and the costa near the apex. The shade, however, is often very ill-defined, and not always symmetrical on both wings. But I have studied this feature very carefully, and conclude that it is characteristic of each species as a whole. The moth is a great rarity in this district, only three specimens having been taken besides those previously mentioned, on Aug. 1st, 1909, and June 5th and 11th, 1910. I saw a specimen bearing a New York label in the American Museum of Natural History which I took to be this species, and so labelled it. One in the Rutgers collection, labelled "New Windsor, N. J., May 27th, 1892, Emily L. Morton," appeared to be this, but had ochreous-tinted secondaries, differing in this respect from any previously seen.

359. *C. asteroides* Guen?—I was quite wrong in listing this species as *postera*. I have a manuscript name for it, and have several times been on the point of describing it, but shall not do

so until I have seen this and typical *asteroides* from the same locality and can distinguish them. The type of *asteroides* is from New York, and is well figured by Hampson. In it the ill-defined discoidal spots are pale fulvous, and slightly paler than the rest of the fulvous shade, which extends longitudinally through the upper portion of the wing. The secondaries are clear pearly-white, with dusky veins and outer border, though the border sometimes covers nearly half the wing. I have specimens of the typical form from New York, Rhode Island, Ohio, Pennsylvania, Illinois and Denver, Colorado. I gave the name to a Montreal specimen for Mr. Winn, on the strength of which it is entered in the Quebec list. The only other named species with which I am likely to have confused it is *montanæ*, as mentioned under that head. In the Calgary form the primaries differ but little, but are generally darker blue gray and more even, with the discoidals even less evident. But the chief difference is that the secondaries are smoky throughout, though darker outwardly. This form is the "*postera*" of the B. C. list, and I have specimens from Windermere and Nelson. Some from Manitoba are the darkest of the series, and differ most from true *asteroides*. The dark secondaries contrast strongly with the pearly whiteness of the typical form, and gives the insect a very different appearance, and the primaries of the dark series seem slightly broader and more rounded on the costa. But I must admit that with the primaries alone I might fail to distinguish between some of the specimens. I have not taken it at Calgary for several years.

360. *C. postera* Guen.—This is the "*florea*" of my original list. The Calgary form is figured by Hampson as *florea*, but seems to me darker and more strongly marked only than the type of *postera* from New York. The chief distinctive character between this and *florea* I have pointed out under the latter heading. Judging from the number I have seen, this species is, with the possible exception of *intermedia*, the commonest of the genus in Canada, though I have not seen it from west of the Rockies. I have named a Montreal specimen for Mr. Winn, which seemed to me about typical. In Prof. Smith's collection, the only specimen which stood under this name was a male from Liberty, N. Y. This was like the

Calgary form, except that it was ochreous-tinted throughout, which made me doubt its identity.

361. *C. speyeri* Lint.—Another female. July 10, 1900, formerly in my *intermedia* series, appears to be this species, but was labelled *intermedia* on Smith's authority. The error was excusable, as it is duller and less black-streaked than *speyeri* usually is, but has all the other characters of that closely allied species, including the pearly white though dusky-margined secondaries. I have the species from Illinois, Volga (S.D.), Colorado, and Aweme, Manitoba.

362. *C. intermedia* Speyer.—I consider that this is the correct name for the form occurring here. Any attempt to separate it from *intermedia* from the east is hopeless, though eastern specimens, as a whole, are a trifle darker, due to their being more suffused with brown shades. I have Calgary and eastern specimens matching exactly. Hampson figures a Calgary example as *cinderella*. The latter was described from a single Colorado male collected by David Bruce. I saw it in the Washington collection, and it has the transverse maculation almost obsolete. A Colorado female in the same collection certainly suggested a faintly marked *intermedia*. The validity of *cinderella* as a species is open to much doubt.

365. *Tapinostola variana* Morr.?—I had listed this species as *orientalis* Grt., but that, according to the description, has a t. p. line of blackish dots, and the subcostal and median nervures are finely lined within the cell with black. This sounds like the species figured by Sir George Hampson, from Renfrew County, Ontario, as *inquinata* Guen., of which he has the type from New York, and of which he makes *orientalis* a synonym. My notes on *inquinata* type do not mention a black streak immediately above the median vein, nor does Hampson mention it in his description in the Catalogue. His synonymy, however, is probably correct. Sir George's description is all I have of *variana*, besides a reference thereto by Grote, and the only difference mentioned is the absence of the t. p. line. Holland figures as *variana* a Winnipeg male from the Washington Museum. I compared this specimen and concluded that the Calgary species was distinct and also probably

distinct from type *inquinata*. But the Winnipeg specimen in question has an obvious t. p. line, which the type *variana* lacked, so that its identity is open to doubt. It is at any rate probably a species not at present in my collection and may be a pale *inquinata*. My Calgary specimens are about the colour of *inquinata* type, but lack all traces of a t. p. line, though some show traces of blackish in the cell. Besides the two before mentioned I have two males taken at light on Sept. 8th, 1906.

366. *Hydrœcia nictitans* Bork.—I feel bound to follow Hampson in treating the North American species as identical with the European *nictitans*. Smith himself referred "Var. *americana* Speyer" to his *atlantica*, so that the former name should have preference in any case. A female type of *atlantica* from Ithaca, N. Y., is in the Washington Museum. No clear differences are pointed out, in fact the impossibility of distinguishing it from the European form except by male genitalia is admitted. Its range is given as "Nova Scotia, Hudson's Bay, Southward to Virginia, West to Colorado". *Interoceanica* was described from three specimens from Winnipeg only. I have none from there exactly, but have seen a pair of types. It was characterized as small and very dark in colour, with the ordinary markings almost blackish, and reniform white. The latter character is of course variable in *nictitans*. I compared Smith's types and did not consider them distinct, nor did they strike me as variations worthy of remark. *Pacifica* was stated to range from California to Vancouver and to be more compactly built than *atlantica* or *nictitans*, and a little more lightly shaded, "the secondaries yellowish or purplish red and somewhat silky, quite different from the eastern examples". I have no Californian examples, but numbers from Vancouver Island, and their variation is much like that of eastern specimens. Concerning his three new names Smith writes in his Revision; "These three species I could hardly have dared to separate from *nictitans* had it not been for the differences in structure in the male genitalia; but these are so radical that specific identity is out of the question". Four genitalic species are claimed for the British Isles, some of which are said to be locally constant in some superficial characters. Hampson unites them all as one species, but quotes six names as

aberrations, including two European, one Asiatic, and three North American; viz:

"*americana.*—Fore wing rather more orange-red. Eastern States and Canada.

interoceanica.—Fore wing browner. Western Canada.

pacifica.—Fore wing grayer, California." This latter is not in accord with Smith's diagnosis.

367. *H. pallescens* Smith.—The male and female types in the Washington Museum are from Calgary. They differ from *medialis* from Colorado in lacking the reddish tints, and in having the entire ground colour washed with white. Some Calgary specimens are a good deal darker than the type, but scarcely reddish. The form was not even recognised as a variety in Dyar's catalogue, though the difference in colour is somewhat striking. Hampson treats them as species. I think it likely that the differences are merely varietal, but have not seen enough Colarado material to enable me to form a fair judgment, and have none from there in my collection. Mr. Baird has taken *pallescens* at High River, and I have it from Cranbrook, B.C.

368. *Papaipema* sp.? The type of *impecuniosa* is a male from Massachusetts, and is in the British Museum. Sir George Hampson figures a specimen like it. There are two Red Deer River specimens from me in the same series. But, like the rest I have seen from that locality, they differ from *impecuniosa* in the form and course of the central shade, which is more like that in *purpurifascia*, *rigida* and *verona*, the latter being a much paler thing from Winnipeg. The form of the t.p. line is something between that in *impecuniosa* and *purpurifascia*. The colour and maculation otherwise is much like that of *impecuniosa*, but the orbicular and claviform may be either yellow or white. Unfortunately I have only three specimens now left in my collection, not having visited the locality for some years.

(To be continued.)

AN ENTOMOLOGIST WANTED FOR ARCADIA

"The Agassiz Association's ArcAdiA is for study and research and for giving information upon any phase of nature to any person who desires to know."

ArcAdiA is well equipped with every facility for studying nature, and especially so in entomology. Within the adjacent territory, especially in Nymphalia, which is a part of ArcAdiA, there are facilities for studying various kinds of aquatic and marsh insects. The laboratory is well equipped with apparatus for classifying, examining, photographing, etc. There are breeding cages for studying the insects in their transformations, and whatever further equipment may be necessary will be made to suit the needs of a student. We want an adult entomologist, preferably a married man, to come to ArcAdiA, lease a building site, erect a small cottage, and live near to nature in the spirit of the Institution. He shall have the freedom of the Institution without expense, but for his services no salary will be paid. We are looking for some one who has retired from the active duties of life, and expects to spend the rest of his days in close proximity to the entomological world.

The experiment has been successfully made in the Department of Botany. Some three years ago a lady in Wisconsin desiring to devote the rest of her life to the study of plants, became a member of The Agassiz Association, at the cost of only three dollars for the first year and only a dollar and a half each year thereafter. She leased a building site and erected, at her own expense, a portable cottage, in which and in the surroundings she leads the ideal ArcAdiAn life in nearness to nature. She devotes all her spare time to the Botanical Department, collecting plants, studying them in their habitat, planting them in her little yard, and studying them under the microscope. A pleasurable part of her occupation is to show the results to the admiring visitors at ArcAdiA. The Agassiz Association remunerates her for her services in giving her all the facilities of the equipment, such as may be needed in her botanical pursuits. In return for her services she receives the best pay in the world—the joy of doing and the joy of helping.

Her attractive little cottage is known as Botany Bungalow. We want some entomologist to make his home in the "Entomologist's Eyrie" or "The Ant Hill," or some similarly named cottage in ArcAdiA. Full particulars as to what the AA is and what is its ArcAdiA, what it has done and what it is trying to do, and including a copy of "The Guide to Nature," will be sent upon application.

On the other hand, full particulars will be required of the personality, skill, experience, plans, etc., of the applicant who would come here and take charge of our Entomological Department. We would prefer someone who has retired from active business life and has means to devote the rest of his days to his favorite pursuits, but such entire devotion of time is not necessary. Arrangements could be made for some income for services, if desired. Employment of various kinds can be obtained in the vicinity, but, as previously stated, the ideal would be one who has retired and intends to devote all the rest of his time to the interests and beauties of entomological nature.

For further particulars, apply to The Agassiz Association—Edward F. Bigelow, President—ArcAdiA, Sound Beach, Connecticut.

THE COTTON MOTH, *ALABAMA ARGILLACEA* HBN.

The photograph from which the accompanying illustration was made, was sent to me by Mr. J. F. Calvart, of London, Ont. These moths were noticed in very large numbers this autumn in Western Ontario. At London, they appeared suddenly either late in the

FIG. 2

evening of Oct. 10, or early in the morning of Oct. 11. The characteristic habit of the moth of resting with its head downward is well shown in the illustration. An account of the occurrence of this moth in eastern Canada in 1912 will appear in an early issue of the Ottawa Naturalist.—ARTHUR GIBSON, Div. of Ent., Ottawa.

ON SEVERAL NEW GENERA AND SPECIES OF AUSTRALIAN HYMENOPTERA CHALCIDOIDEA.

BY A. A. GIRAULT, BRISBANE, AUSTRALIA.

The following genera and species were included within three small collections of this super-family, loaned to me for study, but do not include all of the material from them. Two of these collections were from Queensland, the third partly from Victoria and partly from New South Wales.

Family Chalcididæ.
Subfamily Chalcidinæ.
Tribe Chalcidini.
Genus *Tumidicoxa* Girault.

1. *Tumidicoxa rufiventris* new species.

Female: Length, 5 mm.

Opaque black, the abdomen rufous or orange red, as are also the antennal flagellum, the posterior coxæ, tibiæ (except at tip) and femora (except at apex, lateral), the cephalic tibiæ, except at base, and all the tarsi (somewhat diluted with yellowish); cephalic and intermediate coxæ black or very dark, the proximal half or more of the cephalic and intermediate femora black, their distal half or less honey yellow. Scape dark fuscous, the pedicel somewhat lighter. Tegulæ, a rounded spot at apex of posterior femur laterad, a distinct oval spot near tip of posterior tibiæ and the knees more or less lemon yellow. Wings very slightly stained throughout, the venation smoky. Pubescence not conspicuous, with a reddish tinge.

Scrobicular cavity nearly smooth, shining; head and thorax rugoso-punctate, the propodeum mesad (dorsal aspect) foveate, the abdomen glabrous, but the distal segments finely, polygonally sculptured. Lateral ocelli distinctly more than their own diameter from the eye margin. Plate at apex of scutellum distinctly bilobed. Propodeum in the dorso-lateral aspect, with at least one tooth-like projection, its lateral aspect moderately hairy, but not conspicuously so. Posterior femora with one moderately large tooth, followed by about ten others, which are smaller distad, all the ten much smaller than the first. Antennæ 12-jointed, the single ring-joint large, the pedicel as long as the first funicle joint, which is somewhat wider than long, the remaining joints all wider than long, except the distal club joint, which is longest, conical; distal funicle joints transverse; flagellum clavate, somewhat compressed.

(From two specimens, two-thirds of an inch objection, 1-inch optic, Bausch and Lomb.)

Male: Not known.

Described from two female specimens mounted on pins, labelled "Warburton, Victoria." This species differs from the South American forms strikingly in coloration; also the funicle joints of the antennæ are shorter, the pedicel longer in relation to them, the lateral ocelli farther away from the eyes, the plate of the scutellum more deeply lobed and the stigmal vein not sessile, yet short. Otherwise, it is similar in all details, with the possible exception of the ventral plate on the thorax, which I was unable to see in these specimens because of the manner in which they were mounted.

Habitat: Australia—Warburton, Victoria.

Types: No. *Hy 1178*, Queensland Museum, Brisbane, the above specimens (2 pins) plus a slide bearing an antenna and a posterior leg.

2. *Tumidicoxa flavipes* new species.

Female: Length, about 5 mm.

Like the South American species, but the plate at the apex of the scutellum is not emarginate at the meson, or barely so. Opaque black marked with lemon yellow as follows: The tegulæ except at extreme base (cephalad), tibiæ and tarsi, except the brownish base of posterior tibia and parts of the distal tarsal joint, distal half of cephalic femora, distal third of intermediate femora and the tip of the posterior femora. Legs otherwise black, reddish black on femora and tibiæ on first two legs. Venation and antennæ brownish black, the latter really black, brownish toward tip. Wings perfectly clear. Body rugoso-punctate, the second abdominal segment shining. Pubescence not conspicuous. Antennæ 12-jointed, cylindrical, with one ring-joint, the pedicel small, wider than long, not half the length of the proximal funicle joint, which is the longest joint of the flagellum. Distal club joint longer than the other, obliquely truncate at tip. Distal funicle joint slightly wider than long, the middle joints subquadrate. Scape short, simple. Posterior femora beneath the others, the last three distinctly smaller in succession, the last very small. The teeth are black, and they rather increase in size at the middle (Nos. 5, 6 and 7 from proximal end). Posterior femur minutely punctulate and clothed with soft, greyish pubescence. Agreeing with the generic description, except as may have been noted. The lateral ocelli are somewhat farther away from the eyes.

(From a single specimen, the same magnification.)

Male: Not known.

Described from a single pinned female labelled "Dandenong Range, Victoria."

Habitat: Australia—Victoria (Dandenong Mountains).

Type: No. *Hy 1179*, Queensland Museum, Brisbane, the above specimen on a card; an antenna on a slide.

3. *Tumidicoxa victoria* new species.

Male: Length, 6.1 mm.. Rather large.

Like the preceding species, but larger, more robust, the plate at the apex of the scutellum plainly bidentate, the scape longer, the black parts of the legs darker, and specifically the cephalic tibiæ are brown in the middle, the intermediate ones along the proximal half, except at base and the posterior ones, black in the middle, their tips pale yellowish. The posterior fermora beneath bear nearly the same arrangement of teeth, but there are only eleven, followed by a minute tubercle; numbers 3 and 4 are larger than 2 and those distad of them. The posterior coxæ have a flabellate enlargement at the apex above. The antennæ 12-jointed, the distal funicle joints more transverse than with *flavipes*. Wings hyaline.

(From a single specimen, the same magnification.)

Female: Not known.

Described from a single male, minutien-mounted and labelled "Dandenong Ranges, Victoria."

Habitat: Australia—Victoria (Dandenong Mountains).

Type: No. *Hy 1180*, Queensland Museum, Brisbane, the above specimen plus an antenna on a slide.

4. *Tumidicoxa regina* new species.

Male: Length, 4.95 mm.

Like *flavipes*, but more robust, the scutellum terminating in a distinctly bidentate plate and the postmarginal vein longer. The scape is also longer. Posterior femur armed with twelve teeth, the first large, the next two very small, followed by seven larger ones, (of which numbers six to nine are largest) and two shorter ones, the last broad, its flat upper edge at apex thus emarginate; excluding the first tooth, numbers 6 to 9 are largest. In *flavipes*, teeth Nos. 2 and 3 are not distinctly smaller than the ones immediately following (distad).

(From one specimen, the same magnification.)

Female: Not known.

Described from a single male specimen on a pin, from the collections of the Queensland Museum, labelled "Brisbane, H. Hacker. 3-7-11."

Habitat: Australia—Brisbane, Queensland.

Type: No. *Hy 1181*, Queensland Museum, Brisbane, the fore-noted specimen on a pin, plus one slide bearing antennæ and a posterior leg.

Pseudepitelia new genus.

Female: Resembling *Epitelia* of Kirby, but the abdomen not produced into a stylus distad, the posterior femora without depressed punctures and armed beneath with more teeth, there being six moderately large, more or less, subequal teeth (but the first largest), followed distad by four others, which shorten in succession. The antennæ are 13-jointed, with one ring-joint, inserted nearly on a line with the ventral ends of the eyes. the scrobicular cavity reaching the cephalic ocellus, the lateral ocelli plainly more than their own diameter from the eye margin. The postmarginal vein about half the length of the marginal, slender, the stigmal very short, yet with more or less of a distinct neck. The second abdominal segment occupying more than a third of the abdomen. Propodeum with two small, acute projections in the middle of the dorso-lateral line (seen from ventro-laterad). Body nonmetallic, punctate. Abdomen as in Chalcis. The scutellum terminates in a short, bidentate plate.

Male: Not known.

Type: The following species.

1. *Pseudepitelia rubrifemur* new species.

Female: Length, 5.10 mm.

Opaque black, the second abdominal segment glabrous black, the posterior femur dark reddish, the venter of the abdomen at the meson more or less suffused with dark reddish or yellowish; tegulæ pallid, the wings hyaline, the venation black, face and distal half of the abdomen pubescent. Tarsi more or less brownish. Intermediate and cephalic knees and two distinct elongate spots on each end of the posterior tibiæ exteriorly and not at tip, pale yellowish. Antennae wholly black.

Body moderately finely, densely punctate, the spaces between the punctures lined. First abdominal segment, with very minute punctures, which vary in size, the following segments pubescent and transversely wrinkled, the penultimate segment rougher. Posterior femora densely punctulate the punctures very minute; antennæ, with the distal joint very short, truncate, only about twice the length of the ring-joint; scape very long, narrowing distad; pedicel much longer than the ring-joint, but only half the length of the proximal funicle joint, which is longest of the funicle, twice the length (or nearly) of the subquadrate distal funicle joint. Proximal two joints of the club subequal, the distal joint flat, very short.

(From one specimen, the same magnification.)

Male: Not known.

Described from a single cardmounted female, labelled "Cheltenham, Victoria."

Habitat: Australia—Victoria (Cheltenham).

Type: No. *Hy 1182*, Queensland Museum, Brisbane, the above specimen, plus a slide bearing an antenna and a posterior leg.

2. *Pseudepitelia tricolor* new species.

Female: Length, 5.00 mm.

The same as *rubrifemur*, but the postmarginal vein shorter and stouter, the second (distal) elongate, pale yellowish spot exteriorly on posterior tibia absent, but the two proximal tarsal joints (and less so, the third) of posterior legs white, suffused with yellowish; the scutellum has not a small patch of greyish pubescence at apex, just above the terminal plate as with the type specimen (*rubrifemur*) and in *tricolor* the third abdominal segment is more roughly finely sculptured. There are eleven distinct teeth on the posterior femur instead of the ten of the type species. Intermediate and cephalic tarsi white or whitish. (Antennæ missing; scape black.) Wings hyaline.

(From a single specimen, the same magnification.)

Male: Not known.

Described from a single cardmounted specimen from the collections of the Queensland Museum, labelled "Q.M. Tambourine. H. Hacker, April 2, 1911."

Habitat: Australia—Tambourine, Queensland.

Type: No. *Hy 1183*. Queensland Museum, Brisbane, the above female on a card.

Brachepitelia new genus.

Female: The same as the preceding genus, *Pseudepitelia*, but the antennæ 12-jointed, the scutellum terminating in a short plate, whose distal margins are straight, the plate barely differentiated, The submarginal vein is shorter and stouter. Propodeum without noticeable lateral projections. Second abdominal segment occupying nearly half of the abdomen.

Male: Not known.

Type: The following species.

1, *Brachepitelia rubripes* new species.

Female: Length, 3.70 mm.

Opaque black, marked with dark red as follows: Posterior legs except coxæ; cephalic knees, tibiæ and tarsi; intermediate knees and tarsi (mixed with brownish) and the ends of the tibiæ. Venation dark, the wings hyaline. Head and thorax rugoso-punctate. Posterior femur with ten distinct teeth, the first twice the largest, the distal teeth smaller in succession.

Male: Not known.

Described from a cardmounted female, labelled "Larva of Various Moths, Melbourne."

Habitat: Australia—Melbourne, Victoria.

Type: No. *Hy 1184*, Queensland Museum, Brisbane, the above specimen; an antenna on a slide.

(To be continued.)

REMARKS ON THE DISTRIBUTION OF HETEROPTERA.

BY J. R. DE LA TORRE BUENO, WHITE PLAINS, N. Y.

Among the many problems of nature that engage the attention of the biologist there is one that to me has always been of the utmost interest. It is that of the occurrence of the same species in widely separated regions or through extensive and seemingly dissimilar areas or in isolated and restricted habitats. The classic example of the last, familiar to all entomologists, is the peculiar subarctic and alpine butterfly *Oeneis* or *Chionobas semidea* which from the wilds of Labrador jumps to the high peaks of the Presidential Range of the White Mountains and again is not found till we come to the Rockies in Colorado. Here, however, we have a tenable explanation for this great and peculiar range, in the fact that this is an arctic genus which spread during the ice-age throughout its vast territory, and which, with the recession polewards of the ice cap and the frigid temperatures it caused, travelled northward in its wake. Some, however, followed the receding line of perpetual snow up the mountain sides, and where these were of sufficient altitude, they have contrived to maintain themselves to this late date in the geological history of the earth.

In this paper the Hemiptera only are to be considered, more especially the Heteropterous forms supposed to be common to America and Europe. At the outset we are confronted with a difficulty, which arises from the mistaken reference of American species to European forms. This troublesome condition is directly due to the meagre descriptions of the older authors, who availed themselves principally of colour for specific distinctions and put the structural differences in the generic characterizations. In part, however, our native entomologists are at fault, since much of this confusion can be traced to their neglect of the study of the cognate European species, which, even though they are of the same colour patterns as ours, in so far as any written description can go, are nevertheless sufficiently different in form and structure to be readily distinguishable by the trained eye. This condition in the Hemiptera is being rapidly adjusted, due almost entirely to the labours of the Europeans. In fact, our own present lack of sufficient acquaintance with their writings leads some of us to the perpetuation of errors long since dispelled. It is to the labors of those eminent scientists, Prof. Momtandon and Dr. Horvath, but mainly to the latter, that we owe what has been done towards correcting these mistakes. Horvath's visit to the United States in 1907 and the collecting he then did, enabled him to make the necessary comparisons, and it is his results which form the groundwork for this discussion.

In the writings of the fathers of American Hemipterology, we find much of this erroneous work, in their case most unfortunately unavoidable owing, as already pointed out, to too great reliance on colour characters alone. Thomas Say, who needs no praise to establish his position as the greatest of American entomologists, had indeed a keen and discriminating eye, and nearly without exception his species and genera have withstood the most rigid tests. His successors however, have not been so uniformly successful, so we have for America a list of species of supposedly European Heteroptera (to which in the heat and haste of Hemipterological youth I have added my mite), which includes such species as:

Reduvius personatus L.
Sciocoris lectularius L.
Sciocoris microphthalmus Flor.
Nezara viridula L.
Zicrona cærulea L.
Corizus crassicornis L.
Corizus hyalinus Fabr.
Nysius thymi Wolff.
Nysius ericæ Schill.
Stygnocoris rusticus Fall.
Sphragisticus nebulosus Fall.
Scolopostethus thomsoni Reut.
Aradus crenatus Say.
Aradus lugubris Fall.
Aradus cinnamomeus Panz.
Harpactor leucosdilus Stal.
Gerris rufoscutellatus Latr.
Acanthia pallipes Fabr.
Acanthia xanthochila Fieb.
Corixa germari Fieb.
Corixa præusta Fieb.

These do not include the ones subsequently recognized as undescribed, such for example as *Cymus claviculus* Fall., which turned out to be new and was described by Horvath as *discors*; *Emblethis arenarius*, which comparison with European material showed to be different and which is now known as *vicarius* Horvath; *Pentatoma juniperina*, which is restricted to the other side of the Atlantic, ours being described as new under the name of *persimilis* Horvath.

Returning to the larger aspect of the question, a consideration of the hemipterous forms common to the two continents discloses the fact that in preponderating numbers these are phytophagous and parasitic, the majority being Homoptera of families notoriously injurious to vegetation, namely, the Jassidæ and Aphididæ. The total number of species of this order found on both sides of the Atlantic is in the neighborhood of 160 to 170, a very small number as compared with the Coleoptera.

How are we to account for this dispersal? There are two chief means, the one natural, by migration of the living beings of their own impulse, and the other artificial, through the agency of man. A large proportion apparently belong in the first category. The small remainder, (including therein the classic examples of the unsavory bedbug and other obnoxious personal parasites), owe their distribution undoubtedly to the more or less involuntary agency of man. To-day the constant importation of nursery stock is bringing with it a constant transfer to this continent of various plant pests. Fortunately, the strict surveillance on plants brought from abroad has thus far held in check the spread of these insects to any great extent. On the other hand, sometimes the good perish with the bad, and important predators are fumigated out of existence together with their prey.

An examination of the forms which evidently owe their distribution to natural agencies has shown that the great majority belong to Palæarctic genera and are in the main Palæarctic species of the most widespread character. Take for example *Gerris rufoscutellatus* Latreille, which is without doubt the Hemipteron of widest actual distribution next to *Nezara viridula*. It is known across Northern Europe through Siberia, thence to British Columbia and Oregon, ranging East to the Northern

Atlantic region. Here we have a form undeniably Palæarctic in origin, which has migrated from its native source and travelled 15,000 miles to found its colonies throughout the North Temperate Zone. Its route has certainly been via Bering Straits into Alaska and thence east and south. Its habitat and its predaceous nature have both contributed largely to its fitness for this long voyage. It is furnished with good wings, sucks any insect it can overcome and lives on the surface of the water. It has therefore had an unimpeded and favourable route from the land of its nativity eastward until stopped at the impassable barrier of the Atlantic ocean. Thus also must have migrated the two Corixas, *germari* and *præusta*, out from the Palæarctic region.

This also is the route followed by many of the land bugs, but they indeed must have met the great obstacles, saving only the semi-aquatic strong-flying and predaceous Acanthiidæ, to whom the waters can have no terrors. A number of these terrestrial forms are cannibals and live on other insects, their only requirement being that their prey be not encased in impenetrable armor or too large to be overcome. *Zicrona cærulea* may serve as an example of these carnivores, and here we see how much slower has been its progress than that of the aquatic forms, and seemingly it has met with an unsurmountable boundary in the Rocky Mountains. The advent of the phytophagous forms is similarly explained for the majority of cases, in view of the adaptability of the Hemiptera to any vegetable food other than their native food plants, especially when pressed by hunger. The dispersal of one land group, however, is a subject for interesting speculation. I refer to the three species of Aradids common to the Eastern United States aud Western Europe. Is this their native home? The genus Aradus is boreal in its origin. This much is reasonably certain. But are these three species themselves of Palæarctic or Nearctic origin? And if of Palæarctic origin, how did they get there? And if not, how did they cross Europe?

Aradus crenatus was described by Say in 1832; subsequently Herrich-Schaefer described it and figured it in Wanzenartigen Insekten (IX., fig. 538, p. 90), under the misnomer *corticalis;* and in 1860 Leon Dufour described it as new, and called it *dilatatus*.

So far as I know, it is confined to the Atlantic States and Western Europe. *Aradus cinnamomeus* is in the same case; and *Aradus lugubris* of Fallen was independently recognized by Say also in 1832, who called it *rectus*, and by Kirby in 1837. It appears to extend throughout the northern part of this continent, from east to west and through Siberia into Western Europe. Seemingly, then, *lugubris* has come in by way of Bering Strait, and has travelled eastward. As to the other two, their dispersal might seem to indicate human agency. It is conceivable that they have travelled east into Europe, or west out of Europe concealed in crevices in logs and planks or under loose bark. The earlier discovery of *crenatus* in this country might appear to indicate its American origin, while the fact that *cinnamomeus* is first recognized in Europe might perhaps lead to the inference that that was its native soil, but possibly erroneously, since being a dweller in pine trees it may conceivably have been exported in such timber from this Continent.

There is another small group with a most remarkable distribution. The type of these may be considered to be *Nezara viridula*, which occurs with us commonly in Florida, and thence down into tropic America, across the ocean into Africa, throughout Europe and thence into Asia. Its home is said to be in Africa, whence it has spread so widely. How? No explanation seems to have been offered of its wanderings, but certainly there is no question of the identity of the species, even though the examples come from many lands. In this class, also, belongs *Corizus hyalinus*, which has spread even unto the distant isles of the Pacific Ocean.

It has not been the intention in these remarks to go deeply into the subject or to expound a theory, but simply to set forth a peculiar biological phenomenon and one well worthy of serious consideration and study. A few forms in a restricted group have been referred to, but all orders of insects present the same problem. Where the migration is over extensive land areas with a more or less homogeneous character of vegetation, or when one certain food-plant is widespread, the question presents no difficulties, but where large bodies of water intervene, it becomes more complex, and is a fit subject for scientific inquiry of a high order.

ON THE GENUS *LAMENIA* STAL.

BY F. MUIR, H. S. P. A. STATION, HONOLULU.

Stal founded the genus *Lamenia* in 1859 (Eugenies Resa Zoo., 277, Pl. IV., f. 5), for *caliginea* from Tahiti, and the genus *Herpis* in 1861 (K. Vet. Ak. Hanal., III., No. 6, p. 8), for *fuscovittata* and four other species from Brazil; in 1866 in a footnote on page 193 of Hemiptera Africana he sank *Herpis* and *Lamenia*. Uhler in 1889 (Stand. Nat. Hist., II., 233), placed *Pæciloptera vulgaris* Fitch into *Lamenia* and since then several North American species have been placed in this genus, all congeneric with *vulgaris*. Fowler's *Cedusa funesta* is congeneric with *vulgaris* and (according to Melichar, 1905, Wien. Ent. Zeit., 285), *Attalia*=*Herpis*.

Stal's figure of *caliginea* is very clear, and shows the narrow, parallel-sided form of the tegmen with the subcosta and radia amalgamated to near their apices, and the subcostal cell small, a tegmen typical of *Thyrocephalus* Kirkaldy, whereas *vulgaris* and its allies have the tegmen much broader, the subcosta and radia separate from near the base and the subcostal cell large. For these reasons I do not consider it advisable to keep *vulgaris* and *caliginea* in the same genus. All the specimens I have seen from Central and South America are congeneric with *vulgaris* so that it appears best to place that species along with all its allies under *Herpis* and to have *Lamenia* with its type only, or to place all the eleven known species of *Thyrocephalus* under the latter genus.

Cenchrea dorsalis appears to differ from *Herpis* in having no subantennal keel across the gena, the antennal chamber being entirely pronotal (Westwood's figure of the tegmen also shows differences, which I do not like to emphasize until I can examine a specimen); from *Syntames* it differs by the absence of a central longitudinal keel on face, and from *Basileocephalus* and *Phaciocephalus* by the presence of a transverse keel between vertex and face.

OBITUARY.

Mr. L. E. Ricksecker, well-known collector of California insects, died in San Diego in that State, January 30, 1913. He was especially devoted to the collection of Coleoptera, and distributed amongst his correspondents in the east many interesting specimens.

A NEW SPECIES OF CORIXIDÆ.

BY J. F. ABBOTT, WASHINGTON UNIVERSITY, ST. LOUIS.

Palmacorixa buenoi, new species.

With the general facies of *P. gillettei* Abbott, from which it differs in the coarser texture of the tegmina, the character of the lineations and in the palæ of the male and the first femora of the female. The discovery of a second species of this genus necessitates a revision of the generic diagnosis given with the original description (Ent. News, XXIII, 337). The genus may be characterized as follows: Elongate, tegmina tapered posteriorly, with vermiculate markings. Male palæ thin, plate-like, pegs variable. Large stridular area on femur. Metathoracic wings aborted in both sexes. Male asymmetry and strigil dextral ; fifth tergite entire, sixth divided.

Description: Similar to *P. gillettei* in size and appearance, in the flattened short pronotum, and large head, with prominent posterior angles. Dark yellow to smoky brown, and much darker than *gillettei*. The tegminal lineations are complete, more or less inosculated and confused, but without a marked tendency to longitudinal seriation. Lineations of clavus complete—i.e., not effaced on the inner anterior area as in *gillettei*. Head smoky brown; its length 1¾ in the width in the male, 2¼ in the female; interorbital width twice in the head length in the male, 1¼ in the female. Male fovea more prominent than in *gillettei*, reaching the middle of the eye and clothed with delicate depressed hairs. Pronotum flattened, margined, lenticular in outline, evenly rounded posteriorly, dull and minutely rastrate, with 7–8 approximately parallel lineations, which are more or less broken. the lineations about as wide as the yellow interspaces. Posterior margin brown. Claval lineations delicate, vermiculate and inosculate, covering the whole clavus, fused externally to form a more or less definite oblique line parallel to the corio-claval suture. Clavus rather infuscated and clouded across the middle third. Markings of corium similar to those of clavus, running without interruption over the membrane; inosculated, but scarcely interrupted, sometimes fused into one or two rather indefinite longitudinal lines, which do not extend beyond the embolium. Surface of clavus and corium rather dull and rough, the clavus usually rastrate, the corium merely punctate. Margins of embolium and of clavus elevated. Lower surface and legs pale; posterior tibia fringed with brown hairs. Metaxyphus very short, acuminate. Strigil rounded, 5 striæ, diameter 0.1 mm.

Male palæ cultrate, somewhat produced at the base, the length three times the greatest height. Pegs blunt, elongate, 24–33 in number. The distal ones are somewhat longer and crowded, and may be displaced into two irregular rows; the main row begins midway the base and rises in a curve after the first half dozen pegs; then follows the upper margin, but at some distance from it. A second row of peg-like spines along the lower margin, about 1½ to 2 times the length of the pegs. Tibia subglobular, about as high as the pala. Femur oblong, a little less than twice as long as wide, the stridular area covering the proximal half and consisting of short spines set in transverse rows. Female palæ cultrate, not produced at base, slightly more than three times as long as wide, broadly joined to the tibia. Tibia rounded oblong, tapered proximally, twice as long as high. Femur oblong, 2½ times as long as wide (the width at base in *P. gillettei* is two-thirds the length) with stridular (?) spines on the surface as in *P. gillettei*. Second leg: Femur 2½ times the length of the tibia, the latter equal to the claws,* and 1⅓ the length of the tarsus. Length, 5½–6 mm.; width across pronotum, 1½ mm.

Types 2 ♂ and 2 ♀ from White Plains, New York, collected in August and September by J. R. de la T. Bueno. Other specimens have been examined from Washington, D.C. (coll. W. L. McAtee) Oglethorp, Georgia (coll. T. C. Bradley) Hadley, Mass. (coll. C. A. Frost) and Valhalla, N.Y. (coll. Bueno). The species, therefore, appears to be distributed pretty widely up and down the Atlantic Coast of the United States.

Variation.—Some twenty specimens have been examined in addition to the described types. These individuals show a wide range of variation, such that the extremes would seem to belong to different species were it not for the intergradation. The writer has been unable to find any constant character, however, which would serve as a basis for discrimination. The smallest (White

*Through a *lapsus calami* these are called "spines" in the description of *P. Gillettei* (l. c., p. 339).

Plains) measures but 4¼ mm., the largest (same locality) 6½ mm. The tegminal surface may be smooth and polished, or dull and rastrate, the lineations varying from the regular complete lines of the type to interrupted and confused markings, resembling those of *P. gillettei*; the inner angle of the clavus, however, is never bare of lineations. Pronotal lines 6–9, either entire or much broken and confused. The index of pronotal width divided by pronotal length ranges from 2.22 to 2.60 in the ♀, and 1.79 to 2.73 in the ♂; that of the head width divided by the interorbital width ranges from 2.87 to 3.57 in the ♀ and from 3.60 to 4.20 in the ♂; that of the head width divided by the head length from 2.07 to 2.60 in the ♀ and from 1.68 to 2.33 in the ♂. In the male the palar pegs are sometimes crowded into two rows at both ends of the series. The absence of functional wings in both sexes in this genus certainly interferes with the rapid dispersal or mixing of individuals from adjacent localities, and thus brings about a partial segregation which would preserve and intensify aberrant variations. This possibly explains the very unusual range of variability above described.

ENTOMOLOGICAL MEETING IN CALIFORNIA, 1915.

The Entomological Society of America has received an invitation from the Panama-Pacific International Exposition to hold a meeting in some Californian locality in the summer of 1915. This gathering may be at either of the Universities or on the Exposition Grounds. It has received the enthusiastic support of western entomologists. These latter have attended many eastern meetings, and this is an excellent chance for us to return the compliment. It may be possible for a number to go out with a party, stopping off at one or more interesting points en route. As chairman of a special committee to consider this matter and report at the next meeting of the Association, the undersigned would welcome suggestions in regard to this meeting, and also expressions relative to the support it would probably receive from eastern entomologists.

E. P. FELT, State Museum, Albany, N.Y.

BUMBLE BEES AND WASPS WANTED.

Mr. F. W. L. Sladen, Assistant Entomologist for Apiculture, Division of Entomology, Department of Agriculture, Ottawa, is making a special study of the Bumble Bees (genera *Bombus* and *Psithyrus*) and the Social Wasps (genera *Vespa, Polistes* and *Polybia*). He would be glad if anyone who finds a bumble bee's nest would send him a few specimens of the bees, without destroying the nest, so that he may determine the species. He would also like to receive specimens caught on flowers, especially in out-of-the-way districts.

Bumble bees and wasps are best killed with cyanide of potassium. Crushed tissue paper should be placed in the killing bottle to absorb any moisture, which otherwise mat and spoil the coats of the specimens. Wasps should not be allowed to remain in cyanide fumes for long, or their yellow markings will turn red. The specimens should be packed in soft tissue paper, or mounted on entomological pins and labelled with the date and *locality* of *capture*, and *also the collector's name*, and sent in a strong box, by mail, to the "Dominion Entomologist, Central Experimental Farm, Ottawa." Postage is free.

Specimens should be in good condition, not faded or damaged by exposure, and should include the large queens that are to be found chiefly in May and June, as well as the smaller workers that occur in abundance in July and the males that are common in August and September. Any notes on their habits and about the flowers they frequent and pollinate would be valued.

OBITUARY.

WE regret to record the death of William Greenwood Wright at San Bernardino, California, at the age of 83. Mr. Wright travelled extensively on the North American continent, collecting chiefly Lepidoptera. He assisted W. H. Edwards in the preparation of his "Butterflies of North America", but his most important work was "The Butterflies of the West Coast" published in 1905. He contributed to this Journal.

THE DISASTROUS OCCURRENCE OF *VANESSA CALIFORNICA* IN CALIFORNIA AND OREGON DURING THE YEARS 1911–1912.

BY F. M. WEBSTER, BUREAU OF ENTOMOLOGY, WASHINGTON, D.C.

The interesting note of Mr. J. B. Wallis on the occurrence of this species at Peachland, British Columbia, in 1912, as given in the "Canadian·Entomologist" for December, 1912, comes in very appropriately with the notes and observations made by correspondents of this Bureau, at Lakeview and Waldo, Oregon, and Willow Ranch, California. As the Bureau of Entomology is not likely to publish on this species in the near future, the information here given may be useful in case there should be a re-occurrence of these caterpillars during the summer of the present year.

Our first report of injuries by these caterpillars came from Mr. T. V. Hall, of Lakeview, Oregon, under date of July 27, 1911. Mr. Hall states that there had suddenly appeared in his neighborhood a worm which had taken almost the entire alfalfa crop. "Also has entirely destroyed the prospects for seed, which usually brings in to the farmers of this neighbourhood about $40,000 annually. The worm is from one-half to one inch in length and slender; perhaps 1-8 to 1-12 in thickness, brownish color and sleek appearing surface. It destroys the small tender alfalfa entire. The more mature growth it takes all but the fibre. This worm travels in vast armies. It almost seems as though the ground were in motion when they are in motion. The oldest settlers here state that nothing of the kind has ever appeared here before. This history reaches back at least forty years. We would like well to learn of some method for their destruction, or some way of preventing a repetition of the past, for they have caused a total loss to the year's crop."

The next report came from the same locality, under date of August 25, 1911, from Mr. A. J. Swift, who sent two specimens of these butterflies, which, he says, had been produced in his locality in enormous numbers during that month. Mr. Swift's further statements relative to this occurrence are given in his own words. "So far as known, this butterfly has never occurred here before, or at least in such small numbers as to have escaped comment.

This year, during July, various sections of the country have been covered with a worm of various sizes, but sometimes as large as 1½ inches long and near a ¼-inch in diameter. As I remember it now, it had two pairs of legs forward and three pairs aft, and varied in color with its food supply, some specimens being a bright green and grading from that to nearly black. The worm did immense damage to growing alfalfa and grasses, but so far as I am advised, did not trouble the trees. After the passing of the worm, this butterfly developed, which in its original swarming filled the air with myriads of them, and at this place the entire swarm was headed in one general direction, west, in very rapid flight."

Our next report for 1911, by a coincidence, was of the same date—August 25—from Mr. J. J. Monroe, of Willow Ranch, California, whose letter appears to be of sufficient interest to give in full.

"About June 1, of the present year, an old gardener told me that he noticed many of the specimens of butterfly I send you flitting about his garden and alfalfa fields. About six weeks later many of the destructive larvæ were noticed in the alfalfa fields and in gardens. Thousands of the larvæ left the hay (alfalfa) that I hauled into my barn and attacked one of my gardens which was nearby—i.e., 30 or 40 feet from the barn. They ate any kind of green vegetation—potato tops, peach tree leaves, garden weeds of any and all kinds, gooseberry leaves; in fact, apparently any and all kinds of green vegetation except death-weed. The larvæ have very much the appearance of the ordinary cutworm in the earlier stages of its growth, but it grows to be larger and much longer than the ordinary cutworm, and in the latter stages of the larval growth is of a light green color. Many of the larvæ attain a length of at least two inches, and some a length of probably as much as two and one-half inches. These larvæ while in my garden worked at night—i.e., during the darkness. Looking in the daytime, it was remarkable to find in sight even one larva, but they could be found in abundance in the ground about one inch from the surface. The domestic hen and the ordinary blackbird are very fond of both the larvæ and of the butterfly. The larvæ have destroyed quite an amount of alfalfa that was to be cut for seed, and also some alfalfa that would have been a second cutting for hay. A U.S. forest ranger told me this morning that the first damage he noticed from the larvæ in the forest was the leaves that were eaten from the Snow Brush, in many places the leaves being entirely stripped, eaten off,—i.e. consumed. Yesterday I saw many of these butterflies flitting among the branches and above the tops of the tall pine trees. Sunday, the 20th, I saw millions of these butterflies coming from the direction of the timber and flying on in the direction of Goose Lake. In other words, they were flying just about due west, and at the time there was quite a stiff, constant north wind blowing. These butterflies seem to congregate and alight on the willows, green-growing alfalfa, and in wet, muddy places. At other times—at least, during the day time—they are mostly on the wing. Now, the larvæ haven't done any remarkably great amount of damage yet, but there are butterflies in sufficient numbers now to produce a crop of larvæ next year to entirely destroy all the vegetation that would be produced here next year—i.e., if they are of the kind that comes every year."

Under date of June 12, 1912, Mr. Louis R. Webb, of Waldo, Oregon, wrote us of the appearance there of these caterpillars as follows: "There has appeared in this section of Josephine County a sort of army worm that resembles somewhat the caterpillar, and different from anything I ever saw. It has attacked the grease wood and mountain lilac mostly, and there are many acres in this locality and South River County, Del Norte County, Cal., that have been completely stripped of their foliage, and it has begun to attack fruit trees. It builds no web like the army worm of previous years, and its colour is black, with light streaks along its back. The worm at present is about an inch long and about one-eighth of an inch in diameter. So far, I have failed to find the moth that deposits the eggs."

"Also, under date of July 4, same year, specimens of the larvæ were submitted, and we quote from this letter as follows: "I wish I could send you photos of vegetation destroyed by these caterpillars. When they had eaten all the foliage off grease wood and mountain lilac, they started a sort of exodus and took possession of everything—even our homes could not exclude them. The streams and river were black with them, and tons of them went down the Illinois River, tons of them starved to death, and the brush and trees are covered now with their pupas like the enclosed. They seemed to care for nothing to speak of but grease-wood and lilac, but did eat some on willows, ferns, currant bushes, and very few young apples were gnawed by them like the sample enclosed. I think the enclosed samples will tell you a true story of what they did."

He also wrote us farther, under date of July 23—this time including pupæ of the insect and also specimens of some parasites. The butterfly accompanying this letter was determined by Dr. Dyar as belonging to the species under consideration. The hymenopterous parasites accompanying this letter were determined by Mr. Viereck as *Theronia americana*; while the supposed dipterous parasite was determined by Mr. Walton as *Helicobia helicis*. This latter was more likely to have been a scavenger than a parasite, although both species were reared from material submitted. In this letter Mr. Webb states that the butterflies seem to migrate after they emerged, and that fully half of the chrysalids were destroyed by parasites.

NEW LIFE-HISTORIES IN PAPAIPEMA SM. (LEPID.)

BY HENRY BIRD, RYE, N. Y.

(Continued from Vol. XLIII., p. 47.)

Papaipema moeseri Bird

The larval history accords with the usual routine experienced in *Papaipema*. As it is such a distinct species and so well distributed and accessible when the facts are known, it may be excusable to give some details of its discovery. The root-boring habit of the larva, its superficial resemblance to *P. impecuniosa* Grt., and the fact of its often occurring in the same locality, though in a different food plant, served, through a peculiar chain of circumstances, to retard its apprehension for several years at least.

The first intimation of the species came from Mr. A. F. Winn, of Montreal, whose query as to what *Papaipema* was boring *Chelone glabra*, Turtle-head, had to go unanswered. None of the few plants occurring about Rye gave evidence of being bored, and Mr. Winn was advised to look into the question another year, for we were glad to delegate the matter to such able hands. In due course, the next season, he reported finding a freshly emerged moth of *P. impecuniosa* crawling up the stem of a *Chelone* plant when he was examining it to find the pupa of its borer, and naturally concluded he had found a new food plant for the Grote species.

The following year the writer was at Montreal in mid-July, and happened upon a large colony of Turtle-head borers, working, as it chanced, in a damp area where *Aster puniceus* was flourishing plentifully. The Aster was being bored by numerous *impecuniosa* larvæ, whose identity was beyond question, and a careful comparison of them with the larvæ from the Turtle-head failed to note the slightest difference It was conceded Mr. Winn was doubtless correct in his surmise—a mere case of substitution of food-plants was occurring. Larvæ were then in the fourth stage, and, knowing the trouble it would be to carry them through, the Montreal colony were in no way depleted by accessions in my behalf.

About this time the *Papaipema* investigations of Mr. F. E. Moeser, at Buffalo, prompted a recurrence of the question, what species bores Turtle-head ? And the writer replied with considerable assurance that it was without doubt *impecuniosa.* But when later in the season Mr. Moeser went to get the pupæ from the borings, as can be readily done with the normal workings of the species in Aster, he found they had pupated elsewhere. Even then we were not convinced, for it was recalled when working in *Helenium*, *impecuniosa* usually forsakes its gallery to change. The next year Mr. Moeser decides to settle the matter to his own satisfaction, and scores the breeding of a new species. My own dull eyes had by this time seen a mature larva and had awakened to a relization it could not be the Grote species.

The stations for *moeseri*, doubtless, long endure. Turtle-head is a tenacious perennial in those wet locations that are congenial, and indications point to the well-established colonies existing many years at a given spot. Such a one on Staten Island, N.Y., is called to our attention. Here, almost in sight of the former home of the late A. R. Grote, a woodland rill meanders through the undergrowth, edged with a fringe of *Chelone* that takes root in its very bed. This station for the plant has long been a botanical record for Mr. W. T. Davis, and under his guidance in July, 1911, the place is visited to see if *moeseri* can be found there. Numerous larvæ are located, and in 1912 the colony is found to be still flourishing. There is considerable difference apparently in the time at which the hibernated ova hatch, due to the very moist conditions they endure. While the egg may withstand inundation very well, the young larvæ cannot, and, as with *marginidens* working in *Cicuta* and *Sium*, both water-loving plants, many tardy larvæ occur. Though neither the ova nor the first stages were observed, the first week of June can be figured as their date of general emergence. The stems are entered several inches above ground, and a more or less extended tunnel drilled upward. As they become larger, the boring of necessity becomes small for them, and they turn downward in the underground portion of the stem or root. The stems are often weakened so as to fall, and there are several openings made whereby the frass is thrown out. These castings form in little whitish mounds and become a conspicuous clue to the hidden host. Thus far, parasitism seems abnormally low, but one *Hemiteles* attack having been noticed.

As *moeseri* is so clearly a denizen of the wild woodland or swamp, it seems a coincidence to have been first met within the confines or immediate vicinity of such large cities as Montreal, Buffalo or New York. In southern West Chester Co., N.Y., and on the opposite shore of Long Island no infestations have been found, though it is true no stations of the plant were met that could be expected to support flourishing colonies. The following larval stages were observed:

Stage IV.—Head normal for group, polished, pale brown, marked with a black line at the ocelli, which extends posteriorly oblique across the epicranium, labrum and mouth parts black, seta at tubercle VIII seems longest. Body cylindrical, thoracic joints have the skin puckered, colour is a livid cast of umber brown, which shows on joints four to seven inclusive as a dark band or girdle, the remaining joints relieved by the white longitudinal lines; the dorsal line is unbroken, but its continuation across four to seven is by the merest thread; subdorsal line wider, but breaks abruptly at joints four to seven; subspiracular shows on thoracic joints, on eight to twelve is fused with the white of the ventral area. Tubercles well shown, brownish black; on joint one the cephalic plate forms a complete covering dorsally, being wider than the head, of similar texture, and edged at the side with black; on joint two an elongate plate occurs anterior to Ia and Ib, the fusion of Xa and Xb apparently, and is about twice the length of a spiracle; Ia, Ib and IIa show as mere dots; IIb, III and IV are much larger, being greater than a spiracle; VII of similar size; on joint three tubercles similar, except the elongate plate is absent; on the abdominal joints IV slightly exceeds the spiracle, and on ten is low down; on joint eleven III and IIIa are well separated, and I and II assume their usual large proportions; anal plates well developed; spiracles black.

Stage V.—Similar; on joint ten there is indication of tubercle IVa, but it is not stable for this nor succeeding stages.

Stage VI.—Body colour much lighter; otherwise no change.

Stage VII.—Head has lost oblique lateral marking, body colour fades to whitish translucence at maturity; the fused tubercle Xa and Xb is less prominent; otherwise similar. Larva measures 21, 27, 35, 40 mm. for the stages respectively.

Maturity is reached August 8th to 15th, and the gallery is left for pupation. The pupa is shorter and chunkier than usual, of chestnut brown colour and shows no unusual developments; the cremaster is two sharp, curved hooks; length, 15 to 16 mm.

The emergence dates for thirty specimens include August 26th to September 19th.

Moeseri larvæ in early stages are almost identical with *impecuniosa*, in the last two stages its larger size and middle girth, together with a slight difference of tubercle delineation, readily separate them.

Papaipema stenocelis Dyar.

This species, represented by a unique type from Baltimore, Md., was described in 1907. A second specimen was taken at light at Lakehurst, N.J., by Mr. O. Buchholz, in September, 1910. A relationship is apparent to *P. inquaesita* G & R., and more closely still to *speciosissima* G & R.

It seemed clear that so distinctive a species must have remote haunts and be restricted to a more southern range, else collectors would have cognizance of it long ago. Believing this second capture had bred at Lakehurst, since the habits of this group controvert an assumption of migration, to which the appearance of many late-flying, southern Noctuids is often assigned, led the writer to make an extended search for its larva in the pine barren flora of Lakehurst, in 1911.

The results were negative, and subsequent studies of lists of more southern flora, gave little intimation what particular plant was likely to shelter the *stenocelis* larval tunnel in its stem or root. The larger perennials, with which we are wont to associate these borers, are strikingly absent from pine barrens, and we finally conceived the notion it must bore some fern.

On July 28, 1912, we again invaded the Lakehurst region, with the idea of investigating the unfamiliar ferns, and in a half hour's time had discovered the desideratum. Some orange-coloured frass, similar, yet a little different from that thrown out by *inquaesita* when in the root of *Onoclea*, was noticed about the stipes of *Woodwardia virginica*, and gave intimation that this was the species of which we were in search. Upon uncovering the larva, which was working in the long running rootstock, we became more certain of the determination, as it transpires the tubercles on joint eleven accords with the unique departure shown in *inquaesita*, except that it is more pronounced. Confirmation of the matter occurs on September 13, following, when the first beautiful male moth appears.

The life cycle clearly follows the usual course, the hibernated ova placed in September hatch forth about the first week of June. The normal larval period will likely cover sixty to sixty-five days, and the pupal condition lasts about thirty days.

The newly emerged larva enters the stipe near the base and works down to the running rootstock, where it finds an ample opportunity to mine an extended burrow. Communication with the original entrance is discontinued after a while, and more convenient openings for disposing the frass are made as the tunnel progresses. An Hemiteles parasite, which hibernated in its cocoon, had claimed 85% of the larvæ encountered. Larvæ were observed in the following stages.

Stage V.—The general features as given for the preceding species apply. The colour is a warm shade of brown, the longitudinal lines are narrow and not so contrastingly white, the dorsal alone continuous. On joint eleven tubercles III and IIIa are fused into a plate about three times the size of the spiracle, IIIa occurs distinctly on the preceding abdominal joints; on joint twelve the plates are stronger than with the compared form.

Stage VI.—Little change, colour paler and more of a sienna tint, tubercles appear with better prominence.

Stage VII.—General characteristics normal; colour a pale, dull, pinkish hue, fading to translucence at the sutures; lines indistinct; the blackish tubercles stand out in greater prominence, of the lateral ones, III on joint two, and the fused III and IIIa on joint eleven are most conspicuous, the latter constituting the chief specific character. This plate is four or five times the size of the spiracle. Setæ are so weak as to be unnoticeable without a lens. The first pair of prolegs on joint six are aborted in early stages, and never develop so fully as the succeeding three pairs; crochets here number twelve, while for joint nine the number is eighteen. Larval length, 30, 38, 46 mm., for the stages respectively.

A feature of individuality with *stenocelis* is the prominence of tubercle III on joint eleven, which has evidently taken in IIIa. While these plates often coalesce in other species of the genus, there is not the comparative enlargement as in this case. *Cerussata* and *cataphracta* are examples of large tubercle development, III and IIIa fuse, in this instance, into a large plate with them. But it does not reach the proportions attained in *stenocelis*. With *inquaesita* this plate is of unusual size, since the remaining tubercles are so very weak. Tubercle IIIa on the abdominal joints of Noctuid larvæ seems always obscure and generally wanting, especially is this so on joint eleven. With such a well tubercled larva as that of *Achatodes zeae*, IIIa seems normally wanting on joint eleven, though it occurs conspicuously on the preceding joints. One specimen was observed that had it on joint eleven, but it occurred on one side only.

When ready to pupate, the larva leaves the burrow and changes in the ground. The pupa is of normal appearance, and the period is of usual duration—about a month.

Stenocelis was placed in *Hydroecia* by Dr. Dyar, his type being imperfect in the characteristic tufting that is a feature in differentiating these moths. It is a conventional *Papaipema*, however, and was so referred by Hampson, perfect material having the typical tufting present, while the genitalia conform to the unusual pattern of this group. What is really a better characteristic exists in the larval appearance which accords with the unique pattern disclosed in *Papaipema*—at least, as occurs with thirty of the species whose larvæ are known. One very notable departure happens with *frigida*, whose larva approximates *Hydroecia* characteristics, and is evidently a relic of the stem species, whence both these groups sprung.

The genitalia have not been discussed. These male characters show little to distinguish them from the general type. The broad, heavy side-piece, or clasp is tipped with an irregularly formed cucullus, shaped somewhat like a foot with an over-developed heel, and having the toe, which is the anal angle of the corona, pointing ventrally. This area is set with spine-like setæ that point anteriorly. The harpe is a stout, sharp-pointed hook, curved like a cow's horn. It is shorter than with most species, and is toothed slightly on the outer edge. These teeth, too, are of less prominence. The clavus is marked only by a slight prominence, which is covered with fine setæ. The uncus is the usual finger-like appendage, widened a little near the point.

From our studies of southern flora, now that the food-plant is known, we might predict *stenocelis* may find its principal metropolis in the Dismal Swamp region of Virginia, where Woodwardia reaches a prolific development.

BOOK NOTICES

A CONTRIBUTION TO THE MORPHOLOGY AND BIOLOGY OF INSECT GALLS. By A. Cosens. (Reprinted from the Transactions of the Canadian Institute, Vol. IX., pp. 297-387, 13 pls., 1912.)

That aspect of cecidology which treats of the causes that are operative in the formation of insect galls and the manner in which the plant tissues react to the stimulus is one that has been much neglected, particularly by American students of the subject. Mr. Cosens' work throws considerable light on these interesting problems and is one of the most important contributions to our knowledge of the morphology of galls that has ever been published.

The greater part of the work is devoted to descriptions of the anatomy of sixty-eight kinds of American insect and phytoptid galls. The descriptions are arranged in the order in which the producers are classified, most of the gall-producing families, except those of the Coleoptera, being represented.

Although dealing mainly with matters that are chiefly of interest to the botanist, the author has also cleared up some important difficulties concerning the feeding habits of various gall-producing insects. Cynipid larvæ were found to secrete an enzyme which converts the starch in the nutritive layer of cells surrounding the larval chamber into sugar, which is taken up by the larva through the mouth. The cells of the larval chamber thus remain unbroken, and their inner surfaces present a marked contrast to the ragged cell-layer lining the cavities inhabited by inquiline larvæ. This view is confirmed by the discovery that though, contrary to current views, the intestinal tract in Cynipid larvæ is complete, an anus being present, no frass is expelled, as would be the case were the entire cells devoured, as they are in sawfly galls.

It is suggested that this ferment "may indirectly stimulate cell proliferation by storing the nutritive zone with an unusually large quantity of available nourishment, which can diffuse to all parts of the gall."

Adler's discovery that the gall of *Nematus vallisnierii* is partly formed while the larva is still within the egg, was confirmed in the case of several species of Pontania. It is suggested that the curious power of the excrement of such sawfly larvæ to induce cell proliferation is possibly due to their having swallowed tissues still containing these enzymes, which have retained their stimulating power, even after having passed through the intestinal tract of the larva.

The work, which should be in the hands of every student of insect galls, is beautifully illustrated by thirteen heliotype plates from photomicrographs of sections of the various galls described in the text. There are also a few good text figures.

CONTRIBUTIONS TO THE NATURAL HISTORY OF THE LEPIDOPTERA OF NORTH AMERICA. Parts IV., V. and VI. By Dr. Wm. Barnes and Dr. J. H. McDunnough.

Three more parts of this valuable publication, by Dr. Barnes and Dr. McDunnough, have appeared, bearing dates of July, 1912. Part IV. is entitled, "Illustrations of Rare and Typical Lepidoptera," and contains 27 plates, reproduced by half-tone process from photographs, which present in all 506 figures. Most of these are of moths which have not previously been figured, and a large percentage are the actual types, so the usefulness of the work to students will be realized. The text, 54 pages, and index is mostly an explanation of the figures, with locality of the specimens shown, but in some cases additional notes are given.

Part V.—"Fifty New Species: Notes on the Genus Alpheias"—contains 44 pages of text, three half-tone plates showing 62 figures of types and cotypes of the species described, one plate of genitalia and one of venation. The new species are from Arizona, California, New Mexico, Texas and Utah.

Part VI. is of 13 pages "On the Generic Types of N.A. Diurnal Lepidoptera," and deals with one of the many phases of the vexatious muddles which entomological nomenclature, at present, is in, but it seems probable that the International Congress of Entomology will be able before long to overcome many of the difficulties that make it so easy to keep generic and specific names in a constant state of chaos.

A. F. WINN.

Mailed April 16th, 1913.

The Canadian Entomologist.

VOL. XLV. LONDON, MAY, 1913 No. 5

FURTHER NOTES ON ALBERTA LEPIDOPTERA.

BY F. H. WOLLEY DOD, MIDNAPORE, ALTA.

(Continued from page 98.)

369. *Pyrrhia exprimens* Walk.—I have compared this form with Walker's type from Orillia, Ontario, and consider it correctly named. *Angulata* Grote, from Buffalo, N.Y., is the same species. It is the one with almost blackish central shade and t.p. line, and blackish bordered secondaries. It stands wrongly in Dyar's Catalogue as a variety of *umbra* Hufn., a European species which should be struck out from our lists altogether, and place given to *cilisca* Guen. I have not, as a matter of fact, seen the type of *cilisca*, but Sir George Hampson has, and gives the locality as "Brazil" in the Catalogue. It is in Mons. Oberthür's collection at Rennes. In the British Museum is a Kansas specimen from the Snow collection, marked "*cilisca* Guen," which I must assume has been compared with this type. This is the species figured by Holland as *umbra* Hufn., and is the *umbra* in error of all North American authors.

In *cilisca* the primaries have a crimson irroration which seems to be lacking in *exprimens*. The cross lines are finer and not blackish, and the central shade is less acutely angled in the cell. The secondaries are crimson-bordered, and not blackish. All my specimens of *cilisca* are from the Eastern States, and I have both this and *exprimens* from Milwaukee Co., Wisconsin. Both seem to occur all across the continent in the south, but I have not yet seen *cilisca* from Western Canada.

The European *umbra* combines some of the characters of the two, but I carefully examined the British Museum material, and all the British literature and figures in my possession, and it seems to me to be easily separable from both. It has the dark-bordered secondaries more like *exprimens* than *cilisca*, but the transverse lines are like those of the latter, though it usually lacks the pink irroration. I happen to have but a single example of *umbra* in my own European collection, so I am unable to make further comparisons at present. Sir George Hampson unites them all as one species, under *umbra*, citing:

Ab. 1. *exprimens.*—Fore wing with the postmedial and terminal areas suffused with brown.—Canada and U.S.A.

Ab. 2. *cilisca.*—Hind wing paler yellow, the postmedial band pale crimson.—U.S.A. and Brazil.

373. *Cosmia decolor* Walk. (not *discolor*).—The type is a rather dark, smoky-suffused orange male from Orillia, Ontario, and is probably the one figured by Hampson, but the black streak shown near the inner margin is presumably an artist's error. *Discolor* of our lists was merely a mis-spelling, and the name unfortunately should stand as a synonym.

374. *C. infumata* Grt.=*punctirena* Smith.—Grote's type, a female from Malawqua Co. (?Chautauqua), New York, is a very dark fuscous-brown specimen. Of *punctirena* I have seen two types, male and female, from Yellowstone Park, Wyoming, in the Washington Museum. There is no type there from Cartwright as I previously stated. The types are a trifle reddish, even, and have the t.a. line angled rather than curved, but are certainly the same as *infumata*, and my tentative synonymy of this and *decolor* has proved correct. Sir George Hampson correctly keeps European *paleacea* distinct, but fails to recognize two North American species. I do not blame him. I have no modification to make of my former notes, and nothing to add, but I must admit that I should probably never have suspected, or, at any rate, been able to separate the two species if I had not had the opportunity of studying them in nature. As it is, I cannot always place specimens with certainty.

Hampson places them in the genus *Enargia* Hubn., and, as is his rule, changes the gender of the specific name to concord with that of the genus, thus making the name *decolora* Walk. He makes *infumata* ab.1. "Head, thorax and fore wing thickly irrorated with fuscous." "Ab. 2. Fore wing yellowish white, with slight dark irroration." This is a male from Lower Klamath River, California, and is a very pale whitish *decolor*. I have seen other similar specimens, and Dr. Barnes has such a female from Victoria, B. C., bearing a manuscript name. "Ab. 3. Fore wing pale yellow, irrorated with red, the markings reddish." This specimen, from New York, is *decolor* also, and I have similar ones in my collection.

375. *Orthosia verberata* Smith.—I am perfectly satisfied as to the distinctness of this species from *ferrugineoides*. To my former notes I would add that this species generally has a more or less distinct claviform, which *ferrugineoides* lacks. I have both species from both Cartwright and Miniota, Manitoba. *Verberata* occurs at Kaslo, and on Vancouver Island, but I have not yet seen *ferrugineoides* fom west of the Rockies in Canada, though Hampson lists a specimen from Glenwood Springs, Colo. European *circellaris* Hufn., (*ferruginea* Schiff.) falls, as Hampson correctly places it, between the two. I have eleven British specimens, and have examined more at the British Museum. With a few of the specimens alone I should never have thought of separating it from *ferrugineoides*, and the secondaries in all are more evenly dark, with slightly darker veins and pale costal region, thus resembling *verberata*. There are vague traces of a claviform in a few specimens. In most the general coloration is nearest that of *verberata*—viz., interspersed with varying shades of brown and rufous. The transverse lines are more distinct than is generally the case in *ferrugineoides*, but less so than in *verberata*. Hampson finds that *verberata* has the frons black at sides, and separates it from the other two in the tables by this character. Brown, perhaps, describes it better, but the character is by no means an obvious one, some of my *verberata* having frons scarcely brown at sides at all, whereas some *circellaris* distinctly have.

It is interesting to note that in the present paper there are presented three instances in which a European species has two apparently distinct North American representatives.

EUROPEAN	N. AMERICAN
circellaris Hufn.	*ferrugineoides* Guen. and *verberata* Smith.
paleacea Esp.	*decolor* Walk. and *infumata* Grt.
umbra Hufn.	*cilisca* Guen. and *exprimens* Walk.

In two of these cases, however, I appear at present to be unsupported by other opinions.

376. *O. euroa* G. and R.—Grote changed the name to *puta*, to which Sir George Hampson gives preference. The two names therefore apply to the same type, but I have not discovered where that type is to be found. Presumably it ought to be in the collection of the American Entomological Society at Philadelphia. Smith described *dusca* in Ann. N. Y. Acad. Sci., xviii., 117, Jan. 1908, from Cartwright, Miniota and Winnipeg, Man., and Kaslo, B.C. I have seen a male and female type from Brandon and Miniota, and the type labels bear this name. But in an earlier paper (Trans. Am. Ent. Soc., xxxiii., pp. 350, 360), he makes reference to the form as *duscata*. Compared with *euroa*, it was stated to be "smaller, darker, with more diffuse maculation, and with shorter, broader primaries." Genitalic differences were referred to in the "Transactions." Calgary specimens do not differ from those from Manitoba, and Smith would obviously have called them *dusca*. As a whole, the species is perhaps usually smaller and darker in the west, but not constantly so, and I can see no reason whatsoever for treating the western form as distinct, and must refer *dusca* to the synonymy. Dr. Barnes told me some time ago that he was of the same opinion.

377. *Agroperina lineosa* Smith. (Journ. N. Y. Ent. Soc., xviii, 145, Sept. 1910).—Described from thirty specimens from Calgary and several Manitoba points. *Pendina* Smith, described as a species from the same localities in the same paper, is unquestionably a variety of the same thing, and is almost that form I referred to as "dark crimson." I have such an extreme form, but "deep luteous red-brown," as Smith describes it, is a more common variation and this is the "*morna*, ab. 2, deep rufous," of Hampson's Catalogue, vii., 405, his "ab. 1" being a pale rufous form, intermediate between the more common lüteous *lineosa* and var. *pendina*. The actual specimen figured by Hampson as *morna*, from Yellowstone Park, Wyoming, is of the pale uniform, slightly marked form described by Smith in the same paper, also as a species, as *indela*, from various localities in Wyoming, Idaho, Colorado, Montana and Washington. The *morna* of Strecker, as I have pointed out under my No. 155 (xliii, 230, July, 1911), is not allied to this group at all. By Smith's own admission, *indela* and *lineosa* were very difficult to separate, and from the material I studied in his collec-

tion and elsewhere, it never occurred to me that two species existed at all. But as I happen to possess no specimens from any of the localities given for *indela*, I must give the form the benefit of the doubt, whilst expressing the belief that all the above names, with the exception, of course, of *morna* Streck., are probably only forms of *conradi*, which both Hampson and Smith claim to have from Calgary, and Smith also from Winnipeg. The type of *conradi* is a female in the British Museum, from Colorado, and is, as Grote describes it, "faded ocher brown, . . . the darker specimens having base and subterminal space a little paler, . . . s.t. line preceded by a diffuse darker shade." *Citima* Grote, type a female from Arizona, in the Neumœgen collection at Brooklyn, is like it, but darker and more strongly marked. It is correctly referred as a synonym by Hampson, and Smith accepted the reference. Hampson separated *conradi* from his "*morna*" in the tables by the pale s.t. area. Smith adds, "a rough powdery appearance." Both these characters hold in my only southern specimen, from Las Vegas Range, New Mexico, which I labelled as *conradi* after comparison in the British Museum. Some of the more strongly marked and contrasting Calgary specimens have been cited as *conradi* by both Sir George Hampson and Smith. I have no fault to find with that, except to say that none that I have yet seen from here are quite like the types of either *conradi* or *citima*. But I have entirely failed, after repeated attempts extending over twenty years, to recognize two species amongst my local material, either on treacled posts, flying round a lamp, feeding on flowers, or in the collection. Smith claimed genitalic differences for most of the above named forms, though admitting that they were so slight as to be scarcely noticeable.

Belangeri Morr., found locally in the Province of Quebec, is most suspiciously like a rather suffused fuscous race of *conradi*. I am indebted to Mr. Winn for a nice series, and am able to match more than one of the specimens almost exactly with some of my local *lineosa*. The type is probably in the Tepper collection, but I do not know its origin. Sir George Hampson makes it a synonym of *inficita* Walk., apparently correctly, though that is an unusually even red specimen. It is the specimen figured fairly well, but is a male, not a female. It was described from an unknown locality.

Agroperina is a new generic name used by Sir George Hampson for the foregoing group, and a few other species hitherto standing with *Orthosia*.

(To be continued.)

ENTOMOLOGICAL SOCIETY OF ONTARIO.—FIFTIETH ANNIVERSARY

At the regular meeting of the Society, held on Friday, March 14th, it was decided to hold the Annual Meeting of the Entomological Society of Ontario and the celebration of the 50th anniversary of its formation at the Ontario Agricultural College, Guelph, on Wednesday, Thursday and Friday, August 27th, 28th and 29th. A committee to make all necessary arrangements was appointed, consisting of the following: Professors Bethune, Howitt and Jarvis, and Messrs. Caesar, Baker and Spencer, of Guelph; Dr. E. M. Walker and Mr. A. Cosens, of Toronto; Dr. C. Gordon Hewitt and Mr. Arthur Gibson, of Ottawa, and Messrs. H. H. Lyman and A. F. Winn, of Montreal, with power to add to their number.

As the time of the meeting will be during the first week of the Toronto Exhibition, reduced railway rates will be available as far as that city. The committee will welcome suggestions from any of the members of the Society regarding the best methods of carrying out this celebration.

At a previous meeting a resolution was adopted expressing sympathy with the Royal Geographical Society of London, England, in the loss sustained by them and by the scientific world in general through the death from cold and exhaustion of Captain Scott and his brave companions.

The habitat of *Rhogas indicus* Cam.—In Wien.Ent. Zeit., 1910, p. 3, Cameron has described a new species under the above name, giving as habitat "Sitka on the Ganges (Mannerh?)" There may possibly be some locality Sitka on the Ganges river, but "Mannerh" is apparently an abbreviation of Mannerheim, who possessed many insects from Sitka in Alaska, and this is, I think, certainly the true habitat of the species.

E. Bergroth.

ROBBER-FLY AND TIGER-BEETLE.

On August 21st, 1909, while walking across a young orchard in Peachland, B. C., I flushed a tiger-beetle which flew a few yards. Seeing that it was of a species new to me, I promptly followed it.

Again it flew, but was at once pounced upon by a large robber-fly, *Proctacanthus milberti* Macq., which had been poised on a weed near by.

As the fly flew heavily away with its prey, I netted both. The robber refused to be parted from its dinner, and both were put in the cyanide bottle. Although but a few seconds had elapsed from the seizing of the tiger by its enemy, the poor thing was quite dead, the robber's proboscis having pierced its body exactly between the elytra and about one-quarter of the length of the body from its base.

The beetle proved to be *Cicindela purpurea*, and, strange to say, is the only one I have seen during three visits, each of several weeks, to the valley.

J. B. WALLIS.

NOTES ON THE DEATH FEINT OF CALANDRA ORYZÆ LINN.

BY HARRY B. WEISS, NEW BRUNSWICK, N.J.

In the course of some fumigation work against this insect, which is the common widely distributed "rice weevil," it was noticed that the duration of its death feint was exceedingly brief —so brief, in fact, as to cause one to wonder of just what value such a brief feint was to the weevil. The duration of each feint was ascertained in a number of weevils, and the following table gives the length of time in seconds of the first twenty-five feints in six different weevils. The temperature during these operations was 75°F., and the feint was induced by blowing upon the insect's ventral side or by dropping it through the space of one inch. When dropped from a height of six or eight inches, or more, no feint was produced, the weevils in all cases becoming immediately active.

Duration in seconds of first 25 death feints in 6 weevils

	A	B	C	D	E	F
	3	1	1	1	5	15
	1	8	1	3	5	13
	1	5	3	5	7	15
	8	5	1	1	10	12
	1	4	1	2	2	3
	20	7	4	5	4	3
	1	7	8	5	2	5
	10	5	7	7	2	7
	20	10	5	5	2	8
	25	10	2	5	10	3
	1	7	5	8	3	2
	3	17	2	9	4	1
	1	7	10	2	3	1
	1	7	1	1	1	1
	8	3	8	1	5	10
	15	5	13	1	3	2
	20	3	7	2	1	4
	10	2	8	7	1	1
	23	1	15	5	2	1
	1	7	3	5	8	3
	5	10	2	1	4	2
	5	13	1	3	2	1
	1	17	8	2	6	8
	13	18	2	10	8	2
	10	8	2	5	1	6
Averages . .	8.2	7.4	4.8	4.0	4.0	5.1

From these figures one can see what a wide variation in duration occurs in different individuals, and even in one individual. Twenty-five seconds was the longest feint, and the average ran from four to eight seconds. Thirty-five was the highest number of successive feints it was possible to produce in one individual. After the thirty-fifth, they became only partial—that is, one or two of the legs would stick out from the body as if fatigued.

During every feint the insect was placed on its dorsum, as only in this manner was a successful feint produced. When the insect was placed on its ventral side in almost evey case, the feint would last only a second. On account of the shortness of the feint, it was almost impossible to try the effect of gases, etc. Upon subjecting individuals feigning death to the fumes of carbon bisulphide and chloroform, they instantly became active. Upon placing other feigning weevils upon blocks of ice, they slowly assumed the death attitude, the femurs taking a position at right angles to the body, with the tibiæ and tarsæ loosely folded upon each other—all tending somewhat to bunch together. Individuals starved to death assumed a similar attitude.

The death feigning attitude is quite unlike that of death. The distal ends of the femurs of the first pair of legs extend forward, being pressed against the base of the snout. The femurs of the second pair of legs also extend forward, and are held close to the body. The third pair assume a position similar to the second, except that the distal ends point toward the posterior end of the body. The femur, tibia and tarsus are in all cases folded upon each other and drawn close to the body, while the antennæ take a position parallel to and close against the snout. The entire attitude, however, does not seem to be as rigid as that assumed in the death feint of the plum curculio, but is apparently easily and instantly relaxed.

The value of this brief death feint to the weevil is hardly apparent. Probably on account of its somewhat concealed method of feeding, it has little occasion to feign death, and as a result the duration is correspondingly short.

ERRATA.

Page 2, line 31.—For narrower, read narrow.
Page 3, line 12.—For Manse, read Manee.
Page 7, line 23.—For ♂, read ♂ ♀
Page 9, line 3.—For *annulicomis*, read *annulicornis*.

ON SEVERAL NEW GENERA AND SPECIES OF AUSTRALIAN HYMENOPTERA CHALCIDOIDEA.

BY A. A. GIRAULT, BRISBANE, AUSTRALIA

(Continued from page 106.)

Tribe Haltichellini.

Genus *Stomatoceras* Kirby.

1. *Stomatoceras victoria* new species.

Female: Length, 4.25 mm.

Black, somewhat shining; tegulæ, legs and basal half of abdomen ventrad (also latero-proximad), red, on the abdomen the reddish mixed with yellowish; scape (rest of antenna missing) black; fore wing with a smoky fascia across it at the stigmal vein (accented at the vein) and a rounded smoky spot farther distad nearer the costal wing margin and about half way to the wing apex from the stigmal vein, otherwise both wings hyaline.

Body rather finely rugoso-punctate, the spaces between the punctures smooth; lateral ocelli their own diameter from the eye margin or slightly more; scutellum terminating in two tooth-like plates, one on each side of the meson; abdomen finely reticulated; propodeum in the middle of the dorso-lateral aspect, with one distinct plate-like projection, another broader one indicated cephalad of it. Propodeum punctured like the rest of the thorax. Scape very long, bent at extreme tip, reaching to the cephalic ocellus, which is at the apex of the channel-like scrobicular cavity. Body finely pubescent. Posterior femur without a large tooth ventrad, its ventral margin straight but pubescent and along the distal two-thirds armed with a uniform series of minute, black, comblike teeth. Stylus of abdomen short. Postmarginal vein long.

(From a single specimen, the same magnification.)

Male: Unknown.

Described from a single card-mounted female specimen, labelled "*Cheltenham*, Victoria."

Habitat: Australia—Cheltenham, Victoria.

Type: No. *Hy 1185*, Queensland Museum, Brisbane, the above specimen; a fore wing and an antenna on a slide.

This species closely resembles *S. fasciatipennis* Bingham (1906), described from North Queensland and should be compared with it. However, the second abdominal segment is plainly shorter than the remainder of the abdomen.

Later, among a small collection of *Chalcidoidea* given to me by Mr. F. P. Dodd, I found a species of Stomatoceras which agrees with the description of *S. fasciatipennis* Bingham. Also, it was mounted on a card containing a flat lepidopterous cocoon, in general outline shaped like a spool, from which projected an empty pupal case and also a number of small ants. This card was labelled "Townsville, Qld., 20, 5, 02. F. P. Dodd." Thus, this specimen (a female) is from the type locality of the Binghamian species, agrees with the description and appears to be a part of the same material, since its insect associates agree with those denoted by Bingham. Comparing this specimen (which I have identified as *fasciatipennis* and deposited in the Queensland Museum at Brisbane) with *victoria*, the difference between them becomes more apparent, since in the former the marginal vein is plainly longer and both the subfascia distinctly larger, especially the distal one, which extends distad half way to the apex. Also, the second abdominal segment is somewhat longer in *fasciatipennis*, the third and following segments short, but (segments 3–5) nearly twice the length of the corresponding segments in *victoria*. Otherwise, the two are much alike. A balsam slide bearing an antenna and a posterior leg goes with the cardmount.

2. *Stomatoceras hackeri* new species.

Female: Length, 4.50 mm.

The same as the preceding species (*victoria*), but the scape is also dark red, including also the long pedicel and the first two funicle joints (and a part of the third); the abdomen is reddish, only along the median line of the venter; the fore wings have the same general pattern (as regards fuscation), but they are more irregularly fumated, the two fumated areas less distinctly separated, especially caudad. The posterior femora beneath are toothed less farther proximad and the apical emargination (a convexity) is more pronounced (this crenulation of the margin should not be confused with the first tooth in the family which is usually large); also from between the fine black teeth arise series of solitary, erect, stiff, but short bristles. The postmarginal vein is long.

(From a single specimen, the same magnification.)

Male: Not known.

Described from a single female specimen, minutien-mounted, from the collections of the Queensland Museum, Brisbane, labelled "Brisbane.—H. Hacker.—8-8-11."

Habitat: Australia—Brisbane, Queensland.

Type: No. *Hy 1187*, Queensland Museum, Brisbane, the above specimen, minutien (abdomen separated), plus a slide bearing a fore wing and antennæ.

Stomatoceroides new genus.

Female: Similar to *Stomatoceras* Kirby, but the postmarginal vein well developed, longer than the short marginal vein, four times the length of the stigmal and slender. Antennæ 11-jointed, inserted below the ventral ends of the eyes, the club solid, only slightly shorter than the long proximal funicle joint (a third shorter), the scape simple, long, the pedicel short, the flagellum cylindrical and a single ring joint present. Posterior femora without large teeth beneath, but their ventral margin crenulate or wavy, there being three sloping convexities, the distal two bearing a continuous series of minute, black comblike teeth (along the distal half of the margin). Scutellum terminating in a small, bidentate plate. Metathorax with no dorsolateral projections. Vertex very thin, the cephalic ocellus within the scrobicular cavity, the lateral ocelli distinct from the eye margins. Pronotum thin mesad, broadening laterad. Abdomen not produced distad, normal, the second segment largest.

Type: The following species.

1. *Stomatoceroides bicolor* new species.

Female: Length, 4.10 mm.

Opaque black, the legs dark reddish excepting nearly the whole of the upper margin of the posterior femur, the coxæ, the proximal halves of the tibiæ and the same portions of the cephalic and intermediate femora, all of which are black. Venation brown, the fore wings with a distinct, rounded brownish spot under the marginal vein (against it) and with a larger stain distad more or less obscure and cephalad. Head and thorax rugoso-punctate, the spaces between the punctures with fine grooves, the abdomen, finely densely polygonally reticulated, but the second segment smooth and shining. Antenna wholly black, the distal two funicle segments subequal, each slightly less than half the length of the proximal funicle joint.

(From a single specimen, the same magnification.)

Male: Not known.

Described from a single cardmounted female specimen labelled "Dandenong Ranges, Victoria."

Habitat: Australia—Victoria (Dandenong Mountains).

Type: No. *Hy 1186*, Queensland Museum, Brisbane, the above specimen; also a slide bearing an antenna and a second one, an antenna and a posterior leg.

The following species were thought to represent a new genus, but are all components of this one. Their generic characters are given herewith.

The same as *Stomatoceras* Kirby, the antennæ 11-jointed, the pedicel very small; the scrobicular cavity extends nearly to the occipital margin; thus the vertex acute or like a transverse carina along the occipital margin; the lateral ocelli are not within the scrobicular groove, but between its lateral margin, the eye and the true occipital margin, meso-caudad of the eye; the cephalic ocellus, however, just at the apex of the cavity. Postmarginal vein longer than the moderately long marginal, the stigmal vein very short, sessile and oval, small; submarginal vein more than four times the length of the marginal. Scutellum terminating in a small, bidentate plate. Posterior femora beneath simple—that is, without one or two large teeth, with the black, comblike teeth along distal two-thirds or more of the margin, and hairy; ventral margin of the femur straight. Propodeum with at least one dorso-lateral tooth. Antennæ long, cylindrical, without a ring-joint. Abdomen ovate.

The genus *Stomatoceroides* is more like *Hippota* Walker, but the flagellar joints are much longer, the pedicel smaller, the posterior femora armed and straight beneath, the vertex carinate, the propodeal tooth not prominent, the stigmal vein sessile, the wings clouded.

2. *Stomatoceroides nigricornis* new species.

Male: Length, 4.1 mm. Slender.

Opaque black, the base of the abdomen shining. Marked with dark red (Garnet) as follows: The tegulæ, tarsi, knees, tips of tibiæ and a spot at base of posterior femur (ventrad and exteriorly). Fore wings with two obscure brownish cross-bands—one at the marginal vein and the other nearly half from there to apex; the first accented under the marginal vein, the second more noticeable a short distance out from the costal margin. Venation dark. Rugoso-punctate, the abdomen distad with fine polygonal reticulations.

(From one specimen, the same magnification.)

Described from a single male specimen, minutien-mounted, labelled "Brisbane, 12–5–11." From the Queensland Museum.

Habitat: Australia—Brisbane, Queensland.

Type: No. *Hy 1188*, Queensland Museum, Brisbane, the above specimen; antenna and posterior leg on a slide.

3. *Stomatoceroides versicolor* new species.

Female: Length, 4.0 mm. More robust than the preceding.

Opaque black, the proximal half of the abdomen and the caudal coxa and femur contrasting, bright orange yellow, with some reddish mixed in; legs otherwise black, the knees brownish; antennæ black; tegulæ black. Wings opaque, the venation dark, the marginal vein with a very distinct, sub-elongate dark brown spot under it, which does not tend to cross the wing, but is wider (proximo-distad) than long (cephalo-caudad).

Structurally agreeing with the type species, but the stigmal vein is curved slightly cephalad, the body is more robust, the antennæ very much the same, but the posterior femora beneath with the fine, black, comblike teeth only along the distal third.

(From one specimen, the same magnification.)

Male: Not known.

Described from a single female, minutien-mounted, from the collections of the Queensland Museum, labelled "Hacker, Brisbane.—6–4–11."

Habitat: Australia—Brisbane, Queensland.

Type: No. *Hy 1189*, Queensland Museum, Brisbane, the forenoted female on minutien mount, plus the flagellum on a slide in xylol-balsam.

4. *Stomatoceroides nigripes* new species.

Female: Length, 5.00 mm.

Opaque black, the tarsi fuscous, the wings hyaline, the venation dark, with only a trace of staining under them. Like *versicolor*, but the teeth of the posterior femur along as much as the distal two-thirds of the ventral margin. Antennæ as in the other two species, but the funicle joints are longer.

(From a single specimen, similarly magnified.)

Male: Not known.

Described from a single female, cardmounted, kindly given to me by Mr. F. P. Dodd, of Kuranda, North Queensland. The specimen was labelled "From pupa of the red ant moth, Townsville, 7-11-03.—F. P. Dodd."

Habitat: Australia—Townsville, Queensland.

Type: No. *Hy 1190*, Queensland Museum, Brisbane, the foregoing specimen on a card, plus female antenna and posterior femur on a slide together in xylol-balsam.

Family Callimomidæ.

Podagrionini.

Pachytomoides new genus.

Female: Somewhat similar to *Pachytomus* Westwood and *Podagrion* Spinola, but the antennæ lack the ring-joint and the club is enlarged, as compared with the slender filiform funicle. The second and third tarsal joints are slender. The stigmal vein has a very short neck. Ovipositor very long. Wings infuscated. Propodeum with a semicircular carina at apex around the insertion of the abdomen.

Male: Probably the same.

Type: The following species (*mirus*).

1. *Pachytomoides mirus* new species.

Female: Length, 5 mm., excluding the long, slender and curled ovipositor, which is fully 7 mm. long.

Bright metallic green, the propodeum and head metallic bluish, the abdomen red, except broadly at base above; the fore and intermediate legs reddish brown at the knees, tarsi, tips of tibiæ, proximal third of the swollen femur and distal third of the long subtriquetrous posterior coxa. Ovipositor very thin, fuscous, its valves black. Fore wings irregularly, lightly stained with brownish, the venation black. Eyes red, the ocelli darker red. Antennæ with the scape and pedicel brown, the remaining joints black.

Head and thorax very finely reticulately punctate; abdomen tapering at base, but not petiolate, strongly compressed. Ocelli distant from the eyes. Propodeum with larger reticulate punctures, its dorsum rounded, without a median carina. Postmarginal vein twice the length of the stigmal, the marginal very long, not much shorter than the submarginal. Distal third of scutellum and the mesopostscutellum smooth, but finely, closely, polygonally reticulated. Proximal abdominal segments, with very minute pin-punctures, the distal segments glabrous. Posterior coxae sculptured like the postscutellum, the posterior femur beneath armed with nine large, black, unequal teeth; the first (proximal), eighth and ninth largest; the latter stoutest, triangular, tooth 8 longest, columnar; the seventh next to the shortest, paired—that is, a bidentate, erect plate; the two dentations here counted as separate teeth, though united at base; the two teeth equal; teeth 4 and 5 unequal, also more or less united at base; tooth 2 shortest, obtuse, nipplelike.

Antennæ inserted in the middle of the face, 13-jointed, the funicle filiform, but its distal joint widening somewhat, becoming wider than long; scape simple, not as long as the club; pedicel somewhat longer than the first funicle joint; joint 2 of funicle longest, joint 3 next, the distal joint shortest; joint 5 subequal in length to the pedicel, the following funicle joints all shorter, club joints nearly equal, the distal one slightly the longest.

(From a single specimen, the same magnification.)

Male: Not known.

Described from a single female, minutien-mounted, in the collections of the Queensland Museum, labelled "Q. M. Brisbane. H. Hacker.—20-5-1911."

Habitat: Australia—Brisbane, Queensland.

Type: No. *Hy 1191*, Queensland Museum, Brisbane, the foredescribed female on a minutien mount, plus one slide of xylol-balsam bearing the antennæ and a posterior femur.

2. *Pachytomoides greeni* (Crawford).

Podagrion greeni Crawford, 1912,* pp. 3-4; fig. 1.

This Cingalese species reared from the eggs of a mantid must be referred to this genus, though the female bears an abdominal petiole. Otherwise, it agrees with the species generically.

NEW ICHNEUMONOIDEA PARASITIC ON LEAF-MINING DIPTERA.

BY A. B. GAHAN, MARYLAND AGRICULTURAL EXPERIMENT STATION.

With a single exception the type specimens of the seven supposed new species described in the following paper were furnished by Prof. F. M. Webster, of the United States Department of Agriculture, and the designated hosts are on his authority. The types of one species were reared by the writer.

Family BRACONIDÆ.

Sub-family Opiinæ.

Opius utahensis, n. sp.

Female.—Length, 2.25 mm. Head transverse; vertex, temples, cheeks and occiput smooth and polished with sparse whitish hairs, the frons bare except along the eye margins; face with distinct round punctures and moderately hairy; clypeus fitting closely to the mandibles; mandibles without a notch on the ventral margin; antennæ longer than the body, pubescent, 32-jointed in the type, the first flagellar joint one-third longer than the second. Propleuræ with very fine reticulate sculpture; mesonotum with a median dimple-like impression before the scutellar fovea, parapsidal furrows deeply impressed at the anterior lateral angles, but entirely effaced on the disc; mesopleuræ reticulately sculptured on the disc, with a broad, rugose or foveolate furrow along the dorsal and anterior borders joining a similar furrow which separates the mesopleuræ from the mesosternum; propodeum and metapleuræ strongly rugose. Wings hyaline, stigma lanceolate emitting the radius at about the basal one-third; the radius strongly angulated at the second cubital cross vein, attaining the margin of the wing some distance above the extreme wing apex, its first abscissa less than

*Proc. U. S., National Museum, Vol. 42.

half as long as the width of stigma; second discoidal cell closed at the apex or nearly so. Abdomen broadly oval, the first dorsal plate rather thick, with precipitous edges and finely wrinkled, slightly wider at apex than at base and distinctly longer than broad; second segment two times as wide at apex as at base, smooth like the following segments; ovipositor slightly exserted.

Clypeus, mandibles, palpi, scape, tegulæ, base of wings, legs except apical joint of the tarsi, and abdomen except the first dorsal plate pale testaceous; apical joints of all tarsi, and the flagellum brown-black; wing veins and stigma brownish; remainder of the body black.

Male.—Essentially like the female, but with the antennæ 33-jointed in type.

Type locality.—Salt Lake, Utah.

Host.—*Agromyza parvicornis.*

Type No. 15591, United States National Museum.

One female and five male specimens from the type locality, labelled Webster, No. 8819.—C. N. Ainslie, collector.

Probably closest to *O. bruneiventris* Cr. of the described species, but readily separated from that species by the fact that in *bruneiventris* there is a distinct opening between the clypeus and mandibles, and the mesopleuræ are smooth and polished except for the oblique, foveolated furrow below the middle.

Opius suturalis, n. sp.

Male.—Length, 1.25 mm. Head transverse, smooth, with few hairs above; the face only slightly hairy; clypeus arcuate, leaving a transverse elliptical opening between it and the mandibles; antennæ pubescent, twice as long as the body, 22-jointed in the type. Thorax smooth and shining; mesonotum without a median depression posterior y, the parapsidal furrows indicated only at the anterior lateral angles of the mesonotum; mesopleuræ smooth, with a shallow, ovate, non-foveolated impression below the middle; propodeum smooth and polished. Wings thickly ciliated; the stigma lanceolate, emitting the radius before the middle. The first abscissa of radius short, third abscissa attaining the wing margin far before the extreme wing apex; second discoidal cell not completely closed at the apex. Abdomen spatulate, as long as the thorax, the first dorsal segment very finely but distinctly rugulose; second segment with a distinct transverse suture before the middle, which does not extend quite to the margins; the surface before the suture and for one-third of the distance beyond distinctly rugulose; segments beyond the second smooth. General color shining black; mandibles and palpi slightly fuscous; tegulæ testaceous; wing veins and stigma brownish; legs testaceous, their coxæ piceus. Abdomen wholly black.

Type locality.—Tempe, Arizona.

Host.—*Agromyza pusilla.*

Type No. 15592, United States National Museum. Two male specimens from the type locality, labelled Webster, No. 7215.—V. L. Wildermuth, collector.

Distinguished from *O. aridus* by the presence of a distinct transverse furrow on the second segment and by the rugulose sculpture of that segment. May possibly be the male of *O. nanus* Prov., from the type of which it differs, however, in the smooth propodeum and the wholly black abdomen.

Opius aridis, n. sp.

Female.—Length, 1.25 mm. Head perfectly smooth and polished, the face moderately hairy; vertex, temples and occiput with sparse, inconspicuous hairs; clypeus arcuated apically, leaving a distinct opening between it and the mandibles; antennæ somewhat longer than the body, 18-jointed in the type (varying from 18-jointed to 23-jointed in other specimens of the series), the first joint of the flagellum slightly the longest. Thorax smooth and polished; mesonotum without a median depression or furrow before the scutellar fovea, parapsidal furrows impressed at the anterior lateral angles of the mesonotum, but not attaining to the disc; mesopleuræ smooth, with a shallow, ovate, non-foveolated impression below the middle; propodeum moderately hairy, nearly smooth; the apical margin very slightly roughened. Wings densely ciliated, giving them a brownish tinge; stigma lanceolate, the radius arising before the middle, and attaining the wing margin above the extreme wing apex; its first abscissa shorter than the width of stigma; second discoidal cell closed. Abdomen not longer than the head and thorax; the first dorsal segment longer than wide at apex, smooth and polished like the following segments: ovipositor sheath extending slightly beyond the tip of the abdomen. General colour, black; clypeus, mandibles and palpi stramineous; tips of mandibles brown; legs in the type stramineous, the apices of posterior tibiæ, their tarsi and the median tarsi fuscous (in other specimens of the type series the legs vary from pale stramineous to wholly dark brown); tegulæ and wing base brownish testaceous; 2nd segment of the abdomen more or less stramineous, first segment and those beyond the second piceous to black; ovipositor sheath black.

Male essentially like the female, but with the antennæ 20- 23-jointed.

Type locality.—Tempe, Arizona.

Host.—*Agromyza pusilla.*

Type No. 15593, United States National Museum. The type series contains 10 females and 10 males, labelled Webster, No. 7215.—V. L. Wildermuth, collector.

This species in general appearance closely resembles *Opius* (*Eutrichopsis*) *agromyzae* Vier., which is parasitic on the same host. It may be distinguished from that species, however, by the non-foveolated impression on the mesopleuræ and the smooth first abdominal segment.

Opius bruneipes, n. sp.

Female.—Length, 1.25 mm Head perfectly smooth and highly polished; face sparse'y hairy; vertex, temples and occiput with a few scattering and inconspicuous hairs; clypeus arcuated on the anterior margin, leaving a transverse, elliptical opening between it and the mandibles; antennæ longer than the body, pubescent 21-jointed in the type, the first joint of flagellum slightly longer than the second. Thorax smooth and highly polished, robust, without a median dimple-like depression on the mesonotum, parapsidal furrows wholly effaced or represented by only a few indistinct punctures at anterior lateral angles; mesopleuræ without a trace of an impressed furrow above the coxæ; propodeum entirely smooth and polished, with very few hairs; metapleuræ also smooth.

Wings densely ciliated; the stigma lanceolate and rather broad, much broader than the first abscissa of radius is long; radius arising much before the middle and attaining the wing margin far above the extreme wing apex; its first abscissa very short; second discoidal cell open below at the apex. Abdomen ovate, about as long as the thorax; its first dorsal segment smooth and polished, or nearly so; following segments also smooth; second segment more than twice as wide at apex as at base; following segments tapering to the apex; ovipositor slightly exserted. General color black; mandibles brownish, tips black; palpi fuscous; scape dark brown, flagellum brown-black; tegulæ black; wing veins and stigma brownish; legs including coxæ, dark brown; the anterior pair slightly paler; first and second abdominal segments brownish; following segments black; ovipositor sheath black.

Male.—Essentially as in female.

Type locality.—Lakeland, Florida.

Host.—*Agromyza pusilla.*

Type No. 15594, United States National Museum. The type series consists of three females and three males from the type locality, labelled Webster, No. 9489.—G. G. Ainslie, collector.

The species is distinguished from *O. aridus* by the total absence of the mesopleural furrow, and the open second discoidal cell as well as by the dark brown legs and fuscous palpi.

Opius succineus, n. sp.

Female.—Length, 2 mm. Head transverse, smooth and polished, sparsely hairy; the face moderately hairy, impunctate, with a rather more distinct median ridge than usual; clypeus arcuate, leaving a transverse, elliptical opening between it and the mandibles; antennæ longer than the body, pubescent, 29-jointed in the type. Propleuræ smooth; mesonotum with a median dimple-like impression before the scutellar fovea, the parapsidal furrows distinctly impressed anteriorly for nearly one-third the length of the mesonotum and faintly traceable as shallow impressed lines to the median dimple; mesopleuræ smooth, but with a strongly oblique foveolate furrow below the middle; propodeum indefinitely sculptured. faintly rugulose, with a sinuous, transverse raised line or carina near the middle. Wings hyaline, the stigma lanceolate;

the radius arising near the basal one-third of the stigma and attaining the wing margin only slightly above the wing apex; its first abscissa about as long as half the width of stigma; second discoidal cell closed. Abdomen broadly oval; the first dorsal plate distinctly longer than broad, abruptly narrowed before the middle, indefinitely rugulose; segments beyond the first smooth; ovipositor slightly exserted. General colour brownish yellow; vertex, occiput and temples black; cheeks and face reddish testaceous; ovipositor black; wing veins and stigma brownish; the dorsal abdominal segments beyond the second brownish; scape and legs pale amber.

A male paratype is like the female in sculpture but much darker in colour; the thorax above and at sides strongly tinged with brownish.

Type Locality.—Lafayette, Indiana.

Host.—*Agromyza* sp., mining leaves of Panicum.

Type No. 15595, United States National Museum. The female type is labelled Webster, No. 3814, W. J. Phillips, collector. The male bears the same number, but was collected by P. Luginbill. Another male specimen, labelled Webster, No. 9302—J. J. Davis, collector—was reared from the same source at Danville, Illinois.

This species superficially resembles *Opius diastatae* Ashm., a parasite of the corn leaf-miner, which was described by Ashmead under *Bracon* (Proc. U. S. Nat. Mus., 1888, p. 617). It may be distinguished from that species by the foveolate mesopleural furrow and the dimple-like median impression on the mesonotum.

Family ALYSIIDÆ.

Subfamily Dacnusinæ.

Dacnusa scaptomyzae, n. sp.

Female.—Length, approximately 2 mm. Head transverse, nearly twice as broad as long; above perfectly smooth and highly polished, with a very few scattered whitish hairs on the vertex and occiput; occiput concave; temples broad and slightly rounded; vertex divided by a shallow median groove, running from the anterior ocellus to the occiput; eyes bare, ovate; face with moderately dense whitish pubescence, smooth or nearly so, the punctures being very minute, a rather distinct median carina on the upper half; maxillary palpi 6-jointed, the two basal joints about equal in length and together scarcely longer than the third; labial palpi 4-jointed; mandibles 3-toothed, the median tooth longest and acute; the two laterals short and blunt; antennæ pubescent, 23-24 jointed, a little longer than the body; first joint of the flagellum longer than the second, following joints decreasing in length to the tip.

Thorax smooth and shining; prothorax short, mostly concealed from above; mesonotum gibbous, polished, without pubescence, except for four or five hairs on each lobe opposite the base of the wings; parapsidal furrows impressed anteriorly for about one-third the length of the mesoscutum, a short longitudinal incision on the median line just before the scutellum, varying somewhat in length, but never extending more than half the length of the mesonotum; scutellar fovea broad and deep, with several carinæ crossing it at the bottom; mesopleuræ smooth, polished, glabrous, except for a few hairs at the posterior angle, just above the median coxæ, and with a shallow longitudinal smooth depression below the middle; metapleuræ moderately hairy and mostly rugulose, the disc smooth; propodeum finely rugose, more strongly so posteriorly, not conspicuously pubescent, but with a few scattering hairs most abundant laterally. Wings hyaline, iridescent; stigma long, lanceolate, rather broad, extending half the length of the radial cell; radius arising at about the basal one-third of the stigma and attaining the wing margin about half way between the apex of stigma and the extreme wing apex, its first abscissa nearly perpendicular and slightly longer than the width of stigma; second abscissa slightly straightened toward the wing margin, but not concave beneath, radial cell broad; cubital cross-vein oblique, somewhat longer than the first abscissa of radius; recurrent nervure oblique, joining the first cubital cell before the cubital crossvein, a distance equal to about half the length of the cubital crossvein; first discoidal cell smaller than the first cubital, submedian cell slightly longer than the median, the second discoidal completely closed.

Posterior legs longer than the body, the two trochanter joints together about as long as their coxæ, tibiæ as long as the femora and trochanters combined, tarsi as long as the tibiæ, the first joint nearly twice the length of the second.

Abdomen subsessile, as long as the thorax, ovate, squarely cut off at the apex, the apical segments retracted; first segment rugose, broader at the apex than at the base, as long as the posterior coxæ, its spiracles about midway of the segment and prominent, basally the segment is bicarinate, the carinæ originating at the lateral and angles and meeting before the spiracles, back of the triangular area enclosed by the carinæ, the surface is convex, the posterior lateral angles depressed; segments beyond the first smooth and polished; ovipositor sheath about one-fourth the length of the abdomen.

Colour.—Shining black; palpi, labrum, scape and legs, including the coxæ, testaceous; mandibles slightly darker; flagellum brown-black, the basal oints paler; first segment of the abdomen black, the following dorsal segments very dark brown, the second segment often somewhat testaceous on the disk. Wing veins and stigma brownish testaceous.

Male.—Like the female in every respect, except that the antennæ are 24-25-jointed; the stigma is broader than the length of the first abscissa of radius, considerably broader than in the female; the abdomen is slightly longer than the thorax and attains its greatest width just before the apex, therefore not ovate, but spatulate.

Type locality.—College Park, Md.

Host.—*Scaptomyza flaveola* Meig.

Type Cat. No. 15596, U. S. National Museum. Paratypes in the United States National Museum and the Collection of the Maryland Agricultural Experiment Station.

During the season of 1912 the dipterous leaf miner *Scaptomyza flaveola* Meig. was collected by the writer in three different localities and on as many different dates. June 3rd, at Hyattsville, Md., it was found infesting the leaves of turnips in a small garden plot. Both larvæ and puparia were present in large numbers. The puparia were found either in the original larval mines or beneath wilted and fallen leaves on the ground. The majority seemed to have pupated in the leaves, and none seemed to have entered the soil to transform. Many leaves were collected and taken to the laboratory, and from these were reared during the month of June a large number of the flies, and about an equal number of *Dacnusa scaptomyzae.*

July 1st, at Hancock, Md., mined leaves of radish were collected, from which were reared the same fly, as well as several specimens of the parasite.

July 30th, at College Park, Md., several cabbage plants growing in a box, where they had been seeded for transplanting, were found severely mined. Here again the same fly and many specimens of the parasite were reared during the month of August.

Dacnusa agromyzæ, n. sp.

Female.—Length, approximately 2 mm. Head twice as broad as long, smooth, with a very few scattering hairs on the occiput, vertex and cheeks; the face moderately hairy, with a slight median carina on the upper half; vertex not divided by a median furrow; eyes bare, ovate; maxillary palpi 6-jointed, the two basal joints together not as long as the third, the fourth joint as long as 1, 2 and 3 combined; labial palpi 4-jointed; mandibles with the two lateral teeth acute, the median tooth longer, with a distinct notch on its ventral margin near the base making the mandible appear four-toothed; antennæ 33-36-jointed, nearly or quite twice as long as the body; the first joint of the flagellum about equal to the scape and pedicle combined; following joints shorter and decreasing in length toward the tip.

Prothorax mostly concealed from above; mesonotum slightly bilobed owing to a broad depression extending from base to apex along the median longitudinal line, its surface anteriorly and medially punctate and covered with white hairs, the broad posterior angles opposite the tegulæ smooth and glabrous, parapsidal furrows not at all impressed; scutellar fovea deep, with several cross ridges at the bottom; mesopleuræ polished and glabrous except for a few hairs just above the median coxæ, with a shallow, longitudinal, smooth depression below the middle; metapleuræ covered with a dense, short, white pile, completely concealing its sculpture; propodeum high and broad, abruptly truncate posteriorly, rugose and covered with white pile, which is not as dense as that on the metapleuræ. St'gma linear and extending nearly two-thirds the length of the radial cell; radius arising at about the basal one-fourth of the stigma; its first abscissa not quite perpendicular and slightly shorter than the cubital crossvein, second abscissa curving very slightly into the radial cell toward the apex and attaining the wing margin far above the extreme wing apex; recurrent nervure interstitial with the cubital crossvein; submedial cell longer than the median; second discoidal cell open beneath.

Posterior legs longer than the whole body, their tibiæ scarcely as long as the femora and two joints of the trochanter combined; coxæ equal to the first abdominal segment, first tarsal joint twice the length of the second.

First abdominal segment convex, rugose, wider at apex than at base; bicarinate at base, the carinæ originating at the lateral angles and converging posteriorly, but fading out before meeting; spiracles not prominent and placed slightly before the middle of the segment; sides of the segment parallel beyond the spiracles; the posterior lateral angles somewhat flattened; whole abdomen slightly longer than the thorax; the segments beyond the first smooth and but little wider than the first segment at apex, their sides parallel. Ovipositor sheath less than one-fourth the length of the abdomen.

Colour as in the preceding species, except that the legs are reddish testaceous and the abdomen, including the first segment, is brownish testaceous.

The male is like the female.

Type locality.—Lafayette, Indiana.

Host.—*Agromyza angulata.*

Type No. 15597, United States National Museum.

Four specimens received from Prof. F. M. Webster, reared by P. Luginbill, and bearing Webster's number, 9700.

This species would apparently fall in Foerster's genus Mesora, which genus is believed to be untenable.

NOTES ON SOME SPECIES OF THE GENUS PROSOPIS.

BY J. C. CRAWFORD, WASHINGTON, D. C.

Prosopis mesillæ Cockerell.

This is a valid species, and not a form of *P. cressoni* as it is given by Metz. Externally the two are easily separable. *P.*

cressoni has the propodeum coarsely sculptured, while *mesillæ* has it very finely wrinkled. The eighth ventral plates of the males are quite different, and are therefore figured. In *cressoni* the apical lobes are much shorter than the pedicel attaching them to the plate; in *mesillæ* they are longer than the pedicel.

Prosopis nelumbonis Robertson

Synonym *P. fossata* Metz.

The characters which Metz gives as distinguishing this species from all others—namely, the "coarse, dense, pit-like punctures over the entire head and thorax"— are almost the identical words used by Robertson in his original description of the species. The type of *fossata* is in the U. S. Nat. Museum, and I have carefully compared it with specimens of *nelumbonis* from Illinois.

Prosopis stevensi, new species.

Male.—Length, about 4.25 mm. Black, face below insertion of antennæ old ivory colour, with sparse punctures and silky from minute vertical striatulations, supra-clypeal mark extending upward between antennæ, truncate at tip; lateral face marks extending above insertion of antennæ, dilated above; slightly extending over antennæ and very slightly away from eye margin (fig. 5),

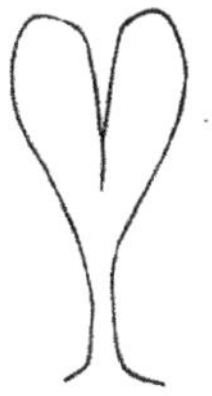

Fig. 3. *P. mesillæ*, male.—Part of 8th ventral plate.

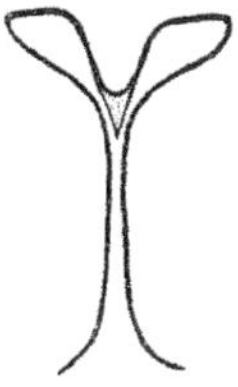

Fig. 4. *P. cressoni*, male.—Part of 8th ventral plate

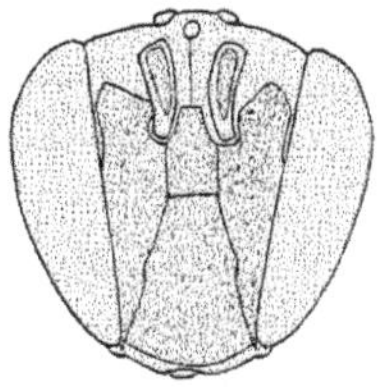

Fig. 5. *P. stevensi*, male.—Face.

face above insertion of antennæ, with rather close and coarse punctures; scape with an ivory stripe in front; flagellum reddish, dusky above; mesonotum with punctures similar to those on vertex, separated from each other by slightly less than a puncture width, surface between punctures lineolate; metanotum rugosopunctate; propodeum with the area not well defined, very coarsely rugose; laterad of it more finely rugose; propodeum sharply truncate behind, truncation surrounded by a salient rim; pronotum with two spots, tubercles, tegulæ with a spot, fore tibiæ with a stripe, mid and hind tibiæ at bases and apices, and basal joints of all tarsi, ivory colour; mesopleuræ more coarsely punctured than dorsum; wings dusky; first abdominal segment finely sparsely punctured, punctures closer towards apex, second and following segments

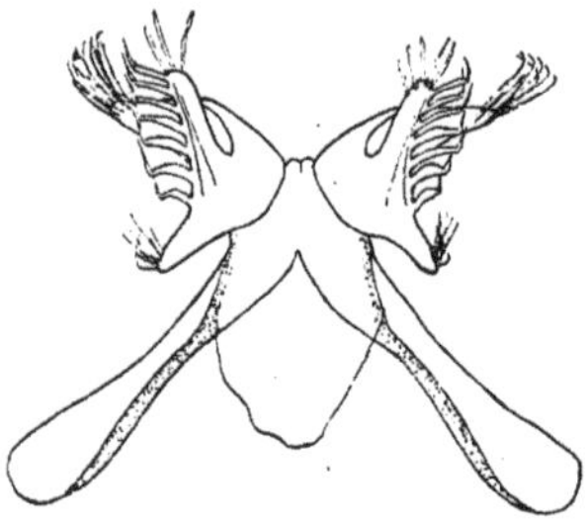

FIG. 6.
P. stevensi, male.—
Seventh ventral plate.

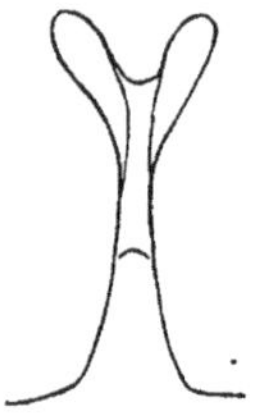

FIG. 7.
P. stevensi, male.—
Part of 8th ventral plate.

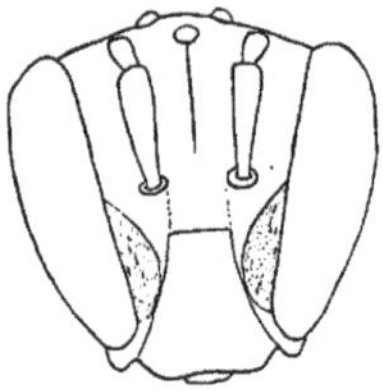

FIG. 8.
P. stevensi, female.—
Face.

more closely punctured. See figures 6 and 7 for structure of seventh and eighth ventral plates.

Female.—Length, about 5.25 mm. Similar to the male; face below antennæ more distinctly striatulate and more silky; a large mark on each side of face (see fig. 8), a spot on each side of pronotum, tubercles, a spot on tegulæ, a spot on fore and mid tibiæ at base and a broad annulus on base of hind tibiæ, ivory colour.

Type locality.—Fargo, N.D.

Type male collected Sept. 6, 1912 (Stevens No. 4154); allotype female the same date (Stevens No. 4152); paratype female, Sept. 8, 1912 (Stevens No. 4194); paratype female, Aug. 26, 1912 (Stevens No. 3947). All four specimens taken on *Melilotus alba* by Mr. O. A. Stevens, after whom the species is named.

Type Cat. No. 15530, U. S. N. M.

Two paratype females in collection Mr. Stevens.

In the classification of the genus by Metz, this species belongs to his *cressonii* division and to the *tridentulus-grossicornis* group and differs from these two species in the male having fewer teeth on the outer margin of each wing of the seventh ventral plate, in these teeth being stouter and more markedly turned up at end. Neither of the older species has the propodeum strongly rugulose nor so abruptly truncate, and the face markings are quite distinct, etc.

OBITUARY.

We record with much regret the death of Miss Mary E. Murtfeldt, which took place at her residence in Kirkwood, Mo., on the 23rd of February last. She was a contributor from time to time to the pages of this magazine and a subscriber for a long series of years. When the late Dr. C. V. Riley was State Entomologist of Missouri she gave him much material assistance, studying and recording the life-histories of many species of insects in the preparation of his series of reports on the Insects of Missouri, which are amongst the most valuable of his writings. After his appointment to be Chief of the Bureau of Entomology at Washington, she continued her interest in entomology. Her contributions were always of much value, as she was very painstaking and accurate in her observations. She belonged to several scientific societies, and was highly esteemed by all who had the pleasure of her acquaintance.

A NOTE ON *GRAPTA J-ALBUM.*

During the month of August, 1908, whilst camped in a mountain valley, engaged in collecting insects, I was interested to observe this butterfly attracted to a piece of bacon hanging in a small tree near the tent. It fluttered round for a few minutes and settled several times upon the bacon. Shortly it was joined by two other individuals, both of which alighted as did the first comer. They must have been attracted as are some other Lepidoptera, by the scent, perhaps. That the drawing power of the bacon was powerful was evident from the fact that during our stay of three days at this spot these butterflies were always to be seen during the warm part of the day hovering round what we called the bacon tree, and constantly alighting on and round the attractive board.

E. P. VENABLES.

BOOK NOTICES.

"INJURIOUS INSECTS: How to recognize and control them," by Prof. Walter C. O'Kane. The Macmillan Company, New York. 414 pages, 606 figures; $2.

The reviewer's pen has hardly dried after noticing Prof. Sanderson's manual of injurious insects when his successor, as Entomologist to the New Hampshire Experiment Station, adds another to the existing number of general works on injurious insects. The two outstanding features of the book are—first, the large number (over six hundred) of illustrations from the author's own photographs; and, second, the arrangement of his subject matter.

In regard to the illustrations: While the author is to be congratulated in his endeavor to provide entirely original illustrations, the preparation of which must have involved an enormous amount of labour, we must admit that in very many instances he would have been more successful in his representation of the insects had he given us line drawings or reproduced some of the really good available cuts. Those who have attempted it realize the difficulties of insect portraiture. The purpose of such illustrations is to facilitate the identification of the insects, but it must be confessed that a considerable proportion of the illustrations are not such as will provide a good means of recognition, especially in the case of larvæ. On the other hand, the author has in some cases given us excellent figures The illustrations would have been more valuable had the magnification been given when the insects are enlarged.

As a means of assisting in the identification of the insect pests of garden and field crops, of orchard and small fruits, all of the chief species of which are described, the author has arranged the insects belonging to these two groups according to the place where they are found at work. Insects working in the soil are considered first, then the borers within the stem, trunk or branch. These are followed by those feeding upon the surfaces of the same. Finally, he deals with the insects feeding on the leaves, flower and fruit in the order named. The leaf-feeders are also grouped. It is hoped by the author that this method of grouping will prevent the usual duplication which sometimes cannot be avoided if the insects are grouped according to host plants, owing to many of the common species feeding on several species of plants. Insect pests of the household and of stored products are also described.

In chapters of varying lengths the morphology, internal structure, senses and behaviour, metamorphoses, classification and means of dissemination are described in language devoid of technicalities that might confuse the general reader. The preventive measures are well discussed, and the chapters on insecticides and fumigants and the methods of applying them form a useful section of the book and increase its value as a book of reference for those who have to deal with insect pests. A list is given of references to bulletins and reports containing detailed descriptions of the insects described in the book. An idea of the large number of insects which the author considers may be gathered from the fact that the index to the book covers twenty-four pages.

Covering the large field that it does, it is not surprising that inaccuracies occur, and space forbids a detailed reference to the same. In compiling information of so varied a character, greater care is necessary than when the information is the result of personal knowledge A work of this character is an enormous undertaking nowadays, and we cannot but feel that the author would have produced a better book had he spent more time in its preparation. Nevertheless it will be a useful book, and the author deserves our thanks.

C. G. H.

Dolichopodidæ in Lundbeck's "Diptera danica."

Diptera Danica.—Genera and species of flies hitherto found in Denmark. Part IV., Dolichopodidæ. By William Lundbeck; 416 pp., 130 figs. (Copenhagen, G. E. C. Gad; London, William Wesley & Son.) Dec., 1912. $4.25.

After a lapse of two years since the previous part of this work appeared (reviewed in this journal, Vol. 43, April, 1911), the author gives us the fourth part, which treats of the single family Dolichopodidæ. Most entomologists know these small, usually metallic golden-green, flies, which appear, as Verrall said, "to be standing on tiptoe," being raised on their front legs. Their wing venation is very characteristic. The most remarkable thing about the family, however, is the sexual dimorphism of the male, which reaches a higher degree of development in this family than in any of the other family of the Diptera. These secondary sexual characters occur primarily on the legs, but they are also found on wings, antennæ and facial region. Associated, as is usually the case, with these secondary sexual characters in the male Dolichopodids are remarkable "courting" habits, which not infrequently strongly recall the analogous amatory preliminaries on higher animals. The flies are all predaceous, feeding on other insects and small invertebrates, and are usually found on bushes, on low herbage and grass in woods and outside, generally in damp localities and more or less near water. In North America we have little information as to their life-histories; the larvæ occur in earth rich in vegetation and under the bark of trees. The species are distributed all over the world, two species of *Dolichopus* being found in Greenland. From North America about 526 species are known, from the palæarctic region about 586 species are known, and ten species are recorded as common to both regions.

Aldrich divides the family in North America into twelve subfamilies, and although the former worker has given no diagnoses, the author of the present work believes them to be good and natural. As he has only examined the Danish species closely, he follows the arrangement of the "Katalog der Paläarktischen Diptera," and divides the family into four subfamilies, at the same time admitting the heterogeneous nature of some of them.

As in the previous parts of this excellent work, the author treats each species fully; where they are known, larval characteristics and habits are given, and the presence of one hundred and thirty figures, chiefly of the antennæ and wings, enhances the value of this further and most welcome addition to our dipterological literature. We look forward to the succeeding parts of this monumental work, in the preparation of which the author has our good wishes.

C. G. H.

Mailed May 17th, 1813.

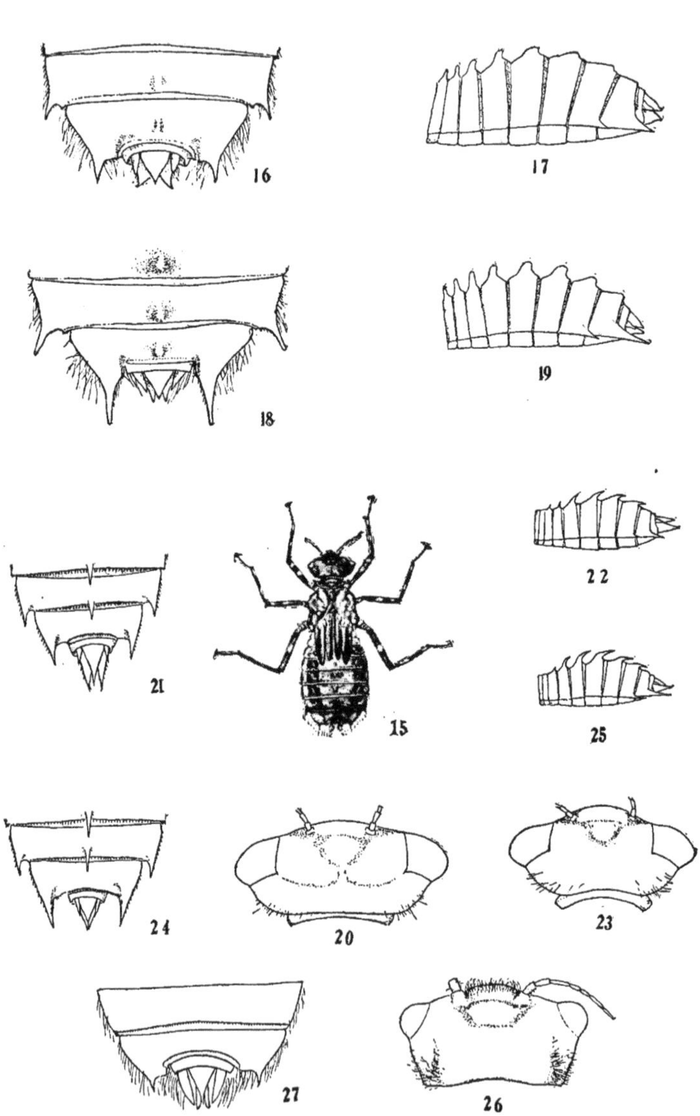

NEW NYMPHS OF CANADIAN ODONATA.

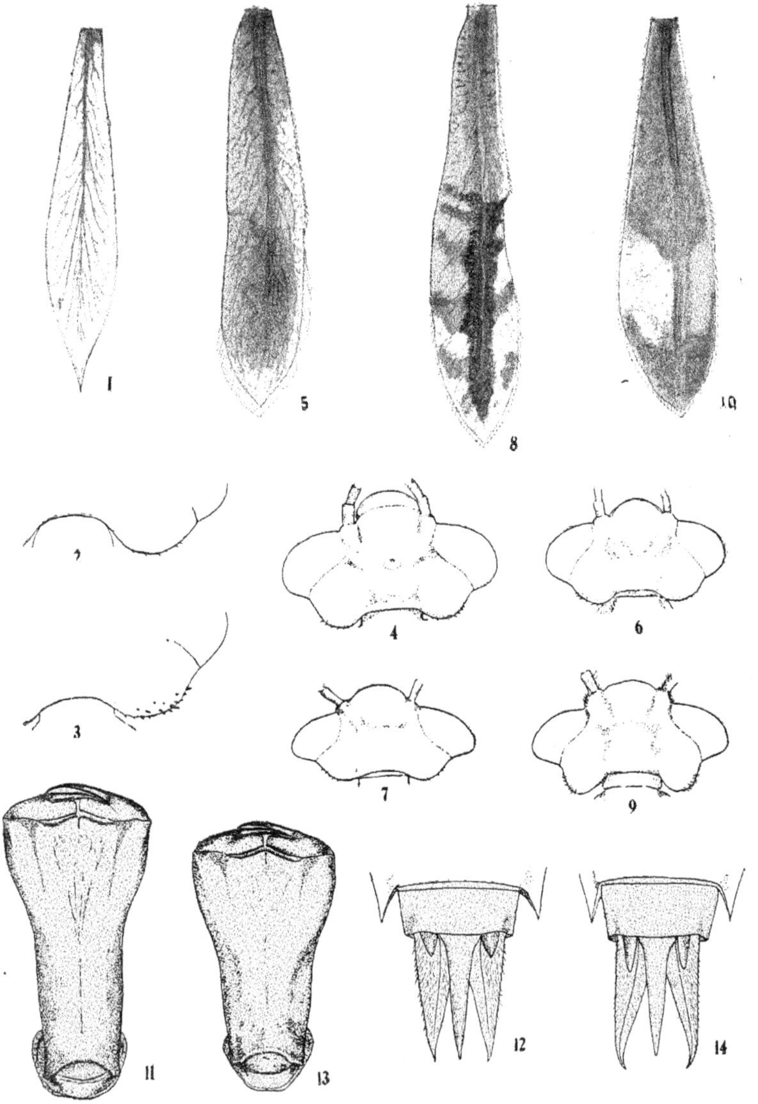

NEW NYMPHS OF CANADIAN ODONATA.

The Canadian Entomologist.

Vol. XLV. LONDON, JUNE, 1913 No. 6

NEW NYMPHS OF CANADIAN ODONATA

BY E. M. WALKER, TORONTO, ONT.

During the summers of 1907, 1908 and 1912, the writer spent much of his time at the Great Lakes Biological Station, Go Home Bay (Georgian Bay), Ont., in collecting and rearing dragonflies (Odonata). A full account of this work will appear in the forthcoming report of the Marine Biological Stations of Canada; but, as this report will not be issued in the immediate future and is comparatively unknown to entomologists, it is thought best to publish in advance the descriptions of the new nymphs obtained.

Nymphs of certain species of Aeshna, which were reared for the first time at Go Home Bay, have already been described in the writer's memoir on this genus,* and are omitted from the present account. In addition to the species described from Go Home Bay, the nymph of *Somatochlora semicircularis* (Selys) from Vancouver Island is also included.

Nehalennia gracilis Morse

A few nymphs of this species were found in floating sphagnum bogs, some distance back from the open water. Several imagoes emerged in the laboratory during July.

I have compared these nymphs carefully with a few specimens of *N. irene* (Hagen) from Toronto, and the only differences I can find are the smaller size, less spinulose hind margin of the head and entire absence of spots on the gills. It is not improbable that none of these characters are constant, as I had but few specimens of either species for comparison.

In *N. gracilis* the convex posterior margin of the head has only 4-6 slender inconspicuous colourless spinules; in *N. irene* there are a dozen or more spinules, which are somewhat coarser and blackish at base (Figs. 2, 3); gills very slender, widest in the distal third, tapering somewhat more gradually than in *N. irene*, with no indication of spots (Fig. 1).

*The North American Dragonflies of the genius Æshna. University of Toronto Studies, Biological Series, No. 11, 1912.

Length of body 8.25-9; 'gills 3-3.75 additional; hind wing-case 2.2-2.7; hind femur 2-2.33; width of head 2.33-2-4.

Enallagma cyathigerum calverti (Morse)

Full-grown nymphs were taken early in the season of 1912, several emerging in the laboratory on June 3 and 4. Mature adults were flying in numbers on June 1, and had about disappeared by the middle of the month. Nymphs were also reared at Lake Simcoe in 1909, adults emerging on June 4.

The nymph (Figs. 4, 5) is very similar in form to that of *E. hageni* (Walsh), but is considerably larger, with much darker gills. Eyes as in *hageni*, less prominent than in *E. signatum* and *E. pollutum* (Figs. 6, 7, 9), the curve of the posterior median excavation somewhat more flattened than that of the rather strongly convex margins on each side, the latter with a dozen or more spinules. Labium with 4 mental setæ and 6 (occasionally 5) lateral setæ; end-hook of lateral lobe preceded by 3 teeth of moderate size, which are preceded by 3 or 4 smaller, somewhat incurved denticles. Gills lanceolate, widest a little beyond the middle, .ventral margin straight at base, dorsal margin convexly curved; apices bluntly pointed, with convexly curved margins or rounded. Across the middle of the gill is a distinct joint, proximad of which the margins are spinulose, the spinules of the ventral margin stronger than those of the dorsal; distad of the joint the margins are beset with a fringe of delicate hairs, much longer than those of *E. hageni*. Color dark brownish (probably olivaceous in life, each abdominal segment, except 10, with a dark lateral blotch, not seen in the exuviæ; femora with a pale ring just before the apex, preceded by a dark ring. Gills dark greyish brown, deepening just beyond the median joint.

Length of body, 15.5(exuvia)-21.5; gills 6.5-8; hind wing 4.5-5; hind femur 4; width of head 3.5-3.7.

Enallagma pollutum (Hagen) Selys.

Among a number of Odonate nymphs, taken by Mr. R. P. Wodehouse at Waubaushene and Fitzwilliam Island, Georgian Bay, in 1912, are numerous specimens of an undescribed form, which is so obviously nearly related to *E. signatum* that we have little hesitation in ascribing it to *E. pollutum*. This species is, moreover, a common one here and is the only Enallagma of the region whose nymph has not been reared (except the rare *E. ebrium*).

Nymph (Figs. 9, 10), long and slender, eyes very prominent laterally, their postero-lateral margins forming with the sides of the head a marked excavation. Hind angles of head with numerous slender setæ, rounded, but very prominent and narrower than the median concavity. Labium with 3 mental setæ: lateral setæ 5; lateral lobes, before the end-hook, with three well-marked teeth, preceded by a feebly denticulate, almost truncate, margin. Abdominal segments 2-7, with prominent postero-lateral angles. Gills large, broad lanceolate, widest at the distal third, with a transverse median joint; basal half dark, except at the base: apical half whitish or grey, except a broad, dark anteapical band.

Colour brown (alcoholic, probably greenish in life), sides of head and thorax with a pale longitudinal band between two dark bands, the most ventral of which passes dorso-caudad to the bases of the front wing-cases. There are usually also a few dark spots on the head and thorax. Abdomen rather dark brown, almost uniform. Legs pale, femora with a very narrow, but usually well-defined, dark ring at the distal fourth.

Length of body 13 (contracted) to 18 (extended); gills 5-6.5; hind-wing 4.3-5; hind femur 3.5; width of head 3.25-3.4.

Boyeria grafiana Williamson.

The dark-coloured nymphs of this species are found rather commonly under stones, along more or less wave-beaten shores or wherever there is a perceptible current. Full-grown specimens were collected on and after June 4, 1912, the first adult emerging in the laboratory on July 14, followed by several others during the succeeding fortnight.

As the nymph of *Boyeria vinosa* was described before *B. grafiana* had been recognized as a distinct species, it is impossible to be certain whether the descriptions all refer to *B. vinosa* or not, but Needham's description* belongs, with scarcely a doubt, to that species.

We have reared a number of nymphs of *B. grafiana* and collected many exuviæ as well as nymphs in several localities. We have also received a series of exuviæ of a Boyeria from the Shawa-

*Bull. 47, N. Y. State Museum, p. 465, 1901.

naga River, collected by Mr. Paul Hahn, which differ very slightly from those of *B. grafiana*. The latter were also found on the same river. As *vinosa* and *grafiana* are the only North American species of Boyeria, and are both common in this region, there can be no doubt that the species not yet reared is *B. vinosa*.

The nymphs of these two forms may be separated as follows:

Mentum of labium, 5.5 mm. long, its middle breadth scarcely less than half its length; fourth abdominal segment without lateral spines; lateral abdominal appendages of female one-fourth to one-third as long as the inferior appendages, and usually about as long as the dorsum of segment 10 . . . . *B. vinosa*.

Mentum of labium 6.5-7 mm. long, its middle breadth distinctly less than half its length; fourth abdominal segment generally with distinct though very small lateral spines; lateral abdominal appendages of female, one-fifth to one-fourth as long as the inferior appendages, and one-half to three-fifths as long as the dorsum of segment 10 . . . . *B. grafiana*.

B. grafiana also differs from *B. vinosa* in the slightly stouter inferior abdominal appendages, which are less incurved at the tips, and in the slightly larger size as shown by the following measurements:

B. vinosa.—Length of body 34-36.5; hind wing 6-7.5; hind femur 5-6; width of head 7.5-8.

B. grafiana.—Length of body 37-39; hind wing 7.5-8; hind femur 6-6.5; width of head 8-8.5.

In coloration the nymphs of these two species are quite similar, except that the pale, wavy, dorso-lateral streak on each side of the abdomen is usually quite distinct in *grafiana*, but more or less obscure in *vinosa*. In both species the depth of coloration varies considerably, usually being a rather dark brown. All the nymphs from the Go Home Bay district are very dark in colour, but the pale bands of the abdomen and legs are quite sharply defined. The most characteristic mark of Boyeria nymphs is a pale oval or diamond-shaped median blotch in the dorsum of segment 8.

Neurocordulia yamaskanensis (Prov.) Selys.

The nymphs of this interesting species are common at Go Home Bay and in the Muskoka Lakes district. They cling to the under sides of boulders, along the more exposed rocky shores or near rapids. Exuviæ are often found on precipitous rocks rising out of the water from a depth of 8 or 10 feet.

Nymph (figs. 15–17) short-legged and somewhat stouter than most Corduliine nymphs. Head broadly convex above and on the sides, eyes not very prominent, frontal ridge with a scurfy pubescence, the anterior margin convexly curved, hind angles of head prominent, distance between them a little greater than half the greatest width of the head; hind margin distinctly excavate.

Labium extending very slightly behind the bases of the front legs; mentum somewhat broader at the distal margin than long, the middle lobe somewhat abruptly deflexed, bluntly obtusangulate; mental setæ 9-11, the innermost 3 or 4 much smaller than the others; lateral lobes triangular, their distal margins produced into seven semi-elliptical teeth; lateral setæ 6; movable hook very slightly arcuate.

Marginal ridge of pronotum produced on each side behind the posterior angles of the head as a prominent process, which is somewhat smaller than the very prominent supra-coxal processes.

Legs short, the length of hind femora being slightly less than the width of the head.

Abdomen ovate, its greatest breadth at segs. 6 or 7, slightly greater than two-thirds of its length; curve of the lateral margins somewhat stronger in the distal than in the proximal half; lateral spines on 8 and 9, in each case about one-third to one-half as long as the corresponding segment, those on 8 strongly divergent, on 9 parallel and extending caudad scarcely or not at all beyond the tips of the appendages.

Dorsal surface rather strongly convex, dorsal hooks present on 1–9, those of the basal segments slender, nearly erect and slightly hooked, becoming gradually broader and lower caudad, and, on 7–9, reduced to scarcely more than a short ridge. Superior appendage triangular, equilateral, very slightly shorter than the somewhat divergent inferior appendages and somewhat longer than the lateral appendages.

Colour yellowish or orange-brown, variegated with dark-brown. Head dark brown above, generally somewhat paler in the centre and on the frontal ridge. Thorax and wing-cases variegated with pale and dark markings, femora and tibiæ dark, with two pale rings—a median and an anteapical. Abdomen yellowish brown, more or less distinctly blotched with dark brown, especially on the dorsal hooks, the lateral margins and spines and the dorso-lateral scars.

Measurements.—Length of body 22-24.5; mentum of labium 4; hind wing 6-7; hind femur 5.5-6; width of abdomen 9-10; width of head 6.5.

The nymph of this species shows the following differences from that of *N. obsoleta* (Figs. 18-19), two exuviæ of which I have from Lake Hopatkong, Pa., received from Professor P. P. Calvert.

Somewhat larger, more elongate, and less depressed; eyes somewhat less prominent, mentum of labium a little longer and more narrowed at base; middle and hind legs somewhat less widely separated at their bases; abdomen narrower, the sides less strongly curved on the middle segments; lateral spines on seg. 9 much shorter than those of *obsoleta*, in which they are fully as long as the segment and extend far beyond the tips of the appendages; dorsal hooks also less developed than in *obsoleta*, in which they form quite prominent tubercles on segs. 7-9.

Tetragoneuria spinigera Selys.

We have reared only two females of this species, these emerging on June 2, 1912, at a time when the period for transformation was about over. We also found a teneral male with its exuvia on June 1 and a large number of similar exuvia, which must belong to *T. spinigera* as *T. cynosura simulans*, the only other species resident in the Go Home Bay district, does not appear until a little later in the season.

A careful comparison was made between the exuviæ of these two species, but no differences could be detected between them. except that in *spinigera* the lateral abdominal appendages average slightly longer than those of *cynosura*. This difference, however, does not appear to be constant. Prof. Needham, who referred certain nymphs to this species by supposition, employed as differential characters the length and amount of divergence of the lateral spines of segment 9. The two species discussed here are quite alike in respect to these features, which vary considerably among individuals of the same species.

Somatochlora semicircularis (Selys).

I have a teneral imago of this species with the exuvia, taken by Dr. A. G. Huntsman from a pond on Mt. Benson, near Nanaimo, B.C., on July 21, 1909. Dr. Huntsman states that both nymphs and imagoes were common here.

Nymphs (Figs. 26, 27): Eyes rather small, but fairly prominent; frontal ridge with numerous coarse hairs, its front margin gently convexly curved; antennæ with the basal segment slightly shorter and stouter than the second, the third slender and nearly as long as segs. 1 and 2, equal in length to 4.5 and to 6, 7 slightly shorter.

Head but little narrowed behind the eyes, the sides nearly straight and about one-third as long as the posterior margin, which is but little excavated; postero-lateral angles subrectangulate, rather prominent, with numerous hairs.

Labium extending back barely to the bases of the middle legs, apical width of mentum about equal to its length, sides slightly flaring distally; middle lobe moderately deflexed, obtusangulate, the margin minutely crenulate and spinulose; lateral lobes concave within, their inner margin with minute spinules of two or three sizes, distal margin with 9 or 10 well-marked, obliquely-cut, apically rounded teeth, each bearing a tuft of 3 or 4 spinules, of which the innermost is the largest. Mental setæ 10 or 11; lateral setæ 6; end-hook scarcely longer than the second antennal segment.

Lateral margin of pronotum and supracoxal process hairy, the former somewhat produced but rounded, the latter not very prominent.

Legs decidedly short, the hind femora being no longer than the hind wing-cases, rather stout and fringed with moderately long hairs.

Abdomen elongate-ovate, about as wide as the thorax, but little flattened, expanding but little from the broad base to seg. 5, narrowing very slightly on 6 and 7, more rapidly on 8 and 9; lateral margin fringed with hairs, which are long and dense, on 8 and 9; much shorter on the other segments. Lateral spines present only on seg. 9, somewhat less than one-third the length of the segment, slender, subparallel, their sharp tips slightly incurved. Dorsal hooks wholly absent. Superior appendages triangular, slightly longer than its basal breadth, acuminate, apex slender-pointed; lateral appendages scarcely longer, flattened, their basal breadth nearly half that of the superior appendage, tapering to a point, outer margin gently curved; inferior appendage slender, slightly divergent, extending a little beyond the laterals.

No trace of a colour-pattern is visible in the exuvia.

Measurements.—Length of body 32; abdomen 13.5; hind wing-case 6.3; hind femur 6.3; width of head 6; width of abdomen 8.

The nymphs of this species differ from the other known nymphs of this genus in the absence of any trace of dorsal hooks. The head is much less narrowed behind the eyes, the postero-lateral angles more prominent and angulate and the legs decidedly shorter than in *S. williamsoni* Walk. and *S. metallica* Vand., the only other species of this genus whose nymphs I possess. It differs in the same characters from the nymph of *Cordulia shurtleffi*, to which it bears a considerable resemblance. There seem to be no very good generic characters for the separation of the nymphs of *Somatochlora*, *Cordulia* and *Dorocordulia*.

Leucorrhinia frigida Hagen.

This species is exceedingly abundant in all swamp waters in the Go Home Bay region, particularly in sphagnum-bogs. We have found the nymphs in large numbers and have reared many specimens.

Needham's description* of the nymph of this dragonfly belongs to another species (probably *L. proxima*). In a letter to the writer he stated that the species had not been reared, but that tenerals of *L. frigida* had been found at the spot where the exuviæ were gathered. The nymph of *frigida*, unlike Needham's species, possesses large dorsal hooks like the other species of Leucorrhinia that have been reared.

Nymph (Figs. 21-23) very similar to that of *L. intacta*, but somewhat smaller and the legs slightly slenderer. Head similar to that of *intacta*, except in the somewhat more prominent eyes; Labium of similar size and form, the lateral lobes somewhat more

*Bull. 24, New York State Museum, Ent. 23, p. 196, 1908.

deeply concave within, the teeth on the distal margin obsolescent. crenate, each with a single spinule; lateral setæ 9 or 10; mental setæ 10-13, the fourth or fifth from the outside longest, the inner four smaller than the others.

Abdomen broadest at seg. 6, scarcely narrowing on 7, slightly on 8, more abruptly on 9; lateral spines on 8 one-half to three-fifths as long as the segment, subparallel, those on 9 reaching about to the tips of the inferior appendages, their inner margins straight and parallel. Superior appendages somewhat less elongate than in *intacta*, acuminate, about twice as long as the lateral appendages. Dorsal hooks on segs. 3-8, larger on 3 and 4 than in *intacta*, less erect and more curved, very slender; those on 5-7 of about the same size as in *intacta* or somewhat larger and slightly more elevated, the curve of the upper margins much stronger proximally, the apices sharp and directed straight back, reaching about the middle of the following segment; on 8 similar to those of the preceding segments but less elevated, directed straight back.

The coloration when well marked is so exactly similar to that of *intacta* that it seems unneccessary to describe it. It is usually, however, rather obscure, though the legs are always distinctly banded.

Measurements. — Length of body 15-16; abdomen 9-10.6; hind wing 4-6-4.75; hind femur 4; width of abdomen 6-6.8; width of head 4.7-4.8.

EXPLANATION OF PLATES I. AND II.

Fig. 1. *Nehalennia gracilis.*—Lateral gill.
Fig. 2. *Nehalennia gracilis.*—Hind margin of head.
Fig. 3. *Nehalennia irene.*—Hind margin of head.
Fig. 4. *Enallagma calverti.*—Dorsal view of head.
Fig. 5. *Enallagma calverti.*—Lateral gill..
Fig. 6. *Enallagma hageni.*—Dorsal view of head.
Fig. 7. *Enallagma signatum.*—Dorsal view of head.
Fig. 8. *Enallagma signatum.*—Lateral gill.
Fig. 9. *Enallagma pollutum.*—Dorsal view of head
Fig. 10. *Enallagma pollutum.*—Lateral gill.
Fig. 11. *Boyeria grafiana.*—Labium.

Fig. 12. *Boyeria grafiana.*—Abdominal appendages of female nymph.

Fig. 13. *Boyeria vinosa.*—Labium.

Fig. 14. *Boyeria vinosa.*—Abdominal appendages of female nymph.

Fig. 15. *Neurocordulia yamaskanensis.*—Nymph.

Fig. 16. *Neurocordulia yamaskanensis.*—Terminal abdominal segments of female nymph.

Fig. 17. *Neurocordulia yamaskanensis.*—Lateral view of abdomen.

Fig. 18. *Neurocordulia obsoleta.* — Terminal abdominal segments of female nymph.

Fig. 19. *Neurocordulia obsoleta.*—Lateral view of abdomen.

Fig. 20. *Leucorrhinia intacta.*—Dorsal view of head.

Fig. 21. *Leucorrhinia intacta.*—Terminal abdominal segments of female nymph.

Fig. 22. *Leucorrhinia intacta.*—Lateral view of abdomen.

Fig. 23. *Leucorrhinia frigida.*—Dorsal view of head.

Fig. 24. *Leucorrhinia frigida.*—Terminal abdominal segments of female nymph.

Fig. 25. *Leucorrhinia frigida.*—Lateral view of abdomen.

Fig. 26. *Somatochlora semicircularis.*—Dorsal view of head.

Fig. 27. *Somatochlora semicircularis.*—Terminal abdominal segments of male nymph.

OBITUARY.

Mr. Franklin A. Merrick died December 16th, 1912, at New Brighton, Pa., at the age of 68 years. He was well known as a diligent and successful collector of Lepidoptera, of which he accumulated a large number of species. Though engaged in business pursuits for a great many years, he found time to devote himself to this Department of Entomology, and maintained a correspondence with others of similar tastes in many parts of the continent. His collection, which was large and valuable, is now in the possession of Dr. Barnes, of Decatur, Ill.

THE IMPERIAL BUREAU OF ENTOMOLOGY

BY C. GORDON HEWITT, D.SC., DOMINION ENTOMOLOGIST, OTTAWA.

As the question of international effort and co-operation in the matter of controlling and preventing the spread of insects, which in various ways affect human activity, is occupying the attention not only of entomologists, sanitarians and workers directly occupied in studying these many-sided problems, but also of statesmen and administrators, the formation in connection with the British Imperial Service of an Imperial Bureau of Entomology at the beginning of the present year will undoubtedly interest all concerned in these problems by whom the progress and work will be watched.

This organization is not a sudden development, but a gradual outgrowth of efforts along similar lines which began in the spring of 1909. In March of that year a meeting was called by the Secretary of State for the Colonies at the Colonial Office in London, in which the present writer had the honour to take part, to discuss the formation of an Entomological Research Committee for the purpose of furthering entomological research in the British possessions in tropical and sub-tropical Africa. The chief insects which it was considered desirable to study were those associated with the transmission of disease. In 1909 an Entomological Research Committee of the Colonial Office was appointed by Lord Crewe, then Secretary of State for the Colonies, and it consisted of the chief experts in entomology and tropical medicine in Great Britain and Ireland, with Lord Cromer as Chairman. Its work fell under three divisions—namely, the carrying on of investigations and entomological surveys in tropical Africa, for the purpose of which two travelling entomologists were employed; the determination of entomological material and the publication of the work so accomplished, for which purpose the "*Bulletin of Entomological Research*," a quarterly journal, was started. Through the generosity of Mr. Andrew Carnegie, the Committee was able also to undertake the training of Entomologists for service in the Dominions and Colonies.

On account of the valuable service which was being rendered by the Committee to the African Crown-Colonies and Protectorates, suggestions were made for the enlargement of the scope of the work of the Committee. Accordingly, in June 1911, advantage was taken of the presence in England of the Prime Ministers of the self-governing Dominions, and a Conference was called by the Secretary of State for the Colonies to consider the desirability of further extending the work already begun by securing the co-operation and financial support of the self-governing Dominions and Colonies. By this means mutual assistance could be rendered by the various countries within the British Empire through the medium of a central bureau which would be engaged in the collection and interchange of information in regard to noxious insects. It was unanimously agreed that the establishment of such a central bureau was desirable, as it was realized what valuable assistance it could render in the way of disseminating information and rendering assistance in other ways. Accordingly, a tentative scheme was submitted to the governments of the various self-governing Dominions and Colonies for their consideration.

After due consideration, a further Conference was held at the Colonial Office in August, 1912, to which the government entomologists of the self-governing Dominions and Colonies, and others similarly interested, were invited to discuss and work out a scheme for Imperial co-operation in preventing the spread and furthering the investigation of noxious insects. At this Conference the whole subject was thoroughly discussed, and a proposal was evolved for the establishment of an Imperial Bureau of Entomology to be financially supported by the various Dominions and Colonies and the British Government.

It was proposed that the functions of the Imperial Bureau of Entomology be as follows:

1. A general survey of the noxious insects of the world and the collection and co-ordination of information relating thereto, so that any British country may learn by inquiry what insect pests it is likely to import from other countries and the best methods of preventing their introduction and spread.

2. The authoritative identification of insects of economic importance submitted by the officials of the Departments of Agriculture and Public Health throughout the Empire.

3. The publication of a monthly journal giving concise and useful summaries of all the current literature which has a practical bearing on the investigation and control of noxious insects.

The scheme was accepted by the various self-governing Dominions and Colonies which were invited to co-operate, and the Crown-Colonies and British Protectorates will also participate in the advantages of the Imperial Bureau of Entomology which has now been established. The former Entomological Research Committee has become the Honorary Committee of Management, with the eminent administrator, the Earl of Cromer, as President, and the Scientific Secretary of the Committee, Mr. Guy A. K. Marshall, has been made Director of the Bureau and Editor of the Journal. The Government Entomologists of the Dominions are *ex-officio members* of the Committee of Management.

The publication of the Bureau's journal, which is entitled "*The Review of Applied Entomology*," was commenced in January. It is being published in two parts: Series A, Agricultural; and Series B, Medical and Veterinary. As the organization and library of the Bureau becomes perfected, the value of this journal to entomological workers cannot be overestimated, when it is remembered that there are no less than 1700 periodicals—scientific, agricultural and medical—which may contain articles dealing with entomology, but a small proportion of which widely scattered entomologists have the opportunity of seeing or the time to consult.

An idea of one aspect of the three years' work of the original Entomological Research Committee will be gathered from the fact that the collections received from collectors in tropical Africa and other parts of the world during that time amounted to about 190,000 insects, of which no less than 56,000 were actual or potential disease carriers. The value of this function of the Bureau to entomologists situated in portions of the Empire where there are no collections and little literature to aid them in identification will be realized by their more fortunate fellow-workers.

It has been stated that the Imperial Bureau of Entomology will serve the needs of the British Empire in a manner similar to that in which the United States Bureau of Entomology serves those of the United States. This statement, however, is not correct. Its primary function will be that of an intelligence Bureau, collecting information for the use of British countries supporting it and assisting entomologists and other officials in those countries in the identification of their material. By these methods which have been mentioned and by the publication of "*The Review of Applied Entomology*," it will furnish a means of assistance and of co-ordination of effort in the war against noxious insects, which will undoubtedly soon make its services invaluable in the further development of the countries, and especially the tropical and sub-tropical countries, of the British Empire. International, as the scope of its inquiries are, the work of the Bureau cannot but prove to be one of the most potent factors in enabling us to develop the agricultural and other resources of the Empire, and our fellow-workers in non-British countries can avail themselves, throug this journal, of some of the fruits of the Bureau's work.

GEOMETRID NOTES.—A NEW VARIETY.

BY L. W. SWETT, BOSTON, MASS.

Therina fiscellaria Gn., var. *Johnsoni*, Nov.—Expands 30 m.m Fore wings smoky ochreous instead of being yellow as in the normal form. The fore wings are smoky to the basal band, which shows as a bright ochreous line, crossing from costa to inner margin in a regular curve. The mesial space is smoky, with black discal dot showing faintly through. Extra-discal line bright ochreous, curved from costa to median vein, then back sharply in a deep curve to inner margin, as in normal *fiscellaria*. Beyond the wing is smoky black to outer margin. Fringe short and smoky ochreous. Antennæ and head ochreous; body of the same colour. Hind wings smoky ochreous to extra-discal line, which rounds out to a point opposite the black discal spot, the line being ochreous as on the fore wing. Beyond the extra-discal line the wing is smoky to outer margin. The insect, on the whole, seems rather semi-hyaline in appearance, and is no doubt a melanic form. Beneath the fore wings are much lighter than above, with the markings showing through. Hind wings lighter than above, almost a dark fawn colour, with lines showing through from above.

This seems to be a rare form and quite distinct from any described variety. I take pleasure in naming this variety after my kind friend, Mr. C. W. Johnson, who has rendered me valuable help and suggestions.

Type.—1 ♂ from Dr. C. S. Minot, North East Harbor, Me., Sept. 24, 1909, in the collection of the Boston Society of Natural History.

NOTES ON THE SYNONYMY OF SOME GENERA AND SPECIES IN THE CHLOROPIDÆ (DIPTERA).*

BY J. R. MALLOCH, BUREAU OF ENTOMOLOGY, WASHINGTON, D.C.

Williston, in his "Manual of North American Diptera," 1908, gives to this family the name Oscinidæ. Unfortunately, the generic name *Oscinis* is a synonym of the earlier name *Chlorops*, as indicated in the following synonymy, so that, even had the name of the family not previously been Chloropidæ, the name Oscinidæ could not be retained. Coquillett, in his paper on "The Type-Species of the North American Genera of Diptera," 1910, made some alterations in the status of certain genera in the family, but some of his conclusions are incorrect. Most European authors refuse to accept Lioy's genera, and of those who have dealt with this family in recent years only Enderlein has recognized any of Lioy's genera as valid. While many of Lioy's genera are synonyms of older genera, and his identifications often obviously wrong, it must be apparent to an unbiased person that wherever it is possible to decide definitely what his genera are, and in all cases he cites species, they must be accepted, provided they are in other respects valid. It seems to me that the acceptance of Meigen's genera included in the 1800 paper, and those of his 1803 paper which had no species included in them, by European writers and their wholesale disregard of Lioy's genera savors slightly of inconsistency. Enderlein, in a paper on the subfamily Oscinosominæ (Sitz. d. Ges. Naturf. Freu, 1911), evidently was unaware of the fact that Coquillett had made use of Lioy's genera in 1910 and retained the generic name *Oscinosoma*, which Coquillett sunk as a synonym of *Botanobia*, and reversed the order as given by that writer. Possibly his reason for using the generic name *Oscinosoma* was to retain as the name of the subfamily one which had as near an approach to the old one (*Oscinis*) as possible. This position might be tenable, even though *Botanobia* has line priority, but for the fact that Coquillett had previously indicated *Botanobia* as the generic name to replace *Oscinis* and gave *Oscinosoma* as a synonym. It is regrettable that these questions of nomenclature occur so often, and that they cause such confusion; but, when they do crop up, it is advisable that they should be settled, and when one under-

*Published by permission of the Chief of the Bureau of Entomology.

takes to decide a matter of this kind, it is always best to give the reasons why such decisions are arrived at. Having been engaged upon some work on the American species in this family, I find that it is necessary for me first of all to decide upon the correct nomenclature, both of genera and species, before I can publish any descriptive matter or give identifications of species to be used in the publications of the Bureau of Entomology. Thus, I have undertaken the rather unwelcome task of revising the nomenclature of the group, in so far as the American genera are concerned, in the hope that such revision may be of use to other students of the family.

Family Chloropidæ
Subfamily Chloropinæ.

ELLIPONEURA Loew, Berl. Ent. Zeitschr., Vol. 13, 1869, p. 44.
Type: *Elliponeura debilis* Loew.

MEROMYZA Meigen, Syst. Beschr. Zweifl. Ins., Vol. 6, 1830, p. 163.
Type: *Musca saltatrix* Linné.

CETEMA Hendel, Wien. Ent. Zeit., Vol. 26, 1907, p. 98.
Centor Loew, Zeit. Ent. Breslau, Vol. 15, 1866, p. 7 (pre-occ.)
Type: *Oscinis cerceris* Fallen.

ANTHRACOPHAGA Loew, Zeit. Ent. Breslau, Vol. 15, 1866, p. 15.
Type: *Musca strigula* Fabricius.

HAPLEGIS Loew, Zeitschr. Ent. Breslau, Vol. 15, 1866, p. 22.
Type: *Chlorops diadema* Meigen.

DIPLOTOXA Loew, Zeitschr. Ent. Breslau, Vol. 15, 1866, p. 31.
Type: *Chlorops versicolor* Loew.

CHLOROPS Meigen, Illig. Mag., Vol. 2, 1803, p. 278.
Type: *Chlorops laeta* Meigen.

Coquillett gives this genus as synonym of *Titania* Meigen, 1800, presumably on the strength of Hendel's representations in his paper dealing with Meigen's genera (Verh. Zool. bot. Ges. Wien., Vol 58, 1908, p. 43), but, as Hendel afterwards points out (Wien. Ent. Zeit., Vol. 29, 1910, p. 312), *Titania* is more probably synonymous with *Gaurax* Loew.

CHLOROPISCA Loew, Zeitschr. Entom. Breslau, Vol. 15, 1866, p. 79.
Type: *Chlorops glabra* Meigen.

EURINA Meigen, Syst. Beschr. Zweifl. Ins., Vol. 6, 1830, p. 3.

This genus has been recorded as occurring in America, but the species included in it from this country is not congeneric with the type, *Eurina lurida* Meigen. In the National Museum collection the series of *exilis* Coquillett stands among the members of the genus *Chlorops*, having been removed to that genus by Coquillett.

ECTECEPHALA Macquart, Dipt. Exot. Supp. 4, pt. 2, 1851, p. 280.
Type: *Ectecephala albistylum* Macquart.

Subfamily BOTANOBINÆ.

CERATOBARYS Coquillett, Jour. N. Y. Ent. Soc., Vol. 6, 1898, p. 45.
Type: *Hippelates eulophus* Loew.

HIPPELATES Loew, Berl. Ent. Zeitschr., Vol. 7, 1863, p. 36.
Type: *Hippelates plebejus.* Loew.
Opetiphora Loew, Dipt. Amer. Sept. Indig. Cent. 10, 1872.
Siphomyia Williston, Trans. Ent. Soc. Lond., 1896, p. 418.

CRASSISETA von Roser, Corres. Landw. Ver. Wurtemb., 1840, p. 63.
Type: *Oscinis cornuta* Fallen.

The genus *Elachiptera*, which has as its type *brevipennis* Meigen, does not occur in America, so far as our present information goes.

GAURAX Loew, Dipt. Amer. Sept. Indig. Cent. 3, 1863, p. 35.
Type: *Gaurax festivus* Loew.
Titania Meigen, Nouv. Class. Mouch., 1800, p. 35 (Nom. Nud.)
Macrostyla Lioy, Atti Istit. Veneto, ser. 3, Vol 9, 1864, p. 1126.

Titania has never had any species placed in it and must be considered as "nomen nudum," though there is a possibility that it may have been a species of *Gaurax* Meigen had before him when he wrote the description in his 1800 paper.

MADIZA Fallen, Dipt. Suec, Oscinid., 1820, 8.
Type: *Oscinis oscinina* Fallen.
Siphonella Macquart, Hist. Nat., Dipt., Vol. 2, 1835, p. 584.
Eurinella Meunier, Bull. Soc. Ent. France, 1893, p. 193.

Coquillett gives *Siphunculina* Rondani, as a synonym also, but this is really the genus afterwards described as *Microneurum* by Becker, which does not at present find a place in the American list.

BOTANOBIA Lioy, Atti Istit. Veneto, ser. 3, Vol. 9, 1864, p. 1125.
Type: *Oscinis dubia* Macquart.
Oscinis authors, not Latreille.

Oscinosoma Lioy, Atti Instit. Veneto, ser. 3, Vol. 9, 1864, p. 1125.

? *Strobliola* Czerny, Verh. Zool. Bot. Ges. Wien., Vol. 59, 1909, p. 289.

Oscinella Becker (Bull. Mus d' Hist. Nat. Paris, 1909, p. 119), Arch. Zool. Budapest, I, 1910, p. 150.

Coquillett accepted *Botanobia* as the name to substitute *Oscinis* Latreille, which had been erroneously used by authors as the generic name for that group, the type of which he indicates as given above. Presumably, he did so because the name appears first in Lioy's paper, though it has only line priority over the one adopted in 1911 by Enderlein, as indicated at the beginning of these notes.

Botanobia frit, Linné, Fauna Suec., 1761, p. 1851 (*Musca*).

Musca hordei, Bjerk., Vetinsk. Akad. Handl. 34, 1777 (*Musca*).

Carbonaria Loew, Dipt. Amer. Sept. Indig. Cent., 7, 1866 (*Oscinis*).

The above synonymy is in accordance with facts ascertained from a comparison of American and European material.

TRICIMBA Lioy, Atti Istit. Veneto, ser. 3, vol. 9, 1864, p. 1125.

Type: *Tricimba linella* Fallen.

Notonaulax Becker, Mitth. Zool. Mus. Berlin, 1903, p. 153.

Through Becker disregarding Lioy's work, he did not recognize the fact that that author had clearly defined this genus, and cited as the type of his genus *Notonaulax* one of the two species Lioy included in *Tricimba*.

This genus occurs in America. The species described as *trisulcata* by Adams (Ent. News, Vol. 16, 1905, p. 111) belongs here.

A NEW GENUS AND ONE NEW SPECIES OF CHALCIDOIDEA.

BY A. B. GAHAN, MARYLAND EXPERIMENTAL STATION, COLLEGE PARK, MD.

During the summer of 1912 a series of specimens of a Pteromalid were reared by the writer from cocoons of *Cladius pectinicornis* Fourcr. They were found to run readily to the genus *Cœlopisthia* Foerst. in Dr. Ashmead's "Classification of the Chalcid

Flies." Upon comparison with a specimen in the United States National Museum, of *Cœlopisthia vitripennis* Thoms., one of the two European species of the genus (not the genotype species), they were found to differ materially. Unfortunately, specimens of the genotype species, *C. cephalotes* Thoms., are not available for comparison, but there seems no reason to doubt that this species and *C. vitripennis* are congeneric. A new genus is therefore erected for the reception of the parasite of *Cladius pectinicornis*, which appears to be undescribed.

Cœlopisthia nematicida (Pack) Hewitt and *C. diacrisiæ* Crawf., being congeneric with the new species, are also included. *C. fumosipennis* Gahan is a true *Cœlopisthia* and the only described North American representative of that genus. *C. smithii* Ashm. (manuscript name in Smith's Insects of New Jersey, 1900, p. 559) does not belong in the tribe Pteromalini since one mandible is 3-toothed and the other 4-toothed. It therefore falls in the tribe Eutilini and does not appear to fit any genus in that tribe.

The new genus is distinguished from all except *Cœlopisthia* in the tribe Pteromalini by the immargined occiput, non-produced propodeum, subequal stigmal and postmarginal veins, and the long antennal pedicel. From *Cœlopisthia* it may be distinguished as follows:

Both antennal ring-joints elongate, as long or longer than broad; discal cilia of the anterior wings reduced to mere dots or punctures, the hairs obsolete; marginal vein nearly three times as long as the stigmal; abdomen short, rotund . *Cœlopisthia* Fœrster.

First ring-joint strongly transverse, the second as long or nearly as long as broad; discal cilia developed on the apical two-thirds of the wing at least; marginal vein scarcely twice the length of the stigmal; abdomen ovate or conic ovate. . . . *Cœlopisthoidea*, n. g.

Cœlopisthoidea, new genus.

Head large, much wider than thorax, broad anterio-posteriorly, occiput concave, the occipital forminal depression angularly defined but immargined. Antennæ 13-jointed, inserted on a line with the lower extremities of eyes; scape slender; pedicel longer than the first joint of funicle; two ring-joints, the first transverse, the second elongate, much longer than the first; funicle 6-jointed cylindrical; club 3-jointed, acuminate. Face below the antennæ receding;

mandibles both four-toothed. Parapsidal furrows subobsolete on the posterior half of the mesonotum, impressed anteriorly; scutellum large, moderately convex; propodeum not prolonged into a neck, the median longitudinal carina and lateral folds present, spiracles prominent long-ovate. Wings hyaline, the marginal vein about twice as long as the stigmal, the postmarginal and stigmal subequal, marginal cilia present but short. Posterior tibiæ with one spur. Abdomen sessile, ovate or conic ovate.

Type of genus—*Cœlopisthoidea cladiæ*, n. sp.

KEY TO THE SPECIES OF CŒLOPISTHOIDEA.

1. Postmarginal vein slightly shorter than the stigmal vein. Anterior wings slightly dusky............*diacrisiæ* Crawf.
   Postmarginal vein not shorter than the stigmal; anterior wings hyaline.
2. Propodeum with a deep circular fovea on either side of the short apical neck; lateral ocelli as far from the eye margin as from each other; anterior wing apparently without marginal cilia..................................*nematicida* Pack.
   Propodeum without deep fovea either side of the apex; lateral ocelli as close to the eye margin as to the anterior ocellus, much closer to the eye than to each other; anterior wings with weak marginal cilia on the posterior margin toward the apex.............. ....*cladiæ*, n. sp.

Cœlopisthoidea cladiæ, n. sp.

Female.—Length about 2.5 mm. Head and thorax æneous, closely reticulate-punctate; scape reddish testaceous, the pedicel and flagellum dark brown, pedicel longer than the two ring-joints and first joint of funicle combined, first ring-joint transverse, second as long as broad; funicle joints not longer than broad, the apical ones not as long as broad; club acuminate, three-jointed, the joints about as long as the funicle joints. Ocelli in an obtuse angled triangle, the lateral ocelli nearer the eye margin than to the anterior ocellus.

Punctures of the mesoscutum somewhat smaller and deeper than those of the head, apical portion of the scutellum differently sculptured from the anterior portion, giving the appearance of a transverse line before the apex; true metanotum punctate; propodeum punctate, the lateral folds distinct and complete, median carina also well defined, spiracles long-ovate and prominent; neck of the propodeum short, smooth and shining and without a distinct circular depression either side. Wings hyaline, the postmarginal as long as the marginal, marginal cilia of the anterior wings absent, except for a very few weak cilia on the posterior margin toward the apex; basal portion of the anterior wing to the apex of the costal cell hairless, except for a single row of hairs in the costal cell, remainder of the wing ciliate but with the hairs very short. Anterior and posterior coxæ more or less metallic on the outer side; median pair brownish; all trochanters, femora, tibiæ, and tarsi pale testaceous, the femora and tibiæ tinged with brownish. Abdomen ovate, pointed at the apex, smooth and shining, dark brown, with the basal segment metallic.

Male.—Coloured like the female, but a brighter green, with stronger reflections; antennæ shorter than in the female, the joints of the funicle not as long as broad, the club short and compact.

Type locality.—Upper Marlboro, Prince George County, Md.

Host.—Cladius pectinicornis.

Type.—Cat. No. 15,506, United States National Museum.

Thirty females and three males in the type series. The type and several paratypes deposited in the United States National Museum. Remaining paratypes in the collection of the Maryland Experiment Station, College Park, Md.

Mr. E. N. Cory, of this Department, brought me several pupæ of the sawfly which he had secured on rose bushes at the farm of Mr. R. S. Hill, Upper Marlboro, Md., August 6, 1912. At the same time he turned over to me a single live female of the parasite which he had taken crawling over the sawfly cocoons. This parasite and the sawfly cocoons were placed together in a vial on my desk. The parasite died and was pinned August 12, without having been observed to oviposit. August 19 there emerged in the vial thirteen specimens of the parasite. Examination of the cocoons on this date showed that all these parasites had come from a single sawfly pupa. One of the remaining cocoons was found to be packed full of the naked pupæ of the parasite, which at this time were pale-yellowish, with the eyes dark-red, and measured a little over 2 mm. in length. August 27th, adults to the number of 20 emerged from this lot of pupæ. While proof is lacking, it seems probable that this last lot of parasites were from eggs deposited by the captured female, either just before or shortly after her capture. *C. nematicida* is said by Hewitt* to be able to develop from egg to adult within a period of twenty-three days.

SPECIES OF LEPIDOPTERA NEW TO OUR FAUNA, WITH SYNONYMICAL NOTES

BY WM. BARNES AND J. MCDUNNOUGH, PH.D., DECATUR, ILL.

In working over some material in the Barnes Collection we have come across several species unrecorded from the United States. As the localities are authentic, we think it wise to note their occurence. We are indebted to Dr. Skinner for several of the determinations.

Diurnals.

Synchloe endeis G. & S.

Synchloe endeis Godman & Salvin, Ann. Mag. Nat. Hist., (6) XIV., p. 97; id., Biol. Cent. Am. Rhop., II., 673, Pl. 108, figs. 5 and 6 (1901).

We have before us 1♂ labelled "Texas" and 1♀ much worn from Edwards Co., Texas, May 1902, received from Mr. H. Lacey, of Kerrville.

Myscelia ethusa Bdv.

Cybdelis ethusa Boisduval in Cuv. Rig. An. Ins. Atl. II., t. 138, fig. 3.

Myscelia cyanecula Felder, Reise Nov. Lep. 408, t. 53, f. 5.

Myscelia ethusa Godman & Salvin, Biol. Cent. Am. Rhop. I., p. 232 (1883).

One very perfect ♂ specimen from Brownsville, Texas, captured Oct. 15th.

Lasaia agesilas narses Staud.

Lasaia narses Staudinger, Exot. Schmett. I., p. 257 (1888): Stichel, Berl. Ent. Zeitsch. 55, p. 48 (1910); id. Gen. Insect. Riod., p. 187 (1911).

Two specimens from Brownsville, Texas, April 11th and June 11th (G. Dorner). We have not seen the original description of this species; but, according to Stichel's short diagnosis, they would seem to be best placed under this name. They certainly do not

*CANAD. ENT., XLIII., 1911, p 302.

agree very well with the figure in the Genera Insectorum of *sula* Stdr. which Stichel records from Texas.

Thecla pastor Butl. & Dru.

Strymon pastor Butler & Druce, Cist. Ent. I., p. 105.

Thecla pastor Godman & Salvin, Biol. Cent. Am. Rhop. II., 34, Pl. 52, figs. 8–10.

Five ♂s and one ♀ from Brownsville, Texas, (May–June). The brown marginal lunules of secondaries on under side point to this species, but the subterminal white broken line is stronger in our specimens than in the figure in the Biologia.

Thecla azia Hew.

Thecla azia Hewitson, Ill. Diur. Lep. 144, Pl. 57, figs 357-8; Godman & Salvin, Biol. Cent. Am., Rhop. II., 91.

One ♀ from Paint Creek, Edwards Co., Texas, received through Mr. H. Lacey. The red marginal line of under side of both wings is characteristic; the species is related to *clytie* Edw., but is without doubt distinct.

Thecla cestri Reak.

Thecla cestri Reakirt, Proc. Acad. Phil. 1866, p. 338. Godman & Salvin, Biol. Cent. Am. Rhop., p. 96, Pl. 58, figs. 12-13.

One ♀ from Brownsville, Texas, Oct. 15th (G. Dorner). The maculation of the underside of secondaries is brown, rather than black, as given by Godman & Salvin; but more material is needed before deciding whether the Texan form represents a geographical race or good species.

Cogia calchas H. S.

Eudamus calchas Herrich Schaeffer, Prodr. III., p. 68 (1868).

Cogia calchas Godman & Salvin, Biol. Cent. Am. Rhop. II., p. 340.

A series of both sexes from Brownsville, Texas, and San Benito, Texas, taken in July and October, is before us.

Xenophanes tryxus Cram.

Papilio tryxus Cramer, Pap. Exot. Pl. 334, figs. G. H.

Xenophanus tryxus Godman & Salvin, Biol. Cent. Am. Rhop. II., p. 387.

Two ♂s and two ♀s are before us, collected at Brownsville, Texas, in May and July.

Noctuidæ.

Oxycnemis dunbari Harv.

Hadena dunbari, Harvey, Can. Ent. VIII., 52 (1876).

Litholomia dunbari, Smith, Bull. 44, U.S.N.M., 226 (1893).

Oxycnemia definita Barnes and McDunnough, Cont. Nat. Hist. N. Am. Lep., I, (5) 17, Pl. II., fig. 3 (1912).

Mr. Wolley-Dod recently called our attention to the fact that our *definita* was probably synonymous with *dunbari* Harv. An examination of Harvey's type in the Edward's Coll. by Dr. Barnes showed a strong claw on fore tibia, which had evidently been overlooked by Smith, and confirmed Dod's suspicions. The species certainly is no *Litholomia*; we place it doubtfully in *Oxycnemis* for the present. Hampson's figure, which misled us, (Cat. Lep. Phal., pl. 100, fig. 31) does not represent this species at all, but is a strongly marked form of *napaea* Morr.

Ozarba fannia Dru.

Eustrotia fannia Druce, Biol. Cent. Am. Het. I., 313, Pl. 29, fig. 12 (1889).

Ozarba fannia Hampson, Cat. Lep. Phal. X., 451 (1910).

Numerous specimens before us from San Antonio, Texas, Black Jack Spgs., Texas, and Kerrville, Texas, agree with Hampson's generic definition and correspond fairly with Druce's figure. The species had been misidentified for us by J. B. Smith as *Thalpochares ætheria* Grt., which, according to Hampson, who has the type before him, is generically distinct.

Eustrotia catilina Dru.

Eustrotia catilina Druce, Biol. Cent. Am. Het. I., 312, Pl. 29, fig. 5 (1889); Hampson, Cat. Lep. Phal. X., p. 598 (1910).

One ♂ and two ♀s from Shovel Mt., Texas; San Benito, Texas; Texas (Rauterberg).

Palindia micca Dru.

Palindia micca Druce, Biol. Cent. Am. Het. I., 319, Pl. 29, fig. 5 (1889).

A single ♂ in very fresh condition from San Benito, Texas, corresponds well with Druce's figure; shows, however, only traces of the terminal dark shading.

Megalopygidæ.

Megalopyge lapena Schaus.

Megalopyge lapena Schaus, Jour. N.Y. Ent. Soc., IV., 58 (1896).

Gasina lapena Druce, Biol. Cent. Am. Het. II., p. 432, Pl. 86, fig. 13 (1897).

Three ♂s and one ♀ of this species, taken in Chiricahua Mts., Ariz., and Palmerlee, Ariz. (Aug.), are before us.

Pyraustinæ.

Edia semiluna Sm.

Lythrodes semiluna Smith, Can. Ent. 37, p. 67 (1905).

Cynaeda bidentalis Barnes & McDunnough, Cont. Nat. Hist. N. Am. Lep. I, (5) 33 (1912).

Edia microstagma Dyar, Proc. U.S. Nat. Mus. 44, p. 320 (1913).

A recent study of the unique type of *semiluna* Sm. proves conclusively that Smith's generic reference was faulty, as the species is a Pyraustid and identical with *bidentalis* B. & McD. There seems but little doubt that Dyar has redescribed the same species, creating the new genus *Edia* for its reception; as a new genus is probably necessary, the synonymy will be as stated above.

Noctuelia castanealis Hlst.

Orobena castanealis Hulst, Tr. Am. Ent. Soc. XIII., 157 (1886).

Thalpochares jativa Barnes, Can. Ent., 37, p. 213 (1905).

An examination of the type specimen of *jativa* shows it to be a Pyraustid and, without much doubt, identical with *castanealis* Hlst., although we have not seen the type of this latter species.

MR. NORMAN CRIDDLE, of Treesbank, Manitoba, has been appointed a Field Officer of the Division of Entomology, Ottawa, to carry on investigations in Southern Manitoba.

MR. L. S. MCLAINE, M. Sc., has been appointed a Field Officer of the Division and is now engaged, through the courtesy of Dr. Howard, in the rearing and collection of the parasites and predaceous enemies of the Brown-tail and Gipsy Moths in Massachusetts, in connection with the work of establishing the same in New Brunswick.

FURTHER NOTES ON ALBERTA LEPIDOPTERA.

BY F. H. WOLLEY DOD, MIDNAPORE, ALTA.

(Continued from page 134.)

378. *Parastichtis discivaria* Walk.—This species is correctly named, and Sir George Hampson changed his opinion as to the distinctness of *gentilis* before publishing. Walker's type is from St. Martin's Falls, Hudson's Bay Territory, and is the strongly marked contrasting form, with pale luteous inner and postmedial areas. Type *perbellis*, from Evans Centre, N.Y., which Hampson makes "ab. 1." is similarly strongly marked, but more even in shade, and lacks the contrastingly pale areas. This is the form figured by Holland. *Gentilis*, from the same locality, is even red-brown, with indistinct maculation. All three forms occur here, and intergrade.

381. *Homoglæa hircina* Morr.—This has been rather common in recent years. I have never seen it in the fall, but it appeared in some numbers in the end of March, 1910, which I thought unusually early. This year however a few were seen at light during a mild spell on the 4th or 5th of March. A fortnight later the thermometer fell to about 15° below zero. It is a strikingly variable species, some of the forms being very pretty. The colour varies from a rather pale reddish luteous to dark chocolate brown. A handsome grey irroration is variably present or absent. Some are practically immaculate; others have the usual geminate cross lines of darker shades filled in with the ground colour, or with grey, the spots also sometimes outlined with grey. Sometimes most of the veins are grey lined. A rare form has black punctiform spots in the s.t., and still more rarely in the t.p. line also. A well defined median transverse shade sometimes exists, and generally runs through the middle of the reniform.

383. *Ipimorpha pleonectusa* Grt.—The type in the British Museum is a male from Evans Centre, N. Y. according to the Catalogue, and the eastern form seems to have reddish brown tints not possessed by specimens from Manitoba and Alberta, which Hampson makes "Ab. 1. Paler, and less red." Dr. Dyar, in the Kootenai List, says that both forms occur at Kaslo, and calls the light clay-coloured one "var. *æquilinea* Smith." Smith refers to

this as a mere synonym in his Catalogue, but I have seen neither type nor description.

384. *Dasyspoudœa meadii* Grt.—High River (Baird).

385. *Copablepharon sp.*—This is not *absidum*, nor apparently any described species. I have seen a few other specimens besides the two previously referred to, but have only a single female in my collection, and do not care to describe from it. Mr. Baird has taken it at High River, and I have also seen a female taken by Mr. C. Garrett in Calgary on Aug. 1st, 1907. The primaries are very pale green, with slight fuscous irroration, and secondaries white, fuscous clouded centrally.

388. *Melaphorphyria oregonica* Hy. Edw.—*Oregona* of our lists is apparently a mis-spelling, as was also my previous rendering (Dr. Dyar's) of the generic name.

389. *Melicleptria septentrionalis* Hy. Edw.—Hampson makes this synonymous with European *ononis* Schiff. The type of *septentrionalis* is a male in the Neumœgen collection at Brooklyn, and is labelled "N.W.B.C."

390. *Heliaca nexilis* Morr.—Rather common at timber line in the mountains. My records are: Brobokton Creek, Wilcox Pass, and Sheep Mountain, July 10th to 22nd (Mrs. Nicholl); Mt. St. Piran, near Agnes Lake, Laggan, 72-75,000 feet, July 17th and 18th. This is the *nexilis* of the British Museum, Rutgers College, and Washington collections.

Var. *elaborata* Hy. Edw.—One male, Head of Pine Creek, June 9th, 1897. High River, June 10th, 1909, two ♀s (Baird). I have seen other specimens taken by Mr. Baird. This form is the *diminutiva* of my former notes on Smith's authority. The error was excusable, as the two are not unlike. But *diminutiva* has truncate frontal prominence, which this has not, and differs in colour and maculation as well. Holland's figure under *diminutina* appears to be *Melicleptria persimilis* Grote, a species with rounded frontal prominence and spined tibiæ. Superficially *persimilis* happens to bear a much closer resembalnce to *elaborata* than to *diminutiva*, but has an additional white spot near base of secondaries.

One of my *elaborata* I have compared with the type, a Colorado female in the Henry Edwards collection. The main features in which it differs from what I take to be the true *nexilis* is that the head, thorax and primaries are strongly overlaid with yellowish, giving the impression, against the black ground, almost of bronzy green. In my three specimens, in addition to the yellowish or whitish band on primaries, there is a small yellowish mark in the cell before the orbicular. There is a trace of this in one only out of my ten mountain *nexilis*, and I notice it exists in Hampson's wood-cut of *nexilis*. After describing *nexilis*, Sir George Hampson gives "Ab. 1. *elaborata*, fore wing without the white spot in cell before the reniform." This is not in accordance with my notes on the type in the New York Museum, but I may possibly have overlooked this difference, which my specimens do not have. Several of my mountain *nexilis* lack this spot. In the British Museum collection an *elaborata* label is placed beside a Washington Forest Reserve specimen, which I should have called typical *nexilis*.

For a long time I was inclined to consider *elaborata* distinct, as I found it hard to believe that a species should occur here on the plains, and in the mountains, in so far as I had observed, at the timber line only. All the B.C. records I can find appear to be from mountains, elevation not given. So closely does *elaborata* resemble *persimilis* that I suspected the existence of tibial spines in the former. But I recently removed, bleached and mounted all the legs of my whole series of *nexilis* and *elaborata* without succeeding in finding a single spine on any tibia. I must admit that the differences between these two latter, such as they are, are very slight, and the observed variation suggests that with more material the forms may be found to overlap. What has made the matter still more interesting is that Mr. A. F. Winn and others have recently discovered *nexilis* at St. Hilaire, which is close to sea level in Quebec. I am indebted to Mr. Winn for a specimen, and except that it has rather less pink on secondaries beneath, it is practically a dead mate for one of my Mt. St. Piran timber line specimens. In Quebec, Mr. Winn says that the species flies in the middle of May. That is two months earlier than the mountain dates, but is probably easily explained by the altitude.

391. *Polychrisia trabea* Smith.—During 1910 I took this species in some numbers at flowers of wild larkspur, on which the larva in all probability feeds.

392. *P. purpurigera* Walk.—Edmonton, July 14th, 1910 (F. G. Carr.)

395. *Euchalcia putnami* Grote.—The type is a female in the British Museum. There I found North American and European specimens associated as one species under the name *festucæ* Linn, and my notes say that the reference is apparently justified, as the European species varies to *putnami*. As a rule, the European form is darker and richer in colour and has a golden metallic spot at the base of the costa which *putnami* generally lacks. Another character not usually found in *putnami* is a metallic outer edging to the t.a. line below the median vein. In *putnami* the two central metallic spots are sometimes joined. I am not sure whether this is ever the case with *festucæ*. At any rate, such variation is rare in Europe. Vancouver Island specimens vary very much nearer to typical *festucæ* than do my local series. Some have the rich dark coloration, the metallic marks at base, and on the t.a. line; but the inner one of the two central spots less frequently extends a little above the median vein than it does in Alberta specimens, or than appears to be the case in *festucæ*.

398. *Autographa californica* Speyer.—The most important distinctive mark between this species and *pseudogamma* I had overlooked in my previous paper. *Californica* has a fine black longitudinal streak anterior to the subterminal line near the apex, which usually reaches, or very nearly reaches, the t.p. line. In *pseudogamma* this streak is non-existent. It exists in *ou*, which resembles *californica* rather closely in pattern, though unquestionably distinct. As *ou* has quite recently been added to the Canadian list, on the strength of a specimen taken at Aweme by Mr. Criddle, a comparison with *californica* may be of special interest. In *californica* the t.p. line is somewhat deeply sinuate near the inner margin. If viewed with the outer margin of the wing upwards, that portion of the line below vein 2 has the shape of a written "n" with the top of the first stroke rather pointed. In *ou* this portion of the line is very slightly waved only. In *californica* the sign is usually

of the well-known Greek "gamma" fcrm, but is sometimes formed of two separate marks. The inner one is roughly V-shaped, with the strokes out-curved. The outer mark is a lobe-shaped dot, which joins or tends to join the V at its apex—that is, at the point nearest the inner margin of the wing. In *ou* the inner mark is more U-shaped, and the outer spot is approximate to, and sometimes joins, it at a point nearest the outer margin. There are other differences in colour and maculation, but these are the most obvious. The difference in the sign is well shown in Ottolengui's figures. As a matter of fact, I find the signs in many Autographas much more variable than I had been led to suppose from first perusals of Ottolengui's paper. The most obvious structural differences between these two species is that *ou* has hind tibiæ strongly spined, whilst *californica* has not.

Holland has his figures of the above three species badly mixed. On plate XXVIII., fig. 25, as *rogationis*, represents *ou*, whilst fig. 33, called *ou*, is of *pseudogamma*, and fig. 35, called *pseudogamma* is obviously *californica*.

The question as to the true status of closely allied forms separated by wide stretches of ocean will probably always give rise to controversy. The best way of dealing with the matter is probably to treat such forms as distinct, unless exactly similar specimens can be found on both continents. I am not aware that similar specimens have been found of our *californica* and European *gamma*, and therefore prefer to treat them as distinct. One difference in pattern appears to be that the upper portion of the t.p. line is more crenate in *californica*. All the maculation in our species is more clearly written, and shows greater contrasts. *Gamma* has the black streaks near the apex, but it is less developed than in *californica*. The sign is about similar in the two, and both have unarmed tibiæ. But in general color of primaries *gamma* is darker and more even, and much more like normally coloured *ou*.

Grote in Can. Ent. XXXV., p. 238, Aug. 1903, states that *ou* and *fratella* are distinct species, and that any confusion between them arose from misidentification of *ou*. In his 1905 list he places *californica* and *russea* as varieties of *ou*. The type of *russea* from

Colorado is in the Henry Edwards collection, and is a reddish *californica*. That of *fratella*, as well as *ou*, is in the British Museum, and I was satisfied that they were one species. *Fratella* is undersized. The type of *ou* had either no hind legs or they were so tucked up in the vestiture that I had no chance of finding spines.

401. *A. rubidus* Ottol.—I have six more local captures of this species in my collection, dated June 1st to July 5th, 1909 and 1910. It comes to light and treacle, and I have taken it on the wing after dark flying over vetches. I took six specimens during 1909 alone, five of them at treacle. The tail of the sign is not always produced to a point as in Ottolengui's figure of the type. It sometimes widens out into a slight lobe, rather like that of *californica*. On the other hand, I have *californica* in which the tail is much like that of the type of *rubidus*. The nearest well-known relation to *rubidus* is *precationis*.

402. *A. alias* Ottol.—I have only four Alberta specimens in my collection which I feel quite certain are this species. The Waghorn (Blackfalds) specimen previously referred to, a ♀ with ♂ abdomen attached, July 25th, 1902. A ♂ and two ♀s, Head of Pine Creek, Aug. 7th, 9th and 16th, 1897 and 1903. They agree with Ottolengui's figures, and have the sign nearer to that of *rectangula* than any other species, but not as heavy. In fact, *alias* is the nearest ally that *rectangula* has. Also taken at Banff, Aug. 4th, 1908 (Sanson).

I long hesitated in separating from this a form which I have been calling *octoscripta*. I have a local female, dated Aug. 21st, 1903, and another from Mr. Wallis, Winnipeg Beach, Man., Aug. 23rd, 1910. The latter specimen is almost the exact counterpart of Ottolengui's figures. Mr. Wallis showed me another female taken at the same place, Aug. 22nd. I have also given this name to a Banff male, one of Mr. Sanson's captures, August 1910. This has a more spider-like sign than any of the others. I have a male from Cowichan Bay, near Duncans, Vancouver Island, which is brighter coloured and has heavier sign, but which I think is the same. It resembles *alias* very closely in colour and general pattern, but is rather darker. It has the irregular dentate terminal line of that species and *rectangula*, and the short blackish streak between that and the s.t. line opposite the reniform, of which there sometimes seems to be a trace in the other two species. The sign seems to be a modification of that in *alias*, which fact for long caused me to associate the two. As Grote describes it, it is "incompletely S-shaped, open superiorly." The outer portion of the S, however, seems sometimes to be a solid dot. The inner portion opens wide like that in *alias*, but is more thread-like. Other differences perhaps distinctive, appear to be that the t.a. and t.p. lines are more direct and less distinct. The t.p. line is, as Grote puts it, "waved or trembled, and appearing thus a distinguishing feature from Guenée's species" (*mortuorum*, *rectangula*). In *rectangula* and *alias* this line is scarcely crenate, but rather obviously waved. In *octoscripta* it is minutely but distinctly crenate, and but very slightly waved. I have carefully studied Grote's description, and the series standing under the name in the British Museum, and must for the present consider this Western form as dark and heavily marked *octoscripta*.

The British Museum series consists of six specimens—all very much alike, and looking somewhat bleached. There are two poor males, Nova Scotia (Redman), one of them badly rubbed. A pair, Grote collection, the male "Can." (This specimen was still unset when I saw it.) Two males, Hudson's Bay. The Grote collection females have two blue-bordered labels in Grote's handwriting—both "*Plusia 8-scripta* Sanb.," the upper label with "M.S." after the name. Whether this is really the type or not I cannot say. They are smaller than mine, as well as paler, and have the sign very thread-like, and similar in the whole series.

The description was published only by Grote, and the name should therefore be credited to him, though he used Sanbourne's Mss. name. The type specimens, number not stated, came from "Anticosti Island (Couper); Racine (O. Meske); Mass. (Prof. Packard)."

(To be continued.)

APPERCEPTIONAL EXPECTANCY AS A FACTOR IN PROTECTIVE COLORATION.

BY HARRY B. WEISS, NEW BRUNSWICK, N. J.

It is a matter of common knowledge that many insects escape detection by reason of their resemblance to certain surroundings. *Conotrachelus nenuphar* Hbst., which drops to the ground when disturbed, resembles bits of soil so closely that it often escapes observation. Certain moths resemble the bark upon which they rest; many caterpillars resemble the foliage upon which they feed; in fact, such resemblances are almost too numerous to mention.

This phenomenon is known as protective coloration, and is usually dismissed without further thought. Upon analyzing it further, however, it is evident that other factors contribute toward the result obtained by the perception of such an insect amid such surroundings, or, in fact, any surroundings. The perception of an insect is modified by associated perceptions from adjoining surroundings. A perception of a colour received, for instance, from a butterfly's wing will depend in part upon other perceptions received at the same time from adjoining surroundings or adjoining parts of the butterfly.

In addition to sensation, which is the result of stimulation upon the organ of sight by the object in question, perception is also determined by apperception, which is the contribution of the mind from previous experience. In other words, the mind also contributes something which helps to form the complete mental content. A red background, for instance, arouses an apperceptional expectancy for red, a green background an apperceptional expectancy for green, and so on. Many green insects are rendered less conspicuous and sometimes inconspicuous against a green background by reason of this expectancy on the part of the observer. Without this expectancy factor such insects would be more conspicuous than they are. When a protectively coloured insect is removed from its surroundings, and both surroundings and insect viewed separately, the sensations are quite distinct.

Many trained observers, and, in fact, numerous birds, are able to overcome this expectancy, and as a result discriminate such insects from their surroundings, although such discrimination may be due in part to an ability to perceive form. By reason of this apperceptional expectancy many insects also appear more conspicuous amid certain surroundings than others.

TABLE SHOWING THE EFFECT OF COLOURED BACKGROUNDS UPON INSECTS, THE COLOUR NEAREST THE NAME BEING THAT AGAINST WHICH IT IS MOST CONSPICUOUS, THE FOLLOWING COLOURS BEING ARRANGED IN THE ORDER APPROACHING INCONSPICUOUSNESS.:

Colias philodice Gdt.	blue	red	green	yellow
Danais plexippus Linn.	yellow	green	blue	red
Pyrameis atalanta Linn.	yellow	green	blue	red
Plathypena scabra Fab.	yellow	green	red	blue
Mamestra trifolli Rott.	yellow	red	blue	green
Anasa tristis DeG.	yellow	green	red	blue
Murgantia histrionica Hahn.	yellow	green	red	blue
Arilus cristatus Linn.	yellow	green	red	blue
Cyllene robiniæ Forst.	yellow	green	red	blue
Rhynchites bicolor Fab.	yellow	green	red	blue
Polistes variatus Cress.	yellow	green	red	blue
Eristalis tenax Linn	yellow	green	red	blue

With most of the above species, at least, it is noticed that yellow is more or less a fatal background, as far as inconspicuousness is concerned. Green is less fatal, but it is only in the red and blue that anything like protection is gained.

The more the coloration of an insect approaches that of its surroundings, the less conspicuous it becomes, but in all cases apperceptional expectancy tends to make this inconspicuousness more complete, and, as a result, more protective.

DR. C. GORDON HEWITT, Dominion Entomologist, was elected a Fellow of the Royal Society of Canada at the recent meeting. He was also chosen to represent the Royal Society at the forthcoming Jubilee of the Entomological Society of Ontario.

DURING the months of June, July and August the editor will be away from the city. Manuscripts for publication may be sent to Mr. A. F. Winn, 32 Springfield Ave., Westmount, Que.

BOOK NOTICE.

"The Importation Into the United States of the Parasites of the Gipsy Moth and the Brown-Tail Moth: A Report of Progress, with some consideration of previous and concurrent efforts of this kind." By L. O. Howard and W. F. Fiske. Bull. 91, Bureau of Entomology, U.S. Dept. of Agriculture, 344 pp., 27 plates, 74 text figs., 3 maps. July, 1911.

Perhaps no recent entomological undertaking has been watched with greater interest by American and Canadian entomologists than the attempt to establish on the American Continent the natural enemies of the Gipsy and Brown Tail Moths. This interest is due to two things: the immense destruction caused by the two insects in Massachusetts and the southern portion of New Hampshire and Maine, and to the fact that it is the first serious attempt to introduce all the effective insect enemies of a Lepidopterous host from one country, or series of countries, into another that has been made in the history of entomology.

The story of the work of introducing these insect enemies, together with that of previous and concurrent efforts of the same nature, is, as the title indicates, told in the Bulletin under consideration.

The first part consists of a discussion of previous work in the practical handling of natural enemies of injurious insects. It is an able discussion including many original and valuable observations. It presents, for the first time, a comprehensive view of the results that have attended the artificial transportation of insect parasites of various hosts in different quarters of the globe.

The second part tells the story of the introduction into the United States of the natural insect enemies of the Gipsy and Brown Tail Moths. The reasons for attempting the work are given at length, and the main issues of the experiment are fully discussed. Biological and other notes on a small army of parasites are recorded. Although the discussion is primarily that of the parasites of two Lepidopterous hosts, yet, on account of the fact that the author brings to bear upon the subject a splendid grasp of the broad subject of insect parasitism, it has a wide biological significance.

The sections on "Studies in the Parasitism of Native Insects" and "Parasitism as a Factor in Insect Control" are particularly interesting. To attempt, however, to pick out the most interesting and valuable portions of the work would be fruitless, as there is scarcely a paragraph that is not well worth reading.

A limited supply of the Bulletin was distributed in July, 1911. A general distribution has, however, only recently been made.

Since its publication a short article on "The Gipsy Moth as a Forest Insect," by the junior author, has appeared as Circular No. 164, U.S. Bureau of Ent. Speaking of the results of parasite importation, Mr. Fiske says: "On the whole, the results are decidedly satisfying, and the State of Massachusetts and the United States Department of Agriculture have no cause to regret having undertaken the unexpectedly formidable task of parasite importation. Within a territory entering a little to the northward of Boston, it may be conservatively stated that fully 50 per cent. of the eggs, caterpillars, or pupæ of the Gipsy Moth, in the aggregate, were destroyed by imported parasites in 1912." It is Mr. Fiske's opinion that this present rate of mortality in the central portion of the infested territory will eventually be considerably increased and will extend itself over the entire area of infectation.

In speaking of the amount of additional control necessary to check the increase of the Gipsy Moth in America, it is stated in Bull 91, p. 117, l. 11, that "An aggregate parasitism of 85% will almost certainly be sufficient, and it may well be that 80%, or even 75%, will answer equally well. Much less than 75% will probably not be effective."

In conclusion, it may be said that the Bulletin contains a wealth of information on a subject that has hitherto been little understood. It treats of a strictly scientific subject in a scientific way, and has the merit of being written in a particularly attractive style. It will be indispensable to any entomologist interested in natural control of insects. The excellent illustrations, of which the majority are original, materially enhance the permanent value of the work. J. D. TOTHILL.

Mailed June 7th, 1913.

The Canadian Entomologist.

Vol. XLV. LONDON, JULY, 1913 No. 7

REPORT ON A COLLECTION OF JAPANESE CRANE-FLIES (TIPULIDÆ), WITH A KEY TO THE SPECIES OF PTYCHOPTERA.

BY CHARLES P. ALEXANDER, ITHACA, N.Y.*

An extensive collection of Japanese crane-flies, taken by Dr. S. I. Kuwana and assistant entomologists in the vicinity of Nishigahara, Tokio, Japan, during the season of 1912, was forwarded to me for examination. The material, alcoholic, is contained in 62 vials, very carefully prepared and with complete data. I express my sincere thanks to Dr. Kuwana and his assistants for this fine representation of Japanese Tipulidæ and Ptychopteridæ.

Family Ptychopteridæ
Genus *Ptychoptera* Meigen.

Key to the species of Ptychoptera.

1. Wings with a distinct brown cross-band along the cord 2
   Wings hyaline or subhyaline without a distinct brown cross-band along the cord.
2. Radial sector more than twice as long as the radio-median cross-vein. (Europe) ....*contaminata L.*
   Radial sector rarely longer than the radio-median cross-vein...3
3. Posterior metatarsus conspicuously white. (Europe)...................................*albimana* Fabr.
   Posterior metatarsus not white.
4. Pleuræ reddish yellow; a short brown cross-band near the middle of the radial cell. (East. U.S.).........*rufocincta* O.S.
   Pleuræ black; no brown cross-band near the middle of the radial cell...5
5. All coxæ yellow or reddish-yellow; scape of antennæ brownish-yellow or yellow...6

* Contribution from the Entomological Laboratory of Cornell University.

Fore coxæ yellowish, other coxæ black; scape of antennæ black. (Japan)................................*japonica*, sp. n.

6. Scutellum yellow. (Europe)................*lacustris* Meig.
Scutellum black....................................7

7. Abdomen with the basal third of the second segment and the basal one-half of the third segment reddish orange. (India)...............................*distincta* Brun.
Abdomen entirely black. (Europe)...........*paludosa* Meig.

8. Femora and tibiæ bright orange-yellow, tarsi coal-black. (Abdomen orange-yellow, tergites with blackish borders to the segments; sternites orange-yellow.) (India)...............................*atritarsis* Brun.
Femora and tibiæ more or less black or brown; tarsi not coal-black.....................................9

9. Pleuræ silvery-white..............................10
Pleuræ not white. Thorax different in colour in the two sexes; femora bright yellow, hind pair black on the basal two-thirds except the extreme base. (India)...........*tibialis* Brun.

10. Hind coxæ black except at tip; femora brown at tip; scutellum reddish; hypopygium large, reddish; first segment of the antennæ reddish. (West. U.S.)................*lenis* O. S.
Coxæ and femora yellow, the latter black at the tip; scutellum yellow; hypopygium small, mostly blackish; antennal scape black. (Europe).......................*scutellaris* Meig.

Ptychoptera japonica, sp. n.

Wings banded; radial sector very short; antennæ of the male very long, about as long as the body; abdomen with little reddish or yellowish colour.

Male.—Length, 8.5 mm; wing, 8.9 mm; antennæ, 8.4 mm.; fore leg, femur, 5.4 mm.; tibia, 5.4 mm.; tarsus, 8 mm.; middle leg, femur, 5.4 mm.; tibia, 5.1 mm.; tarsus, 7 mm. Hind leg, femur, 6.1 mm.; tibia, 6.8 mm.; tarsus, 6.3 mm.

Female.—Length, 11.5–13.5 mm.; wing, 10.7–10.8 mm. Fore leg, femur, 5.6–5.8 mm.; tibia, 5.1–5.4 mm.; tarsus, 7 mm. Middle leg, femur, 5.4 mm.; tibia, 5.4 mm; tarsus, 6.8 mm. Hind leg, femur, 6.2 mm.; tibia, 6.8 mm.; tarsus, 6.2 mm.

Male.—Rostrum and palpi light brownish-yellow; front and vertex very dark coloured, occiput similar. Antennæ, segment one black, segment two black at base, brown apically, segment three yellowish on basal half, black apically, remainder of antennæ black; antennæ very long, as long as the body ; segments one and two short, the third segment very long, segments 4 to 15 long, gradually shortening, terminal segment very short.

Thoracic pronotum deep bluish-black; mesonotum, including the pleuræ similar. Halteres rather pale dull whitish. Fore legs with yellow coxa, dark at base, yellow trochanter, yellow femur broadly tipped with blackish, yellow tibia narrowly tipped with blackish, metatarsus yellowish-brown—darkened into brownish-black at the tip, remaining tarsal segments brownish black; middle and hind legs similar, but their coxæ blackish and the black femoral tips narrower. Wings with cell C yellowish brown, Sc and R more yellowish, remainder of wing hyaline or nearly so, a brown mark at the base of the wing in the neighbourhood of cross-vein h, a cross-band at the cord, often irregular, often a rounded brown spot on vein $Cu._1$ midway between cross-vein m-cu and the tip of the vein, brown marks at end of vein R_1, fork of R_{4+5} and fork of M. Venation (see plate III., fig. 7); Rs very short, much shorter than cross-vein r-m, basal deflection of R_{4+5} short but distinct, about one-half as long as Rs, cross-vein m-cu long, curved, longer than the basal deflection of $Cu._1$, placed opposite or very slightly beyond cross-vein r-m.

Abdomen, 1st segment very short, 2nd a little longer than the 4th, 3rd very long, as long as the succeeding 4 segments combined, segments 4 - 8 successively shorter. Abdomen dark brownish black, basal half of segment 4 orange. Hypopygium, 8th tergite narrow, short, widely separated from the somewhat broader 8th sternite, 9th tergite viewed from above very deeply incised, this incision rectangular, the caudad projecting lateral lobes are somewhat swollen basally, narrowed behind, slightly enlarged at the tips, densely clothed with long black hairs, between the lateral arms is a small rounded lobe, directed caudad; the 9th pleurite reaches the 8th tergite, the 9th tergite and 9th sternite being more widely separated; the 9th tergite is triangular, its apex rounded, bearing a long slender appendage at its tip on the inner side, this appendage long, slender and curved proximad so that each touches its mate of the opposite side, these appendages thickly clothed with long black hairs. The 9th sternite is very high at its base, extending up beyond the ventral level of the 8th tergite, its caudal ventral margin strongly chitinized, produced caudad and dorsad into a long slender arm, just dorsad of which is a shorter, strongly chitinized arm, with five or six blunt teeth on the ventral face. The guards of the penis are separated except at the base, divergent, chitinized, slender, rather blunt at the end, but the outer angle produced distad into a long slender arm. (See pl. IV., fig. 12-16).

Female.—Similar to the male, with the following exceptions: Antennæ short; black on tips of femora even more extensive, in fore femur covering almost one-half of the segment; tibiæ almost uniformly brown. Abdomen, tergites 1 to 6 dark brown; segment 7 brown, apical third white; 8th tergite mostly whitish; sternum lighter brown. 9th tergite, blade-like, pointed; 9th sternite short, produced into a short lobe on its dorsocaudal angle; ovipositor chestnut-brown. (See pl. IV.; fig. 11.)

Vial No. 29, Tokyo, Japan; May 7, 1912. 1 ♂, 5 ♀.

Holotype.—Male, Tokyo, Japan; May 7, 1912.

Allotype.—Female, with the type.

Paratypes.—Four females, with the type.

Types in the author's collection.

Paratypes in the U. S. National Museum and Cornell University collections.

Family Tipulidæ

Tribe Limnobini.

Genus Dicranomyia Stephens.

DICRANOMYIA JAPONICA, sp. n.

Subcosta long; wings with a distinct stigma and faint clouds along the cord; femora tipped with brown.

Male.—Length, 9–9.4 mm; wing, 9.4–10 mm; antennæ 3.2 mm. Female: Length, 10.2–11.4 mm; wing, 9.3–10.6 mm.

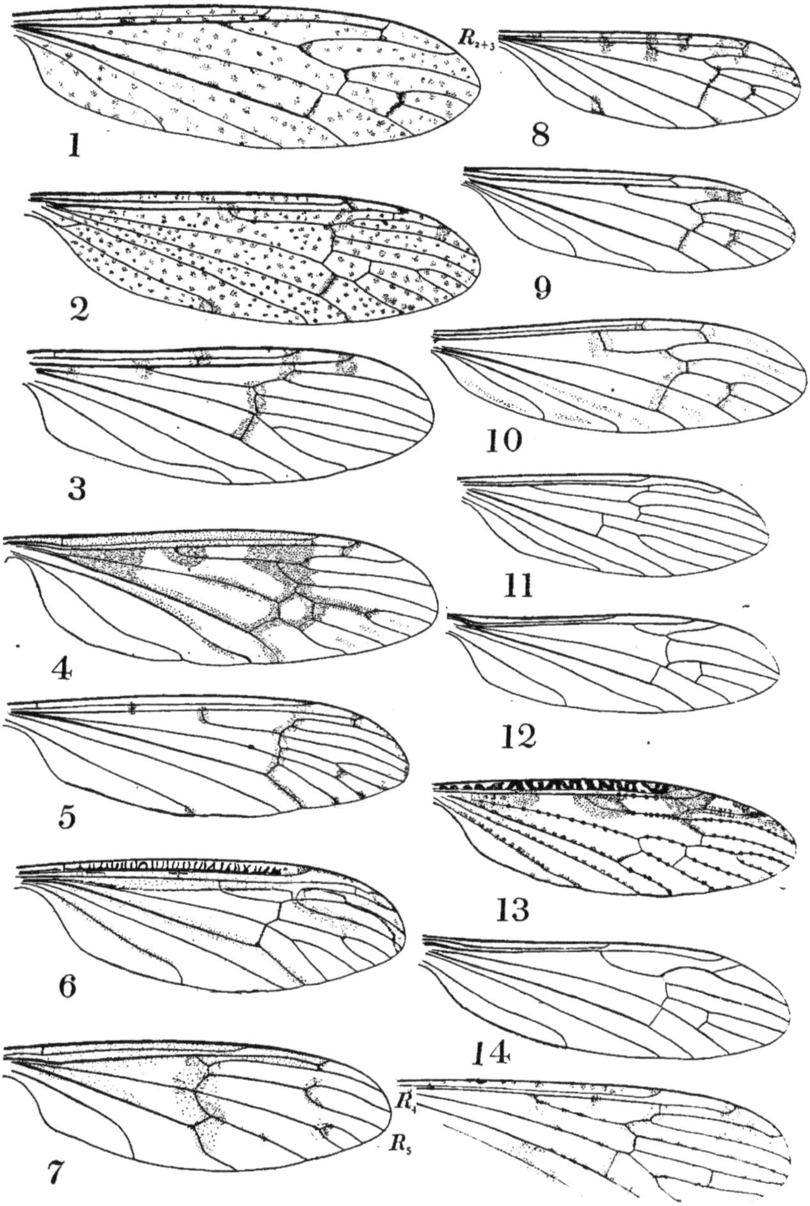
1
2
3
4
5
6
7
8
9
10
11
12
13
14
R_{2+3}
R_4
R_5

Male.—Rostrum and palpi brown; antennæ, segments 1 and 2 pale whitish yellow; segment 3 yellowish basally, brown at tip; remainder of antennæ dark brown. Antennæ long; flagellar segments long, cylindrical, subequal in length. Front, vertex and occiput, dark brown; genæ lighter colored, more yellowish.

Pronotum dark brown medially, yellowish on sides; mesonotal præscutum light yellow, with a broad, conspicuous median brown stripe; scutum with the lobes dark brown, paler medially; scutellum dark brown, except the narrow median incision on the anterior margin; post-notum largely dark brown. Pleuræ dull light yellow, the mesopleuræ suffused with brownish. Halteres rather long, pale, knob a little darker. Legs, coxæ and trochanters light yellow, femora dull yellow, the tip brown; tibiæ dull yellow, tip scarcely darker; tarsi, segment 1 dull brownish yellow basally, darkening to brown on apical third; remainder of tarsi brown. Wing pale brownish yellow, costal and subcostal cells rather clearer yellowish; veins brown; a conspicuous brown stigma; very pale grey clouds along the cord, outer end of cell 1st M_2, and at origin of Rs. Venation see fig; Sc long, ending before fork of Rs, Sc_2 longer than Sc_1, at the tip; Rs long arcuated at origin sometimes with a spur. (See pl. III.; fig. 9.)

Abdomen, tergites largely brown, usually with a yellow triangle on the anterior portion of the sides of the sclerites; sternite yellow; 8th and 9th, brown; 8th tergite, with caudal margin pale, straight; 9th tergite, with caudal margin strongly convex; with a brown median mark. Pleural pieces short, triangular, very broad at base, narrowed apically; dorsal apical appendage short, cylindrical, narrowed at tip, its inner or caudal margin provided with 4–5 rounded teeth. Ventral arm a small, rounded, little chitinzed lobe, covered with long hairs; guard of the penis very long, pale, projecting beyond the apical appendages, bifid at tip with 2 slightly chitinzed divergent horns, these horns directed ventrad; 2nd gonapophyses, slender, much shorter than the penis guard, scarcely enlarged at end, but inner face produced into a short, indistinct tooth. (See pl. IV.; fig. 10.)

Female about as in the male; valves of the ovipositor rather long, the tergal valves much longer than the sternal valves.

Variations: In some specimens the basal 4 or 5 segments are pale; yellow triangles on sides of abdominal tergites vary in distinctness.

Vial No. 4.—Tokyo, Japan; April 25, 1912. 1 ♂.
Vial No. 14.—Tokyo, Japan; April 25, 1912. 1 ♀.
Vial No. 15.—Tokyo, Japan; April 25, 1912. 4 ♂s, 2 ♀s.
Vial No. 24.—Tokyo, Japan; April 27, 1912. 7 ♂s, 6 ♀s.
Vial No. 32.—Tokyo, Japan; May 7, 1912. 3 ♀s.
Vial No. 33.—Tokyo, Japan; May 7, 1912. 2 ♂s.
Vial No. 37.—Tokyo, Japan; May 7, 1912. 1 ♂.
Vial No. 38.—Tokyo, Japan; May 7, 1912. 1 ♀.

Holotype.—♂, Tokyo, Japan; April 27, 1912. (Vial 24).

Allotype.—♀, with the type (Vial 24).

Paratypes.—14 ♂s; 12 ♀s; Tokyo, Japan; April 25–May 7, 1912.

Types in author's collection.

Paratypes in U.S. National Museum and Cornell University collections.

D. japonica resembles *umbrata* Meij. from Java (1) but the legs are much paler, wing-pattern and venation different, and it is a much larger species (wing, 9–10 mm.; in *umbrata*, 5 mm.).

Dicranomyia nebulosa, sp. n.

Subcosta long; wings clouded with grey; femora pale apically, with a dark subterminal ring.

Male.—Length, 5.4 mm.; wing, 5.8 mm.

Male.—Rostrum and palpi dark brown; antennæ, 1st segment brown at base, more yellowish at the tip, succeeding segments brown; flagellar segments rounded, short-pedicallate, these pedicels being whitish; front, vertex and occiput, very dark blackish.

Pronotum brownish-yellow, darker brown medially above. Mesonotum rather gibbous, brown, a narrow, darker brown, median line on the præscutum; lateral margin of this sclerite with a rounded dark brown spot which is connected with short lateral stripes nearer to the median vitta; scutum light brown, lobes margined with dark brown; scutellum with a dark brown median mark; postnotum brown. Pleuræ brown, almost uniform, paler near the

(1) (Tijd voor Entomol.; Vol. 44; p. 25; pl. 1, f. 7.; 1911.)

sternum. Halteres pale. Legs, coxæ and trochanters light yellow; femora light brown, becoming light yellow on the apical sixth and with a conspicuous, dark-brown, subapical ring; tibiæ dark brown; tarsi broken. Wings, whitish or subhyaline; costal cell slightly more yellowish; grey clouds as follows: At origin of Rs (largest), at stigma, at tip of Sc, along cord, along outer end of cell 1st M_2 and in the center of most of the cells. Venation, (see pl. III.; fig.10); Sc long, extending far beyond the origin of Rs, Sc_2 at the tip of Sc_1, Rs almost square at its origin and spurred (in the types), cell R_2 almost as far proximad as cell 1st M_2 (as in *F. stulta* O.S.), cell 1st M_2 long, longer than the veins issuing from it, basal deflection of $Cu._1$ at the fork of M.

Abdomen, tergum dark brown; caudal margins of the 7th, 8th and 9th segments more yellowish; sternum dull yellow. Hypopygium (see figs. 8, 9; pl. IV.); 9th tergite short, its cephalic and caudal margin convex, its caudal half provided with a number of long hairs. Pleuræ very long, cylindrical, the tips produced into a slender lobe on the ventral side; two apical appendages, which are very short and inconspicuous, being scarcely one-third as long as the plura; dorsal appendage simple, short, slender and subchitinized, not exceeding the ventral appendage; ventral appendage double, its dorsal arm being small, triangular and with the caudal or outer face bearing a chitinized tooth, its tip produced entad and cephalad into a blunt lobe; the ventral arm is produced entad into a small lobe, with the tip evenly rounded. Viewed from the side, the pleura is broad, its ventral margin rounded at the base, at the middle of its length produced into a spatulate fleshy lobe which is directed caudad. The guard of the penis is long (extending about to the extreme tip of the pleura), and slender. broad at the base, narrowed toward the tip, the end little, if any, enlarged; the apex is very slightly notched; viewed from the side, it is seen that the extreme tip is bent ventrad; viewed from above, the guard seems to be concave, its lateral margins being more strongly chitinized. The second gonapophyses are rather long, dark brown, subrounded or scarcely pointed at the apex; at their base they are about as broad as the base of the penis guard; the lateral margin of the apophyse is produced dorsad into an incurved, chitinized flap or margin, which, on the sides, protects the short, slightly emarginate anal tube.

Vial No. H.—Tokyo, Japan; Aug. 1912. 1 ♂.

Holotype, ♂.—Tokyo, Japan; Aug. 1912.

Type in the author's collection.

D. *nebulosa* resembles *unibrata* Meij. (Java), but the leg-pattern and venation are quite different.

Genus Geranomyia Haliday.

Geranomyia avocetta, sp. n.

Wings spotted; thoracic dorsum brown, the humeral portions of the præscutum yellow; tibial apices not blackened.

Male.—Length, excluding the proboscis, 7.5–7.7 mm.; proboscis, 3–3.6 mm.; wing, 7.8–7.9 mm.

Male.—Proboscis and palpi dark brown, the former more yellowish basally; antennæ, basal segments dark brown, flagellar segments somewhat lighter brown, segments rounded-oval; front, vertex and occiput dark-colored, almost black.

Pronotum dark brown; in the paratypical specimen, the caudal margin of the scutum and the scutellum, yellowish. Mesonotal præscutum with a broad, dark brown, median line, widened behind; humeral angles conspicuously light yellow, behind darkening into brown of a lighter shade than the broad median vitta; scutum with the lobes dark brown, median line paler; scutellum and postnotum brown. Pleuræ dull brownish-yellow, clearer below. Halteres pale, knob a little browner. Legs: Coxæ and trochanters light yellow, the latter margined with black at the tip; femora and tibiæ light brown, scarcely darkened at their tips; terminal tarsal segments darker brown. Wings, hyaline or nearly so, the costal cells and veins more tawny; veins light brown, darker brown where traversed by dark markings; seven brown marks along the costal margin, the third at the origin of Rs extending down almost to vein M; the fourth at the tip of Sc extending down into cell 1st R_1; the 5th (stigmal) spot, largest, rectangular; the sixth and seventh spots at ends of veins R_{2+3} and R_{4+5}; cord and outer end of cell 1st M_2 seamed with brown; a brown spot at ends of most of the veins, most distinct and largest at the 2nd anal vein. Venation (see pl. III.; fig. 8): Sc long, ending nearer to the fork of Rs than to its origin; Sc_2 at tip of Sc_1; Rs long, nearly three times as long as the basal deflection of R_{4+5}; basal deflection of $Cu_{.1}$ at fork of M.

Abdominal tergum brown, anterior margins of the basal segments somewhat more yellowish; sternum pale whitish-yellow. Hypopygium (see figs. 5–7; pl. IV.): 8th tergite short, consisting only of a narrow ring, almost straight on its cephalic margin, concave on the caudal margin; 9th tergite convex anteriorly, concave on caudal margin. Pleural pieces very short, cylindrical, not more than twice as long as wide, bearing two apical appendages. The dorsal appendage is a short, slender, strongly curved hook, sharp pointed and more chitinized at its tip; it is directed entad, cephalad and dorsad. The ventral lobes are long, fleshy, between two and three times as long as the pleura and much thicker; at their base, on the inner side, is a short, fleshy tooth, more chitinized at its tip, directed cephalad and dorsad and meeting its mate of the opposite side on the median line; near the tip, on the outer or caudal face, are two, long, slender, subequal bristles, directed caudad. The ventral side of the pleura is produced into a lobe, enlarged apically and directed entad and slightly caudad. The guard of the penis is short, extending slightly beyond the most caudad-projecting portion of the pleura; it is swollen at the base, less so in the middle of its length, its tip small, chitinized, bifid at apex, the tip directed slightly ventrad. The second gonapophyses are very short, and, viewed from above, barely project beyond the fleshy lobe lieing between them.

Vial No. 8.—Tokyo, Japan; April 25, 1912. 1 ♂.

Vial No. 49.—Tokyo, Japan; August, 1912. 1 ♂.

Holotype, ♂.—Vial No. 8.

Paratype, ♂.—Vial No. 49.

Types in the author's collection.

G. avocetta, compared with the four Javan species described by de Meijere, agrees most closely with *G. montana*, which, however, has the wing-pattern much less distinct. From the North American *G. rostrata* Say, it differs conspicuously in its unicolorous tibiæ.

Genus Rhipidia Meigen.

Rhipidia pulchra septentrionis, subsp. n.

This subspecies differs from typical *pulchra* Meij.* (Java) in

*Neue und bekannte sudasiatische Dipteren; p. 92, fig. 7. Bijdragen tot de Dierkunde, vol. 17, 1904.

antennal coloration, the flagellar segments being alternately dark and light-coloured; segments, 4, 6, 8, 10 and 12 are whitish, the remainder of the antennæ brown. The wings have a large spot at the base of Cu and the venation is not as figured by de Meijere. (Compare fig. 1; pl. III.)

Female.—Length, 7.6–8.6 mm.; wing, 7.4 mm.

Vial No. 10.—Tokyo, Japan; April 25, 1912. 2 ♀s.

Holotype and Paratype in author's collection.

In Tijd Voor Entomol., Vol. 44, p. 27, figs. 14–16, de Meijere refers this to *Dicranomyia*. However, I believe his original reference of the species to be the correct one—this belief based on venational hypopygial characters.

Tribe Antochini

Genus Rhamphidia Meigen.

Rhamphidia nipponensis, sp. n.

Rostrum short; palpi pale; wings hyaline without darker marks.

Female.—Length, 8.9 mm.; wing, 7.8 mm.; middle leg, femur, 6.6 mm.; tibia, 7 mm.; tarsus, 6.7 mm.

Female.—Rostrum light brown; labrum light yellow; palpi light brownish-yellow; antennæ brown, flagellar segments cylindrical with short black bristles not exceeding the segment in length, the outer segments not conspicuously narrowed; front, vertex, occiput and genæ dark brown.

Pronotum dark brown, mesonotal præscutum light brown, with three broad, darker brown stripes, the median one longest, broadest, very dark brown in front; the lateral stripes begin behind the pseudosutural fovea and cross the suture, suffusing the lobes of the scutum; scutum medially light brown, on margins yellowish-brown; scutellum brown, margined with yellowish; postnotum brown. Pleuræ brownish-yellow, suffused with brown on portions of the mesopleuræ; mesosternum brown. Halteres light yellow, knob slightly darker, brown. Legs: coxæ light yellow, tipped with pale brown; trochanters yellow; femora yellowish-brown, rather clearer yellowish basally; tibiæ brown, tarsi brown, terminal segments rather darker. Wings, hyaline or nearly so; veins brownish

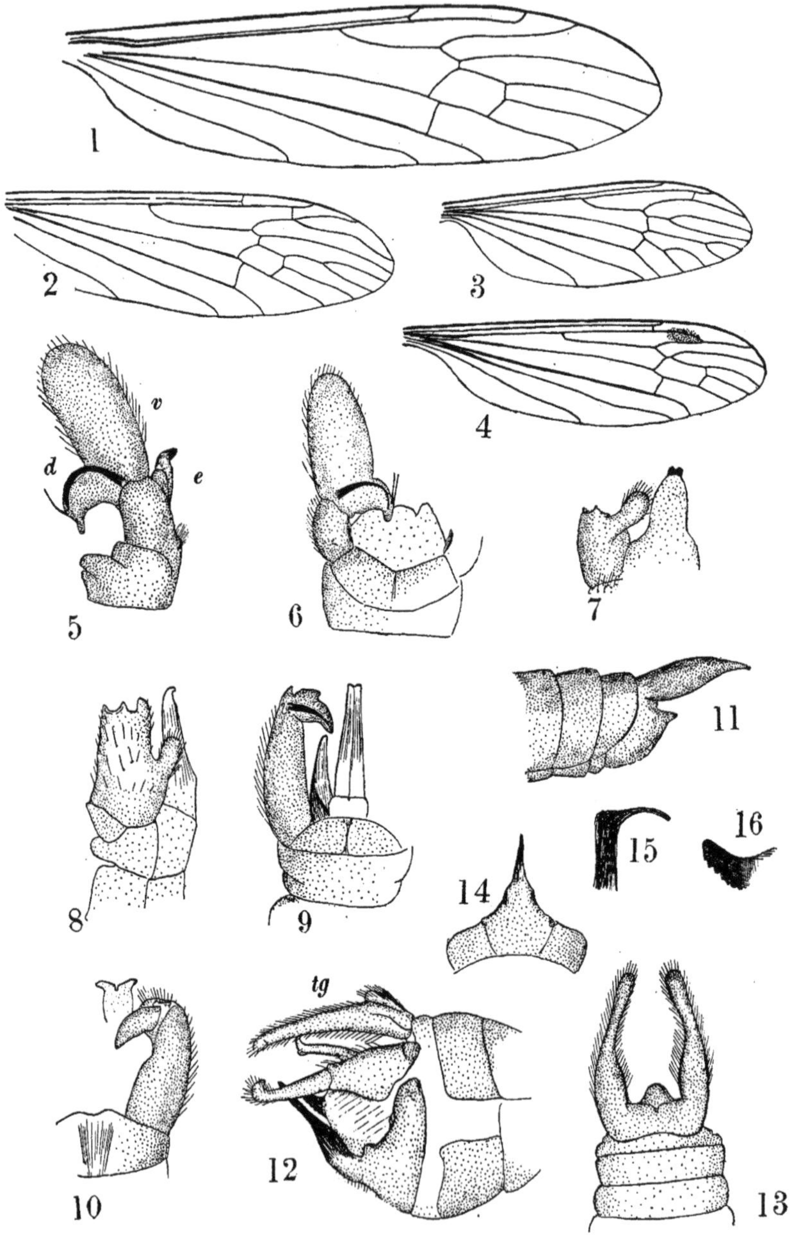

JAPANESE CRANE-FLIES (ALEXANDER).

yellow, stigma not indicated. Venation (see fig. 1; plate IV.); cross-vein r-m distinct; basal deflection of Cu_7 beyond the fork of M.

Abdomen, tergum and sternum dark brown; ovipositor light yellow.

Vial No. 28.—Tokyo, Japan; April 26, 1912. 1 ♀.

Holotype, ♀.—Vial No. 28, in author's collection.

This species differs from the European *R. longirostris* by its shorter rostrum, cylindrical flagellar segments with short bristles; pale maxillary palpi and other colorational differences, which may, of course, vary in series.

EXPLANATION OF THE PLATES.

Plate III.

Fig. 1. Wing of *Rhipidia pulchra septentrionis*, sub sp. n.
Fig. 2. Wing of *Limnophila japonica*, sp. n..
Fig. 3. Wing of *Erioptera elegantula*, sp. n.
Fig. 4. Wing of *Limnophila satsuma* Westwood.
Fig. 5. Wing of *Tricyphona vetusta*, sp. n.
Fig. 6. Wing of T. *kuwanai*, sp. n.
Fig. 7. Wing of *Ptychoptera japonica*, sp. n.
Fig. 8. Wing of *Geranomyia avocetta*, sp. n.
Fig. 9. Wing of *Dicranomyia japonica*, sp. n.
Fig. 10. Wing of D. *nebulosa*, sp. n.
Fig. 11. Wing of *Molophilus pegasus*, sp. n.
Fig. 12. Wing of *Gonomyia insulensis*, sp. n.
Fig. 13. Wing of *Conosia irrorata* Wiedemann.
Fig. 14. Wing of *Gonomyia superba*, sp. n.
Fig. 15. Wing of *Erioptera asymmetrica*, sp. n.

Plate IV.

Fig. 1. Wing of *Rhamphidia nipponensis*, sp. n.
Fig. 2. Wing of *Limnophila inconcussa*, sp. n.
Fig. 3. Wing of *Tricyphona insulana*, sp. n.
Fig. 4. *Liogma kuwanai*, sp. n.

Fig. 5. Hypopygium of *Geranomyia avocetta;* lateral aspect. e—penis guard; d—dorsal apical appendage; v—ventral apical appendage.

Fig. 6. Hypopygium of *Geranomyia avocetta;* dorsal aspect.

Fig. 7. Hypópygium of *Geranomyia avocetta;* ventral aspect, showing a portion of the *hypopygium.*

Fig. 8. Hypopygium of *Dicranomyia nebulosa;* lateral aspect. The apical appendages are not included.

Fig. 9. Hypopygium of *Dicranomyia nebulosa;* dorsal aspect.

Fig. 10. Hypopygium of *Dicranomyia japonica;* dorsal aspect.

Fig. 11. Ovipositor of *Ptychoptera japonica;* lateral aspect.

Fig. 12. Hypopygium of *Ptychoptera japonica;* lateral aspect. tg—9th tergite.

Fig. 13. Hypopygium of *Ptychoptera japonica;* 9th tergite, dorsal aspect.

Fig. 14. Hypopygium of *Ptychoptera japonica;* 9th sternite, ventral aspect.

Fig. 15. Hypopygium of *Ptychoptera japonica;* guard of the penis (?).

Fig. 16. Hypopygium of *Ptychoptera japonica;* ventral appendage.

(TO BE CONTINUED.)

DONACIA EMARGINATA KIRBY (COLEOPTERA.)

A Biographic Note.

BY L. B. WOODRUFF, NEW YORK CITY.

Donacia emarginata Kirby may gain its sustenance from various water-loving plants, but that which it seems to find superlatively to its taste near New York City is the Marsh-marigold, *Caltha palustris.* In a certain wooded swamp just outside the city limits, always wet under foot and in April excessively "soft," grow and bloom great masses of these glorious golden flowers; and when they reach the zenith of their splendor, in almost every clump, half buried under their stamens, are from one to several of these graceful metallic beetles. The sturdy crowfoot cup gives them secure support, and in them throughout the flowering period they are to be found in breeding pairs. On the stems just above the roots the pupal cocoons are attached, sometimes several in a row; but when the swollen buds expand the beetles emerge, leave their lowly dwellings, and, climbing up the stems, attain the scene of their ensuing revels. When not too much engrossed, they display the instinct shared with so many other strongly flying members of their order, and, on the approach of danger, clamber to the petal's edge and seek safety by dropping to the cover that lies below.

The Caltha seems to be an unrecorded food-plant for the genus, though hardly a surprising one in view of its evident adaptability and its environmental association with the skunk cabbage, the resort of certain others of its component species.

So far as they have come under the writer's observation, the males of *D. emarginata* in this neighborhood are uniformly purplish or bluish-black, while the females are never like them in colour, but vary through shining olivaceous green, the shade most commonly occurring, to brassy and rich bronze. If these colour distinctions hold constant with the beetles from other localities, we have here secondary sexual characters which are worthy of note.

A NEW BRACONID OF THE GENUS MICRODUS FROM CANADA.

BY C. H. RICHARDSON, JR., FOREST HILLS, MASS.

Among a number of parasitic hymenoptera reared from the Bud Moth (Tmetocera ocellana Schiff.), at the Dominion Entomological Laboratory, Bridgetown, Nova Scotia, by Mr. G. E. Sanders, there is a Braconid belonging to the genus Microdus which appears to be new. Since it is desired to refer to this species in the near future, Dr. Hewitt, Dominion Entomologist has asked me to describe it at the present time.

Microdus ocellanæ, sp. nov.

Description of the type (female): Length 5 mm. Wing 4 mm. Ovipositor about 5 mm. Head, thorax and abdomen black, refulgent; palpi pale fulvous; fore and middle legs pale fulvous, with the apical joints black; hind legs pale fulvous except for the black coxæ, the black apical annuli on the tibiæ, the darkened distal ends of the first tarsal joints and the complete darkening of the succeeding joints. A large fulvous spot covering the first and second abdominal segments ventrally. Pubescence light. Wings slightly infuscated, iridescent; stigma black. Head slightly wider than thorax, less than three times as wide as thick; clypeus slightly produced; clypeal foveæ large, each equaling an ocellus in size; face punctulate, pubescent; vertex, occiput and genæ sparsely punctulate and pubescent. Antennæ 42-jointed, stape and pedicel longer than the first joint of the flagellum; joints of flagellum subequal. Mesonotum punctulate, with deep punctate parapsidal grooves which meet posteriorly. Scutellum punctulate flatly convex; anterior depression of scutellum with four deep umbilicate punctures. Metanotum rugose-punctate; metathoracic spiracles oval, slightly longer than wide. Mesopleuræ sparsely punctulate and pubescent with a curved punctured line just below the tegulæ; a single post-median fovea and a longitudinal row of umbilicate punctures below this. Metapleuræ more densely punctulate and pubescent. First segment of the abdomen deeply striated longitudinally; the second segment weakly and irregularly aciculated with a median transverse depression; remaining segments smooth, shining.

Type ♀ No. 4001d, July 28, 1912; in Coll. Div. Ent., Ottawa.

Type locality.—*Kentville, Nova Scotia, Canada.*

Paratype (♀) agrees essentially with the type.

This species is related to *Microdus earinoides* Cresson resembling it in size and colour, but differing in the sculpture of the abdomen,the possession of black hind coxæ and the extent of the black on the hind tibiæ as well as the quite distinctly infuscated wings. It is also very similar to *Microdus nigricoxis* Provancher, but only the hind coxæ are black and the basal segment of the abdomen is striated, not rugose. Acknowledgments are due Mr. C. T. Brues for aid in looking up the literature.

A SUCCESSFUL MOVE

Recently I had occasion to move my entire collection of over 200 well-filled boxes of Hemiptera from Buffalo, N. Y., to San Diego, Calif., and on unpacking them here was surprised to find that not a single specimen had been damaged. The boxes were packed in straw in two large willow pottery crates and were shipped by freight through one of the household shipping agencies. However, they had to go through two storage warehouses and be twice reshipped before starting on their long ride which speaks well for the packing. I received my instructions for packing from Dr. E. D. Ball and will gladly pass it on to any one contemplating a similar move.—E. P. Van Duzee, 4020 Ivy St., San Diego, Calif.

DESCRIPTION OF TWO NEW SPECIES OF OCHTERUS LATR. (HEMIPTERA) WITH AN ARRANGEMENT OF THE NORTH AMERICAN SPECIES.

BY H. G. BARBER, ROSELLE PARK, N.J.

The genus Ochterus Latr. (Pelogonus Latr.) is represented in North America by five species—four from Mexico, Central America or the Antilles, and only one has been described from the United States—*O. americanus* Uhl. (Bull. U.S. Geol. and Geogr. Surv. I, 335, 1876). The Mexican and Central American species are well characterized and figured by Champion, in Biol. Cent. Amer. Hem.-Het. II., 344–346, 1901.

I herewith add two more species to the list—one from the collection of Mr. Nathan Banks, who collected several specimens at Glencarlyn, Va., in June, and the other represented by a single specimen from Mrs. Slosson's collection, taken by her at Ormond, Florida, in the spring. The former of these must be closely related to the Palæarctic *O. marginatus* Latr., the latter is more closely allied to *O. americanus* Uhl.

The following synopsis of the North American species of Ochterus is adapted from Champion's key:

Anterior angles of the pronotum acute; humeri rounded; face not at all or obsoletely carinate between the eyes . . . . *perbosci* Guér. (Mex., Antilles.)

Anterior angles of the pronotum obtuse or rounded.

- Humeri rounded.
  - Face not carinate between the eyes . . . . *ænifrons* Champ. (Mex., Cent. Am., Antilles)
  - Face distinctly carinate between the eyes.
    - Clavus entirely yellow . . . . *flaviclavus*, n. sp. (Florida).
    - Clavus concolorous.
      - Entire lateral pronotal margins broadly pale . . . . *banksi*, n. sp. (Virginia)
      - Lateral pronotal margins, with only a pale spot anteriorly . . . . *americanus* Uhl. (U.S.)

Humeri subacute. Face carinate and closely rugulose between the eyes....................*viridifrons* Champ. (Cent.Amer.)

Humeri acute. Face carinate, but almost smooth between the eyes....................*acutangulus* Champ. (Cent. Amer.)

Ochterus banksi, n. sp.

Broad ovate, brownish black. The head, behind the vertex, opaque, from there anteriorly, shining and obilquely, finely regulose and tricarinate; one carina next each eye and a median one, continuous from vertex to apex; transversely sulcate midway between ocelli and base of head. Pronotum with anterior margin almost truncated, with the anterior angles next the eyes rounded and not projecting forwards or outwardly beyond the exterior margin of the eyes; entire lateral margins gently rounding posteriorly; humeral angle rounded, not very prominent; lateral margins broadly expanded, pale; this mark broadest about the middle, more abruptly rounded anteriorly and tapering posteriorly to occupy the entire margin; the remainder of the surface brownish black, elevated and transversely, but not very deeply, sulcate a very little behind the middle; posterior lobe, middle and anterior part of first lobe more coarsely punctate, the latter with two or three transverse weak furrows. Scutellum almost equilateral, rather coarsely punctate and transversely furrowed; anteriorly with a transverse elevated ridge, behind which it is depressed. Corium not demarked from membrane, broadest across the middle, with lateral margin gently rounded to just beyond middle, where it more abruptly rounds off to the rather narrow apical part of membrane; the external margins either broadly pale throughout or in part suffused with fuscous and reflexed. without the usual series of pale marginal spots

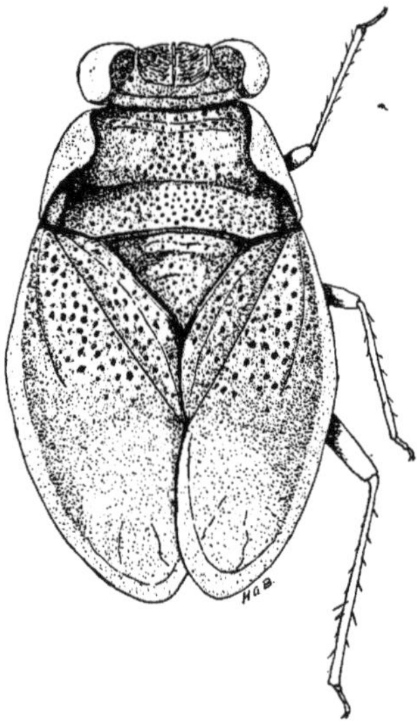

FIG. 9.
Ochterus banksi, n. sp.

which occur in *O. americanus.* Clavus and corium, anteriorly, with coarse scattered punctures. Nervures of the membrane indistinct. General surface with indications of the customary bluish grey markings, unless denuded, when the whole upper surface is smooth and shining. Beneath on sternum and venter paler, with rostrum, acetabulæ, coxæ, legs and external angle of metathorax pale yellow. Prosternum rather coarsely punctate.

Length, 4 mm. Width of pronotum, 2 mm.

Described from three males and one female collected by Mr. Nathan Banks at Glencarlyn, Virginia, in June. Judging from the meagre descriptions and indifferent illustrations at hand, I am led to the opinion that this species is most nearly related to *O. marginatus* Latr., of Europe. But having no specimens of that species for comparison, I am, at this time, unable to settle the point. *O. banksi* can readily be separated from *americanus* by its difference in color markings, and the character of the pronotum. Apex of membrane is more narrow than in *americanum.*

Ochterus flaviclavus, n. sp.

Brownish-black. Very much the appearance of *O. americanus*, to which it is closely related, having the usual carinate and rugulose face. However, somewhat smaller than that species with the clavus entirely yellow. The pronotum with the lateral margins gently rounded, more converging anteriorly, the anterior margin being narrower than the width across the eyes; the anterior angle of the pronotum sharply rounded and not projecting anteriorly as in *americanus;* the expanded part of lateral margins narrower, with a small yellowish spot just posterior to the anterior angle; the humeral angle almost rectangular, projecting but a trifle beyond margin of corium. Extreme edge of corium very narrowly pale, but the usual pale marginal spots are lacking. Surface with the usual pearl grey spots. Beneath, with the sternum slate grey; the acetabula, posterior and lateral flange of the prosternum, elytral flange anteriorly, posterior margin of metasternum, legs and venter, pale; legs lightly infuscated. Prosternum, mesosternum externally and metasternum before the posterior angle distinctly punctate.

Length, 3½ mm.; width of pronotum, about 2 mm.

Described from a single male in the collection of Mrs. Annie Trumbull Slosson, taken by her at Ormond, Florida.

A SECOND ADDITION TO THE AUSTRALIAN HYMENOPTERA MYMARIDÆ.

BY A. A. GIRAULT, BRISBANE, AUSTRALIA.

The following species have recently been captured by Mr. Alan P. Dodd and very kindly given to me. They are the eighteenth and nineteenth species of *Gonatocerus* and the sixth, seventh and eighth of *Polynema*. All in normal position.

1. *Gonatocerus bicolor*, new species.

Female.—Length 1.65 mm. Large for the genus.

Black, the abdomen contrasting orange reddish, dorsad with faint duskiness, the scape and pedicel lemon yellow, as are also the legs and coxæ; tibiæ fuscous. Ovipositor not exserted. Fore wings of the narrower type, yet moderately broad, bearing about thirty longitudinal lines of very fine discal cilia, lightly fumated throughout, the marginal cilia short, the longest not more than a fifth of the greatest wing width. Proximal tarsal joints very long. First funicle joint longer than either the pedicel or the second joint of the funicle, subequal to funicle joint 3, joints 4 and 5 each somewhat shorter than 3, joint 5 shorter than 4, 6 still shorter than 5, while 7 lengthens slightly, subequal to 2; distal funicle joint shortest, subequal in length to the pedicel. Of the general habitus and structure of *spinozai* Girault and belonging to the group of species with graceful fore wings and usually golden bodies (*e.g.*, *comptei*, *cingulatus*). Marginal vein very long. Caudal wings with an incomplete, more or less variable, paired line of midlongitudinal discal ciliation. Club long.

(From one specimen, 2-3-inch objective, 1-inch optic, Bausch and Lomb).

Male.—Not known.

Described from a single female captured by sweeping jungle growths along forest streamlet, near Nelson, North Queensland, December 6, 1912 (A. P. Dodd).

Habitat: Australia—Nelson (Cairns), Queensland.

Type: No. *Hy 1293*, Queensland Museum, Brisbane; the above specimen on a slide of xylol-balsam.

This beautifully coloured species may be distinguished with ease by the great contrast between the black of the thorax and the orange of the abdomen, by the clouded wings and long venation and by the long first joint of the antennal funicle. It is allied to *spinozai*, but could not be confused with that characteristic species.

2. *Gonatocerus spinozai* Girault.

At the same time that the above new species was captured Mr. Dodd obtained a pair of this species. The male was unknown and, since it differs considerably in coloration from the female, I briefly point out its characteristics. In structure, similar to the female, but the antennæ are 13-jointed and filiform, the pedicel very small and sublobate, funicle joints 2-4 and 9-10 subequal, longest, about thrice longer than wide; joints 1 and 5 subequal; somewhat shorter than the others; joints 13, 6 and 7 subequal, still somewhat shorter; joint 8 shortest, a third shorter than joint 2. Abdomen subpetiolate, declivous from above at base, ovate, striped dorsad with black, transversely (6 stripes counting the broadest at extreme base). Propodeum purplish black, its spiracle very minute, round, the surface finely reticulated, a median carina present (its exact shape not seen, probably paired). Tip of dorsal abdomen black. Otherwise coloured as in female. When mounted in balsam, the tip or apex of the declivous part of the base of the abdomen closed up to the thorax, partially concealing the real nature of the segmentation; this apex is projected or heeled, stopper-shaped and appearing as if it was intended to fit against the thorax.

3. *Gonatocerus fasciativentris*, new species.

Male.—Length 1.15.mm.

Black, the abdomen golden yellow, conspicuously striped transversely with black above and below, the intervening yellow stripes much narrower, the lateral line yellow (about six black stripes). Legs yellowish brown, the coxæ black. Wings hyaline, the fore wings of the less graceful type, the marginal vein moderately long; fore wing with about twenty-five lines of discal cilia; posterior wings narrow. Scape, pedicel and first funicle joint more or less suffused with yellowish. Antennæ strongly longitudinally striated, the funicle joints short and subequal, each about one and a half times longer than broad. Allied with *coethei*, but, besides the differences in coloration, the fore wings are broader. Pedicel only half the length of the first funicle joint.

(From one specimen, similarly magnified).

Female.—Not known.

Described from a single male captured with the preceding two species.

Habitat: Australia—Nelson (Cairns), N. Q.

Type:Hy 1294, Queensland Museum, Brisbane, the above specimen (mounted with the types of *G. brunoi lyelli* and *Polynema devriesi* both described beyond).

4. *Gonatocerus brunoi lyelli*, new variety.

Male: Like the typical forms, but the abdomen at its distal half dorsad distinctly banded by narrow golden yellowish stripes (two or three), the wings very dark.

(From one specimen, enlarged as with preceding species).

Respectfully dedicated to the late Sir Charles Lyell, the author of the "Principles of Geology."

Described from a male captured with the preceding species.

Habitat: Australia—Nelson (Cairns), Queensland.

Type: No. *Hy 1295*, Queensland Museum, Brisbane, the above specimen (mounted with the type of *Gonotocerus fasciativentris* Girault and a *Polynema*).

Genus *Polynema* Haliday.

1. *Polynema devriesi*, new species.

Male: Length, 1.2 mm.

Somewhat similar to both *draperi* and *romanesi* Girault, but differing from the former in having the discal cilia of the fore wing much coarser, from the latter in the same point, from both in general coloration being ferruginous, the distal third of the abdomen black. Scape and pedicel concolorous, the flagellum black, its joints very long, as are also the proximal tarsal joints. About nine lines of rather coarse discal cilia, the marginal cilia longer than the wing's greatest width. Distal tarsal joints black. Wings obscurely fumated, the posterior ones very narrow, the fore wings narrowing proximad before venation.

(From one specimen, similarly magnified).

Female: Not known.

Described from a single male captured with the species of *Gonatocerus* noted above. Respectfully dedicated to Hugo De Vries, the author of the mutation theory in biology.

Habitat: Australia—Nelson (Cairns), N. Q.

Type: No. *Hy 1296*, Queensland Museum, Brisbane, the above specimen in balsam (mounted with the types of *Gonatocerus fasciativentris* Girault).

2. *Polynema mendeli*, new species.

Male: Length 1.20 mm.

Like *devriesi*, but the discal cilia of the fore wing is finer, the marginal cilia shorter, not quite as long as the greatest width of the blade, subfuscous, not as slender proximad before venation; in this species the proximal funicle joint is much shorter than the next joint, not half its length, while also joints 5 distad of the flagellum are all short, more or less subequal to 1, flagellar joint 2 longest, 3 and 4 next in succession. This antennal structure easily separates this species from *draperi* and *romanesi*. Ferrugineous, the abdomen (exclusive of pedicel) black, as are also the distal tarsal joints and the flagellum; proximal funicle joint yellowish, head blackish. Fore wings with about 10 lines of fine but rather long discal cilia.

(From one specimen, enlarged as in previous descriptions.)

Female: Not known.

Described from one male, captured with the preceding species. Dedicated to Abbé Gregor Mendel, who established the Mendelian principle of inheritance.

Habitat: Australia—Nelson (Cairns), Queensland.

Type: No. *Hy 1297*, Queensland Museum, Brisbane, the above specimen in balsam (mounted with specimens of *Gonatocerus spinozai* and the type of *Polynema nordaui*, described beyond).

3. *Polynema nordaui*, new species.

Female: Length 0.60 mm. Small for the genus.

Black, the first three antennal joints, abdominal pedicel, legs except distal half of posterior femur and distal tarsal joints, orange yellow. Like the North American *longipes* Ashmead, being about the same size and habitus, but differing in that the wings of *longipes* are much narrower and slender and the antennal segmentation entirely different, since in this Australian species the second and third funicle joints are long and subequal. Very much like *draperi* in wing structure, but the legs are brighter and orange. Funicle joint 1 longer than the pedicel, joints 2 and 3 longest, subequal, elongate, one and a half times longer than 1, joint 4 a fourth shorter, 5 shorter, somewhat enlarged, somewhat longer than 1. Scape moderate in length.

(From one specimen, enlarged as in preceding.)

Male: Not known.

Described from one female, captured with the preceding species.

Habitat: Australia—Nelson (Cairns), Queensland.

Type: No. *Hy 1298*, Queensland Museum, Brisbane, the above female in balsam (mounted with *Gonatocerus spinozai* and the type of *Polynema mendeli*).

Respectfully dedicated to Max Nordau.

This species may be the female of *draperi*, which it resembles closely, but there are differences which make me doubt it, especially in the shape of the fore wings, the relative length of the cephalic marginal cilia of those wings and the differences in colour.

SOME NEW AUSTRALIAN GENERA IN THE HYMENOPTEROUS FAMILIES EURYTOMIDÆ, PERILAMPIDÆ, EUCHARIDÆ AND CLEONYMIDÆ.

BY A. A. GIRAULT, NELSON (CAIRNS) N. QUEENSLAND.

Family Eurytomidæ,

Eurytomini.

Xanthosomoides, new genus.

Female.—Non-metallic, yellow, body not umbilicately punctate, fore wing with a stigmate spot at the stigmal vein. Head normal, the antenna inserted in the middle of the face, 11-jointed, the club solid, the funicle 7-jointed, cylindrical, its joints not much longer than wide, the single ring-joint rather stout, the pedicel nearly as long as the first funicle joint, the scape rather long, simple. Wings large, the marginal vein long and slender, at least two-thirds the length of the long submarginal vein, thrice or more the length of the rather short stigmal vein, the postmarginal vein also very long, nearly as long as the marginal or quite equal to it (or slightly longer than it), tapering distad. Wings normally ciliate, the marginal fringes short. Abdomen as long as the thorax, the ovipositor and its valves exserted, curved upward, fully as long as the rest of the body. Abdomen sessile but narrowed at base, triangular from lateral aspect. Propodeum slightly shorter than the scutellum or the prothorax and simple, without carinæ. Parapsidal furrows complete. Eyes ovate, the ocelli in a triangle in the centre of the vertex, the lateral ones distant from the eye margins. Propodeal spiracle elliptical. Face subquadrate, wide.

Male.—Not known.

A genus related to *Xanthosoma* Ashmead, from which it differs in bearing the much longer marginal and postmarginal veins and non-moniliform funicle joints.

Type.—The following species:

1. *Xanthosomoides maculatipennis*, new species.

Female.—Length, variable, 2.50 mm. exclusive of ovipositor, the exserted portion of the latter about the same length.

Pale cadmium yellow, the head, pronotum, legs and a transverse spot laterad of the mesopostscutellum (the spot cephalad of the propodeal spiracle) contrasting yellow, lighter, lemon yellow; also more or less, the produced part of the ventral abdomen. Lateral suture of scutellum, the visible (dorsal, lateral) portions of the occiput and the cephalic margin of the propodeum, black. Dorsal aspect of abdomen suffused irregularly with brown. Venation black. Fore wings hyaline, but with a conspicuous, rather large, black globe-like stigmal spot, regularly oval in shape, obscuring the curved stigmal vein and appearing as if suspended by a short pedicel from the end of the marginal vein. Ciliation normal and dense, the marginal fringes very short. Antennæ yellow suffused with much black. Ovipositor brown, the valves black. Thorax delicately, transversely wrinkled. Club solid, first funicle joint widening distad, nearly twice longer than broad at apex. Scape yellow, black above.

(From three specimens, the same magnification.)

Male.—Not known.

Described from three females on cards from the collections of the Queensland Museum, labelled "Bred out of Gall 5 A. Brisbane. H. Hacker."

Habitat: Australia—Queensland (Brisbane).

Type: No. *Hy 1192*, Queensland Museum, Brisbane, one female on a card.

2. *Xanthosomoides fulvipes*, new species.

Female.—Length, 4.2 mm. excluding the ovipositor, the latter exserted for a length nearly equal to that of the body. The same as *maculatipennis*, but much more robust; also the pronotum is nearly as dark as the scutum, its caudal margin contrasting lemon yellow followed by a narrow black stripe running across the cephalic margin of the scutum; the propodeum is wholly black or very dark, the ovipositor fuscous, the meson of the thoracic venter is black and in the venter of the prothorax there is a distinct triangular black marking like the Greek letter Delta of the capital case. The postmarginal vein is slightly longer than in the first species, the apparent petiole of the stigmal spot in the fore wing also longer. Vertex dark ochreous, the face lemon yellow. The wings are large. First funicle joint longer, cylindrical, more than twice longer than broad at apex.

(From a single specimen, the same magnification.)

Male.—Not known.

Described from a single female specimen from the collections of the Queensland Museum, labelled "Brisbane, H. Hacker. 3-7-1911."

Habitat: Australia—Brisbane, Queensland.

Type in the Queensland Museum, Brisbane, the foregoing specimen on a card.

Melanosomellini, new tribe.

Antennæ 12-jointed, with one ring-joint, the club 3-jointed, the male antennæ different and bearing long ramii; otherwise as in the Eurytomini and Rileyini as limited by Ashmead. The marginal vein two and a half times its own width, but shorter than either the stigmal or postmarginal veins. Probably differing totally in habits from the Rileyini, since the latter appear to be egg parasites of the Orthoptera. The following genus:

Melanosomella, new genus.

Female.—Head (cephalic aspect) slightly wider than long, the antennæ inserted slightly below the middle of the face, the scrobes short and not deep, the lateral ocelli far distant from the eyes; pronotum not long, the parapsidal furrows complete, the head and thorax smooth. Antennæ 12-jointed, the ring-joint large, nearly as long as wide, the funicle apparently compressed, the joints all transverse and lamellate or produced toward one side, the distal joint much less so and largest of the funicle; pedicel subquadrate, much longer than the proximal funicle joints; club long-ovate, longer than the cylindrical, simple scape, its joints obliquely truncate, the distal joint short and conic. Fore wings normal, the marginal cilia sparse and short. Propodeum with a bright median carina, its spiracle large and nearly round. Abdomen short and stout, no longer than the thorax, its second segment occupying half of the surface. Scutellum longer than the propodeum. Parapsidal furrows complete. Posterior tibiæ apparently with but one apical spur.

Male.—The same, the abdomen more depressed and cylindrical; antennæ entirely different, the scape much shorter, dilated ventrally, the antennæ 12-jointed, the pedicel not much longer than thick, the ring-joint like a ring, the first funicle joint very transverse and lamellate; following five funicle joints very transverse and increasing in length, each bearing a long, curved, cylindrical ramus from its disto-lateral margin, joint 2 no longer than the diameter of its ramus and practically forming a continuation of it; joint 3 slightly larger than wide; joint 6 much longer than wide; the ramii longer proximad, the shortest and distal one distinctly longer than any single joint of the antenna. Proximal joint of club elongate, obconic, forming half of the club and longer than the distal funicle joint; the other two club joints subequal. Funicle and ramii with sparse, long fine hairs.

Type.—The following species (flavipes).

Female.—Length, 3 mm.

Black and shining, the face, genæ, legs (except coxæ, the tarsi more brownish), scape (except at tip, where it is blackish) the margin of the eyes dorsad and caudad more or less obscurely, lemon yellow; the black of the vertex at the meson projects obtusely into the yellow of the face, some distance directly cephalo-ventrad of the cephalic ocellus. Tegula brownish. Fore wing with a distinct brownish band nearly across it from the apex of the submarginal vein; this stripe is interrupted. Venation black. Antennal flagellum brownish, subfuscous. Face with thimble punctures; remainder of body apparently simple and shining more or less.

(From one specimen, the same magnification.)

Male.—The same, but the fuscous stripe on the wing subobsolete.

(From one specimen, the same magnification.)

Described from a single pair received for study from the Acting Government Entomologist of Victoria, cardmounted and labelled "From unknown galls on Eucalyptus, N.S.W."

Habitat.—Australia—New South Wales.

Types: No. *Hy 1193*, Queensland Museum, Brisbane, the above specimens (2 pins) plus a slide bearing male and female antennæ.

Family Perilampidæ.

Epiperilampus, new genus.

Female.—The same as *Perilampus* Latreille, but the thorax not coarsely punctate but only with scattered thimble-punctures and transversely wrinkled, the antennæ with two large ring-joints and a well-defined, 3-jointed club, the joints of the flagellum distad transverse, the pedicel larger than the first funicle joint. Marginal, stigmal and postmarginal veins shortened but still moderately long, yet the postmarginal is somewhat shorter than the other two, which are subequal; the stigmal vein with a slender neck. Fore wings with a fuscous blotch under the end of the submarginal vein. Antennæ inserted in the middle of the face, the head more or less lenticular from cephalic aspect. With an encyrtine habitus. Scutellum simple. Axillæ separated. Discal ciliation of the fore wing not quite normal. Second segment of abdomen nearly half the latter's length, the third short.

Male.—Not known.

A genus resembling *Perilampus*.

Type: The following species:

1. *Epiperilampus xanthocephalus*, new species.

Female.—Length, 2.5 mm.

Orange yellow, the parapsides cephalo-mesad marked with metallic bluish, the propodeum and abdomen shining blackish or dark metallic bluish, but the latter in the dorsal aspect of the base of its distal half with a conspicuous yellow marking, incised medially from behind (caudad). Legs nearly all dark metallic bluish, but with brownish markings at the knees and tarsi, the cephalic tibiæ nearly all brown. Venation brownish, the marginal, stigmal and postmarginal veins lemon yellow; the fore wings lightly embrowned throughout and with a distinct, smoky brown cloud under the apex of the submarginal vein, extending across the wing, but interrupted caudad of its middle by a clear longitudinal streak; its proximo-cephalic margin is accented and another shorter clear streak enters it from proximad nearer the caudal wing margin. Marginal fringes extremely short, as is also the discal ciliation, which is speckled over the wing surface like minute pin-points, quite irregular but not dense. Scape yellow, dark above and at tip, the remainder of the antenna brownish yellow, sometimes bluish, proximal joint of club subequal to distal funicle joint, both wider than long; funicle joints 2 and 3 subquadrate, subequal. Thorax finely polygonally sculptured, the scutum with obscure punctures.

(From many specimens, the same magnification.)

Male.—Not known.

Described from a number of specimens in the Queensland Museum, mounted on cards labelled respectively: "Gall, No. 6 Brisbane. H. Hacker. 19-7-11." 4 ♀'s. Types:"Gall No. 6; 3 ♀'s": "Gall No. 6," three cards 5 ♀'s, 5 ♀'s and 6 ♀'s; and "Gall No. 6. Brisbane. H. Hacker. 19-7-11." 3 ♀'s. Evidently reared from galls.

Habitat: Australia—Brisbane, Queensland.

Types: No. *Hy 1194*, Queensland Museum, Brisbane, the four females on a single card as above noted, plus a slide of xylol-balsam bearing an antenna and a pair of wings.

Family Eucharidæ.

Epimetagea, new genus.

Female.—The same as *Metagea* Kirby, but the antennæ not moniliform and only 10-jointed. Also agreeing somewhat with *Pseudochalcura* Ashmead, but differing again in lacking one antennal joint. Head thin, triangular, the antennæ inserted slightly below the middle of the face, 10-jointed, the club solid and ovate, longer than any of the funicle joints, but slightly shorter than the simple, cylindrical scape; pedicel obconic, short, subequal to joint 4 of the funicle, bearing from one side of its apical margin a single, very long, slender but stiff bristle-like seta, which reaches distad nearly to the apex of joint 3 of the funicle. Proximal funicle joint longest, nearly twice the length of the pedicel, all the funicle joints obconic, widening distinctly distad, all more or less prolonged obtusely from one apical corner, the distal joints more so. None of the joints petiolate or subpetiolate; no ring-joint. Mandibles long and falcate, acute at apex, the right with two large triangular teeth within, the left one which is larger than either of those of the right; also exteriorly at base each with a large tooth. From beneath the clypeus there projects a flat, palmate (9-digitate) brownish plate, above and between the mandibles; clypeus convex along the distal margin, the latter with two teeth on each side of its end, the first very obtuse, the second more tooth-like, but not large. Ocelli nearly in a straight line across the vertex, the cephalic one within and at the apex of the short scrobicular cavity. Parapsidal furrows complete, with deep punctures. Scutellum normal, terminating in a short plate whose distal margin is entire though convex. Thorax elevated convexly in places, but the convexities obtuse. A rather large, tooth-like plate from the lateral aspect of the thorax some distance beneath the axilla. Thorax with large, irregular reticulations or narrow carinate lines, but not punctate excepting the large punctures in sutures. Abdomen with a distinct petiole, which is moderate in length, depressed, diamond-shaped from dorsal aspect, opaque. Proximal tarsal joints of all the legs long and slender. Venation obscure, the stigmal and postmarginal veins short, much shorter than the marginal, the stigmal the longer of the two, curved or bent like a boomerang. Wings hyaline, all ciliation nearly absent; a trace of marginal cilia disto-caudad. From lateral aspect, scutellum appearing as if terminating in a short, acute tooth.

Male.—Not known.

Type: The following species (*purpurea*).

1. *Epimetagea purpurea*, new species.

Female.—Length, 3.5 mm.

Metallic purple, the abdomen with metallic green reflections; knees, tibiæ, tarsi (except distal dark part of distal joint) and the antennæ, brown, the latter suffused with purplish distad. Venation nearly invisible, but the stigmal vein brownish. Head impunctate, but with very fine circular striæ; lateral ocelli very distant from the eye margins; scrobicular cavity with its lateral margins noncarinate; a tubercle at latero-cephalic aspect of pronotum. Scutellum between and behind the axillæ (at the meson) sunken. Abdominal petiole longitudinally striate. Cephalic part of thorax dorsad (cephalad of the middle of the scutum) coarsely reticulate, as is also much of the scutellum. Base of propodeum with deep, transverse foveæ.

(From three specimens, the same magnification.)

Male.—Not known.

Described from three female specimens kindly given to me by Mr. F. P. Dodd, mounted together on a card labelled "From ant pupæ. Townsville, July 1902."

Habitat: Australia—Townsville, Queensland.

Types: No. *Hy 1195*, Queensland Museum, the above specimens (two more or less mutilated) on a single card, plus a slide of xylol-balsam bearing female head and antennæ.

Family Cleonymidæ.

Chalcedectinæ.

Calosetroides, new genus.

Female.—Allied to *Amotura* Cameron, but the front femora are swollen, compressed and excised beneath at apex, the posterior femora unarmed beneath. Legs unarmed otherwise; cephalic tibiæ somewhat compressed; caudal coxa compressed, flat interiorly, the caudal femur enlarged but unarmed; caudal tibiæ with two unequal spurs, both rather large. Tarsi five-jointed. Antennæ inserted distinctly below the ventral ends of the eyes; very near the clypeus, the scape obclavate and long, the flagellum 9-jointed, no ring-joint. Scrobicular cavity long, but not including the cephalic ocellus, the lateral ocelli separated from the eye margin, the three in a small triangle in the center of the vertex; eyes somewhat convergent above, long-ovate, naked. Bulbs separated by a long, acutely triangular raised area in the scrobicular cavity. Genæ long, genal suture distinct. Pronotum incised at meson. Parpasidal furrows complete, the axillæ rather widely separated. Scutellum simple, its caudal margin carinate and preceded by a line of deep punctures separated by narrow, short carinæ. Propodeum with a short, solid acutely margined median carina, which is V-shaped and margined on each side by a broad sulcus; the spiracle cephalad, large, elliptical. Abdomen sessile, the ovipositor not exserted, the abdomen not any longer than the head and thorax combined, flat above, acutely conic-ovate, its second segment smooth, forming nearly half of the surface. Wings infuscated; marginal vein long, only slightly shorter than the submarginal, the stigmal and postmarginal veins also long, the former curved, only half the length of the postmarginal, which is three-fourths the length of the marginal. Metallic, large.

Male.—Not known.

Type: The species *australica*, described forthwith.

1. *Caloseteroides australica*, new species.

Female.—Length, 5.65 mm.

Metallic purplish with aeneous tinges, the face metallic green; legs reddish brown, the coxæ, the posterior femora (exteriorly only) concolorous, the intermediate tibiæ promixad and exteriorly and the cephalic tibiæ exteriorly or along the outer margin, black. Wings with a distinct, large embrowned subsagittate cloud in its middle, longitudinally, the area appearing as if hung by one of the lateral angles from the apex of the stigmal vein; also there is an elliptical spot suspended from the apex of the submarginal vein. Antennæ black, the scape concolorous. Head and thorax granulately punctate.

Male.—Not known.

Described from a pinned female received from the Acting Government Entomologist of Victoria, labelled "Millbrook, Victoria."

Habitat: Australia—Victoria (Millbrook).

Type: No. *Hy 1196*, Queensland Museum, Brisbane, the above specimen, plus a slide, bearing fore wing, the legs and antenna.

SOME FOSSIL INSECTS FROM FLORISSANT, COLORADO.

BY T. D. A. COCKERELL, UNIVERSITY OF COLORADO, BOULDER, COL.

The insects now described have a very modern aspect. The anal cell of *Venallites*, taken by itself, may be thought of as primitive, but the fly is otherwise a specialized type. Certainly there has been little advance in insect evolution since the Miocene, but many genera have become extinct.

HOMOPTERA

Echinaphis new genus (Aphididæ)

Stout, with long antennæ; the two basal joints short as usual; the first somewhat gibbous at apex on inner side; front broad; abdomen with six longitudinal rows (the outer was lateral) of about six very strong black spines; the apex of abdomen, which is broad, with a transverse (marginal) row of six still larger and stronger spines; cornicles not evident, probably small; hind wing of rather coriaceous texture, the venation essentially as in *Chaitophorus*. Anterior wings not preserved in type.

Echinaphis rohweri, n. sp.

Length, 3 mm.; width of abdomen, 1.75 mm.; length of hind wing a little over 2 mm.; dark colored, with the anterior legs clear ferruginous; wings reddish; front and sides of thorax without hairs. The following measurements are in microns: Width of front between eyes 320; length of first antennal joint 128; of second 80; antenna, from base of third joint to apex, 1665; length of a dorsal spine about 160; of a caudal one about 270; distance between the wing-veins (Cu and M.) at base (separation from Rs) about 112. The veins are nearer together at base, and less parallel than in *Chaitophorus populicola*.

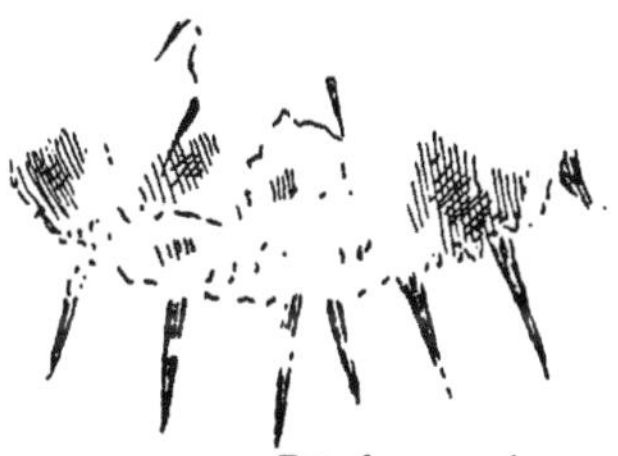

FIG. 9
Echinaphis rohweri. Ckll.

Miocene shales of Florissant, Station 13 (*S. A. Rohwer*).

This singular species is quite unlike any of the fossil Aphids previously described from Florissant. In the development of spines, it has a certain resemblance to the living *Chaitophorus spinosus* Oestl., found on oak in Minnesota. *Sipha glyceriæ* (Koch), which is also spiny, has much shorter antennæ. Close to the type of *Echinaphis rohweri* is an elongated, minutely reticulated object 690 microns long, shaped like the egg of *Aedes grossbecki*, but broader, with the reticulations considerably more minute and rather transverse than longitudinal, in the manner of *Aedes colopus*. It is, I think, a mosquito egg, and is the first fossil from Florissant I have been able to refer to the Culicidæ.

DIPTERA.

Asilus peritulus Cockerell.

Two wings from Station 14 (Geo. N. Rohwer).

Verrallites, new genus (Bombyliidæ).

A genus of slender-bodied Bombyliidæ, with clavate but not much elongated abdomen, characterized especially by the anal cell being very widely open, its width on margin in the typical species 720 microns, which is a slight fraction more than the width of the third position cell on the margin. Head and thorax apparently bare; abdomen sparsely minutely hairy; costa minutely bristly; auxiliary vein longitudinal, reaching costa near (apparently a little before) middle of wing (practically as in *Lordotus*); marginal cell long and narrow, its lower side gently concave, its apex broadly rounded, the second vein turned basad before reaching the costa (the cell practically as in *Lomatia lateralis*, except that the outer angle with costa is more acute in the fossil); two submarginal cells the second elongate, widened apically (about as in *Phthiria pulicaria* except that the upper nervure curves upwards apically, more as in *Geron*); four posterior cells, the first nearly parallel-sided throughout (in the manner of *Phthiria*), the others widely open, the third very broadly open (much as in *Ploas virescens*, only much longer); fourth posterior narrowed basally and extremely widely open apically (*Phthiria*-style, only more elongated); anterior cross-vein far beyond middle of discal cell, beyond the beginning of its last third.

Verrallites cladurus, n. sp.

Length, about 7 mm., with the abdomen gently curved; abdomen with a depth of 2 mm. near apex; wings 5.75 mm. long. Head and thorax probably black in life; abdomen apparently brown, the sutures broadly colourless; wings clear hyaline.

Miocene shales of Florissant, Colorado, Station 13 B (Univ. of Colorado Exped.)

This remarkable genus is dedicated to G. H. Verrall, whose writings are invaluable to students of fossil Diptera, although he studied only living forms. In Williston's table (N. Am. Diptera, 3rd. Edit.) it runs to 29, and the wings, except for the anal cell, show a rather close general resemblance to those of *Lepidophora*. In *Legnotomyia* the anal cell is as widely open on the wing margin as the third posterior, but these cells are not nearly as wide as in *Verrallites*; the discal cell in *Legnotomyia* is also much shorter than in the fossil, and there are other important differences. From all the genera of fossil Bomlyliidæ from Florissant, *Venallites* is easily known by the form of the anal cell.

We are still without a single Tachinid or Muscid s. str. from the Florissant shales. *Glossina* alone (two species) represents the whole series of Calyptrate Muscoids! In Coleoptera, we are still without a Histerid or Cicindelid. A *Cypris* is, so far, the sole representative of the Crustacea. The total number of species described is now so great that these blanks become significant. In the Neuropteroid series we have plenty of Ephemerids and Termites; numerous Raphidiids, Chrysopids and Hemerobiids; a Nemopterid and an Embiid; but as yet not a single Perlid. The Panorpids are represented by three species. We have no less than five species of the Dipterous family Nemestrinidæ, now so rare in this country. The quite numerous Bombyliidæ, as well as the very numerous Aphididæ, *all* belong to extinct genera; but the Phoridæ, Syrphidæ, Therevidæ, Leptidæ, etc., are referable to genera still living.

HYMENOPTERA.

Alysia ruskii, n. sp.

♀—Robust, length almost 5 mm.; anterior wings broad, broadly rounded at apex, nearly 4 mm. long; expanse about 9 mm.; head and thorax black; base of abdomen (apparently two segments) clear ferruginous, the rest black or dark brown; antennæ nearly 3 mm. long, dark, thick, the joints just before the end about as broad as long, with a diameter of about 110 microns; legs ferruginous, the hind femora incrassate, suffused with dark brown, the base broadly and apex more narrowly pallid; hind tibial spur long and sharp; head and thorax apparently closely but shallowly punctured;

parapsidal grooves of mesothorax distinct, entire; width of abdomen nearly 1½ mm.; wings hyaline, slightly dusky because minutely hairy all over, the hairs dark; nervures ferruginous, very distinct; costa not bristly; stigma large, about 720 microns long and 320 deep; a linear, hardly noticeable, costal cell; basal nervure leaving costa very obliquely near base of stigma, its lower part very strongly arched, its lower end only about 320 microns in a straight line from subcosta; marginal cell subtriangular, sharply pointed, about 930 microns long, its lower side beyond the submarginal cells faintly concave (bulging inward); first s.m. diamond-shaped except for the large part cut off by the stigma, its basal end only a short distance down nasal nervure; first section of radial or marginal nervure having stigma beyond middle, nearly at right angles; second section nearly obsolete, but marked by the bend in the nervure; second t.c. wholly obsolete, but marked at each end by an angle in the nervure where it should arise; recurrent nervure exactly meeting first t.c.; lower end of b.n. basad of t.m. a distance equal to rather more than half of latter; t.m. very oblique; second discoidal complete.

Florissant, in the Miocene shales (*Willard Rusk*). Type U. of Colorado Museum, 4903. Easily known from the two species described by Brues from the Florissant shales by the obsolete second t.c. Except for this the venation is nearly as in *A. petrina* Brues, except that the first section of radius is about as long as second, the marginal cell is narrower apically, the b.n. is strongly bent (straight in *petrina*), and the second s.m. has its apical corner more produced. The linear costal cell is not different from that seen in other forms in which this cell is described as "absent," because it is not readily seen without a microscope. According to Ashmead's tables, the absence of the second t.c. would throw it in Dacnusinæ; but, as Marshall observes, in true Dacnusinæ the radius beyond the first section presents an unbroken curve, without any angle where the second t.c. should be inserted. In the meeting of the a.n. and first t.c., *A. ruskii* resembles *Alysia* (*Goniarcha*) *atra* Hal., but that species has the first s.m. with a broad side on b.n. In the shape of the first discoidal cell, the fossil is suggestive of *Dacnusa* (*Phænolexis*) *petiolata* Nees.

Alysia ruskii should perhaps form a new genus near *Alysia*, but it seems better to leave it in *Alysia* sens. latiss.

Heriades saxosus, n. sp.

♂—Length about 7¼ mm., in a rather contracted state, the abdomen strongly convex dorsally in profile; head and thorax dark brown, probably black in life; abdomen lighter and redder; wings hyaline, very faintly dusky; anterior wings 4 mm. long; venation as in *H. sauteri* from Formosa, except that lower section of basal nervure is more arched, the marginal cell is considerably longer and more pointed, and the bend in the second t.c. is less distinct. As in *H. sauteri*, the second a.n. squarely meets the second t.c. The following measurements are in microns: Length of marginal cell 1152; depth of marginal cell 304; greatest (diagonal) length of first s.m.. 768; second s.m. on marginal, 240; lower side of second s.m. 544; second s.m. on first discoidal 80; greatest (diagonal) length of first discoidal about 976. The basal nervure practically meets the transversomedial, which, as usual in Heriadines, is oblique, the lower end most basad.

Florissant, Colorado, in the Miocene shales; Station 14 (*W. P. Cockerell.*)

Among the fossil bees hitherto found at Florissant, this comes nearest to *Heriades laminarum* Ckll., but is smaller, with the second r.n. meeting second t.c., and the b.n. hardly falling short of the t.m. The apex of the marginal cell is pointed, if rather obtusely, not rounded. The first r.n. joins the second s.m. at a distance from its base equal to a little over a third of the length of the first t.c., the latter being about 224 microns long. The stigma is well developed.

CONCERNING THE REPUTED DISASTROUS OCCURRENCE OF *VANESSA CALIFORNICA* IN OREGON AND CALIFORNIA

BY J. MCDUNNOUGH, DECATUR, ILL.

In the April number of the CANADIAN ENTOMOLOGIST, Prof. F. M. Webster of the Bureau of Entomology, Washington, D. C., recounts several instances of devastation of crops and foliage which he attributes to the larvæ of *Vanessa californica*. A careful study of the various letters quoted convinces us that in all but the last instance the author is in error in determining the larvæ as belonging to this species.

In the Proceedings of the California Acadamy of Science, June 7th, 1875, Hy. Edwards gave a detailed account of the larva of *Californica*, citing the food plant as *Ceanothus;* according to this account the larva is jet black, *strongly spined* (a characteristic of all *Vanessa* larvæ) with five branched spines on each segment, the middle spine being bright-yellow at the base; at the bases of the spines are bright, steel-blue tubercles and between them numerous circular, whitish-yellow dots, giving the appearance of a yellow dorsal line. It is a well-known fact that the larvæ of the various *Vanessa* species are restricted to one or two food plants and it would be a most extraordinary proceeding if a Vanessid larva, normally restricted to *Ceanothus* as a food plant, should suddenly be found devastating alfalfa and garden truck.

Taking the various reports in order, we note from that of Mr. T. V. Hall of Lakeview, Oregon, that the "worm" which had destroyed the alfalfa crop was brownish colcur, with *sleek* appearing surface. This description could hardly, even by the most ignorant, be drawn up from the jet black, heavily spined Vanessid larva; it could, however, easily apply to any one of the "cut-worm" species.

The next letter, from Mr. A. J. Swift of the the same locality, reports the occurrance of vast swarms of *californica* a month after the crops had been ravaged by a "worm" varying from bright green to nearly black, according to its food supply. There is nothing, except the imagination of the writer and the appearance of the butterfly at a later date than the larvæ, to connect the two. The swarms of the butterfly, which doubtless was *californica*, may be accounted for either as due to imaginary instincts or to the fact that the larvæ had actually bred in numbers on *Ceanothns* in the high valleys, a feature which would naturally not be observed by farmers, who are principally interested in their crops.

In the report from Mr. J. J. Monroe of Willow Ranch, California, we note one feature that would absolutely preclude the determination of the destructive larvæ as *californica*, i. e., the fact that they burrowed in the ground during the day, feeding by night. This is characteristic of "cut-worms" but unknown in Vanessid larvæ, which remain on their food plants continually, usually feeding gregariously by day.

Mr. Webb's report from Waldo, Oregon, actually does deal with *californica*. He cites the larvæ as completely stripping the foliage off grease-wood and mountain lilac. We do not know just what is meant by this latter plant, but believe that *Ceanothus* is often locally called grease-wood. From this report it would seem that there is some danger, when vast numbers of the larvæ are present, of fruit trees being attacked, but it is apparent that only when the natural food supply is exhausted would this occur. We note that it is distinctly stated that "they seemed to care for nothing to speak of but grease-wood and lilac", and the fact that "tons" of them perished on water and land in their vain search for a further food supply only goes to support our previous statement that *californica* is very restricted in its choice for food plants and the idea of its being held responsible for damage to alfalfa and other crops may be banished as so improbable as to be almost ridiculous.

ANNUAL MEETING OF MONTREAL BRANCH

The fortieth annual meeting of the Montreal Branch of the Entomological Society of Ontario was held at the residence of Mr. Henry H. Lyman on Saturday evening, May 17th. Mr. G. A. Southee, President, occupied the chair, seven members being present.

After the reading of the minutes and election of Mr. G. M. Henderson as a member the reports of the council and of the treasurer were read and adopted. The president delivered his annual address dealing with the good work accomplished by members of the branch in spite of the exceptionally unfavourable weather conditions, several new species and varieties of moths having been discovered as well as some rare captures, notably *Hepialus auratus*, the second Canadian specimen.

The election of officers resulted as follows: President, A. F. Winn; Vice-President and Librarian, G. Chagnon; Secretary, Geo. A. Moore; Treasurer and Curator, Henry H. Lyman; Members of Council, G. A. Southee, E. C. Barwick, G. H. Clayson.

GEO. A. MOORE, SEC., 850 St. Hubert Street.

FURTHER NOTES ON ALBERTA LEPIDOPTERA.

BY F. H. WOLLEY DOD, MIDNAPORE, ALTA.

(Continued from page 192.)

403. *A. excelsa* Ottol.—I have no local captures in my collection, but several from Banff, July 30th–August 19th (Sanson). Under the description Dr. Ottolengui mentions having three specimens from Laggan, and claims to have seen many more from there. I have one from Field, B.C., and a few from Kaslo. My series are all much alike, and one or two agree concisely with Ottolengui's figure. I have *angulidens* from Colorado, and, though closely allied, I believe they are distinct. The difference was pointed out by Ottolengui. I would say, in addition, that whilst in *angulidens* the outer stroke of the U portion of the sign is evenly out-curved, the outer stroke of the V in *excelsa* is either direct or in-curved for the lower two-thirds of its length. In both it has generally a slight inward hook at the tip. *Vaccinii* appears to be another very close ally. There was a series of that in the Washington collection from the White Mountains, N.H., in which the sign seemed to me very variable. Also associated with them, justly as far as I could judge, was an unset Kaslo specimen, recorded by the name in the Kaslo list. I have suspicions that this specimen was really *excelsa.*

I have what I feel sure is another slightly larger closely allied species from Kaslo and Nelson, B.C., which was recorded in the Kaslo list as "*u-aureum* Guen." but which was sent me subsequently by Mr. Cockle as "*u-aureum* of the Kaslo list, but *excelsa* by Dr. Barnes, compared with Ottolengui's naming." I feel sure that *excelsa* is wrong for this form, and I am by no means satisfied that it is *u-aureum.* Compared with both *excelsa* and *angulidens,* it has a wider open sign; in fact, more rectangular than V or U-shaped. I may call it a more *octoscripta*-like sign, more resembling that of *arctica* than of any other of Ottolengui's figures. The outer spot is in every specimen larger than in *excelsa,* and sometimes hollow—that is to say, dark filled centrally, and more often touches the outer line of the larger sign at varying points. It is a slightly larger species, but as regards the rest of the maculation and color of the primaries, there is really very little difference. The secondaries differ, however. In *excelsa* the secondaries may be described as dull fuscous, with a broad but ill-defined yellowish white median band. The outer fuscous border is rather narrow, and the pale median band is fuscous suffused. My specimens of *excelsa* all agree exactly with Ottolengui's figure in this respect. In the other species the secondaries are better described as yellowish white, slightly fuscous at the base, and with a broad fuscous outer band, occupying the outer third of the wing. The central portion of the wing is thus much dirtier in *excelsa*, but the outer border narrower. On the underside *excelsa* is more suffused with gray and fuscous than the unknown species. The discal spot on secondaries beneath in *excelsa* is scarcely more than a point. In the unknown species it is obviously V-shaped. In both species spines are usually, but apparently not always, present on the hind tibiæ.

If the Kaslo specimens formerly recorded as *u-aureum* were subsequently named *excelsa* after a comparison with a co-type of that from Jefferson, New Hampshire, which my notes tell me I saw in the Washington collection, then it is possible that the co-type in question is not *excelsa*. Of course, Dr. Ottolengui may have mixed these two species in his description, but I am taking it for granted that his figure represents the type.

The Kaslo and Nelson specimens in question have a most remarkable resemblance to Mr South's most excellent photo-lithograph figures of *interrogationis* Lin., Pl. 26, figs. 4,5, of his "Moths of the British Isles,"Series ii.,though I appear to have overlooked the resemblance in the British Museum, if, indeed, I noticed the Linnean species at all. I had a Nelson specimen with me, and it did not satisfy me as agreeing with the *u-aureum* of that collection. I noticed several similar B.C. specimens there however, standing under more than one name. It differs most obviously from what I have listed as *octoscripta*, which it sometimes nearly resembles in the sign, by a totally different arrangement of color, the less crenate t.p. line, and the absence of blackish dashes both before and after the s.t. line.

I have seen specimens of it standing under *celsa*, described, I believe, from Oregon, and have two from Duncans, Vancouver Island, which appear to be the same species, though slightly larger and with sharper contrasts, one of which agrees with Ottolengui's figure of *celsa* in every detail except the sign. In this one specimen, as in the figure, the inner part of the sign is V-shaped. Mine has, instead of a tail, a large round outer dot touching the lower angle of the V. The fact that one of my eight specimens has an almost V-shaped sign, and the rest have it nearer rectangular, does not indicate greater variation than exists in *californica* and other speciesin my collection. On the strength of this Duncans specimen, which I may remark bears some resemblance to a small *viridisignata*, it seems not unlikely that *celsa* may turn out to be at least one of the correct names for my Kaslo and Nelson species. I quite expect ultimately to find at least a close relationship to *interrogationis*.

As regards the great variation known to exist in the signs of some species of this genus, the late Mr. Tutt's remarks concerning *interrogationis* in the British Isles are interesting. "The great character in this species is the endless variation which the central silvery marks or characters undergo. Truly no two are alike, and to look down a long series at this mark, is something like looking at a series of Chinese characters. Some are like the normal mark in *iota* and *pulchrina*, composed of a V and a dot; others have them united as in *gamma*; others again are like the Greek ϵ; one forms a tiny solid blotch as in *bractea*, and so on." (British Noctuæ and their Varieties, IV. 36, 1892.)

As to *u-aureum*, which Ottolengui claimed was not a North American species at all, and further remarked that the description associated it with *interrogationis* (Journ, N.Y. Ent. Soc. X. 69, June 1902), it may be observed that the only localities given for it in Staudinger's Catalogue are Greenland, Labrador, and North America. He also places "*Interrogationis* var. *grœnlandica* Staud." as a synonym. The types of *u-aureum* are probably in Mr. Oberthur's collection at Rennes. which by an unfortunate chance I just missed seeing in March 1912. Under the name in the British Museum were three specimens supposed to be North American. One had label "United States" at side. Sir George Hampson wrote me concerning the species: "Our specimens are from the Grote collection without exact locality. It is considered that the types really came from Labrador, and not from Dalecarlia, Sweden, as described." Concerning the Grote collection specimens, during my first visit to the British Museum early in 1909, I wrote,

"I can't distinguish them from *vaccinii*,"of which there was a ♀ from Mt. Washington. About this specimen I wrote: "Darker than the *u-aureum* series, but seems to me exactly like it." My sketch of the sign of the three specimens shows that it was exactly like some in my unknown species, which I call "*u-aureum* of the Koctenai list." But in February, 1912, I compared a Nelson, B.C., specimen with them and do not seem to have found that they matched it. This time my notes read: "The *u-aureum* of this collection is not improbably *zeta* Ottol., judging by the figure of the type of that, though the t.a. (in *u-aureum*) seems less even, and in none does the outer spot join top of sign. Secondaries are alike exactly, but basal area of primaries seems paler in *zeta*." The description of the latter was made from a single ♀ from"North West Territory" and came from Mr. Jacob Doll's collection. I have nothing to match it exactly, but it appears to be of this group.

406. *A. falcifera* Kirby—Dr. Ottolengui's remarks on this species appear correct. Kirby described the grey form from Nova Scotia, and *simplex*, of which the type is a female in the British Museum, from Trenton Falls, N.Y., is a very dark brown specimen. I have tried hard to recognize two species in these forms, noticing that most *falcifera* seemed to have a smaller and more slender sign. This difference is not constant, however, and I must admit that I can discover no other means whatsoever of separating them except by color, in which they grade easily through.

It seems hard to believe that *simplicima* Ottol., described from a single female from the State of Washington, is anything more than an unusually small *simplex*, with a sharp-pointed sign. His remark that the sign is "always knobbed in *falcifera* and *simplex*" is not correct. I have a Calgary *falcifera* in which it is sharp, though not quite as sharp as in the right wing of his figure of *simplicima*.

407. *A. orophila* Hamps.—Sir George Hampson, in CAN. ENT. XL. 105, March 1908, thus named the Rocky Mountain form previously passing as *diasema*. The description was made from six males and a female from Brobokton Creek, Alberta Rockies, and one male from Early Winter Creek, Washington Forest Reserve, all taken by Mr. Nicholl. The type is a male from the former locality, and is marked as taken at 5,500 feet, on July 10th, 1907.

Its describer remarks: "*Diasema* Bdv., ... which is found in N. Europe and Asia, and in America from Greenland to Labrador, has the head, thorax and fore wing much more strongly tinged with red-brown, the last with the antemedial line excurved below the cell, the stigma more V-shaped, with a slight tail or point beyond its lower extremity; the hind wing with the terminal area, reddish-brown."

On my first visit to the British Museum, in January 1909, I found two Hudson Bay specimens and three others—one marked Lapland, standing under *diasema*. From notes I took on them I concluded on my return home that the Banff specimen I had recorded under the name was correct. Three years later I actually compared this specimen with the *diasema* series, and concluded that it fitted *orophila* better, and that, moreover, I had never seen true *diasema* from the Canadian Rockies at all. My series at present consists of a male and four females from Brobokton Creek, August 13th, 1907 (Mrs. Nicholl), Banff, August 13th, 1900, and August 1st, 1910 (Mr. N. B. Sanson and the author), and a pair from Kaslo, B.C. (Cockle), the female dated September 10th, 1907. I have also seen a Banff male from Mr. Sanson, dated September 1st, 1909, as well as more Kootenai specimens in Mr. Cockle's collection. The course of the t.a. line varies somewhat, and so does the size and shape of the sign. Both strokes of the latter vary considerably in their course, as well as in the amount of grey space which they define. The lower stroke may be almost direct, or slightly curved, or even almost obtusely angled at about its middle. The inner one may bend outwardly or inwardly, or both ways, and may so connect with the outer as to form either an even curve, an obtuse or a right angle, or a decided tail or point. Any specimens, however, which may have been named *diasema* by me have been so named erroneously.

409. The species referred to under this heading is not *snowi* nor does it bear any close resemblance thereto. It is *microgramma* Hbn., a European species not previously recorded from North America. I have compared a local specimen with a series in the British Museum. I referred to this in 40th Rep. Ent. Soc. Ont.,

p. 118, 1909. I have only two poor specimens left in my collection. On several occasions I have made special trips to the locality at about the time for its appearance, but have not been fortunate enough to meet with the species again. It is the size of *alticola* and *devergens*, in color and maculation not unlike *orophila*, except as to the sign, which much resembles that of *californica*. It is very distinct from anything else North American.

410. *Syngrapha alticola* Walk.—Walker's type from the Canadian Rockies (Lord Derby) is in the British Museum, and my specimens agree with it. They are labelled Laggan, July 17th, 1904, and Wilcox Pass, Rockies, Alta., July 26th to Aug. 13th, 1907, Mrs. Nicholl. It flies at low altitudes (5,000 ft.), but I do not know how high it goes. Sir George Hampson, in CAN. ENT. XL. 106, March, 1908, records more of Mrs. Nicholl's captures on Mt. Assiniboine, Brobokton and Brazeau Creeks, Alta., and Kicking Horse Pass, B.C., and states that the species is quite distinct from European *devergens* Hbn. I have two specimens of the latter from the Swiss Alps, and have examined others, and believe his statement to be correct, though they are very close allies. *Devergens* has been recorded from Labrador, but I have not the literature by which to investigate either the record or the correct spelling of the name. Holland's figure of *devergens* is *parilis*.

411. *S. ignea* Grt.—I have seen neither description nor type of this species. Smith's Catalogue states that the type should be at Philadelphia, and Grote makes the same assertion in his 1895 list. Smith's reference of *ignea* to *alticola* is after Grote, who admitted that he had never seen Walker's type, and many have mistaken his species. My *ignea* is the same as that of the British Museum from the Grote collection, and the same as Holland's figure of *hohenwarthi*, misspelt, as elsewhere, *hochenwarthi*. The latter stands as distinct in all our lists. I have specimens from Alberta, Colorado, Utah, and several European localities, and am unable to recognize two species. *Divergens* Fabr. is given as a synonym by Smith and Grote, but the name is attributed to Hübner by Staudinger. I am unable to discover whether the latter name has any connection with *devergens*.

414. *Therasea angustipennis* Grt.—I have not seen the description of this species, but Hampson figures the type, a female from Bosque County, Texas. That has fewer whitish areas than any of my series, which are nearly all from Alberta, but is evidently the same species. In common with most species in this and allied genera, the males have usually much more white than the females. Some of my females have the olive brown shading on the costa from the base to the t.a. line, and in one it continues with scarcely a break to the t.p. In some the costa is almost clear except for three or four patches, some or all of which usually join the extensive brown region below the median vein. In males, the costa is on the average much cleaner, and the patches are much reduced, sometimes almost entirely lacking. Their position is sometimes indicated by distinct yellowish shades, which may extend faintly all along the costal area. Specimens with the yellow shades are var. *flavicosta* Smith, which was described as a species from five males and two females from Hot Springs, New Mexico; Colorado, and Montana. I have compared one of my specimens with all, or nearly all, the type material. A male type from Colorado in the Washington collection has the costa clear nearly to the apex, with very little yellow, indeed. The variation appears to be more common in the male sex. The species is by no means rare on the Alberta Prairies.

415–416. The specimens formerly referred to by me under these two headings appear to be all one species, *tortricina* Zeller, by the British Museum collection, which Hampson places in his genus *Tarachidia*. The typical form appears to be ochre yellow, which is my No. 416. Hampson mentions three varieties as aberrations. "Ab.I., with the markings almost obsolete," is *obsoleta* Grt., though Grote's type, from Illinois, happens to be itself obsolete, all except the left hind wing. "Ab. 2, *modesta*, grey brown, slightly suffused with yellowish white." This form occurs here, and is one described by Henry Edwards. "Ab. 3, *deleta*, dark brown, suffused with olive yellow scales, leaving the termen and cilia dark, almost without markings." I seem to have this form from here also, and it was likewise described by Henry Edwards. *Inorata* Grt. stands as a synonym. I have a series of eight specimens, taken on Pine Creek and on the Red Deer River prairie.

One is creamy whitish, as mentioned in my former notes. The series shows a gradation through.

Fasciatella Grt. is entirely distinct, and I have no authentic Canadian record. Hampson places it by itself in *Fruva* Grt. I have an Arizona specimen compared with the type in the British Museum, from Texas.

417. *Drasteria erechtea* Cram.—The species I have listed under this name is apparently that of which Holland figures both sexes on Pl. XXX., figs. 14 and 15, the latter figure as *crassiuscula*. Of local captures I have at present twenty-five males and three females.

418. *D. crassiuscula* Haw.—I have taken no more females than the one I previously referred to. Males, of course, I am uncertain about.

419. *D. distincta* Neum.—Under this heading in my previous notes, Vol. XXXVIII., p. 47, line 8 of the note, instead of "for these species," read "for three species." It was a printer's error, and the correction is an important one, as the point I wished to emphasize was not that I had gone to the trouble of verifying the names, as far as that was possible, but that I was under the impression that I had taken three allied species in Alberta. I have recently spent some hours studying the group again with the aid of material from other localities, and have found no reason to alter my opinion. Separation into three species in Alberta is quite easy, excepting, of course, with males of *erechtea* and *crassiuscula*, but I have much difficulty in coming to a decision about some outside material. For instance, I have males from the eastern coast which are superficially inseparable from my local males of *distincta*, but no females at all like mine, which differ very little from the males. From Vancouver Island I have females of *crassiuscula* and *erechtea* and a series of thirteen good males, which probably includes both. Another species from there is about the size of Alberta *distincta*, but shows very much stronger sexual dimorphism. The males are like dark and ochreous *distincta*, but the females are not unlike very small *crassiuscula*, though the subapical black marks are usually lacking. It seems not unlikely that we have a fourth

species in this group excluding *cærulea* and *conspicua*. Slingerland gives us to understand that there are some very marked differences in the male genitalia. Careful examination of numbers of these might give enlightenment, and, in addition to breeding, the forms require to be studied almost by the hundreds from various localities.

420. *D. annexa* Hy. Edw., syn *conspicua* Smith.—Edwards' type is a male in the British Museum, labelled "West U.S.A., Walsingham," and is the *conspicua* of Smith. It appears to agree structurally with *distincta*, and has all the tibiæ spined. It differs in several points of structure from *cuspidea*. My series has been reduced to two pairs, and I have no recent captures, though I occasionally notice it in the spring. I have seen it from Similkameen River in the collection of Mr. E. M. Skinner, of Duncans, B.C., and there is a specimen in the British Museum, taken by Mrs Nicholl in the Upper Keremeos. Both of these localities are in Southern British Columbia, near the border of Washington. It occurs at Banff.

421. *Euclidia cuspidea* Hbn.—I have a specimen from Edmonton, taken by Mr. F. S. Carr.

422. *Syneda hudsonica* G. & R.—The species is not *limbolaris*, which is correctly figured by Holland. No. 422 stands correctly named in the Neumœgen and Henry Edwards collections. It was described from the Hudson Bay Territory. I have not seen any type, but both sexes are figured with the description, and appear to be this species. This is not the form figured by Holland as *hudsonica*, which is referred to under No. 424. The female is quite unlike the male, having the primaries much more evenly grey, sometimes quite a blue-grey, with the maculation blurred, indistinct. In this respect it differs strikingly from No. 424, formerly listed as *hudsonica*, and in which the sexes are superficially alike. It is not uncommon on the prairie, and occurs in Manitoba, but I do not seem to have met with it here in the hills. A day flier.

(To be continued.)

Mailed July 8th, 1913.

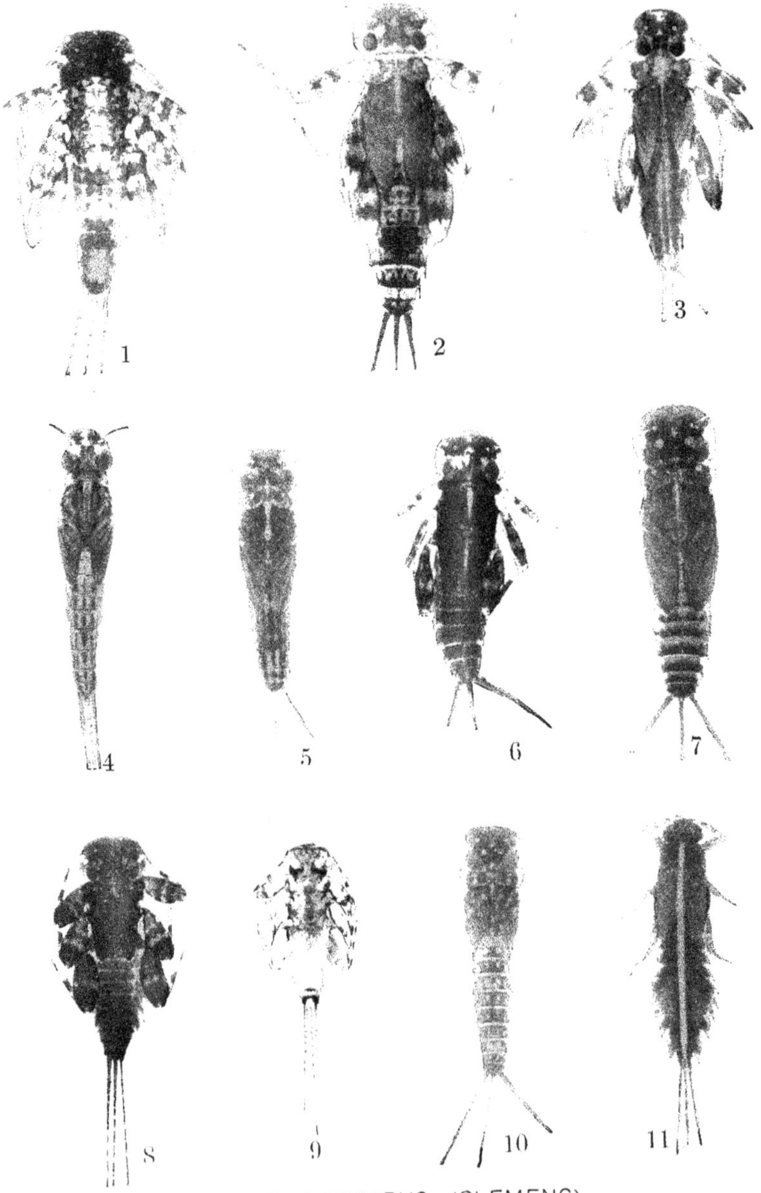

MAYFLY NYMPHS (CLEMENS).

The Canadian Entomologist.

VOL. XLV. LONDON, AUGUST, 1913 No. 8

THE ENTOMOLOGICAL SOCIETY OF ONTARIO.

The Ontario Agricultural College, Guelph, is the scene from year to year of many notable gatherings. At the end of August there is to be a somewhat remarkable assembly, when the Entomological Society of Ontario holds its fiftieth annual meeting, at which a number of eminent entomologists are expected in celebration of the Jubilee. Scientific societies and institutions in many parts of the United States are sending delegates, and some are also coming from Great Britain. The Universities of Edinburgh and Manchester are to be represented, and also the Natural History Department of the British Museum, the venerable Linnæan Society of England and the Entomological Society of London and the Entomological Society of South London. Also delegates from entomological societies in various places. It is somewhat remarkable that the two men by whose efforts the Society was formed half a century ago are still in active work, and will be present at the meeting, namely, Dr. William Saunders, who established the experimental farms of the Dominion, and was for twenty-five years director, and Professor C. J. S. Bethune, of the Agricultural College, who is President of the Society for the current year. In addition to those referred to, there will, of course, be a large representation of Canadian entomologists from all parts of the Dominion. The meeting will begin on Wednesday, August 27th. During that afternoon the delegates will present their congratulatory addresses, and Dr. Saunders and President Bethune will give some reminiscences of the formation of the Society and its early days. That evening a social reception will be given to the visitors by President and Mrs. Creelman at their residence in the College. Thursday morning will be taken up with the reading of papers and addresses, and the afternoon in motor excursions in the neighbourhood or visits to the College buildings. On Thursday evening a public lecture will be given in Massey Hall.

On Friday, the 29th, the visitors will be taken to Grimsby and given an opportunity of seeing the results of economic work in the Niagara fruit district. As the Toronto National Exhibition will be going on that week, reduced railway fares will be available from many points to that city.—C. J. S. B.

NEW SPECIES AND NEW LIFE HISTORIES OF EPHEMERIDÆ OR MAYFLIES.

BY W. A. CLEMENS, TORONTO, ONT.

While at the Go Home Bay Biological Station on Georgian Bay, during the summer of 1912, I made a special study of the Ephemeridæ of that district, under the direction of Dr. E. M. Walker, to whom I am much indebted for advice and kindly criticism. A full account of the investigations will appear in the report of the Marine Biological Stations of Canada, this paper being confined chiefly to new species and new additions to the life-histories of several forms.

The work was carried on from May 25 to September 10, and consisted chiefly in the collecting and rearing of nymphs or larvæ. Collections were made in as varied localities as possible, as there are nymphs for almost every condition of fresh water. The nymphs were taken to the laboratory in jars or bottles of water, where they were examined under the binocular microscope and the species separated. A number of each species were then transferred to breeding jars, which consisted of glass vessels, fitted up as nearly as possible to the conditions in which the nymphs were found, and supplied with running water. Over the jars, wire cages were placed to catch the subimagos as they emerged. As the subimagos appeared, they were transferred to other vessels, where they were kept until the final moult, which usually took place in a day or two. The imagos were killed with potassium cyanide and then preserved dry or in alcohol. The subimago exuvial and final nymph sloughs were also preserved for future reference. In this way about 180 specimens were bred out during the summer. The following is a list of the forms taken:

Subfamily: Ephemerinæ	1.	*Hexagenia bilineata* Say.
	2.	*Ephemera simulans* Walker.
Subfamily: Heptageninæ	1.	*Heptagenia flavescens* Walsh.
	2.	" *lutea* sp. nov.
	3.	" *fusca* sp. nov.
	4.	" *tripunctata* Banks
	5.	" *rubromaculata* sp. nov.
	6.	" *luridipennis* Burm.
	7.	" *canadensis* Walker
	8.	" *frontalis* Banks.
	9.	" sp.? (nymphs only).
	10.	*Ecdyurus maculipennis* Walsh.
	11.	" *lucidipennis* sp. nov.
	12.	" *grandis* sp. nov.
Subfamily: Baetinæ	1.	*Baetisca obesa* Walsh.
	2.	*Leptophlebia*, sp. ? (nymph only).
	3.	*Blasturus cupidus* Say.
	4.	*Blasturus nebulosus* Say.
	5.	*Choroterpes (?) basalis* Banks.
	6.	*Ephemerella lutulenta* sp. nov.
	7.	*Ephemerella lineata* sp. nov.
	8.	*Ephemerella bicolor* sp. nov.
	9.	*Drunella* sp. ? (nymph only).
	10.	*Cænis diminuta* Walker.
	11.	*Tricorythus allectus* Needham.
	12.	*Chirotenetes albomanicatus* Needham.
	13.	*Siphlurus flexus* sp. nov.
	14.	*Baetis propinquus* Walsh.
	15.	*Clœon dubium* Walsh.
	16.	*Callibætis ferrugineus* Walsh.

Descriptions and Notes.

Genus Heptagenia

Special attention was given to this genus on account of its abundance and the comparatively large number of species. The nymphs of eight species were taken and imagos reared, three of which are new species and the nymphs of the other five have not previously been described. The Heptagenia nymphs were the dominant forms in the swift waters and along the exposed shore. Their bodies are very much flattened, legs spreading, femora flattened, claws pectinated, gills placed dorsally in an overlapping series, and eyes on dorsal surface of head, and so are adapted to a life in the swiftest water. They are able to cling very tightly, for when they are lifted from a stone, quite a resistance can be felt. The clinging habit is very strong, for if a number are placed in a vessel of water without anything else to cling to, they begin clinging to each other and are soon all in a mass. They are quite active and are able to scurry over the surface of a stone, even going sideways and backwards. Their food consists of the various algal forms on the stones to which they cling.

A Heptagenia probably completes its life cycle in a year. It spends all its life in the water except for four or five days as subimago and imago. The egg hatches in about 40 days. This calculation is based upon the fact that about two months after the appearance of the imagos of *H. tripunctata* the small nymphs of the next generation were found, and this is the time required for the eggs of *Hexagenia bilineata*. The nymphs moult about once every two weeks, and as the time of emergence approaches, they probably migrate into quieter water. I have not observed the emergence of a Heptagenia subimago in the open, but in the laboratory the nymphs would crawl up the sticks placed in the jars for the purpose and transform just above the water level. The subimago stage usually lasts a day, but occasionally only a few hours and in the early part of the season it frequently lasted three or four days. Temperature and humidity seemed to be important factors. The imagos commenced their flight shortly after sundown along the lake shore, dancing in their rhythmic up and down manner at a height of from 12 to 20 feet. The females deposited their eggs by flying over the surface of the water and brushing off the eggs into the water as they appeared from the openings of the oviducts. Of the eight species the first to appear was *H. tripunctata* about June 1, and the last, *H. luridipennis*, September 2.

There are two distinct groups. In the first, consisting of *H. tripunctata*, *H. luridipennis*, *H. flavescens*, *H. lutea*, *H. fusca*, *H. rubromaculata*, the nymphs are characterized by having the lamellæ of the gills oblong, claws usually pectinated, distal segment of maxillary palpus thickest about its middle and with a small tuft of bristles near its distal end. The body is much flattened and the colour olive brown or greenish yellow. The male imagos have the penis lobes rather L-shaped and the second and third tarsal segments of the fore legs are equal, while the fourth is about four-fifths the length of the second. In the other group, consisting of *H. canadensis*, *H. frontalis*, and a third undetermined species represented by the nymph only, the nymphs have the lamellæ of the gills oval and produced distally into a sharp point; the claws are not pectinated, the distal segment of the maxillary palpus thickest towards the distal end and the tuft of bristles larger than in group 1. The body is less flattened, more reddish or yellowish, and has the appearance of being striped longitudinally on dorsal surface of abdomen. The male imagos have the penis lobes oblong instead of L-shaped and the second and third tarsal segments are not quite equal, while the fourth segment is about half the length of the second.

The following keys will serve to separate these eight species:

Key to Male Imagos:

A. No black spots or bands on face below antennæ. Group 1.

B. Very pale species.

C. Notum ferruginous, stigmal dots distinct........................ *H. flavescens.*

CC. Notum lighter, no stigmal dots......*H. lutea.*

BB. Dark species.

D. Large, entirely brown species.

E. Thorax with a broad dark median stripe or two narrow stripes close together.............. *H. verticis.*

EE. Thorax without dark median stripe..................*H. fusca.*

DD. Not entirely brown.

F. Two very small dots on median carina between antennæ.........*H. tripunctata.*

FF. No dots on median carina; thorax and top of abdomen dark.

G. Reddish area in pterostigmatic space of wing...*H. rubromaculata.*

GG. Without reddish area in wing....*H. luridipennis.*

AA. Two black spots or bands on face below antennæ. Group 2.

H. A black band on face below antennæ, a dark dash in wing, abdomen dark..*H. canadensis.*

HH. A black spot on face below antennæ, no dash in wing, abdomen lighter..*H. frontalis.*

Key to Nymphs :

A. Gills oblong. Group 1.

B. Nymphs entirely brown, without a distinct dorsal colour pattern.

C. An inverted dark U-shaped mark on ventral surface of 9th segment and a dark spot on ventral surface of the 8th. Dorsal surface of body has a smooth appearance.............*H. flavescens.*

CC. A row of dark mushroom-shaped marks along ventral surface and a rectangular dark mark on 9th. Dorsal surface has a rather granular appearance and lateral margins of body quite hairy.....................*H. rubromaculata.*

BB. Nymphs not entirely dark brown and have a distinct colour pattern.

D. Ventral surface of abdominal segments banded with dark bands along posterior margins.

E. Broad dark bands at posterior margin of each segment on dorsal surface................*H. fusca.*

EE. Dark bands at posterior margins of segments 7, 8, 9 and 10; not as broad as preceding species and a more elaborate colour pattern. . . . . *H. lutea.*

DD. Ventral surface not banded.

F. Two rows of black dots along ventral surface of abdomen. . . . . *H. tripunctata.*

FF. No dots. . . . . *H. luridipennis.*

AA. Gills oval and pointed. Group 2.

G. Two light longitudinal stripes on dorsal surface of abdomen close to median line.

H. Stripes fairly uniform for entire length. Reddish species . . . *H. canadensis.*

HH. The stripes not of uniform width, very wide on 8th segment, very narrow on 5, 6 and 7, so that darker intermediate parts have oval shapes. Lighter species. *H. frontalis.*

GG. Dorsal surface of abdomen has appearance of three longitudinal dark stripes. Colour greenish yellow. *H.*, sp. undetermined.

Mr. Nathan Banks kindly identified the imagos for me and loaned me specimens of *Heptagenia verticis*, *H. luridipennis* and *H. terminata*, for comparison.

Heptagenia flavescens, Walsh.

Nymph.

Measurements: Body 8-9 mm.; setæ 10-13 mm. Head brown, very slightly covered with light dots; a light spot above each ocellus; a small light dot on each side of median ocellus; an irregular light area anterior and lateral to each eye. Pronotum brown, with two light spots on each side. Mesothorax similar in colour to prothorax. Abdomen of a uniform brown colour dorsally, having a smooth appearance; lighter ventrally, with a semicircular brown band on 9th segment and a median brown spot on 8th. Spines of lateral edge short. Setæ banded, usually three segments dark and one light, sparsely fringed, usually only at base of light segment. Femora much flattened, brown and dotted with light spots, and having three irregular light bands; covered dorsally with small spines and posterior margin fringed with hairs and spines. Tibia with median and distal light bands. Tarsus tipped with white. Claws with two pectinations.

The nymphs of this species were taken up the Go-Home River on June 16, 1912, immediately above Flat Rock Falls, where the water was flowing swiftly but smoothly. They were clinging to stones in water one to one-and-a-half feet deep not far from the shore. On the same date they were found just below Sandy Gray Falls, two miles farther up the river. Here the water was swift and rough. I was successful in rearing only two specimens, the dates being June 27 and July 3.

Heptagenia lutea, sp. nov.

Male imago.

Measurements: Body, 9-10.5 mm.; wing, 10.5 mm.; setæ, 20; fore leg, 10.

This is a light-coloured species, slightly reddish on face below antennæ; reddish brown between ocelli and eyes. Thorax almost whitish yellow dorsally, light yellowish brown laterally; a dark area on each side of pronotum, slight red and brown markings below bases of fore and hind wings. Each abdominal segment 1-8 banded dorsally at posterior margin, remaining part of these segments being almost white; segments 9 and 10 entirely reddish brown; stigmal dots not marked; wings clouded in pterostigmatic

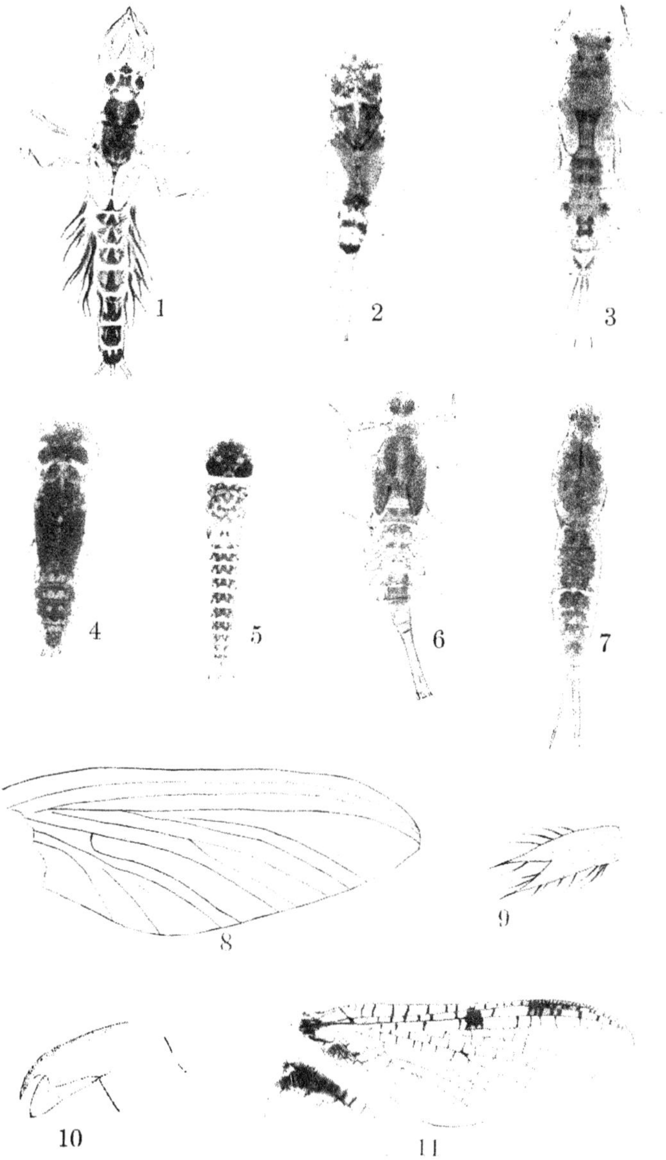

MAYFLY NYMPHS (CLEMENS).

space, a few cells reddish. Femora with median and apical bands; tibia-tarsal and tarsal joints black; fifth tarsus and ungues dark.

Female imago.

Measurements: Body, 11 mm.; wing, 12; setæ, 22; abdomen more yellowish than male.

Nymph.

Measurements: Body, 10 mm.; setæ, 13-16 mm. Head, light brown in colour and dotted with light dots; light areas over ocelli; another at posterior margin of head in median line and a larger one lateral to each eye. Pronotum with a broad, colourless lateral margin; remainder light brown, with numerous irregular light spots. Abdomen darker dorsally and with a rather complicated colour pattern. First segment light, with two brown areas at side; second with a narrow brown band along posterior margin and five brown areas and four light ones placed alternately; third almost entirely dark, with a few light dots; fourth with two dark spots in posterior lateral angles of segment, also a large dark area in centre of segment with a light area within it; fifth with a dark spot in each posterior lateral angle as in preceding segment, a dark band along posterior margin, two light areas surrounded with brown and a dark spot in centre of each; sixth almost entirely brown except for two light areas in anterior lateral angles; seventh with two large light areas, with a brown dot in each toward inner side; eighth an irregularly light and dark coloured segment; ninth has a narrow brown band along posterior margin and a dark longitudinal stripe in median line; tenth almost entirely dark. Ventrally, the lateral and posterior margins of segments 2-8 dark; segment nine with two large brown spots. Setæ greenish; basal half well fringed at joints, distal half with each two segments alternately light and dark and few hairs at joinings. Femora with alternately light and dark irregular bands and covered with minute spines dorsally; posterior margins fringed with hairs, anterior margins also fringed, but hairs shorter. Proximal ends of tibiæ dark and have dark bands slightly beyond middle. Tarsi with reddish-brown bands very near proximal ends. Claws with two pectinations.

These nymphs were very abundant along the open shore of Station Island and west of it, my collection dating from June 3 to July 2. A few were taken in a rapid on the Muskosh River on June

30 and several small specimens from Sandy Gray Falls, August 23. Imagos were reared from June 27 to July 3.

Heptagenia fusca, sp. nov.

Male imago.

Measurements: Body, 10 mm.; wing, 13; setæ, 26; No markings on face; ocelli almost in a straight line, the middle one the smallest. Pronotum brown, slightly darker along the median line; mesothorax uniformly brown. Abdomen with posterior one-third of each segment of same brown colour as thorax and projections from this band anteriorly in the median line, almost forming a continuous longitudinal stripe on the abdomen; the band widens laterally also; remaining portions of each segment somewhat light brown; ventrally very slightly banded. Forceps and penis lobes of usual form. Femur banded in middle and at distal end. Wings large; costa, subcosta and radius light in colour, while remainder of longitudinal and the cross veins brown. No cloud in pterostigmatic space.

Female imago.

Measurements: Body, 10-12 mm.; wing, 14 mm.; setæ, 18; Quite similar to male, except that abdomen is considerably darker.

Nymph.

Measurements: Body, 12-14 mm.; setæ, 15-20; antennæ, 3. Head brown, dotted with light spots; usually three light areas at posterior margin between eyes and two lateral to each eye; anterior margin well fringed with hairs. A light longitudinal median line on pronotum; two light areas on each side and lateral margin colourless; remainder of pronotum brown, with small light dots. Posterior one-third of each abdominal segment 6-10 almost black; segments 1-6 brown; the remainder of each segment varying from light brown to greenish yellow; ventrally posterior one-fourth of each segment 2-8 brown; ninth segment has two dark areas laterally. Femur light brown on upper surface, with a few lighter areas and covered with minute spines dorsally; posterior margin fringed with hairs; proximal end of tibia dark brown and its third quarter dark; proximal half of tarsus dark. Setæ well fringed with hairs at the joinings.

While on a canoe trip up the Go-Home River, June 16th, I collected a number of the nymphs of this spaces just below Sandy

Gray Falls. The only imagos I have are those bred from this collection. The dates of emergence are June 23rd and 24th.

This species is close to *H. verticis*, but lacks the dark median stripe on the thorax, and does not show the slightest trace of a dash in the wing under the bulla.

Heptagenia tripunctata, Banks.

Nymph.

Measurements: Body, 11-14 mm.; setæ, 12-16. Head deep brown, occasionally almost dark dotted with light spots; three light spots; three light areas along anterior margin of head and one at posterior margin between eyes. Pronotum similar in colour to head, with light dots and about five larger light areas on each side; lateral with a light area which extends inwards some distance. A light area in antero-lateral angle of mesothorax. Femur stout, with five irregular light areas; small spines very numerous; posterior margin fringed with hairs. Tibia with two dark and two light areas, arranged alternately. Abdomen similar in colour to head and thorax; a light area on segments 4 and 5 containing a small triangular dark area at anterior margin of segment 5, lateral to which are two dark dots; another light area on segments 7, 8, 9 and 10 containing two dark dots on 8th and two on 9th segments; usually three small dark dots at posterior margin of each segment. Ventrally two longitudinal rows of dark dots, increasing slightly in size toward posterior end; segment 9 usually with two pairs, the anterior pair small, posterior pair larger. Setæ with alternate dark and light areas. Gills have the lamellæ slightly rounded at distal end.

The nymphs of this species were seldom found in swift water, but were everywhere abundant about Go-Home Bay, in quiet bays, along open shores and in quiet streams. They could be found at any time during the summer. The first bred specimens emerged May 31, but the first capture was not made until June 11. On this date a small swarm of about 20 individuals was discovered about 8.15 p.m. flying from 10 to 20 feet high along the shore of Station Island, facing north. One female and several males were taken. Soon after this they became very abundant and remained so until about July 5th. The last bred specimen is dated Aug. 13.

Heptagenia rubromaculata, sp. nov.

Male imago.

Measurements: Body 8 mm.; wing 8; setæ 17; fore leg 7. No markings on face; darker spot at posterior margin of head between eyes. Thorax dark; median longitudinal dark stripe on pronotum; dark brown stripe on coxa of fore leg and extending up the side of prothorax. Abdominal segments 1-7 light; 8-10 dark, similar to thorax; each segment banded at posterior margin; stigmal dots distinct; wing has a reddish area in pterostigmatic space.

Female imago.

Measurements: Body 9-9.5 mm.; wing 13-14; setæ 15-22; often slightly reddish on face beneath antennæ. Dark brown on dorsal surface of head behind ocelli. Abdomen varies from reddish to a yellowish colour in dried specimens.

Nymph.

Measurements: Body 9-10 mm.; setæ 10. Head dark brown, dotted with minute light spots. Pronotum similar in colour to head; two light areas on each side, the outer one sometimes joined to the light margin. Abdomen dark brown, with a granular appearance; sometimes a faint, broad, dark, longitudinal streak can be made out with two dots on each side of it on each segment excepting 9 and 10; ventral surface lighter, with a median row of irregular dark spots and lateral rows of small dots or lines; the median dots are sometimes broken up so that only four or five small dots remain in its place; on segment 9 the markings are usually joined, forming roughly three sides of a square. Femur with four irregular dark bands; both posterior and anterior margins very hairy; claws pectinated. A very hairy species, having anterior margin of head, sides of thorax and abdomen very hairy.

This nymph was first taken on June 15 in what is commonly called the Narrows, near the mouth of the Go-Home River. The water here had a well-marked current, but not swift. On June 30 I found them very numerous in the very swift water of a rapids near the mouth of the Muskosh River. Nearly a month after this, on July 20th and 22nd, I discovered mature nymphs at an old lumber chute on the Go-Home River in fairly swift water. Imagos were bred from the nymphs taken at the Narrows on June 22nd and

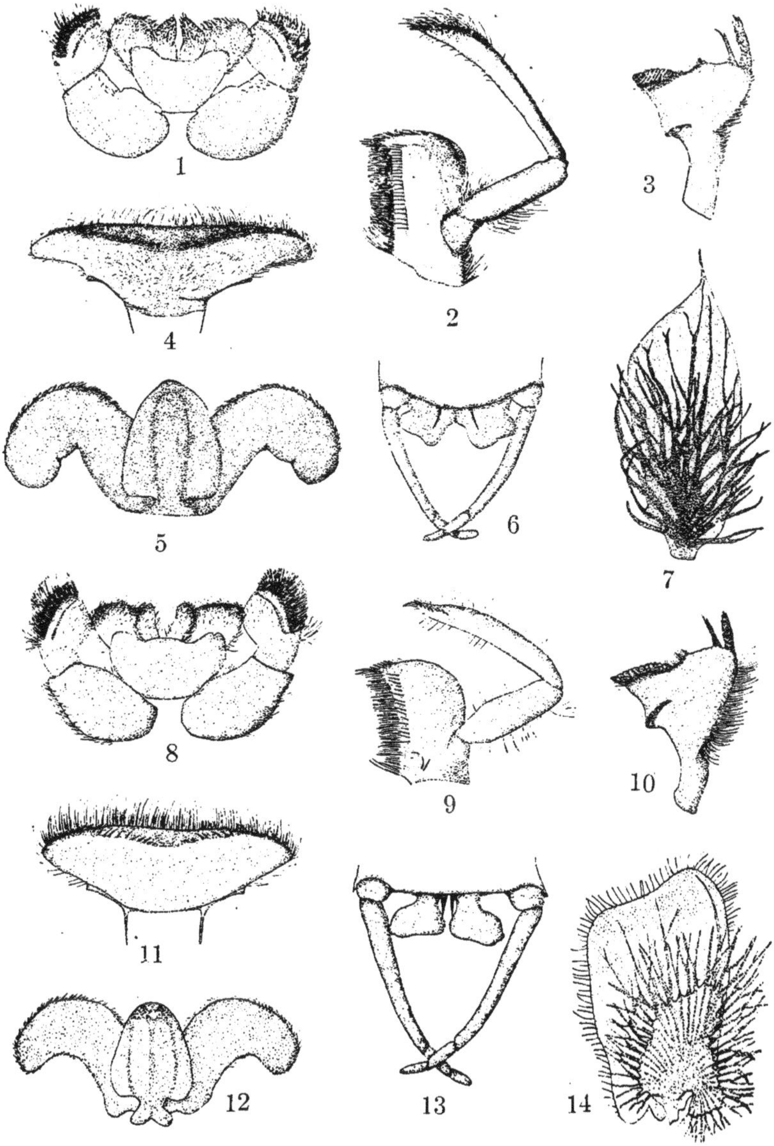

MAYFLY STRUCTURE (CLEMENS).

25th; in the Muskosh Rapids from July 3rd to 5th, and at the Chutes, July 24-29th.

Heptagenia luridipennis, Burm.

Nymph.

Measurements: Body 7-8.5 mm.; setæ 10-14. Head brown, with light dots; anterior margin fringed with hairs. Prothorax similar in colour to head; on pronotum a light spot on each side of median line; lateral to this another larger one, and lateral to this another which extends to the lateral margin. Abdomen similar in colour to prothorax; a row of black dots on each side corresponding to the stigmal dots of imago; segment 3 for the most part light, with a round brown spot in the median line and with two short projections laterally; segment 4 with a small triangular brown spot in median line with base to anterior margin, while apex meets a large brown area, leaving a small light area on each side of triangle; lateral to the brown area is a light one, and lateral to this again is a triangular dark spot in the posterior angle of the segment; segment 5 much like the 4th; segment 6 entirely dark, except for two small spots at anterior margin and two toward lateral margin; segment 7 with a triangular dark spot in median line, with base to anterior margin and apex reaching about middle of segment; on each side of triangle two dark spots; segment 8 similar to the 6th; segment 9 irregularly marked; roughly, it is dark, with a darker median longitudinal line, two light spots on each side and another at lateral margin; segment 10 entirely dark; ventrally there are two dark spots at lateral margins of 9th segment, just beside the lateral spines of that segment; sometimes a triangular spot in median line also. Setæ with basal half fringed with hairs.

This was the last species to be taken. On August 23rd I found them in a rapid just above Sandy Gray Falls, where the river flows through a small gorge. I was successful in rearing quite a number of imagos, dating from August 28th to September 1st. These are slightly smaller than those Mr. Banks sent me and considerably smaller than the measurements given in various descriptions.

Heptagenia canadensis, Walker.

Nymph.

Measurements: Body 11 mm.; setæ 15; antennæ 3.5. Head reddish brown in colour; a small dark area immediately in front of each antenna, and another about the same size in front of each eye, a black dot behind each lateral ocellus; a light area in front of median ocellus, and a larger light area between each lateral ocellus and eye; another lateral to each eye along margin of head. Mouth parts of the type belonging to group 2. Pronotum reddish brown, with a dark and an approximate light area in each lateral half; margin colourless. Abdomen darker than thorax; each segment with four light longitudinal streaks, two near median line and the other two near lateral margin; black dots, corresponding to the stigmal dots just inside the lateral light streaks. Ventrally the abdomen is almost white, each segment has two light brown lateral streaks, while the 9th has its lateral and posterior margins margined with light brown. Short lateral spines at posterior lateral angles of segments 8 and 9. Setæ of equal length; light brown in colour; joinings fringed with hair. Gills oval and pointed. Femur of fore leg light brown, with four light areas; two small ones toward anterior margin and two large ones toward posterior; distal end light coloured. Femora of hind legs with fewer pale markings. Tibiæ alternately banded with brown and white; tarsi have very broad median bands; legs slightly hairy along posterior margin.

This species was very abundant at Go-Home Bay, being next in numbers to *H. tripunctata*. The nymphs were taken from May 25th to July 10th in various localities, but never in swift water, the usual place being quiet bays. Small nymphs of the next generation were found on September 5th. The first bred specimen is dated June 1st and the last July 4th. Imagos were most abundant at Station Island from June 25th to July 15th.

Heptagenia frontalis, Banks.

Nymph.

Measurements: Body 9-10 mm.; setæ 9-10; Head yellowish brown in colour; three almost round light spots along anterior margin of head; usually a light area in front of each ocellus, and another along median line between eyes and two smaller ones lateral to this along posterior margin of head. A black dot below each antenna, in front of each eye and near inner margin of each eye. Thorax lighter in colour than head; on each side of pronotum, near median line, is a small light spot; just lateral to this is a triangular dark spot, and lateral to this again is another light area; in anterior angle of pronotum is an oval light spot; along posterior margin extending some distance on either side of median line is a broad light band, which is connected by a light longitudinal stripe along median line of mesonotum to a large irregular light area on the mesonotum. Abdomen usually a light yellowish brown; the colour pattern roughly has the appearance of a broad light band along median line, in which in segments 5, 6 and 7 are oval dark areas; in 8, a narrow stripe, and in 9 a round, dark area in each segment; on either side of this broad light band is a short light stripe; ventral surface almost white, with two lateral light brown longitudinal stripes on segments 1-9; a broad band across 9th along posterior margin, joining the two lateral stripes. Segments of setæ alternately light and brown. Legs pale, colour pattern similar to *H. canadensis*.

This species was not nearly so abundant as *H. canadensis*. The nymphs were taken in similar localities, but were not so widespread or plentiful. They were found from July 15th to July 2nd and imagos were reared from June 26th to July 4th.

Heptagenia................? (undetermined).

Nymph.

Measurements: Body, 10-11 mm.; setæ, 12-13.; head, light brown; sometimes three light areas along anterior margin, but frequently the middle one is lacking and the two lateral ones are connected with the light margins lateral to the eyes. An almost black spot in centre of each half of pronotum; around this is an irregular light area, exterior to which is a brown area. Abdomen whitish yellow, with five longitudinal yellowish brown stripes in each segment 1-8. Setæ light greenish yellow; joints abundantly fringed with hairs. Legs yellowish brown in colour; pattern similar to the two preceding species.

These nymphs were collected along the east shore of Manitoulin Island, on June 26th, 1912, by Mr. R. P. Wodehouse, who kindly handed them over to me. As imagos were not reared, the species cannot be determined at present.

Explanation of Plates.

Plate V.

Photographs of Mayfly Nymphs.

Fig. 1. *Heptagenia tripunctata* Banks.
Fig. 2. *Heptagenia lutea* sp. nov.
Fig. 3. *Heptagenia canadensis* Walker.
Fig. 4. *Heptagenia* sp ?
Fig. 5. *Heptagenia frontalis* Banks.
Fig. 6. *Heptagenia rubromaculata* sp. nov.
Fig. 7. *Heptagenia fusca*: sp. nov.
Fig. 8. *Heptagenia flavescens* Walsh.
Fig. 9. *Heptagenia flavescens* (ventral view).
Fig. 10. *Ecdyurus pullus* sp. nov.
Fig. 11. *Ephemerella lineata* sp. nov.

Plate VI.

Fig. 1. *Nymph Hexagenia bilineata* Say.
Fig. 2. *Nymph Heptagenia luridipennis* Burm.
Fig. 3. *Nymph Ephemerella bicolor* sp. nov.
Fig. 4. *Ecdyurus maculipennis* Walsh.
Fig. 5. *Nymph Ecdyurus lucidipennis*, sp. nov.
Fig. 6. *Nymph Bætis propinquus* Walsh.
Fig. 7. *Nymph Cloeon dubium* Walsh.
Fig. 8. Venation of wing pad of nymph of *Siphlurus flexus* sp. nov.
Fig. 9. Fore claw of nymph of *S. flexus* sp. nov.
Fig. 10. Fore claws of imago of *S. flexus* sp. nov.
Fig. 11. Wings of *Siphlurus flexus* sp. nov.

Plate VII.

Mouth parts, gills and genitalia of *Heptagenia canadensis* and *H. tripunctata*.

Fig. 1. Labium and 2nd maxilla, *H. canadensis*.
Fig. 2. First maxilla, *H. canadensis*.
Fig. 3. Mandible, *H. canadensis*.
Fig. 4. Labrum, *H. canadensis*.
Fig. 5. Hypopharynx, *H. canadensis*.
Fig. 6. Genitalia, *H. canadensis*.

Fig. 7. Gill, *H. canadensis.*
Fig. 8. Labium and 2nd maxilla, *H. tripunctata.*
Fig. 9. First maxilla, *H. tripunctata.*
Fig. 10. Mandible, *H. tripunctata.*
Fig. 11. Labrum, *H. tripunctata.*
Fig. 12. *Hypopharynx*, *H. tripunctata.*
Fig. 13. Genitalia, *H. tripunctata.*
Fig. 14. Gill, *H. tripunctata.*

(TO BE CONTINUED.)

A JUMPING MAGGOT WHICH LIVES IN CACTUS BLOOMS (*ACUCULA SALTANS*, GEN. ET SP. NOV.).

BY CHARLES H. T. TOWNSEND,
Director of Entomological Stations, Lima, Peru.

On January 25, 1913, the writer was exploring a rocky draw among the bare hills in the western base of the Andes, above Santa Ana ranch house, about forty miles inland from Lima, and at about 4,000 feet elevation above sea. In this draw a columnar cactus was found growing in bunches, probably *Cereus* sp., which at that date showed few blooms opened, but many unopened buds. One large bud evidently past opening time, and in reality a bloom whose opening had been prevented by the shrivelling of the petals which effectually closed it, was cut open and disclosed five maggots that possess the power of jumping six or eight inches high from a hard surface. The cactus buds were all numerously attended by a large brown ant, specimens of which have been sent to Dr. Wheeler for determination, and the closed bloom containing the maggots was simply massed with the ants on the outside, much more so than the buds in general, yet no entrance had been effected by them into this bloom. The bloom was cut open with the idea that the ants were inhabiting it, and thus the discovery of the maggots was purely accidental. The maggots were found to be boring among the clotted mass of stamens and anthers. Fermentation of the mass was evident from the sour odour, but no actual putrefaction had taken place. The maggots had not penetrated the septum covering the ovarial chamber, and the developing seeds appeared to be in normal condition.

Description of Third-stage Maggot.—Length, extended, 9 to 10 mm. Pale yellowish or straw-colored, anal plates and cephalopharyngeal skeleton black. Mandibular hook double, not coalesced. Anal plate in one transverse piece of chitin, with a sharp spine pointed upward from each end. Anal stigmata situated one on each side in end of anal plate next to and just inside of the spine. At outer end of anal plate on each side is a chitinous black ocellus. Ventral surface of body has spinose areas at junction of segments, being eleven half rings of microscopic spines, the front one faint and situated opposite the pharyngeal sclerites, the hind one on the subanal proleg-like hump or tubercle. Dorsum of body without spines or spine areas. Thirteen segments appear marked by integumental divisions, and counting the apparent second segment as II. and III. the total is fourteen, XIII carrying the subanal tubercle and XIV. the anal plate, though the last is small and ill-defined.

The maggot jumps by curling the body until the head and anal plate meet, the mandibular hook being appressed ventrally to the dorsal surface of the anal plate, whose lateral hooks are dorsally directed, the anal plate being then forcibly thrust free from the mandibular hook by a sudden and rigid straightening of the body from the anal end, while the mandibular hook is maintained continuously at resistant tension. This produces the leap, probably after the same manner as in the maggot of *Piophila*. While the body is curled, the ventral surface represents the concavity and the dorsal the convexity of the curve assumed. Probably this jumping power of the maggot has been developed for the purpose of escaping the ants or other enemies when the flower is abandoned for pupation in the soil.

On January 26 the maggots were found to have issued from the bloom. Soil was supplied to three of them, into which two of them immediately entered, but the third had already begun to contract for pupation and remained on the surface. Issuance had not taken place up to some fifteen or twenty days after, but on February 27 the three flies were found issued, perfectly transformed, and dead. The pupational period is evidently close to three weeks. The fly is of unusual interest on account of the long and extremely needle-like ovipositor of the female.

ACUCULA, gen. nov.

Head flattened or shortened-subhemispherical, in form approaching that of *Milichia*, but longitudinal axis less. Front of male about three-fifths of eye-width, that of female about eye-width. A pair of reclinate and slightly convergent vertical bristles, a pair of reclinate orbitals in front of ocelli, a pair of proclinate ocellar bristles, these all being equal in strength and length; rest of parafrontals with fine hairs half the length of the bristles. Antennæ inserted below eye-middle, reaching about three-fifths way to oral margin, third joint somewhat elongate, arista short-pubescent. Peristomalia with five or six equal bristles, the vibrissæ not differentiated. Eyes descending to lower margin of head in profile, the cheeks narrow. Proboscis and palpi short, not exserted, the oral cavity rather pronounced.

Mesoscutum with bristles near posterior border only, short hairs on rest; scutellum subtriangular but rounded apically, bearing a pair of apical and a pair of lateral bristles slightly longer than those of mesoscutum. Abdomen broad in both sexes, as broad as, or slightly broader than, the thorax, suddenly narrowed at base; oblong and flattened in male, slightly arched in female, but also flattened and shortened-subrounded rapidly tapering apically. Male hypopygium rather small. Ovipositor three-jointed; the basal joint widened and flattened, about as long as basal width; second and third joints equal and twice as long as basal or nearly that, the second a little wider than thick, the third filiform needle-like, with microscopically sharply-pointed tip and evidently telescoping within second joint; whole ovipositor conspicuously longer than female abdomen, but about as long as abdomen of male. Legs short, normal in both sexes; the hind metatarsi a little elongate, middle and front ones successively less so; middle tibiæ with very weak short apical bristle. Auxiliary vein coalesced with first vein throughout, latter ending a little before small crossvein; apical cell not narrowed, second basal and anal cells distinct; hind crossvein about half way between small crossvein and point where fourth vein reaches wing margin; a slight emargination of costa at end of first vein.

Type: *Acucula saltans*, n. sp.

Acucula saltans, n. sp.

Length of body of male, 5 mm.; body of female to end of extended ovipositor (axis of abdomen and ovipositor flexed to axis of thorax), 7 mm.; ovipositor, 3 mm.; wing, 4 to 4.5 mm.. Two males and one female reared from maggots found in cactus bloom at Santa Ana, Rio Rimac Valley, Peru, about 4,000 ft.

Wholly bluish-greenish black, polished, metallic; eyes, face and antennæ brown; face slightly cinerous in oblique lights; legs brown, tibiæ tawny or obscure yellowish. Wings clear, tawny whitish at base.

The eggs are evidently deposited within the cactus bud at a certain stage of development of the latter, the elongated needle-like ovipositor being used for piercing the wall of the bud. The maggots evidently feed on the fermenting juices of the flower mass, whose development is arrested by their presence.

This fly appears to be intermediate between the Milichiidæ and the *Sepsidæ*, partaking largely of the characters of both. The head, abdomen, wings, legs and vibrissæ are more like *Milichia;* while the frontal characters and larval habits are more like *Sepsis*. The larval saltatory habit finds its only known counterpart in *Piophila*. The fly is probably to be considered an aberrant member of the *Sepsidæ*, certainly so if the saltatory habit signifies anything.

OVIPOSITION HABITS OF *CULEX ABOMINATOR* DYAR AND KNAB.

BY B. R. COAD, WASHINGTON, D.C.

To the best of the writer's knowledge, the oviposition habits of *Culex abominator* have not been published, and, as they are unique for a species of Culex, they are perhaps worthy of note.

The larvæ of this species are indigenous to the beds of aquatic vegetation which frequently form in the rivers and lakes of the north-central states. These beds are composed of Ceratophyllum, Potamogeton, Lemna and similar aquatic plants. This growth is more or less impervious to fish, but provides sufficient open water surface to allow the breeding of great numbers of mosquitoes.

The eggs of *abominator* are laid on the upper surface of Lemna fronds in rather large masses. In only one instance were the eggs found on any other plant, and in this case they were on the edge of a Potamogeton leaf which was floating on the water. They are quite firmly attached to the frond and to each other. The base of

FIG. 10.—Egg masses *Culex abominator*.

the egg is truncate, facilitating a firm attachment. The eggs are very black and the masses show up distinctly in contrast with the green of the frond. They are always near the margin of the frond and, upon hatching, the young larvæ immediately wriggle off into the water.

The usual appearance of the mass is shown in the accompanying photograph, thanks for which are due Mr. H. P. Wood.

The writer made these observations while working on the mosquitoes at Havana, Illinois, in the employ of Dr. S. A. Forbes, and this note is published with his kind permission.

NOTES ON SOME COLEOPTERA OF THE OKANAGAN VALLEY.

BY E. P. VENABLES, VERNON, B.C.

In preparing the following list, I am fully aware that many of the observations and records may not be new. But as some years have been spent by the writer in collecting and recording, as opportunity has offered, insects of the Vernon district, there will no doubt be found among the species given some records of interest from the standpoint of geographical distribution. The Okanagan Valley, at Vernon, has an elevation of 1250 feet and it is at this point that all the material has been taken. Vernon is situated in what is known as the dry belt of B.C.

The summers are, as a rule, somewhat dry and irrigation is a necessity. During March and April there is, as a general rule, no rain to speak of beyond a few showers, and the months of May and June are what may be called in this district the wet months. From the middle of June to the middle of September the weather is very warm and bright. The snowfall begins at the end of November, and there is in most seasons a foot or more on the ground, which does not disappear until early in March.

The dates when given are those on the labels of the specimens and are not meant to show the period of activity of the species. This can, of course, only be done after many seasons of careful observation and accurate note-keeping.

I have not worked to any extent in the Carabidæ; hence the small number of species given. It is hoped to present lists of other families from time to time.

CICINDELIDÆ.

Cicindela longilabris Say.—IV. 04, V. 05, VIII. 08.

purpurea Oliv.—Rare, I. IV. 04.

vulgaris Say.—Very common in sandy places.

oregona Lec.—Another abundant species found on the shores of Long Lake.

imperfecta Lec.—Not commonly observed, V. 04.

CARABIDÆ.

Cychrus marginatus Fisch.—Found in rotten stumps, fairly numerous.

angusticollis Fisch.—Only one specimen.

Carabus tædatus Fab.—A common species to be found in the spring under dead leaves, etc.
serratus Say.—Numerous at all times.

Calosoma calidum Fab.—Very common.
subæneum Chd.

Elaphrus riparius Linn.—Taken in numbers on damp beaches, VIII.

Opisthius richardsoni Kby.—Taken under logs in August.

Nebria gebleri Dej.
sahlbergi Fisch.

Bembidium bifossulatum Lec.—III. 07; IV 06.
planatum Lec.—VIII. 05.

Bembidium transversale Dei.—Taken under leaves in spring.
lucidum Lec.—Numerous at all times in loose soil.
pictum Lec.—Fairly common.
nigripes Kirby.
4-fossulatum Mann.

Patrobus longicornis Say.—Taken under stones near water.

Pterostichus validus Dej.—Common under stones.
riparius Dej.—V.

Amara latior Kirby.—An abundant species during the whole year.
californica Dej.—Fairly common.

Badister pulchellus Lec.—Found under leaves in woods.

Platynus cupripennis Say.—Not common, taken in November and April.
subsericeus Lec.
obsoletus Say.
sordens Kirby.
gemellus Lec.

Lebia virdis Say.—Taken on low bushes in May.

Cimindis planipennis Lec.—Found under stones.

Chlænius sericeus Forst.—Under stones near water.
pennsylvanicus Say.—Taken in May and November.

Harpalus computatus Say.—Not numerous.
fraternus Lec.

Stenolopus conjunctus Say.—March and April.

Tachycellus badiipennis Hald.

SOME BEES FROM NEW BRUNSWICK, WITH DESCRIPTION OF A NEW SPECIES OF *HERIADES*.

BY J. C. CRAWFORD, WASHINGTON, D.C.

While collecting Ichneumonoidea in New Brunswick, Mr. A. Gordon Leavitt also collected a number of other Hymenoptera, and below is given a report on most of the Apoidea. Some of the material, mostly Megachilinæ, Sphecodes and male Halictus, has not been identified, and is not included. A few of the determinations were made by Mr. H. L. Viereck, and credit is given in the proper places.

Owing to our scanty knowledge of the bees of Canada in general, the exact dates, sexes and number of individuals determined have been recorded.

Bombus fervidus Fabricus.

Nerepis, July 19, 22, two females; July 19, 22, 24, Aug. 22, five workers; Aug. 19, two males.

St. John, July 14, two females; July 14, 18, Sept. 23, three workers; Sept. 23, one male.

Red Head, St. John, Sept. 1, five workers, one male.

Bombus ternarius Say.

St. John, July 14, one female; Sept. 23, two workers; Aug. 19, Sept. 23, three males.

Nerepis, Aug. 18, 19, 22, Sept. 8, 9, twenty-seven workers.

Douglas Harbor, Grand Lake, Aug. 14, one worker.

Red Head, St. John, Sept. 1, one worker.

Bombus terricola Kirby.

St. John, Oct. 3, one female; July 14, one worker; Sept. 23, one male.

Nerepis, July 24, Aug. 18, 19, five workers; Aug. 18, one male.

Red Head, St. John, Sept. 1, one worker.

Bombus vagans Smith.

St. John, July 14, one female; Oct. 3, one worker; Sept. 23, to Oct. 3, three males

Nerepis, July 22, one female; Aug. 18, 19, Sept. 9, five workers; Sept. 8, one male.

Psithyrus ashtoni Cresson.

St. John, Oct. 2, one male.

Red Head, St. John, Sept. 1, one male.

Psithyrus insularis Smith.

St. John, Oct. 3, one female.

Nerepis, July 19, Sept. 8, two females.

Red Head, St. John, Sept. 1, one female.

Psithyrus laboriosus Fabricius.

Nerepis, Aug. 18, 19, Sept. 9, five males.

Clisodon terminalis Cresson.

Nerepis, Aug. 19, one female.

Stelis foederalis Cresson.

Nerepis, July 18, Aug. 18, two females.

Macropis morsei Robertson.

Nerepis, July 24, Aug. 19, six males; July 24, two females.

Alcidamea producta Cresson.

Nerepis, July 18, 22, two males; July 18, 24, Aug. 18, 20, 22, seven females.

Heriades carinatum Cresson.

Nerepis, Aug. 19, one male; Aug. 22, one female.

St. John, Sept. 9, one female.

Heriades leavitti, new species.

Male.—Length, about 5 mm. Black, with white pubescence, head and thorax closely and very coarsely punctured; face rugoso-punctate, the sides of face and clypeus almost concealed by the pubescence; second joint of antennæ subquadrate, the third, along its shortest side, not as long as broad; antennæ beneath dull reddish; wings brown; legs dark, obscurely reddish; abdomen coarsely and closely punctured, the punctures finer than those on thorax; the punctures on segments 1-3 hardly half a puncture width apart; segment four apically and 5 and 6 rugoso-punctured; 1st ventral segment elongate, medially at apex pointed, at base without a median elevation.

Habitat: Nerepis, New Brunswick.

Described from two specimens collected Aug. 22, by Mr. A. Gordon Leavitt, after whom it is named.

Type Cat. No. 16069 U. S. N. M.

This species resembles H. carinatum, but is smaller; lacks the projection at the base of the first ventral segment, has this sclerite elongate instead of short and truncate at apex medially, and has the first three dorsal abdominal segments more closely punctured, etc.

Osmia atriventris Cresson.

Nerepis, Aug. 18, 20, 22, seven females.

Osmia melanotricha Lovell & Cockerell.

Nerepis, Aug. 20, 22, four females.

As represented by this series, in this region the present species is larger than the preceding.

Megachile infragilis Cresson.

Nerepis, July 19, one female. Det. by Mr. Viereck.

Megachile melanophæa Smith.

Nerepis, July 19, 22, Aug. 18, 22, fifteen males. Det. by Mr. Viereck.

Megachile vidua Smith.

Nerepis, July 19, 22, Aug. 22, five females; July 22, one male.

St. John, July 14, four females; July 14, one male. Det. by Mr. Viereck.

Perdita octomaculata Say.

St. John, Sept. 9, one female.

Panurginus asteris Robertson.

St. John, Sept. 9, three males; Sept. 8, 9, ten females.

Nerepis, Aug. 18, 19, 22, twenty males; Aug. 18, Sept. 8, four females. Det. by Mr. Viereck.

Calliopsis andreniformis Smith.

Nerepis, Aug. 20, 22, four females; July 24, Aug. 20, 22, five males.

Augochlora confusa Robertson.

Nerepis, July 24, Aug. 19, two females; Aug. 22, one male.

St. John, Sept. 9, six females; Sept. 8, 9, four males.

Halictus albipennis Robertson.

St. John, Sept. 9, three females.

Nerepis, July 22, one female.

Halictus arcuatus parisus Lovell.

St. John, Sept. 8, two females; Sept. 8, 9, 18, twenty males.

Nerepis, Sept. 9, one female; Sept. 8, 9, six males.

Halictus coriaceus Smith.

St. John, Sept. 9, Oct. 3, two males.

Nerepis, Sept. 8, one female; Sept. 8, 9, four males.

Halictus craterus Lovell..

Nerepis, July 18, Aug. 18, 19, three females; Aug. 19, one male.

St. John, July 14, Sept. 8, 15, 18; Oct. 3, six females; Sept. 8, 9, 18, Oct. 2, 3, ten males.

Halictus cressoni Robertson.

St. John, Sept. 9, Oct. 3, two females; Sept. 9, one male.

Nerepis, July 18, one female.

Halictus lerouxii Lepelletier.

St. John, Sept. 8, Oct. 3, two males.

Nerepis, Aug. 18, one male

Halictus oblongus Lovell.

St. John, July 18, Sept. 8, 15, Oct. 3, six females.

Nerepis, July 22, Aug. 19, two females.

Halictus pilosus leucocomus Lovell.

St. John, Sept. 8, one female.

Nerepis, July 22, Aug. 19, two females; Aug. 18, three males.

Halictus provancheri Dalla Torre.

St. John, Sept. 8, 9, nineteen females; Sept. 8, 9, five males.

Nerepis, July 18, Aug. 18, three females; July 22, Aug. 18, 20, 22, ten males.

Halictus versans Lovell.

St. John, July 14, 18, seven females.

Nerepis, Aug. 19, one female.

Andrena canadensis Dalla Torre.

Nerepis, Sept. 8, 9, two females. Det. by Mr. Viereck.

Andrena cratægi Robertson.

Nerepis, July 22, 24, three females. Det. by Mr. Viereck.

Prosopis basalis Smith.

Nerepis, Aug. 18, one female.

Prosopis cressoni Cockerell.

St. John, Sept. 8, one male.

Prosopis modestus Say.

With yellow spots on collar: St. John, July 18, two males.

Nerepis, July 22, two males.

Without yellow on collar: St. John, July 14, 18, six males.

Nerepis, July 11, 22, 24, Aug. 18, nine males.

Prosopis varifrons Cresson.

St. John, July 18, two males.

Nerepis, Aug. 18, 19, two females.

Prosopis ziziæ Robertson.

Nerepis, July 22, Aug. 19, two males.

THREE NEW NORTH AMERICAN DIPTERA.

BY J. R. MALLOCH, BUREAU OF ENTOMOLOGY, WASHINGTON, D. C.

Chætoneurophora macateei, new species.

Male: Black, shining, but not glossy. Palpi and legs more or less brownish. Halteres black.

Frons about one and two-thirds as wide as its length at centre; lower row of bristles convex; surface with numerous short hairs; antennæ normal; proboscis normal; palpi of moderate size, numerously bristled; one large downwardly directed bristle on posterior margin of cheek, besides the numerous smaller cheek bristles. Mesonotum with one pair of dorso-centrals; scutellum with four equal sized bristles. Second and fifth segments of abdomen elongated; hypopygium large, knob-like, anal protuberance slightly projecting, with several short bristles. Legs strong; fore tibia with one strong bristle at about middle, mid tibia with two at before basal third, one antero-dorsal and one almost dorsal, and one anterior bristle at near apex; hind tibia with one dorsal bristle at about one-third from base, one antero-dorsal at about same distance from base, and an antero-dorsal one at near to apex. Wings clear, veins yellowish; third vein bristled to fork; fork of third vein acute; first costal division equal to 2-3 together; fourth vein leaving at beyond fork of third with a decided curve.

Length, 4 mm.

Locality: Plummer's Island, Maryland, April 23, 1913 (A. K. Fisher), one specimen.

Near to *curvinervis* Becker, but differing in the bristling of the hind tibia and some minor particulars.

Female similar to male, except in form of abdomen. The sixth segment in this sex is distinctly the longest and the apex of abdomen is rather pointed. Same data as male.

This species is dedicated to W. L. Macatee of the Bureau of Biological Survey, in whose collection the types are.

Botanobia (=*Oscinis*) *varihalterata*, new species.

Female: Glossy black. Frons with triangle glossy, frontal stripe with a silky lustre, anterior margin obscurely brownish; antennæ reddish, third joint brown on upper surface; arista brown; face brown; cheeks glossy black-brown; proboscis brown, apical portion pale yellowish; palpi reddish. Thorax without any indication of dusting either on disk or pleuræ. Abdomen glossy black on dorsum, especially on apical half, venter opaque brown. Legs yellow, almost white, mid and hind femora except narrowly at bases, and basal third of hind tibiæ glossy black. Halteres pale yellow, with an elongate, glossy black streak on outer side of knob. Wings clear, veins brownish yellow. Hairs pale, bristles yellowish brown.

Frons slightly over one-third the width of head; triangle occupying two-thirds the width of vertex, and extending slightly over two-thirds to anterior margin of frons; surface hairs sparse and distinct; antennæ of moderate size, third joint rounded; arista pubescent, tapering, in length at least equal to width of frons; cheeks almost linear, especially at anterior margin; marginal hairs distinct, the upper strongest; proboscis and palpi normal; eyes distinctly higher than long, bare. Mesonotum without sulci, the hairs very closely placed and, though in rows, not easily distinguishable as such, the base of each hair in a very minute puncture; scutellum short, rounded in outline, the two apical bristles of good length, the two subapical bristles much shorter, disk haired as mesonotum. Legs normal; surfaces pale haired. Wings with third costal division three-fifths as long as second; veins 3-4 slightly divergent; penultimate sections of veins 3-4 subequal; outer cross vein at one and one-half times its own length from inner and two and one-half times its own length from end of fifth.

Length, 1.75 mm.

Type locality: District of Columbia, June 5, 1913 (J. D. Hood, and J R. Malloch).

Readily distinguished from any other previously described species by the peculiar mark on the haltere. In this respect the species resembles one in *Agromyza* which occurs in North America.

Apocephalus antennata, new species.

Female: Yellow, subshining. A small black spot on pleura

below wing base and another on the posterior surface of mid coxa. Posterior margin of abdominal segments narrowly browned; ovipositor glossy brown black.

Frons with two rows of four bristles and in front of those a single pair situated close together on middle of frons; lower pair of bristles, represented in *A. wheeleri* Brues, absent; third antennal joint very large, reaching as high as vertex, conical; arista brown, terminal, pubescent, slightly shorter than third antennal joint; palpi large, the surface bristles very minute. Mesopleura bare; mesonotum with one pair of dorso-centrals; scutellum with two marginal bristles. Second abdominal segment elongated, 2-3 bristles on lateral margins; sixth segment elongated and with a few short posterior hairs; ovipositor elongate-conical, four times as long as its basal width. Legs normal; hind tibia with 10-11 postero-dorsal setulæ. Wings with costa to middle; first division as long as 2+3; third about one-third as long as second; angle at fork of third vein obtuse; costal fringe twice as long as diameter of costal vein; fourth vein gently arcuated, leaving at fork of third, and ending distinctly in front of wing tip.

Length, 1.5 mm.

Locality: Plummer's Island, Maryland, June 8, 1913 (W. L. Macatee.)

CARNEGIE SCHOLARSHIP IN ENTOMOLOGY.

Mr. John D. Tothill, B. S. A., a graduate of the Ontario Agricultural College, Guelph, has been awarded the Carnegie Scholarship in Entomology in order to enable him to take a year's post graduate course at Cornell University. The value of the scholarship is $625.00 and includes travelling expenses. These scholarships are somewhat similar in character to the Rhodes scholarships at Oxford and are intended to enable qualified young men in various parts of the British Empire to spend a year in study at some University in the United States. Mr. Tothill is a field agent of the Division of Entomology at Ottawa, and is at present carrying on investigations under the direction of Dr. Hewitt, in the work of parasites of the Brown-Tailed Moth in N. B., his headquarters being at Fredericton.

A SECOND NEW GENUS OF CHALCIDOID HYMENOPTERA OF THE FAMILY MYMARIDÆ FROM AUSTRALIA.

BY A. A. GIRAULT, NELSON (CAIRNS) QUEENSLAND.

The following genus has the general appearance of certain Entedoninæ and resembles also species of *Gonatocerus*, but it is very small. It is allied with *Gonatocerus*. Its many-jointed antennæ are unique for the family.

Agonatocerus, new genus.

Normal position.

Female: Like *Gonatocerus* Nees, but the antennæ 13-jointed, the body much smaller. Proximal four funicle joints small, subequal, the distal six subequal, each over four times longer than any ot the proximal four, subequal in length to the pedicel. Fore wings with short marginal fringes. Club solid, not long. Abdomen subsessile, the phragma absent. Scutum with a median grooved line.

Male: Not known.

Type: *A. humboldti*, described herewith.

1. *Agonatocerus humboldti*, new species.

Female: Length, 0.65 mm.

Dusky brown, the base of the abdomen golden yellow, the wings hyaline. Antennæ and legs somewhat darker, dusky, the proximal half of the scape pallid. Fore wings without discal cilia under the venation or for some distance beyond, distad bearing about eighteen lines. Mid-longitudinal line of posterior wing without discal cilia.

(From one specimen, two-third-inch objective, one-inch optic, Bausch and Lomb.)

Male: Not known.

Described from a single female captured in the first week of December, 1912, by Mr. Alan P. Dodd by sweeping in a forest.

Habitat: Australia—Nelson(Cairns), Queensland.

Type: In the Queensland Museum, Brisbane, the above specimen in xylol-balsam.

Respectfully dedicated to Alexander von Humboldt.

Mailed August 5th, 1913

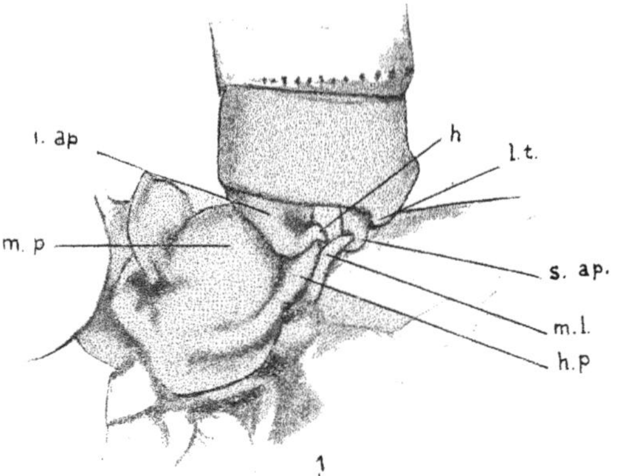

1

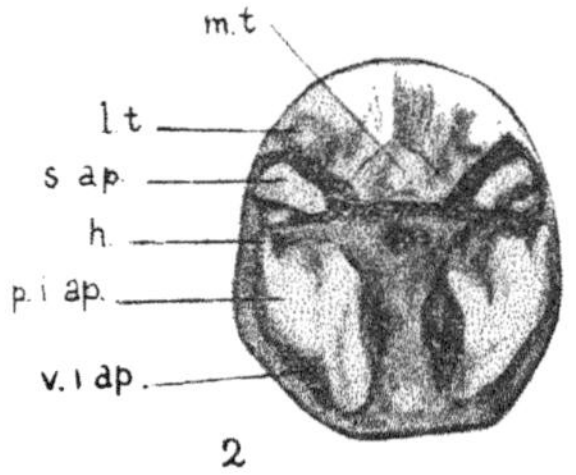

2

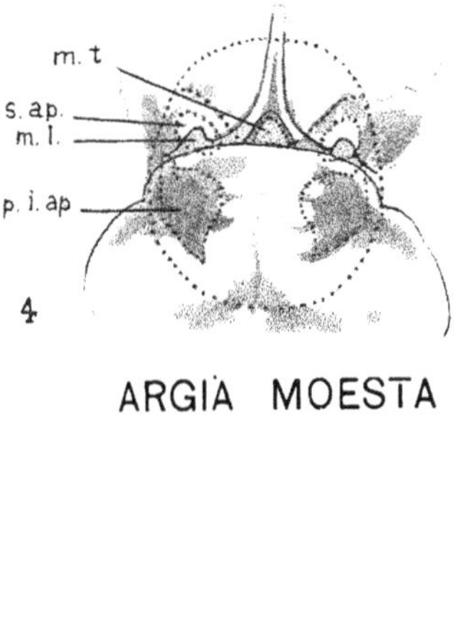

4

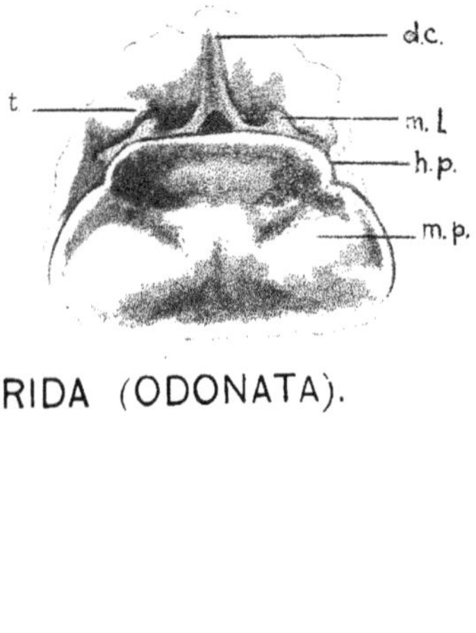

3

ARGIA MOESTA PUTRIDA (ODONATA).

The Canadian Entomologist.

Vol. XLV. LONDON, SEPTEMBER, 1913 No. 9

MUTUAL ADAPTATION OF THE SEXES IN *ARGIA MOESTA PUTRIDA*.

BY E. M. WALKER, TORONTO.

On August 1st, 1912, I captured a pair of *Argia moesta putrida* at Go Home Bay, Ont., and by killing them suddenly with gasoline, prevented the separation of the abdominal appendages of the male from the parts of the female with which they were in contact. I noted carefully the relations of the structures forming the connection, but unfortunately made no drawings at the time, as the specimens remained in their natural position after drying and the connection was apparently permanent. In carrying the specimens to Toronto, however, they separated, so that I have had to rely upon my original observations and a close scrutiny of the structures concerned in my further study of the method by which coupling in this species is effected. Some difficulties as to the precise position of the inferior appendages of the male in relation to the pronotum of the female were readily solved by making plasticine models of the parts of both sexes and fitting them together.

The only published account of the process of coupling in the genus Argia is given by E. B. Williamson in an article entitled "Copulation in Odonata."* In this paper a classification of the methods of coupling in a number of zygopterous genera is given, and the following extract gives all that is known in regard to this process in the genus Argia, the observations having been made upontwo species—A. moesta putrida and A. apicalis.

> "BB. Inferior appendages forming two jaws which grasp the anterior surface of the hind lobe of the pronotum of the female, the superior appendages resting in cups formed by depressions in the mesostigmal laminæ and the rear surface of the hind lobe of the pronotum and, depending on their

*Ent. News, XIII., pp. 143-148, 1906.

form, grasping the mesostigmal laminal or not. The female, by drawing the hind lobe of the pronotum closely against the mesostigmal laminal, prevents the escape of the male.

D. Dorsum of apex of segment 10 of male modified to form a brace against the middorsal carina or its fork or the cavity in the fork. *Anomalagrion*, *Ischnura*, *Enallagma*.

DD. Dorsum of apex of segment 10 of male with a viscid pruinose tubercle on either side which attaches itself to the mesoepisternum of the female on either side of the fork of the middorsal carina, the tubercle which corresponds to the inferior appendage of Anisoptera engaging the cavity in the fork between the mesostigmal laminæ. *Argia* (*putrida* and *apicalis*)."

In my specimens the inferior appendages of the male were in contact with the dorsum of the pronotum of the female in the position shown in fig. 1. The posterior prominence of the inferior appendages fits into the depression between the middle and hind lobes of the pronotum, which is deepened on each side to receive it (figs. 3 and 4). Thus the posterior surface of the inferior appendages (p. i. ap.) is applied to the anterior surface of the hind lobe of the pronotum of the female and the postero-ventral surface of the appendages (v. i. ap.) rests upon the postero-dorsal surface of the middle lobe of the pronotum. The upper and outer angle of each inferior appendage bears a small slightly hooked process (n), which clasps the posterior margin of the hind lobe of the pronotum. The superior appendages do not "rest in caps formed by depressions in the mesostigmal laminæ, etc." but the reverse is the case. The "ear" of each mesostigmal lamina is received into the concavity of the corresponding male superior appendage (figs. 1 and 4), which apparently rests in the pit on the mesoepisternum just beneath the former. The mesoepisternal tubercles of the female, which are but slightly developed in this species, do not seem to play an important part in this process, except perhaps in forming the outer boundary of the pit just referred to. I have not actually verified the statements made by Williamson in Section DD of his analysis, but, in fitting my model together, I found that the parts mentioned by him here must necessarily be related in exactly the manner described.

As stated by Williamson in Section BB, "The female, by drawing the hind lobe of the pronotum against the mesostigmal laminæ, prevents the escape of the male." In the case of Argia moesta putrida, the result of this action is that the two pairs of appendages of the male are drawn together and it can readily be seen by examining the figures that in such a position these appendages are incapable of being shifted in any direction, and hence escape of the male is impossible unless permitted by the female.

The mutual adaptation of these structures in the two sexes is so precise that it seems improbable that copulation could take place between different species of Argia, even though very closely related.

Explanation of Plate VIII.

Fig.1.—Position of the abdominal appendages of the male in relation to the thorax of the female in copulation.

Fig. 2.—Posterior view of end of male abdomen.

Fig. 3.—Dorsal view of parts of the pronotum and mesoepisternum of the female.

Fig. 4.—Semi-diagrammatic combination of figures 2 and 3, showing the relative positions of the parts in coupling. The parts of the male are indicated by dotted lines.

S. ap., superior appendage of male; i. ap., inferior appendage of male; p. i. ap., posterior surface of same; v. i. ap., postero-ventral surface of same; h. terminal hook of same; m. t., tubercle which engages the cavity between the forks of the dorsal thoracic carina; v. t., viscid lateral tubercle; m. p., middle lobe of pronotum of female; h. p., hind lobe of pronotum of female; d. c., middorsal thoracic carina; ms. l., mesostigmal lamina of female; ms. t., mesoepisternal tubercle of female.

A NEW SPECIES OF NEUROTERUS FROM WASHINGTON.

BY WILLIAM BEUTENMULLER, NEW YORK.

Neuroterus washingtonensis, sp. nov.

Male.—Head black, mouth parts pitchy brown; front shining and indistinctly rugose; eyes very large and conspicuously reticulated. Antennæ 14-jointed, basal joints pale yellowish brown, terminal joints darker. Thorax dark brown, with the whole surface finely crackled and with minute whitish hairs; parapsidal grooves very fine and line-like and extending to beyond the middle; median line also fine and almost reaching the scutellum. All the lines may be seen by transmitted light; sides of thorax yellowish brown. Scutellum blackish brown, large, rounded and obtusely pointed at the apex; it is more distinctly crackled than the thorax, with a fine transverse line at base, and covered with a few scattered whitish hairs. Abdomen small, smooth and shining; petiole long and yellowish brown. Legs pale yellowish brown. Wings hyaline; veins brown, radial area closed; cubitus not extending to the first cross-vein; areolet large and triangular; anal vein broken. Length, 1.75 to 2 mm.

Female.—Wholly black, abdomen robust and large, petiole very short. Antennæ darker and shorter than the male, 13-jointed. Legs shorter and stouter than those of the male, yellowish-brown, with all the femora dark brown to nearly the tip. Ovipositor very long.

Length: 1.50–2 mm.

Gall: On the leaves of white oak (*Quercus garryana*), singly or in numbers on the mid-rib and principal veins, sometimes deforming the entire leaf. Rounded, irregularly rounded, oval or elongate and often forming a shapeless mass when confluent. Green and fleshy when fresh; brown, hard and woody when old and dry. Inside it is solid, whitish and filled with numerous larval cells.

Length, 10–35 mm. Width, 10–15 mm.

Habitat: Friday Harbor, Puget Sound, Washington. Galls, July 2nd, 1911. Flies, July 30th, 1911. Lewis H. Weld, collector.

The species is allied to *Neuroterus batatus* and *N. noxiosus*, and it is probably double brooded like these two species. The gall somewhat resembles that of *N. noxiosus*. Hundreds of specimens

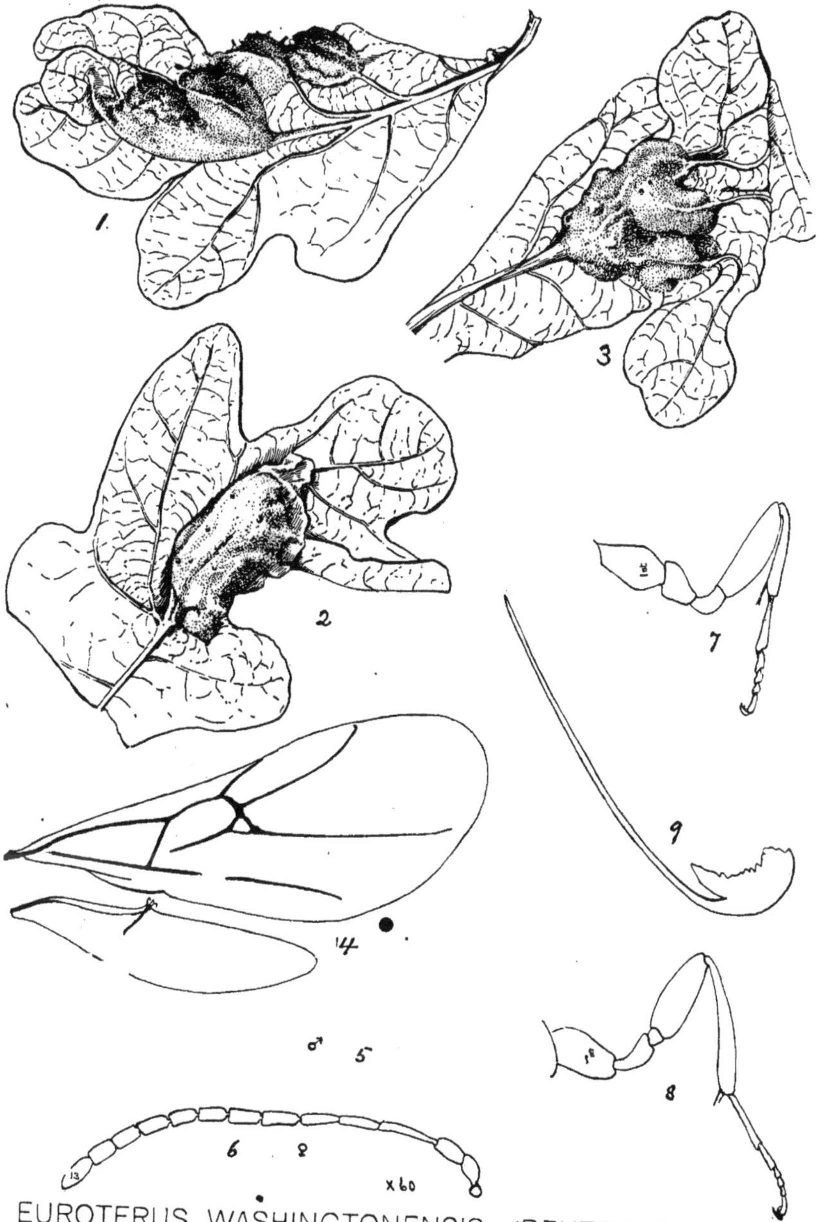

EUROTERUS WASHINGTONENSIS (BEUTENMULLER).

of *N. washingtonensis* were bred from the galls by Mr. Weld, and they are all essentially the same as color, form and sculpture. The types are in my collection and paratypes were deposited with Mr. Weld, and the following institutions: United States National Museum, Museum Comparative Zoology, American Entomological Society, Cornell University and British Museum.

EXPLANATION OF PLATE IX.

Figs. 1, 2, 3.—Galls, natural size.
Fig. 4.—Wings, greatly enlarged.
Fig. 5.—Antennæ of male, greatly enlarged.
Fig. 6.—Antennæ of female, greatly enlarged.
Fig. 7.—Anterior leg, greatly enlarged.
Fig. 8.—Posterior leg, greatly enlarged.
Fig. 9.—Ovipositor of female, greatly enlarged.

Figures 1–3 were made by Mrs. E. L. Beutenmuller, and figures 4–9 by Mr. Lewis H. Weld.

NEW NORTH AMERICAN DIPTERA.

BY J. R. MALLOCH, WASHINGTON, D.C.

The two species of Chloropidæ described herewith I had purposed including in a larger paper dealing with the whole genus to which they belong, but the carrying out of this project is at present not possible, and I thus present them in their present form pending the possible completion of my work on the different genera of Chloropidæ at some future time. The species of *Milichiella* I took after my paper dealing with the Agromyzidæ had gone to the press. The characters given in the description should readily separate it from any described American species of this genus.

Madiza (=*Siphonella*) *nigripalpis*, new species.

Male and Female: Glossy black. Antennæ sometimes brownish on the inner surface of third joint; palpi generally black, but sometimes brown. Legs black; fore tibiæ entirely, and apical third of mid and hind tibiæ as well as entire tarsi on all legs, except apical joint, clear yellow. Halteres whitish. Wings clear, veins yellow.

Frons about one-third wider than either eye, whole surface glossy; triangle poorly defined, occupying more than two-thirds the width of vertex and reaching slightly over midway to anterior margin of frons; surface hairs on frons weak, those on triangle confined to margins; antennæ of moderate size, third joint rounded; arista short, not as long as width of frons, basal joints elongated and slightly swollen, pubescence indistinct; face concave; cheek distinctly higher than width of third antennal joint, the anterior angle produced slightly; lower half of cheek bristled, the usual strong hair at anterior angle; eye pubescent, distinctly higher than long. Mesonotum thickly covered with short, pale hairs, at the base of each of which there is a small puncture; there are traces of a central line of punctures, and, more distinctly, of two lateral, broader lines; scutellum with disk haired as of disk of mesonotum, rounded in outline; two apical marginal bristles and two weaker bristles anterior to them, at the base of each of the long bristles there is a slight tubercule. Wings with third costal division distinctly more than half as long as second; veins 3–4 slightly divergent at apices; outer cross vein oblique.

Length, 1.5 mm.

Type in author's collection.

Locality: Beltsville, Maryland, swept in hay field, June 15, 1913 (J. R. Malloch, J. D. Hood).

Madiza (=*Siphonella*) *projecta*, new species.

Female: Yellow, slightly shining. Ocellar spot black; proboscis glossy brown at joint. Mesonotum with three black-brown stripes, which are not clearly defined and sometimes indistinct, the side stripes generally shortened posteriorly; pleuræ with a dark, brownish spot on the middle; scutellum sometimes with a discal brown spot. Abdomen with three longitudinal rows of brown spots, or with fore margins of segments brownish. Legs entirely yellow, or with the femora slightly darkened at middle, and the bases of hind tibiæ darkened. Wings clear. Halteres yellow.

Head in profile distinctly produced at mouth margin; frons about one-half the head width, surface hairs pale; proboscis very long, either half as high as head; arista bare, short, not as long as breadth of frons; cheek as high as breadth of third antennal joint;

eyes pubescent. Mesonotum with a more or less distinct central line of punctures and the disk covered with short pale hairs; scutellum rounded, four distinct marginal bristles present, the disk with short hairs. Legs normal. Wings with third costal division two-thirds as long as second; veins 3–4 subparallel; outer cross vein slightly oblique; last section of fifth vein nearly twice as long as penultimate section of fourth.

Length, 1-5 mm.

Type: Cat. No. 16002, U.S.N.M.

Locality: Las Cruces, New Mexico, May 20, 1896, on *Yucca angustifolia* (T. D. A. Cockerell).

Paratypes: Same data, and one specimen, Mesilla Park, May 7 (T. D. A. Cockerell). Five specimens.

Milichiella urbana, new species.

Female: Black-brown, slightly shining. Mesonotum opaque brown. Abdomen slightly shining and almost black in color. Legs with tarsi brownish, the remainder black. Wings clear, veins brown. Halteres brown.

Frons at least one-half the head width, opaque, orbits very narrow, grayish pollinose, bristles distinct, and extending to the base of antennæ; center rows slightly divergent towards ocelli; antennæ of normal size; arista slightly pubescent, hairlike, a little longer than length of frons; three bristles above level of mouth; cheek linear, marginal bristles distinct; palpi and proboscis normal; incision on posterior eye margin distinct. Mesonotum with disk covered with short black setulæ and with one pair of prescutellar dorso-centrals; scutellar bristles subequal. Abdominal segments with short discal hairs, apex of last segment setulose. Legs without bristles, their surfaces with short hairs. Wings with veins 2-3 slightly divergent at apices, veins 3-4 slightly convergent at apices; last section of fifth vein slightly longer than penultimate section of fourth.

Length, 1.25 mm.

Type in author's collection.

Locality: Washington, District of Columbia, June 23, 1913, at an open window in center of city (J. R. Malloch).

REPORT ON A COLLECTION OF JAPANESE CRANE-FLIES (TIPULIDÆ), WITH A KEY TO THE SPECIES OF PTYCHOPTERA.

BY CHARLES P. ALEXANDER, ITHACA, N.Y.

(Continued from Page 210.)

Gonomyia (*Gonomyia*) *superba*, sp. n.

Antennæ, brown; color, brown and yellow; vein, *Sc* ends slightly beyond the origin of *Rs*.

Male.—Length, 5–5.5 mm.; wing, 4.9 mm.

Female.—Length, 5.9 mm; wing, 5.2–5.5 mm.

Male.—Rostrum yellow, palpi brown; antennæ brown, including the basal segments; front, vertex and occiput dull yellow, the vertex clearer yellow behind.

Pronotum, clear light yellow above; on the sides, a short, dull brown stripe from the cervical sclerites down to above the fore coxa. Mesonotum, præscutum very light yellowish brown, with rich chestnut-brown stripes, a median stripe, broad and dark in front, narrow behind, and again enlarged at its end divided by a pale, narrow, median stripe; lateral stripes short, beginning behind the pseudosutural pits crossing the transverse suture and suffusing the lobes of the scutum; lateral edge of the præscutum, in front, yellowish; behind, brown; scutellum pale, whitish; the base and lateral edges tinged with brownish, post notum brown. Pleuræ clear yellowish white, an irregular dark brown mark behind and above the base of the coxa; sternum yellow, the sides of the mesosternum, between the fore and middle legs, brown, separated by a broad median pale mark; the propleural stripe begins on the prosternum as a rounded mark which sends out a narrow caudal prolongation. Halteres light yellow. Legs: coxæ and trochanters light yellow, margins of the segments more or less brown; femora and tibiæ light brown; tarsi somewhat darker brown. Wings, hyaline or nearly so; veins brown, costa more yellowish. Venation (see fig. 14, pl. III): Sc ending slightly beyond the origin of Rs; basal deflection of Cu^1 about at the fork of M.

Abdomen, tergum, light yellow, each segment with a large brown mark on basal half, the caudal margin of this mark much rounded; sternum light yellow. Hypopygium (see fig. 1 and 2, plate X). Pleurites short and broad, the caudal end produced into one fleshy and three chitinized appendages, as follows: Viewed from above, a fleshy lobe in front, the inner dorsal margin produced entad and dorsad into a slightly curved slender spine; behind the fleshy lobe arises a stout hook, very strong at the base, constricted before the middle, the tip slender and pointed, this hook directed entad and caudad; from the outer ventral angle of the pleurite arises a long, straight chitinized appendage, directed entad and caudad, narrow basally and more enlarged apically. The guard of the penis is long, pale, ending in a long, slender, tube-like point. On either side of the penis guard arises an elongate, very slender, chitinized hook, which is straight for about three-fifths its length and then bent strongly inward; viewed from the side, these hoops are bent very strongly ventrad and then caudad. Summarized, the hypopygium bears eight chitinized slender arms, all except two (which are probably homologous with the second gonapophyses) being borne by the pleurites.

Female.—Very similar to the male, but larger.

Vial No. 1.—Tokio, Japan; Aug. 1912. One ♂.

Vial No. 5.—Nishigahara, Japan; Apr. 25, 1912; 5 ♂, 4 ♀.

Holotype, ♂; Vial No. 1.

Allotype, ♀; Vial No. 5.

Paratypes, 5 ♂, 3 ♀; Vial No. 5.

Types in author's collection; Paratypes in U.S. National Museum and Cornell University Collections.

G. superba differs from *nubeculosa* Meij. (Java). (Tijd. voor Entomol., vol. 44, p. 48, 49; fig. 36, 1911) in the unspotted wings; from *metatarsata*, (l.c., p. 48, fig. 35) in its closed cell 1st M_2, etc.

Gonomyia (*Leiponeura*) *insulensis*, sp. n.

Pleuræ without longitudinal stripes; vein *Sc* ends far before the origin of *Rs*.

Female.—Length, 3.9–4 mm.; abdomen, 2.6 mm.; wing, 4 mm.

Female.—Rostrum yellow, palpi brown; antennæ, segment one yellowish, remainder dark brown; front, vertex and occiput yellow, the vertex suffused with dark colored.

Mesonotal præscutum yellowish, with three brown stripes, the median one broad, not divided by a pale median vitta, extending to the suture, the lateral stripes are broad, narrow, uniform in width until they cross the suture (not expanded behind), lateral margin of the sclerite dull yellow, the ground color between the brown stripes is very reduced; scutum, lobes dark brown, median line yellowish; scutellum yellow, a brown median spot in front; postnotum brown. Pleuræ, mesopleuræ brown in front, extending from the lateral margin of the præscutum down to and suffusing the mesosternum on the sides; metasternum pale brown. Halteres dull yellow. Legs: coxæ and trochanters yellow, suffused with brown in front; femora, tibiæ and tarsi brown, a little darker toward the tip. Wings subhyaline, veins brown. Venation (see fig. 12, plate III); Sc. ending far before the origin of Rs; R^{2+3} almost parallel to R^1.

Abdominal tergites yellowish-brown; sternites light yellow.

Vial No. F.—Tokio, Japan; August, 1912; 1 ♀.

Holotype, ♀; in Vial F.

Type in author's collection.

The three species of *Gonomyia* described by de Meijere as *Atarbæ* (Tijd. voor Entomol.; vol. 44, 1911) are all members of the subgenus *Leiponeura* Skuse. These species are *Gonomyia nebulosa* (l.c., p. 42, fig. 25); *pilifera* (l.c.; p. 43, fig. 26) and *diffusa* (l.c.; p. 43, 44). They have nothing in common with *Atarba* and are quite distinct from any members of the *Leiponeura* group, that I know of, in their clouded wings. *G. insulensis* differs from all of the above species in its unmarked wings.

Genus Erioptera Meigen.,
Subgenus Acyphona Osten-Sacken.

Of this subgenus, two species were included, both of which are herein characterized as new. The only described Palæarctic species, *Acyphona maculata* Meigen, of Europe, differs from the Japanese species, as follows: Wing pattern, in *maculata* large, rounded brown markings mostly with grey centers; the body-shade is much lighter in *maculata* and there are several important differences in hypopygial characters, these being shown by the following key:

1. 9th tergite broad and thin, at its apex deeply notched; two chitinized teeth at the base of the pleura on the ventral side. . . . . .2.

9th tergite provided with two chitinized hooks at its apex; no chitinized teeth at the base of the pleura on the ventral side; [horns of the second gonapophyses long, widely separated at the base] (Japan). . . . . *asymmetrica*, sp. n.

2. Base of pleura on sternal side provided with a chitinized plate which is bidentate, the proximal tooth free, the distal one joined to the pleura; 2nd gonapophyses short, chitinized at tip and on sides; apex merely notched. (Europe). . . . . *maculata* Meigen.

Base of pleura on sternal side provided with a small chitinized tooth, minutely denticulate; 2nd gonapophyses long, the tips long and widely separated (Japan). . . . . *incongruens*, sp. n.

Erioptera (*Acyphona*) *incongruens*, sp. n.

Small species; light brown, with narrow dark brown pleural stripes; wings thickly spotted with brown.

Male.—Length, 5 mm.

Male.—Rostrum and palpi brown. Antennæ long, segment one brownish-yellow; segments two to eight light yellow; remainder with increasing amounts of brown at their tips, the apical segments all brownish. Front, vertex and occiput dark brown.

Thoracic pronotum brownish-yellow, brown on the sides. Præscutum reddish-brown, with a double median brown stripe; humeral region brighter yellow; sides of the sclerite darkened; scutum, scutellum and postnotum brown. Pleuræ reddish-brown with narrow dark-brown lines, the most dorsal one continuing from behind the fore coxa underneath the wing to the postnotum; the second beginning on the mesosternum running above the middle coxa, becoming very narrow and indistinct before the root of the halter; the last stripe on the metasternum over the hind coxa. Halteres light yellow. Legs: coxa brown; trochanters brownish-yellow. (The legs are all detached and loose in the vials; most of these have the femora largely brown, basal third mostly paler, yellowish; a post median yellow ring, tip usually pale; tibiæ and tarsi clear light yellow, sometimes infuscated at the tips; tibiæ often with a sub-basal aunulus. In the vial were several specimens of *E. asymmetrica*, a closely allied form, and most of the legs evidently belong to that species. Two legs in the vial are very different and may belong to this little species, this being rendered probable by the size; in these the entire legs are clear, light yellow, the femora with a rather narrow subapical dark brown ring).

Wings spotted with brown.

Abdomen: Tergum dull brownish yellow, apex and lateral margins of the sclerites brown. Hypopygium unsymmetrical as in the genus, the 9th abdominal segment being twisted one-half around. Suture between the 9th tergite and the 9th sternite not indicated. The 9th tergite is broad and long, its hind margin produced caudad in a wide, thin plate which is broadly and rather deeply notched at its middle; no chitinized hooks at its apex. The pleurites are convex outerly (produced into two apical appendages), the base (dorsal) produced entad and cephalad in a long, chitinized hook; the ventral edge of the pleura near the sternum possesses a small chitinized organ which is directed caudad and is provided with two or three denticulæ; of the two apical appendages, the ventral one is chitinized, the dorsal one is fleshy, the second gonapophyses are close together, the chitinized tips rather long and deeply divided. (See plate X, figs. 5 and 6).

Holotype, ♂. Vial 6, April 25, 1912; Tokio, Japan.

Erioptera (*Acyphona*) *asymmetrica*, sp. n.

Resembles *incongruens* closely, but is larger, the coloration darker, especially on the pleuræ and usually on the abdomen. Wings hyaline, spotted with brown, varying considerably in the intensity and size of the markings; in some the dots are small, not confluent, in the darker specimens the spots on the costal half of the wing tend to flow together to form large blotches. The male genitalia of the two species is remarkably different. (See plate III, fig. 15, wing.)

The hypopygium is, as in the genus, asymmetrical, the usual dorsal portions of the 9th sclerites being switched around on a level with the pleural sutures of the remaining segments. (See fig. 7-9, plate X), suture between 9th tergite and sternite obliterated, 9th tergite broad and long with a cross-shaped mark; near its tip set with two small, semicircular, chitinized pieces which are produced into sharp points on the proximal ends. Pleurites short and stout, at the base on the dorsal side, produced into a long, slender, chitinized arm which is directed entad, two apical appendages, the more ventrad being chitinized, especially at the tips, the dorsal apical appendage fleshy. Between the tergite and the unarmed sternite, nearly in the median plate, is a rectangular, subchitinized organ, bearing at its outer angles chitinized hooks, bent ventrad and inward, these hooks minutely denticulated at tip.

♂.—Length, 5.8 mm.; wing, 6.3 mm.

♀.—Length, 6.4–7.1 mm.

Holotype.—Vial 6, April 25, 1912; Tokio, Japan.

Allotype.—Vial 6, April 25, 1912; Tokio, Japan.

Paratypes.—Vial 6 and L; 4 ♀, 2 ♂, April 25, 1912; Aug. 1912, Tokio, Japan.

Subgenus Erioptera, Meigen.

Erioptera (*Erioptera*) *elegantula*, sp. n.

Wings with brown spots.

Male.—Length, 5.4 mm.; wing, 7.7–7.9 mm.

Female.—Length, 6–6.5 mm.; wing, 7–8.3 mm.

Male.—Rostrum and palpi dark brown, antennæ with basal segments brown, flagellar segments short, dark brown; front, vertex and occiput dark brown.

Pronotum dark brown above, lighter colored on the sides. Mesonotum dark brown, the region before the pseudosutural pits more yellowish; scutum, scutellum and postnotum dark brown. Pleuræ dark brown. Halteres pale. Legs: coxæ dark brown; trochanters brown; femora dark brown; tibiæ dark brown, a little paler at the extreme base; tarsi dark brown. Wings subhyaline with greyish-brown marks, as follows: A large rounded spot at origin of Rs, a second at Sc^2, a third at end of Sc^1 running down over cross-vein r; a fourth spot at tip of R^1 and a smaller one at tip of R^2; cord broadly margined with the same color; less distinct clouds at ends of the other veins and along most of these veins. Venation, (see fig. 3, plate III.)

Abdomen dark brown, densely clothed with long whitish hairs. Hypopygium. 9th tergite broad at base, narrowed at the middle, the tip rather expanded with a deep V-shaped incision, the lobes rounded. Pleurites long, cylindrical, not very convex on outer face; three apical appendages, the more dorsal being somewhat fleshy, brown, elongate-cylindrical, narrowed basally, provided with long hairs, and, at its tip, with a slender hook directed cephalad; the median apical appendage is longest, chitinized, very strongly so at its tip; tip broadly expanded and concave, this concavity provided with minute denticulæ; the ventral apical appendage is shorter than the median one, fleshy, cylindrical, narrowed at base. Viewed from beneath, the 9th sternite is straight on its caudal margin, pleurites very broad at base, produced entad and almost meeting on the median line on the sternum; second gonapophyses long, slender, acicular, the tips barely projecting beyond the caudal level of the 9th sternite.

Female.—Similar, but averages larger in size.

Vial No. 1.—Tokio, Japan; 2 ♂, 2 ♀.

Vial No. 16.—Tokio, Japan; 2 ♀ (small, but apparently of the same species.)

Holotype.—♂. Vial No. 1, I.

Allotype.—♀. Vial No. I.

Paratypes.—1 ♂, 3 ♀, Vials I and 16.

Types in author's collection.

E. elegantula differs from *E. javensis* Meij. (Tijd voor Entomol., vol. 44, p. 45, 46, fig. 28, 1911) and *E. notata* Meij. (l.c., p. 46, figs. 29–31) in its spotted wings.

Genus Molophilus Curtis.

Molophilus pegasus, sp. n.

Antennæ of the male short; color of body brown.

Male.—Length, 4.2 mm.; wing, 4.3 mm.

Female.—Length, 4.9 mm.; wing, 5.1 mm.

Male.—Rostrum and palpi dark brown; antennæ light yellow, the flagellar segments with the exception of the first, a little more brownish; antennæ short, extending about to the base of the wings, segments of flagellum cylindrical; front, vertex and occiput brown.

Pronotum above, light yellow, darker on the sides. Mesonotal præscutum reddish-brown, with a broad, dark brown median stripe, and less distinct but broader lateral stripes, which begin behind the pseudosuture, broaden out behind and fuse with the median stripe near the transverse suture; scutum, lobes brown, median line paler; scutellum lighter colored, yellowish medially, brown on the sides; postnotum brown. Pleuræ brown except dorsally, where there is a pale band running from the pronotum back to the wing basis. Halteres light yellow. Legs: coxæ and trochanters pale yellow, femora short, incrassated beyond the base, brown, paler basally; tibiæ and tarsi brown. Wings slightly tinged with yellowish-grey; veins yellow. Venation (see fig. 11, plate III).

Abdomen, tergites dark brown; sternites rather lighter brown, extreme apices of the sclerites pale. Hypopygium (see figs. 3 and 4, plate X); 9th tergite and sternite completely fused so that no pleural suture remains; viewed from beneath, the 9th sternite projects backward, its caudal margin rather squarely truncated; the outer ventral pleural arm is straight, fleshy, rather thickly covered with long hairs; just entad of the outer arm and nearer to the base of the pleurite, arises the inner ventral pleural arm, which is elongate, slender, its tip strongly chitinized and denticulated at the extreme end and bent inward; the guard of the penis is a pointed, chitinized organ, nearly as long as the outer pleural arm. Viewed from the side, outer ventral arm of the pleurite directed caudad; inner ventral arm with the tips conspicuously arcuated and bent ventrad; just above the base of the inner arm arises the dorsal pleural appendage, very broad at the base, its tip chitinized and directed slightly dorsad, on the dorsum of the pleurite are two protuberences clothed with long hairs. Viewed from above, the pleurites are very broad, so that the space between them on the median line is narrow; about midway of their length, on the inner face, is a strong protuberance, directed inward; it is strongly chitinized and almost touches its mate of the opposite side.

Female.—Similar, but larger; the abdomen is dark brown, the genital segment much brighter, yellowish-brown.

Vial No. 19.—Tokio, Japan; June 25, 1912; 1 ♀.
Vial No. 20.—Tokio, Japan; June 25, 1912; 1 ♀.
Vial No. K.—Tokio, Japan; Aug. 1912; 1 ♂.
Holotype.—1 ♂, Vial K.
Allotype.—1 ♀, Vial 20.
Paratype.—1 ♀ Vial 19.

Types in author's collection; paratype in U.S. National Museum collection.

M. pegasus differs from *bicolor* Meij. (Java) (Tijd. voor Entomol.; vol. 44, p. 45, fig. 27) in its darker brown body-color and darker legs.

Genus Conosia Van der Wulp.

Conosia irrorata Wiedmann.

The following papers since Kertesz (1902) may be cited:

1904.—*Conosia irrorata* de Meij; Bijdragen tot de Dierkunde; p. 92.

1911.—*Conosia irrorata* de Meij; Tijdschrift voor Entomologie; vol. 44, p. 51.

1911.—*Conosia irrorata* Brun.; Rec. Indian Museum; vol. 6, part. 5, p. 283.

1912.—*Conosia irrorata* Brun.; Fauna Brit. India, Dipt. Nemat., p. 497.

One female in vial 47; Tokio, Japan. The wing pattern is figured on pl. III; fig. 13.

Explanation of Plate X.

Fig. 1.—Hypopygium of *gonomyia superba;* dorsal aspect; x, y, z = chitinized pleural appendages.

Fig. 2.—Hypopygium of *gonomyia superba;* lateral aspect, sternum uppermost; lettering as in fig. 1.

Fig. 3.—Hypopygium of *Molophilus pegasus;* lateral aspect; t = 9th tergite; s = 9th sternite.

Fig. 4.—Hypopygium of *Molophilus pegasus;* dorsal aspect; p = pleura.

Fig. 5.—Hypopygium of *Erioptera (Acyphona) incongruens*, sp. n.; dorsal aspect.

Fig. 6.—Hypopygium of *Erioptera (Acyphona) incongruens;* 9th tergite, dorsal aspect.

Fig. 7.—Hypopygium of *Erioptera (Acyphona) asymmetrica;* 9th tergite, dorsal aspect.

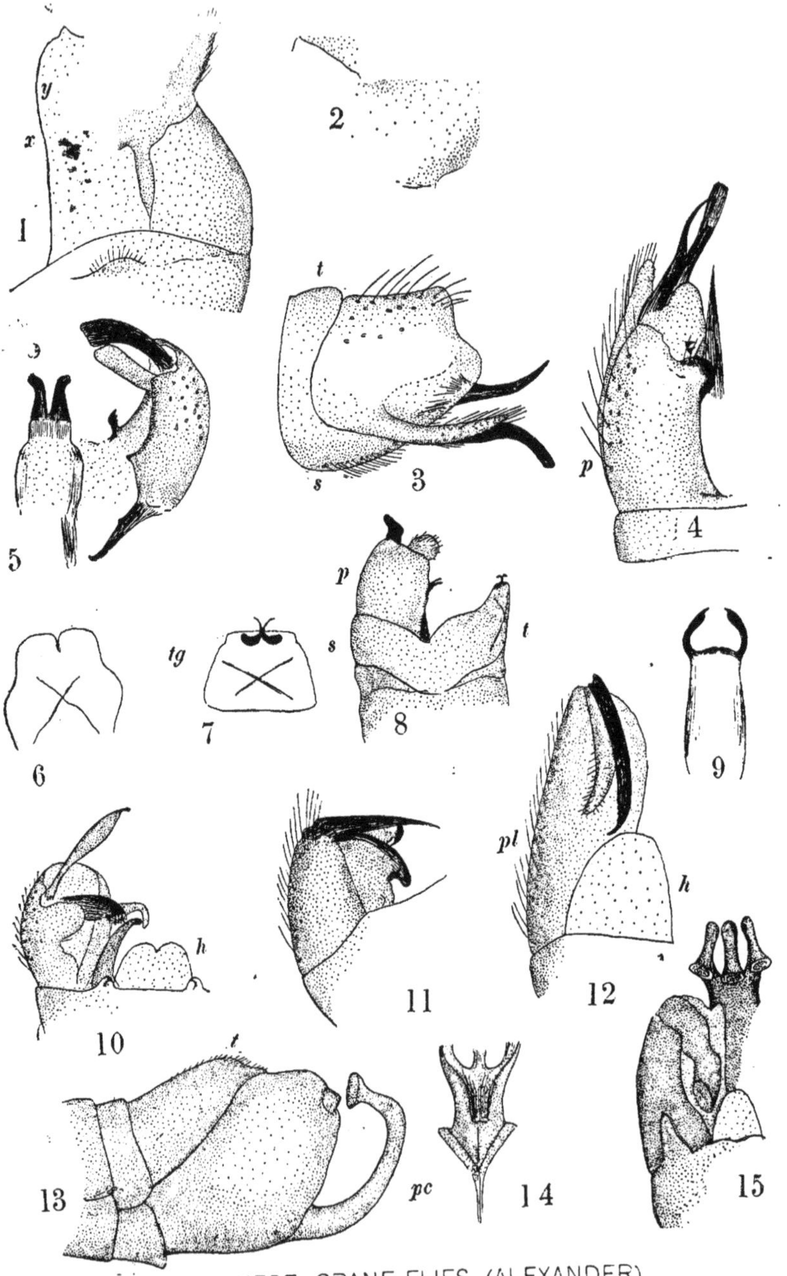

JAPANESE CRANE-FLIES (ALEXANDER)

EXPLANATION OF PLATE X.—Continued.

Fig. 8.—Hypopygium of *Erioptera* (*Acyphona*) *asymmetrica;* lateral aspect; p = pleura; s = 9th sternite; t = 9th tergite.

Fig. 9.—Hypopygium of *Erioptera* (*Acyphona*) *asymmetrica;* dorsal aspect; gonapophyse.

Fig. 10.—Hypopygium of *Limnophila japonica;* dorsal aspect; h = anal tube.

Fig. 11.—Hypopygium of *Limnophila satsuma ;* ventral aspect.

Fig. 12.—Hypopygium of *Limnophila inconcussa;* dorsal aspect; h = anal tube; pl = pleura.

Fig. 13.—Hypopygium of *Liogma kuuanai;* lateral aspect; t = 9th tergite; pc = penis-guard.

Fig. 14.—Hypopygium of *Liogma kuwanai;* ventral aspect of the base of the tripartite penis-guard.

Fig. 15.—Hypopygium of *Liogma kuwanai,;* dorsal aspect.

(TO BE CONTINUED.)

A NEW PYROMORPHID FROM TEXAS.

BY WM. BARNES, M.D., AND J. McDUNNOUGH, PH.D., DECATUR, ILL.

Acoloithus novaricus, sp. nov.

Very similar to *falsarius* Clem., having the wings of the same dull black colour. The distinguishing feature is that the collar is unbroken reddish-orange, whereas in *falsarius* this colour is confined to the lateral areas, the centro-dorsal portion being black. Expanse, 14 mm.

Habitat: Kerrville, Texas; Shovel, Mt. Texas (July), 2 ♂'s. Type and cotype coll. Barnes. 4 ♂'s (Texas). Cotypes, Tring Museum, England.

Dr. K. Jordan, with whom we have recently had some correspondence concerning this group, has called our attention to this species and expressed the desire that we describe it. We take pleasure in doing so, as the characteristic feature seems very constant.

FURTHER NOTES ON ALBERTA LEPIDOPTERA.

BY F. H. WOLLEY DOD, MIDNAPORE, ALTA.

(Continued from page 244.)

423. *S. athabasca* Neum.—The only locality given for this species in Smith's Catalogue is "British Columbia," presumably on the strength of the description, which I have not seen. But I have seen the type, a male, in the Neumœgen collection, and it is labelled "Belly River," which is in Southern Alberta, and no portion of it in B.C. I have seen the species fairly swarming around Gleichen, and on the Blackfoot Indian Reserve near there. It is almost or quite exclusively a day flier, and revels in hot sunshine, usually accompanied, in far fewer numbers, by *Melicliptera septentrionalis* and *Melaporphyria oregonica*. The Laggan specimens I referred to as having orange secondaries are *petricola* Walker, described from Rocky Mountain specimens taken by Lord Derby's collectors. A prairie and a mountain series of these respectively might easily give every impression of two species, especially if the series were short ones. Mountain specimens are usually a trifle more robust and larger, have yellowish or orange secondaries and ochreous tinted primaries, the depth of this tint varying as the depth of color of the secondaries. In size, my prairie specimens vary from about 28 to 31 mm., smaller specimens being uncommon. Mountain specimens seen to average scarcely more than 1 mm. larger, but my largest specimen, a handsome female from Field, B.C., expands very nearly 35 mm. My darkest and most richly coloured example is from Windermere, also in B.C. But an orange-tinted form is rare on the prairie, and a form with creamy white ground is equally rare in the mountains. Each of these grades through to the predominating form in their respective districts, and the extremes in each overlap those in the other. I regret to say that my entire series of these at present consists only of twenty-five specimens, but I have examined a good many more, both dried and in nature, and after years of deliberation have come to the conclusion that the balance of evidence is strongly in favor of there being only one species.

Petricola may stand for the form with orange secondaries. Walker's type happens to be paler than average mountain specimens. The names of course should be reversed, thus:

S. petricola Walk.—Ground color of primaries creamy ochreous, secondaries orange. Alberta Rocky Mountains. Rare on the prairie.

Var. *athabasca* Neum.—Ground color of all wings whitish, sometimes pure white. Alberta prairies. Rare in the mountains.

It usually flies on the prairies from the middle of June to the middle of July. My mountain specimens are dated from June 30th to July 27th, but I don't know that I was ever in the mountains before the earlier date. It flies most freely at low levels, but my palest mountain specimen comes from 7,500 feet on Mt. Saskatchewan, and was taken by Mrs. Nicholl. I have a female from Eureka, Utah, which is probably *petricola*. *Athabasca* is well figured by Holland on Pl. XXX., f. 29. What his fig. 35 is I don't know. It may possibly be a worn specimen of this species, but shows more reddish in the secondaries than any of mine. It certainly is not *alleni*. Grote's type of *alleni* in the British Museum appeared to me to be *adumbrata*, as did also the *alleni* of the Henry Edwards collection.

An analogous case to this amongst Alberta butterflies is perhaps found in *Argynnis electra*, which is the predominating form in the mountains where *lais* is rare; whereas *lais* is the normal prairie form, where *electra* is seldom met with.

424. *S.* sp.—I had listed this as *hudsonica*, and it is the *hudsonica* of the Kootenai List. I do not feel quite confident that it is not a variation of that species (No. 422), but I have so far been unable to connect them. I have never taken it on the prairie, nor nearer to Calgary than Laggan, but have seen it on the wing in the daytime in abundance all through the mountains during July, and Dr. Dyar records it apparently as somewhat common in the Kootenai from May 29th to July 13th. So far as the males are concerned, the maculation and color scheme is practically identical with *hudsonica*, but the tone is very much darker, there being more inoration with black scales, and all the black shades blacker.

Both species have usually a fuscous transverse bar or shade through the outer portion of the median pale band on the primaries in both sexes. This is normally very distinct, and I have no specimens in which it is not at least traceable. In the males only of both, this bar or shade is sometimes chestnut brown, and when best developed this is very conspicuous, as it is the only portion of the fore wing, and that almost of the palest ground, in which this color appears. In *hudsonica* this takes somewhat the form of a bar with well-defined edges, and may consist of a broadly geminate waved brown line, with the included space paler brown. In No. 424, though often broader, it has ill-defined edges, coalescing gradually with the pale ground.

But the most obvious superficial difference between the forms is seen in the females. Whereas in *hudsonica* this sex has the primaries much greyer and more even, with the maculation partially obsolete or ill-defined, in No. 424 the sexes are alike, with the exception that in neither case is the brown-barred variation found in the females, at least so far as I have yet observed.

In the Kootenai List, Dr. Dyar referred to the brown-barred form as var. *seposita* Hy. Edw., and I found a series of such forms separated under that name in the Washington Museum. I saw a male type of *seposita*, from Colorado, in the Henry Edwards collection. It certainly seemed near this form, and may be the same, though I did not compare a specimen, and do not feel at all convinced of its identity. There is another male type in the Neumœgen collection.

The ground color of the secondaries of *hudsonica* is pale creamy white. In No. 424 it is darker, a trifle ochreous, sometimes slightly orange. Holland's Plate XXX., fig. 31, under *hudsonica*, is apparently this form, and I can match it very closely in my collection.

It is quite possible that this is a somewhat similar example of what I believe to be racial variation in *petricola* and *athabasca*, and that *hudsonica* is easily influenced by environment and has developed strongly marked races in localities no great distance apart.

430. *Philometra metonalis* Walk.—The type is a male from St. Martin's Falls, Hudson Bay Territory. It is a very even specimen, and has scarcely any trace of transverse lines. *Gaosalis*

Walk. (not *goasalis*), from Nova Scotia, a male from Lieut. Redman, is better marked. They appeared to me to be the same species, and the one here listed.

THYATIRIDÆ

433. *Pseudothyatira expultrix* Grt.—I have seen a specimen from Edmonton, taken by Mr. F. S. Carr, on June 10th, 1910. The same collector also submitted to me a specimen of *cymatophoroides*, taken on July 17th of the same year, which is now in my collection. These forms appear to me to be distinct species, as the transverse lines differ. In *cymatophoroides* the posterior edge of the transverse anterior bar is much less waved or crenate than in *expultrix*, and the same is true of the t.p. line, which in *expultrix* is generally deeply crenate. This diflerence is most obvious near the anal angle, where is the dark-brown patch generally so conspicuous in *cymatophoroides*. This patch is sometimes much reduced, whilst in one of my Vancouver specimens of *expultrix* there is a small patch of dark brown after the t.p. line at this point, though the line itself is deeply crenate. The same specimen has no trace of dark brown in the t.a. bar.

In the Kootenai List, Dr. Dyar suggested that these forms might be distinct, as he found a slight, though apparently not constant, difference in the larvæ.

NOTODONTIDÆ.

436. *Melalopha albosigma* Fitch.

437. *M. brucei* Hy. Edw.—I have occasionally taken both these species at light, not uncommonly, at the end of May and early June.

439. *Notodonta simplaria* Graef.—High River (Baird). Also at head of Pine Creek, May 29th, 1910.

440. *Pheosia dimidiata* H.-S.—One at light at head of Pine Creek, June 5th, 1910. High River, May 7th, 1910 (Baird).

441. *Harpyia scolopendrina* Bdv.—I have about twenty local specimens under this name, and have taken more, May 11th to June 5th. Also one from Banff (Sanson), June 25th. I have not been able to verify the name. It is doubtfully the same as the species figured by Holland on Plate XXXIX., Fig. 11., being paler in ground colour, with much more distinct lines in the postmedial area, and

none of my specimens have as wide a band. In all my series the median band is entire.

442. *H. modesta* Huds.—Besides the three previously referred to, I have only taken four more specimens, on June 7th and 17th, 1910. In three out of this series of seven the median band is broken centrally and forms two quite distinct blotches. In a fourth it is so constricted that the margins nearly meet below the median vein. The others are not unlike Holland's *scolopendrina*, referred to above, and have a similar band on secondaries, but the ground colour of my specimens is more ochreous, the transverse postmedial lines more distinct, and the discal dot on all wings heavier. I query the name, as I find I have made a note to the effect that, "Packard's figure of *borealis* is the '*modesta*' of my Calgary list." If the note and Packard's figure are correct, it remains to be discovered what is the correct name for the species—not in my collection—figured by Holland as *borealis*, and standing as such in the British Museum.

442a. *H.* (? var.) *albicoma* Strk.—I have thirteen Alberta specimens in my collection, from Red Deer River and head of Pine Creek, collected by myself, and from High River, from Mr. Thomas Baird. May 30th to July 7th. I have the same form from Wellington, Vancouver Island, and it is that figured by Holland under this name. It differs from what I hold as *scolopendrina* in the slightly paler and less smoky ground, in the narrower median band, which is often much constricted and sometimes divided into two blotches, in the greater preponderance of fulvous scales, especially in the band, and in having the discal spot on primaries more usually punctiform than linear. In my former notes, the words, "has no fulvous scales," were a grave error, as were also "the two inner lines of the three beyond the cell are obsolete" (Vol. XXXVIII., p. 52, Feb., 1906). The form is hard to separate from what I hold as *scolopendrina*, and may not improbably prove a variety of that.

443. *Gluphisia septentrionalis* Walk.—High River (Baird). Two females, June 30th and July 7th.

444. *G. lintneri* Grt.—A few more specimens have come to hand, April 19th, 1906, May 26th, 1907, and one in 1911. The first of these was taken flying in sunshine.

LIPARIDÆ.

• 446. *Notolophus antiqua* Lin.—I have occasionally seen the male of this species on the wing in sunshine during September in some numbers, and have bred it from Salix.

447. *Olene vagans* Barnes & McDunnough, var. *grisea* B. & McD. (Contributions, Vol. II., No. 2, pp. 60 to 63, pl. III., April, 1913).—This is the species standing as *Olene plagiata* Walk. in our lists, and was supposed to be the *Acyphas plagiata* of Walker. When at the British Museum early in 1912 I saw Walker's type, and recognized it as the well-known *leucostigma* of Abbott and Smith. This is referred to by Messrs. Barnes & McDunnough on page 50 of their "Contributions" quoted above, and Walker's type is figured on Plate VII., Fig. 1. Those authors considered the present species unnamed, and described it accordingly as *vagans*. They divided it into three sub-species: (a) *vagans*, types from St. Johns and Montreal, Que., and Yaphang, L. I.; (b) *grisea*, types from Eureka and Provo, Utah; (c) *willingi*, types from Humboldt, Sask. I have seven Alberta specimens, six males and a female, from Red Deer River, head of Pine Creek, and High River, the latter from Mr. Baird, June 17th to July 27th. All of these seem referable to *grisea*. I have the same form from Cartwright, Man., Wellington, V. I., and Eureka, Utah.

GEOMETRIDÆ.

452. *Rachela bruceata.*—The larvæ of this species have again denuded hundreds of acres of *Populus tremuloides* in this district during the present year (1913). The denudation is greater in extent than it was ten years ago, though none has been observed during the intervening period. After starving themselves out on a patch of poplars, they spread to neighboring species of *Salix*.

453. *Talledega montanata* Pack.—Red Deer River, July 8th, 1905; Banff, June 28th—July 1st, 1907. The Banff specimens, like a series from Field, B. C., are rather darker than those from the prairie, and have duller secondaries.

486. *Mesoleuca hersiliata* Guen.—I have taken two more specimens, on August 24th, 1907, and July 24th, 1909.

490. *Hydriomena multiferata* Walk.—Lake Louise, Laggan, July 18th, 1907, one specimen.

491. *H. custodiata* Grt.—Three more specimens. Red Deer River, July 21st and 24th, and head of Pine Creek, July 27th, all in 1907.

493. *Cœnocalpe magnoliata* Guen.—Two more specimens, at Head of Pine Creek, on July 15th, 1906, and July 16th, 1911; one at Banff, July 1st, and another at Laggan, July 18th, 1907.

495. *C. topazata* Grt.—Head of Pine Creek, May 26th, 1907.

498. *Xanthorhoe abrasaria* H.-S.—One at Head of Pine Creek, July 3rd, 1904.

501. *X. turbata* Hbn., syn. *circumvallaria* Taylor.—I quoted Mr. Prout's MS. reference in a footnote to my previous notes. The same reference is made by Messrs. Barnes & McDunnough in Vol. XLIV., p. 274, Sept., 1912. Taylor described it from Laggan specimens only.

502. *X. fossaria* Taylor—Described from three males and a female from Laggan, Alta., and two males from Mt. Cheam, B. C. One from the latter locality is the type (Can. Ent., XXXVIII., 401, Dec., 1906). In March, 1906, I visited Mr. Taylor, and received from him, amongst other species, a series of six Wellington specimens as *fossaria*, bearing dates of June 15th to 30th, 1902 and 1905. They certainly look to me the same species, but it is strange that no mention is made of its occurrence at Wellington in the description.

503. *Synelys enucleata* Grt.—Two more Red Deer River specimens, July 24th, 1907.

519. *Deilinia borealis* Hulst.—I have a female taken by Mrs. Nicholl at Banff on June 24th, 1907.

(To be continued.)

ODOUR PREFERENCES OF INSECTS.

BY HARRY B. WEISS, NEW BRUNSWICK, N.J.

Moths, butterflies, bees, flies and other insects feed upon the nectar of flowers, being guided to them presumably by the senses of smell and sight. Various investigators have differed in this, Lubbock claiming that bees, for instance, recognize at a distance and prefer certain colours; while Plateau found that neither form nor colour played any part in attracting insects and that they were guided entirely by a sense of smell.

This sense is defined by Forel as "a special sense which allows the animal to recognize at a distance by some specialized energy the (chemical) nature of a certain body." Our scientific knowledge of odours is rather meagre. Some are known vaguely as pleasant or unpleasant and for many we have no definite names whatever, and are forced to liken them to the few odours with which we are familiar and for which we have definite names. Moreover, some smells are exceedingly complex experiences involving elements of taste, touch and vision. The most satisfactory classification of smells is that adapted by Zwaardemaker from the classification of Linnæus, which groups natural objects according to similarities, but does not aim to itemize all smells. This list is as follows:

1.—Ethereal smells, including all fruit odours.

2.—Aromatic smells; for example, those of camphor, spices, lemon, rose.

3.—Fragrant smells, those of most flowers.

4.—Ambrosiac smells—all musk odours.

5.—Alliaceous smells—those of garlic, asafœtida, fish, chlorine.

6.—Empyreumatic smells—those of tobacco, toast.

7.—Hircine smells—those of cheese, rancid fat.

8.—Virulent smells—those of opium.

9.—Nauseating smells—those of decaying animal matter.

In the Lepidoptera practically all members are attracted by fragrant smells. The Coleoptera have a somewhat wider range. Dermestidæ are attracted by fragrant and also hircine odours; *Dermestes lardarius*, for instance, the larva of which feeds on bacon, cheese, meat and feathers. The bumble flower beetle, *Euphoria inda*, finds ethereal and fragrant odours to its liking, being found feeding on peaches, grapes, apples and the pollen of flowers. Locust borers and soldier beetles are plentiful on goldenrod and various Buprestids also visit flowers, while the cigarette beetle has an empyreumatic taste. The Silphidæ, however, are drawn to nauseating odours, feeding, as they do, on decaying flesh.

With the exception of the ants, nearly all Hymenoptera are attracted by fragrant odours and also ethereal odours, the Vespidæ and bees being very fond of nectar and fruit juices. Ants have a wider range, ethereal, alliaceous, hircine and nauseating odours all being more or less attractive.

The range of the Diptera is exceptionally wide, embracing ethereal, fragrant, alliaceous, hircine and nauseating odours. Certain species of mosquitoes, bee flies, and syrphus flies are found feeding on nectar. *Eristalis tenax* visits cesspools, dung-pits and decaying vegetable matter in addition to different flowers. Drosophilidæ visit decaying fruits both for food and egg deposition, and *Piophila casei* is drawn toward cheese, ham and partly spoiled vegetable matter; while the house fly, as everyone knows, shuns nothing except aromatic and virulent odours.

Robertson's records show clearly that the Hymenoptera and Diptera are especially fond of fragrant odours. He found that Pastinaca sativa was visited in twenty-six days by 173 Hymenoptera, 72 Diptera, 14 Coleoptera, 9 Lepidoptera, 6 Hemiptera and 1 Neuropteron; also that Asclepias verticillata was visited by 52 Hymenoptera, 42 Diptera, 16 Lepidoptera and 3 Coleoptera.

It would be extremely interesting to find the effect of exhaustion upon the end organs of smell. A bee, for instance, visiting innumerable flowers of the honeysuckle must have its organ fatigued by the continuous smelling of this one odor. How, then, would it react to other odours? Does its physiological mechanism of smell consist of distinct parts, one of which might be put temporarily out of commission without impairing the others, or does it consist entirely of one part?

THREE NEW GALL MIDGES (DIPTERA).

BY E. P. FELT, ALBANY, N.Y.

The following descriptions are of species which have been reared and of one concerning which we possess some exceptionally interesting data. There is much to be learned about our tropical or subtropical midge fauna. There must be hundreds of interesting and undescribed species existing in the West Indies and adjacent countries.

Karschomyia cocci, n. sp.

The midges described below were reared from a sugar-cane mealy bug, Pseudococcus sacchari (?) collected at Central Providencia, Patillas, P.R., January 30, 1913, by Mr. D. L. Van Dine.

The type species of this genus, *K. viburni* Felt is easily distinguished by the almost trinodose character of the flagellate antennal segments of the male; while the only other known species, the Peruvian *K. townsendi* Felt, has much more slender flagellate antennal segments in the male, the stems in this latter each having a length 3½ times their diameter. Described from specimen in alcohol.

Male.—Length, 1 mm. Antennæ, ¼ longer than the body, rather thickly haired, yellowish brown; 14 segments, the fifth having the stems with a length 1½ and two times their diameters, respectively; circumfili and setæ well developed. Palpi: First segment short, subquadrate; the second with a length three times its diameter, the third ½ longer, more slender; the fourth ½ longer than the third. Mesonotum dark brown, the submedian lines fuscous yellowish. Scutellum and postscutellum yellowish. Abdomen yellowish white, the dorsal sclerites and genitalia somewhat fuscous. Halteres pale yellowish. Coxæ and femora mostly pale yellowish; tibiæ and tarsi fuscous yellowish. Claws slender, strongly curved, the anterior unidentate, the pulvilli rudimentary. Genitalia: basal clasp segment moderately stout, the posterior external angles somewhat produced and bearing a group of three or four stout setæ; terminal clasp segment subapical, swollen near the middle, curved; dorsal plate long, deeply and narrowly emarginate, the lobes broad, narrowly rounded.

Female.—Length, 1.5 mm. Antennæ extending to the third abdominal segment, sparsely haired, dark brown; 14 segments, the fifth with a stem about ¼ the length of the cylindric basal enlargement, which latter has a length about twice its diameter; terminal segment produced, the enlargement with a length three times its diameter and apically a broad, knoblike appendage. Mesonotum dark brown, submedian lines indistinct. Scutellum and postscutellum yellowish. Abdomen yellowish orange, the dorsal sclerites somewhat fuscous. Ovipositor short, yellowish, the terminal lobes narrowly oval. Halteres: Coxæ and femora mostly pale yellowish; tibiæ and tarsi light straw. Type, Cecid a2415.

Mycodiplosis insularis, n. sp.

This midge was reared from a vial containing leaves of Leonotis nepetæfolia abundantly infested with red spider. There were also small, white coccons among the colonies of red spider collected at Pio Piedras, P.R., August 6, 1913, by Thomas H. Jones. This species appears to be allied to *M. reducta* Felt, from which it is most easily separated by its larger size and the somewhat longer distal stem of the 5th antennal segment. Described from specimens in alcohol.

Larva.—Length, 1.3 mm.; moderately stout, pale yellowish. Head apparently with a length nearly twice its diameter, broadly rounded anteriorly. Antennæ long, with a length fully 10 times the diameter, slender, curving, posterior extremity subtruncate and irregularly papillate.

Male.—Length, 1 mm. Antennæ, ¼ longer than the body, sparsely haired, light brown; 14 segments, the 5th having the stems with a length 1½ and 2½ times their diameters, respectively; circumfili well developed. Palpi: the first segment short, subquadrate, the second with a length twice its diameter, the third a little longer, more slender, the fourth ½ longer than the third. Mesonotum dark reddish brown, the narrow submedian lines yellowish. Scutellum and postscutellum pale yellowish. Abdomen mostly pale yellowish, the dorsal sclerites slightly fuscous. Halteres, coxæ and femora mostly pale yellowish, tibiæ and tarsi light straw. Claws slender, strongly curved, the pulvilli as long as the claws. Genitalia: basal clasp segment moderately stout, terminal clasp segment swollen basally, slightly curved; dorsal plate moderately long, deeply and roundly emarginate, the lobes narrowly rounded.

Female.—Length, 1.25 mm. Antennæ extending to the second abdominal segment, sparsely haired, light brown; 14 segments, the fifth with a stem ¼ the length of the cylindric basal enlargement, which latter has a length 2½ times its diameter; terminal segment slightly produced, with a length three times its diameter, broadly rounded apically. Face yellowish brown. Ovipositor short, yellowish, the lobes narrowly oval. Type, Cecid a2413.

Clinodiplosis examinis, n. sp.

The midges described below were present by hundreds, if not thousands, upon a screen door, or hanging from cobwebs attached thereto at Nassau, N.Y., June 19, 1913. The insects were so numerous as to fairly dot the surface of the screen here and there, and where spider webs occurred it was not uncommon to see 5 or 6 in a line usually about a quarter of an inch apart. They hung lightly from the web, were easily disturbed and frequently returned to their fragile supports.* The insects were so numerous that it was comparatively easy to capture some 50 with an ordinary collecting bottle by simply placing it over groups of three or four here and there on the screen. The midges had not been observed previously and presumably represent the emergence of a brood from some nearby food plant or food material, possibly plant lice inhabiting adjacent maple or elm trees.

Male.—Length, 1 mm. Antennæ fully ½ longer than the body, thickly haired, light brown; 14 segments, the fifth having the two portions of the stems 3 and 3½ times their diameters, respectively; the distal enlargement with a length ¾ greater than its diameter and a slight constriction near the basal third. Palpi: first segment subquadrate, with a length fully twice its diameter; the second a little longer, more slender; the third nearly twice the length of the second, somewhat dilated; the fourth a little longer than the third, more slender. Mesonotum yellowish brown. Scutellum brownish yellow, post scutellum fuscous yellowish. Abdomen mostly reddish brown, the genitalia reddish yellow. Costa fuscous straw. Halteres mostly yellow transparent, slightly fuscous subapically; Caxæ and femora mostly pale yellowish, the tibiæ and tarsi fuscous straw; claws slender, strongly curved, the anterior unidentate, the pulvilli about half the length of the claws. Genitalia: basal clasp segment moderately stout; terminal clasp segment long, slightly swollen basally; dorsal plate short, broad, deeply and triangularly emarginate, the lobes broadly rounded; ventral plate long, rather broad, broadly and roundly emarginate, the lobes short and with a few coarse setæ apically; style long, stout, tapering.

Female: Length, 1.5 mm. Antennæ nearly as long as the body, sparsely haired, reddish brown; 14 segments, the fifth with

*Mr. Frederick Knab (N.Y. Ent. Soc. Journ 20: 143—46) records a number of Diptera as habitually occuring on spider's webs In this connection it is worthy of note that Mr D B. Young found last June at Albany N. Y. a Tipulid hanging on cobwebs, leaving and returning thereto at will. The species appears identical with a specimen in the state collection determined by Mr C. P. Alexander as Oropega obscura.

a stem nearly ¾ the length of the cylindric basal enlargement, which latter has a length about twice its diameter; terminal segment produced, the basal enlargement cylindric, with a length more than three times its diameter and apically a finger-like process. Palpi: first segment subquadrate, with a length more than twice its diameter, the second twice the length of the first, the third a little longer, somewhat dilated; the fourth a little longer and more slender than the third. Mesonotum slaty brown, the submedian lines indistinct. Scutellum yellowish, postscutellum fuscous yellowish. Abdomen brownish red, the dorsal sclerites somewhat fuscous. Ovipositor reddish yellow. Halteres yellowish transparent, fuscous subapically. Coxæ pale yellowish, femora light straw, tibiæ and tarsi fuscous straw. Ovipositor stout, nearly as long as the abdomen, the terminal lobes lanceolate, sparsely setose. Type, Cecid a2411.

Described from a number of males and females taken together and presumably specifically identical.

A NEW SPECIES OF HELIOTHRIPS (THYSANOPTERA) FROM MARYLAND AND ILLINOIS.

BY J. DOUGLAS HOOD,
United States Biological Survey, Washington, D. C.

The systematist's interest in the genus *Heliothrips* Haliday is enhanced by the fact that it includes several of the best known and most troublesome species of the order. *Hæmorrhoidalis* and *femoralis* are cosmopolitan greenhouse pests ; *rubrocinctus*, a widely distributed tropical species, injurious to cacao, has lately appeared in Florida as an enemy of the mango and avocado; *fasciatus* often proves troublesome to beans and other crops in California ; while the recently described *phaseoli* is an important bean pest in southernmost Texas.

In a recent paper on the genus,* Dr. Karny unites *Dictyothrips* and *Parthenothrips* with *Heliothrips*, recognizing them as subgenera only, and erects a fourth subgenus, *Selenothrips*, for a new species which he calls *decolor* and for *rubrocinctus* Giard. While admitting that *Selenothrips* is a well-founded subgenus, I can not follow Dr. Karny in his treatment of *Dictyothrips* and *Parthenothrips*. In the

*Revision der Gattung *Heliothrips* Haliday, Ent. Rundsch., 28 Jhg., no. 23, pp. 179-182, 5 figs.; 1911.

sinking of the former, Dr. Karny was led into error by Dr. Hinds' description of *Heliothrips fasciatus* Pergande,** in which the number of segments in the maxillary palpi is erroneously given as three, instead of two—a mistake which was copied also by Moulton.*** *Dictyothrips* is thus readily separable by the three-segmented maxillary palpi; while the number of antennal segments and the decidedly anomalous character of the fore wings marks *Parthenothrips*, in my opinion, as one of the most distinct genera of the entire family.

The new species described below is the tenth one of the genus to be recorded from North America; and as the entomological fauna of the tropical and sub-tropical south becomes better known, this number will doubtless be greatly increased.

Heliothrips striatus, sp. nov.—Figs. 11 and 12.

Female.—Length about 1.1 mm. General color, dark blackish brown (nearly black); head and thorax paler than abdomen, the former with a yellow spot each side of the ocelli; legs dark brown, with the femora and tibiæ paler at extremities; tarsi pale; abdomen slightly paler at tip.

Head about 1.6 times as wide as long and about equal in length to prothorax; cheeks rounded to eyes, narrowed to base; dorsal

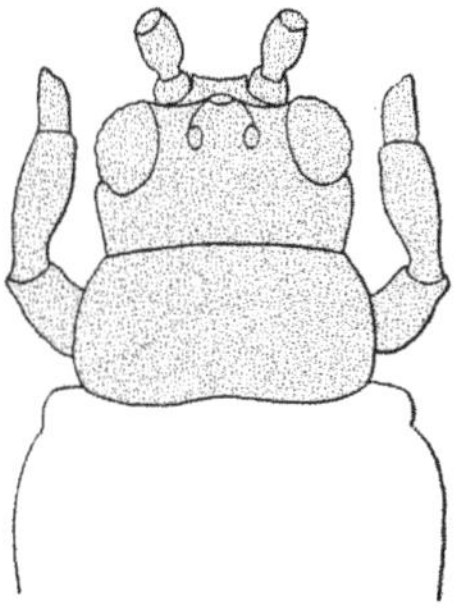

FIG. 11.—Head and prothorax.

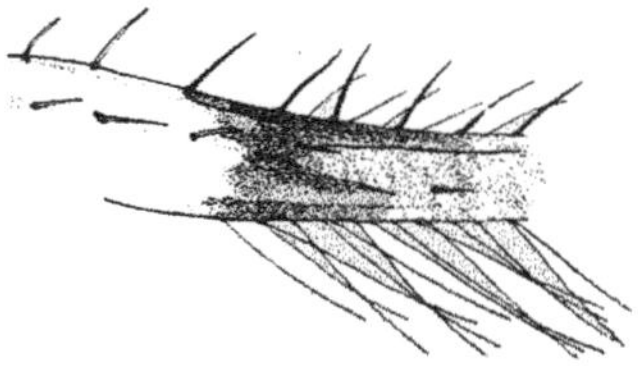

FIG. 12.—Portion of fore wing.

Heliothrips striatus Hood.

surface reticulate, roughened between the lines of reticulation; frontal costa broad, much wider than first antennal segment; vertex subcarinate in front of ocelli. Eyes less than twice as long as their distance from posterior margin of head, slightly protruding, setose. Ocelli approximate, the posterior pair opposite center of eyes. Antennæ about 2.3 times as long as head; segments 1 and 2 light brown; 3, dark gray, paler in second fifth; 4, dark gray, pale in basal half; 5, pale grayish white, slightly darker apically; 6–8, dark gray. Maxillary palpi two segmented.

Prothorax twice as wide as long, about equal in length to head and with similar reticulation. Pterothorax somewhat broader than prothorax, dark brown in color. Wings long, surpassing the abdomen; fore wing about thirteen times as long as width at middle and with two veins nearly or quite attaining tip; basal vein with four spines, of which the distal is much stouter, black, and situated at the fork (see figure 12); anterior vein usually with one spine (rarely two) at base and two near apex of wing, all black, the basal one unusually stout; posterior vein with five or six equidistant spines at middle, of which three or four are black; fore wings slightly darkened with brown at extreme base, clear white in basal fourth, nearly black in second and third fourths (darkest toward base), clear white again in seventh eighth, and nearly black again in apical eighth, where it is margined with darker.

Abdomen broadly ovate, pointed at tip; *notum of segments 1-8 closely striate laterally*, the striæ transverse toward middle of segment and longitudinal at sides. Segment 10 without longitudinal dorsal suture, though irregularly weakened toward tip.

Measurements of holotype: Length, 1.07 mm.; head, length .120 mm., width .192 mm.; prothorax, length .114 mm., width .228 mm.; mesothorax, width .324 mm.; abdomen, width .372 mm. Antennal segments: 1, 21μ; 2, 45μ; 3, 50μ; 4, 45μ; 5, 41μ; 6, 32μ; 7, 15μ; 8, 34μ; total length of antenna, .28 mm., width at segment 4, .027 mm.

Male.—Length about .84 mm. Sternum of abdominal segments 3-7 each, with a large, pale, transverse area about nine times as wide as long. Segment 9 with two pairs of dorsal spines, of which the basal is much shorter and stouter than the apical.

**Proc. U. S. Nat. Mus., vol. XXXVI, p. 174; 1902.
***Tech. Ser. 21, Bur. Ent., U. S. Dept. Agr., p. 14 ; 1911.

Measurements of allotype: Length .84 mm.; head, length .102 mm., width .168 mm.; prothorax, length .090 mm., width .196 mm.; mesothorax, width .252 mm.; abdomen, width .228 mm.

Described from three females and one male, taken near Chevy Chase Lake, Maryland, July 6, 1913, by W. L. McAtee, on the under surface of leaves of a tulip tree (*Liriodendron tulipifera* L.); and from one female collected at Parker, Illinois, July 14, 1909, by C. A. Hart, on the same food plant.

Type locality: Chevy Chase Lake, Maryland.

The abdominal sculpture is almost identical with that of *H. phaseoli*, figured by the writer in Psyche, Vol. xix, No. 4, plate 8, fig. c, August, 1912. From that species it may be known by the broader head, the much darker color of the body, the details of wing coloration, and the stout black spines on the fore wings at the junction of the two principal veins.

ANOTHER RED SPECIES OF THE GENUS *OLIGOSITA*.

BY J. C. CRAWFORD, WASHINGTON, D. C.

Oligosita giraulti, new species.

Female.—Length about 0.5 mm. Brilliant vermilion, including marginal and stigmal veins of fore wings and marginal vein of hind wings; the femora and hind tibiæ red, the red color decreasing apicad on legs and the rest of legs testaceous; submarginal vein with a bristle at middle, and one at apex of vein, near base of marginal vein a short one followed by two longer ones, then one or two shorter ones and a long at apex of vein, fore wings with no discal cilia; marginal cilia at apex of wing slightly longer than width of wing; fumated spot in under stigmal knob distinct; stigmal knob almost circular but with a projection apicad; pedicel about as long as the funicle joint, about as long as middle segment of club which is longer than either the first or third joints; base of abdomen with a whittish band, more or less suffused with reddish and occupying about one-fourth of the abdomen.

Type locality: St. Clair Experiment Station, Trinidad.

The type slide has two specimens reared from the eggs of *Tomaspis varia* by Mr. P. Lachmere-Guppy.

There is also a paratypic slide with one specimen with the record "Reared from grass, Verdant Vale, Jan. 30, 1913, and oviposited in frog hopper eggs" (same date). F. W. Urich, collector.

Type Cat. No. 15971. U. S. N. M.

Easily distinguished by the red color and coloring of the veins; *sanguinea* Girault lacks the red in the veins and *neosanguinea* Girault has a white band occupying about one-third of the abdomen, and the basal joint of the club is at the second.

The species is named in honor of Mr. A. A. Girault who has done most of the work in this family.

Mr. Lachmere-Guppy writes that "this insect in life is vermillion with jet black eyes, and acts like a mymarid".

Mailed September 12th, 1913.

The Canadian Entomologist.

Vol. XLV. LONDON, OCTOBER, 1913 No. 10

REPORT ON A COLLECTION OF JAPANESE CRANE-FLIES (TIPULIDÆ), WITH A KEY TO THE SPECIES OF PTYCHOPTERA.

BY CHARLES P. ALEXANDER, ITHACA, N. Y.*

(Continued from page 295).

Tribe *Limnophilini*.
Genus *Limnophila* Macquart.

KEY TO THE JAPANESE LIMNOPHILÆ.

1. Wings unspotted (subgen. *Limnophila*) ......*inconcussa*, sp. n.
   Wings marked with brown (subgen. *Pœcilostola*).............2.
2. Large species (male, length 22-25 mm.; wing over 15 mm.); wings with a few large seams or blotches....*satsuma* Westw.
   Small species (male, length 10-13 mm.; wing under 12 mm.) wings with abundant dots in the cells.................3.
3. Legs and abdomen yellow throughout; petiole of cell M_1 as long as cell 1st M_2.......................*varicornis* Coq.
   Legs with segments tipped with brown; abdomen yellow and brown; petiole of cell M_1 longer than cell 1st M_2..............................*japonica*, sp. n.

Limnophila inconcussa, sp. n.

Wings unspotted; cross-vein r far from tip of R_1.

Rostrum brownish yellow beneath, brown above, palpi brown; antennæ dark brown, the third segment more yellowish at its base; antennæ short, reaching about to the wing basis; front, vertex and occiput dark brown, dusted with grey.

Mesonotum greyish with a median brown stripe; pseudosutural fovea and tuberculate pits very distinct, black; scutum, scutellum and postnotum brown, pleuræ dark brown (probable that the body, in dried specimens, is grey). Halteres pale. Legs: coxæ and trochanters dull yellow; femora yellow, a little darkened

*Contribution from the Entomological Laboratory of Cornell University.

before the tip; tibiæ yellow, brown at tip; first tarsal segment yellow, brown at tip, remainder of tarsi brown. Wings with a brownish yellow tinge; stigma indistinct, brown; veins Sc and R yellow, remainder brown. Venation (see fig. 2, pl. II): R_{2+3} arcuated, long, cross-vein r almost at its fork; Rs long; cross-vein r-m more distad than fork of cell; basal deflection of Cu_1 at or slightly beyond the fork of M.

Abdomen: tergites light brown; sternites much paler, yellowish. Hypopygium (see fig. 12, pl. X): pleurites elongate, slender, cylindrical, clothed with long hairs; two apical appendages, elongated, the outermost longest, more slender, chitinized, directed cephalad, its tip produced into a slender spine and its inner or cephalic edge near the tip armed with blunt denticulæ; inner appendage shorter, a little stouter and more fleshy, clothed with long hairs, especially on the inner face; the anal tube prominent; oval.

Vial No. 2.—Tokyo, Japan; 1 ♂, 1 ♀.

Vial No. 9.—Tokyo, Japan; April 25, 1912; 2 ♂, 2 ♀.

Vial No. 17.—Tokyo, Japan; April 25, 1912; 3 ♂, 7 ♀.

Vial No. 27.—Tokyo, Japan; April 25, 1912; 4 ♀.

Holotype.—♂, Vial 2.

Allotype.—♀, Vial 2.

Paratypes.—5 ♂, 13 ♀, in Vials 9, 17 and 27.

Types in author's collection; paratypes in U. S. National Museum and Cornell University collections.

Of the American species, *inconcussa* is most like *toxoneura* O. S. (East. U. S.), but the cross-vein r is removed from the tip of R_1, fusion of R_{2+3} is longer, etc.; the coloration of the two species is quite different. In Verrall's key to the British species (Ent. Mo. Mag., April, 1887, p. 264, 265), it runs down to *lucorum* Meig., which has a dark brown abdomen, brown legs, etc.

Limnophila (*Pœcilostola*) *satsuma* Westwood.

1876.—*Limnobia satsuma* Westwood, Trans. Ent. Soc. Lond., p. 504, pl. 3, fig. 5a, 5b.

1881—*Limnobia satsuma* Westwood, Trans. Ent. Soc. Lond., p. 383.

1888.—?*Epiphragma satsuma* Bergroth, Ent. Tidskrift, p. 138.

1902.—*Limnobia satsuma* Kertesz, Cat. Dipt., Vol. 2, p. 177.

Male—Length 22.6 mm.; wing 16.8 mm.; hind leg, femur 14.1 mm.; tibia 12.2 mm.

Male—Rostrum and palpi brown, the apical segment of the latter darker; antennæ, segments one and two dark brown, flagellum yellow except the two last segments which are brown; front dark brown, vertex and occiput reddish brown, a narrow median streak continued back from the front.

Pronotum with the scutum dark brown, scutellum yellowish. Mesonotal præscutum rich reddish brown, the lateral margins of the sclerite more greyish, a darker brown median triangle, broadest in front, narrowed to a point at the suture, lateral stripes similar in colour to the median stripe; scutum, lobes dark brown, median line yellowish, dark brown on caudal portion; scutellum and postnotum dark brown. Pleuræ light brownish yellow; propleuræ and dorsal portions of the mesopleuræ up to the wing; root dark brown; mesostigma very large, conspicuous, situated just behind and under the pronotal scutellum. Halteres short, stem yellow, knob brown. Legs: coxæ light yellow; trochanters reddish yellow; femora yellow, tip brown, with a still darker subapical ring; tibiæ slightly darkened at the extreme base, a whitish sub-basal annulus, tip narrowly dark brown: tarsi brownish yellow, tips of the segments darker; legs conspicuously hairy. Wings (see fig. 4, pl. III.): cephalic third deep yellow, caudal portions yellowish grey; surface with conspicuous brown marks: a large blotch at base of M; at origin of Rs; at the cord; a narrow seam to cross-vein r; paler crown margins to Cu and the veins in the vicinity of cell 1st M_2 (discal). Venation (See fig. 4).

Abdomen: tergites rich yellow, extreme apical margin of the sclerites darker; a brown lateral line; sternites lighter yellow, apices, especially of the terminal segments, darker. Hypopygium (See fig. 11, plate X.): viewed from beneath, 9th sternite with caudal margin straight, the sides oblique; pleuræ very short, stout; dorsal apical appendage directed inward, cylindrical, chitinized, its tip with a sharp recurved hook; ventral apical appendages, two, the outermost chitinized, broad at base, rapidly tapering to a sharp point, directed inward, the lower appendage is fleshy at the base, more chitinized at the tip, its caudal or outer margin grooved to receive the outer appendage. Viewed from above, 9th tergite concave, with a projecting median lobe; anal tube conspicuous, more pointed at upper end than in *japonica*

One male (Vial No. C; Tokyo, Japan; August, 1912); I give the above description to supplement Westwood's brief characterization. The species agree with *barbipes* Meigen (Europe) in its conspicuously hairy legs.

Limnophila (*Pœcilostola*) *japonica*, sp. n.

Wings spotted; tibiæ and femora tipped with brown.

Male.—Length 10-13 mm.; wing 9.8 mm. Female, length 15 mm.; wing, 11-12.3 mm.

Male.—Rostrum and palpi dark brown; antennæ dark brownish black, except segment three, which is pale yellow basally, the tip brown; antennæ short, if extended backward it would barely reach the wing basis; segment one elongate, as long as the succeeding three combined; segments 2-5 broad, oval-pyriform, gradually becoming more cylindrical; segments 6-16 cylindrical more elongate toward the end; front, vertex and occiput dark brown.

Pronotum and mesonotum dark brown: Pleuræ dark brown. Halteres long, stem yellow, knob brown. Legs: coxæ and trochanters dull brownish yellow; femora light yellowish brown, the tip broadly brownish black; tibiæ with base narrowly dark brown, remainder yellow, except the broad dark brown tip; tarsi dark brownish black. Wings tinged with brownish, cells C and Sc rather brighter; veins yellowish brown; wing spotted with brown, varying greatly in the size of the markings; in one (♀, Vial A), there are large brown spots at origin of Rs, tip of Sc and fork of R_{2+3}, and abundant pale brown dots over the wing surface; in a second specimen (♂) the wing disk is heavily marked with brown, a series of brown marks in the costal cell, a large square blotch at origin of Rs, another at tip of Sc_1 extending partly down across the cord; others at tips of R_1, R_2 and R_3; large, paler brown dots in all the cells of the wing. Venation (see fig. 2, pl. III).

Abdomen: tergites brownish; sternites dull yellow, apical third of each sclerite brown. Hypopygium (see fig. 10, pl. X): viewed from above, 9th tergite, caudal margin almost straight with a little rounded knob or hook on either side of the median line; pleurites very short and stout, with three apical appendages, the more dorsal being the longest, slender at base, swollen subapically, the extreme tip slightly hooked and strongly chitinized, this appendage directed caudad and entad; two ventral appendages, the more dorsal being short, blunt, very strongly chitinized at its tip and with numerous, triangular denticulæ, closely and regularly set; ventral appendage slender, curved at a right angle, its tip directed cephalad. Anal tube very conspicuous, pale whitish, slightly notched at its tip. Second gonapophyses rather slender, tips expanded, the organs directed caudad. Viewed from beneath, the 9th sternite has a rectangular median protuberence.

Female.—Similar, larger; the dark apices of the abdominal sternites not well marked.

Vial No. A.—Tokyo, Japan; April 25, 1912; 1 ♀.

Vial No. 7.—Tokyo, Japan; April 25, 1912; 2 ♂, 1 ♀.

Vial No. 18.—Tokyo, Japan; June 26, 1912; 4 ♀.

Vial No. 23.—Tokyo, Japan; June 25, 1912; 3 ♂.

Vial No. 48.—Tokyo, Japan; August, 1912; 1 ♂.

Holotype.—♀. Vial 48.

Allotype.—♀. Vial A.

Paratypes.—5 ♂, 5 ♀. Vials 7, 18, 23.

Types in author's collection; paratypes in U. S. National Museum and Cornell University Collections,

This species differs from *L. varicornis* Cog. (Japan)* in its shorter antennæ; legs not all yellow, but the segments conspicuously tipped with darker; abdomen not yellow; wings with petiole of cell M_1 much longer than cell 1st M_2, etc. *L. varicornis* also, is probably a *Pœcilostola*.

Tribe Pedicini.

Genus Tricyphona Zetterstedt.

KEY TO THE JAPANESE TRICYPHONÆ.

1. Wings hyaline or nearly so, not spotted or striped; cross-veins r-m connected with vein R_{4+5} beyond the fork of Rs................................,*insulana*, sp. n.

*Proc. U.S. Nat. Mus. vol. 21, p. 301 (1893).

Wings spotted or striped with brown or yellow; cross-vein r-m connected with the radial sector at or before its fork.....2.

2. Wings with a broad, yellow subcostal streak, extending from the base of the wing to the apex; median cross-vein absent..............................*kuwanai*, sp. n.

Wings with a narrow brown seam along the cord, and rounded brown spots on most of the cross-veins and at the ends of most of the longitudinal veins; median cross-vein present..............................*vetusta*, sp. n.

Tricyphona kuwanai, sp. n.

Color yellow; mesonotum with black markings; wings with a conspicuous yellow longitudinal streak.

Female.—Length 15.8 mm.; wing 12.2 mm.; abdomen 12.4 mm.

Female.—Rostrum and palpi brown; antennæ, segment 1 brown, segments 2 to 16 light yellow, the terminal flagellar segments more brown; front and vertex brown, the hind part of the vertex, the occiput and the genæ clearer reddish brown.

Pronotum light yellow, brown medially. Mesonotal præscutum light brownish yellow, darkest medially, the sclerite with four rounded, velvety-dark brownish black spots as follows: a small rounded spot on either side of the median line, about mid-length of the sclerite; an oval spot on the sides of the sclerite, about midway between the pseudosuture and the transverse suture; a small triangular black spot on the middle of the transverse suture; scutum light yellow, with velvety-black marks as follows: a double, semilunar transverse mark on the cephalic portions of the sclerite, caudad of these marks are four small dots, the outermost larger, rounded, occupying the middle of the scutal lobes, the inner small and oval, on either side of the median line; a small elongate black mark on the suture, between the scutum and scutellum; scutellum and postnotum brown. Pleuræ light brownish yellow. Halteres light yellow. Legs: coxæ and trochanters light yellow; femora and tibiæ yellow, tip of the latter narrowly dark brown; first three tarsal segments light yellow, narrowly tipped with dark brown; segments 4 and 5 dark brown. Wings hyaline or nearly so, a broad yellow streak running from the base of the wing around to beyond the apex, embracing the caudal portion of cell C, cell Sc, cephalic portion of cell R and 1st R_1, caudal portion of cell 2nd R_1 and outer half of cell R_2; cell C is hyaline with small, rather evenly spaced dark brown cross stripes; the margin of the wing from the end of cell C around to end of cell R_5 is light brown; the caudal margin of the longitudinal yellow streak above described is narrowly brown at the deflection of R_2, a slender brown streak runs caudad and outward along R_{4+5}, ending opposite the fork of R_{1+2}; Cu and 2nd anal margined with bright yellow. Venation (see fig. 6, plate III), Rs beyond the cross-vein r-m short, a little shorter than r-m; R_2, R_3 and R_{4+5} all originate at a common point; R_2 at origin is perpendicular; cross-vein m lacking; basal deflection of Cu_1 at fork of M.

Abdomen: tergites light brownish yellow, with numerous slender black hairs; segment 2 with a short black sub-basal streak on the margin; segments 3 to 6 with longer marginal streaks, which cover almost the basal half of the sclerite; sternites light yellow, with black marks on the sides remote from the margin of sclerite, that on the second oblique, meeting its mate on the venter, the others longitudinal.

Vial No. 31.—Tokyo, Japan; May 7, 1912; 1 ♀.

Holotype.—♀, in Vial 31.

Type in author's collection.

Tricyphona insulana, sp. n.

Brown; wings hyaline without a stigma; no median cross-vein; legs largely yellow.

Female, length 9.6 mm.; wing 9.4 mm.

Female.—Rostrum and palpi dark brown apices of the palpal segments a little paler: antennæ, basal segments pale brown, flagellum dark brown; front, vertex and occiput dark brown, probably with a grey bloom in dry specimens.

Pronotum dark brown. Mesonotum dark brown with indications of stripes near the median line; it is probable that the thorax is covered with a grey bloom; scutum dark brown; scutellum brownish yellow; postnotum brown. Pleuræ brown. Halteres light yellow. Legs: coxæ yellow, more brown basally;

trochanters yellow; femora yellow, darkening to light brown at the tip; tibia light yellow, brown at tip; tarsi brown. Wings hyaline; veins light brown. Venation (see fig. 3, plate IV) cross-vein r-m connects R_{4+5}; no cross-vein m.

Abdomen: tergum reddish brown, segments with a dark brown apical ring; pleural line yellow; sternites brown, ovipositor with yellow valve.

Vial No. 27.—Tokyo, Japan; April 25, 1912; 1 ♀.

Holotype, ♀, Vial No. 27.

Type in author's collection.

Related to *T. vitripennis* Doane (West. U.S.) but lacks a brown stigma, has no median cross-vein, etc. From *T. immaculata* Meigen (Europe) it differs in having cross-vein r farther removed from the tip of R_1, cross-vein r-m far beyond the fork of Rs, not at it; the legs much more yellow, not mostly brown; ovipositor of the female yellow, not patch brown, etc.

Tricyphona vetusta, sp. n.

Wings spotted with brown; cross-vein m-cu of the wings present; cross-vein m present.

Female.—Length 16 mm.; wing 14.8 mm.; hind legs femora 8.4 mm.; tibia 10.3 mm.; tarsus 8.9 mm.

Female.—Rostrum light brownish yellow; palpi with segments dark brown, the apical ones with bases yellow; antennæ, base light brown, flagellum dark brown; front, vertex and occiput dark brown.

Pronotum dark brown. Mesonotum, præscutum, greyish with four brown stripes, the median one double, narrowed behind; scutum dark brown, the lobes paler brown; scutellum dark brown, much lighter on the sides; postnotum dark brown with a large oval spot behind on either side of the median line. Pleuræ dark brown, indistinctly variegated with darker. Halteres light yellow, the knob light brown. Legs: Coxæ, especially the fore and middle, brownish at the base, remainder light yellow; trochanters yellowish; femora yellow, darkening into brown at the tip; tibia yellowish brown, rather darker apically; tarsi dark brown. Wings, tinged with light yellow, cells C and Sc a little brighter, with brown marks as follows: a rounded spot at Sc_2, a larger one at origin of Rs, a crossband on the cord running from the tip of Sc_1, down to fork of Cu and thence to the wing-margin along Cu_2; a round spot at cross-vein r, apical margin of the wing brown, a brown seam on cross-vein m, brown dots at ends of all the longitudinal veins; veins yellowish brown. Venation (see fig. 5, plate III); Rs in a line with R_{4+5}; R_{2+3} short, gently arcuated; crossvein r very far distad so that R_1 beyond it is about equal to it in length; cross-vein m present, connecting M_2 with M_3; cross-vein m-cu present.

Abdomen: Tergites, segment one brown, segments two and six dull yellow, an indistinct median brown stripe becoming more plainly defined behind until on the 8th and 9th tergites it abruptly suffuses the entire sclerites; pleural stripe broad, dark brown, extending the length of the abdomen as a conspicuous lateral line; sternites light yellow, a rounded ill-defined brown mark on the 8th sternite.

Vial No. 26.—Tokyo, Japan; April 25, 1912; 1 ♀.

Holotype, ♀, in Vial 26.

Type in author's collection.

Related to *T. constans* Doane (West. U.S.) but is much smaller with a very different wing pattern. In venation, suggesting *T. vernalis* Osten-Sacken of the Eastern United States.

Tribe Cylindrotomini.

Genus Liogma Osten-Sacken.

Liogma kuwanai, sp. n.

Resembles *L. nodicornis* O. S., of the United States, but the tripartite penis-guard is very much longer and directed dorsad.

Male.—Length 15.9 mm.; wing 11.4 mm.; antennæ 3.8-3.9 mm.

Male.—Rostrum and palpi light brown, remaining segments dark brown; flagellar segments slender at base, the inner face produced into a subtriangular tooth, making the flagellum strongly serrate; front, vertex and occiput dull dark brown, very rugulose, the vertex broad.

Mesonotum dark brown, a lighter brown line extending from the median line of the scutum, branching Y-shaped and extending to the pseudosuture, this pale line being somewhat impressed; scutum, scutellum and postnotum brown, the latter rather darker. Pleuræ, propleuræ and cephalic and dorsal portions of the mesopleuræ, up to the wing-root, yellowish; remainder of the pleuræ brownish. Halteres pale, yellow. Legs: coxæ suffused with brown; trochanters light yellow; femora yellow basally, becoming brown at the tip. Wings tinged with grey, stigma elongate-oval, brown, distinct. Venation (see fig. 4, plate IV).

Abdomen light brownish yellow, the caudal half of the 7th, 8th and 9th tergites brown. Hypopygium (see figs. 13-15, plate X): 9th tergite, viewed from above, with the lateral ears or lobes prominent, the interval between them almost straight, not deeply notched as in *nodicornis;* 9th sternite and its pleurite fused, massive, as in the genus, the apical appendate stout, directed cephalad, flattened at its apex. Viewed from the side, the penis-guard is conspicuous, tripartite as in the tribe, it is very long, arising from the ventral wall, directed caudad and thence dorsad, almost attaining the level of the dorsal edge of the 9th sternite, toward their end, directed cephalad, the tip flattened; anal tube conspicuous. Viewed from beneath, the massive sterno-pleurites meet in a straight median suture, which is membranaceous; the tripartite penis-guard is deeply concave below the forking.

Vial No. E.—Tokyo, Japan; Aug., 1912; 1 ♂.

Holotype.—♂, Vial E.

Type in author's collection.

The difference between the American and Japanese species are shown by the following key:

1. Abdomen brown; ♂ hypopygium, 9th tergite with a deep median notch; guard of the penis short, directed caudad. (East. U. S.) . . . . *nodicornis* O. S.

   Abdomen reddish brown; ♂ hypopygium, 9th tergite without a deep median notch between the prominent lateral ears; guard of the penis elongate, conspicuous, directed caudad and dorsad, almost attaining the dorsal level of the abdomen. (Japan) . . . . *kuwanai*, sp. n.

The succeeding parts dealing the Tipulinæ will conclude the Tipulidæ.

THE CADDIS-FLIES (*TRICHOPTERA*) OF JAPAN. I.—FAMILY PHRYGANEIDÆ.

BY WARO NAKAHARA, HONGOKU, TOKIO, JAPAN.

Through the kindness of Mr. Miyake, who has generously permitted the free use of the valuable literature and collections in his possession, and has given me much valuable advice, I have recently had the opportunity of studying Japanese caddis-flies or Trichoptera. The purpose of the present study is to record the species known from Japan, offering such notes as may suggest themselves, and to describe any forms that appear to be unknown.

The present paper deals with the family Phryganeidæ, which includes some of the most beautiful caddis-flies in the world.

FAM. PHRYGANEIDÆ.

Genus *Neuronia* Leach.

1. *Neuronia regina* Maclachlan.

Holostomis regina MacLachlan—Journ, Linn. Soc. London, Zool., XI, p. 104 (1871); Matsumura, Thous. Ins. Jap., i, p. 165, Pl. XII, fig. 11, ♀ (1904).

Neuronia regina Ulmer—Cat. Coll. Selys, Fasc. VI (1), p. 6, figs. 1, 2 and 3, Pl. i, fig. 1 (1907) ; Ulmer, Gen. Insect., Pl. XXIX, fig. 3 (1907).

This magnificent species, which is common in China, as well as in India, is not rare in Japan.

The manner of flight of this species resembles that of a certain moth. Occasionally they are found on the bark of trees closely resembling the colour of the forewings, which always cover the abdomen and beautiful hind wings, when they are at rest.

Already reported from Hokkaido and Honto.

2. *Neuronia reginella* sp. nov.

Head blackish, clothed with brownish hairs, especially on the face; vertex wholly fuscous; ocelli brown; labial palpus consisting of four joints, brown; maxillar palpus four-jointed, the last joint more slender than others and fuscous; all the others mostly brown and each thickened at extremity. Antennæ lost, except two basal joints. Prothorax light brown with a median longitudinal impression, clothed with long, fuscous hairs. Meso- and metathorax fuscous, the former with stout fuscous hairs, the latter with long and weak gray hairs. Underside of thorax mostly brownish. Legs brownish, tibiæ and all the tarsal joints fuscous; spurs on tibiæ 2, 4, 4; hind femora somewhat dark, with a brown ring near the extremity. Fore wing light fuscous yellow, with fuscous markings, as shown in figure, with some stout blackish hairs at the base;

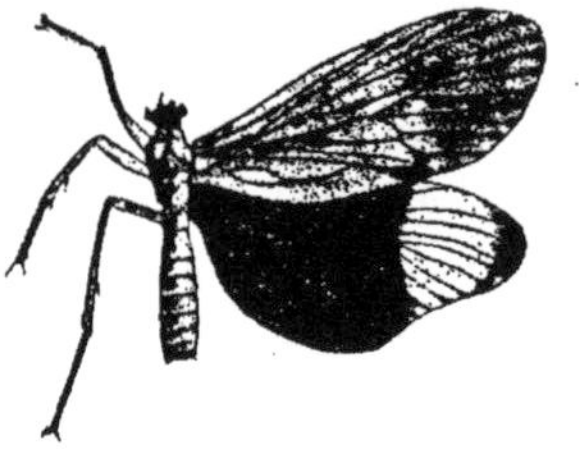

Fig. 13.—*Neuronia reginella* n.sp (Male).

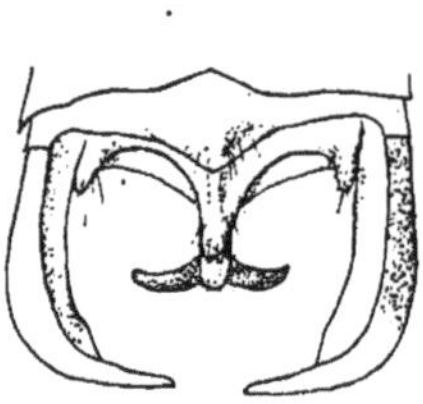

Fig. 14.—*Neuronia reginella* (Male), genitalia, dorsal view.

veins yellowish. Inner half of hind wing violet black, forming a broad, beautiful orange band between this and the fuscous black apical spot. Abdominal segments blackish, fuscous on both dorsal and ventral sides, each segment with hind margin narrowly brown. Suranal plate in the genitalia of male individuals rather broad and little produced in its hind margin; superior appendage, with two small projections and many hairs on the end; intermediate one separated at the end of a small lobe curving little upward; inferior claspers stout and long, suddenly becoming much more slender near its end.

Length of body 20 mm.; length of fore wing 33 mm.; length of hind wing 28 mm.

The type is a single male specimen in the collection of the Imperial Agricultural Experiment Station at Nishigahara. The specimen was captured by Mr. Murata at Nikko, on July 28th.

This species is closely allied to *Neuronia regina*, which is a little larger, but the differences in the genitalia, mouth parts and wing markings seem to me to warrant the specific separation of the two forms.

3. *Neuronia clathrata* Kolenati.

Anobolia (*Oligostomis*) *clathrata* Kolenati, Gen. et spec. Trichoptera, i. p. 82 (1848).

Neuronia clathrata Walker—Cat. Neuropt, Brit. Mus., Pt. I, p. 7 (1852); Matsumura, Journ. Coll. Agr. Tohoku Imp. Univ., IV, p. 16 (1911).

4. *Neuronia phalænoides* Linné.

Phryganea phalænoides Linné—Faun. Suec., p. 378 (1761).

Holostomis phalænoides Walker—Cat. Neuropt. Brit. Mus., Pt. I, p. 6 (1852).

Neuronia phalænoides Matsumura—Journ. Coll. Agr. Tohoku Imp. Univ., IV, p. 15 (1911).

The above two species occur in Europe and Siberia, and have been described by Matsumura (l. c.) from Saghalien. It is said that a few specimens have been obtained at Solowiyofka, Chipsani and Galkinowraskoe, on that island. No specimen before me.

5. *Neuronia apicalis* Matsumura.

Neuronia apicalis Matsumura—Thous. Ins. Jap., I, p. 172, Pl. XII, fig. 11 (1904); Matsumura—Journ. Coll. Agr. Tohoku, Imp. Univ., IV, p. 15-16 (1911).

6. *Neuronia fluvipes* Matsumura.

Neuronia fluvipes Matsumura—Thous. Ins. Jap., I, p. 172, Pl. XII, fig. 12 (1904).

Unfortunately, neither *N. apicalis* nor *N. fluvipes*, both of which were described by Matsumura about ten years ago from Hokkaido and Honto, are represented in the material before me. The same professor recorded the former from Saghalien, also.

7. *Neuronia melaleuca* MacLachlan.

Phryganea melaleuca MacLachlan—Journ. Linn. Soc. Lond., Zool., XI, p. 106 (1871).

Holostomis melaleuca Matsumura—Thous. Ins. Jap., I, p. 166, Pl. XII, fig. 2, ♀ (1904).

Neuronia melaleuca Ulmer—Doutch. Ent. Zeit., p. 339 (1908).

The specimens in hand, which I believe to be true *N. melaleuca*, differ to a certain extent from that species as described and figured by Matsumura in his "Nippon-Senchu-Zukai" (Thousand Insects of Japan), though the too meagre description does not enable me to satisfactorily determine it. It may be doubted whether Matsumura's identification of his specimen with *N. melaleuca* be justified.

In any case it may safely be said that there is another form in Japan besides *N. melaleuca*, closely allied to this species. It is said that the habits of this species resemble those of *N. regina*.

Habitat—Hokkaido, Honto.

Genus *Phryganea* Linne.

8. *Phryganea japonica* MacLachlan.

Phryganea japonica MacLachlan—Trans. Ent. Soc. Lond., (3) V, p. 248 (1866); Matsumura—Thous. Ins. Jap., I, p. 167, Pl. XII, fig. 3, ♂ (1904); Ulmer—Cat. Coll. Selys, Fasc. VI (1), p. 10, figs. 11, 12 and 13, Pl. I, fig. 2 (1907); Ulmer—Gen. Insect., Pl. XXX, fig. 1 (1907).

The markings of the fore wing of this species are subject to variation, and the material before me can be separated into two types:

(i) Those that have conspicuous fuscous lines along the cubital and the fourth apical veins.

(ii) Those that have faint and obscure fuscous lines along the cubital and the fourth apical veins.

Though there are some other minute differences in the markings of the fore wing between types i and ii, I think they are not worthy of specific rank, since I could not recognize any difference in the genitalia, nor in any other respects, that appear to be specific. Until a more comprehensive study of these two forms is published I shall have to include them in one species, *Phryganea japonica*. It would be very interesting if their life-histories were known.

It seems to me that Ulmer's figure in the Selys Catalogue represents type i and his figure in Genera Insectorum type ii Matsumura's figure seems to represent type i.

This is one of the most common caddisflies of the family in the Main Island of Japan, occurring also in Hokkaido.

9. *Phryganea sordida* MacLachlan.

Phryganea sordida MacLachlan—Journ. Linn. Soc. Lond., Zool., XI, p. 106 (1871); Ulmer—Cat. Coll. Selys, Fasc. VI (1), p. 8, figs. 6-10 (1907).

A single female specimen in the collection of the Imperial Agricultural Experiments Station, from Gifu, labeled "Hayafumiyama."

10. *Phryganea latipennis* Banks.

Phryganea latipennis Banks—Proc. Ent. Soc., Wash., VII, p. 107 (1906); Ulmer—Cat. Coll. Selys, Fasc. VI (1), p. 10, figs. 14-20, Pl. I, fig.3 (1907).

A single male specimen in the collection of the Agricultural Experiments Station from Gifu, where the type specimen of this species was obtained.

The above two species seem to be uncommon.

Genus *Limnoceutropus* Ulmer.

11. *Limnoceutropus insolitus* Ulmer.

Limnoceutropus insolitus Ulmer—Cat. Coll. Selys, Fasc. VI (1), p. 14, figs. 21-23 (1907).

This is the single species of the genus *Limnoceutropus*, and is known only from the female. I have not seen specimens of it. Taken at "Nikko, 600-2000 m."

Komagome-Higashikatamachi, Tokyo, Japan.

THE OCCURRENCE OF THE MYMARID GENUS *COSMOCOMOIDEA* HOWARD IN AUSTRALIA (*HYMENOPTERA*).

BY A. A. GIRAULT, NELSON, N. Q., AUSTRALIA.

The following remarkable mymarid represents the fifteenth genus of the group known to occur in Australia. The original description of the genus is not accessible to me just at present, but I should call attention to the fact that the tarsi are *five-jointed*, not as in *Polynema*, as the name would lead one to infer. I have a specimen of the type of the genus, one of the series on which the species was founded, but not a type.

Genus Cosmocomoidea Howard.

1. *Cosmocomoidea renani* new species.

Normal position.

Female.—Length 2.00 mm. Large for the family. Shining black, the bullæ of the scape, cephalic legs, trochanters, knees, proximal four tarsal joints and tips of tibia, rich brown. Wings conspicuously infuscated at tip (about distal fourth), the proximal margin of the fuscation convex; slightly distad of the middle, where a rather broad band crosses, not quite its own length from the end of the stigmal vein, and obscurely under the marginal vein. Scape more or less brownish along proximal half. Coxæ black. Venation brown. Posterior wings clear. Head and thorax with a scaly, polygonal reticulation, the propodeum less scaly, smooth and shiny between the median carinæ.

Differs from the type of the genus (*morrilli* Howard) in being black, in having the flagellum uniformly black, the wings more conspicuously and differently fumated, the greater size, and in having joints 4 and 5 of the funicle longest of that region; also, the abdomen is not distinctly petiolate, but only tapers at base—slender there. The following important structural characters are noted: The thorax is rather peculiar, for there is a mesopræscutum present at the meson cephalad of the scutum, and which is moderately large and subquadrate; the pronotum is short at the meson, but dorso-laterally long, extending broadly halfway down the scutum (but not by far to the tegulæ), then curving off; the axillæ are small, but distinct, not advanced into the parapsides and widely separated. Scutellum subquadrate, as long as the scutum, the latter with a median grooved line. Parapsidal furrows complete, short, curved, the parapsides short and wedge-shaped, with the base of the wedge mesad. Propodeum with a carina on each side of the meson, the two rather widely separated; the spiracle minute and round, near postscutellar margin. Tarsi 5-jointed. Ovipositor not exserted.

(From one specimen, ⅔-inch objective, 1-inch optic, Bausch and Lomb.)

Male.—Not known.

Described from a single female specimen captured by sweeping grass and foliage in a forest at Nelson, N. Q., December 13, 1912 (A. P. Dodd). Other specimens were captured a few weeks later in the same place.

Habitat.—Australia, Nelson (Cairns), Queensland.

Type.—In the Queensland Museum, Brisbane, the above female in xylol-balsam.

[Dedicated to Ernest Renan.]

NEW SPECIES AND NEW LIFE HISTORIES OF EPHEMERIDÆ OR MAYFLIES.

BY WILBERT A CLEMENS, TORONTO, ONT.

(Continued from page 262.)

Subfamily—*Heptageninæ.*

Ecdyurus maculipennis Walsh. (Pl. VI, fig. 4, Nymph.)

Only a few imagos of this species were taken, although the nymphs were abundant along open stony shores and in rapids. My collections of nymphs date from July 2nd to August 23rd, and rearings from July 6th to August 30th.

Ecdyurus lucidipennis sp. nov.

This was not a very abundant species, but nymphs were collected July 1st and 14th, and imagos reared July 4th and 17th, respectively.

Male imago:

Measurements—Body 6 mm.; wing 7mm.; fore leg 6.5 mm. Face very slightly obfuscated; dorsal surface of head dark brown or reddish. Notum dark brown; sides of thorax and ventral surface light yellow. Dorsum of abdomen a blackish brown; venter considerably lighter. Penis lobes and bases of forceps yellow; forceps tinged with black. Setæ with basal halves slightly tinged with black and minutely hairy. Fore femora dark, middle and hind yellowish. Wings hyaline; longitudinal veins slightly dusky, especially costa and subcosta; cross-veins entirely colourless.

Female imago:

Measurements—Body 6 mm.; wing 7.5 mm.; fore leg 4 mm. Thorax and abdomen lighter in colour than male.

Nymph: (Pl. VI, fig 5.)

Measurements—Body 7-8 mm.; setæ 3-4 mm. Head brown, with numerous light spots, chief of which are 6 along anterior margin, 2 lateral to each antenna, 4 small, elongated ones between antennæ, and 2 small round spots anterior to these latter. Thorax above lighter brown, with numerous light areas. Anterior part of each abdominal segment brown; four light spots along anterior margin, one large one at each lateral margin, and 3 along posterior margin. Setæ of about equal length and fringed with hairs; middle one slightly smaller in size than lateral ones. Femora flattened, fringed with spines along anterior margin and with hairs along posterior; rather light in colour, with 2 zigzag brown marks about middle and brown areas at distal and proximal ends. Tibiæ banded about the middle with brown. Tarsi with distal and proximal ends dark.

Ecdyurus pullus, sp. nov.*

This is a large form, compared with the two previous species. The nymphs were found along the very stony, exposed shores of small islands three and four miles out in the open bay. The collections are dated June 23rd and July 6th, and the rearings July 2nd. A few imagos were captured June 27th.

Male imago:

Measurements—Body 10-11 mm.; wing 11 mm.; setæ 22 mm.; fore leg 11-12 mm. Face pale, slightly tinged with brown along the carina. Dark brown on dorsal surface of head between eyes. Pronotum dark brown; mesonotum lighter; a dark brown line on each side of prothorax extending forward from base of fore wing; other dark brown marks at bases of wings and legs. Dorsal surface of abdomen dark brown, somewhat lighter laterally toward anterior margin; ventral surface light in colour. Genitalia of the usual *Ecdyurus* type. Legs light in colour, dark at joints. Tarsi of fore legs in order of increasing lengths, 1, 5, 4 (3 and 2) equal. Wings with longitudinal and cross veins brown and very slightly darkened in apical costal region.

Nymph: (Pl. V, fig. 10.)

Measurements—Body 12 mm.; setæ 15 mm. Head brown, with a colourless area on each side from eye to lateral margin of head, and three light dots between eyes; slightly fringed with hairs along anterior and lateral margins, and a light area about the middle of each half of pronotum. Mesonotum darker, with numerous light spots. Each segment of abdomen brown; 1-8 have six light spots, and 4-8 have the two spots near the median line fused, forming a large rectangular area; segment 9 with only four light spots; segment 10 entirely brown. Gills comparatively small, lamellæ oval. Setæ of about equal size, with each two alternate segments brown; sparsely fringed at joints, and outer

*This species was listed on p. 247 as *Ecdyurus grandis*, sp. nov.

margins of lateral ones not fringed. Femora stout and flattened, brown in colour, lighter at distal and proximal ends, and two or three irregular light areas toward middle; covered with minute spines and fringed along posterior margin with hairs. Tibiæ alternately light and dark banded; fringed along both anterior and posterior margins. Tarsi brown, with proximal tips colourless. Ungues double on each leg, the large one well covered, the other small and lateral to the large one.

Subfamily—*Ephemerinæ.*

Hexagenia bilineata Say.

This was a very common species at Go-Home Bay. The nymphs were first taken on June 6th, by dredging in water 15 to 45 feet deep. The bottom was very muddy. When the nymphs were placed in jars containing about 4 inches of mud, they immediately began to burrow, and were able to bury themselves in a very short time. At first the gills were left partly exposed, and the position of the nymphs could be detectcd by the waving motion in the thin mud. Later on they completely buried themselves, and only the round openings of the burrows could be seen. The first of these nymphs to emerge was on July 3rd, and others followed in July and August, while one was still alive in the breeding jar on September 9th, when the Station was closed. On June 13th the first subimago was captured at large, but not till June 28th did imagos appear in large numbers. They would commence their flight shortly after sunset, flying in large swarms about the tree tops. The hum of their wings could be heard up to a distance of 125 feet or more. The females deposited their eggs by flying up and down the shore, brushing off the eggs as they appeared in two small, rather compact columns from the openings of the oviducts, by dipping to the surface of the water. On July 12th a female was caught just after copulation, and she deposited a large number of eggs by being held by the wings and touching her abdomen frequently to some water in a jar. These eggs hatched in thirty-six days.

Nymph: (Pl. VI, fig. 1.)

Measurements—Body 30-35 mm.; setæ 13-15 mm.; antennæ 5-6 mm. Head rather yellowish, with dorsal surface between ocelli and between eyes entirely brown or in some cases lighter along median line and posterior margin. Antennæ very hairy at joints of basal halves, while apical halves are entirely bare and become very slender. Margin and base of frontal piece hairy. Clumps of hairs between eyes and at bases of antennæ, in front of lateral ocelli and posterior to eyes. Mandibular tusks ¾ length of antennæ, upcurved, brown at tips, and with three longitudinal rows of hairs. Prothorax has a broad longitudinal band of brown on each side of middle line on dorsal surface, and is very hairy along lateral margins. Mesothorax brown for the most part, dorsally. Each abdominal segment has a large, almost triangular brown area with two light areas within it; these light areas often reduced to mere stripes. Ventrally on segments 6 to 8 there is a faint median longitudinal dark streak, while on ninth segment are two lateral streaks. Setæ of about equal length, and very hairy at joinings for entire length. Gills and legs of usual *Hexagenia* type.

Ephemera simulans Walker.

The imagos of this species appeared from June 5th to July 27th, but were most abundant during the first two weeks in July. The nymphs were not taken at Go-Home Bay, although diligent search was made. The male imagos would appear shortly before 8 o'clock in the evening, and were often noticed in the morning, also, as late as 10 o'clock. They would dance in swarms of a couple of hundred individuals, usually at a height of from 10 to 35 feet. When a female appeared, several males would take after her. The successful male, flying up beneath the female, would seize her around the prothorax with his long fore legs, and bending up his abdomen would grasp her abdomen with his forceps, and his penis could then be inserted in the oviducts. His setæ usually aided him in securing and maintaining his hold by being bent up over the female's body. The couple would then go off on a gradual downward slant toward the water, before reaching which the male would disengage himself and fly back to the swarm, while the female would fly out over the water and soon begin depositing her eggs by skimming the surface of the water with her abdomen. A peculiar thing was noticed, namely, that the male *Ephemera* frequently attempted copulation with the male *Hexagenia*, apparently being deceived by the colour.

Subfamily—*Baetinæ*.

Baetisca obesa Walsh.

The very interesting nymphs of this species were quite abundant along the north-east shore of Giant's Tomb Island, on May 26th. The shore is rather sandy, with numerous small stones, and deepens very gradually. The nymphs were clinging to the stones in water 3 to 15 inches deep. Imagos did not emerge from this collection until July 13th.

Leptophlebia sp. ?

A single almost mature nymph was taken on July 21st in quiet water at the side of an old lumber chute, but it died before time of emergence.

Blasturus cupidus Say.

This is an early species, mature nymphs being found May 25th, and subimagos appearing May 31st. A small nymph, collected May 31st, was observed to be filled with small, oval, brownish bodies. Upon dissection by Mr. A. R. Cooper, these were found to be the eggs of a trematode, and in the midst of them was the trematode itself, which belonged to the genus *Halicometra*. Another nymph, taken some time afterwards, was also discovered to be parasitized.

Blasturus nebulosus Walker.

The nymphs and imagos of this species were first taken June 9th, on a small, bare, granite island a short distance out in the open bay. On top of this island were numerous pot-holes of all sizes, and in these, under loose pieces of rock and some rubbish the nymphs were very abundant, having tadpoles, chironomid larvæ and water beetles for associates. Many nymphs were covered with *Vorticella*. Several nymphs were obs rved to crawl out of the water and transform on the rock just above the surface of the water. Subimagos were clinging to the sides of the rock in sheltered places, and a few imagos were flying above the pools. This species was again observed on June 27th, on an island five miles from the mainland. The island had an area of about three acres, and was almost smooth, bare granite. On top was a pretty lagoon, margined with water plants, shrubs and a few small trees. Imagos of *Blasturus nebulosus* were dancing over this pond in the sunlight, about 3 p.m., matings frequently occurring.

Nymph:

Measurements—Body 9.5-10 mm.; setæ 7-10 mm. General colour blackish brown. Head brown, with a dark area behind middle ocellus and between the lateral ones; black, scroll like markings between the eyes. Pronotum has a small light spot on each side, close to median line and near anterior margin; posterior to this and farther from median line is another larger oval light spot, and lateral to this again is an elongated light area; the rounded lateral margin is colourless. Abdomen is blackish brown, with light brown markings; segments 5 or 6 to 10 have a light median longitudinal stripe; on each segment is a slightly elongated, incurved small light spot on each side of median line toward the anterior margin of the segment; posterior and more lateral is a larger round light area, which usually disappears on segments 8, 9 and 10. Ventral surface is light brown, with three faint dark longitudinal lines, one median and two lateral; on each side of the median line in each segment is a very small white oblique line near anterior margin, and posterior to this is a small light dot. Median seta shorter, slenderer and lighter in colour than lateral ones. All fringed with hairs at joints. Legs light brown; posterior margins of tibiæ and tarsi fringed with hairs, and anterior margins covered with serrated teeth; inner margins of claws with rows of teeth for their entire lengths.

Up to the present time I have not been able to find any apparent differences between the nymphs of these two species of *Blasturus*.

Choroterpes (?) basalis Banks.

Large numbers of the nymphs of this species were found in a small stream July 30th, clinging to the lower sides of stones in the quiet water. The next day several subimagos emerged from this collection. As late as September 5th mature nymphs could be found here.

Ephemerella lutulenta sp. nov.

Male imago:

Measurements—Body 8-9 mm.; wing 10 mm.; setæ 12-14 mm.; fore leg 8 mm. Face dark brown; a spotted reddish-gray streak down carina, and two similar lateral streaks from it to the bases of antennæ. Thorax dark reddish brown. Abdomen blackish brown; segments 9 and 10 slightly lighter in colour; venter pale; posterior lateral margins of 9th segment produced into spines. Forceps pale, with tips brown. Setæ reddish brown toward bases, but becoming pale toward tips; articulations brown. Legs greenish yellow, claws brown. Segments of fore tarsi in order of increasing lengths, 1, 5, 4, 3, 2; 1 very small; fore femur about five-sixths length of fore tibia. Wings entirely clear.

Female imago:

Measurements—Body 9-10 mm.; wing 10 mm.; setæ 10-12 mm.; fore leg 5 mm. Quite similar to male. Posterior lateral projection of 9th abdominal segment not as long as in male.

Nymph:

Measurements—Body 10-11 mm.; setæ 6-7 mm. A large species, with colour varying from a dirty brown to a deep blackish brown, often of a granular appearance. Body and legs hairy. Head with a pair of occipital tubercles of varying size; in the male these are often obscured by the developing eyes of the imago. Pronotum rectangular. Abdominal segments 2-9 produced laterally into flat spines; none on segment 1, minute on 2, increasing in size to the 9th, none on the 10th. A double row of spines on dorsal surface, very minute on segments 8-10, large on 1-7. On venter, six small black dots on each segment, sometimes very faint. Rudimentary gills on segment 1; gills on segments 4-7, covered by a large jointed elytroid gill cover 1.5 mm. in length. Femora stout, brown in colour, with numerous round white dots and several irregular light areas. Tibiæ with median brown band, distal ends light, proximal ends dark. Tarsi about same length as tibiæ, and with proximal half dark and distal half light. Claw with numerous pectinations. Setæ well fringed with hairs along middle, almost bare at base and tip; each two alternate segments brown.

The nymphs were taken almost everywhere about Go-Home Bay from May 29th to June 19th, in quiet water. Mr. R. P. Wodehouse gave me specimens from various places around Georgian Bay, including Shawanaga Bay, Penticost Island, French River and Sturgeon Bay.

Ephemerella lineata sp. nov.

Female imago:

Measurements—Body 9 mm.; setæ 14 mm.; wing 10.5 mm. Very similar to female of *E. lutulenta*, but has a rusty brown median longitudinal stripe on dorsal surface of abdomen. In a fresh specimen the stripe would probably extend over the thorax, and thus correspond to the stripe of the nymph. No male specimens were reared.

Nymph (Pl. V, fig. II):

Measurements—Body 10 mm.; setæ 6 mm. Slightly smaller than *E. lutulenta*, but very similar in colour, except that there is a dorsal median longitudinal white stripe from the anterior margin of the pronotum to the posterior margin of the 10th abdominal segment. This stripe lies between the double row of spines on the abdomen. Occipital tubercles slightly longer than those of preceding species.

The nymphs were not very abundant, but were found in about the same localities as *E. lutulenta*, from June 3rd to July 9th. My bred specimens are dated June 14th and June 15th.

Ephemerella bicolor sp. nov.

Male imago:

Measurements—Body 5-6 mm.; wing 6 mm.; setæ 8-9 mm.; fore leg 6 mm. A small brown species, very similar to *E. lutulenta* in form and structure, but very much smaller. The size, apparently, is the only character by which to distinguish it.

Female imago:

Slightly larger than male.

Nymph: (Pl. VI, fig. 3.)

Measurements—Body 6-6.5 mm.; setæ 3 mm. These nymphs show a great variation in colour pattern. The light-coloured specimens are of a dirty white colour, with brown markings. Head for the most part brown, slightly paler towards posterior margin. Pronotum brown laterally; anterior margin of mesonotum brown, and a brown area at posterior margin between the wing pads. Anterior halves of abdominal segments 2 and 3 brown, and slight marks on the 4th segment; brown areas on 6 and 7 about the median line, and on segment 9 there are two small brown dots at anterior margin, and a rather semicircular brown band posteriorly. Some specimens are almost entirely brown, and between these two extremes the amount of brown and white varies. Many specimens, especially females, show slight indications of tubercles on the head, but they are never large, as in the preceding species. A double row of spines on abdominal segments 1-7; postero-lateral margins of segments 3-9 produced into broad, flat spines. Gills on segments 4-7 covered by a large jointed elytra. Setæ light brown basally, becoming paler distally; well fringed with hairs; joints brown. Legs rather small; femora stout; colour for the most part brown, divided into two areas, the proximal one large and contains a rectangular white spot, the distal one smaller and contains a perfectly round white dot. Tibiæ brown at proximal end and a brown band near distal end. Tarsi with a brown band toward proximal end. Claws dark and pectinated.

The nymphs were everywhere abundant, in exposed as well as sheltered places. Imagos were captured and reared from July 1st to July 12th.

Genus *Drunella* Needham.

Several nymphs of this genus were taken, but no imagos were reared.

Cænis diminuta Walker.

This little nocturnal species came to the lamp in the reading-room for the first time on July 2nd, and was taken as late as August 12th. The nymphs were common in ponds and lagoons from June 5th to July 30th.

Tricorythus allectus Needham.

Imagos were captured on July 3rd and 9th, but none were reared. Nymphs which apparently belong to this species were dredged up from a slightly sandy bottom in water 5 to 15 feet deep, on September 3rd.

Chirotenetes albomanicatus Needham.

On July 16th I found a nymph slough at Sandy Gray Falls, on the Go-Home River, but was unable to find either nymphs or imagos. I did not get up to the falls again until August 23rd, and then found the numerous small nymphs of the next generation.

Siphlurus flexus sp. nov.

Two beautiful *Siphlurus* nymphs were taken early in the season, but both died before time of emergence. The first was found May 25th in the bottom of a canoe, when some water was being emptied from it; the other was found June 3rd beneath a stone in about 1½ feet of water along the open, exposed shore of Station Island. Quite a number of imagos, apparently *Siphlurus*, were captured about this time, and it seemed quite probable that they were the same species as the nymphs. I think I have proved this quite conclusively by the wing venation. The wing of the imago has a very characteristic bend in cubitus 2 at the base, and the wing pad of the nymph shows this bend very distinctly. Again, the imago has claws like *Ameletus*, the two on each leg being unlike. These two unlike claws can be made out in one of the nymphs, due to the nymph dying just when about to transform. Imagos were captured on May 23rd, May 26th, and June 12th. On the latter date a swarm of 12 or 15 were observed flying off the west point of Station Island, about 5.30 p.m., at a distance of from 12 to 20 feet from the surface of the water. They faced the west, and had the characteristic fluttering rise and leisurely fall.

Male imago: (Pl. VI, figs. 10, 11.)

Measurements—Body 13-14 mm.; wing 12-13 mm.; setæ 23-24; fore leg 12-13. Head blackish brown, except lower part of face, which is tinged with brown; eyes large, meeting dorsally. Notum blackish brown. Sides of thorax marked irregularly with white. Abdominal segments 1, 8, 9 and 10 dark, segments 2-6 lighter in colour; these latter are light toward anterior margin and brown toward posterior; in the median line the brown is dark and forms a triangular area, the apex extending almost to the anterior margin; from the anterior margin in the median line two bands arise, composed of black dots, passing backwards, curving outward, end near the base of the triangular brown area; between this line and the triangular area is a light brown oval area; segments 7-10 almost entirely blackish brown dorsally, but 7 and 8 have triangular white areas on sides, and 9 a slight indication only; segment 10 has the sides of dorsum white. Ventrally segment 1 is dark brown, and remainder white with brown markings; segment 2 has two brown spots; 3 with two smaller brown spots and a slightly reddish area at anterior margin in median line; on 4 and 5 the brown spots become smaller and the reddish areas larger; on segment 6 the reddish area is elongated to the posterior margin; on 7 and 8 there is a median longitudinal brown line, thickened about the middle, and two dots of unequal sizes on each side of it; segment 10 brown, except for a lateral white streak on each side. Forceps white and 4-jointed. Setæ white, with brown joints, and minutely pubescent. Fore legs brown; femur with a light area near distal end, next to which is a dark brown band; tarsal joints 1, 2 and 3 about equal in length, 4 slightly shorter, and 5 about half the length of 4th. Hind legs lighter in colour than fore; a brown band on femur in distal half; tibia with a brown band about middle; tarsus light in colour, but brown at joints; joint between tibia and tarsus 1 not distinct. Claws unlike. Wings with brown neuration; costal cross veins and others toward base of wing margined more or less with brown; a slight brown cloud in apical costal area; a heavy brown cloud at bulla; often a small cloud at bifurcation of median vein; cubitus 2 strongly bent at base; hind wing with a large brown cloud at base.

Female imago:

Quite similar to male.

Nymph: (Pl. VI. figs. 8, 9.)

Measurements—Body 15 mm.; setæ 5 mm. The two nymphs collected proved to be a male and female, both mature, but, unfortunately, both died when just about to transform. On this account it is difficult to describe the colour pattern, as the body of the subimago shows through the nymph skin.

Head vertical; body curved. Posterior lateral margins of abdominal segments 1-9 produced into spines. Dorsal colour pattern distinct on segments 9 and 10 only; 9th pale with a short median longitudinal brown stripe commencing at anterior margin; on each side of this is a short stripe of about same length, but placed more posteriorly; lateral to this, again, is a large brown area, roughly triangular, apex at posterior margin, base at anterior; at lateral margin, slightly below middle line, is a small brown spot; on the 10th segment is a median brown longitudinal stripe, with two dots on each side of it. Ventral surface of abdomen white, with three longitudinal brown stripes, one median and two lateral. Gills on segments 1-7, double on 1, 2 and 3. Three setæ of equal length; lateral ones fringed with hairs on inner margins only, except tips; all banded toward distal end with brown. Legs pale; femur with proximal end brown and a brown band beyond middle; tibia with a brown band about the middle; tarsus with brown band toward proximal end; fore tarsus much longer than fore tibia; hind tarsi only slightly longer than hind tibiæ; fore claw rather short, broad and bifid at tip; hind claws about twice length of fore, and very pointed.

Baetis propinquus Walsh.

The imago is described in Eaton's Monograph, but my specimens do not show the subopaque area between the two nervures of the hind wing. Nymphs were taken at Go-Home Bay from June 14th to July 22nd; on August 19th large numbers of them were discovered in a little bay of a small, bare island about three miles out in the open. This rock was the home of numerous gulls, and hence is commonly called "Rookery" Island. The nymphs were mature, and imagos emerged on August 21st and 22nd.

Nymph: (Pl. VI, fig. 6.)

Measurement—Body 6 mm.; setæ 2 mm. Face vertical, mostly brown in colour; on dorsal surface of head on each side of median line is a row of irregularly-shaped light spots. Notum brown with various light areas. Dorsum of abdomen for the most part brown; segments 2-4 brown, with a light area in each half of each segment, and margins colourless; on segment 4 there is also a light area in median line; segment 5 quite light in colour; segment 6 brown, with a light area along anterior margin and two faint ones posterior to it; segments 7 and 8 each with two rather large pale areas in posterior half; segment 9 almost entirely pale; segment 10 slightly brown, especially along posterior margin; on each of the brown segments there are two small faint pale oblique, slightly-curved streaks, and a pale dot posterior to each. Ventrally, the joining of segments brown. Setæ slightly tinged with brown, with tips darker brown and a brown band beyond the middle; lateral setæ fringed on inner sides only. Legs pale; femora banded with brown about middle; tibiæ and tarsi darker toward distal ends; each claw with a lateral row of sectinations.

Cloeon dubium Walsh.

The imagos I have agree with the description in Eaton, except that the intercalar veins are single, not in pairs. Adults were numerous at Station Island about July 10th, flying in small swarms along the shore at a height of from 10 to 15 feet. They appeared about 7.45 in the evening. Not many nymphs were taken, collections dating July 30th to August 12th. Imagos were reared July 30th and August 2nd.

Nymph: (Pl. VI, fig. 7).

Measurements—Body 4-4.5 mm.; setæ 1.5 mm. Face vertical, with two large pale areas above antennæ; between eyes a large pale area, partly divided into two parts, and containing two brown stripes. Notum brown, with irregular light areas. Dorsum of abdomen brown, except lateral margins, which are colourless; on each segment there are two small, oblique, pale streaks and two round dots posterior to the streaks. Setæ pale, with brown band toward distal end; lateral setæ fringed on inner sides only. Gills double apparently on segments 1 and 2 only; broader than gills of *Bætis*; a main trachea in each slightly to outer side, and branchlets on inner side only. Legs pale; femora banded with brown in distal half; tibiæ and tarsi brown toward proximal ends; claws comparatively long, sharp-pointed, not pectinated.

Callibætis ferruginea Walsh.

Imagos, subimagos and nymph skins of this species were collected in Toronto by Dr. E. M. Walker, who kindly handed them over to me. The date would be about August 20th. None were taken at Go-Home Bay.

I am very grateful to Dr. Anna H. Morgan, Mount Holyoke College, So. Hadley, Mass., for the identification of a number of species for me.

VANESSA CALIFORNICA AGAIN.

BY F. M. WEBSTER, WASHINGTON, D. C.

I note with interest the criticism of Dr. J. McDunnough of Decatur, Illonois, in the July Number of the Canadian Entomologist, on my note in the preceding April number relating to the above-named species.

The trouble seems to be that Dr. McDunnough is looking at the matter solely from the viewpoint of what the writers of the letters, quoted by me, claimed to have seen, while I used these letters in their entirety, together with the identifications as they came to me, for the purpose of complete record, not only with the object of showing what these people stated that they saw, but also what they actually "produced in court," thus the better enabling everyone to draw his own conclusions from all of the evidence presented.

The problem of the exact larval food habits of the species is not susceptible of solution, either in Washington, D. C., or in Decatur, Illinois.

A PARASITE OF THE CHINCH BUG EGG.*

BY JAMES W. MCCOLLOCH,
Assistant Entomologist, Kansas State Agricultural College and Experiment Station.

In the experiments conducted this year to determine the time of the first appearance of young chinch bugs and the mortality of the eggs, a large number of eggs were collected in the field for examination. The eggs, which were collected at different intervals and in different localities, were examined daily. While thus examining the eggs it was noticed that some of them became dark in colour instead of assuming the usual red colouring. These eggs were isolated, and on May 19 there emerged from them three parasites. With these three parasites as a basis, the life-history was carried through four generations, running up to July 5. Since this was the first time between the two broods of the chinch bug,

*Mr. A. B. Gahan, Entomological Assistant of the Bureau of Entomology, U. S. Dept. of Agric., to whom specimens of the parasite were sent for determination, says: "I have made a partial examination of these parasites, and find them to belong to the family *Proctotrypidæ*, and they probably fell close to the genus *Telenomus*. It will require further study for me to determine definitely regarding them. It seems probable that they represent not only a new species, but possibly a new genus."

it became impossible to obtain additional chinch bug eggs with which to continue the work. From July 5 to July 23 only an occasional parasitized egg was found in the field, but beginning with the latter date, parasitized eggs were found in large numbers in the cornfields, and the second generation was obtained by August 10. Up to the present date this year over 275 individual parasites have been bred out. The length of the life cycle has been found to vary from ten to eighteen days, depending on the climatic conditions.

The parasite has been found in every wheat- and cornfield examined around Manhattan. Of 3,101 eggs collected between April 28 and June 10, the average per cent. of parasitism was 20.8, and of 116 eggs collected at Crawford (Central Kansas), 19 eggs, or 16.3, were parasitized.

The work is still under way, and a full description of the parasite, together with notes on its life-history and efficiency, will be published later.

A NEW SPECIES OF PHENGODES FROM CALIFORNIA (COLEOPTERA).

BY HERBERT S. BARBER, BUREAU OF ENTOMOLOGY, WASHINGTON, D.C.

With regret the writer feels forced to offer the description of the following species in advance of its publication in a monographic revision of the Phengodids now in manuscript, the appearance of which has been delayed far beyond contemplation.

Phengodes bellus, n. sp.

Large, strongly bicoloured. Antennæ (except two basal joints), palpi, elytra and dorsum of last two abdominal segments (except lateral margin) black; wings creamy white; all other parts luteous.

Length 20 mm.; width across humeri 3.8 mm. Habitat, California.

Occiput coarsely strigose; eyes separated above by slightly more than twice the width of one eye as seen from above, below by about one and one-fourth times the width of one eye as seen from below. Antennæ extending to about third abdominal segment, black, except two basal joints, which are luteous; rami about ten times as long as supporting joint, black with pale hairs. Maxillary palpi black, last three joints subequal (the penultimate slightly shorter), apex of terminal joint very obliquely truncate. Mandibles strong, upper side nearly flat. Under side of head sparsely pubescent, the hairs arising from moderately fine, strigose punctures; gular suture very strong anteriorly, fossa elongate, very narrow, nearly closed. Pronotum half as long as wide, as wide as body at humeri; disc smooth, shining, very minutely punctulate, with small impression before scutellum; sides broadly explanate, the dilated margins each about one-eighth of the entire width, base almost straight, feebly trisinuate; hind angles rounded; side margins straight, very slightly convergent anteriorly; front angles obtuse, broadly rounded; front margin strongly arcuate at middle, nearly straight on each side. Elytra black, one-third longer than width across humeri, feebly bicostate, surface shining, scabrose, punctulate, with fine dark brown pubescence, apices attenuate strongly divergent. Wings creamy white, costa and media brown, other veins pale, a cross-vein between the forks of the cubitus. Abdomen pale, except large black spot on dorsum of last three segments. Legs pale, except tarsi, which are black with dark pubescence, fourth tarsal joint with an elongate, whitish membranous lobe projecting under base of fifth joint; claws with very obtuse tooth at base on inner edge.

Type in the Carnegie Museum. Paratype, No. 16332, U. S. National Museum.

Two specimens collected in June or early July, 1904, by the late Dr. W. Miller, in San Bernardino Co., Cal. (exact locality unknown), and kindly loaned to the writer by Mr. H. G. Klages, who has generously placed the paratype in the U.S. National Collection. This is certainly the handsomest species of the genus known in our fauna. It is distinguished from any other species in the United States by the black elytra and whitish wings. A variety (?) of *P. bipennifera* Gorh. is figured in the Biologia Centrali-Americana as having black elytra but specimens have not been seen by the writer.

A WORM THAT CARES.

BY XIMENA MCGLASHAN, TRUCKEE, CALIFORNIA.

Does the worm have care or thought for the adult it is to produce? Many writers assert that there are no signs of sentiment in any of the stages of moth or butterfly existence. They say the mother fly lays her eggs because of natural law, the eggs hatch because they must, the larvæ simply live to eat, and the chrysalis, however wonderful, is only a part of the process. That is all very interesting, but the mother never sees nor cares for her progeny, nor does the offspring care for anything but itself. If one were to cross pens in a friendly tilt with these writers, the best illustrations of loving care would doubtless be sought in the pains and trouble which the mother fly manifests in depositing her eggs, or in the solicitude of the larva for the protection of its pupa.

In my home at Truckee, California, there is a species of *Cossus*, which Barnes and McDunnough say is "probably *Cossus angrezi* Bailey," which lays its eggs under the bark and in the wood of the cotton-wood tree in August. The female will oviposit if confined in a paper bag, and lays more than a hundred eggs; but, if allowed to have her own way, she hides each egg in the wood or bark of the tree. The larvæ burrow into the interior of the trunk, and up to the time when they wish to pupate they are entirely hidden from view. They pupate in the bottom of their burrow, and if they only plan for themselves there would seem to be no reason why they should delay the transformation when the time arrives. As a matter of fact, however, they seem to know that the adult must have access to the open air which they themselves have never breathed. Just before pupation they carry their burrow to the surface and smooth the jagged ends of the bark and wood of the opening so that nothing will retard the egress of the moth. They do one thing more which shows a high order of instinct, if it be not reason. The diameter of the opening, just at the surface, is made a trifle less than that of the burrow itself. A little thin ledge projects inward all around the edges of the hole. When the adult is ready to emerge, with the large pupa-case around its body, it arrives at the projecting ledge on the inner side of the opening, and the case itself is a trifle too large to slip through. It is held fast by the ledge while the adult pulls itself out. When the moth has escaped, bits of the end of the pupa case project outside the burrow, and the empty case may be forcibly extracted before it dries. If this *Cossus* larva pupated in the earth at the foot of the tree there would be a good reason why it should have carried the burrow to the surface. As it does not pupate outside the tree, and as it remains in the open air only long enough to shape and smooth the opening, may we not conclude that here is a worm which cares for its adult?

A REMARKABLE NEW PLATYGASTERID GENUS FROM AUSTRALIA.

BY ALAN P. DODD, NELSON, N. Q. AUSTRALIA

Platygastoides nov. gen.

Female (?).—Head transverse, as wide as the thorax; ocelli far apart, the lateral ones touching the eye margins. Antennæ 10-jointed; scape extraordinarily dilated, scarcely longer than wide, half as wide as the head; when in the normal position the rest of the antennæ lies back along the scape; pedicel slender, twice as long as wide; 1st funicle joint as long as the pedicel and narrower; 2nd as long as wide; 3rd and 4th wider than long; club 4-jointed; 1st joint very short, transverse; club joints 2-4 large, wide.

Thorax short, scarcely longer than wide; pronotum scarcely visible from above; mesonotum wide, with the parapsidal furrows present, wide apart; outside the parapsidal furrows are two parallel groove lines; scutellum semicircular, with a median groove line ; metanotum with two deep sulci, separated by a median carina; lateral edges of the sulci carinate.

Fore wings rather short, broad, without veins. Abdomen sessile, as wide as the thorax, and longer than the head and thorax united; 2nd segment equal to one-half the abdominal length.

Legs rather short; tarsi 5-jointed.

Type.—The following species:

Platygastoides mirabilis sp. nov.

Female (?).—Length, 1.50 mm. Black; legs, except coxæ, reddish yellow; antennæ reddish yellow, the scape and club suffused with black. Head and thorax finely sculptured; abdomen very finely reticulately rugulose. Fore wings infuscated, opaque; marginal cilia very short; discal cilia very fine and dense.

(From 4 specimens, 2-3 inch objective, 1 inch optic, Bausch and Lomb.)

Male.—Unknown.

Described from two ♀ specimens caught while sweeping the forest slopes of Mount Pyramid, 1,500-2,500 feet, near Cairns; one ♀ caught while sweeping in a jungle, Goondi (Innisfail), N. Q.; and one ♀ received from the South Australian Museum, and labelled, "Cairns district, N. Q., A. M. Lea."

Habitat.—North Queensland (Mount Pyramid, near Cairns, Innisfail).

Type.—South Australian Museum, Adelaide, a ♀ tagmounted plus a slide bearing head, antennæ and forewings.

STRANGE ACTION OF *BOMBUS OCCIDENTALIS.*

BY J. WM. COCKLE. KASLO, B. C.

Whilst walking across my garden to-day I observed a number of bees disporting themselves on the flowers of some Chinese Cabbage that were running to seed.

On closer inspection I found that they were all *Bombus occidentalis* workers, with the exception of a very few *A. mellifica.* The *Bombus* were there in thousands, and their actions caused me to stop and watch them. Instead of settling and inserting their tongues amongst the pistils of the flower, they tumbled in every direction over the flower, and seemed to be looking for hidden treasure at the base of the corolla. Being unable to see what they were so assiduously hunting for, I sat down in the middle of the patch in order to get a closer observation.

They inserted their tongues in small holes at the base of the corolla and between the folds at the base of the petals. In many cases they seemed to have considerable difficulty in forcing an entrance, raising their bodies and thrusting the tongue down with force.

An examination of the flowers showed that all the open flowers were punctured at the top of the bulb which forms the base of the corolla. I made a minute examination of the flowers to find out if the punctures were the work of other insects, but could find no other insect on or in any of the flowers, and also that none of the unopened flower buds showed any sign of puncture. Eventually, by the aid of a glass I found that the puncture was made from the outside and was ragged and torn, and ultimately I was fortunate in seeing one bee actually pierce the base of the corolla whilst I was observing it. As stated before, there were no other bees there except *Apis mellifica*; these were acting in a normal manner, seeking the honey through the centre of the flower, and in no case did I see one attempting to follow the example of the *occidentalis*.

The reason for this (to me, at least) strange action of *B. occidentalis* may possibly be explained by the fact that the tongue of *occidentalis* when fully extended is not nearly long enough to reach the honey sac, but the fact of the folds at the base of the petals being easily pried apart gave them ready access, and it is also probable that when they found a freshly-opened bud on which the folds of the petals had not yet commenced to separate that they found easy access by puncturing the corolla; they most assiduously hunted for a puncture and invariably thrust their tongue into it. Some of the flowers I examined had been punctured in several places. It would be interesting to know if this action of puncture is shared by any of the other bees, or if it is an invariable practice of *occidentalis* when attacking a flower having a deep-seated honey sac.

[NOTE.—*B. occidentalis* belongs to the same group of *Bombus* as the European species *terrestris*, which, it is well known, punctures with its mandibles the base of such flowers as Snapdragon and Broad-bean to obtain the nectar, thereby sometimes damaging the seed vessels. I have seen the workers of *B. terricola*, the representative of this group in Eastern Canada, puncturing the spur of *Impatiens biflora* and sucking the nectar through the wound thus made, though *B. vagans* and *fervidus* were observed obtaining the nectar by entering the flower in the legitimate way.—F. W. L. Sladen, Central Experimental Farm, Ottawa.]

Mailed October 13th, 1913

ENTOMOLOGICAL SOCIETY OF ONTARIO—FIFTIETH ANNUAL MEETING,

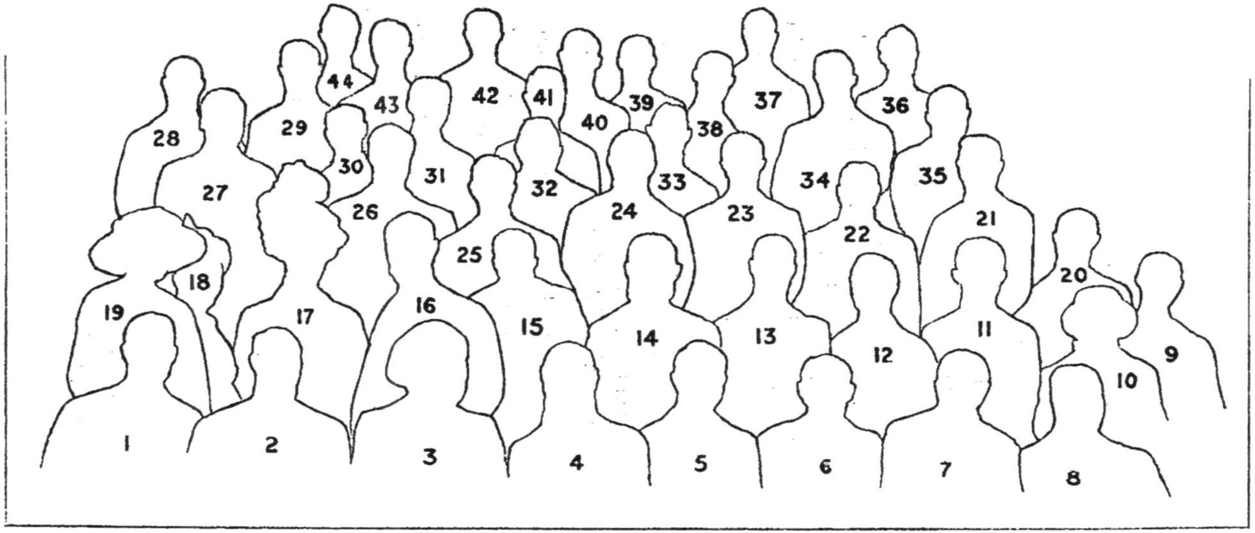

1. Prof. W. M. Wheeler. 2. Rev. T. W. Fyles. 3. Mrs. Comstock. 4. Rev. Prof. C. J. S. Bethune. 5. Dr. C. G. Hewitt. 6. Prof. J. H. Comstock. 7. Prof. F. M. Webster. 8. Pres. G. C. Creelman. 9. Prof. T. D. Jarvis. 10. Mrs. Jarvis. 11. Dr. E. P. Felt. 12. Prof. P. J. Parrott. 13. J. F. Brimley. 14. J. D. Evans. 15. G. Meade-Waldo. 16. H. H. Lyman. 17. Mrs. Lyman. 18. Miss Bethune. 19. Mrs. C. G. Hewitt. 20. A. Gibson. 21. F. W. L. Sladen. 22. Dr. R. S. MacDougall. 23. Prof. E. M. Walker. 24. Prof. W. Lochhead. 25. Prof. J. Dearness. 26. Mr. W. E. Saunders. 27. Dr. A. Cosens. 28. Mr. J. J. DeVyver. 29. H. Curran. 30. L. Caesar. 31. R. S. Hamilton. 32. J. C. Chapais. 33. C. E. Petch. 34. R. C. Treherne. 35. Prof. T. J. Headlee. 36. W. A. Ross. 37. A. W. Baker. 38. J. D. Tothill. 39. L. S. McLaine. 40. G. J. Spencer. 41. W. A. Clemens. 42. J. F. Hudson. 43. A. Burrows. 44. G. E Sanders.

1. Prof. W. M. Wheeler. 2. Rev. T. W. Fyle 3. Mrs. Comstock 4. Rev. Prof. C. J. S. Bethune. 5. Dr. C. G. Hewitt. 6. Prof. J. H. Comstock. 7. Prof. M. Webster. 8. [illegible] C. C. [illegible]man. 9. Prof. T. D. Jarvis. 10. Mrs. Jarvis. 11. Dr. E. P. Felt. 12. Prof. P. J. Parrott. 13. [illegible] Brimley. 14. J. D. Evans. 15. G. Meade-Waldo. 16. H. H. Lyman. 17. Mrs. Lyman. 18. [illegible] 19. Mrs. C. G. Hewitt. 20. A. Gibson. 21. F. W. L. Sladen. 22. Dr. R. S. MacDougall. 23. Fr[illegible]ker. 24. Prof. W. Lochhead. 25. Prof. J. Dearness. 26. Mr. W. E. Saunders. 27. Dr. A. Cosens. 28. Mr. J. J. DeVyver. 29. H. Curran. 30. L. Caesar. 31. R. S. Hamilton. 32. J. C. Chapais. 33. C. E. Petch. 34. R. C. Treherne. 35. Prof. T. J. Headlee. 36. W. A. Ross 37. A. W. Baker. 38. J. D. Tothill. 39. L. S. McLaine. 40. C. J. Spencer. 41. W. A. Clemens. 42. J. F. Hudson. 43. A. Burrows. 44. C. E. Sanders

The Canadian Entomologist.

VOL. XLV. LONDON, NOVEMBER, 1913 No. 11

THE FIFTIETH ANNUAL MEETING OF THE ENTOMOLOGICAL SOCIETY OF ONTARIO.

One of the most important and interesting events in the history of the Entomological Society of Ontario, the celebration of its Fiftieth Annual Meeting, took place on Wednesday, Thursday and Friday, August 27-29, 1913, at the Ontario Agricultural College, Guelph. Wednesday and Thursday were devoted to the reading of papers and presentation of addresses by the representatives of other societies and institutions and to the routine business of the society, while on Friday an excursion was made to Grimsby and the Niagara Fruit District.

The President, Rev. Dr. Bethune, was unfortunately unable to act in his official capacity on account of defective eye-sight, but he was nevertheless present at all the meetings, and his place in the chair was ably filled by the Vice-President, Dr. Hewitt.

The meetings were attended not only by a large number of the Society's members and visitors from the town, but also by a goodly representation of distinguished entomologists from the United States and Great Britain. Keen regret was felt by all in the absence of Dr. Wm. Saunders, whose serious illness prevented him from being present, as had been hoped for, but, apart from the disappointment caused by his absence and that of several other prominent entomologists who had been expected, the meeting was a most successful and memorable occasion and was much enjoyed by everyone present. Much of the success of the meeting is to be credited to the excellent arrangements of the committee in charge for the comfort and accommodation of the members and visitors.

The following were present at the meeting:

Prof. J. H. Comstock, Hon. Member, Entomological Department, Cornell University, Ithaca, N.Y.; Mrs. J. H. Comstock; Prof. F. M. Webster, Hon. Member, Bureau of Entomology,

Washington, D.C.; Dr. E. P. Felt, Hon. Member, State Entomologist, Albany, N.Y.; Dr. R. Stewart MacDougall, University of Edinburgh; Mr. Geoffrey Meade-Waldo, British Museum, London, Eng.; Prof. W. M. Wheeler, Harvard University, Cambridge, Mass; Prof. T. J. Headlee, State Entomologist, New Brunswick, N.J.; Prof. A. D. MacGillivray, Urbana, Ill.; Prof. P. J. Parrott, Geneva, N.Y.; Prof. J. J. de Vyver, Entomological Society of N.Y.; Mr. W. A. Clemens, Ithaca, N.Y.; Dr. C. Gordon Hewitt, Dominion Entomologist, Ottawa; Mrs. C. Gordon Hewitt; The Rev. T. W. Fyles, D.C.L., Ottawa; Mr. and Mrs. Henry H. Lyman, Montreal; Prof. W. Lochhead, MacDonald College, P. Q.; Mr. J. C. Chapais, St. Denis (en bas), Quebec; Mr. John D. Evans, Trenton; Ontario; Mr. F. J. A. Morris, Peterboro; Dr. E. M. Walker, University of Toronto, Toronto; Mr. J. B. Williams, University Museum, Toronto; Dr. A. Cosens, Parkdale Collegiate Institute, Toronto; Mr. R. S. Hamilton, Galt; Prof. J. Dearness, London, Ont.; Mr. J. F. Brimley, Grimsby, Ont.; Messrs. Arthur Gibson and F. W. L. Sladen, Division of Entomology, Ottawa; the following Field Agents of the Dominion Division of Entomology: Messrs. Sanders, Tothill, Petch, Ross, Hudson, McLaine and R. C. Treherne, Vancouver, B.C.

The Ontario Agricultural College was represented by the following: President Creelman, Prof. C. A. Zavitz, Prof. C. J. S. Bethune, Prof. T. D. Jarvis, Prof. Hutt and Prof. Crow; Mr. L. Caesar, Mr. A. W. Baker, Dr. R. E. Stone, Prof. E. J. Zavitz, Mr. Wright, Mr. G. J. Spencer; Messrs. Burrows, Curran, Good, Hart and others.

On Wednesday evening a meeting of the Council was held in the Biological Lecture Room, at which, among other matters, certain proposed changes in the constitution of the Society were discussed. These changes, which were afterwards adopted at the General Meeting, will be given in full in the December number.

In the afternoon the members and delegates met in the Massey Hall Auditorium, the proceedings commencing with an address of welcome by President Creelman of the College, which was delivered in his usual genial manner and vigorous style. Congratulatory addresses were then presented by the following representatives of other societies and institutions:

Prof. E. M. Walker, University of Toronto; Prof. Wm. Lochhead, University of McGill; Dr. C. Gordon Hewitt, University of Manchester, Royal Society of Canada, Academy of National Sciences of Philadelphia and the Canadian Department of Agriculture; Dr. R. Stewart MacDougall, University of Edinburgh and Imperial Bureau of Entomology; Prof. W. M. Wheeler, Harvard University and Boston Society of Natural History; Prof. J. H. Comstock, Cornell University and the Entomological Society of London; Prof. A. D. MacGillivray, Entomological Society of America; Prof. P. J. Parrott, American Association of Economic Entomologists; Dr. E. P. Felt, New York Entomological Society; Prof. F. M. Webster, Entomological Society of Washington and the Bureau of Entomology, U. S. Department Agriculture; Mr. Arthur Gibson, Ottawa Field Naturalists' Club; Mr. J. C. Chapais, Quebec Society for the Protection of Plants; Mr. R. C. Treherne, Entomological Society of British Columbia; Mr. A. F. Winn, Montreal Branch, Ent. Society Ont.; Dr. A. Cosens, Toronto Branch, Ent. Soc. Ont.; Mr. Geoffrey Meade-Waldo, British Museum, Natural History Department, London, England.

A message of congratulation from Dr. William Saunders, who was too ill to be present, was conveyed to the Society by his son, Mr. W. E. Saunders. It was prefaced by a few remarks of appreciation by the Chairman.

Letters of congratulation were also read from the following:

The Imperial Academy of Natural Sciences, St. Petersburg, Russia (by cable); The Vice-Chancellor of the University of Oxford; The Vice-Chancellor of the University of Cambridge; The President, Laval University, Que.; Dr. Walther Horn, Director of the German Entomological Museum, Berlin, Germany; J. P. Moore, Secretary of the Academy of Natural Sciences of Philadelphia; Geo. A. Dean, Kansas State Agricultural College, Manhattan, Kansas; E. Baynes Reed, Dominion Meteorological Station, Victoria, B. C.; N. H. Cowdry, Esq., Chicago; Dr. L. O. Howard, Bureau of Entomology, Washington, D. C.; The University of Chicago, Chicago, Ill.; The Trustees of the British Museum (Natural History), London, Eng.; A. Ross, Sec. Natural History Soc. of Glasgow, Glasgow, Scotland; Guy A. Marshall, Imperial Bureau of Entomology, London, Eng.; Geo. Francis Dow, Secretary, Essex Institute, Salem, Mass.; State Commission of Horticulture, Sacramento, Cal.; Prof. T. D. A. Cockerell, Boulder, Col.; Prof. Harrison Garman, Lexington, Ky.; Prof. H. F. Wickham, University of Iowa, Iowa City.

Letters expressing regret at their inability to attend the meeting or to send a representative were received from the following: Mr. E. T. Cresson, American Entomological Society, Philadelphia; S. A. Rohwer, Secretary, The Entomological Society of Washington, Washington, D.C.; Prof. H. F. Wickham, State University of Iowa, Iowa City; Prof. Francis John Lewis, Edmonton, Alta, representing the Linnean Society of London, Eng.; Stanley Edwards, Hon. Secretary, South London Entomological and Natural History Society, London, Eng.; Dr. Walcott, Secretary, Smithsonian Institution and the United States National Museum, Washington, D.C.; Louisiana State University, Baton Rouge, La.; The Central Museum, Brooklyn, N.Y.; The Director of the Missouri Botanical Garden, St. Louis, Mo.; Professor V. Kellogg, Leland Stanford University, California; President W. O. Thompson, Ohio State University, Columbus, O.; President Stephen A. Forbes, Director, Ill. State Laboratory of Natural History, Urbana, Ill.; President Robert J. Aley, University of Maine, Orono, Maine; Yale University, New Haven, Conn.; F. J. Skiff, Director, Field Museum of Natural History, Chicago; Dr. L. O. Howard, Director Bureau of Entomology, Washington, D.C.; Professor R. Matheson, Agricultural College, Truro, N.S.; Mr. A. F. Burgess, Secretary American Association of Economic Entomologists, Melrose Highlands, Mass.; Hon. Jas. S. Duff, Minister of Agriculture for Ontario, Toronto; Mr. W. Bert Roadhouse, Deputy Minister of Agriculture for Ontario, Toronto; Dr. C. C. James, Ex-Deputy Minister of Agriculture for Ontario, Toronto; Mr. C. E. Grant, Orillia, Ont.; John Bland, Esq., Secretary, Mo. State Board of Horticulture, Columbia, Mo.; Prof. A. L. Melander, State College of Washington, Pullman, Wash.; The President, University of Montana, Missoula, Montana; Prof. Geo. A. Dean, Kansas State Agricultural College,

Manhattan, Kansas; Joseph H. Kastle, Director, Kentucky Agricultural Experiment Station, Lexington, Ky.; E. Davenport, Director, University of Illinois, Champaign, Ill.; Geo. F. Dow, Secretary, Essex Institute, Salem, Mass.; W. W. Atwood, Secretary Chicago Academy of Sciences, Chicago, Ill.; Harry Piers, Secretary N. S. Institute of Sciences, Halifax; A. F. Winn, Secretary, Montreal Branch of the Entomological Society of Ontario, Montreal; Prof. H. Garman, Agricultural Experiment Station, Lexington, Ky.; Prof. J. G. Needham, Cornell University, Ithaca, N. Y.

The evening was marked by one of the most enjoyable features of the meeting—a reception given to the members by President and Mrs. Creelman at their residence.

On Thursday morning a business meeting was held in the Biological Lecture Room, at which the officers for the ensuing year were elected and several matters of interest to the members were discussed. Of these reference has already been made to the revised constitution of the Society. Among other matters, a resolution was passed recommending that the various entomological societies be properly represented at the International Congress of Entomology. The Rev. Dr. T. W. Fyles was elected a life menber of the Society. Mr. J. M. Swaine and Dr. E. M. Walker were appointed to represent the Entomological Society of Ontario on the American Committee of Nomenclature. It was decided to hold the next annual meeting at Toronto, the date to be chosen on a later occasion.

The remainder of the day's session was occupied by the reading of addresses and papers, commencing with the Presidential Address by Dr. Bethune, an extremely interesting review of the Society's early history. An abstract of this address is given below, together with the other papers presented.

The feature of the evening meeting was a most interesting and instructive address on Ants by Prof. W. M. Wheeler, an abstract of which is also given below. The lantern slides, with which the lecture was richly illustrated, were of quite exceptional excellence and beauty.

Address of President.

The President, Dr. Bethune, stated that, owing to defective eye-sight, he was unable to prepare a formal written address and would, therefore, endeavour to give some account of the origin of the Society and the proceedings that led to its formation.

When a student at Trinity College, Toronto, he began the collection and study of insects. At that time there were no available books on the subject. The first work that gave him any assistance in naming specimens was Gosse's "Canadian Naturalist," a delightful work giving an account of observations made in various departments of natural history during each month of the year in the eastern townships of the Province of Quebec. In the Canadian Journal there were published excellent short descriptions of the more conspicuous beetles found in the neighbourhood of Toronto by Wm. Couper, a printer by trade. These were supplemented by lists furnished by Prof. Croft, of the University of Toronto. Kirby and Spence's "Entomology" and Westwood's "Modern Classification of Insects" were published about that time and afforded the first scientific aids to the knowledge of insects. Through the kindness of Prof. Croft, the speaker had access to the library of the University of Toronto, which contained several rare works on entomology. He was also permitted to consult the books in the library of Parliament, which, at that time, was located in Toronto. In these libraries he spent much of his leisure time in laboriously transcribing descriptions of Canadian insects, which, for the most part, had to be translated from Latin and French, and also in making copies of illustrations. These difficulties can hardly be realized by students at the present day who have such an abundance of literature upon every department of natural history. Such works as Comstock's "Manual for the Study of Insects" and Mrs. Comstock's "How to Know the Butterflies" would, at that time, have been treasures indeed. However, there is no doubt that the difficulties encountered helped one to build upon a sound foundation and to acquire a more complete knowledge than could be attained by attempting to hastily read a superabundance of publications.

At the suggestion of Prof. Croft, the speaker made the acquaintance of Mr. Wm. Saunders. of London, who carried on at the time the business of chemist and druggist on a moderate scale. The acquaintance thus formed soon ripened into a mutual friendship and esteem which has continued unbroken to the present time.

In the Canadian Naturalist for June 1862 there appeared a list of persons residing in Canada, all interested in the collection and study of insects, which contained no less than thirty-six names. This was prepared by the speaker, with the assistance of Mr. Saunders. It was then proposed that a meeting should be held for the purpose of bringing together as many as possible of those interested, and to form some kind of club or society which would be of general benefit to those concerned. The result of this publication was the holding of a meeting in Toronto at the residence of Prof. Croft in September, 1862. As there were only ten persons present, it was thought inadvisable to form a society at that time, but a draft of a constitution was drawn up and it was decided to hold another meeting during the coming year. On the 16th of April, 1863, a meeting was held in the library of the Canadian Institute and the formation of the Entomological Society of Canada was then decided upon and its constitution drawn up and adopted. The attendance was small, but several who were unable to be present had given in their adhesion to the movement. Dr. Henry Croft, Professor of Chemistry in the University of Toronto, was elected President; Mr. Saunders, of London, Secretary-Treasurer, and the Rev. Jas. Hubbert, Curator. The others present were: The Rev. Wm. Hincks, Professor of Botany and Zoology at the University of Toronto; Dr. Sangster, Principal of the Normal School, Toronto; Dr. Beverley R. Morris, an Englishman who not long after returned to England and there became editor of a popular magazine on natural history, Dr. Cowdry and his son, Mr. N. H. Cowdry, of York Mills, and Messrs. Saunders and Bethune. The following gentlemen were unable to be present, but became original members of the society: Mr. E. Baynes Reed, Barrister, London, Mr. E. Billings, editor of the Canadian Naturalist and Geologist, for many years attached to the Geological Survey. Mr. R. V. Rogers, Barrister, Kingston; Mr. T. Reynolds, Engineer of the Great Western Railway, now part of the Grand Trunk system, Hamilton; Mr. B. Billings, Prescott, who subsequently lived in the neighborhood of Ottawa and formed a large collection of Coleoptera; Rev. V. Clementi, Peterborough, an English Church clergyman, who was greatly interested in the various aspects of natural history. Mr. Wm. Saunders was appointed by the Dominion Government in 1886 to establish and superintend a series of Experimental Farms extending from Nova Scotia to British Columbia. For twenty-five years Dr. Saunders conducted this work in a most able and successful manner, and his name is well known not only throughout Canada, but also in the United States and Great Britain.

The Society thus formed began its career of active usefulness and it has steadily grown and prospered to the present time. In 1868 the publication of the Canadian Entomologist was begun, the first number consisting of only eight pages. It is now in its 45th year of publication and is sent to all parts of the world. In 1870 the first Annual Report of the Society on Noxious, Beneficial and other Insects was published, the three contributors being Dr. Saunders, Mr. Baynes Reed and the speaker. What really made the fortunes of the Society was the invasion of Ontario by the Colorado Potato Beetle. The Board of Agriculture for the Province requested the Society to report on the insect and to advise as to the best methods of checking or controlling its ravages. An admirable report was prepared by Messrs. Saunders and Reed, the former being a practical chemist was able to experiment with various poisons and to discover that Paris green was the most convenient and reliable substance for the destruction of the beetle. The result of this report was a grant from the Department of $400 per annum, which was afterwards increased to $1,000 and the incorporation of the Society under the name of the Entomological Society of Ontario. For a few years the Canadian Institute in Toronto gave the Society the privilege of using its library and museum for its meetings and collections. After a few years, however, the headquarters were removed to London and continued there until 1906, when a change was made to the Ontario Agricultural College, Guelph.

The speaker expressed the great pleasure which it gave him and his colleagues to find that so many friends had come from long distances to join in the celebration of the Jubilee Meeting of the Society. He joined in giving them all the most hearty welcome and expressed his hope that they would fully enjoy their visit.

GREEN LANES AND BYWAYS.

BY REV. THOMAS W. FYLES, D.C.L., OTTAWA.

I.

Old Country Lanes.

"Through the green lanes of England, a long summer day,
When we wandered at will in our youth's merry May;
When we gathered the blooms o'er the hedge-rows that hung,
Or mocked the sweet song that the nightingale sung.

In the autumn we knew where the blackberries grew,
And the shy hazel-nuts hidden deep in the shade;
And with shouting and cheer, when the Christmas drew near,
In search of the ripe ruddy holly we stray'd."

These lines appeared in the "Illustrated London News" for January the 24th, 1852. They are dear to my remembrance, for they were sung to me by a much-loved companion—long gone to his rest—as we strolled along an English lane, one day in the summer, after their appearance. From this friend* I received my first lessons in Entomology.

The enclosures in the rural parts of England, by which the road-ways pass, have been from times immemorial, and for the most part they are known each by its proper name, as "Nether lea," "Ea-side," "Haly-well Croft," Twenty acres," "Basket lot," etc. The boundaries of the fields are quickset hedges, with ditches on the outer sides. Six feet from the roots of a hedge was allowed for the ditch.

The original growth of the hedges was Hawthorn (*Cratægus oxycantha* L), but, as time passed on, birds and other agents dropped seeds of many plants among the thorns. The most noteworthy of the intruding growths are: Blackthorn (*Prunus spinosa*), Dog-rose (*Rosa canina*), Honeysuckle (*Caprifolium perfoliatum*), Holly (*Ilex aquifolium*), Traveller's Joy (*Clematis vitalba*), Elder (*Sambucus nigra*) and Bindweed (*Convolvulus sepium*).

The mud from the ditches—washings from the roads and fields—is thrown up periodically to the hedge-bottoms, and the fresh soil maintains the varied growth in constant vigour.

*Mr Edwin Tearle, in after years Rector of Stocton, in the Diocese of Norwich.

Some of the byroads of England were formerly important highways. In a tour I made, in my youth, to Tennyson's country in North Lincolnshire, I came one day to a little place that, I was told by a countryman, was "Spittle-in-the-Street." After a little thought I understood the name. "Spittle" was '*Spital*, a contraction of *Hospital*, and the "Street" stood for the *Stratum*, the Roman way from Lincoln (*Lindum Colonia*—the *Colony-in-the-Marsh*) to the Humber. Yes, along that way, centuries ago, marched the legionaries of the Cæsars, in stern array, while the woad-stained Cortiani peeped out upon them from their coverts, in hatred and fear.

In after and pre-reformation days, a religious house of entertainment for travellers was erected beside the ancient roads, and this was the Hospital-in-the-Street. There remained of it a farmhouse and the chapel. In the latter a clergyman from a neighbouring parish held services at stated intervals.

In some parts of England where the country is of rolling surface, and the soil light—the lanes being frequently cut up by heavy farm waggons, and but little cared for—the soil is constantly washed by the rains to lower levels, and hollow ways are formed, such as those spoken of by Kirke White in one of his sonnets:

"God help thee, traveller, on the journey far,
The wind is bitter keen, the snow o'erlays
The hidden pits and dangerous hollow ways,
And darkness will involve thee."

In that powerful description of the Battle of Waterloo, given by Victor Hugo in *Les Miserables*, we are told of a grand charge made by three thousand five hundred French cuirassiers upon the English centre. At full speed, in the fury of the charge, the warriors came to the hollow way of Ohain, twelve feet deep, of which they were unaware. Unable to check their steeds, they plunged in, one upon another, and piled up—a writhing mass, crushed and broken. "One-third of Dubois' brigade"—says Hugo—"fell into that abyss." "This," he says, "began the loss of the battle."

But let us quit the contemplation of disasters and consider the delights of English lanes. And, truly, those lanes are delightful

—with their hedgerows gay with blossoms, diffusing sweet perfumes and jubilant with the song of birds!

English hedges are famous nesting-places for many of the feathered tribes. I can recall the pleasure of my first inspection of the nest of the Long-Tailed Tit (*Parus caudatus*). It was a seemingly compact ball of the finest and greenest moss; but it had on one side a small round entrance, closed with a feather. The tit lays many tiny white eggs, spotted with lilac.

Another nest that attracted my attention in my early days was that of the Red-backed Shrike (*Lanius collurio* L). The mother bird was sitting on her pretty, cream-coloured, richly spotted eggs. Meanwhile her mate was busy attending to her wants. He kept her larder well supplied. On the thorns around her were impaled little blind mice and callow birds, shewing that the common name of *Butcher-bird* was justly given to this feathered pillager. But—as an Eastern Township housewife said in praise of her husband, so we may say of the Shrike—"He is a good provider."

It is said* that the English ornithologist, Gould, dated his interest in bird life from the time when, in his childhood, he was lifted up to see the pretty blue eggs in a hedge-sparrow's nest.

Here and there, in the South of England, a lane leaves the enclosures and traverses a piece of common land covered with bushes of the Furze (*Ulex europæus*). This strange plant, which has spines instead of leaves, is, in its season, gorgeous in its wealth of golden bloom. Linnæus, on first beholding it upon Wandsworth Common, fell upon his knees and thanked God who had created a thing so beautiful.

Elsewhere the lane enters, it may be, a stretch of woodland, the game preserve of the lord of the surrounding Manor; and there, truly, the wayfarer is in the midst of charming sights and sounds. In early spring the woods around him are ankle-deep with blue-bells, anemones and primroses. Later in the year the stately foxglove (*Digitalis purpurea* L.) rears its shafts of purple bloom, and "lords and ladies" look out from their stalls.

Many beautiful butterflies sport around. I can mention but a few of them. The pretty Speckled Wood (*Lasiommata ægeria*)

*Country Walks of a Naturalist with his Children, p. 109.

is everywhere in evidence. The lovely Peacock (*Vanessa io*) and the Brimstone (*Gonepteryx rhamni*) shew well against the surrounding foliage. The Silver-washed Fritillary (*Argynnis paphia*) flits over the brambles, on which its larvæ feed. Once in an age a Queen of Spain Fritillary (*Argynnis lathonia*) makes its appearance—blown over, it may be, from France. The Bath White (*Pieris daplidice*) sometimes shews itself, and formerly the Black Veined White (*Aporia cratægi*) could often be seen.

Years ago, in such a wood, I saw what English entomologists seldom see—a specimen of the Camberwell Beauty (*Vanessa antiopa*). It came sailing over the tree-tops and lit upon an oak sapling immediately before me, and then opened its lovely wings. A moment—and it was gone! And I saw it again no more.

Where oak trees are plentiful in the forest, the monarch of English butterflies, the stately Purple Emperor, may sometimes be seen, and there the Purple Hairstreak will surely be found.

Remarkable instances of insect mimicry will engage the attention in such a wood. Here by the road-side is a bush of Broom—the *Planta genista* of olden times, from which the great Plantagenets of English History derived their surname:

"That name Count Geoffrey did assume
When, riding to the chase,
He wore in his casque, instead of plume,
A nodding crest of the yellow Broom,
In its fresh and fragrant grace."

As the traveller approaches the shrub, he will be surprised to see a number of supposed *leaves* of the plant detach themselves from the twigs and flutter away. They are specimens of the tiny Green Hairstreak (*Thecla rubi*).

At another time, noticing the long cylindrical catkins of the Birch, he will be astonished to see that which he had taken to be one of them move away with alternate loops and strides. It is a larva of the Large Emerald Moth (*Geometra papilionaria* L.).

In the woodland lane the ear is—"charmed with concord of sweet sounds." Suppose yourselves in such a lane—call to your imagination its sights and sounds, and—

Let us recline beneath this tree,
  So ragged with lichens—ragged and gray;
Its fretwork of leaves shall our canopy be,
  Our carpet the moss where the sunbeams play.

And we'll list to the pipes of the robin and wren,
  To the flute of the merule so loud and clear,
To the trumpet call of the cuckoo, and, then,
  To the deep bassoon of the stock-dove near.

See you the black-cap 'mid the leaves!
With his glad song his bosom heaves;
His efforts rouse to rivalry
The pride of all Pan's company.

Of choristers, sweet Philomel,
And now soft cadence and rich swell,
  And hurried note and note prolonged,
Echo the glades and thickets through;
As oft, when Sol is borne from view,
In his car of crimson clouds they do,
  Till heaven with listening stars is thronged.

—T. W. F.

The linnet, the goldfinch, the bullfinch, the greenfinch, the whitethroat, the yellowhammer, the thrush, the misselthrush, and other birds, do their best to render the concert of the feathered tribes effective.

Here and there in the road-side hedges a crab-tree may be seen, and here and there a holly.

The holly is sometimes grown as an ornamental hedge. John Evelyn had such a hedge, and he tells how the Czar of Muscovy (Peter the Great) and his outlandish crew amused themselves by trundling one another in a wheel-barrow, backwards and forwards through the prickly barrier. Evelyn had lent his house and grounds for the accommodation of the Muscovites. When the foreigners retired, they left a *muss* behind them.

II.

Canadian Lanes.

Doubtless, in olden times, when men were few and land grants under the feudal system extensive, hedging and ditching were ready means for enclosing and draining the land, and they have been enduring means.

In Canada the roads that remind one of English lanes, though in truth they are very different, are such as lead through parts of the country in which the old-fashioned snake-fences still enclose the farms and in which brush has been allowed to grow freely in the angles of the fences. In such localities, old roads abandoned for new ones, concession roads leading to a few homesteads off the main lines of travel, roads through sugar-woods and the uncleared forest—these, in their quietude and freedom from dust, are suggestive of English lanes—though they lack much of their beauty.

I will speak briefly of a few such roads:

The Caledonia Road.— Skirting a tract well known to the naturalists of Ottawa, by the name of "The Beaver Meadow," is a lane connecting the Aylmer Road with the Chelsea Road. It was originally a "Corduroy road," and it still ends in the remains of a swamp, in which *Typha latifolia* grows freely. Improvements in the neighbourhood have altered its appearance: the logs are gone, and the bed-rock is seen in much of its length; and this, in summer, is carpeted with Stone-crop (*Sedum acre* L.).

Alas! the Beaver Meadow has now been cleared, drained and laid out into building lots. The city naturalists will have to go farther afield for their investigations, and the Caledonia Road will soon become a city street. When I lived in Hull, however, I spent many tranquil hours within its quiet limits.

Muddy spots in the road were much frequented by butterflies. In bright days in April hibernated specimens of *Aglais milberti* Godart might be seen there. The spring larvæ of this species may be found feeding upon the young shoots of the Stinging Nettle (*Urtica dioica* L.). I raised two batches of them in 1911. They went into chrysalis in the first week of June. Sixty per cent. of them were parasitised by *Protopanteles atalantæ* Packard. The grubs of this fly issued from the larvæ of the butterfly—not through the spiny upper parts, but—through the tender ventral portions.

They spun their white, compact cocoons in clusters attached to the skins of their victims. The first imagos of *milberti* appeared in my breeding-cage on the 13th of June.

The large Skipper (*Eudamus tityrus* Fab.) might be seen on the Caledonia Road. I had become acquainted with this insect on Mount Royal, where its larvæ fed on the Hog-peanut (*Amphicarpœa monoica* Nutt), but there I had seen it in its short flights only, as it skipped from bush to bush. When there I witnessed its rapid flight through the open for the first time, I was puzzled. Its direct course; the peculiar motion of its wings; the flashes, in the sunshine, of the large, heart-shaped, silvery patches on the under side of the hind wings—all were new to me. I had to catch the insect to make sure of its identity. In the neighbourhood of Hull its larvæ feed on *Robinia pseudacacia* L. It gathers several leaflets of the tree together, binds them, and feeds under their cover.

A stream, the outlet of Fairy Lake, crossed the Caledonia Road, and over it a rude wooden bridge was thrown. At this point the Turtle-head (*Chelone glabra* L.), the Vervain (*Verbena verticillata* H. B. K.), the lovely Swamp Loosestrife (*Decodon verticillata* H.B.K.) and the Joe Pie Weed (*Eupatorium purpureum* L.) grew in a tangle. On the last named the larvæ of the handsome Tiger Moth (*Arctia caja* L.) fed.

Ought not this specific name to be written and pronounced *Caia?* Linnæus, in naming it, probably had in mind the form of words spoken by the bride in the marriage ceremonies of the ancients: "*Ubi tu Caius, ibi ego Caia.*" We have an instance of the use of the long i, or j, in the last of the numerals representing *four* —iiij. *Halleluiah* was spelt with a j in former times; and I once knew a worthy clergyman whose name was Micaiah, but who always spelt it *Micajah*, with a thought, I doubt not, of the sacred name in the 68th Psalm.*

On the growth spoken of above the pretty Neuropteron *Chauliodes serricornis* Say was often to be seen.

Along the Caledonia Road locusts were numerous. In 1909, particularly, our largest species, *Dissosteira carolina* L., abounded.

* Praise Him in His name Jah and rejoice before Him. Psalm LXVIII, 4.

But a natural check to its undue increase came; many of the insects were affected by *Entomophthora grylli*, and the species has not been so plentiful since.

LEVIS MILITARY ROAD.—A by-way of interest to naturalists is the road connecting the Forts on Levis Heights. The ramparts raised for the defence of this road are now overgrown with brush, and bushes and young trees have sprung up on both sides of it. In the scrub the tall *Diplopappus umbellatus* (Miller) grows abundantly, and upon this the galls of *Gnoremoschema gallæ-diplopappi* Fyles may be found.

What a formidable name "Gnoremoschema" is! It was derived, I suppose, from the Greek, Gnorimos—well known, and Cheima—in winter. The insects that cause the galls, however, do not occupy them in winter. Having escaped their enemies and come to perfection, they quit their dwellings in August, or September at the latest.

But in some instances the galls are not without winter tenants, several kinds of Ichneumon flies, having preyed upon the former inhabitants, spin their cocoons within the galls and remain in them till summer comes around.

The young gregarious larvæ of that lovely butterfly *Melitæa harrisii* Scudder may be found, late in the season, in dingy, closely clinging webs, on the stalks of the Diplopappus. In the spring they disperse and thrive rapidly on the young shoots of the plant.

In this locality the Large-leaved Aster (*Aster macrophyllus* L.) grows plentifully. An insect of remarkable habits feeds upon it, viz., *Tricotaphe levisella* Fyles. The larvæ of this species fasten the edges of the large bottom leaves together and thus form ample tents within which they feed. A full description of the insect in its different stages is given in the 33rd Annual Report of our Society on page 28.

Another insect deserving of notice that may be met with along this military road is the fine ruby-winged locust described by Harris under the name *Locusta corallina*. (See "Insects injurious to Vegetation," p. 176).

OLD ST. HENRY ROAD.—This road, when I lived at South Quebec, was a rich hunting ground for the naturalist. No less than eight species of the Cicindelidæ frequented it, viz., *longilabris, 6-guttata, limbalis, purpurea, vulgaris, 12-guttata, repanda* and *hirticollis.*

I took *Lexis bicolor* Grote on this road. *Thecla titus* Fabr. was plentiful there, and *Debis portlandia* Fabr., *Phyciodes nycteis* Dbl. and *Pamphila paniscus* Fabr. were there to be seen.

Where the road passed through damp woods, a plant that attracted attention was the White Lettuce (*Nabalus altissimus* Hooker). Its stout stems rose like spires, from the wayside, tall as a man, and clothed with long leaves. This plant is a habitation and food-store for *Aulax nabali* Brodie. By slitting its stalks late in the season, the cells or cocoons of the species may be found. The imagos bite their ways of exit from their hibernacula in March.

Eastern Township Lanes.—There are lanes and by-ways in the Eastern Townships that more nearly resemble the green lanes of England than those I have spoken of, and interesting objects appear in them. Riding slowly through one such lane in the year 1867, I witnessed a sight which I had never seen before, and which I do not expect to see again, namely—a small flight of Passenger Pigeons (*Ectopistes migratorius*). There were seven or eight of them. They lit on some second growth maples a few yards in advance of me. They flapped their wings, and flirted their long tails, and preened their fine plumage, greatly to my delight.

Two other kinds of birds, especially worthy of notice that came under my observation in the Eastern Townships' lanes were the Great Grey Owl (*Scotiaspex nebulosa*) and the Barred Owl (*Strix varia*). The former whose big round head seemed too large for his body was greatly disturbed at my appearance. It rolled its head and fidgeted and blinked at me, but seemed to doubt the propriety of taking flight—it may have been recently mobbed by other birds. I left it unmolested to its wise cogitations.

The Barred Owl is a smaller bird—trim and alert.

Green lanes in those parts are frequented by the strangely elusive and tantalizing butterfly *Grapta j-album,* Boisd. & LeC. It is an insect of rich colouring and powerful wing. It rises before you, and you watch its direct and rapid flight, and note the spot where it alights. You hasten thither, and, drawing nigh, walk warily; but, look carefully as you may, you cannot perceive it. Suddenly it starts up, a few yards before you, and dashes away, and so on, till you abandon the pursuit. Its under side is of sober browns, like the fencing on which it usually alights. Gosse took this insect in the "Grove Lane" at Compton, P. Que. He named it the "Compton Tortoise." (See *Canadian Naturalist*, p. 247).

Along a by-road leading to the estate of the late Col. Calvin Hall in East Farnham a row of white elms had been planted. When I took notice of them, they were about fifteen feet high. It was in the Fall of the year, when, from some cause or other, the leaves of the elm curl over, and form rolls, on which the veins of their under sides are very conspicuous.

The trees I speak of had been visited by the Sphinx, *Ceratomia amyntor* Hubner, and I found a number of the larvæ of this insect feeding upon them, Strange to say, the larvæ took positions in which they closely resembled the rolled leaves—the ribbed side-lines of the caterpillars mimicking the veins of the leaves.

As the season advanced, the leaves of the elms changed from green to rusty brown, *and a corresponding change took place in the colour of the larvæ.*

But it is time I brought this paper to a close. It is one of reminiscences—a record of days gone by. I have written it in the hope that some into whose hands it may fall may be led by it to take a deeper interest in Nature Studies, to perceive a little more clearly some of the beauties in God's marvellous works, and to look up with deeper feelings of love and reverence to Him, for whose pleasure all these things are and were created.

THE CANADIAN ENTOMOLOGIST.

Owing to the greatly increased cost of printing of late years, it has been found necessary to raise the price of this magazine from $1.00 to $2.00 per annum, payable in advance. This includes postage to any part of the world. The change will go into effect with the beginning of the volume for 1914. Members of the Society, who are residents within the Dominion of Canada, will receive the magazine and annual report without charge if their subscriptions have been duly paid in advance.

THE IMMATURE STAGES OF THE TENTHREDINOIDEA.

BY ALEX. D. MACGILLIVRAY,
University of Illinois, Urbana, Illinois.

An interest in the study of the adults of the Tenthredinoidea has emphasized the necessity for some knowledge of their immature stages. This opportunity came the past summer through the offer of the Maine Agricultural Experiment Station to spend some time there collecting, breeding, and studying the larvæ of this group.

It is essential in all phylogenetic studies that the most generalized forms should be identified and the lines of specialization from these forms determined. This identification has been made for the adults and it was hoped that a study of their larvæ would throw some light on the validity of this classification.

The eggs are laid by the female within the tissue of the host plant. Where the larvæ are borers, the eggs are laid in holes made in the stems of shrubs or in the hard wood of the limbs or trunks of trees, with a thread-like ovipositor. Where the larvæ are leaf feeders, the eggs are placed in slits made by the female from the under surface, with an ovipositor consisting of two plate-like structures. The number of eggs placed in a single leaf varies greatly among the different species. In some only a single egg is inserted in a leaf; in others a large number, varying from three or four to thirty or forty. The recently laid eggs are difficult to locate, but they become swollen with age and then their location is easily determined.

The method of placing her eggs adopted by the female determines, to a certain extent, the feeding habits of the larvæ, as to whether they are solitary or gregarious feeders. Many species are solitary feeders throughout their entire life, a single larva on a leaf, part of a bush, or entire bush; the others are gregarious through the placing of many eggs in a single leaf or on closely adjacent leaves. Where many eggs are laid in a single leaf, the larvae developed from these eggs may be gregarious throughout their entire larval life or only for a time, when they are half grown they gradually disperse over all parts of the bush and are solitary in their habits for the remainder of their life.

The embryo requires from twelve days to three weeks to complete its development, the length of time varying considerably with the species. While the larva of each species passes through a definite number of instars, the number varies considerable among the different groups of species. The larvæ usually require about fifteen days for the completion of their growth.

There is great variation in their method of feeding. The great majority of the nematids are edge feeders; they cling to the edge of the leaf with the thoracic legs and when disturbed, bend their abdomen back onto the thorax in an S-shaped curve. Many emphytids and selandriids rest with their body stretched out flat on the underside of the leaf and eat holes in the leaf or feed from the edge; others, while feeding in a similar manner, rest with the body curled helix-like, with the anal prolegs forming the apex of the helix. This is also true of the cimbicids, *Cimbex* and *Trichiosoma*, which are edge feeders, resting curled on the upper surface of the leaf. A few nematids and selandriids also feed on the upper surface—some with the body helix-like, others extended. The phyllotomids, including the genus *Caliroa*, are leaf skeletonizers, feeding only on the parenchyma of the under surface. The very young larvæ of many of the groups are also skeletonizers during the first or second instars. The lophyrids, which feed only on conifers, begin at the the free end of the needle, clasping it between the prolegs, and eat towards its base until only a stub is left. The dolerids, or what I believe to belong to this group, for none of the American species have been bred, feed on sedges (*Carex*), feeding on the ends of the blades; the same is true of some of the grass-feeding nematids. The pamphiliids are either gregarious, when they fasten several leaves together into a nest with threads of silk, or solitary, when they roll the edge of the leaf and fasten it together with silk. These larvæ are the only ones, so far as I know, that use silk in this way. The fenusids and scolioneurids are leaf miners.

The body of the larva may be either black, white, green, spotted or banded. A large proportion of the species are white or green. In the green species the colour is due, in great part, to the colour of the blood and the food contained in the alimentary canal. They are also usually banded dorsally (the dorsal blood vessel) and marked on each side of this band and along the pleura by frosted lines (the dorsal pleural air tubes). A few species are marked on the dorsum and some on the pleura adjacent to the spiracles of certain segments by large yellow patches, which are due to lobes of adipose tissue or fat, which show through the cuticle. Many species are marked with pigmented colours; only one so far as observed has a median dorsal line, *Cimbex*. There is great variation in the arrangement and time of appearance of these colours. The entire body may be black or brown, or consist of longitudinal or tranverse rows of spots. Larvæ that are white when they emerge from the egg, may be entirely or almost black or chocolate brown in the latter part of their larval life, while those that are black when they emerge from the egg, may be almost entirely light-coloured when mature.

The larvæ, when fully fed, moult their skins and seek a place for pupation. This last moult may take place before they leave the host plant or after that time. There is no striking difference in the form and appearance of the body of most of the green or white larvæ, but those with prominent spines and black spots loosen the spines and spots and become opaque white; some that are white become distinctly spotted; some that are black or black-spotted become green, white, or glassy green, and still others that are opaque white through all the preceding stages have black spots about where the setæ were located. This diversity in form in the different stages makes it very difficult, until both stages have been recognized for a given species, to determine whether you are working with one or two species. This last larval stage has been designated as the ultimate stage by Dyar, who has done more toward elucidating the life-histories of the American species of sawflies than all the other workers together.

After the assumption of the ultimate stage some larvæ remain quiet for a time, resting upon the food plant; but the great majority leave the host plant and wander about in search of a place to prepare for pupation. The xyelids, pamphiliids, and blennocampids form cells in the ground, the emphytids and selandriids bore into rotten wood; also some nematids, which have the same habit, and so far as observed all these have a striking ultimate stage. The end of the tunnel is plugged with frass and the cavity left unlined with silk except a few species, which make a very thin cocoon. The great majority of the larvæ of the saw-flies form cocoons just beneath the surface of the ground or among the debris on its surface, as the lophyrids, dolerids, phyllotomids, tenthredinids, holocampids, acordulecerids, hylatomids, and most nematids. Some individuals of the first brood of the lophyrids attach their cocoons to the leaves of the host plants. The cocoons of the acordulecerids are white and compact; those of the hylatomids large, white and lace-like; while in all the others the cocoons are dense and black or brownish in colour.

In most insects, when the cocoon has been formed by the larva, it transforms almost immediately to a pupa—at least, within a week or two. A different condition is found among the larvæ of the saw-flies. Some of the species are, so far as known, always single brooded; the larvæ of such species emerge early in June, complete feeding in a few weeks, enter the ground or rotten wood, form their cocoons or cells, but live as larvæ within them until the following spring, when they transform to pupæ, and emerge as adults in May or June. A similar condition is found in those species that are apparently more than one-brooded, the ultimate larval stage is long and the pupal stage short. Writers frequently refer to the time when the larvæ form their cocoons or enter the ground as the beginning of the pupal stage. Such a designation is clearly incorrect.

Any statement as to the number of broods of any given species of saw-fly should be made with reservation. Many insects, probably a large majority, appear at a definite, stated time, usually not exceeding a period of two or three weeks and sometimes less. Such a condition does not exist in this group, for adults may appear over a period of four to eight weeks, so that it is possible to find on the same host plant and even on the same leaf, if the plant be one that has large leaves as a dock (*Rumex*) larvæ that are fully grown, others that have just emerged from the egg, and various sizes between these. Such a condition in June would probably mean that they were larvæ from the eggs of females produced from wintering larvæ, but in July the young larvæ may have been produced from the eggs of females that have matured the same season or from females produced from wintering larvæ that have been very slow in development. The field conditions would warrant considering such a species as single brooded, while the breeding experiments show conclusively that it is two brooded. The determination is further complicated by the fact that in those species producing summer females, not all the larvæ transform the same season, but, with the exception of a few species, the great majority of the females do not appear until the following spring. So that the second so-called brood is in reality only a partial brood; this is a well-established fact in the case of *Lygæonematus erichsonii* and *Pteronus ribesii*. In some species there is evidently a partial third brood.

The pupæ of saw-flies do not differ from those of other Hymenoptera. The antennæ, legs and wings are enclosed in separate cases and lie free on the breast of the insect.

The fact that saw-flies produce only partial second or third broods is also substantiated by the scarcity of adults during July and the following months. This time will vary somewhat, depending upon the altitude and latitude of the location. In most regions the adults are found in greatest abundance in May and June. They should be sought on the leaves of plants along the edge of forests, along fences and roadsides, and on the plants of marshy places.

ADAPTATION IN THE GALL MIDGES.

BY E. P. FELT, ALBANY, N.Y.

Adaptation is defined in the Century Dictionary as an "advantageous variation in animals or plants under changed conditions." This definition is sufficiently broad to include practically every modification resulting in a variation from what might be construed as the normal for a given family, tribe, genus, or even species. It is well known that every animal is exposed to numerous natural hazards during its life. Existing species must be equal to these perils or become extinct. It is convenient to group the forms of adaptation under three heads.

1. *Strength, aggressive and defensive.* We can all recall forms which appear well-nigh invincible because of superior physical development—muscular or defensive. The lion and rattlesnake represent two familiar and diverse types belonging in this category. One is remarkable for its superior muscular development and the other possesses a peculiarly efficient means of defense.

2. *Prolificacy.* There are numerous species with no particular physical efficiency. Some of these latter owe their existence largely to prolificacy. The common river shad, for example, may produce from 60,000 to 156,000 eggs, while a seventy-five pound cod may contain 9,100,000 ova. This extraordinary prolificacy is evidently a provision of nature to offset the numerous perils threatening the fry. Some of our plant-lice attain the same end by producing a number of generations annually. For example, the common hop plant-louse is capable of producing twelve generations in a season, the final progeny amounting to over ten sextillion. The increase in this latter species is by geometrical, not arithmetical, progression.

3. *Evasive adaptations.* There are hosts of species which escape extinction by the exhibition of more or less cunning in avoiding the many natural perils. This may be the result of modifications in the biology, peculiarities in habit, specializations in structure, or even cryptic or other resemblances. We have sometimes wondered if these factors, physical development or strength, prolificacy and evasive adaptations could be assigned sufficiently exact values that, if two were known, the third could be ascertained.

The gall midges exhibit a most interesting condition. The approximately 800 American species known probably represent only one-third to one-fifth of our fauna. Some 450 species have been reared from 183 plant genera representing 65 plant families. The largest of the gall midges is only about one-fourth of an inch in length, while the smallest measures scarely one-fiftieth of an inch. Local in habit, slow of flight, fragile in structure and far from attaining an extraordinary prolificacy in many instances, how do these multitudinous species maintain themselves? Physical development, either aggressive or defensive, is hardly worth mentioning.

Biological adaptations. There are good reasons for believing that gall midges are allied to the fungus gnats or Mycetophilidæ, many of which live as larvæ in decaying organic matter. The inner bark of various trees in incipient decay may contain hosts of Miastor and Oligarces larvæ. These maggots are remarkable because they exhibit a modification of parthenogenesis known as pædogenesis, an adaptation of inestimable value to species living under such conditions and dependent upon weakly organized adults for their establishment in favourable conditions. These midges produce only a few eggs and evidently possess very limited powers of flight. The larvæ are capable of penetrating only the weaker, semi-rotten tissues of bark and sapwood and are preyed upon by voracious maggots belonging to the genera Medeterus, Lonchea and Lestodiplosis. All too frequently the only evidence of Miastor infestation is the abundance of predaceous maggots which have devoured practically every inhabitant of a once populous colony. The ability to produce young in an indefinite series of generations by maggots advancing in unoccupied tissue is a great advantage in avoiding such enemies as those mentioned above. We also have in this series of pædogenetic generations an example of multiplication by geometrical progression such as obtains among our plant-lice.

Certain species like the Hessian-fly, sorghum midge, violet midge and rose midge depend for existence to a considerable extent upon the production of several generations annually; in other words, increase is by geometrical progression. The extraordinary efficiency of this form of adaptation is strikingly illustrated in plant lice as mentioned above. Such species, if able to subsist upon farm crops or other products valuable to man, are potentially serious pests. One generation annually appears to be the normal for many midges, and consequently the ability to produce more in a season must be considered a favourable adaptation to existing conditions.

Midge galls. Recalling the fact that the more ancient type of gall midges appears to be related to the fungus gnats or Mycetophilidæ and that they furthermore exhibit similar preferences in that the larvæ occur in organic matter in various stages of decay, one would expect to find a series of galls showing gradual modifications from this comparatively simple habitus to the more complex type of shelter so frequently observed in this group.

Bud galls. Possibly the simplest type of midge gall is to be seen in the irregular, loosely and various developed bud galls produced by some species of Dasyneura and its allies. The eggs appear to be simply dropped among the developing floral organs or leaves and the larvæ obtain their sustenance by absorbing nutriment from adjacent tissues. The weakening of the latter prevents normal development and, in some instances, at least, we have the conspicuous and rather characteristic rosette galls such as those of species of Rhopalomyia upon Solidago and of Rhabdophaga upon willow.

The growing point of a plant stem, whether it forms a leaf, bud or a flower, affords such ideal conditions for nourishment, that it is not surprising that certain genera should be restricted in large measure to such favourable habitus. This is particularly well marked in Asphondylia and certain of its allies, which not only confine themselves largely to bud galls, but have become so specialized that they are particularly adapted to the production of such deformities.

Leaf galls. The leaf gall, like the bud gall, usually begins as a development upon expanding or tender tissues. The simplest type is probably a marginal leaf roll, and this differs from certain of the bud galls simply by the fact that in the roll only a portion of the leaf is involved, while in the bud gall all of the several leaves may be distorted or have their development arrested. Vein folds are produced simply by the larvæ congregating or restricting their operations to this portion of the leaf rather than to the margin. They vary greatly in character and may be limited to the midvein or to the lateral veins, may be comparatively simple and composed of greatly hypertrophied tissue or ornamented with a conspicuous white pile or other development such as is found in that of *Cecidomyia niveipila* O.S. These leaf rolls and vein folds are usually produced by a number or small colony of larvæ.

Blister leaf galls and the more highly developed globular or conical galls are generally produced by single larvæ hatching from eggs deposited in or upon the buds before the leaves have unfolded. The peculiar blister galls on Solidago and Aster are multilocular, are easily recognized by the typical discoloration and thickening of the leaf, and are produced almost without exception by the genus Asteromyia. These galls represent a slightly more advanced condition than obtains in certain species which live between the upper and lower epidermis and either produce only a slight discoloration as in certain species of the genus Cincticornia, or else excavate a fairly well defined mine, such as that of *Lasioptera excavata* in Cratægus. The globular or lobulate galls of *Cincticornia globosa*

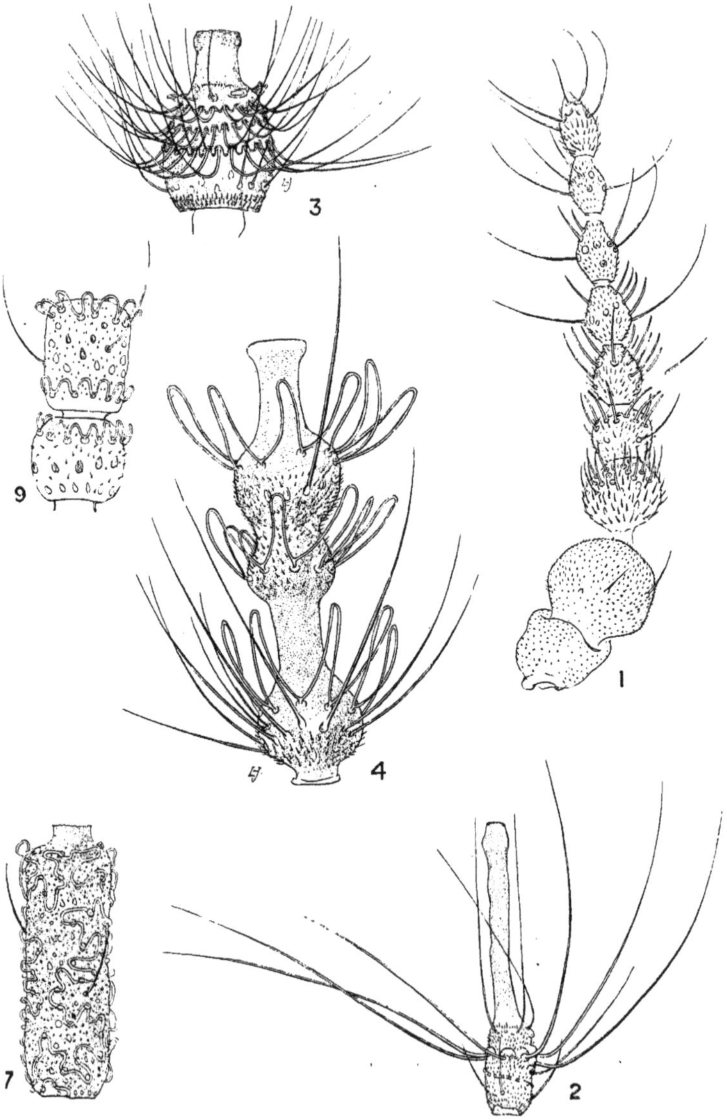

GALL MIDGE STRUCTURES.

and *C. pilulæ*, respectively, as well as the conical and globose enlargements of various species of Caryomyia upon hickory, must be considered as extreme types or modifications of the blister gall.

Stem galls. No part of the plant is exempt from infestation by the small representatives of this large family, be it seed, flower, leaf, stem or root. The stem gall is usually subcortical, and in those produced by midges, development generally begins while the tissues are still in a soft and plastic condition. They are usually polythalamous and are frequently irregular, more or less confluent swellings in the bark.

The medullary, stem or branch galls differ from the preceding in that the larvæ confine their operations to the interior of the affected tissues, frequently restricting themselves to the pith and producing rather characteristic deformities.

Root galls. There are only a few root galls known, probably because of the great difficulty in finding them. There appears to be no marked difference between these and the stem galls, aside from the point of location.

Recalling the fact that gall midge larvæ are small, without defensive armor or apparatus, with masticatory, or boring organs poorly developed or absent, it is obvious that this gall-making habit is one of the most important adaptations in the family. The gall midges have been able to maintain themselves in hosts and in many and varied forms by adaptations which have led to their seeking sustenance and shelter in places comparatively free from invasion by other insects. Not only have these small insects learned to prey upon numerous plants, but some have found it advantageous to wring sustenance from their associates. The species of Lestodiplosis, in particular, may be reared from a great variety of galls, and the larvæ have even been observed preying upon gall midge maggots, especially those of Miastor. Members of this family have also learned the value of other insects as food, and we now have records of a number of species preying upon scale insects, various plant-lice and red spiders.

Intimate relations exist between certain genera of gall-midges and families and species of plants. It is perhaps sufficient to note in this connection that the genus Cincticornia is practically confined to Quercus, Caryomyia to Carya, Rhopalomyia largely to

Solidago and Aster, and Rhabdophaga mostly to Salix. The mere statement of these facts indicates a correlation which has been discussed more fully by the writer elsewhere and need not be dwelt upon at the present tine.

Structural adaptations. It might be thought that this host of gall midges, with its general similarity of habit, would exhibit comparatively slight variations in structure. Modifications in anatomy almost invariably mean variations in habits ,and consequently they are worthy of note, even though they be but signs of unknown facts, in the same way that irregularities in the movement of a celestial body may mean the existence of an unknown planet. We wish for a few minutes to call attention to some of the more structural modifications.

Antennæ:—The antennæ in this family present a most extraordinary range in development, varying from comparatively insignificant and presumably relatively useless organs with but 8 segments in Tritozyga and Microcerata to the rather highly specialized organs with as many as 33 segments in *Lasioptera querciperda.* There is an equally great variation in the form of the antennal segment and their sensory organs. The cylindric antennal segment is undoubtedly the more generalized type, as it is the one found most frequently in the Mycetophilidæ. This may be modified to form a cylindrical larger base and a greatly produced distal stem, in some instances the latter attaining a length three times that of the basal enlargement. The basal portion of the antennal segment may be conical as in many Campylomyzariæ or globose as in Joannisia, while in the Itondidinariæ we have a dumb-bell-shaped structure, the basal distal enlargements being separated by a stem, with a similar constriction at the apex of the segment. This peculiar modification undoubtedly means greater efficiency in the sensory organs, since they are more widely separated, and is characteristic of the males in one large tribe.

The antennæ of the more primitive groups, such as the Campylomyzariæ and the Heteropezinæ, bear a number of peculiar sensory organs, the more remarkable of which are the so-called stemmed disks in the genus Monardia. These are probably olfactory in function.

The Itonididinæ, as limited by us, may be easily recognized

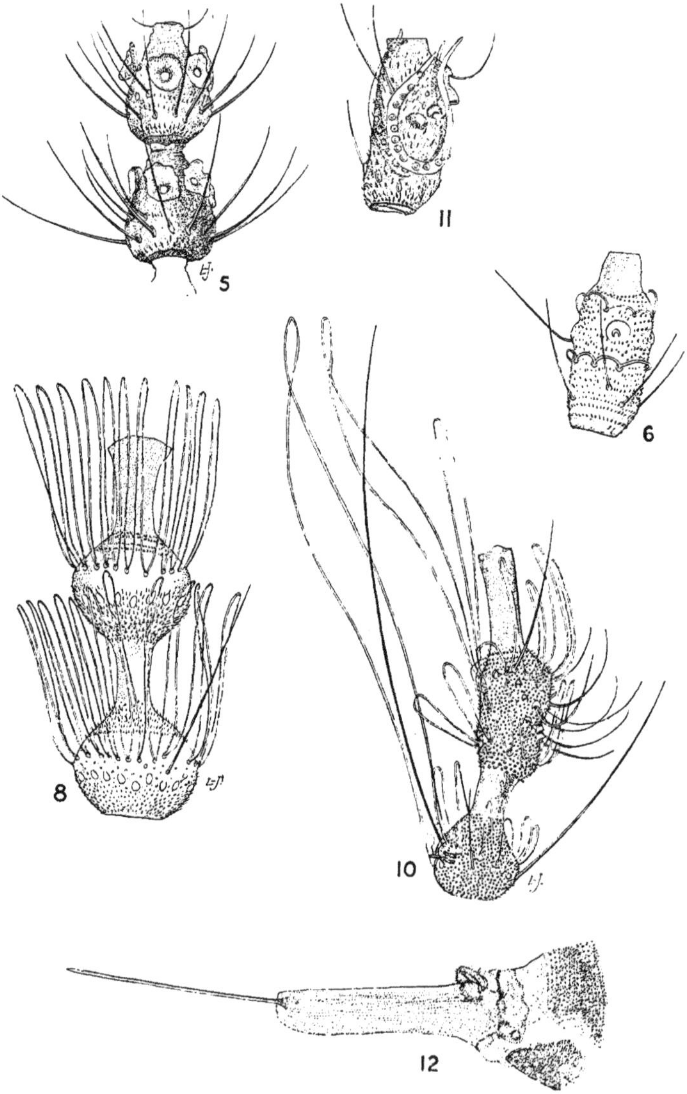

GALL MIDGE STRUCTURES.

by the presence of peculiar, colourless, thread-like, homogeneous, chitinous structures which we have named circumfili because they invariably run around the segment. They originate or arise from the interior of the segments, are presumably auditory in nature and are discussed by Europeans under the names of arched filaments (verticili arcuata and filets arques) and bow whorls (Bogenwirtel), since these common names aptly describe the structures as seen in the males. These organs in the females generally form a slender girdle near the base and distal portion of the enlargement on the flagellate antennal segments, the two being connected by one or two longitudinal threads. In the males the development may be very diverse. In the case of the male Asphondylia the circumfili consist of a more or less variable series of extremely tortuous, slightly elevated threads reaching from the base to the apex of the segment. In the Itonididinariæ the circumfili of the male are frequently prolonged into a series of bow-like loops girdling the basal and apical enlargements of the antennæ; one on each in the bifili and in the trifili with two on the distal enlargement. The loops of the circumfili or bow whorls may be simply conspicuous sinuosities as in Caryomyia or greatly prolonged on one side and having a length equal that of the entire segment as obtains in Aphidoletes and Bremia. A unique form of circumfili occurs in the genus Winnertzia. Here these structures greatly resemble minute, horseshoe-like appendages, one on each face of the segment, the produced free ends extending beyond the apex of the enlargement, while the supporting vertical threads give the appearance of a series of nails.

The peculiar circumfili, quite distinct in structure from auditory setæ, suggest our latest means of communication, the much vaunted "wireless," and present distinct analogies thereto. Both respond to impulses conveyed through air. It is possible the circumfili are "tuned" to vibrations unrecognizable with our finest instruments, and while the devices of men may convey signals several thousand miles, there is no reason for thinking that these unique antennal structures are relatively less efficient.

Palpi:—The normal number of palpal segments appears to be four, though these organs may become greatly reduced in any one of the tribes and in one genus, Oligarces, appear to be wanting.

The development of these organs affords a good systematic character and is correlated in certain instances at least ,with important modifications in habit.

Wings:—The organs of flight are of great value in taxonomic work and, in this family, present satisfactory characters for the delimitation of subfamilies and tribes. There is a cross-vein connecting subcosta and the third vein which occurs in a well-developed condition in the Lestremiinæ and the Epidosariæ, it being rudimentary or absent in the other groups except certain Heteropezinæ. The presence of the fourth vein is limited to the Lestremiinæ, in which subfamily it may be either forked or simple. The fifth vein also presents important modifications in that it may be simple, in which case there is frequently a sixth vein, or forked, in which latter instance the sixth has become partly fused with the fifth. Certain genera in the Heteropezinæ are remarkable because of the weak wings and greatly reduced venation.

Tarsi:—The normal number in segments is five, members of the Itonididinæ invariably having the first segment greatly reduced. Certain genera of the Heteropezinæ have four; others three, and in Oligarces there are but two tarsal segments. The claws may be simple, pectinate or dentate. They vary greatly in development and the same is true of the pulvilli.

There are other structures presenting equally significant modifications. This is particularly true of the generative organs and is especially well shown in the modified ovipositor which reaches an extreme development in the needle-shaped organ of Asphondylia, an instrument evidently designed for the piercing of thick bud tissues so that the egg may be deposited close to the growing point and in a place where conditions are most favourable for the development of the young.

It will be seen from the foregoing that the gall midges can not be counted as particularly strong or prolific forms, yet they have been able to maintain themselves largely by what we term evasive adaptations, which have resulted in their securing a very large degree of protection at the expense of the host plant. This summary is not intended to exhaust the subject, but is presented for the purpose of calling attention to a group exhibiting numerous unsolved and exceedingly interesting biological and morphological problems. There is perhaps no insect family better suited for the study of adaptation in numerous ways than the gall midges, a large group which up to recent years has been almost ignored by students.

EXPLANATION OF PLATES XII AND XIII.

1. Antenna of *Microcerata spinosa* male, showing 9 short segments. This organ is shorter in this species than the palp.

2. Sixth antennal segment of *Colpodia diervillæ* male. Note the greatly produced distal stem.

3. Fourth antennal segment of *Prionellus graminæ* male, showing the conical shape and the peculiar whorls of long setæ arising from distinct crenulate chitinous ridges.

4. Fifth antennal segment of *Karschomyia viburni* male, showing a binodose, almost trinodose structure of the segment and the peculiar circumfili or bow whorls.

5. Seventh and eigth antennal segments of *Monardia toxicodendri* female, showing the general shape of the segments and the characteristic stemmed disks.

6. Fifth antennal segment of a *Rhopalomyia* female, showing the generalized type of segment and the low circumfili commonly occuring in the female *Itonididinariæ*.

7. Sixth antennal segment of *Asphondylia monacha* male, showing the low, very tortuous character of the circumfili.

8. Fifth antennal segment of the pear midge, *Contarinia pyrivora* male, showing the binodose character of the segment and the two well-developed circumfili, the latter characteristic of the bifili.

9. Fifth antennal segment of *Caryomyia caryæ* male, showing the short though plainly sinuous circumfili, the three on a segment being characteristic of the trifili.

10. Fifth antennal segment of *Aphidoletes hamamelidis* male, showing its binodose character, circumfili, and particularly the greatly produced loops and setæ on the dorsal aspect.

11. Sixth antennal segment of *Winnertzia calciequina* female, showing the peculiar horseshoe-like circumfili attached to opposite faces of the subcylindric segment.

12. Extended ovipositor of the nun midge, *Asphondylia monacha* female, showing the basal pouch, the thick eversible basal portion of the ovipositor and the highly developed needle-like terminal part.

INSECT GALLS.

BY A. COSENS. TORONTO

(Abstract of lecture, illustrated by lantern slides.)

In the evolution of the study of galls there are different epochs, each merging gradually into the following. From early historical times these abnormal structures have excited attention. In the first instance, this was in all probability due to the fact that they presented phenomena unusual and out of the ordinary. At this earliest epoch witchcraft and like fanciful explanations were proposed to account for their origin. Gradually, as they were better understood and seen to involve a stimulus by a parasite and a response by a host, the examination of them became more scientific, and the hypotheses concering their causes, as a consequence, more valuable. The problem presented was recognized as one of great scientific interest, since it presented the unique feature of a foreign organism stimulating and controlling for its own benefit the growth of a host. Within the last few years it has been shown that a close relation exists between the structure of the bacterial crown gall and certain malignant animal tumors. Thus the second epoch with the subject of theoretical interest seems gradually to be passing into a third in which it will rank as one of the greatest practical importance.

The term "gall" is applied to any enlargement of plant cells, tissues, or organs induced by the stimulus of a parasitic organism as a regular incident in the life history of the parasite.

Galls are divided into two classes, according to the agent that produces the stimulus—namely, *Phytocecidia*, those owing their origin to parasitic plants and *Zoocecidia*, those produced by animal parasites. The former are caused by many different classes of plants, myxomycetes, bacteria, algæ and fungi. Even the flowering plants are represented among the gall producers, since the witches' brooms and the spherical stem swellings on the black spruce are due to the stimulus of the dwarf mistletoe *Arceuthobium pusillum*. The latter are incited by mites (*Acarina*) and by insects in several different orders as follows: *Hemiptera* (Families *Aphididæ* and *Psyllidæ*), *Diptera* (Families *Cecidomyidæ* and *Trypetidæ*), *Coleoptera* (Families *Buprestidæ*, *Cerambycidæ* and *Curculionidæ*), *Lepidoptera* (Families *Gelechiidæ*, *Sesiidæ*, *Tineidæ*),

Hymenoptera (Families *Cynipidæ* and *Tenthredinidæ*). From the Bryophytes to the Spermatophytes nearly all plants are subject to gall formations of this class.

The type of gall produced by the orders Acarina and Hemiptera is simple in structure, consisting usually of a more or less pronounced folding in the leaf of the host, often accompanied in the former by an abundant production of trichomes. The Coleoptera and Lepidoptera originate galls that show little differentiation of tissues and an entire lack of a well-defined nutritive layer. The Dipterous forms are in some cases as simple in structure as the Acarina pouch galls, but in others are as complicated as any of the highest types of galls. In the order Hymenoptera are two families, Cynipidæ and Tenthredinidæ, the members of which produce galls that are in marked contrast to each other. The sawfly galls are characterized by a very pronounced proliferation of tissue without differentiation into distinct layers except at the very earliest stages of gall production. The Cynipid galls, by way of contrast, have invariably three distinct zones of tissues, and only seldom is a fourth absent. These layers have the following relation to each other. Lining the larval chamber is the nutritive zone with cells oriented usually in a radial direction. Bounding this layer on the outside is situated the protective sheath, the zone that is absent in a few types. Outside of that again the parenchyma or tannin zone is differentiated, passing out to the epidermal layer.

One fundamental and far-reaching principle of gall production by insects is that the stimulus does not endow the protoplasm of the host with power to produce new types of organs, tissues, etc. Structures are in many cases originated that are not found on the same part of the normal host, but invariably their prototypes are present on another part of the plant or a nearly related species. The protoplasm is so stimulated that not only are dominant characteristics strengthened, but also in certain cases latent properties are called into activity, and thus the apparent new type of production appears in the host. This principle can be illustrated in the case of glands, trichomes and aeriferous tissue.

It may be stated, as an unvarying rule, that when glands are present in the normal tissue, they are always more plentiful or larger in the gall originating from that tissue. This is exemplified in the galls produced by *Eurosta solidaginis* Fitch, *Aulacidea nabali* Brodie and numerous other species.

But glands also occur in certain galls on parts of the host that are normally glandless. Thus they are plentiful in the gall produced by *Neolasioptera perfoliata* Felt on *Eupatorium perfoliatum* L, but are not found in the same location in the normal, but are, however, present at the base of the stem. In *E. urticæfolium* Reichard they likewise occur in the transitional region between stem and root, while in *E. purpureum* L they are present in the roots, petioles and flowering axes, as well as in the cortex and pith of the stem. In the case of gland production, it is clear that not only have active characteristics of the protoplasm in that direction been stimulated to an activity greater than the normal maximum, but nearly dormant properties have sometimes been aroused into action.

The trichomes exemplify the principle in a very similar manner to the glands. When the gall produces types different from the normal, these are invariably found on the reproductive axes of the host. The unicellular acicular hairs of *Eriophyes querci* Garman are totally unlike the stellate hairs of the leaf, but their exact counterparts are found on the reproductive axes of the host *Quercus macrocarpa* Michx. The much convoluted type of hair present in the Acarina dimple gall on the leaves of *Acer negundo* L. are found plentifully distributed over the reproductive axes, although the normal leaf hairs are straight. The trichome-producing activity of the protoplasm has thus been stimulated by the foreign organism to a degree reached in the normal only at the time of reproduction.

The production ofaeriferous tissue in certain Salicaceous galls also substantiate the principle in a very striking manner. These galls contain examples of a typical aeriferous tissue, comparable, indeed, to that found in such aquatics as Nymphæa, Potamogeton or Saururus; while in the corresponding parts of the host it does not occur. Indeed, this statement may be extended to include all the species of the host genus. A cross section of the gall originated on *Salix cordata* Mühl. by *Rhabdophaga triticoides* Walsh shows this tissue surrounding each larval cel[1]. It is present in the abnormal stem and extends entirely across the pith. While this tissue is present in the primary cortex of the normal stem of both Salix and Populus, and indicated in the pith of the latter, it is entirely absent from the pith in the corresponding part of the stem of Salix. It is abundant in such primitive regions of Salix as the reproductive axes, nodes and leaf traces. Thus the unexpected appearance of this tissue in the gall cited is readily explainable on the same grounds as in the case of the glands and trichomes—namely, the power to produce this tissue is latent in the protoplasm of the host, and it becomes sufficiently active to reinstate the tissue only when the gall-producing stimulus gives rise to unusual conditions.

A further illustration of this principle is shown in the production of cork in an aphid gall on the leaf of *Passiflora suberosa.* While this tissue is entirely absent from the unstimulated leaf, the stem produces it normally. Also, *Rhodites multispinosus* Gillette stimulates the usually unarmed stem of *Rosa blanda* Ait to the production of an exceedingly spiny gall. The production of spines, however, is a marked characteristic of the genus and a dormant activity has again been aroused.

Concerning the mode of application of the stimulus by the parasite, it may be stated that in none of the orders of insects except the Tenthredinidæ is there any evidence that indicates the beginning of gall formation before the hatching of the larva. In this family the source of the stimulus is in all probability the ovipositor of the insect, since it has been conclusively shown that the gall structure is well advanced while the larva is still within the egg membranes.

From observations on the galls of *Neuroterus læviusculus* and *Biorhiza aptera*, Adler concluded that cell division commenced only after the larva emerged from the egg. Weidel lately has shown that such is the case in the gall produced by *Neuroterus vesicator* Schlecht. It may, as a consequence, be accepted as proven that the source of the stimulus in the galls produced by the Cynipidæ is the larva of the producer.

As already published* the writer has proven by a series of experiments, that the larva of *Amphibolips confluens* Harris

*Transactions of the Canadian Institute, Volume IX., 1912.

secretes an enzyme capable of changing starch to sugar, and has also demonstrated the presence of salivary glands opening externally in *Philonix nigra* Gillette and *Amphibolips confluens* Harris. We may conclude, then, that at least one enzyme is present in the salivary secretion of the larvæ of the Cynipidæ and that this acts as a pre-digestive ferment on the contents of the nutritive zone. By its action, starch is changed into a readily soluble substance, and is consequently readily absorbed by the digestive tract of the larva. On account of this amylalytic ferment in the larval secretion the nutritive zone will become stored with an unusually large amount of available nourishment which can diffuse to all parts of the gall. The material thus prepared supplies nourishment for both the larva and the gall. The protoplasm of the latter is thus rendered unusually active since it receives an abnormal quantity of available food material in a limited area. The hypertrophy and cell proliferation and probably also the appearance of vestigial tissue, or other primary characters, are, in my opinion, the response of the protoplasm of the host to the additional food supply.

CHRYSOMELIANS OF ONTARIO.

BY F. J. A. MORRIS, PORT HOPE, ONT.

The title of my paper may be misleading to some of you, and I should like at the outset to explain my attitude. It is simply that of a nature-lover led(more or less by accident) to collect some of the insects observed by him about trees, flowers and leaves, while roaming about the countryside with what Wordsworth calls "a heart that watches and receives."

Of technical knowledge I have little or none to offer, and my interest in the economics of Entomology is subject to prolonged fits of catalepsy; indeed, I doubt if it has ever shaken off this blanket of suspended animation sufficiently to appear in really stark-naked wide-awakeness. The fact is, an amateur collector is drawn chiefly by the giddy pleasure of the eye; most of the time he goes about craving new specimens, probably those of large size and bright colour; he is an enthusiastic and irresponsible schoolboy, easily pleased, easily deceived. I knew a collector once in England —I should have called him then, in my ignorance, an old man—

he certainly had grey hairs in his head—a respectable married man and a regular church-goer, but alas, gentlemen, a lepidopterist in an advanced stage. He greatly coveted specimens of the swallow-tail butterfly. This is almost extinct in Great Britain, though still occasional in the fens of Cambridgeshire; the made-in-Germany kind that are exported from the continent to English dealers, ready set and pinned, did not satisfy him, and at last he was obliged to compromise matters by rearing some imported larvæ and liberating the imagoes in his back garden, in order to catch them again with his butterfly net. Now, what is that but childish make-believe? Unfortunately, most of us left this faculty of self-deception behind in the nursery and are incapable of hoodwinking ourselves so easily. Yet I confess to a greater liking for my specimens of Asparagus Beetle since I took them on wild plants that were not growing in a garden, and I never really loved the Potato Bug and the Squash Beetle till I caught them on my side of the farmer's fence, the one feeding on the Bittersweet and the others on the blossoms of the Goldenrod.

Moreover, were it not that such a consummation would jeopardize the existence of one of the world's lilies and eventually defeat its own end, I'd sooner see every stalk of asparagus in my own as well as in all my neighbours' gardens devoured by either species of Crioceris(both, perhaps) than invent or discover an insecticide that should prove fatal to so pretty a beetle.

It is, I admit, bearding the lion in his den to appear before an audience largely composed of economic entomologists and talk from so alien a point of view as this about Chrysomelidæ of all insects in the world; for in the whole order of Coleoptera this is probably the one family that most violently flaunts its existence before the public eye, by the invasion of the kitchen garden.

Is there such a thing as a beetle-fancier, I wonder? If there is, that's what I am, and to show you that I have the courage of my opinions, I invite you all as fellow-members of this Society, or as guests interested in insects, to join me in a cross-country tramp north of Port Hope on a fine day about the middle of July. We shall start from our honoured President's old home of Trinity College School, and in order to enjoy the day thoroughly I'll ask each of you for a little while to fancy yourselves back at school once more—throw away the burden of years and the cares of a responsible position; drop the handle from your name, college degree and the rest of it—forget it all. What you want is a little zest for the day's captures and (as we shall be out for the day) a sandwich or two in your pocket against the noontide hour.

We have green lanes and fields right at our door, but as our road will in any case be a long one, we shall condescend to get a lift by boarding the morning train for Peterborough and riding as far as Quay's Crossing, five miles up the track. I am giving myself as well as you a treat, for this is a favourite walk, and I may not have many more opportunities of taking it. But for all the hundreds of times that I have trodden these paths and roamed the woods and fields, I do not think I have ever come out entirely or even primarily as a Coleopterist. The countryside all means far more than beetles to me, so I must ask you to pardon the digressions, which may be many. I hope they will not weary you.

During the few minutes of our train ride, let us briefly review the family of Chrysomelians. There are no less than 18,000 species of these leaf-eating beetles known in the world; the vast majority are tropical: North America can claim only about 1-25th of this number and Ontario about 1-70th. But even Ontario's share, nearly 300 species, makes a long list, the mere detailing of which would take some pages, while anything like systematic treatment, with specific or even generic description would require a volume; it would, besides, be more than tedious—it would be deadly dull. Henshaw's check-list makes about as inspiring reading as the list inspired of Walt Whitman's poems, and for the same reason—it's a mere catalogue. There are purple patches, I grant you, and not a few in LeConte and Horn or in Blatchley as there are in Professor Wickham's papers on the Chrysomelidæ of Ontario and Quebec (contained in volumes 28 and 29 of the Canadian Entomologist, 1896–7). What are these purple patches of interest?—these oases in a desert of dry description? At first sight they seem of varying nature; sometimes a brilliant generalization or an ingenious analogy; at others a quaint observation of habits or a personal experience. But they all resolve themselves, at last, into the personality of the writer. It is the personal element that lends interest to a book or a paper on a technical subject; it is just this that makes the old-fashioned Lexicon of Samuel Johnston or Noah Webster an enthralling romance beside a modern dry-as-dust scientific work-of-a-syndicate like the Standard Dictionary.

It would obviously be impossible to write an interesting account of 264 species of beetles or even of 96 genera, but for the convenience of systematic treatment, this enormous mass of individuals, countless as the sands of the sea, has been marshalled, like the children of Israel, into 12 tribes, and every one of these tribes has several representatives in Ontario. In our day's tramp we shall run across at least one representative of each tribe, from Reuben the first born to little Benjamin, our ruler; in plain terms, from Donacia, the reed beetle, cousin german to the more ancient Cerambycidæ, to Chelymorpha and Coptocycla the little Tortoise. Of these twelve tribes, the most numerous in boreal America, as well as the most important, are the five numbered VI–X. These comprise more than 450 species out of a total (to the family) of less than 600 and more than 70 genera in a total of about 100; i.e., ¾ of the entire genera and species belong to five consecutive tribes out of the twelve. Of these five tribes, again, two are supreme, the 9th and 10th included by LeConte and Horn in the single tribe of Galerucini or Helmet-grub beetles, with a total of more than 200 species and over 40 genera; i.e., nearly half the family.

In the tropics, where vegetation is most luxuriant, these beetles play an important part in checking the too-lavish growth; but in the Temperate Zone, where civilized man has brought the earth under cultivation, these twelve tribes, the chosen people of my paper, are nothing better than one of the plagues of Egypt, a most distinctive pest, and man's best wits are taxed to prevent an annual loss of many million dollars.

The Chrysomelians represent a later development than the Cerambycidae or wood-borers, and their adaptation to succulent herbage and the deciduous foliage of flowering plants pari-passu, with changes in the vegetable kingdom from sporophytes and gymnosperms, presents in its way as wonderful an illustration of adaptive development as more specific examples like symbiosis which has isolated the Yucca and its moth from all creation, till each depends on the other for its very existence and on the other only.

The larvæ of the Chrysomelians are in general soft and helpless; feeding, as they do, in the open and gregariously, they are easily destroyed; but several factors contribute to their notable success in the struggle for existence: their immense numbers, the rapidity of their growth (which enables them to produce more than one brood in a season), and the ability of the mature insect, in most cases, to hibernate.

A few of them retain traces of an earlier condition in being stem-borers, or in tapping the roots of plants, as the Donacias; and it may be a sort of atavism that impels Cryptocephalus and Glyptoscelis to resort to the needles and bark of white pine.

Our train is now slowing down to let us off at Quay's Crossing, and for the rest of the day we'll have to put our best foot forward, for it is going to be Shank's mare with us. First we go a quarter of a mile east to Mose Robinson's mill-pond and Pine Grove School-house. Just after crossing the stream here we turn south down a grassy lane, flanked on the west by an old snake fence and on the east by a still more ancient stump-fence; the snake fence appears to spring from a bed of fern-oak and brittle bladder. The lane is filled with sweet-briar and the stump fence festooned with wild grape-vine; a fortnight ago the briar and the grape-vine were both in bloom and the lane was redolent with two of the most delicious scents on earth. A little way on, at the foot of a sandy slope, we cross a tiny brook of lovely, cool spring water, its surface mantled with water-cress. Here in the early season, as early as April, are nearly always to be found about the grass-blades, some specimens of the Donacia. This is our representative of Tribe I, a small tribe generically, consisting of two members only; the genus Hæmonia has only one species, but the Donacia (Reed-beetle, as the Greek name implies) has more than 20 species in North America. The kind I have found here is much like a Longicorn, and in early days was mistaken by me for a member of that family; it differs from the Chrysomelians in being long and narrow in shape, usually yellowish brown in colour and of a metallic lustre. The larva feeds about the roots and bases of aquatic plants and has acquired the power of living under water by tapping the air-vessels of its food-plant. It has actually a small process on the body which it uses as a probe. When about to pupate, it encloses itself in an air-tight cocoon which is fastened to the roots or stems of the food-plant beneath the surface. The beetle is covered on the under side with a pubesence that acts as a perfect aquifuge shedding the water like oilskin. The species found here in the cool days of April is more or less cylindrical (convex on the upper side) and quite sluggish in habit, but the Donacia of the dog-days in the height of summer is a very different creature. I well remember during my first visit to the Algonquin Park how one day I went over with the late Dr. Brodie to the little land-locked Cranberry Lake in the heart of the hardwood forests. It was a glaring hot day, with the sun at its height and perfectly calm. We rowed a boat down to the Cranberry marsh at the foot of the lake, where all sorts of botanical treasures awaited us. On the way we passed through a patch of water-lilies and flushed a covey of Donacias; there must have been hundreds, leaping and flying from the lily-pads, striking the sides of the boat, sometimes in the water, occasionally on our clothes, darting and glittering in the sun like sparks from the molten surface of the cauldron of heat formed by this woodland lake at high noon beneath an August sun. The activity of movement and extraordinary vitality in the sun's heat are not common among the Chrysomelians, but they are among some of the Longicorns, with which the Donacias have a close affinity. Lords, for the nonce, of all three elements, earth, air and water, they moved easily about all three, perfectly at home and at their ease. On cooler days, or when the breeze blows, they love to sit on their beloved lily-pads, like miniature batrachians, their thorax and head partly raised and their antennæ thrust forward alertly, something like the asparagus beetle when it scents danger.

We shall now stroll south about a mile, along the edge of a wood we call the North Wood, a wood sacred by many memories, rich in flowers, the home of some rare orchids, in and about which I have found more than 20 species of ferns and a wide range of warblers and other birds at the spring migration; it is, besides, the scene of many of my best captures among the Coleoptera. Ten minutes'walk brings us to where the wood narrows close to a division fence, running west across meadow-lands to the railway. Just here stands, on the edge of the wood, a hawthorn, whose blossom, for some reason or other, has proved a beetle-trap or bait for an extraordinary number of species. It was on this blossom that I first captured specimens of the Orsodacna, our representative of Tribe II, and on the top-rail of the snake-fence, beside it I took one of the few specimens I have ever seen of Syneta, another of the four genera contained in this tribe. The Orsodacna (or Bud-gnawer) is said by Blatchley to feed on willow-blossoms, and this season, as early as April, I was on the look-out for it about clumps of willows in bloom, but the only thing new to me that I observed was a small moth dancing up and down in lively zigzag flight over the willow bushes; it was almost as small as a clothes moth, blackish with a cream or white bar near the apex of the wing. From its extremely long hair-like antennæ I should judge it a species of Adela. We have but one species of Orsodacna, and I have always found it in great numbers, once here and once in Lakefield. The specific name is *atra* (black), but it is very variable, and specimens sent by me to Guelph, taken all at the same time off this hawthorn bush some years ago, were returned labelled under no less than four varietal forms. The pigmentation of the elytra, normally black, becomes less heavy and the wing-covers show light brown with darker disks and markings. In some of its forms the blend of colours is very pretty; the beetle is narrow-oblong and the texture of its upper surface is of an oily smoothness.

Let us cross the meadow west to the railway track; near the fence that extends from the hawthorn tree to the railway, on the south side are some sand-drifts where I have captured no less than six species of Tiger-beetle at various times in the season. The meadow to the north is less sandy and springs ooze out from its surface and meander over the grassy slopes. Here in September the meadow is white and fragrant with *Spiranthes cernua*, the nodding Ladies' Tresses, one of our autumn orchids. Just where we strike the railway is an immense patch of that rather rare plant, the Grass of Parnassus, whose green-veined creamy white blossoms in August and September make as brave a show as the anemone in June and July. It is a sure sign of springs in the soil and further south there are traces of an old sphagnum moss swamp; though it is years since the railway hacked away the trees and shrubs, marsh pyrola and the Showy Ladies' Slipper annually rear their upright stalks and unfold their blossoms for gauze-winged visitors to gather nectar from beneath the July sun.

Here, along the right of way, grows wild Asparagus, and on it you will find at least one species of the Asparagus beetle, which we shall take to stand for Tribe III. The first specimens I ever saw of this beetle were in a Kentish garden; they belonged to the species commonly known as the striped asparagus beetle, and at first I did not recognize the insect—all I had by way of guide was an old book of Stevens with coloured illustrations that were several times magnified. The picture showed a gorgeous insect, in rich dark green and cream hues, which to my excited imagination must be nearly as large as a June Bug. I found ,however, to do the old naturalist justice, that though in the dead insect the sutural stripe, the basal marks and the cross-bar on the elytra appear black on a ground colour of opaque straw-yellow—in life these colours are a rich, vivid, dark green, on a ground colour of translucent cream, extremely beautiful when scanned with a lens. The 12-spotted species, which seems the commoner in Ontario and is apparently more hardy, I first found in the late Dr. Brodie's back garden in Toronto. Until five or six years ago neither species had made its way to Port Hope, but the spotted one appeared in several gardens then, followed a season or two later by the striped, and two seasons ago I first found the *Crioceris duodecimpunctata* on wild asparagus. There is only one other genus in this tribe—the Lema, of which there are no less than 16 species in North America, only a few occur in Ontario, and I have only found one—*Lema trilineata*, a beetle which sometimes shares with one of the Blister beetles the title of "the old-fashioned Potato Beetle"; it feeds on various plants of the Potato family, and I have found it in some abundance on the Physalus or Ground Cherry, while searching vainly for specimens of *Coptocycla clavata*, the Rough Tortoise Beetle. Before we leave the asparagus and return to our little brook a mile north, I may mention that it was on some garden Asparagus at Lakefield that I found my reward for a day's umpiring at a cricket match, in the shape of a beetle called *Anomoea laticlavia*. This is the only species in the IVth Tribe known to me; for though North America has seven genera in the tribe and over 20 species, there are but four genera represented in Canada, each by a single species.

It is, for a Chrysomelian, a decidedly large insect, stout and of striking appearance, light-brown in colour, with a black sutural stripe, which is slightly thickened from about midway down the elytra to near the apex. I have never since seen it on asparagus, but more than once I have taken it feeding in large numbers on willow-shrubs about the right of way, a few miles north of our present halting-place on the Peterborough railway. Last year I discovered it very abundant, almost a pest, on wild grape-vines near Sackville's Swamp, on the South Shore of Rice Lake, between Bewdley and Gore's Landing.

We now return to the little brook where our first Donacias were captured. Just over the fence, on our right hand, is a small pine wood, out of which, indeed, it is that our little brook emerges. This wood is a great place for early morels; it has also yielded some very interesting species of Longicorn and Clerid on the occasional windfall of white pine. Towards the north-east side of it, where our way lies, grows a patch of raspberry canes, where I captured once in full flight, with my hand, that most elusive of dodgers, the Oberea. On the leaves of the raspberry once I saw some tiny dark conical galls, as I supposed, and one of these I tried to tear from the leaf; to my surprise, when I had partly wrenched it aside, it distinctly moved and glued itself back on the leaf. This was something new for a gall, and I pulled it away from its fastenings to find that it contained a live larva, whose legs were kicking frantically to get back to the leaf. You have often seen a refractory man-child plucked suddenly up by the nurse from the place where it was playing? Well, that's how this caterpillar kicked. It was Chlamys, one of two genera that represent the Vth tribe. These insects construct a case out of their own excretions, and under cover of this tiny, steeple-crowned brownie's cap of a case they move about and feed securely; when the time comes to pupate, they simply close the door at which they have grazed and behold a ready-made cocoon. The insect itself is dark brownish black, and covered with little warty excrescences; when alarmed it closes its legs and falls to the ground, where it escapes notice entirely or is passed over by warblers and other insectivorous birds as a pebble or a pellet of dirt; one more instance of protective mimicry preserved in this creature through all stages of its existence.

(To be continued.)

APPLIED ENTOMOLOGY FOR THE FARMER.

BY F. M. WEBSTER, WASHINGTON, D.C.

Of all husbandmen, the true farmer, the grower of grains and forage crops for sale or consumption on his premises, has been the last to profit by the applied science of entomology. He in the past has indeed supposed himself as helpless against the inroads of insects upon his crops as the Indian squaw whose only hope of saving her patch of Indian corn was in the effect of charms and incantations in warding off attacks of wireworms, cutworms and perhaps other similar pests.

The beginnings in applied entomology consisted in dusting garden vegetables with soot, lime, ashes, and, somewhat later, with powdered hellebore. But to the farmer these precautions meant practically nothing. Though his farm might not be a large one, the area was usually too wide to render these measures practicable, even if they proved effective in a small way. It is true that the trapping of cutworms under compact bunches of elder sprouts, milkweed, clover and mullen, "placed in every fifth row between every sixth hill," was known as early as 1838, but these constituted only a trap or baits, the worms found under the traps being killed by some sharp instrument. This measure, however, seems to have never become popular.

The spread of the so-called Colorado potato beetle over the country from the west eastward brought the use of the Paris green and London purple as insecticides to the front, but, again, this did not help in the least the troubles of the ordinary farmer.

The work of Riley, Packard and Thomas, on the western migratory locust, was the first important effort made to aid the farmer in devising practical measures of fighting destructive insects over large areas.

The spread of the cabbage butterfly from the east to the westward brought into use as an insecticide the powdered blossoms of Pyrethrum, but the farmer does not raise cabbage as either a grain or a forage crop.

Studies of the cotton-worm, by Riley and others, brought Paris green again into use and developed that useful insecticide, kerosene emulsion, but the farmer cannot make use of these in his cultivation of wheat, oats, corn, rye or barley; neither can he apply them to insect pests on his broad acres of forage crops.

In the same way, fighting the codling moth and San José scale have developed the use of arsenical sprays, as well as those of lime and sulphur, crude petroleum and other sprays and washes. But none of these are of the slightest use to the farmer in his fields, no matter how valuable they may have been to the fruit grower.

The farmer has, therefore, largely occupied the position of a skeptical spectator, who, while seeing clearly the benefits derived from applied entomology by his brother husbandman, the fruit grower, the gardener and even the cotton planter, was seemingly himself debarred from sharing in these benefits, because of the measures being inapplicable to his crops, and, even if this were not the case, his wide areas would render their use impracticable.

Besides all this, the farmer has, himself, held somewhat the position of a critical onlooker as the result of other causes.

Before the advent of experiment stations, and even for some time afterward, letters addressed to the members of university faculties, complaining of the ravages of insects and asking relief, brought the actual farmer little consolation. The replies he received to his appeals for relief were usually couched in terms to which he was unused and much of the text of these replies in a language that he did not understand. Moreover, the replies were usually penned by men who had little or no practical knowledge of agriculture, and thus there grew up between the two not only a continually widening breach, but in many cases an absolutely intolerant feeling on the part of each for the other.

This was approximately the relative positions of the man from the campus and the man from the farm, at the time of the establishing of the Experiment Stations, though there were. of course, some brilliant exceptions. Besides this, many, probably the majority, of those who were afterwards to make the Experiment Stations a success, were yet to be trained and given their practical experience in combining the practice and science of agriculture; and it may be stated that the science of entomology, for reasons previously given, has impressed the farmer the least favourably. Farmers had always looked upon insect depredations precisely as they did other natural phenomena like drouth, storms and floods, fully convinced by ages of experience that nothing could be done to prevent them, and, therefore, they must be endured to the end. Entomological literature, however elementary and popular, they simply would not read. This was, generally speaking, the situation at the time when I was just beginning my entomological work among the farmers of Illinois.

We will now step over the intervening 25 years and look at the situation as it is to-day. It will be an obscure section of the country, indeed, if where there are serious insect depredations going on, we at the Department of Agriculture do not promptly receive a report of it through one or the other of several sources. These reports are received through letters addressed direct to either the Department or Bureau, and are coming each year with increasing frequency, through experiment stations, the press, and last, though not least, through members of Congress.

Perhaps nothing better illustrates the changed condition and rapid growth of agriculture as a science than the immense strides made by economic entomology as applied over and throughout the broad acres of the ordinary farmer. At the present time, instead of receiving a stereotyped reply to his applications for relief, when he applies as an individual, or for his neighbourhood, to the Department of Agriculture, either directly, or, as is becoming every day more frequent, through his representative in Congress, he is very often surprised when, within two or three days after the receipt of his complaint, there appears in his neighbourhood a young man who, in most cases, has grown up a farmer's son on the farm, and, besides this, has had a thorough university training, and, perhaps, is further equipped by having been engaged in the investigation of insects over a wide range of country, including, perhaps, no small number of the United States. Instead of receiving a letter which to him might, perhaps, so far as practical aid is concerned, have been written in a foreign language, he finds that his visitor can go about over his and his neighbours' farms with him and with a clear understanding of the crops cultivated can point out the work of insects and tell then in what manner they might have avoided these injuries and saved their money. He will tell him of things that, though he may have spent a life time in farming, neither the farmer nor his neighbours have ever yet been able to observe. His caller not only fits into their farm life and speaks to him in the language of the farmer, but is able to explain, in a perfectly natural and intelligible way, much of what to him has heretofore been a mystery. The young man points out to him wherein their farm methods have, in many cases, been primarily responsible for their previously sustained losses by insect attack. The farmer is now in a position to read entomological literature intelligently and with pleasure to himself. It does not greatly matter of what State he may be a resident, if his locality is not too inaccessible and the matter is of more than local importance any of the men located at the fifteen different field stations can be wired instructions that will send them to his relief. In this way entomology as applied to the broad acres of the farm has within the last twenty-five years become completely revolutionized. This means much to the growers of grains and forage crops and to the stock breeder. Moreover, it means almost equally as much to the banker, the manufacturer, and the merchant, all of whom are coming to recognize the fact. It has been my own practice to take up only such investigations as involve several States, leaving local matters to State institutions, where such are equipped for the work, and, when called upon to deal with such, I have urged that the State be at least given an opportunity to help itself, while we stood ready to reinforce their efforts if need be. This course has been followed especially with reference to local outbreaks of grasshoppers. Where investigations can be carried out in any State, as a part of an extended plan of work, notably that of wheat sowing in fall to evade the fall attack of Hessian fly, we have carried out such experiments with the co-operation of farmers at whatever points seemed most desirable for obtaining results which would benefit the greatest number of farmers. In many cases these sowings have been also made in co-operation with State institutions. The alfalfa weevil investigations have been carried on in co-operation with the State agricultural college and station at their request.

Besides the field laboratories there are being carried on field experiments, out on the farms, under precisely the same conditions as those with which the farmer has himself to meet. These experiments are conducted in such a way that farmers can see just what is done, how it is done, as well as the object of the experiment itself. They can also see what results are obtained, and what we have done, under their conditions, they, under like conditions, can do for themselves; and the proof thereof is right before their eyes in their own fields. We find that these object lessons and personal contact are primarily worth vastly more than whole volumes of literature, and, gradually the farmer is coming to learn that there is help for him as well as for the horticulturist, in combatting insect pests, even though his acreage may be many times theirs and his crops radically different in nature.

ILLUSTRATED LECTURE ON "ANTS" (ABSTRACT).

BY PROF. W. M. WHEELER.
Bussey Institution, Forest Hills, Mass.

By way of preface the lecturer made some general statements in regard to the 5,000 known species and sub-species of ants, described the development and metamorphosis of the individual ant, the various castes, or polymorphic phases represented by each species and the function of each of these castes in the life of the colony. Then the general behavior of ants was treated from the standpoint of the three basic biological activities, namely reproduction, nutrition and protection.

Special emphasis was placed on the behavior of the female, or queen ant and her methods of establishing the colony in contrast with the behavior of the queen honey-bee and with the male ant, which takes no part in the activities of the colony as such, but functions only as a fecundating agency during the nuptial flight. The queen ant was shown to possess all the instincts of the worker forms in addition to some of her own and thus to represent the most complete embodiment or epitome of the species. This statement requires qualification only in the case of certain parasitic and slave-making species, in which the queen is degenerate like the queen honey-bee and no longer able to establish a colony and bring up the first brood of her offspring without the aid of workers either of her own or of an alien species.

The peculiar structure of the ant's alimentary track was described in some detail, with its "social" and "individual" stomachs, which enable the insects not only to store their liquid food in the most economical manner but also to distribute it equally among the various members of the colony both larval and adult. For the purpose of illustrating this portion of the lecture more fully, the various adaptions of ants to living in very dry regions, such as deserts, were examined, and it was shown that these insects have evolved four very different methods of circumventing the difficulties inseparable from life under conditions that imply a great scarcity of their natural insect food. A certain number of species have exaggerated their primitive predatory instincts and have become rapacious hunters (e.g. the species of *Cataglyphis* in the North African deserts). Others have taken to storing quantities of liquid food in the crops, or social stomachs of certain workers of the colony for the purpose of tiding over the long droughts (e.g. the honey ants of the South-western States and Australia belonging to the genera *Myrmecocystus*, *Melophorus*, *Camponotus*, *Leptomyrmex*, etc.). Other species have become agricultural or harvesting ants (the species of *Messor*, *Pogonomyrmex*, many species of *Meranoplus*, *Pheidole*, *Solenopsis*, etc.), and have therefore become addicted to a vegetable diet. These forms store the seeds of various desert plants in their nests. Lastly, a group of American ants, comprising the species of *Atta* and allied genera, has learned to grow fungi for food on pieces of leaves, caterpillar excrement or other vegetable detritus. Although this habit seems to have originated in the moist woods of South and Central America, several of the species which acquired it were able by its means to invade the deserts of the Mexican plateau and of the South-western States and thus to remain independent of the precarious supply of insect food peculiar to those regions. This represents the most specialized stage of ant dietetics.

The protective instincts of ants, apart from their stinging and biting proclivities, attain their most striking expression in the construction of the nests. The various types of these structures were briefly considered: the small crater nests in the soil, the nests under stones and in wood, the larger mound nests, which are characterized by a superstructure of accumulated vegetable detritus which is used as an incubator for the larvæ and pupæ, the carton nests constructed in trees by various tropical ants of the genera *Crematogaster*, *Azteca*, *Dolichoderus*, and *Polyrhachis*, and the extraordinary silken nests of *Oecophylla smaragdina* and some species of *Polyrhachis* and *Camponotus*, which are woven by the ants using their spinning larvæ as shuttles.

THE EXCURSION TO GRIMSBY.

In accordance with the prearranged programme, the visiting entomologists were all invited to participate in an excursion to the town of Grimsby, which is situated near the centre of the chief peach district of the Province. About thirty-five availed themselves of the opportunity. It had been expected that the party would arrive in Grimsby soon after noon, but owing to a very severe thunderstorm the previous evening, the electric cars were running irregularly and it was not until about 2 p.m. that we arrived there. Lunch was at once served. After lunch there were two or three very interesting, short addresses of appreciation of the pleasant trip and of the entertainment. Immediately afterwards those who were enthusiastic collectors set out in a body to search the flower-clad side of the so-called mountain for their favorite kinds of insects. The remainder, under the guidance of Mr. Caesar, visited the neighboring orchards, especially the peach orchards. Fortunately the peaches were just ready to pick, and the healthy trees, with their luxuriant green foliage and the branches bending down almost to the breaking point with the weight of golden fruit, aroused the enthusiasm and admiration of those who had never before seen an Ontario peach orchard. About two hours were spent driving through or past peach and other orchards, noting at the same time a few of the special insect pests of the locality, and then all returned to the hotel to meet the party of collectors who reported a considerable number of interesting captures. Farewells were given and the convention was at an end.

L. C.

ENTOMOLOGICAL SOCIETY OF ONTARIO.

OFFICERS FOR 1913-1914

PRESIDENT.—Dr. C. Gordon Hewitt, Dominion Entomologist, Central Experimental Farm, Ottawa.

VICE-PRESIDENT.—Mr. A. F. Winn, Westmount, Que.

SECRETARY-TREASURER.—Mr. A. W. Baker, Demonstrator in Entomology, O. A. College, Guelph.

CURATOR.—Mr. G. J. Spencer, Assistant in Entomology, O. A. College, Guelph.

LIBRARIAN.—Prof. C. J. S. Bethune, O. A. College, Guelph.

DIRECTORS.—Division No. 1, Mr. Arthur Gibson, Division of Entomology, Central Experimental Farm, Ottawa; Division No. 2, Mr. C. E. Grant, Orillia; Division No. 3, Mr. A. Cosens, Parkdale Collegiate Institute, Toronto; Division No. 4, Mr. C. W. Nash, East Toronto; Division No. 5, Mr. F. J. A. Morris, Peterborough; Division No. 6, Mr. R. S. Hamilton, Collegiate Institute, Galt; Division No. 7, Mr. W. A. Ross, Jordan Harbour.

DELEGATE TO THE ROYAL SOCIETY.—Mr. H. H. Lyman, Montreal.

AUDITORS.—Messrs. J. E. Howitt and L. Caesar, O. A. College, Guelph.

REV. DR. C. J. S. BETHUNE, who has been suffering for some years from defective eyesight, recently underwent an operation, which has completely restored the sight of his right eye. We offer him our heartiest congratulations, and feel sure that the news of his recovery will be received with pleasure by all readers of THE CANADIAN ENTOMOLOGIST.

Mailed November 18th, 1913.

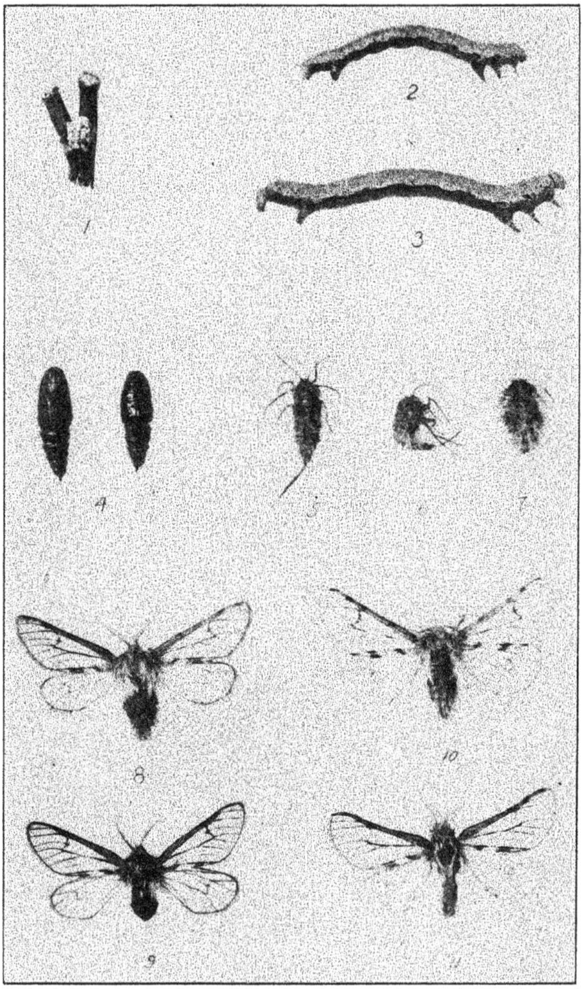

APOCHEIMA RACHELAE, HULST.

The Canadian Entomologist.

VOL. XLV. LONDON, DECEMBER, 1913 No. 12

THE PREPARATORY STAGES OF *APOCHEIMA RACHELÆ* HULST.*

BY ARTHUR GIBSON, DIVISION OF ENTOMOLOGY, OTTAWA.

A cluster of about 200 eggs of this rare moth was received from Mr. Norman Criddle in 1903. Oviposition took place at Aweme, Man., during the nights of April 20, 21 and 22, and the larvæ hatched at Ottawa on May 9 and 10.

To the notes made in 1903, further information has been added from a recent study of inflated larvæ.

Egg.—The eggs were laid in an agglutinated mass and when received (May 1) were yellowish in colour. On May 6-8 they turned pinkish, then black, and before hatching were beautifully iridescent. In shape, the egg is oval; height 0.6 mm.; breadth 0.3 mm.; the whole surface roughened.

The eggs were secured by confining the female moth in a collecting net-bag. Since, however, Mr. Criddle has found eggs under natural conditions, viz., on a twig or poplar (see figure 1, plate XIV) these had been deposited in a tightly compressed mass. On another occasion, a female which the same observer had in confinement laid the eggs in a cluster on her own body.

Larval Stage I.—Length 2.5 mm. Head 0.4 mm. wide, somewhat quadrate in shape, slightly depressed at vertex; dull black in colour, mouth-parts reddish brown, ocelli black. Body velvety black, with five transverse bands of white on abdomen; collar and stigmatal stripe white; thoracic feet black; prolegs concolorous with body.

Stage II.—Length 6 mm. Head 0.7 mm. wide, blackish brown, with conspicuous whitish patches, giving a mottled appearance; the lower half of clypeus and lower portion of epicranium reaching to ocelli, almost wholly whitish in some specimens; in others the front of head is mostly whitish, with a few dark brown spots on clypeus and inner margins of cheeks. The larvæ are now of a grayish-brown colour, with whitish longitudinal lines and

*Contributions from the Division of Entomology, Department of Agriculture, Ottawa.

streaks. Anterior third of thoracic shield mostly white. Transverse whitish bands on dorsum not so conspicuous as before. Tubercles on body blackish; setæ short, blackish.

Stage III—Length 13 mm. after moulting. Head 1.1 to 1.3 mm. wide, which now may be described as white, mottled with conspicuous dark brown, irregularly-shaped spots, mostly present on upper two-thirds epicranium. Body now pale mauve, above spiracles, venter darker, the whole skin marked with many black spots and streaks. The longitudinal stripes on the body indistinct, except three pale yellow stripes on venter, which are more apparent owing to the darker colour of the under surface of body. The transverse bands on dorsum are irregular and broken and in colour pale yellow. Stigmatal band whitish, blotched with yellow, broken, irregular, bordered with a wide blackish band beneath. Spiracles blackish. Thoracic feet dark brown, or black; prolegs concolorous with venter.

Stage IV (Fig. 2, plate XIV).—Length 21 mm. after moulting. Head 2.0 to 2.2 mm. wide. This stage is much the same as stage III, but the longitudinal stripes are now all yellow, and the skin is not so heavily marked with black spots and streaks. Transverse dorsal bands on abdominal segments as before. The lower joint only of each thoracic foot is now wholly black, the other joints being mostly white, banded above with black.

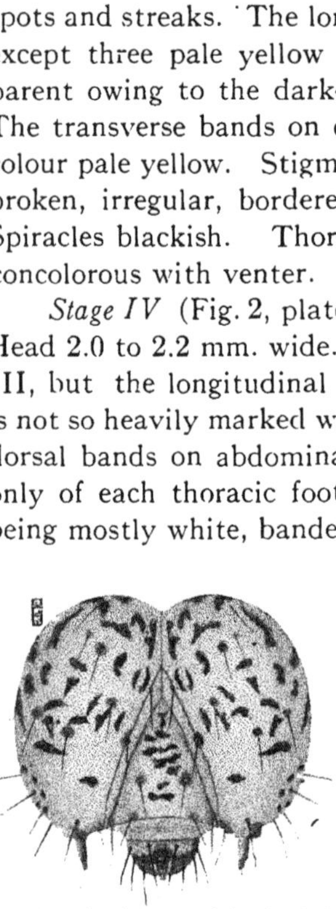

Fig. 15—*Apotheima rachelæ*, head of mature larva (X 33).

Stage V (Fig. 3, plate XIV).—Length 26 mm. after moulting. Head (fig. 15) 2.9 to 3.2 mm. wide, slightly indented at summit, flattened in front; whitish with yellowish tinge; mottled with brown as before; margined behind with black. Larvæ cylindrical in shape, and, in general, the same as Stage IV. Just after moulting the longitudinal stripes are plainly visible; as the larvæ grow these markings become less discernible. When mature and ready to pupate, the larvæ may be described as pale mauve or pale yellow—the colour varying in the specimens—with numerous irregularly-shaped streaks and spots of brown, venter darker; the markings on the dorsum not so heavy as in previous stages. Tubercles black, very small, of equal size; on segments 5 to 9 inclusive, tubercle iii is slightly above, and anterior to, the spiracle; iv is behind and almost in a line with the lower end of spiracle; v is below, and slightly anterior to, spiracle; same distance therefrom as iii. On segments 10 to 12, however, tubercle v is immediately below the anterior side of spiracle. On segment 12, tubercle iv is also below the spiracle. Spiracles black, oval in shape. Stigmatal band whitish, blackish border beneath irregular and not so conspicuous as in previous stage. Thoracic feet and prolegs as in Stage IV.

On June 4 some of the larvæ were mature and entered the earth for pupation. On this date they were 39 mm. in length and 4.5 mm. in width.

Pupa (fig. 4, pl.XIV).—Length 14-16 mm., 4.5-5 mm. in width at widest part; shining, reddish brown, thorax and wing-cases wrinkled. Abdomen moderately punctured, abdominal fold densely, very minutely, punctured in concentric rows. Spiracles black. Cremaster dark brown, bearing two divergent, almost straight, stiff, spines; close to the base of the cremaster there are also two, short, thick, blunt, spiniform protuberances, one on either side.

Most of the moths emerged (in a cool cellar) in the end of April and early in May, 1904. A breeding jar containing some of the pupæ was kept in the laboratory and in this moths emerged in January. Males and females were kept alive and in one instance, on January 9, a male and female mated and remained *in coitu* for 30 hours. In the wooden box in which the moths were confined there was an open crack in the bottom, and when the box was moved on January 11 it was noticed that the female had inserted her long ovipositor through the crack and laid eggs on the underside of the bottom close to the crack, the bottom of the box being slightly elevated above the lower edge of the sides. The ovipositor of the moth (fig. 16) measured in one instance 7.5 mm. in length; near and at the tip it is distinctly pilose, the hairs being slender and of varying lengths.

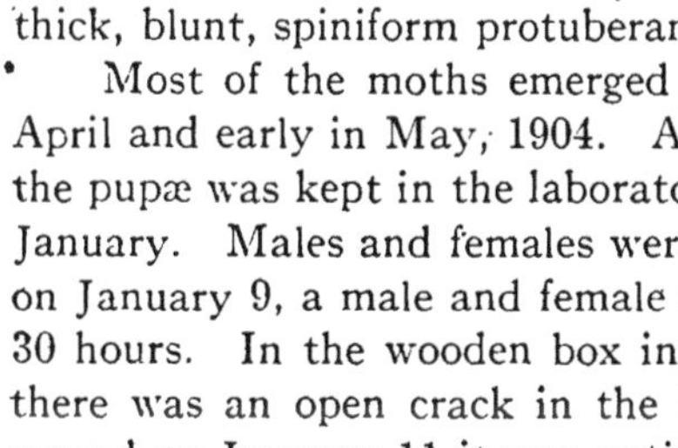

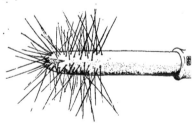

Fig. 16.—*Apocheima rachelæ*, ovipositor, showing arrangement of hairs near the end (X 19).

From the fact that these female moths have such remarkably long ovipositors, it would seem strange that they should deposit their eggs openly, as indicated on the plate herewith. One would naturally expect that the eggs would be laid in crevices, or other recesses. Further observations on the egg-laying habits of the species are most desirable.

The captured females which we have vary in size from 11 mm. in length, by 4 mm. in width (at widest part) to 16 mm. in length by 6 mm. in width. Scales on outer half of rudimentary wings mostly reddish ochreous, remaining scales black. A sprinkling of reddish-ochreous scales also occurs on sides of thorax and abdomen and a conspicuous cluster on summit of head. A distinct dorsal band of these scales is present on thorax and abdomen. Balance of body, black, clothed with long, grey hairs. Antennæ filiform, black, hairy. Legs hairy, tarsal claw brown.

Food Plants.—The larvæ were fed on aspen (*Populus tremuloides*) and cottonwood (*Populus deltoides*). At Aweme, Man., the larvæ have frequently been collected by Mr. Criddle from aspen and also from several species of willow; on one occasion he found a larva about two-thirds grown on elm.

Distribution in Canada.—In Alberta males have been collected at Head of Pine Creek, April and early May (F. H. Wolley-Dod), Millarville, May 9 (A. F. Hudson); High River, March 12 (T. Baird) and at Edmonton, May 12, 1896 (F. C. Clare): in Saskatchewan; Saskatoon, April, 1913, and Regina, April 23 (T. N. Willing): in Manitoba; Rounthwaite, April 15, 25 (L. E. Marmont), Aweme, April 18-27 (N. Criddle); Winnipeg, April 29 (Hanham).

The photograph from which the plate was made was taken by my colleague, Mr. F. W. L. Sladen. Figs. 12 and 13 in the text were drawn by Mr. A. E. Kellet, artist assistant in the Division.

EXPLANATION OF PLATE XIV.

Apocheima rachelæ Hulst.

(All figures natural size).

1. Egg cluster on poplar.
2. Larva, Stage IV.
3. Larva, Stage V, mature.
4. Pupæ.

5. Female moth, bred specimen showing extended ovipositor.
6. Female moth, bred specimen, side view.
7. Female moth, specimen found on poplar.
8. }
9. } Male moths; bred specimens.
10. Male moth, collected at Winnipeg, Man.
11. Male moth, collected at Aweme, Man.

SOME PARASITES OF SIMULIUM LARVÆ AND THEIR POSSIBLE ECONOMIC VALUE.*

BY E. HAROLD STRICKLAND, DIVISION OF ENTOMOLOGY, OTTAWA.

During a year's residence at the Bussey Institution of Harvard University, in 1911, the writer had an opportunity of studying the Simuliid larvæ which abound in the streams of that locality. This study revealed the fact that a large percentage of these larvæ are attacked by parasites, the presence of which appears to result, in all cases, in the death of the host. Simuliid larvæ are most abundant in these streams from early March until May. Isolated specimens are found from then onwards till October, when other species occur in considerable numbers upon the rocks and vegetation in rapidly flowing parts of the streams.

A brief resume of the peculiarities in structure and habits of these interesting larvæ will be of advantage here, as their curious modifications have a very direct bearing upon their liability to parasitism.

The larvæ are to be found either solitarily or gregariously, according to species, attached by means of a caudal sucker to stones or vegetation, only in the fastest flowing water. Silken threads secreted from the salivary glands act as anchor lines, holding the larva in a vertical position, and retain a hold upon the support should the caudal sucker become detached. The cylindrical head bears on its anterior border two fan-like organs carried on elongate pedicels. When expanded, these fans form two very efficient bowl-shaped strainers, through which the water flows. They can be closed at will, and brought over the mouth orifice, carrying with them the small particles of vegetation, and diatoms which constitute the food of the larvæ. Since Simuliid larvæ,

*Contributions from the Division of Entomology, Ottawa.

unlike those of most Nematoceran genera, live entirely, and continually, below the surface of the water, they are apneustic, and all of the oxygen they require has to be extracted from the water through three supra-anal finger-like gills. The relatively small size of these accounts, in all probability, for the necessity of these larvæ living in fast flowing water, for when they are placed in still water they soon die, presumably of asphyxiation.

In maturing larvæ the histoblasts of the pupal and adult organs are well developed and most conspicuous. Thus on each side of the thorax can be seen the three leg-, the wing-, and the halter-histoblasts, as distinctly limited whitish areas. The pupæ of these flies resemble the chrysalids of the Heterocera with the exception of having all the spiracles closed, and the respiratory function being accomplished by a tuft of respiratory filaments situated on each side of the prothoracic region. These project far out of the slipper shaped cocoon in which the pupal stage is passed. The histoblasts of these filaments turn black in the later stages of larval development, and when the latter assumes the chestnut brown colour of maturity they appear as a black triangular area on each side of the prothorax.

The commonest Simuliid around Boston, in Spring, is *Simulium hirtipes*. In the larva of this species two classes of parasites occur. One of these is represented by a nemathelminth worm, belonging to or near the genus *Mermis*. The worm lives either singly, or in considerable numbers, coiled up within the body cavity of its host, where it occupies the ventral portion of the somewhat swollen abdominal region (Pl. XV, fig. 1). When one worm only is present it measures about three centimetres, which is nearly three times the length of its host. The greatest number of worms found in a single larva was twelve. In this case none attained to a greater length than 1 cm. The most striking effect of these parasites upon their larval host is that they so far inhibit the development of the histoblasts that pupation becomes impossible (fig. 2). This suppression of pupal and adult organs is accompanied by a slight increase in the size of the larval tissues, for most parasitised larvæ were from 2 to 3 mm. longer than their healthy companions. This condition is opposite to *Prothetely*, which name Kolbe ('03) ascribed to the several recorded cases in which larvæ of various orders had their histoblasts so accelerated that the pupal, or in some cases adult, organs appeared as external structures in the as yet immature larvæ. Such cases have been recorded by Heymons ('96) in *Tenebrio molitor*, Hagen ('72) in *Bombyx mori*, and others in various Coleopterous larvæ. A study of the cases of prothetely now on record shows that they were all produced under artificial conditions. This would suggest that it is due to some pathological disturbance, which has caused an excessive stimulation of the enzymes, whose action brings about the multiplication of adult tissue forming cells, without appreciably affecting those of the larval tissues. It then follows that there are two sets of enzymes concerned in the maturation of holometabolic insects, one of which may be termed the "larval enzymes" and the other the "adult enzymes." The suppressed growth of the histoblasts in parasitised Simulidæ would then be due to the worm decreasing the stimulating action of the adult enzymes by impoverishing them either in quality or quantity. This subject is discussed more fully in an earlier paper by the writer ('11), where this pathological condition is termed Methetely, in contradistinction to Kolbe's Prothetely.

The parasitised Simuliid larvæ are unable to pupate and are finally killed by the worm, which bores its way out through the "skin" and thus escapes into the water. Here it probably leads a free sexual life, as do the related nemathelminths found in grasshoppers of which it is only the larvæ that are parasitic. It is surmised that the larval worm passes into the body cavity of its host from the alimentary tract, into which it would be readily taken with the food. This worm was found during the spring in varying abundance in most of the streams examined. The largest percentage of infection was 25, equally distributed between the two species of *Simulium* present in that stream. It was never found in the fall, and has probably one generation only per annum.

During the spring there was a very high percentage of parasitism by various Myxosporidia S. L. (Sporozoa). When these were discovered, they had all sporulated, and were therefore at too late a stage in development for their taxonomic position, or life history, to be ascertained, but they were evidently related to the organisms causing the Pebrine disease of silkworms. The body of the parasitised larva becomes enormously distorted and swollen, particularly near the apex of the abdomen (fig. 3). The skin, normally dark green in colour, is rendered almost transparent owing to its great distension, and through it can be seen a white irregular mass of parasitic material. Upon dissection, this parasitic mass is found to consist of countless spores. Four distinct types of spores were found in different localities. These represented, probably four, or even more, species. The simplest type was a plain ovoid, about 5μ by 3μ in size. Another, similar in size, had at one end a flattened disc. In a third, this disc was replaced by a stout flagellum-like organ about 2-3 times the length of the spore, while the fourth resembled the third, but had in addition two raised annulations around its equatorial region. The first type of spore represented, probably, a normal species of the genus *Glugea*, of which three more species were found during the following fall. Unless the spores bearing appendages belong to the Myxosporidia, S.S., which seems to be improbable, they represent entirely different types to anything previously described. Up to 80% parasitism was recorded, every case of which is believed to have been fatal. This could not be proved definitely since all Simulium larvæ kept under observation in captivity died. The following observations tend, however, to confirm this supposition. No pupæ containing parasites could be found, even where 80% of the larvæ had been infested. No reproductive organs were found in parasitised larvæ. There is very little fat body stored up in these larvæ. The voluminous proportions of the parasite would require an enormous rent in the ectoderm in order that it might escape, and were it to pass over into the adult it is inconceivable that the latter would be able to escape from the water when so hampered.

Throughout the summer isolated specimens of *S. bracteatum* were present in the streams. These were casually examined, but no parasites were found. By the beginning of October larvæ of this species were abundant. *S. vittatum* was represented, also, by a few specimens, and by the middle of November *S. hirtipes* was once more hatching out from recently deposited egg masses. The latter species seems to æstivate throughout the Summer, for no signs of it were seen between the end of May and the beginning of October.

The larvæ present in the streams during the fall months were found to be parasitised with species of two distinct Protozoan orders—namely, Gregarinida and Myxosporidia S. L. The former was represented by a single species, the taxonomic position of which was not ascertained. Larvæ parasitised by this species were conspicuous in presenting a white speckling over the entire surface of the body (fig. 4). This, upon closer examination, was seen to be due to innumerable small globular cysts, measuring up to .25 mm. in diameter, which either floated freely in the blood, or were still attached to the original seats of infection. The tissues invaded were the ectodermal epithelium, the cells of the fat body and the layer of pigment cells which cover the nervous system in these larvæ. The sexual organs were never found in parasitised larvæ.

The cysts when sectioned, and stained with iron hæmatoxylin, were seen to be composed of a homogeneous mass of granular protoplasm, in which was situated numerous free masses of chromatic material. In some young cysts there were vacuoles, but these were detected in living specimens only. In other fresh material there seemed to be a distinct ectosarc layer of a perfectly clear fluid. By the end of November a developmental stage was reached in which the protoplasm began to collect around the scattered masses of nuclear material, and the cyst contents were divided up into numerous uninucleate globular bodies. If a cyst were then dissected and allowed to float in water it soon bursts, liberating countless numbers of these minute globules which, within a quarter of an hour of their escape, began to move independently,and were soon actively darting around, in a limited area, in the water. Killed and fixed specimens revealed the fact that each was provided with a flagellum. No further study of these organisms was made. The larvæ thus parasitised had their histoblasts retarded, though to a less extent than those which contained *Mermis* sp. Since this parasite was present in about 50% of the larvæ in streams where it was found, it must have a distinct economic value. The retardation of the histoblasts is sufficiently pronounced to assure the death of the larvæ, which was in all recorded cases that of *S. bracteatum*. No other species of Simulium larvæ were present however, in the streams where this parasite was found.

The order Myxosporidia S.L. was represented by three species of the genus *Glugea*. These species received special attention, and their life histories were worked out. The appearance of parasitised larvæ was similar to that of the specimens, found in Spring, containing other Myxosporidia S.L. The life history of the organisms is, in brief, as follows: a *spore* is taken into the alimentary tract of a very young larva. From this escapes an amœboid *germ*, which passes between the cells of the mesenteric epithelium, and thus gets into the body cavity. Here it attacks, and enters, a cell of the fat body, where it grows with great rapidity, soon bursting the cell and living free life in the body cavity of its host. It is now termed a *trophozoite*, and consists of a multinucleated mass of protoplasm. As the trophozoite matures, a small clearly constricted globule of protoplasm collects around each of the numerous nuclei, to form a spherical *sporont*. The single nucleus of the sporont undergoes three (or more in some species) divisions, thus forming eight nuclei, which in time become the centres of eight small bodies known as *sporoblasts*. Around each of these sporoblasts is secreted a thick shell, which activity is accompanied by a complicated internal development. This converts the sporoblast into a mature *spore*, which is capable of spreading the infection, which liberated in the water by the death and subsequent decay of its host.

It is believed that infection by this class of parasite can be accomplished only in the earliest stages of larval life, before the peritrophic membrane lines the entire surface of the mesenteron. The latter, it would seem, is the only part of the alimentary tract which would not resist the attacks of an unarmed germ. A fuller account of this exceptionally stout peritrophic membrane, and its development, has been published by the writer ('13). In this paper, also, the three Glugeid species discovered as parasites of Simuliid larvæ were described.

It will be seen, from the above descriptions of the various parasites of Simuliid larvæ found around Boston, that they are very conspicuous, and would readily attract the attention of an observer. Notwithstanding this fact, there are no other records of their occurrence in North America. This would appear to indicate that, in those sections of the country where species of Simulium are most abundant, these parasites do not exist, for in these places careful studies of the larval stages have been made by several observers. A Glugeid was described from the European

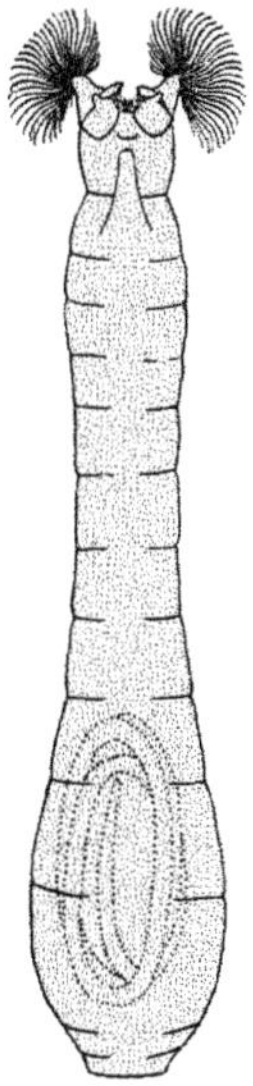

1.

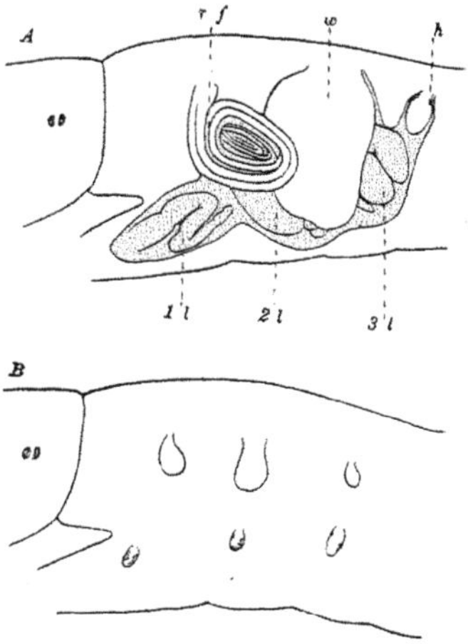

2.

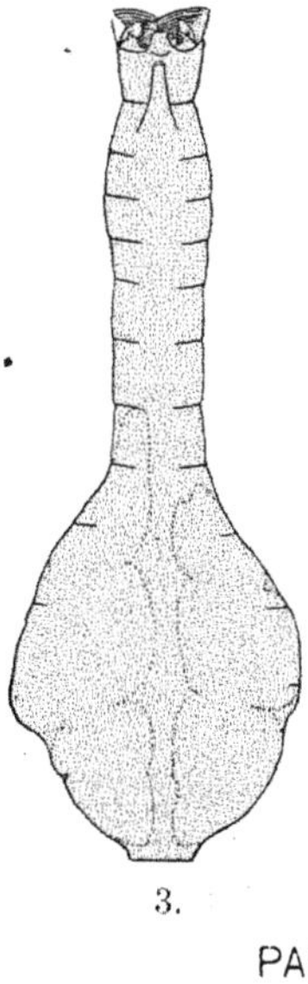

3.

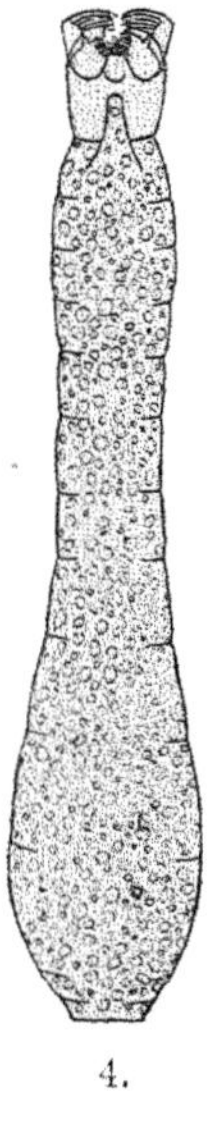

4.

PARASITES OF SIMULIUM.

S. ornatum by Leger ('97), and another from a Brazilian ? species by Luta and Splendore ('04 and '08). Lutz ('09) also records the presence of a nemathelminth in Brazilian Simuliidæ.

The parasites found in Simuliid larvæ around Boston during 1911 may be summarized as follows:

1. Parasites of the Spring brood of Simulium.
   Various Myxosporidia spp. up to 80% mortality.
   Mermis sp. up to 25% mortality.
2. Parasites of the Fall brood of Simulium.
   Glugea spp. up to 10% mortality.
   Gregarine sp. up to 50% mortality.

No experiments have been made upon the possibility of transferring these parasites from one species of Simulium to another, but so far as can be seen there should be no great difficulty in accomplishing this, for in all cases observed the parasites infected all species of larvæ present at the same time in the streams where the former occurred. There is, however, a seasonal variation of parasitism, for the species taken in the spring were not found in the fall, so that it is probable that only those species of Simulium whose life history coincides with parasitised species could be infected with the parasites of the latter.

Explanation of Plate XV.

Fig. 1. Mermis parasites in situ.

Fig. 2. A, normal histoblasts of mature larva. B, histoblasts of full-grown larva containing Mermis, sp.; r f, respiratory filament histoblast; w, wing histoblast; h, halteres histoblast; l (1, 2, 3), leg histoblasts.

Fig. 3. Glugeid parasite in situ.

Fig. 4. Gregarine in situ.

Literature Cited.

Hagen, H. A.—'72. Stettin ent. Zeit., pp. 392-393.

Heymons, R.—'96. Sitzungsber. d. Ges. nat. Fr. Berlin.

Kolbe, H. J.—'03. Allgem. Zeitsch. für. Ent., Bull. 8, No. 1, p. 28.

LUTZ, A.—'09. Memorias do Institute Oswald Cruzo. Tomo I, Faciculo II, pp. 124-146.

LUTZ AND SPLENDORE.—'04. Centralblatt für Bakteriologie, Bd. 42. Id., '08· Centralblatt für Bakteriologie Bd 46, pp 311-315.

STRICKLAND, E. H.—'11. Biological Bulletin, Vol. 21, pp. 302-338. Id., '13· Journal of Morphology. Vol. 24, pp. 43-105.

NOTES ON SOME SPECIES OF CECIDOMYIIDÆ.

BY WILLIAM BEUTENMULLER, NEW YORK.

Dasyneura hirtipes Osten-Sacken.

Cecidomyia hirtipes OSTEN-SACKEN, Mon. Dipt. N. Am., Vol. I, 1862, p. 195; Glover, MSS. Notes, Dipt. 1874, p. 8; Bergenstamm and Loew. Verh. zool. bot. Gesell. Wien., Vol. XXVI, 1876, p. 47.

Dasyneura hirtipes ALDRICH, Cat. N. Am. Dipt. 1905, p. 155.

Male and Female.—Head red, clothed with red scales; antennæ brown. Neck red; thorax above, smooth, polished, blood-red, blackish above with a few erect brown hairs, pleura marked with black; scutellum red. Abdomen bright scarlet red with blackish appressed scales. Legs wholly black; coxæ reddish. Wings with a dense blackish pubescence. Halteres reddish or orange. the club black. Expanse 5.5-6 mm. Length 3.5-4 mm.

Gall.—Polythalamous, large bud-like and solid when immature, with a number of aborted leaves surrounding it. When mature, it is soft inside and filled with a white foam-like substance, in which are a number of long, narrow, larval cells. When the flies are ready to emerge, the gall bursts open and it is then of the appearance of a miniature cauliflower; the large white center surrounded by the dark green leaves, giving it the appearance of that plant. Diameter 50 to 25 mm.

The gall is formed at the tip of stunted stalks of the fragrant golden-rod (*Solidago graminifolia*) in June and July. When old, the white foam-like internal substance decays and the gall is then hard, woody and hollow inside, with a large opening on top. It remains on the bushes in this state over winter. I have collected the gall of *C. hirtipes* at Fort Lee, New Jersey, late in June, from which the adults began to emerge on July 6.

Dasyneura seminivora Beuten.

Cecidomyia (?) *seminivora*, BEUTENMULLER, Bull. Am. Mus. Nat. Hist., Vol. XXIII, 1907, p. 390, pl. XV., figs. 1-4 (larva and gall).

Male and Female.—Head and antennæ black. Thorax deep brown-black, somewhat shining, with erect hairs in the grooves; scutellum deep red. Abdomen deep red, covered with deep brown black scales above; under side wholly deep red with rather long, erect, dark hairs. Legs blackish above, pale brown beneath. Wings blackish hyaline with bluish reflection; extreme base of wings red. Halteres yellowish brown. Expanse 4 mm. Length 2 mm.

This species was heretofore known from the larva and gall only, and the adults are here described for the first time. The flies emerged from May 31 to June 15. The gall is a deformation of the seed-pods of certain species of violets.

Lasioptera podagræ, sp. nov.

Male and Female.—Head clothed with gray scales in front and behind; eyes black; antennæ deep brown, first and second joints dull red. Thorax narrowly banded on each side from the anterior portion of the scutellum, with gray and brown hair-scales. Along the middle of the thorax is a very broad golden brown band from the scutellum to almost the anterior portion. Scutellum marked with gray scales. Abdomen velvety black, with a row of large, broad gray marks on each side, decreasing in size toward the last segment. Under side: Thorax black marked with gray, abdomen wholly gray; legs black above, white on the joints, tarsi banded with white; femora beneath white from the base to about the middle. Wings hyaline with black scales, red at the extreme base and with a very small white mark at the middle of the costa. Halteres red or white. Expanse 2.50 mm. Length 1.50 mm.

Gall.—Polythalamous, consisting of an elongate swelling enlargement of the stems of a species of aster. Length 20 to 50 mm., Width 8 to 10 mm.

Habitat: Fort Lee, New Jersey; Bronx Park, New York City (W.B.).

Types: Collection Am. Mus. Nat. Hist.

Lasioptera linderæ Beuten.

Lasioptera (?) *linderæ* Beutenmuller, Bull. Am. Mus. Nat. Hist., Vol. XXIII, 1907, p. 398, pl. XIV., fig. 3-6 (gall and larva).

Male and Female.—Head black, face and posterior portion snowy white; antennæ black, first, second and third joints white. Thorax velvety black, broadly bordered with white around the anterior portion to the base of the wings. In the grooves are a few scattered pale brown hairs. Scutellum black, with a few long white hairs. Abdomen black, first segment white, second segment with a small white mark on the dorsum posteriorly, the two following segments each with a broad white band posteriorly, the remaining segments wholly black. Under side of abdomen broadly white. Legs pale brown or sordid white; tibiæ marked with black at the middle. Wings hyaline with black scales and a white mark at the middle of the costa and at the base of the wings. Halteres white. Expanse 3 mm.

The adults are here described for the first time. They emerge from the galls in June.

Lasioptera vernoniæ Beuten.

Cecidomyia (?) *vernoniæ* Beutenmuller, Bull. Am. Mus. Nat. Hist., Vol. XXIII, 1907, p. 389, pl. XV., figs. 7-8.

Male and Female.—Head white in front, black posteriorly; antennæ black; first and second joints white. Thorax deep velvety black, with a very broad white margin around the anterior portion to the base of the wings. In the two grooves of the thorax are a few scattered, white hairs. Scutellum scaled with white and with a few long hairs of the same colour. Abdomen deep velvety black, each segment with a very narrow white band posteriorly and not extending to the extreme sides of the abdomen. Legs black, with the joints pure white; coxæ white. Under side of abdomen broadly white. Wings blackish hyaline, with a white mark at the middle on the costa. Halteres white. Expanse 3.5 mm. Length 1.5 mm.

Heretofore, this species was known only from the larva and gall.

Cecidomyia meibomiæ Beuten.

Cecidomyia (?) *meibomiæ* BEUTENMULLER, Bull. Am. Mus. Nat. Hist., Vol. XXIII, 1907, p. 390, pl. XV, figs. 9, 10, 11.

Male and Female.—Eyes large, contiguous at the vertex, black; face semi-translucent, dull orange red. Antennæ yellowish brown with erect black hairs. Thorax dull, semi-translucent red, with rather long, blackish hairs in the grooves on top, forming two parallel lines, and a few hairs at the sides; scutellum dull semi-translucent red. Abdomen dull semi-translucent, orange red sparsely, with brown black hairs; tip of abdomen blunt. Under side of body dull red. Legs fuscous. Expanse 3.50 mm. Length 1.33 mm.

Since my description of the larva and galls of this species I succeeded in breeding the adults from galls collected in the Valley of the Black Mountains, North Carolina, in September 1906, and they are here characterized for the first time.

Cecidomyia clavula, sp. nov.

Female.—Head black, eyes large, contiguous, face pale orange; antennæ pale, semi-translucent, orange yellow, banded with black, hairs gray. Thorax above semi-translucent orange, tinged with red; underside orange. Abdomen orange, with yellow hairs. Legs white, with deep black bands. Wings hyaline yellow, with metallic blue and purplish reflections; a broad, dusky grayish, wavy transverse band a little beyond the middle, followed by a similar close to the outer margin; on the outer margin is a grayish patch almost connected with the preceding transverse band; costa yellow, deep black below the apex; fringes yellow. Halteres semi-translucent orange. Length 2 mm. Expanse 4 mm.

Very closely allied to *Cecidomyia variegata* of Europe. Bred from enlargements of the terminal twigs of Dogwood (*Cornus florida*). It is the gall described by me in the Bulletin Am. Mus. Nat. Hist., Vol. IV, 1892, p. 269.

Rhopalomyia remuscula Beuten.

Cecidomyia (?) *remuscula* BEUTENMULLER, Bull. Am. Mus. Nat. Hist., Vol. XXIII, 1907, p. 392, pl. XVII, figs. 7, 8.

Male.—Head black; antennæ fuscous. Thorax shining black above, with two pale, parallel lines dorsally, pleura yellowish brown; scutellum yellowish. Abdomen dull yellowish, covered with brown hairs. Wings blackish, hyaline. Halteres yellowish, knobs black. Legs fuscous. Length 2 mm.

Hitherto known only from the gall and the larva.

TWO NEW CANADIAN GALL MIDGES.

BY C. P. FELT, ALBANY, N.Y.

Cystiphora canadensis n. sp.,

The interesting midge described below was reared July 10, 1913, by Dr. A. Cosens, Toronto, Canada, from an inconspicuous flat, blister gall on the leaves of white lettuce or rattlesnake root, *Prenanthes altissima* or *P. alba*. This is the second American species of Cystiphora to be discovered and is easily distinguished from *C. viburnifolia* Felt by colorational and structural characters.

Gall.—Circular, diameter 5 mm., dark purplish with a paler center. There is no perceptible thickening of the tissues. The galls are placed irregularly between the veins.

Female.—Length 1.25 mm. Antennæ extending to the third abdominal segment, sparsely haired, dark brown; 13 or 14 segments, the fifth cylindric, with a length twice its diameter; terminal segment either simple or composed of two closely fused segments. Palpi: first segment irregularly capitate, the second subquadrate, the third slender, with a length fully five times its diameter. Mesonotum shining dark brown. Scutellum, postscutellum and abdomen mostly reddish brown, the terminal abdominal segment fuscous, the tip of the ovipositor yellowish. Wings hyaline, the third vein uniting with the anterior margin at the distal ninth. Halteres pale yellowish. Coxæ and femora basally, fuscous yellowish, the distal portion of femora, tibiæ and tarsi mostly fuscous; claws slender, toothed, the pulvilli as long as the claws. Ovipositor with a length nearly half that of the abdomen, the basal half distinctly swollen and rather heavily chitinized, the distal half with a diameter about half that of the basal portion, tapering slightly to a narrowly rounded apex bearing a slender spur. Type Cecid a2441.

Hormomyia helianthi Brodie

1894, Brodie, William—Biol. Rev. of Ont., I, pp. 44-46 (Cecidomyia).

1909, Jarvis, T. D.—Ent. Soc., Ont., 39th Rep't., p. 83 (Cecidomyia).

The axillary galls of this species occur on Helianthus. They are more or less cylindric, occasionally flask-shaped and, according to the describer, have a length of 10 to 25 mm. and a diameter of from 1.5 to 5 mm. There may be 1 to 10 galls in an axil, firmly attached to the stem by an expanded base and projecting in various directions, usually upwards, often at right angles to the stem and occasionally downward. They occur on the upper third of the stems of *Helianthus decapetalus* and *H. divaricatus*, growing in open woods or in shaded situations. This gall occurs about Toronto, Ont., and has been collected at Evanston, Ill., by Mr. L. H. Weld. A species of Torymus has been obtained from this insect. Specimens of this midge were reared June 23, 1907, by Dr. A. Cosens, of the University of Toronto. The following descriptions are drafted from this material.

Male.—Length 2.5 mm. Antennæ probably extending to the fourth abdominal segment, sparsely haired, light yellow; 14 segments, the fifth binodose, the basal stem with a length one-half its diameter, the distal with a length about three-fourths its diameter, the basal enlargement subglobose, the distal subcylindric, slightly expanded apically and with a length nearly twice its diameter; three low, broad circumfili occur on each flagellate segment; terminal segment irregularly binodose, with an indistinct constriction near the basal third, the apex narrowly and irregularly rounded. Mesonotum yellowish brown. Scutellum and postscutellum fuscous yellowish. Abdomen dark yellowish brown, the genitalia fuscous yellowish. Wings hyaline, costa thickly haired, light straw, the third vein uniting therewith well beyond the apex, the fifth at the distal fourth, its branch near the basal half. Halteres, coxæ and femora basally yellowish, the distal portion of femora and tibiæ dark brown, the tarsi a little lighter; claws slender, evenly curved, simple, a little longer than the large pulvilli. Genitalia: basal clasp segment short, broad; terminal clasp segment stout, slightly curved and with a length thrice its diameter; dorsal plate short, very broadly and roundly emarginate, the lobes strongly divergent; ventral plate short, broadly rounded; style short, obtuse.

Female.—Length 3.5 mm. Antennæ nearly as long as the body, sparsely haired, light yellow; 14 segments, the fifth with a stem ¼ the length of the cylindric basal enlargement, which latter has a length 2½ times its diameter and three irregular, anastomosing circumfili on the basal portion of the enlargement, with a fourth

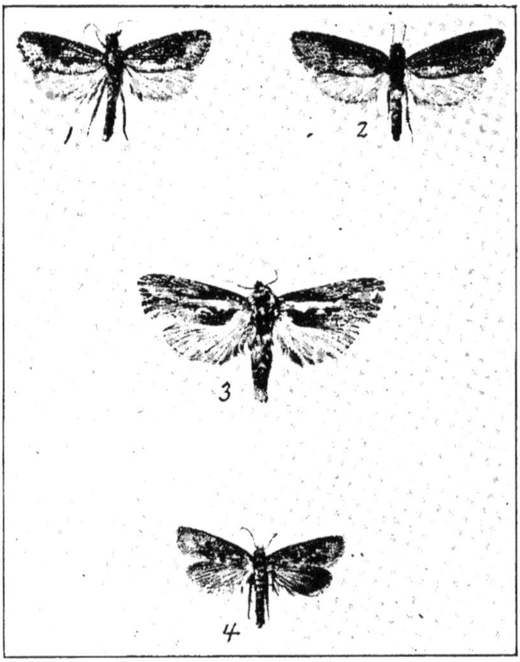

NEW NORTH AMERICAN ANAPHORINAE.

1. Pseudanaphora quadrellus. Type ♂.
2. Pseudanaphora quadrellus. Type ♀.
3. Neolophus antonellus. Type ♂.
4. Eulepiste antonellus. Type ♂.

apical filum; terminal segment slightly reduced, with a length about twice its diameter and tapering suddenly to an obtuse apex; the surface of this segment is nearly covered with very irregular, anastomosing circumfili. Palpi yellowish, basal segment roundly quadrate, the second segment reduced, conical. Mesonotum reddish brown. Scutellum brownish, yellowish apically, postscutellum brownish yellow. Abdomen sparsely haired, dark brown. Halteres and coxæ yellowish. Legs mostly yellowish straw. Ovipositor when extended about as long as the abdomen, the terminal portion moderately stout and with ndistinct, inarrowly rounded apical lobes.

This species departs from the typical Hormomyia in the greatly produced ovipositor of the female and it is possible that it, with related forms, should be referred to a distinct genus.

Type, C. a2453.

SOME NEW NORTH AMERICAN ANAPHORINÆ.

BY WM. BARNES, M.D., AND J. MCDUNNOUGH, PH.D., DECATUR, ILL.

In sorting over a large accumulation of material belonging to the Tineid group called by Walsingham *Anaphorinæ*, we discovered several species which did not agree with the descriptions of any known North American species; these we venture to describe as new. We have followed Walsingham in the generic references, although Busck states that these genera are not tenable; we feel, however, that Walsingham's genera serve at least to define the position of the species in the group rather more clearly than if we had employed a more general term.

Neolophus antonellus, sp. nov. (Fig. 3.)

♂.—Antennæ strongly serrate and fasciculate; palpi rough-haired, upturned to well above front and rather closely appressed, ochreous, deep brown at base; collar brown at base, ochreous apically; thorax brown centrally, this portion rather sharply defined by pale ochreous patagia with light brown center. Primaries light brown with broad, very prominent pale yellowish stripe from base of wing through the fold to approximately the end of cell; upper margin of this stripe is rather even, bordered by a black line which curves slightly upwards and ends in a black discocellular dot;

lower margin irregular, the stripe projecting slightly about center of wing in small quadrate patch towards inner margin, bordered basally with blackish and with brown patch beyond the projection; outer boundaries of stripe diffuse, two faint narrow black lines extending beyond it on each side of vein 4 to termen; slight sprinkling of gray scales above anal angle; costa and inner margin shaded with black-brown basally; indistinct series of marginal dots extending from costa before apex to anal angle; fringes brown, cut by ochreous. Secondaries smoky ochreous at base. Beneath smoky brown tinged with ochreous. Expanse 33 mm.

Habitat: San Antonio, Texas. 1 ♂. Type, Coll. Barnes.

Eulepiste pyramellus, sp. nov. (Fig. 4.)

♂.—Antennæ annulate; palpi rather smooth, upturned to above front, but not appressed; front pale ochreous, thorax darker; primaries an admixture of pale gray and brown scales, maculation very indefinite and indistinct, in well-marked individuals consisting of a brownish blotch in cell near base, another at end of cell and a third midway between these two above inner margin, these latter are at times connected outwardly by a whitish oblique waved line, which is usually more or less obsolete; indistinct costal and terminal dark dots; secondaries and underside unicolorous smoky brown. Expanse 23 mm.

Habitat: Pyramid Lake, Nevada. 4 ♂. Type, Coll. Barnes.

The species appears to be intermediate between *hirsutus* Bsk. and *occidens* Bsk.

Pseudanaphora quadrellus, sp. nov. (Figs. 1, 2.)

♂.—Antennæ very slightly serrate below, palpi upturned, roughly haired, brown; thorax chocolate-brown; primaries chocolate brown shaded with pale ochreous especially along inner margin and termen; costa with alternate striæ of chocolate-brown and ochreous; slight ochreous tinge in cell; dark discocellular dash; inner margin broadly ochreous, more or less striate with brown, upper edge of this ochreous stripe irregular with prominent blunt tooth of ground-colour projecting downward towards middle of inner margin; before and after this tooth the margin is rounded, bent sharply upwards beyond origin of vein 3 as far as vein 7, bending again at right angles and attaining termen below apex, forming a large subquadrate terminal ochreous patch; faint terminal row of dark dots; fringes checkered brown and ochreous with pale basal line. Secondaries pale smoky brown with ochreous terminal line and checkered fringes. Beneath smoky brown, costa of primaries apically ochreous with 3 or 4 brown striæ, narrow terminal ochreous line, secondaries and fringes as above.

♀.—Palpi short, hairy, porrect; primaries more uniform chocolate brown with only faint traces of ochreous along inner margin; a paling of the ground colour represents the quadrate terminal patch so prominent in the ♂. Expanse ♂ 25 mm. ♀ 28 mm.

Habitat: Palmerlee, Ariz. 7 ♂, 3 ♀. Types, Coll. Barnes.

The species is allied to *davisellus* Beut., but should be readily distinguished by the dark apex and subquadrate ochreous terminal patch with sharply defined inner-edge. The males vary in the amount of brown striations on the ochreous area in some there are scarcely any, in others they show a tendency to obscure this area more or less completely.

EUGONIA CALIFORNICA BDV. IN THE PACIFIC NORTHWEST.

During the summer of 1912 there was an unusual occurrence of the caterpillars of *Eugonia californica* in a number of places throughout the states of Washington, Idaho, and also British Columbia. They were reported as being present "by the millions" and defoliating the buckbrush or ceanothus (*Ceanothus velutinus*). It was even reported that the caterpillars were blocking the trains at Clayton, a village about 30 miles north of Spokane, Wash. This story was not exaggerated in the least, as I had occasion to ascertain.

On the 18th of June I visited the field of devastation at Clayton. When I first arrived at the place, I thought the caterpillars to be those of Vanessa antiopa, but such they did not prove to be. I saw that there were a great many of the caterpillars and started to step off the distance across the small area ahead of me, but I soon found that it was not a matter of yards or rods, but of miles. As far as could be seen to the westward, the ceanothus looked as if it had been scorched by fire. All the bright, green glossy leaves had been eaten, and the branches were entirely bare except for the millions of crawling, spiny, dark caterpillars. They were crawling everywhere—on the ground, over sticks, stones, logs, stumps, up in the young pines above one's head, on the fence posts, up and down the timbers of railroad bridges, in the water, and even on the rails of the railroad. Some of the caterpillars were about one-third grown, some were nearly mature, while a few were hanging pendant, like fruit, preparatory to pupation. All about one could hear the incessant rattle of their feeding, and the patter of their falling excrement like the patter of summer rain. The ground was covered with excrement. Thousands of the cast skins were to be seen on the naked branches. Where the railroad train had been blocked countless thousands of the dead and crushed caterpillars were found, their bodies covering the ground for rods. In the midst of this field of devastation stood willows, pines, roses, firs, tamaracks, grasses and lupines all untouched. Just as I was leaving the place, I noticed that the ground was covered with caterpillars, all crawling in the same direction. I found that they had completely defoliated the ceanothus bushes where they had been and were now on their way to find other food.

Two days later, in Spokane, I found a few acres of ceanothus on a hill side completely defoliated as at Clayton. Here many of the caterpillars had transformed to chrysalids. Several times they were noticed to shake themselves violently until the bushes shook from the effect.

On July 7 I found millions of the caterpillars on the south slope of Moscow Mt., Idaho. Some of these were parasitized by an undetermined species of Braconid.

On July 13, when I again visited the place of infestation at Moscow Mt., I found that all caterpillars and all chrysalids were gone. The caterpillars had evidently not migrated, for all around as far as I could see the ceanothus had not been touched. Even had the caterpillars migrated that would not explain the absence of the chrysalids. I think that the total disappearance of these caterpillars and chrysalids was no doubt due to birds. A similar disappearance of all the caterpillars in the other districts visited seems to confirm this opinion.

These caterpillars were reported from the following places: Chelan, Wn.; Brownsville, Wn.; Moscow Mt., Idaho; Nelson, B.C.; Peachland, B.C., and Clayton, Wash. M. A. YOTHERS.

NOTE ON THE OVIPOSITION OF *AEDES CALOPUS* MEIGEN.

BY JAMES ZETEK, ANCON, CANAL ZONE.

The writer was taking the adults from a lot of breeding cages and, by accident, one female *Aedes calopus* escaped. Hardly more than two minutes later he turned his attention to an uncovered cage containing water and larvæ of *Culex coronator* D & K, and discovered a mosquito resting on the side of the jar, the tip of its abdomen extending and touching the water, depositing eggs. This occurred at 4.10 p.m., July 2nd, 1913. The mosquito was *A. calopus*, and most probably was the one that had escaped a moment before. The cage was covered with gauze.

When the cages were examined at 1.00 a. m., July 5th, very young larvæ of the yellow-fever mosquito were seen in this particular jar. The egg instar is about 60 hours. These larvæ were allowed to mature, and from them emerged 14 ♀ and 18 ♂ adults.

The female which had deposited these eggs had no blood meal, nor any other food than which was present in the air and water. In the original cages of *Aedes calopus*, copulation was frequently noted, occurring chiefly in the daytime. In the act the male is underneath, clasping the female, the two mosquitos facing each other. The male clasped the female as frequently in flight as when at rest. In one cage, containing one male and many females, the male copulated three successive times during the half hour under observation.

The rapidity with which the mosquito found water suitable for oviposition after its escape is remarkable, and places emphasis upon the cautions to be taken while working with disease-transmitting insects or pests.

PHLEBOTOMUS AND VERRUGA.

It would appear that an addition is likely to be made to the ever-increasing number of cases of the relation of insects to disease. In certain of the valleys of the Pacific slope of the Peruvian Andes, in South America, an obscure disease, known as *Verruga*, has existed for years. Recently, the possibility of the transmission by some species of insect, or tick, has been seriously entertained, and we now learn from "Science" (August 15th, 1913) that Mr. Charles H. T. Townsend, who was some time ago especially charged by the Peruvian Government with the investigation of the insect transmission of verruga, injected a dog with triturated females of *Phlebotomus* on July 11th, and on July 17th secured as a result an unmistakable case of verruga eruption. The gnats used for the injection were secured on the night of July 9th, in Verrugas Canyon, a noted focus of the disease. This is the first experimental transmission of verruga by means of insects, and adds a notable case to the list of insect-borne diseases. The details of the experiment will appear shortly. Further transmission work in laboratory animals will be pursued at once, both by injections and by causing the gnats to bite. C. G. H.

A NEW LEPTODESMID FROM MONTANA.

BY RALPH V. CHAMBERLIN, CAMBRIDGE, MASS.

The following description is published separately in order that the name may be available for early use.

Leptodesmus (*Chouaphe*) *elrodi*, sp. nov.

Light brown to very deep brown and brownish black, the background sometimes rather obscurely chestnut. Carinal and anal process in darker individuals orange, in paler even more yellowish; the first dorsal plate also paler, yellowish, oblong anterior margin. The metazonites may be be paler caudally.

Head with the median sulcus deep. Vertex moderately finely uneven or coriaceous, bearing several long bristles across vertex and also in clypeal region above those of labial border. Antennæ of moderate length. First or cervical dorsal plate narrower than the second one, anteriorly strongly convex; caudal margin moderately deeply concave mesally; laterally margined. Caudolateral angles with caudal side nearly straight. Dorsum strongly arched; anterolateral corners of plates convexly rounded, in the second to fifth plates a little extended cephalad, but in others more and more sloping off caudad. Caudal corners in about the second to the eighth plates bent moderately forward, then becoming straight in a few and then bent caudad, and in the last few forming a distinct but always distally rounded process; edges of lateral carinæ narrowly elevated; pores opening ectad; nineteenth plate much shortened, its processes reduced; metazonites with transverse furrow distinct; surface to naked eye appearing nearly smooth and shining, under the lens finely coriaceous, more strongly roughened laterally. Sternites smooth, glabrous. A deep transverse sulcus at middle and a weaker median longitudinal one crossing it at right angles. Anal process in dorsal outline subtriangular, distally subcylindric, the tip a little depressed; a transverse row of four bristles near middle of length and toward and on tip about eight more. Anal scale semilunar in outline, but with the anterior margin weakly convex, bearing a bristle on each side a little in front of caudal margin. Anal valves elevated along mesal border, each bearing a bristle a little ectad of mesal edge near middle of length and a second one in line with it farther caudad. Legs clothed with stiff hairs, which proximally are sparse, but distally, and especially on dorsal surface, become more dense and at the same time shorter. Male gonopods consisting of two long prongs, of which the posterior one is distally slender and style-like and curves evenly, first dorsal and somewhat mesad and then back proximad. The anterior branch just proximad of the curve in the first one expands into a subtriangular plate on the dorsomesal side, which along its distal or caudal edge bears several short teeth, of which the one at angle is the longest; the prong continues beyond this on a more slender blade which bends abruptly dorsad or dorso-ectad and bears on its proximal edge and mostly on distal portion a series of long, curved, spine-like processes. (See fig. 17).

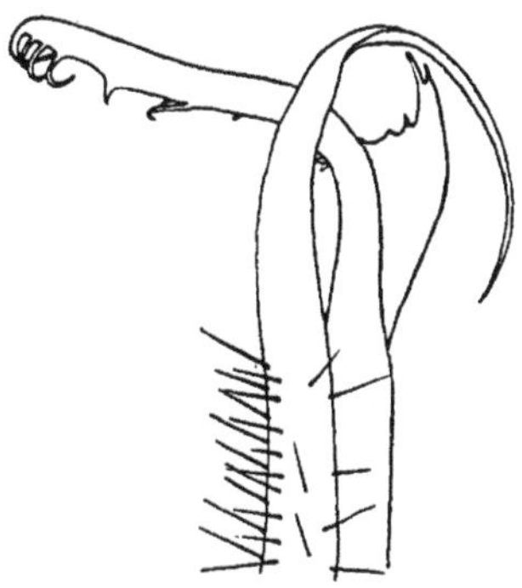
Fig. 17.—*Leptodesmus elrodi*, caudo-ventral view of distal portion of right male gonopod.

Locality.—Montana (Flathead Lake).

The specimens forming the types of this species were collected by Dr. C. C. Adams in the summer of 1912. The species is named in honour of Prof. Morton J. Elrod, of the University of Montana, who is doing much in aiding the advancement of our knowledge of the Natural History of Montana.

A NEW PAMPHILA FROM NEW MEXICO (LEPIDOPTERA).

BY HENRY SKINNER, M.D., SC.D.

Pamphila margarita, n. sp. The male expands 14.5 mm. and the female 15.5 mm., the measurements being taken from the base to the apex of one wing. The colour of the species is tawny olive (Ridgway) and the same colour as *pittacus* Edwards. There is a very faint stigma in the male and on the primaries three vitreous subapical spots; a rectangular spot, constricted in the middle, at the end of the discoidal cell; three spots in an oblique line across the median interspaces, the middle one being the largest and triangular in shape, and the lower one is somewhat linear, with the inner end pointed. The secondaries have a crooked transverse row of four vitreous spots below the middle of the wing; the lower two are small and parallel to the margin, while the upper two are the larger and at right angles to the margin. Fringe dirty white. Underside: Primaries with the spots repeated and also on the secondaries, but larger, and there are in addition a few spots at the base of the wing. The female is like the male, but larger, and the spots are more conspicuous.

This species is allied to *pittacus* Edw. and looks much like it. The transverse row of spots on the upper side of the secondaries of *pittacus* consists of four, straight, distinct rectangular spots, and the two species may be separated by the difference in this row of spots.

Described from a number of specimens of both sexes submitted by Mr. R. C. Williams, the species being named in honour of his wife. They were captured at Jemez Springs, New Mexico, May 26th to June 9th, by Mr. John Woodgate. The type is in the collection of the Academy of Natural Sciences of Philadelphia.

A NEW SPECIES OF ELASMIDÆ OF THE GENUS *EURYISCHIA* HOWARD FROM AUSTRALIA, AND A NEW *PODAGRIONELLA*.

BY A. A. GIRAULT, NELSON, N.Q., AUSTRALIA

The following species has, in my own mind, served to confirm the position of *Euryischia* Howard in the family Elasmidæ. It was captured with the sweeping net January 6th, 1913, from foliage and grass along the Cape River, Capeville, Pentland, Queensland.

Genus *Euryischia* Howard.

1.—*Euryischia sumneri*, new species.

Female.—Length 2mm.

Black-blue, the distal third or more of the fore wing distinctly embrowned (from about distal fourth of marginal vein to the apex); postmarginal vein somewhat longer than the long stigmal, and over half the length of the marginal. Cephalic tibiæ and tarsi brownish, also the tegulæ; caudal coxæ normal for the family, but the caudal femora enlarged and compressed. Scutellum finely alutaceous, the scutum the same but clothed with dense, black, stiff bristles. Antennæ yellowish, club 3-jointed, funicle 3-jointed, the distal joint wider than long; the first subquadrate, the club ovate, larger than the funicle; no ring-joint. Both mandibles tridentate, the inner tooth broad and truncate, shorter. Pedicel longer than any of the funicle joints. Forewings proximad with several very long bristles from the blade. Posterior tibial spurs white.

(From one specimen, ⅔-inch objective, 1-inch optic, Bausch and Lomb.)

Male.—Not known.

Habitat.—Australia—Capeville (Pentland), Queensland.

Type.—In the Queensland Museum, Brisbane, the above specimen on a tag plus a slide bearing the head.

Dedicated with much respect to Charles Sumner for his orations on war.

Genus *Podagrionella* Girault.

1.—*Podagrionella pentlandensis*, new species.

Female.—Length 5.10 mm., exclusive of ovipositor.

Very similar to the type of the genus, but the antennal club darkens at tip, the flagellum reddish brown, the pedicel darker.

The abdominal petiole is shorter, very short, wider than long; the distal two funicle joints are slightly wider than long; the cross dash on the fore wing is the only fuscation present in these wings. The whole body is more robust, the abdomen stouter and longer. Otherwise structurally like *fasciatipennis*, with which I have compared it. Mandibles tridentate.

Male.—Not known.

Described from one female captured by sweeping miscellaneous foliage and grasses along the Cape River, Capeville, Pentland, Q., January 6th, 1913.

Habitat.—Australia—Queensland, Capeville (Pentland).

Type.—In the Queensland Museum, Brisbane, the above specimen on a tag with a slide bearing posterior femur and head.

CONSTITUTION OF THE ENTOMOLOGICAL SOCIETY OF ONTARIO.

Incorporated 1871.

SECTION I.—(OBJECTS AND MEMBERSHIP).

1.—The Society shall be called "THE ENTOMOLOGICAL SOCIETY OF ONTARIO," and is instituted for the investigation of the character and habits of insects, the improvement and advancement of Entomological Science, and more especially its practical bearing on the Agricultural and Horticultural interests of the Province. The Society shall consist of not less than twenty-five members.

2.—The Society shall consist of four classes, viz.—Members, Life Members, Honorary Members and Corresponding Members.

3.—Members shall be persons whose pursuits, or studies, are connected with Entomology, or who are in any way interested in Natural History and who are resident within the Dominion of Canada.

4.—Life Members shall be persons who have made donations to the value of $25 in money, books or specimens (the two latter to be valued by competent persons) or who may be elected as such at the General Meeting of the Society, for important services performed, and after due notice has been given.

5.—Entomologists residing outside Canada may be elected Corresponding Members of the Society, but such membership will not entitle them to the publications of the Society except on payment of the subscription to the Journal of the Society.

6.—Honorary Members shall be members of high standing and eminence for their attainments in Entomology.

7.—The number of Honorary Members shall be limited to twenty-five.

8.—The Officers of the Society shall consist of a President, a Vice-President, a Secretary-Treasurer, and not fewer than three, and not more than five, Directors, to form a Council; all of whom, with two Auditors, shall be elected annually at the Annual General Meeting of the Society, and shall be eligible for re-election. The said Council shall, at their meeting, appoint a Curator.

SECTION II.—(ELECTION OF MEMBERS).

1.—All candidates for Ordinary or Life Membership must be proposed by a member at a regular meeting of the Society and be balloted for; the affirmative vote of three-fourths of the members present shall be necessary for the election of a candidate.

2.—Honorary Members must be recommended by at least three members, who shall certify that the person named is eminent for his Entomological attainments; the election in their case shall be conducted in the same manner as laid down for other members.

3.—Corresponding Members shall be elected in the same manner as Honorary Members.

4.—Whenever any person is elected a member in any class, the Secretary shall immediately inform him of the same by letter, and no person shall be considered a member until he has signified his acquiescence in the election.

5.—Every person elected a member is required to pay his first contribution within one month of the date of his election; otherwise his election shall be null and void.

SECTION III.—(CONTRIBUTIONS).

1.—The annual contribution of members shall be one dollar; all contributions to be due in advance on the first day of January in each year, the payment of which shall entitle the member to a copy of all the publications of the Society during the year. All new members, except those elected at and after the Annual General Meeting and before the following first of January, shall be required to pay the subscription for the year in which they are elected.

2.—Every member shall be considered to belong to the Society, and as such be liable to the payment of his annual contribution, until he has either forfeited his claim or has signified to the Secretary, in writing, his desire to withdraw, when his name shall be erased from the list of members.

3.—Whenever any member shall be one year in arrear in the payment of his annual contribution, the Secretary shall inform him of the fact in writing. Any member continuing two years in arrears shall be considered to have withdrawn from the Society, and his name shall be erased from the list of members.

4.—Life and Honorary Members shall be required to pay an annual contribution.

SECTION IV.—(OFFICERS).

1.—The duties of the President shall be to preside at all the meetings of the Society, to preserve good order and decorum and to regulate debates.

2.—The duties of the Vice-President shall be the same as those of the President during his absence.

3.—The duties of the Secretary-Treasurer shall be to take and preserve correct minutes of the proceedings of the Society, and to present and read all communications addressed to the Society; to notify members of their election, and those in arrear of the amount of their indebtedness; to keep a correct list of the members of the Society, with the dates of their election, resignation or death and their addresses; to maintain the correspondence of the Society, and to acknowledge all donations to it. He shall also take charge of the funds of the Society and keep an accurate account of all the receipts and disbursements, and of the indebtedness of the Members, and render an Annual Report of the same at the Annual General Meeting of the Society, in the manner required by the Act respecting the Board of Agriculture and Arts.

4.—It shall be the duty of the Curator to take charge of all books, specimens, cabinets, and other properties of the Society; to keep and arrange in their proper places all donations of specimens;

to keep a record of all contributions of books and specimens, with a list of the contributors; and to oversee and direct any exchange of specimens. He shall also report annually to the Society on the condition of the specimens and cabinets under his care.

5.—The Officers of the Society shall form a Council who shall have the direction and management of the affairs of the Society. The Council shall meet once in every quarter, the time and place of meeting to be appointed by the President, and notice to be given by the Secretary at least ten days beforehand.

6.—The Council shall draw up a Yearly Report on the state of the Society, in which shall be given an abstract of all the proceedings, and a duly audited account of the receipts and expenditure of the Society during their term of office; and such report shall be read at the Annual General Meeting.

SECTION V.—(MEETINGS).

1.—Ordinary Meetings shall be held once a month, on such days and at such hour as the Society by resolution may from time to time agree upon.

2.—The Annual General Meeting of the Society shall be held at the place and during the same time as the Exhibition of the Agricultural and Arts Association is being held in each year, to receive and deliberate upon the Report of the Council on the state of the Society, to elect Officers and Directors for the ensuing year and to transact any other business of which notice has been given.

3.—Special Meetings of the Society may be called by the President upon the written request of five members of the Society, provided that one week's notice of the meeting be given, and that its object be specified.

SECTION VI.—(BRANCHES OF THE SOCIETY).

1.—Branches of the Society may be formed in any place within the Dominion of Canada on a written application to the Society from at least six persons resident in the locality.

2.—Each Branch shall be required to pay to the Parent Society fifty cents per annum for each paying member on its list.

3.—Every Branch shall be governed by the constitution of the Society, but shall have power to elect its own officers and enact by-laws for itself, provided they be not contrary to the tenor and spirit of the Constitution of the whole Society.

4.—All Members of the Branches shall be Members of the Society and entitled to all the privileges of Members.

5.—No Corresponding or Honorary Member shall be appointed by the Branches, but such members may be proposed at General Meetings of the Society by any Branch as well as by the individual members.

6.—Each Branch shall transmit to the Parent Society, on or before the first of September each year, an Annual Report of its proceedings, such report to be read at the Annual General Meeting.

SECTION VII.—(ALTERATION OF CONSTITUTION).

1.—No article in any section of this Constitution shall be altered or added to, unless notice be first given at an Ordinary Meeting of the Society, or of a Branch, and the alteration or addition be sanctioned by two-thirds of the members present at the next ensuing meeting; the Secretary of the Society, or of the Branch, shall then notify the Secretaries of all the other Branches; when the sanction of all the Branches has been obtained in the same manner, the alteration or addition shall become law.

SECTION VIII.—(SUBSCRIPTION PRICE OF MAGAZINE).

The Annual Subscription Price of the CANADIAN ENTOMOLOGIST shall be two dollars ($2.00), postage included, payable in advance. Members of the Society who have paid their annual dues shall receive the Magazine free of charge, as stated in Section III, Clause I, of the Constitution of the Society.

Mailed December 19th, 1913.

Index to Volume XLV.

Lightning Source UK Ltd.
Milton Keynes UK
UKHW012138180219
337529UK00012B/1344/P

9 780266 043096